Biotechnology: In Context

Biotechnology: In Context

Brenda Wilmoth Lerner & K. Lee Lerner, Editors

VOLUME 1

ADVANCED BIOMASS COOKSTOVES TO GREEN CHEMISTRY

GALE
CENGAGE Learning·

Detroit • New York • San Francisco • New Haven, Conn • Waterville, Maine • London

Biotechnology: In Context

Brenda Wilmoth Lerner and K. Lee Lerner, Editors

Project Editor: Elizabeth Manar

Contributing Editors: Jason Everett, John McCoy

Managing Editor: Debra Kirby

Rights Acquisition and Management: Robyn V. Young

Imaging: John L. Watkins

Product Design: Kristine A. Julien

Composition: Evi Abou-El-Seoud, Mary Beth Trimper

Manufacturing: Wendy Blurton, Dorothy Maki

Product Manager: Douglas A. Dentino

For product information and technology assistance, contact us at **Gale Customer Support, 1-800-877-4253.**
For permission to use material from this text or product, submit all requests online at **www.cengage.com/permissions.**
Further permissions questions can be emailed to **permissionrequest@cengage.com**

Cover photographs: Apoptosis in cells, © Dr. Thomas Deerinck/Visuals Unlimited/Corbis; gloved hand with Petri dish, © Alexander Raths/Shutterstock.com; field of rapeseed, © Rasmus Holmboe Dahl/Shutterstock.com; cell manipulation, © Sebastian Kaulitzki/Shutterstock.com; bioluminescent jellyfish, © John Wollwerth/Shutterstock.com; mouse embryo, © Natalia Sinjushina and Evgeniy Meyke/Shutterstock.com; hazmat training drill, © TFoxFoto/Shutterstock.com; identical sheep, © Jason Bennee/Shutterstock.com; Arabidopsis plant, © Vasiliy Koval/Shutterstock.com; white lab rat, © lculig/Shutterstock.com.

While every effort has been made to ensure the reliability of the information presented in this publication, Gale, a part of Cengage Learning, does not guarantee the accuracy of the data contained herein. Gale accepts no payment for listing; and inclusion in the publication of any organization, agency, institution, publication, service, or individual does not imply endorsement of the editors or publisher. Errors brought to the attention of the publisher and verified to the satisfaction of the publisher will be corrected in future editions.

LIBRARY OF CONGRESS CATALOGING-IN-PUBLICATION DATA

Biotechnology / Brenda Wilmoth Lerner & K. Lee Lerner, editors.
    v. cm. – (In context)
    Includes bibliographical references and index.
    ISBN 978-1-4144-9083-0 (v. 1) ISBN 978-1-4144-9084-7 (v. 2) – ISBN 978-1-4144-9082-3 (set) ISBN 978-1-4144-9085-4 (e-book)
    1. Biotechnology–Popular works. I. Lerner, Brenda Wilmoth. II. Lerner, K. Lee.
    TP248.215.B563 2012
    660.6–dc23   2011045784

Gale
27500 Drake Rd.
Farmington Hills, MI, 48331-3535

ISBN-13: 978-1-4144-9082-3 (set)   ISBN-10: 1-4144-9082-8 (set)
ISBN-13: 978-1-4144-9083-0 (vol. 1)   ISBN-10: 1-4144-9083-6 (vol. 1)
ISBN-13: 978-1-4144-9084-7 (vol. 2)   ISBN-10: 1-4144-9084-4 (vol. 2)

This title is also available as an e-book.
ISBN-13: 978-1-4144-9085-4   ISBN-10: 1-4144-9085-2
Contact your Gale, a part of Cengage Learning, sales representative for ordering information.

Printed in China
1 2 3 4 5 6 7 16 15 14 13 12

Contents

Contents

VOLUME 2

Contents

Advisors and Contributors

While compiling this volume, the editors relied upon the expertise and contributions of the following scientists, legal scholars, and researchers, who served as advisors and contributors for *Biotechnology: In Context*.

Susan Aldridge, Ph.D.

Independent scholar and writer
London, England

Stephen A. Berger, M.D.

Director, Geographic Medicine
Tel Aviv Medical Center
Tel Aviv, Israel

Bryan Davies, J.D.

*Independent scholar
and science writer*
Whitby, Ontario, Canada

Sandra Galeotti, M.S.

Independent scholar
Porto Alegre, Brasil

Brian D. Hoyle, Ph.D.

Microbiologist
Nova Scotia, Canada

Joseph P. Hyder, J.D.

Independent scholar
Jacksonville, Florida

Phillip Koth

Independent scholar
East Peoria, Illinois

**Kenneth La Pensee,
Ph.D., MPH**

*Epidemiologist and medical
policy specialist*
Hampton, New Jersey

Adrienne Wilmoth Lerner, J.D.

Independent scholar
Jacksonville, Florida

Pamela S. Michaels, M.A.

Independent scholar and writer
Santa Fe, New Mexico

Anna Marie Roos, Ph.D.

*Research Fellow in Modern
History*
University of Oxford
Oxford, England

Melanie Barton Zoltán, M.S.

Independent scholar
Amherst, Massachusetts

Acknowledgments

The editors also wish to thank John Krol, whose sound judgments greatly enhanced the quality and readability of the text.

The editors gratefully acknowledge and extend thanks to Julia Furtaw, Douglas Dentino, and Debra Kirby at Cengage Gale for their faith in the project and for their sound content, advice, and guidance. Without the able guidance and efforts of talented teams in technology, imaging, and rights acquisition at Cengage Gale this book would not have been possible.

The editors extend both professional and personal appreciation to Project Manager Elizabeth Manar. Her sharp eye and keen mind greatly enhance every aspect of this book. Her grace and humor enriches our lives, and her skillful editing elevates our work.

Introduction

Life is ancient, grounded in the physics, chemistry, and the evolutionary biology of Earth. In contrast, today's biotechnology marshals new ideas and techniques with the potential to reshape the planet and now, life itself.

The historical development of biomedical technology, the fusion of mathematical, scientific, and engineering principles, is deeply entwined with centuries of advances in medicine. Modern biotechnology is also unquestionably a rapidly expanding field that promises important and rewarding work for generations of students. The problems facing the world in fighting hunger, pollution, and disease cry out for innovative scientific solutions. Accordingly, for many, biotechnology is a beacon toward engineering solutions, but for some, the manipulation of life offers a siren song of peril.

Biotechnology: In Context is devoted to helping high school students and other readers understand an array of biotechnology issues in their historical, social, and ethical contexts. The entries in this book explore fundamental molecular processes and applications ranging across medicine, agriculture, and industry. Entries and readings are designed to provide an introductory overview offering a solid foundation and motivation toward further study.

At the core of the advances in biotechnology lies the science of molecular biology and genetics. Because *Biotechnology: In Context* is designed for younger students and general readers, the book correspondingly includes foundational coverage of central processes such as transcription, translation, and replication. Entries also provide scientifically grounded insights into current controversies, and selected primary source readings deepen reader understanding and help develop critical thinking skills needed to evaluate news and media coverage of biotechnology-related topics.

Written by experts, teachers, and writers in the fields of physics, molecular biology, genetics, microbiology, biotechnology, law, and philosophy, *Biotechnology: In Context* presents complex scientific concepts clearly and simply, without sacrificing fundamental accuracy. Color illustrations and images that stimulate interest and understanding are included throughout the book.

Biotechnology: In Context is specifically designed to enable readers to grasp and articulate key facets of biotechnology issues, construct critical arguments about those issues, and to better understand the viewpoints of others on issues with which they may hold differences.

Such critical thinking and balance are important because biotechnology issues are often multifaceted. As exemplified by the search for an effective way to prevent the mosquito-borne disease dengue fever, biotechnology offers both hope and hazards.

Known as breakbone fever throughout much of the developing world, dengue is endemic in at least 115 countries and spreading globally. There is no cure. Caused by four unique viruses, dengue spreads wherever its carrier, the *Aedes Aegypti* mosquito, thrives due to lack of sanitation or standing water.

Except for the more deadly hemorrhagic form, dengue fever is usually less lethal than influenza. Of the 30,000 to 50,000 people who die from dengue each year, however, children under five are especially vulnerable. Although not especially lethal, breakbone fever earned its name by causing incapacitating joint pain. As outbreaks sweep through villages and towns, stricken farmers watch crops rot in fields and livestock die from lack of care. Dengue perpetuates poverty, robbing millions of their economic health—and in the poorest and most vulnerable regions, dengue steals already slim chances to live a long, productive, and healthy life.

Passions flare over whether vector control, treatment programs, or potential vaccines should receive priority for scant resources. Yet there is consensus that dengue is a looming twenty-first century pandemic.

There is, however, uncertain glory and profit to be gained from tackling dengue, a disease that receives only a fraction of the money and media attention devoted to AIDS, malaria, and tuberculosis. Dengue is officially classified by the World Health Organization as a Neglected Tropical Disease (NTD) and only a handful of biotechnology and pharmaceutical companies are technically and financially capable of tackling the complex problems presented by dengue—especially with the liabilities associated with developing a tetravalent vaccine, effective against all four dengue viruses, that prevents mixing and reversion of live attenuated viruses.

Despite these obstacles and uncertainties, biotechnologists are seeking solutions that range from genetic alteration of the mosquito vector to the development of a chimera vaccine that may offer protection to the billions of people at risk for dengue fever. An increasing number of experts assert that with preventive measures proving ineffective or inefficient, the world's best hope against dengue lies in the development of a tetravalent vaccine. Such a complex vaccine requires pioneering biotechnology that also poses deeply philosophical questions.

To stop dengue, biotechnologists may need to create life—or something very close to it. As of 2011, several vaccine candidates in development are on track to become the first chimera vaccine ever licensed for human use. The chimera-based vaccines contain new genetic molecules, a fusion of elements of dengue and yellow fever viruses. In creating these chimera molecules, however, scientists are giving them a 4.4 billion-year head start on evolution, a process that simultaneously spurs apocalyptic Hollywood fabulist fears, scientific concerns about adequate testing and safeguards, and both legal and ethical concerns about the creation, ownership, and use of life.

As with the use of genetically modified seeds, nanotechnologies, and innovative gene therapies, the use of biotechnology to create chimera-based vaccines will undoubtedly provoke controversies. In the particular case of the pioneering vaccine for dengue, the world will be asked to balance the benefits and risks of a biotechnology-generated chimeric vaccine.

Biotechnology: In Context also tackles the contentious byproducts of biotechnology advances. Even under optimal circumstances, for example, vaccine production is often limited, requiring decisions about who receives limited supplies of vaccine. In the case of a potential dengue vaccine, will children under five years old living in areas most at risk receive the vaccine at a cost that is not prohibitive, or will travelers from wealthy countries willing to pay a high price for protection consume a disproportionate amount of the vaccine?

As with other *In Context* books, the goal of the authors and editors is not to provide definitive answers to these and other complex questions, but rather to provide students and readers the basic knowledge and diverse insights that allow them to critically and meaningfully engage issues in ways that lead to informed conclusions.

With microorganisms we coexist. "They outnumber us. We must use our wits against their genes."

—Nobel Laureate Joshua Lederberg

K. Lee Lerner & Brenda Wilmoth Lerner, Senior Editors

CAMBRIDGE, MASSACHUSETTS, DECEMBER 2011

The Lerner & Lerner / LernerMedia Global portfolio includes award-winning books, media, and film that bring global perspectives to science related issues. Since 1996, they have contributed to more than 60 academic books and served as editors-in-chief for more than 30 books related to science and society. Their book *Infectious Diseases: In Context* was designated an ALA Outstanding Academic title and their book, *Climate Change: In Context*, both published by Cengage Gale, was named an ALA RUSA Outstanding Reference Source for 2009.

About the *In Context* Series

Written by a global array of experts, yet aimed primarily at high school students and an interested general readership, the *In Context* series serves as an authoritative reference guide to essential concepts of science, the impacts of recent changes in scientific consensus, and the influence of science on social, political, and legal issues.

Cross curricular in nature, *In Context* books align with and support national science standards and high school science curriculums across subjects in science and the humanities and facilitate science understanding important to higher achievement in science testing. The inclusion of original essays written by leading experts and primary source documents serve the requirements of an increasing number of high school and international baccalaureate programs and are designed to provide additional insights on leading social issues, as well as spur critical thinking about the profound cultural connections of science.

In Context books also give special coverage to the impact of science on daily life, commerce, travel, and the future of industrialized and impoverished nations.

Each book in the series features entries with extensively developed words-to-know sections designed to facilitate understanding and increase both reading retention and the ability of students to advance reading in context without being overwhelmed by scientific terminology.

Entries are further designed to include standardized subheads that are specifically designed to present information related to the main focus of the book. Entries also include a listing of further resources (books, periodicals, Web sites, audio and visual media) and references to related entries.

Each *In Context* title has approximately 300 topic-related images that visually enrich the content. Each *In Context* title will also contain topic-specific timelines (a chronology of major events), a topic-specific glossary, a bibliography, and an index especially prepared to coordinate with the volume topic.

About This Book

The goal of *Biotechnology: In Context* is to help high school and early college students understand the essential facts and deeper cultural connections of topics and issues related to the scientific study of biotechnology.

In an attempt to enrich the reader's understanding of the mutually impacting relationship between science and culture, as space allows we have included primary sources that enhance the content of *In Context* entries. In keeping with the philosophy that much of the benefit from using primary sources derives from the reader's own process of inquiry, the contextual material introducing each primary source provides an unobtrusive introduction and springboard to critical thought.

General Structure

Biotechnology: In Context is a collection of 236 entries that provide insight into increasingly important and urgent topics associated with the study of biotechnology.

The articles in the book are meant to be understandable by anyone with a curiosity about topics related to biotechnology, and the first edition of *Biotechnology: In Context* has been designed with ready reference in mind.

- Entries are arranged alphabetically, rather than by chronology or scientific subfield.

- The **chronology** (timeline) includes significant events in the history of biotechnology. Where appropriate, related scientific advances are included to offer additional context.

- An extensive glossary section provides readers with a ready reference for content-related terminology. In addition to defining terms within entries, specific Words-to-Know sidebars are placed within each entry.

- A bibliography section (citations of books, periodicals, websites, and audio and visual material) offers additional resources to those resources cited within each entry.

- A **comprehensive general index** guides the reader to topics and persons mentioned in the book.

Entry Structure

In Context entries are designed so that readers may navigate entries with ease. Toward that goal, entries are divided into easy-to-access sections:

- **Introduction**: A opening section designed to clearly identify the topic.

- **Words-to-Know** sidebar: Essential terms that enhance readability and critical understanding of entry content.

- Established yet flexible **rubrics** customize content presentation and identify each section, enabling the reader to navigate entries with ease. The goal of *In Context* entries is a consistent, content-appropriate, and easy-to-follow presentation. Accordingly, inside *Biotechnology: In Context*, most entries contain internal discussions of **Historical Background and Scientific Foundations**, followed by a section articulating **Impacts and Issues**.

- **Bibliography:** Citations of books, periodicals, web sites, and audio and visual material used in preparation of the entry or that provide a stepping stone to further study.

- **"See also" references** clearly identify other content-related entries.

Biotechnology: In Context special style notes

Please note the following with regard to topics and entries included in *Biotechnology: In Context*:

- Primary source selection and the composition of sidebars are not attributed to authors of signed entries to which the sidebars may be associated. In all cases, the sources for sidebars containing external content are clearly indicated.

- The Centers for Disease Control and Prevention (CDC) includes parasitic diseases with infectious diseases, and the editors have adopted this scheme.

- Equations are, of course, often the most accurate and preferred language of science, and are essential to epidemiologists and medical statisticians. To better serve the intended audience of *Biotechnology: In Context*, however, the editors attempted to minimize the inclusion of equations in favor of describing the elegance of thought or the essential results that such equations yield.

- A detailed understanding of biology and chemistry is neither assumed nor required for *Biotechnology: In Context*. Accordingly, students and other readers should not be intimidated or deterred by the occasional complex names of chemical molecules or biological classification. Where necessary, sufficient information regarding chemical structure or species classification is provided. If desired, more information can easily be obtained from any basic chemistry or biology reference.

Bibliography citation format

In Context titles adopt the following citation format:

Books

Davis, Dena S. *Genetic Dilemmas: Reproductive Technology, Parental Choices, and Children's Futures*, 2nd ed. Oxford and New York: Oxford University Press, 2010.

Periodicals

Liras, A. "Future Research and Therapeutic Applications of Human Stem Cells: General, Regulatory, and Bioethical Aspects." *Journal of Translational Medicine* 8 (December 2010): 131–145.

Web Sites

"Svalbard Global Seed Vault." *The Global Crop Diversity Trust*. http://www.croptrust.org/main/arcticseedvault.php?itemid=842 (accessed November 26, 2011).

Using Primary Sources

The definition of what constitutes a primary source is often the subject of scholarly debate and interpretation. Although primary sources come from a wide spectrum of resources, they are united by the fact that they individually provide insight into the historical *milieu* (context and environment) during which they were produced. Primary sources include materials such as newspaper articles, press dispatches, autobiographies, essays, letters, diaries, speeches, song lyrics, posters, works of art—and in the twenty-first century, Web logs—that offer direct, first-hand insight or witness to events of their day.

Categories of primary sources include:

- Documents containing firsthand accounts of historic events by witnesses and participants. This category includes diary or journal entries, letters, email, newspaper articles, interviews, memoirs, and testimony in legal proceedings.

- Documents or works representing the official views of both government leaders and leaders of other organizations. These include primary sources such as policy statements, speeches, interviews, press releases, government reports, and legislation.

- Works of art, including (but certainly not limited to) photographs, poems, and songs, including advertisements and reviews of those works that help establish an understanding of the cultural environment with regard to attitudes and perceptions of events.

- Secondary sources. In some cases, secondary sources or tertiary sources may be treated as primary sources. For example, if an entry written many years after an event, or to summarize an event, includes quotes, recollections, or retrospectives (accounts of the past) written by participants in the earlier event, the source can be considered a primary source.

Analysis of primary sources

The primary material collected in this volume is not intended to provide a comprehensive or balanced overview of a topic or event. Rather, the primary sources are intended to generate interest and lay a foundation for further inquiry and study.

In order to properly analyze a primary source, readers should remain skeptical and develop probing questions about the source. Using historical documents requires that readers analyze them carefully and extract specific information. However, readers must also read "beyond the text" to garner larger clues about the social impact of the primary source.

In addition to providing information about their topics, primary sources may also supply a wealth of insight into their creator's viewpoint. For example, when reading a news article about an outbreak of disease, consider whether the reporter's words also indicate something about his or her origin, bias (an irrational disposition in favor of someone or something), prejudices (an irrational disposition against someone or something), or intended audience.

Students should remember that primary sources often contain information later proven to be false, or contain viewpoints and terms unacceptable to future generations. It is important to view the primary source within the historical and social context existing at its creation. If, for example, a newspaper article is written within hours or days of an event, later developments may reveal some assertions in the original article as false or misleading.

Test new conclusions and ideas

Whatever opinion or working hypothesis the reader forms, it is critical that they then test that hypothesis against other facts and sources related to the incident. For example, it might be wrong to conclude that factual mistakes are deliberate unless evidence can be produced of a pattern and practice of such mistakes with an intent to promote a false idea.

The difference between sound reasoning and preposterous conspiracy theories (or the birth of urban legends) lies in the willingness to test new ideas against other sources, rather than rest on one piece of evidence such as a single primary source that may contain errors. Sound reasoning requires that arguments and assertions guard against argument fallacies that utilize the following:

- false dilemmas (only two choices are given when in fact there are three or more options);

- arguments from ignorance (*argumentum ad ignorantiam*; because something is not known to be true, it is assumed to be false);

- possibilist fallacies (a favorite among conspiracy theorists who attempt to demonstrate that a factual statement is true or false by establishing the possibility of its truth or falsity. An argument where "it could be" is usually followed by an unearned "therefore, it is.");

- slippery slope arguments or fallacies (a series of increasingly dramatic consequences is drawn from an initial fact or idea);

- begging the question (the truth of the conclusion is assumed by the premises);

- straw man arguments (the arguer mischaracterizes an argument or theory and then attacks the merits of their own false representations);

- appeals to pity or force (the argument attempts to persuade people to agree by sympathy or force);

- prejudicial language (values or moral goodness, good and bad, are attached to certain arguments or facts);

- personal attacks (*ad hominem*; an attack on a person's character or circumstances);

- anecdotal or testimonial evidence (stories that are unsupported by impartial observation or data that is not reproducible);

- *post hoc* (after the fact) fallacies (because one thing follows another, it is held to cause the other);

- the fallacy of the appeal to authority (the argument rests upon the credentials of a person, not the evidence).

Despite the fact that some primary sources can contain false information or lead readers to false conclusions based on the facts presented, they remain an invaluable resource regarding past events. Primary sources allow readers and researchers to come as close as possible to understanding the perceptions and context of events and thus to more fully appreciate how and why misconceptions occur.

Primary Source Acknowledgments

The editors wish to thank the copyright holders of the excerpts included in this volume and the permissions managers of many book and magazine publishing companies for assisting us in securing reproduction rights. We are also grateful to the staffs of the Detroit Public Library, the Library of Congress, the University of Detroit Mercy Library, Wayne State University Purdy/Kresge Library Complex, and the University of Michigan Libraries for making their resources available to us. The following is a list of the copyright holders who have granted us permission to reproduce material in *Biotechnology: In Context*. Every effort has been made to trace copyright, but if omissions have been made, please let us know.

COPYRIGHTED EXCERPTS IN *BIOTECHNOLOGY: IN CONTEXT* WERE REPRODUCED FROM THE FOLLOWING PERIODICALS:

Rio Declaration on Environment and Development, June 14, 1992. Copyright © 1992 by United Nations Environment Program. Reprinted with the permission of the publisher. **—*Cartagena Protocol on Biosafety to the Convention on Biological Diversity***, 2006. Copyright © 2006 by United Nations Publications. Reproduced by permission of the publisher. **—CORDIS** for "EU-Funded Study Yields Disease-Resistant Crops," http://cordis.europa.eu/, © European Union, 2010. **—*Genetic Bill of Rights***, 2000. Copyright © 2000 Council for Responsible Genetics. Reproduced by permission of the publisher. **—*Associated Press***, January 1, 2005, for "Seed Company Cracks Down on Recycling Farmers" by Paul Elias; January 14, 2010, for "Technology Aims to Replace Animal Testing" by Matthew Perrone. Copyright © 2005, 2011 by Press Association, Inc. Reproduced by permission of the publisher. **—*President George W. Bush, State of the Union Address***, January 31, 2006. Copyright © 2006 by U.S. Government Printing Office. Reproduced by permission of the publisher. **—*Introductory Remarks of Senator Edward Kennedy (D-MA) on the Genetic Information Nondiscrimination Act***, January 22, 2007. Copyright © 2007 by Genetic Alliance. Reproduced by permission of the publisher. **—*Remarks of President Barack Obama — As Prepared for Delivery Signing of Stem Cell Executive Order and Scientific Integrity Presidential Memorandum***, March 9, 2009. Copyright © 2009 by U.S. Government Printing Office. Reproduced by permission of the publisher. **—*The Age***, April 10, 2006, for "Frost Tolerant Crops in Reach" by Martin Daly. Copyright © Fairfax Media, 2006. Reproduced by permission of the author. **—*Baltimore Sun***, August 19, 2006, for "Baltimore Firm Wins FDA Nod for Mix of Beneficial Viruses" by Jonathan D. Rockoff and Hanah Cho. Copyright © 2006 by *Baltimore Sun*. Reproduced by permission of the publisher. **—*Cornell Chronicle***, July 25, 2006, for "Seven-year Glitch: Cornell Warns that Chinese GM cotton Farmers Are Losing Money Due to 'Secondary' Pests"; May 1, 2007, for "Student Designer and

"Slick Solution: How Microbes Will Clean up the Deepwater Horizon Oil Spill" by David Biello. Copyright © 2010. All rights reserved. —*Svalbard Global Seed Vault: Securing the Future of Agriculture*, February 26, 2008. Copyright © 2008 by Global Crop Diversity Trust. Reproduced by permission of the publisher. —*The Telegraph*, January 31, 2009, for "Extinct Ibex Is Resurrected by Cloning" by Richard Gray. Copyright © 2009 by *The Daily Telegraph*. Reproduced by permission of the publisher. —*Convention on Biological Diversity*, 1992. Copyright © 1992 by United Nations. Reproduced by permission of the publisher. —*Trials of War Criminals before the Nuremberg Military Tribunals under Control Council Law*, February 1949. Copyright © 1949 by U.S. Government Printing Office. Reproduced by permission of the publisher. —*The Convention on the Prohibition of the Development, Production and Stockpiling of Bacteriological (Biological) and Toxin Weapons and on their Destruction*, 1972. Copyright © 1972 by United Nations. Reproduced by permission of the publisher. —*Washington Post*, April 13, 2008, for "Building Baby From the Genes Up" by Ronald M. Green. Copyright © *Washington Post*, 2008. Reprinted with permission of the author.

COPYRIGHTED EXCERPTS IN *BIOTECHNOLOGY: IN CONTEXT* WERE REPRODUCED FROM THE FOLLOWING BOOKS:

—Fleming, Alexander. From *Nobel Lectures, Physiology or Medicine 1942-1962*. Elsevier Publishing Company, 1964. © The Nobel Foundation 1945. Reproduced by permission of the Nobel Foundation.—Lourie, Alan. From *Federal Reporter, Third Series (F.3d)* . West-Thomson, 2010. Reproduced by permission of Thomson Reuters.—Micklos, David A., and Greg A. Freyer. From *DNA Science: A First Course*. Cold Spring Harbor Laboratory Press, 2003. Copyright © 2003, Cold Spring Harbor Laboratory Press. Reproduced by permission of the publisher.—Scalia, Antonin. From "Eli Lilly & Co. v. Medtronic, Inc." in *United States Reports (U.S.)*. U.S. Government Printing Office, 1990. Copyright © 1990 by U.S. Government Printing Office. Reproduced by permission of the publisher.—Scalia, Antonin. From "Medimmune, Inc. v. Genentech, Inc.," in *United States Reports*. 2007. U.S. Government Printing Office, 2007. Copyright © 2007 by U.S. Government Printing Office. Reproduced by permission of the publisher.

Glossary

A

ACETYLSALICYLIC ACID (ASA): The modern pharmaceutical derived from salicylic acid, the chemical found in the bark of willow trees, a substance known to possess pain relief and anti-inflammatory properties since antiquity. Aspirin, the most famous of the ASA-related medications, was first patented in 1900 by the German pharmaceutical company Bayer.

ACT: A statute, rule, or formal lawmaking document enacted by a legislative body or issued by a government.

ADJUVANT THERAPY: The administration of cancer drugs after surgery has removed all visible cancer. The aim is to prevent recurrence by killing off any invisible cancer cells that may remain after surgery and otherwise go on to form a new tumor.

ADMISSIBILITY: Refers to evidence, such as forensic evidence, that is accepted by a court of law.

ADRENAL CORTEX: The part of the adrenal glands (located on each kidney) that synthesizes and secretes cholesterol derivatives, including corticosteroids.

ADULT STEM CELL: Adult stem cells are stem cells found throughout the body that are capable of differentiating into a limited number of cell types. Examples include brain, bone marrow, blood vessels, gut, liver, ovary, testis, skin, teeth, and heart.

ADVANCED BIOFUELS (SECOND GENERATION BIOFUELS): Liquid fuels manufactured from sustainable sources that do not impact the crops otherwise available for human food or feedstock. These fuels include alcohols such as ethanol and butanol converted from biomass.

AEROBIC: Any process or organism that requires the presence of oxygen. An organism requiring oxygen to function is described as aerobic. The term may also be applied to any process taking place in a bioreactor that requires oxygen.

AGAROSE: A specialized chemical used in gel electrophoresis and made from agar, which is extracted from a type of seaweed called red marine algae. Although it is said to be non-toxic, it is best to wear gloves and use appropriate ventilation equipment when handling agarose. Agarose comes in powdered form for laboratory use; it is mixed with buffer solution, heated until it melts, then poured and allowed to set into a gel. The higher the concentration of agarose insulation, the firmer the gel will be.

AGE-RELATED INFERTILITY: Female mammals are born with their lifetime supply of eggs, no more are manufactured after birth. Over time, the viability of healthy eggs decreases, with a dramatic decline in human fertility after the age of 35.

AGREEMENT: An arrangement among parties to do something or not do something in order to achieve a certain goal. It also refers to the document in which the agreement is recorded. It also may be called a contract, protocol, or treaty.

AGREEMENT ON TRADE-RELATED ASPECTS OF IN-TELLECTUAL PROPERTY RIGHTS (TRIPS): Administered by the World Trade Organization (WTO), TRIPS is the 1994 international treaty negotiated to serve as a comprehensive international intellectual property rights (IPR) regime. TRIPS was created to advance the liberalized trade philosophy of the WTO; it has been criticized by numerous indigenous peoples' rights groups as a structure that fails to account for the importance of specific legal protections for traditional resource rights.

AGRIBUSINESS: A large commercial farm operation, often encompassing development, production, marketing, and branding of food or other farmed products.

AGRICULTURAL PATENTS: The legal protection sought by developers of biotechnological applications such as genetically modified (GM) seeds.

AGRONOMIC: Refers to the cultivation characteristics of a plant.

ALBUMIN: A protein produced by the liver that assists in the movement of small molecules through the bloodstream. Albumin is essential in the prevention of blood fluid leaking from its vessels into adjacent tissues.

ALCANIVORAX BORKUMENSIS: The oil-consuming microorganism identified in bioremediation research conducted during the 2010 Gulf oil spill. The microbe has a set of genes that apparently allow it to reduce oil alkanes to a usable food source.

ALLELE: The expression or form of a particular gene.

ALLELOMORPH (ALLELE): The generic term that describes the type of genes found in a biological sample; a common example is the dominant and recessive alleles that contribute to variable human characteristics such as hair color.

ALLOGENEIC: In cell therapy, allogeneic refers to cells taken from a donor, cultured in a lab, and then applied to a patient as therapy.

ALLOGRAFT: An organ or tissue transplant in which the recipient of the transplanted material is the same species as the donor but their genetic structures are different. Allografts may involve living related persons, unrelated persons, or cadaveric sources.

AMERICAN ASSOCIATION OF TISSUE BANKS (AATB): The AATB is the leading North American tissue bank organization. Founded in 1976, it has more than 100 national tissue bank members. The AATB is active in the accreditation of tissue banks; its "Standards for Tissue Banking" is recognized as the definitive global standard. It is also active in the provision of public education services concerning the importance of tissue

and organ donations. The European Association of Tissue Banks (EATB), founded in 1991, is similarly organized and structured.

AMINE GROUP: A chemical formation composed of one nitrogen atom to which two hydrogen atoms are attached. All amino acids have an amine group.

AMINO ACIDS: Simple small peptides (protein units) that serve as the building blocks of more complex proteins. Amino acids also play essential regulatory roles in physiological systems such as basal metabolism, cell functioning, brain chemistry, and functional regulation.

AMMONIUM NITRATE: A fertilizer with high nitrogen content produced through the combination of ammonia and nitric acid. Ammonia nitrate is also a powerful explosive; its sale and distribution are restricted in many countries due to its potential use by terrorists.

AMYOTROPHIC LATERAL SCLEROSIS (ALS): Also referred to as Lou Gehrig's Disease in the United States, ALS is a progressive neurological disease that results in the degeneration of nerve cells in the brain and spinal cord, leading to muscle weakness and atrophy, paralysis, and death.

ANAEROBIC: A process or organism not dependent on the presence of oxygen.

ANAEROBIC DIGESTION: The chemical process in which microorganisms break down organic materials in the absence of oxygen. Biogas (the combination of methane and carbon dioxide) is the usual by-product.

ANAEROBIC DIGESTION OF ORGANIC WASTE: Processes used to speed up the decomposition of organic waste materials without adding oxygen to the system.

ANAEROBIC MICROORGANISMS: Bacteria that cannot grow in the presence of oxygen (obligatory anaerobes) or those that do not require oxygen to produce energy (facultative anaerobes).

ANALGESICS: The group of medications that is associated with the phrase "painkillers"; the term has its roots in the Greek words for "without pain."

ANALYTE: A substance to be measured or analyzed.

ANAPHYLACTIC SHOCK: Violent reaction by the immune system against an allergenic substance, leading to serious systemic complications and even death.

ANEMIA: Deficient levels of red blood cells (erythrocytes), causing poor transport of oxygen to cells in tissues and organs.

ANGIOGENESIS: The development of new blood vessels from existing blood vessels.

ANODE: The fuel cell electrode where oxidation occurs, resulting in a loss of electrons in the particles involved in the chemical reaction. All fuel cells have an anode as the negative terminal.

ANTHRAX: An infectious disease caused by contact with *Bacillus anthracis* bacteria. Anthrax is a particularly feared biological weapon because it can survive for years in harsh environments, and it is transmitted by spores that may be airborne, present on various surfaces that promote human contact, or through ingestion of contaminated foods.

ANTHROPOCENTRIC: Human-centered or from a human source.

ANTIBIOTIC: A substance, often produced by microorganisms, that can kill or inhibit the growth of bacteria or fungi.

ANTIBIOTIC RESISTANCE: This occurs when bacteria and other microbes become immune to the effects of antibiotics that previously killed them or halted their proliferation.

ANTIBODY: A protein molecule produced by white blood cells in response to the presence of a foreign protein or antigen. An antibody binds to the antigen in a very specific way, with each antibody binding its own antigen. Antibodies are used widely in science and medicine because of this very specific binding response, which enables them to target molecules of interest for analysis.

ANTIGEN: Any protein or protein fragment or chemical compound that elicits a response from the immune system.

ANTIPYRETIC DRUGS: Medications that relieve fever.

ANTISENSE: A way of blocking gene expression by exposing messenger RNA to its complementary, or antisense, sequence so it cannot carry out its usual function.

APOPTOSIS: A programmed series of events leading to the death of a cell. Often known as cell suicide, it is often faulty in cancer when damaged cells continue to survive instead of dying, but can be triggered by interfering with the cancer cell's functioning. Apoptosis is distinct from necrosis, a type of cell death caused by oxygen deprivation or other injury.

AQUACULTURE: Commercial breeding of fish.

ARABIDOPSIS: *Arabidopsis thaliana* is a plant that is widely used by research biologists as the model organism in many frost-resistant crop experiments. *Arabidopsis* is a member of the Brassicaceae plant family, and it is genetically similar to radish and cabbage plants. The genetic structure of *Arabidopsis* is well understood. This plant provides researchers with numerous investigative avenues to advance the understanding of plant genetics.

ARRAY: Probes arranged in a pattern on a solid support, typically in a rectangle or square.

ARTIFICIAL INSEMINATION BY DONOR (AID): Also known as donor insemination, this is the insertion of donor sperm into the recipient's vagina to allow fertilization.

ASEXUAL REPRODUCTION: The ability to produce offspring in the absence of a mate. Because there is no exchange of genetic material from two parents, asexual reproduction results in low genetic variation in the specific species as a whole.

ASSISTED REPRODUCTION: Assisted reproduction refers to a number of techniques, including in vitro fertilization, which are used to help achieve fertilization of an egg with a sperm in cases of infertility.

ASSISTED REPRODUCTIVE TECHNOLOGIES (ART): A group of technologies, such as in vitro fertilization, which substitute one or more steps of natural procreation to establish a pregnancy, sometimes using donor gametes.

ATOM ECONOMY: A measure of how much starting material in a chemical reaction is turned into the desired product. Wasteful processes, with by-products, have low atom economy, whereas efficient processes, with high product yield, have high atom economy.

AUTOGENUOUS BONE GRAFT: A bone sample used for substitution that is taken from the patient's own bones.

AUTOGRAFT: Tissue transferred from one part of the body to another. Common examples include skin grafts to treat burns and blood vessel grafts used in cardiac bypass surgery. Also, transplants involving organs or tissue between identical twins.

AUTOLOGOUS: Obtained from the same individual. In an autologous blood transfusion, blood from the recipient is removed and stored, then transfused back into the original donor.

B

***BACILLUS THURINGIENSIS* (Bt):** A naturally occurring bacterium used as an insecticide; also introduced into transgenic crop species by plant geneticists to increase defenses against insects.

BASE PAIR: Two chemical bases bound together by chemical bonds to form a rung of the DNA double helix. The amino acid pairs in DNA are adenine, cytosine, guanine, or thymine. The ways in which the bases are paired identifies and defines the organism.

BENCH TO BEDSIDE: The process of integrating research and practical patient application from start to finish through a specific research study's life cycle.

BIG PHARMA: A collective term for the largest international pharmaceutical companies, especially when they are seen to function as an socially and politically influential business group.

BIO-FILTER: Any of a variety of filters used in waste gas and wastewater treatment technologies in which microorganisms are employed to counteract the effect of pollutants through their biological degradation.

BIO-HACKER: An amateur scientist who experiments with genetic material as a hobby, sometimes referred to as a garage biologist. Not to be confused with the term bioterrorist.

BIO-REPORTERS: Use of a whole organism as a biosensor, able to detect a target substance in the environment and indicate its detection.

BIOACCUMULATION: The toxic buildup of chemicals within the body. Bioaccumulation can occur through the inhalation or ingestion of toxins; it may also result from direct contact between tissues and contaminated water, soil, or materials.

BIOACTIVITY: Bioactivity is the term that describes the specific effect observed when living animal or plant tissue is exposed to a particular substance.

BIOASSAY: A test to appraise the strength or activity of a biological substance by comparing it with an accepted standard.

BIOAUGMENTATION: The addition of specifically chosen, cultured, prepared microorganisms, often bacteria or fungi, to the environment in order to enhance bioremediation of contaminated or polluted areas.

BIOCATALYST: A substance that initiates or accelerates the rate of a biochemical reaction. Most biocatalysts are enzymes, and the two terms are sometimes used interchangeably, but hormones may also be considered as biocatalysts.

BIOCHEMISTRY: Short for biological chemistry; the discipline devoted to investigating the chemical compounds and processes that occur in living organisms.

BIOCOMPATIBLE: A material that does not have a toxic, or otherwise injurious, effect on biological tissue, particularly in the context of the human body.

BIOCOMPATIBLE MATERIALS: Natural or synthetic materials that can be used safely as an interface with living tissues.

BIODEGRADABLE: A material that uses biological processes or natural weathering to degrade rapidly with minimal environmental impact.

BIODEGRADATION: Microbial process of breaking down complex molecules and digesting both organic and inorganic matter through fermentation.

BIODETECTOR: An analytical or sensor device designed to detect either a single analyte (a chemical substance to be analyzed) or multiple discrete analytes. It contains physiochemical and biological detector components.

BIODIESEL: Liquid fuel produced from refined animal fats and vegetable oils.

BIODIVERSITY: In general, biodiversity refers to all of the different types of organisms found in a particular geographic area or ecosystem. Specifically, it can refer also to all varieties in current existence for a specific type of organism; for example, all strains and varieties of soybeans currently being cultivated versus the total number two or three centuries ago.

BIOETHERS: Ethers produced from biomass sources that are used as additives in conventional fuels such as gasoline to enhance engine performance and reduce carbon monoxide emissions. Methyl tert-butyl ether (MTBE) is a commonly used bioether.

BIOFORTIFICATION: The supplementation of a biological material with a compound or compounds that enhance health.

BIOFUELS: A form of fuel created from biomass. Biofuels are produced through the use of renewable resources. The three most common types of biofuel are bioethanol as a replacement for traditional fossil fuels, biodiesel as a "green" replacement for diesel fuel, and biogas used instead of natural gas.

BIOGAS: When organic matter is degraded without the presence of oxygen, a common by-product is biogas, which is combustible and can be used as a biofuel. It contains methane, carbon dioxide, and hydrogen sulfide and can be burned to produce electricity or heat.

BIOGENERICS: Compounds that have the same structure and activity as the originally-developed compounds and can be licensed for sale when the period of patent protection of the original compound has expired.

BIOHAZARD: In the field of biological materials and research, any agent or situation that constitutes a threat to health or environment.

BIOINFORMATICS: The National Institutes for Health Biomedical Information Science and

Technology Initiative Consortium defines bioinformatics as research, development, or application of computational tools and approaches for expanding the use of biological, medical, behavioral, or health data, including those to acquire, store, organize, archive, analyze, or visualize such data.

BIOLOGICAL DATABASE: A compilation of data gathered from laboratory or literature research, experimental outcomes, clinical trials, biotechnology, and computer analyses in the biological sciences.

BIOLOGICAL DETERGENT: A detergent containing enzymes including proteases, amylases, and lipases, which break down protein, carbohydrate, and fat stains respectively.

BIOLOGICAL RECOGNITION SYSTEM: Also called a biodetector or biosensor, this is an instrument that recognizes, records, analyzes, and transmits data in response to specific biochemical, chemical, or physical changes recorded by the system.

BIOLOGICS: Drugs that are made in living cells and include recombinant proteins, vaccines, and monoclonal antibodies.

BIOLUMINESCENT: Light produced by living organisms due to physiological processes.

BIOMARKER: A gene, a protein, or some other measurable characteristic linked to a disease process or response to a drug.

BIOMASS: The biodegradable component found in all agricultural products, their waste and residues, vegetation, forest products, and industrial and municipal waste.

BIOME: An ecosystem containing indigenous fauna and flora integrated into a local food chain, such as lakes, marshes, or deserts.

BIOMECHATRONIC: Incorporating biology, electronics, and mechanics.

BIOMECHATRONICS: A multidisciplinary field that combines biotechnology, electronics, mechanics, biology, neurosciences, and robotics. Biomechatronic developers make devices and prostheses that interface directly with the body, connecting the machine to nerves, muscles, and bone to facilitate improved control and enhance natural movement lost through trauma, disease that necessitates amputation, or genetic malfunction resulting in missing body parts.

BIOMETRICS: Derived from the Greek words *bio* (life) and *metrics* (measure), the science associated with technologies employed to measure and analyze personal human characteristics, including physiological and behavioral traits. Finger, facial, and retinal scans are the best known of these technologies widely used to authenticate identity.

BIOMIMETIC: The study and application of functions and structures found in nature to engineering and technology.

BIONANOSENSORS: The term that describes analytical devices used in biology constructed from biomolecular sources and nanomaterials. A rapidly emerging nanotechnology field is the development of sensors that provide immediate feedback concerning specific chemical levels in the body to prompt immediate adjustments (e.g., glucose levels in a diabetic).

BIOPHARMACEUTICAL: A drug that is manufactured in a living organism, including a cell culture, plant, or animal.

BIOPSY: A term that refers to several techniques that are used to obtain tissue samples.

BIOREMEDIATION: The use of living organisms to degrade noxious compounds that have been accidently released into water or soil.

BIOSAFETY: Applicable to all facilities in which biological hazards are present, biosafety covers the protocols and ongoing process of hazard and risk identification, assessment, and prevention.

BIOSENSOR: A sensor composed of two parts: a biological element and a human-made transducer.

BIOSIMILAR: Also known as a follow-on biologic, a biosimilar is a new version of a biopharmaceutical drug for which the patent has expired. The word similar is used because the version is unlikely to be an exact copy of the original biological molecule. Biotechnology production processes are more variable than chemical production processes. An exact copy can be made of a small molecule drug, such as aspirin, which is known as a generic version.

BIOSTIMULATION: Adding specific, targeted nutrients to a contaminated soil area in order to support and reinforce the activity of naturally occurring, already-present microorganisms in an effort to enable them to break down and digest the waste, converting it to harmless byproducts. No additional microorganisms are added to the soil.

BIOTECHNOLOGY: The integration of technology, engineering, and the life sciences.

BIOTECHNOLOGY INDUSTRY ORGANIZATION: Also known as BIO, this leading international trade organization contains more than 1,110 biotechnology firms, academic institutions, and non-profit organizations.

BIOTERRORISM: The use of microorganisms, toxic agents, or biological diseases as an act of terrorism. Bioterrorism acts fall into three categories: Category A acts pose a large threat to national security due to rapid transmission or dissemination, severity of human/animal symptoms and high mortality rate; Category B acts pose a moderate security risk due to relative ease of dissemination and transmission, leading to moderate symptoms and low mortality rates; and Category C acts involve developing or emerging pathogens capable of causing extremely severe mass effects due to their ease of dissemination and gravity of effects.

BIOTERRORISM PREPAREDNESS AND RESPONSE ACT (2002): United States legislation enacted in wake of the 2001 terrorist attacks. The Act established regulations to govern the security of biological materials that are capable of use by bioterrorists.

BIOTIC STRESS: When plants are negatively affected by the presence of pathogens such as fungi, bacteria, viruses, insects, or weeds, they are said to be impacted by biotic stress.

BIOTRANSFORMATION: Chemical conversion of one substance into another by the action of enzymes, either isolated or within a cell.

BIOTURBATION: A process by which soil is enriched, aerated, and physically restructured by organisms living within it. Those organisms can be mobile, such as ants, worms, and other insects, or relatively stationary plants and other fauna.

BIOVENTING: A bioremediation process used to remove contaminants from groundwater by increasing oxygen or nutrient levels to encourage the contaminant-removing action of naturally occurring mircroorganisms.

BIOWASTE: The vegetal and animal waste resulting from industry, food production and household use, called biowaste, produces greenhouse gases, primarily in the form of methane, and it is toxic to the environment.

"BLACK CARBON": Black carbon, or soot, is particulate air pollution produced from biomass fuel combustion. When released into the atmosphere, these particles reflect sunlight otherwise absorbed by Earth's surface and thereby contribute to increased atmospheric temperatures. For this reason, black carbon is closely linked to global warming and climate change.

BLASTOCYST: The hollow, cellular structure formed in early embryonic development. Its inner cell mass is the basis of the embryo and subsequent tissue development. An outer cell layer is the basis for placenta formation.

BLOOD PATTERN ANALYSIS: Also known as "blood spatter" analysis, the branch of forensic science devoted to the relationship between the patterns left by blood at a crime scene and type of injury sustained or the manner in which a weapon might have been used to produce the particular pattern.

BOLAR PROVISION: A statutory or regulatory provision that exempts the research and development required for generic drug approval from patent infringement during the life of the underlying patent. Bolar provisions take their name from the U.S. court case, *Roche Products, Inc. v. Bolar Pharmaceutical Co.* and have become widely adopted in patent laws around the world.

BONE MARROW: Tissue found inside bones that, in humans, is the site of production of blood cells.

BOOLEAN LOGIC: Also called Boolean algebra; a form of logical calculus derived by George Boole (1779–1848) in the mid-1800s. Boolean logic has two values, used as zero and one in the world of computers.

BOTOX: The trade name used for botulinum toxin type A formulations, one of seven forms of the toxin isolated. Allergan, Inc. is one of several pharmaceutical companies that use this trade name; in all commercial preparations, Botox is a purified form of the toxic bacterium *Clostridium botulinum* commonly found in various soils.

BOVINE SOMATOTROPIN (bST): A natural hormone, the synthesized version of which is used regularly by commercial North American dairy operations to stimulate greater milk production in cows.

Bt: Short for the soil bacterium *Bacillus thuringiensis*, which produces a toxin, Cry1Ab, that kills many insects, including the European corn borer.

BUILDER: A water-softening substance that improves the efficiency of synthetic surfactants by removing excess calcium ions from hard water.

C

C-VALUE: The total amount of DNA in an organism, measured in pictograms.

CADAVERIC MUSCULOSKELETAL TISSUE: The umbrella term that describes any of the muscles, tendons, ligaments, bones, fibers, and all associated tissues necessary for human movement and structure recovered from a deceased person for the purpose of future allographs.

CANCER: Diseases that share the characteristic of uncontrolled cell growth and division.

CAPSID: The repeating unit making up the protein coat that surrounds the genetic material of many viruses. The capsid structures may fit together in icosahedral symmetry to form a nearly spherical protein coat or in helical symmetry to form a helix structure.

CARBOHYDRATES: Sugar molecules found abundantly in nature and used as a source of energy by most organisms. Carbohydrates are also structural features of plants (cellulose fibers) and unicellular microorganisms' walls. Examples of carbohydrates include glucose, ribose, maltose, erythrose, cellulose, sucrose, and glyceraldehyde, among many others.

CARBON CYCLE: The biogeochemical exchanges that ultimately result in the rotation of all existing carbon atoms (such as the carbon dioxide / oxygen exchange that occurs during photosynthesis in terrestrial plants and marine biota). The cycle includes carbon movement between the biosphere, atmosphere, oceans, and geosphere.

CARBON SEQUESTRATION: The intentional removal of carbon from the atmosphere by various geoengineering methods, including the separation of carbon dioxide from industrial flue gases for permanent underground storage. Carbon sequestration alters the function of the carbon cycle, including the biogeochemical exchanges that involve carbon atoms, such as the carbon dioxide / oxygen exchange in photosynthesis that occurs in terrestrial plants and marine biota.

CARBOXYL GROUP: A chemical formation composed of one carbon, one hydrogen, and two oxygen atoms. All amino acids have a carboxyl group.

CASH CROP: A crop grown for sale and profit rather than for use for food or other personal use by the grower.

CATHETER: A hollow, flexible tube inserted into a blood vessel or body cavity to allow passage of fluids in and out of the body.

CATHODE: The fuel cell electrode in which reduction occurs and the particles gain electrons. In a fuel cell, the cathode is the positive terminal.

CBRN ATTACKS: The term used by national security agencies to describe potential terrorist attacks based on chemical, biological, radiological, or nuclear components.

CELL CULTURE: The growth of cells outside the body from an initial small sample of cells.

CELL LINE: A cell line consists of continuing generations of cells that originated from a single cell type.

CELL LINES: Identified cell cultures capable of infinite cell division. In the United States in 2001, research projects on embryonic stem cells that were made eligible for U.S. federal government funding were limited to specified embryonic stem cells lines identified through prior research. Parts of this restriction were lifted in 2009 by a presidential executive order.

CELL MEMBRANE: The outer boundary between a cell and its contents. The cell membrane is semipermeable and composed of protein and lipid molecules. The contents of a microinjection enter the cell through its membrane.

CELL SEEDING: When groups of cells of a specific type are placed on a scaffold for growth or repopulation.

CELLULOSIC BIOMASS: Biomass, including grass and trees, the main carbohydrate component of which is cellulose rather than starch.

CERAVITAL: The registered trademark for the first Bioglass material that was developed for clinical applications in 1973. Ceravital remains the most commonly used Bioglass material in middle ear reconstructive surgery.

CHIMERA: A substance created through genetic engineering using genes or proteins of two or more species.

CHIRAL: A carbon atom in an organic molecule that is attached to four different atoms or groups of atoms is said to be chiral. A molecule with a chiral carbon atom exists in space as two non-superimposable mirror images that may have different biological properties. Enzymes, amino acids, and other biological molecules are chiral.

CIRRHOSIS OF THE LIVER: A group of liver conditions in which liver cells are damaged and replaced by scar tissue, thereby reducing the amount of healthy, functional liver.

CISGENIC ORGANISM: A genetically modified organism that contains genetic sequences from the same, or closely related, species that have been spliced together in a different order.

CISGENICS: A molecular biology technique in which genes associated with a particular trait, such as disease resistance, are identified in an organism. Those genes are reinforced and transferred into other members of the same species with the expectation that the trait will be evinced in the parent plants and magnified in successive generations.

CIVIL LIBERTIES: The fundamental rights of the individual, such as freedom of speech, guaranteed under the law against unwarranted infringement by the government or others.

CLINICAL TRIAL: A research study using human participants that gathers information on the safety and efficacy of a drug or medical procedure.

CLONE: An entire cell, portion, or organism that is genetically identical to the unit or individual from which it was derived.

CLONING: Growing an individual organism from a single cell and producing a genetically identical copy of the organism from which the cell was removed. This has been done successfully with animals. Human cloning is currently illegal.

CLONING VECTOR: The selected DNA molecule that will replicate within a living cell during the recombinant process. Plasmids are a source used for these molecules.

COBALAMIN: The B vitamin (vitamin B12) that is involved in metabolic functions, DNA synthesis and repair, and brain and nervous system functions.

COCHRANE COLLABORATION: An international network and independent non-profit organization that includes physicians, researchers, statisticians, policy makers, and others who focus on systematic reviews of clinical research to evaluate best practices in medicine that are based on evidence and objective standards.

CODEX ALIMENTARIUS: According to the World Health Organization, the Codex Alimentarius was created in 1963 for the purpose of developing food standards, guidelines, and food-related codes of practice. The primary purpose of the Codex is to protect the health and safety of consumers as well as to ensure that global fair trade practices are maintained throughout the food industry.

CODOMINANT TRAITS: Inherited genetic characteristics that have the same degree of dominance.

CODON: A sequence of three of the following four nucleotides: adenine, guanine, cytosine, and thymine, present in a strand of DNA or RNA. Codons encode the genetic information to make a specific amino acid which will be incorporated into a protein chain or serve as a termination signal. Codons are also known as triplets.

COGENERATION: Fuel cell systems generally convert to electricity between 40 to 50 percent of the chemical energy in the hydrogen fuel source. Gasoline powered engines perform at approximately 20 percent efficiency, and conventional electrical battery powered systems achieve 65 percent efficiency.

Cogeneration is the process used to harness system thermal output to improve overall efficiency of the fuel cell system; energy efficiency rates approach 80 percent when the fuel cell systems incorporates thermal cogeneration.

COLLAGEN: A fibrous protein found in tendon, bone, cartilage, skin, and other connective tissue.

COMBUSTION: The chemical destruction of a compound in the presence of oxygen, which produces energy.

COMPANION DIAGNOSTIC: A test given before a drug, including a cancer drug, is prescribed to ascertain, by measurement of a marker molecule such as surface antigen, whether the patient will respond.

COMPARABLES: Worth range of similar or equivalent products sold in the market, used to calculate the value of a new product.

COMPOSITE MATERIAL: A material composed of two or more substances that remain distinct from one another. A low-tech example is straw combined with mud to create bricks (adobe). The resulting composite material exhibits properties different from that of its individual components.

COMPOST: The mixture resulting from combining organic materials such as food scraps, manure, and lawn clippings with wood or other bulk materials in order to speed up decomposition. When the materials have thoroughly mixed and degraded, the compost is typically dark brown and clumpy and smells much like topsoil. It is used to fertilize plants or to cover seedlings.

COMPOSTABLE: Organic and engineered wastes able to be composted; that is, able to be broken down by the aerobic biological various microorganisms.

CONFOCAL MICROSCOPY: Confocal microscopy enables the creation of sharp images from a specimen by excluding much of the light that would otherwise illuminate the microscope slide. The image has superior contrast and shows fine detail. Images from confocal microscopy can be used to build an overall three-dimensional image.

CONNECTIVE TISSUE: Fibrous tissue that forms a network to support organs, joints, muscle, and body tissue. Collagen is a type of connective tissue found in many parts of the human body.

CONSTANT GAZE: Persons who are being surveilled in a panopticon believe they are under the "constant gaze" of an unknown watcher. The constant gaze may be real or it may be an illusion, but the watched does not know whether the watcher is present; sometimes known as "unequal gaze."

CONTAINMENT BOOMS: The physical barriers placed around oil spills that occur on a body of water. At the height of the Gulf of Mexico cleanup operations undertaken in response to the *Deepwater Horizon* disaster, more than 345 miles (550 km) of containment booms were deployed by emergency crews. Containment booms are distinct from the sorbent booms designed to absorb oil directly from the water surface.

CONTROL GROUP: A group of subjects in a medical experiment that receive a placebo or proven intervention so that the response of the control group may be measured against the group receiving experimental treatment.

CONVENTION ON BIOLOGICAL DIVERSITY (CBD): A 1993 international treaty enacted to promote biological diversity, sustainability principles, and the equitable sharing of all benefits derived from global genetic resources.

CORD BLOOD BANKING: The storage of umbilical cord blood so an infant's stem cells can be used in treatment at some point in the future.

COSMECEUTICAL: A cosmetic said to contain biologically active ingredients and to have drug-like properties. Cosmeceuticals are therefore part cosmetic and part pharmaceutical. The term cosmeceutical is used within the cosmetics industry but is not recognized by the U.S. Food and Drug Administration (FDA).

COSMID: A circular piece of DNA (plasmid) that contains DNA sequences from a type of virus called a lambda phage and that is useful in building genome libraries.

COX-2 INHIBITORS: Two cyclo-oxygenase (COX) enzyme systems first identified in the early 1990s that assist in the production of prostanoids that are essential to the anti-inflammation response; also known as non-steroidal anti-inflammatory drugs or NSAIDs.

CROSS POLLINATION: Cross pollination occurs when plants of two different varieties are bred together. The offspring will have a gene pool that combines characteristics of both varieties, creating a new strain. Cross pollination can occur intentionally or through pollination by wind or insects.

CRY PROTEIN: An abbreviation of crystal proteins, such as those produced by the *Bacillus thuringiensis* bacteria, that are activated by an insect's digestive enzymes.

CRYOBIOLOGY: The study of how different components in living organisms react when exposed to low temperatures. The contemporary study of frost tolerance is primarily directed to the cell and protein structures of the crop plant species that make the greatest contribution to global food supplies. These species include wheat and other grain crops.

CRYOGENIC: Relating to freezing and storage of an entity at temperatures far below zero. Cryogenics are typically used for longer-term storage of substances that are fragile or unstable at traditional freezing temperatures.

CRYONICIST: Cryonicists are those who believe cryonic preservation leading to reanimation and the restoration of health is a future certainty based on the current pace of bio- and nanotechnical advances. They actively support and may plan to participate in the cryonic preservation process. Per the January 1983 issue of *Cryonics*, cryonicists believe that individuals in cryostatic preservation are neither alive nor dead, but in a state of "uncertain waiting."

CRYOPRESERVATION: A technique of preserving cells and tissues, by freezing them rapidly, usually in liquid nitrogen, and storing them at a temperature of $-212°F$ ($-100°$ C), or below.

CRYOSTAT: Technically, this can be any device used to maintain cryostatic temperatures around $-320°F$ ($-196°C$). In the context of cryonics, the term refers to a container for storing cryopreserved individuals, organs, or DNA in liquid nitrogen. It is also called a dewar. Individuals are stored head down in cryostats.

CULTIVAR: A plant variety produced by selective breeding. A variety itself is a subdivision of a species, differing from other varieties of the species in minor ways.

CURED FOODS: The general term used to describe any food preservation process that involves salt, sugar, or nitrates either alone or in combination. High quantities of these preservatives have been linked to various health problems, with high blood pressure attributed to excessive sodium intake a prominent example.

CYANOBACTERIA: The phylum of single cell aquatic and photosynthetic bacteria commonly known as blue-green algae. Cyanobacteria tend to thrive in warm, stagnant water that is rich in nitrogen and phosphates.

CYTOPLASM: All the protoplasm in a living cell that is located outside of the nucleus, as distinguished from nucleoplasm, which is the protoplasm in the nucleus.

CYTOTOXIC: Refers to a drug that is directly toxic to a cell, the action of which often leads to the death of the cell.

D

DATA MINING: The process of utilizing statistical methods and relevant algorithms to search for comparisons, correlations, and patterns in large databases.

DEEP VEIN THROMBOSIS (DVT): The sometimes fatal condition associated with thrombophlebitis, in which blood clots form in a major vein, usually in the lower legs of the patient.

DEFORESTATION: The permanent destruction of indigenous forests; the removal of forests that otherwise absorb carbon dioxide (CO_2) from the atmosphere contributes to increased CO_2 levels associated with the "greenhouse effect" prominent in all modern discussions concerning climate change.

DEMATERIALIZED BONE MATRIX (DBM): Glycoprotein that comprises the transforming growth factor family (TGF-β) aids in the development of human tissue and organs. DBM is a commercially available compound that is used to heal problematic fractures. DBM does not contribute to increased structural support in the damaged bone.

DENATURATION: The unfolding of the three-dimensional structure of a biological molecule, such as a protein, on exposure to increased temperature or other unfavorable change in condition. Denaturation leads to loss of biological function.

DENDRITIC CELLS: Part of the immune defense with the function of capturing compounds, proteins, or any other foreign substance and presenting it to T cells or B cells for recognition. If the substance is alien, an immune response is prompted.

DEOXYRIBONUCLEIC ACID (DNA): DNA is a polymeric molecule whose basic units, the nucleotides, are composed of a phosphate backbone, a sugar bridging group, and a nitrogenous base. DNA makes up genes, and the informational content of DNA is carried in the sequence of bases in the DNA molecule.

DERMIS: The inner layer of the skin, it is comprised primarily of water, collagen, elastin, and blood vessels.

DESICCATION: The removal of water from a living organism.

DESIGNER BABY: A baby whose genetic material has been altered, either by adding or removing specific genes. The egg and sperm are combined through in vitro fertilization and the fertilized embryo undergoes germ line genetic manipulation. The genes are altered to make certain that unwanted characteristics are eliminated and desired ones are present. This primarily is done to select healthy embryos free of genetic diseases. Occasionally, this has been used to ensure that the resultant individual has healthy stem cells compatible with an older sibling with a life-threatening genetic condition that potentially could be cured via an infusion of cord blood/stem cells from a healthy sibling.

DESIZING: A sizing material is added to a textile to protect its fibers during weaving. Desizing is the removal of this material from the textile after the weaving process.

DEW POINT: The temperature at which water vapor will condense into liquid form. The condensation is dew; the dew point varies in accordance with the prevailing barometric pressure.

DIABETES: In type 1 diabetes, which a person is either born with or develops during childhood, the islet cells in the pancreas fail to produce insulin. In type 2 diabetes, which tends to start at a later age, cells become resistant to insulin. In both types, failure of control of blood glucose occurs, which may lead to long-term complications such as heart or kidney disease.

DIALYSIS: The separation of particles in a liquid on the basis of differences in their ability to pass through a membrane. This clinical purification of blood thus is a substitute for the normal function of the kidney.

DICOTYLEDON: Tomato, potato, and pepper are examples of dicotyledons, whose seeds contain two cotyledons (the first leaves of seed plants that develop from the embryo).

DIETARY GUIDELINES FOR AMERICANS: Evidence-based guidance compiled by the U.S. Department of Agriculture, most recently in 2010, to promote health and reduce the risk of chronic disease, overweight, and obesity.

DIFFERENTIATION: The process by which a stem cell turns into a more specialized cell such as a blood cell or skin cell.

DIGESTIBLE: Substances that can be broken down by enzymes and gastric acids into simpler molecules. Complex multicellular organisms digest food in their stomachs in order to be able to absorb the simpler molecules into the blood stream.

DILUENT: A diluting agent.

DIPLOID CELL: A cell that contains two sets of chromosomes. In the fertilized ovum, one set of chromosomes comes from each parent.

DISCOUNT RATE: An arbitrary value assumed to reflect all the risks involved in research and development of a new product. Such risk value is then discounted from the estimated final product worth.

DISPERSANTS: The general term to describe all chemical agents used to remove spilled oil from a water body surface. As the name suggests, dispersants are one of the "first response" options available to combat oil spill effects, because these agents act to transfer accumulated spilled surface oil into the water subsurface. In the subsurface, the oil is diluted, and the speed of natural degradation is increased.

DISTRICT ATTORNEY'S OFFICE V. OSBORNE: The 2009 United States Supreme Court decision that states are not required to turn over DNA evidence for convicts to use for post-conviction DNA testing.

DNA: An abbreviation for deoxyribonucleic acid, the genetic material that is the basis of life for most organisms. (See also deoxyribonucleic acid.)

DNA DATABASE: A computerized database that stores DNA profiles of individuals, typically criminals.

DNA FINGERPRINTING: Often used in forensics or paternity testing, DNA fingerprinting involves sequencing an individual's DNA and comparing it to evidentiary DNA. It is also called DNA typing.

DNA PROBE: A short length of DNA tagged with a fluorescent or radioactive atom that binds to a complementary sequence in the genome.

DNA PROFILING: The omnibus term used to describe the forensic processes that determine identity based on the match between a biological sample and a known DNA source.

DNA TESTING: The use of gathered DNA from human tissue (hair, semen, saliva, etc.) to test for confirmed identity or statistical probability of a match between the two or more individuals.

DNA/TISSUE CRYOPRESERVATION: The storage of DNA or tissue samples in special containers suspended in liquid nitrogen, with the intent of future reconstruction.

DOUBLE HELIX: A three-dimensional spiral structure containing two matched linear strands held together in opposite directions by chemical bonds.

DOWNSTREAM PROCESSING: All the process steps involving isolation and purification of a biotechnology product.

DRAGLINE: The silk line used by a spider to construct the scaffolding of its web. Also known as the lifeline used for suspension, it has the highest tensile strength of any silk.

DRUG PRICE COMPETITION AND PATENT TERM RESTORATION ACT: Also known as the Hatch-Waxman Act, in 1984 this legislation amended the Federal Food, Drug, and Cosmetic Act of 1938 by adding laws related to the approval and regulation of generic drugs.

E

E-COMMERCE: Buying, selling, banking, or conducting business or service using electronic means, most often through the Internet.

EBOLA: An extremely infectious virus with a high death rate. Research on Ebola must be carried out in the most stringent biological containment facilities (BSL-4).

ECOSYSTEM: A biological community consisting of a group of interacting organisms and their habitat, or living environment, to include (but not be limited to) animals, plants, insects, trees, and water sources.

EFFLUENT: Liquid waste created as a by-product of organic solid waste treatment processes.

ELECTROMAGNETISM: The branch of physics that deals with electricity and magnetism, and the interrelation between the two.

ELECTRON BEAM LITHOGRAPHY: The nanofabrication process used to imprint fine patterns (widths often smaller than 10 nanometers) essential to the function of modern integrated circuits. The patterns are created when electrons from a scanning microscope are accelerated through long chain polymers (resists) into a silicon surface. The secondary electrons generated in the process destroy the polymer chain bonds to produce the pattern on the circuit surface.

ELECTRON MICROSCOPY: A type of microscopy in which a beam of electrons is used instead of visible light, allowing much higher magnification.

ELECTROPHORESIS: A technique used to separate various molecules, such as proteins (RNA or DNA segments), according to size and molecular weight by moving them across a field, usually a gel, carrying an electrical charge.

EMBRYO: An organism in the early stages of development. For the human embryo, this refers to the period within the first eight weeks.

EMBRYOID BODY: An aggregate of embryonic stem cells that tends to form spontaneously.

EMULSIFIER: A substance added to an emulsion to stop it from separating into oil and water phases.

EMULSION: A mixture consisting of droplets of oil in water or droplets of water in oil.

ENDANGERED SPECIES: A species at risk of extinction throughout all or most of its habitat. There are two risks with endangered species: first, that the species will disappear completely from the planet;

second, that the surviving gene pool will be weakened or threatened by the low numbers of organisms available for breeding.

ENDOSPERM: Plant tissue; the endosperm of all grains and flowering plants contains starch.

ENGINEERING: The discipline of applying principles found in mathematics and physics to the design, production, and operation of various devices, machines, systems, and processes.

ENVIRONMENTAL DEGRADATION: The process by which the health and integrity of an environment declines as a result of human action.

ENZYME: A protein that allows a chemical reaction to proceed more easily than it otherwise would and is not altered by the reaction.

ENZYMOLOGY: The study of enzymes, their different types and functions in biological systems, and applications in therapeutics or in industrial chemical reactions.

EPIDERMIS: The upper layer of skin composed of epithelial tissue containing pigmented cells called keratinocytes.

ERADICATION: The worldwide reduction in the prevalence of a specific disease to zero.

ESCHERICHIA COLI: From the family Enterobacteriaceae, genus *Escherichia*, and species coli, *E. coli* is the widely recognized short form for the bacterium that is an abundant microorganism in the lower intestines of most mammals, including humans. First isolated in 1885, *E. coli* is the most commonly used bacterium as the medium for the production of recombinant proteins.

ETHICS: As compared and contrasted with morality, the set of rules, principles, or intellectual approaches to reasoning and problem solving common to an identified group. Morality is generally regarded as a concept rooted in the distinction between "right" or "wrong" when a specific action, attitude, or intention is evaluated.

EUGENICS: A social theory, popular in the early decades of the twentieth century, that sought to remove so-called socially undesirable hereditary traits from the population through institutionalization, segregation, and limits on childbearing, including sterilization, of people who were members of specific races or had certain medical conditions.

EUKARYOTE: A cell in which genetic material is contained within a specialized nuclear membrane.

EUROPEAN MEDICINES AGENCY (EMA): The organization that regulates drug approvals in the European Union. The EMA tries to work with the U.S. Food and Drug Administration (FDA) to streamline regulatory processes for pharmaceutical companies, but they are separate organizations with different legislative frameworks.

EUTHANASIA: The painless killing of a patient suffering from an incurable disease or condition.

EUTROPHICATION: The process that leads to the depletion of dissolved oxygen in bodies of water (hypoxia). Excessive nutrient levels in water promote the growth of algae, reduce water quality, and eliminate habitats for aquatic species that require colder, oxygen-rich waters. Fertilizer runoff is a primary contributor to eutrophication.

EVIDENCE-BASED MEDICINE (EBM): By combining a physician's clinical expertise, patient experience and values, and the latest and best research evidence for a particular condition or set of symptoms, EBM works to produce optimal outcomes for clinical practice.

EXXON VALDEZ: The name of an oil tanker that ruptured in Prince William Sound, Alaska, in March 1989, spilling upwards of 750,000 barrels of oil. Part of the cleanup involved the use of a genetically-modified bacterium that was the first patented living organism.

F

FABRY DISEASE: A lysosomal storage disease caused by deficiency of the enzyme alpha-galactosidase A, which is essential to breakdown of molecules known as glycosphingolipids. Without the enzyme, these molecules accumulate in the nerves, kidneys, and heart.

FASCICLE: A bundle or cluster; in human anatomy, the term is used to describe how nerve fibers (axons) or skeletal muscle tissue is arranged.

FEDERAL FOOD, DRUG, AND COSMETIC ACT: The Food, Drug, and Cosmetic Act (FFDCA or FD&C Act) is a set of laws passed by Congress in 1938 and amended numerous times that grants the U.S. Food and Drug Administration (FDA) the authority to regulate food, drugs, cosmetics, medical devices, food additives, dietary supplements, and related products.

FEEDER CELLS: Mouse cells added in a layer to a human stem cell culture, thereby providing a sticky layer to which the human cells can attach to grow. The feeder cells also secrete various nutrients into the growth medium.

FERMENTATION: The process by which yeast turns the sugar in fruits and grains into alcohol and carbon. The term is often used more generally to describe any microbial biotransformation taking place in a bioreactor.

FIBROBLAST: A connective tissue cell, found in the structural parts of the body such as bone, skin, and cartilage, and also in between other tissues, playing a supportive role.

FIBROSIS: The condition caused when fibrous connective tissue accumulates on the surface of an organ or tissue. Fibrosis may result from the surgical repair of an injured area; it is commonly observed at the point of amputation in a limb. It also occurs when the tissues react to an implanted object such as a pacemaker or sensors.

FLAVIVIRUS: Positive-sense single-strand RNA viruses transmitted by insects such as mosquitoes and ticks.

FLOCCULATION: Colloids are materials evenly dispersed in a fluid. Flocculation is the process that removes colloids from their suspended state to form floc, the clumps or flakes of material that can be readily removed from the fluid.

FLUE GAS: The generic term that describes the exhaust gas released into the atmosphere when combustible materials are burned. The flue is the pipe, chimney, or vent through which exhaust gas is released. Typical flue gas components are the greenhouse gases carbon dioxide, nitrogen oxides, and carbon monoxide, as well as particulates.

FLUORESCENCE: The emission of radiation, typically in the wavelength spectrum of visible light, of one wavelength by a molecule following the absorption of light of a different wavelength and excitation of the absorbing molecule.

FLUORESCENCE MICROSCOPY: In fluorescence microscopy, a fluorescent tag molecule is attached to an antibody that targets the cell structure or molecule to be imaged. The sample is then exposed to light, which excites the fluorescent tag molecule so that it emits light of a longer wavelength that is detected by the microscope, creating an image of the part of the cell where the antibody is attached.

FLUX: The flow of energy through a particular system or electronic medium. In metabolic engineering, flux is defined as the rate of turnover of molecules in a metabolic pathway. Enzymes regulate the rate of flux.

FOLATE: Folate, the naturally occurring form of folic acid, is a B vitamin (vitamin B9) that is essential to numerous bodily functions, including the synthesis and repair of DNA and cell division and growth.

FOLLICULAR ASPIRATION: The method by which eggs are removed from the ovaries by a fine needle.

FOOD INSECURITY: The phrase used to describe any set of conditions in which a human population is threatened by hunger or starvation. Climatic conditions such as drought and frost can reduce harvests, as can insect infestations and crop diseases. Human conditions such as war or political instability in a region will also often contribute to food insecurity.

FORENSICS: The application of science-based techniques and tests to the investigation of crime scenes.

FOSSIL FUEL: Hydrocarbons (molecules containing only hydrogen and carbon atoms) derived from ancient organisms that typically lived millions of years ago. Petroleum, coal, and natural gas are all examples of fossil fuels.

FROST PROTECTION: Many crops are protected from frost by artificial means; the objective is to maintain the crop ambient air temperature above 32°F (0°C). In some instances, wind turbines are used to redirect warm air from the atmospheric inversion layer to the crop. Water sprinkler systems are also employed to moisten the soil surface. This procedure tends to prevent the surface temperature of the crop and soil from falling below freezing.

FUEL CELL STACK: The stack is composed of individual fuel cells that are arranged in circuits to increase voltage.

FUNCTIONAL FOODS: Foods that are enhanced by an additive or in which the quantity of a present bioactive compound is increased to enhance specific physiological benefits.

FUNCTIONAL GENOMICS: The branch of genomics concerned with the function of genes.

FUSARIUM: A common fungal disease that destroys wheat prior to the wheat reaching maturity. *Fusarium* is a specialized strain of mycotoxins that not only impair the healthy development of wheat, they also pose a significant threat to human and animal health when introduced into the food chain.

FUSION PROTEIN: A type of biologic drug created from genes coding for two different proteins. This is usually done to improve the property of the protein produced. In etanercept, for instance, one protein binds to TNF while the other makes the drug easier to deliver in the body.

G

GAIT: The style or pattern of motion a person displays while walking.

GAMETE: Sperm or egg cell.

GASIFICATION: The conversion of a carbon-based solid (e.g., wood) into carbon monoxide, hydrogen, carbon dioxide, and methane gases. These products generally are known as syngas, used as a fuel or in the manufacture of chemicals such as the propellant dimethyl ether.

GASTROENTERITIS: Stomach and intestinal infection caused by bacteria or virus present in contaminated food, soil, or water.

GENBANK: GenBank is a database containing real-time data on all publicly available nucleotide sequences, their protein translations, and all relevant bibliographic data. It is updated daily.

GENE: Each DNA molecule contains many genes, the basic physical and functional units of heredity. A gene is a specific sequence of nucleotide bases, whose sequences carry the information required for constructing proteins that, in turn, provide the structural components of cells and tissues as well as enzymes for essential biochemical reactions.

GENE EXPRESSION PROFILE: Genes vary in their level of activity, or expression, and the pattern of expression of a group of genes is known as a gene expression profile. Different tissues and disease states have different gene expression profiles.

GENE GUN: A device that shoots tungsten or gold particles coated with DNA into plant cell tissue in genetic engineering experiments.

GENE MAPPING: Creation of a genetic map by identifying and locating the genes in an organism's chromosomes or plasmid.

GENE PROBE: Each location, often known as a spot, on a microarray, has a fragment of a specific gene, known as a gene probe, chemically attached to it.

GENE TRANSFER: Gene transfer occurs when a DNA sequence is transferred from one organism to another, unrelated organism. This may involve two members of the same family or organism, or transfer between different types of organisms.

GENERIC DRUG: A product that is pharmacologically similar to a brand name drug, including similar safety and efficacy, which is released following the expiration of the patent on the name brand drug.

GENETIC ENGINEERING: When recombinant DNA or other biotechnology techniques are used to intentionally alter the genetic material of an organism in order to introduce a new trait or characteristic that can then be transmitted to future generations.

GENETIC MODIFICATION: A term often used interchangeably with genetic engineering, which refers to transfer of a gene from one organism to another by experimentation rather than natural means.

GENETIC POLYMORPHISMS: Enzyme variations attributed to different amino acid sequences.

GENETICALLY MODIFIED (GM) FOODS: GM foods include crop plants in which the original genetic traits are altered to improve a specific crop feature, such as drought resistance, herbicide tolerance, enhanced nutritional capacity, or protections against insects.

GENETICALLY MODIFIED ORGANISM (GMO): Also called genetically engineered organism (GEO). An entity in which the genetic make-up has been permanently altered by the insertion of a gene or genes from a different variety of the same species or a completely different species through the use of genetic engineering. Such modifications are likely to be heritable, i.e., they are passed on to offspring. Modification is usually in order to produce a more disease-resistant, higher yield version of the organism.

GENETICS: The biological science of heredity and genes that often explores the similarities and differences of organisms.

GENOME: The complete set of genes containing the genetic instructions for making an organism. An organism's genome contains the master blueprint for all cellular structures and activities for the lifetime of the cell or organism. All human cells, with the exception of red blood cells, contain the complete human genome. The simplest bacterial organisms have genomes containing about 600,000 base pairs; the human and mouse genomes each contain more than three billion base pairs.

GENOMICS: The study of the complete DNA sequencing of organisms, aimed at gathering data about the structure and function of genes, as well as the mapping of genomes.

GENOTYPE: The physical expression of the genetic makeup of an individual. An individual's genetic identity is physically and biochemically expressed through his or her entire set of genes, called the genotype.

GERM LINE: The lineage of cells resulting in eggs and sperm.

GERM LINE GENE THERAPY: Modification of genetic material at the pre-implantation level. This process involves making genetic changes in fertilized embryos by injecting new genes into an embryonic cellular nucleus. The embryo, with the new genetic material, is implanted into the female host with the hope that a healthy, full-term pregnancy will ensue and result in a live birth of a genetically enhanced/altered individual.

GERM THEORY OF DISEASE: This is the theory that bacteria and microorganisms cause disease and that the spread of disease can be at least partially mitigated by hand washing and maintaining basic hygiene and environmental cleanliness.

GERMPLASM: Refers to germ cells containing heritable resources for a single organism. It incorporates the genetic material needed for propagation or reproduction.

GESTATIONAL CARRIER: A woman who completes pregnancy and gives birth to a baby for another woman. A gestational carrier is implanted with a zygote created from the ovum of another woman, not her own.

GLOBAL CROP DIVERSITY TRUST: A partnership between the private and public sectors, dedicated to the assurance of funding for important food crop genetic preservation in perpetuity. The Trust seeks to create and guide efficient and effective sustainable conservation methods in order to provide a means for preserving and protecting diverse food crop genetic material.

GLOBAL WARMING: The warming of Earth's atmosphere that is occurring due to the accumulating of compounds that increase the retention of heat.

GLYCOSAMINOGLYCAN: A polysaccharide found in connective tissue.

GOLDEN RICE: A GM rice that contains genes that make vitamin A, thereby increasing its nutritional value.

GRAM NEGATIVE: Gram's stain is used to distinguish between different bacterial species. The stain is not retained by Gram negative bacteria because they have a thin permeable cell wall from which the purple stain is readily washed.

GRAM POSITIVE: A Gram positive bacterium retains the purple Gram stain in its relatively thick cell wall.

GREENHOUSE GAS: Any gas in Earth's atmosphere that contributes to the greenhouse effect, wherein solar radiation (such as visible light) largely passes through such gases, which block infrared radiation from Earth or the atmosphere from escaping into space.

GREENPEACE: A leading organization campaigning on environmental issues.

H

HAPLOTYPE MAP: Individuals with common SNP alleles may have the same variations in alleles at other sites, creating what researchers call a "haplotype map" that drives the research into finding patterns of genetic variation.

HARD WATER: Water that contains a higher-than-average amount of the minerals calcium and magnesium. Clothes washed in hard water have duller or grayer appearing colors and tend to feel rough and somewhat dense or hard.

HEAT TRANSFER EFFICIENCY: This expression is used to describe the efficiency of any process in which heat is generated to achieve a desired result. Expressed as a ratio, efficiency is defined as the amount of energy available for use versus the amount of energy expended in the process.

HEMATOPOIESIS: Blood formation within the body.

HEMATOPOIETIC STEM CELL: Hematopoietic stem cells are multipotent stem cells, originating in the bone marrow, which give rise to the whole range of blood cells.

HEMOPHILIA: A recessive hereditary disorder affecting only males, which causes a bleeding disorder associated with deficiencies of clotting factors.

HEMOSTATIS: The process in which blood loss is reduced and ultimately stopped through the combined action of a platelet and fibrin, a fibrous protein. Fibrin creates a nanoparticle-sized mesh that forms over the wound where blood is lost.

HEPATITIS C: A chronic liver infection with the hepatitis C virus that can lead to cirrhosis. Hepatitis C has become the leading cause for liver transplantation in the United States.

HERBICIDE: Chemical or biological products used in agriculture to control weeds in crops.

HERD IMMUNITY: Also referred to as community immunity, this occurs when the disease-resistant portion of a community provides a degree of protection for susceptible individuals.

HERITABILITY: The capability of a genetically determined characteristic, such as eye color, to be passed from generation to generation.

HIV: The human immunodeficiency virus (HIV) causes acquired immune deficiency syndrome (AIDS) by attacking immune cells so the body is vulnerable to infections and cancers.

HOLOCENE EXTINCTIONS: The widespread animal species extinctions observed in the last 11,500 years since the end of the last Ice Age. This period is known as the Holocene epoch. Human activity is the primary cause of many of the extinctions recorded during this period, as forests and range land that supported many species have been converted to human use.

HOMOCYSTEINE: An amino acid. High levels of homocystein have been linked to cardiovascular disease and indicate vitamin B deficiencies.

HORMONE: A protein that produces a certain physiological response.

HOST: The cell that receives the added genetic material.

HUMAN EMBRYONIC STEM CELL: A human embryonic stem cell is one source of stem cells. They are found, as the name suggests, in the human embryo and have the capacity to develop into all the cells and tissues of the human body.

HUMAN GENOME PROJECT (HGP): The international genomics initiative was established in 1990 with two primary objectives. The project sought to identify all of the genes present in human DNA. Once identified, the HGP assembled a map of the entire genome and the sequences of the estimated three billion DNA base pairs. The HGP was completed in 2003 with these objectives achieved.

HUMECTANT: A substance that helps water retention in the skin by attracting water molecules to itself.

HURDLES (COMBINATION PROCESSING): The use of different food preservation methods in combination or sequences. The harmful microorganisms present in food that are not destroyed by a single preservation method will not withstand the combined effects of several methods. A typical hurdle may include pasteurization, refrigeration, and the use of preservatives.

HYBRID SEED: Seed that results from the cross pollination of plants.

HYBRIDIZATION: The reproductive crossing of two organisms with different genotypes. Hybridization can involve two different varieties of a species, or it can result from the crossing of two different species.

HYBRIDS: Animals or plants bearing genes of two different species.

HYDROPHOBIC: Water insoluble; oil is an example of a substance with a chemical composition that repels water.

HYGROZYME: A proprietary enzyme technology developed for the United States Postal Service after 2001 to detect anthrax in the wake of mail terror attacks directed at various U.S. Senate members.

HYPERACCUMULATOR: A plant species that takes up a pollutant such as a heavy metal to a concentration level that would be toxic to a non-accumulator.

HYPOXIA: Reduction of oxygen levels to below those required to sustain animal life.

I

IATROGENIC: From the Greek *iatros*, meaning physician; and genic, meaning origin. Iatrogenic diseases are transmitted by physicians and other health professionals.

IMMUNITY: The ability by the host's immune system to recognize and kill invading microorganisms that cause infectious, contagious diseases.

IMMUNIZATION: Immunization or vaccination is the method of exposing the host's immune system to a dead or attenuated virus or bacterium or to a specific antigen in order to prompt recognition and attack by the host's immune cells against those infectious agents.

IMMUNOASSAY: Laboratorial assays that use proteins or biological molecules to cause specific responses from cells of the immune system.

IMMUNOGENIC: Any foreign substance that can cause an immune reaction, inducing inflammation or allergies.

IMMUNOLOGICAL TOLERANCE: The extent to which the immune system can ignore a threat such as an invading pathogen, cancer cells, or organ transplant. Prior to the full development of the immune system, a fetus has immunological tolerance. Antirejection drugs induce a state of immunological tolerance in organ transplant recipients.

IMMUNOSUPPRESSION: Reduced or absent normal immune response to disease. Immune deficiency disease and treatment with anticancer drugs or antirejection drugs are major causes of immunosuppression.

IN NATURA: In its natural state, not processed by cooking, smoking, or other methods of alteration.

IN VITRO FERTILIZATION (IVF): An assisted reproduction technology process involving the fertilization of ova by sperm outside of the body in a laboratory setting.

INBREEDING: When two closely related individuals, whether plant or animal, reproduce. As part of a selective breeding program, this is often continued over many generations.

INDIGENOUS PEOPLES: Two international law definitions are often associated with this term: (1) tribal peoples whose status within a specific national community is determined with reference to its customs or traditions; (2) any people who are descended from a specific indigenous group that claim indigenous status through self-identification.

INDUCED PLURIPOTENT CELL (iPSCs): Adult stem cells that have been reprogrammed to enter an embryonic stem cell-like state by being forced to express specific genes that have this effect.

INDUSTRIAL BIOTECHNOLOGY: The use of bio-products (such as enzymes) and biomass feedstock in industrial products that are normally made from petroleum-based products such as fuel, polymers, pet-rochemicals, bulk chemicals, and specialty chemicals.

INFORMED CONSENT: The permission granted by the subject of a medical procedure or clinical trial after the individual has been apprised of all potential benefits and risks.

INHALATION: Breathing in of particles, including microorganisms.

INNOCENCE PROJECT: A nonprofit organization devoted to pushing for laws and policies to permit post-conviction DNA testing for convicted prisoners.

INSECTICIDAL TOXINS: Naturally produced or artificially synthesized substances that are lethal to insects. They are commonly used as active ingredients in agricultural and household pesticides.

INSULIN: A peptide hormone with 51 amino acids in all, arranged in two chains. Insulin regulates the level of glucose in the blood, storing any excess produced after a meal in the liver as glycogen.

INTELLECTUAL PROPERTY: A creative or engineered idea, product, or work that has commercial value. It can also refer to the rights and legal instruments that protect such property, such as patents or copyrights, which are also valuable commodities.

INTELLECTUAL PROPERTY PATENT: Proprietary protection of a process, product, plan, or method of doing something that is novel and useful, usually granted by a government. When creators want to protect their rights to maintain ownership or proprietorship of a product or idea, they protect it with a patent for a designated number of years.

INTELLECTUAL PROPERTY RIGHTS (IPR): The legal protections extended to creators of new innovations. IPR includes copyright, patents, trademarks, some forms of trade secrets, and industrial designs. Stem cell IPR has often involved efforts to secure patent protection.

INTERSPECIES NUCLEAR TRANSFER: Nuclear transfer experiments involving donor cells and egg cells from different species, such as human skin cells and cow eggs.

INTERVENTIONAL STUDY: A type of research protocol in which individuals are assigned to groups in which they will receive some sort of treatment or clinical (or other) intervention.

INTRA-CYTOPLASMIC SPERM INJECTION (ICSI): An assisted reproductive process in which sperm are injected directly into an egg in order to facilitate fertilization as part of an IVF process.

INVASIVE PROCEDURE: Any medical diagnostic or therapeutic technique that requires entry into the body or interruption of normal body functions to perform. Invasive procedures are typically more painful and require greater patient recovery time than noninvasive procedures.

IONIC LIQUID: An ionic compound that is liquid below 212°F (100°C), the boiling point of water. Conventional ionic compounds have much higher melting points. Many ionic liquids are powerful solvents, without the toxic properties of organic solvents, and so play a role in green chemistry.

ISCHEMIA: The death of organ or tissue cells due to lack of oxygen; living organs must be transplanted shortly after their removal from the donor for this reason.

ISCHEMIA AND REPERFUSION DAMAGE: Refers to the potential for damage done to cells when deprived of oxygen and nutrients. Reperfusion damage occurs when blood is recirculated after a prolonged period of ischemia. The combination of ischemia and reperfusion can cause permanent, severe damage to cells and organs.

K

KERATINOCYTE: A skin cell found in the outer layers of the epidermis that helps form a tough protective layer after it migrates up from the basal layer where it is formed. Keratinocytes produce a protein called keratin.

KNOCK-IN MOUSE: A genetically-modified mouse that has a gene with a specific mutation inserted into its genome in order to study its function.

KNOCKOUT MOUSE: A genetically-modified mouse created by gene targeting by homologous recombination, which has the effect of inactivating or deleting a specific gene, so its function can be studied.

L

LACTIC ACID BACTERIA (LAB): Natural organisms, including various organic acids, that promote carbohydrate fermentation and produce lactic acid. LAB possess specific antimicrobial properties that promote food preservation. In addition, LAB contribute to unique food flavor, such as the sharpness often noted in fermented milk products such as cheese and yogurt.

LAMARCKISM: The belief that acquired characteristics can be inherited, that is, that changes to an organism that happen during its life can be passed on to offspring.

LARVA: An immature form of an insect.

LATITUDINAL GRADIENT: The relationship between biodiversity and the latitudinal position of a geographic region on Earth's surface. As a general rule, greater genetic and species diversity is observed the closer an ecosystem is located to the equator.

LIGAND: A molecule, such as a charged atom (ion), that is bound to another atom.

LIGNIN: A rigid polymer found in the cell walls of many types of plants.

LIGNOCELLULOSIC MATERIALS: Any materials made from wood; when gasified, the components are used to manufacture various types of biosynthetic natural gas (syngas).

LIMBUS: The edge of the cornea, where it joins the sclera or white of the eye.

LIPOSOME: A microscopic particle composed of an outer layer of lipid molecules around an aqueous center. Liposomes can encapsulate drug molecules to make them have a longer lifetime in the body.

***LISTERIA MONOCYTOGENES* (LISTERIA):** Named for Joseph Lister (1827–1912), an early proponent of sterile surgical practices, the bacterium is commonly found in milk, soft cheese, undercooked meat, and processed meats. Listeriosis, in the form of septicemia, meningitis, or encephalitis, is a common and sometimes fatal form of food poisoning.

LIVING BARRIER: A form of genetic engineering in which bacteria are inserted into seed genomes. The bacteria grow with the seedlings' roots, protecting against numerous varieties of nematodes that feed on soybean seedlings.

LIVING MODIFIED ORGANISM (LMO): Any living organism that has been genetically modified through the use of biotechnology.

LOCUS: The location of a gene, or other DNA sequence, on a specific chromosome within the genome (plural, loci).

LOOSE FILL: Any form of packaging material used to cushion items shipped in boxes or larger containers. The fill is placed in the spaces between the shipped item and the container wall.

LYSENKOISM: A type of pseudoscience that arose in the Soviet Union in the 1930s and destroyed Soviet biology for decades. Lysenkoists denounced modern evolutionary biology and genetics.

LYSOSOME: Organelles that are involved in the degradation of proteins and other substances in the cell. Lysosomal storage disorders lead to accumulation of lysosomes because there is a lack of an enzyme involved in normal lysosomal functioning.

M

MACRONUTRIENTS: The essential elements required for healthy crop plant development. Nitrogen, sulfur, and phosphorous are macronutrients typically found in most fertilizer formulations.

MAGNETIC FIELD: A region that arises due to the presence of a magnet(s) or electric currents, and in which a measurable magnetic force exists at each point within the region.

MAGNETISM: The response of a particle to a magnetic field.

MARKER-ASSISTED SELECTION (MAS): A technique that combines molecular biology/biotechnology with classic genetics to select genes controlling specific desired traits and to systematically strengthen them across successive generations through various selective breeding processes.

MASS SPECTROMETRY: An analytical technique that utilizes magnetic and/or electric fields to separate ions (such as charged molecules) in order to identify the different molecular types.

MEDICAL DEVICE: A medical device is a product used for the diagnosis or treatment of patients.

MEDIUM: A liquid or gel containing all the components, such as glucose and vitamins, needed to support the growth of cells.

MEMBRANE OXYGENATION: The artificial process by which carbon dioxide is removed and oxygen simultaneously added to blood. Oxygenation treatments are used if the patient's lungs are so damaged by disease or injury that they do not function. Whereas the current techniques involve external devices that are connected to the body, ongoing artificial lung research seeks to build on membrane oxygenation principles to create an implantable artificial organ.

MENHADEN: A herring fish species found in North America Atlantic coastal regions. Too oily for regular human consumption, menhaden are used to produce fish oils, animal feed, and organic fertilizer.

MESENCHYMAL STEM CELL: Mesenchymal stem cells are multipotent stem cells originating in the bone marrow that can differentiate into many cell types, including bone, cartilage, and fat cells.

METABOLISM: Conversion of one substance to another in the cell, usually by enzyme action. Drugs are metabolized mainly in the liver and are broken down in the series of metabolic reactions before being excreted.

METABOLITES: Molecules resulting from enzymatic breakdown of other more complex substances, such as medications, nutrients, and hormones. Some examples of metabolites are activated drug molecules and nutrients, as well as inactivated toxins.

METHICILLIN-RESISTANT *STAPHYLOCOCCUS AUREUS* (MRSA): MRSA is used to describe any strain of *S. aureus* that has become resistant to any beta-lactam antibiotic, including methicillin.

MICROBIAL GENOMICS: The study of the genetic structure of microorganisms. This field includes important research into the genetic structure of the pathogens such as *Salmonella* that cause food poisoning.

MICROELECTRONICS: A subset of electronic technology specializing in the creation, design, development, and utilization of compact, very small, and energy efficient components.

MICROGRAVITY: An environment in which the gravitational forces are small compared to those on the surface of Earth.

MICRONUTRIENT: The trace elements required for healthy crop plant development. Boron, chlorine, and zinc are important fertilizer micronutrients.

MICROORGANISM: Any organism that is too small to be seen unless magnified, as with the use of a microscope.

MIGRAINE HEADACHES: Serious and often debilitating headaches that afflict up to one in 10 North American adults. The causes of migraines are not entirely understood; research has established likely connections between migraine onset and hormonal changes (in women), as well as chemical imbalances in the brain, most notably the level of the pain regulator serotonin.

MIMIC: To simulate, or resemble closely.

MINISATELLITE DNA: Short repetitive stretches of DNA within the genome that are used as a basis for DNA fingerprinting.

MITOCHONDRIAL DNA: The DNA located within the mitochondria, the center for energy production located within all human cells. Mitochondrial DNA is the smallest unit of the genome, the hereditary information stored throughout the 22 chromosomes found in every human cell.

MODELING: The use of mathematical means to describe the aspects of a system and/or a process, thereby focusing on the factors believed to be important.

MOLD: Molds are a form of fungus that grow on organic matter. They aid in decomposition of organic substances and often secrete mycotoxins that inhibit the growth of bacteria. Penicillin is derived from the *Penicillium* mold.

MOLECULAR ARCHAEOLOGY: The study of DNA extracted from archaeological specimens, such as bones and mummies.

MOLECULAR BIOLOGY: A branch of biology concerned with the structure and function of key biological molecules, including nucleic acids and proteins.

MOLECULAR CLONING: A technique that assembles and replicates recombinant DNA (rDNA) molecules with a host.

MOLECULAR FARMING: The use of plants to produce a target compound.

MOLECULAR WEIGHT: The weight of a molecule, which is the sum of the weights of all of its component atoms. The unit of weight is the dalton, which equals one-twelfth of the weight of a carbon 12 atom.

MONOCLONAL ANTIBODY: An antibody is an immune protein produced in response to the introduction of an antigen protein into the body, usually on the surface of a pathogen or cancer cell. The antibody binds to the antigen as part of the immune response. A monoclonal antibody is a type of biologic drug consisting of a population of identical antibodies that can bind to a target antigen involved in a disease process. The antibodies are produced by a clone or genetically homogenous population of hybrid cells. Hybrid cells are cloned to establish cell lines producing a specific antibody that is chemically and immunologically homogeneous.

MONOCOTYLEDON: The embryonic seed leaf, which stores food for the growing plant, is known as a cotyledon. A monocotyledon, or monocot, such as rice, contains a single cotyledon.

MONOCULTURE: The practice of cultivating only one crop species in the same land, season after season.

MONOSODIUM GLUTAMATE: Known by the acronym MSG, this chemical additive is a popular food supplement that is promoted as a flavor enhancer and preservative. MSG mimics the flavor of sodium chloride (salt).

MORBIDITY: Disease incidence.

MULTIPOTENT: Able to differentiate into a limited range of cell types.

MUNICIPAL SOLID WASTES: Solid refuse, including household and business trash, that is disposed of by local sanitation agencies.

MUTATION: A variation in the DNA sequence of a gene that may lead to a defect in or even absence of the protein for which the gene codes. Mutations range from single base changes to more extensive changes in the DNA sequence. Mutation may or may not lead to disease.

MYCO-DIESEL: Based on the interaction of the fungus species *Gliocladium roseum* and cellulose materials to produce hydrocarbon molecules that mimic those found in conventional diesel fuels.

MYCOREMEDIATION: A bioremediation process that uses the mycelia, or vegetative part, of fungi to remove contaminants from the soil or groundwater.

MYOELECTRIC: Using electrical signals from muscles to operate devices, especially prosthetics.

MYOSITIS: The inflammation of muscle fibers.

N

NANO: Scale of measurement in units of 1 to 100 billionths. In 1960 the nanoscale was confirmed as an accepted international standard of measurement.

NANOBIOTECHNOLOGY: The technological use of microorganisms to achieve results at the molecular level.

NANOFIBERS: Fibers that have a diameter of less than 1,000 nm; a nanometer is the unit of measure defined as 1 billionth of 1 meter.

NANOTECHNOLOGY: Nanotechnology involves the engineering of molecules and atoms at sizes of less than 100 nanometers. One nanometer is equal to one billionth of a meter in length.

NATURAL LAW: A concept distinct from the laws of nature described by science, natural law is constructed on two principles: (1) moral standards ultimately govern human behavior; (2) as all humans are inherently rational, people will behave in accordance with their rational nature.

NATURAL SELECTION: Also known as survival of the fittest; the natural process by which those organisms best adapted to their environment survive and pass their traits to offspring.

NEURAL INTERFACE: Also known as a brain-computer interface, a neural interface is a path of communication between the brain and an external computer system.

NEUROIMAGING: The production of images of the brain through noninvasive techniques.

NEUROTOXIN: A biological or chemical complex that damages or destroys tissues in the nervous system.

NEUTROPENIA: A decrease in neutrophil count in the blood. Neutrophils are immune cells that are vital in defending the body against infection.

NITRIFICATION: The biological oxidation of ammonia into nitrates. Ammonia is produced when organic sources of nitrogen break down through natural processes. Municipal wastewater has a high level of organic waste, and the ammonia levels in wastewater are correspondingly high. *Nitrosomona* bacteria are commonly used to treat municipal wastewater to facilitate the ammonia oxidation process.

NON-STEROIDAL ANTI-INFLAMMATORY AGENT (NSAID): Any of a group of drugs that act locally, at the site of pain and inflammation, to reduce symptoms. They do not contain steroids, which can also reduce inflammation.

NUCLEAR ENVELOPE: The double membrane that separates the nucleus and its contents from the cytoplasm of the cell. The nuclear envelope is perforated by a number of pores, which allows penetration of certain substances, including DNA delivered by microinjection.

NUCLEOBASE: Any of a group of nitrogen-based molecules—adenine, cytosine, guanine, thymine, and uracil—that form nucleotides. The first four are the building blocks of DNA. In RNA, the nucleobase thymine is replaced by uracil.

NUCLEOTIDES: The three-part molecules that combine to form DNA. The parts are pentose (a sugar with five carbon atoms) and four nitrogenous bases (organic compounds essential to molecular structure): adenine (A), guanine (G), thymine (T), and cytosine (C).

NUCLEUS: The part of a living cell that is enclosed within a membrane and that contains all the genetic information in the form of DNA.

NUREMBERG CODE: In 1947, in Nuremberg, Germany, a panel of American judges presided over the Doctors' Trial, involving 23 individuals charged with murder and torture occurring during medical experimentation performed on non-voluntary human subjects housed in concentration camps during World War II (1939–1945). As a result, a set of 10 criteria for designing ethical human subjects research was created, called the Nuremberg Code.

NUTRACEUTICALS: The generic term that describes any food or food component that is proven to contribute to disease prevention or treatment.

O

OBESITY: The condition of having a body mass index of 30 or more, whereas overweight is defined as having a body mass index between 25 and 30. Body mass index is an individual's body weight divided by the square of his or her height.

OLIGODENDROCYTE: A type of cell found in the central nervous system, different from a neuron. The role of oligodendrocytes is to insulate the axons along which nerve impulses are propagated.

OLIGONUCLEOTIDE: A short polymer of nucleic acid that typically contains 50 or fewer bases.

ONCOGENE: A mutated form of a gene known as a proto-oncogene, which is generally involved in important functions such as cell division. Once an oncogene is formed, it may go on to contribute to the development of a cancer.

ONCOLOGY: The field of medicine devoted to the diagnosis and treatment of cancer.

ONCOMOUSE: Also known as the Harvard Mouse, the oncomouse was the genetically-altered mouse designed at Harvard University for specific types of cancer research. Its creators sought to patent the life form; the Supreme Court of Canada ruled in 2003 that such patents were unlawful in that jurisdiction, though this decision was later modified.

OOCYTE: The female mammalian germ (reproductive) cell, which matures into an ovum (egg) by meiosis. The words oocyte and ovum are often used interchangeably, although this is not strictly accurate: An ovum can be fertilized, an oocyte cannot.

OOPLASM: Cytoplasm of an ovum.

OPIOIDS: Opioids can be defined as drugs that are derived directly from the opium poppy, also called opiates, or synthetic or semi-synthetic compounds that simulate the effects of opiates.

ORGAN PROCUREMENT AND TRANSPLANTATION NETWORK (OPTN): The OPTN is a public-private network linking all the professionals involved in the organ donation and transplantation system. It was established by Congress in 1984 under the National Organ Transplant Act.

ORGANIC COMPOUND: Generally-speaking, any compound that incorporates carbon into its molecular composition.

ORGANISM: An individual living thing.

ORPHAN DRUG: A drug, including a vaccine or medical device or medical food, that is developed to treat diseases so rare that companies are reluctant to develop them under current market conditions as their investment is unlikely to be repaid by sales revenues.

OSMOTIC PRESSURE: Pressure applied to prevent the flow of water into a cell to equalize the salt concentration on the inside and outside of the cell.

OSSEOINTEGRATION: A direct interface formed between an implant (artificial or by autograft) and an original bone structure.

OSTEOARTHRITIS: An inflammatory condition of the joints caused by aging and wear and tear, rather than autoimmunity.

OUTCROSSING: Outcrossing occurs when two organisms from different varieties of the same species are bred. Outcrossing serves as a means of introducing new genetic material into the breeding pool.

OUTSOURCE: When research or development organizations opt to contract with a third party to carry out specific functions that they might have otherwise done in-house, such as a clinical trials process.

OVUM: An ovum, or egg, is the female gamete (reproductive cell) of animals, including humans. A human ovum is capable of developing only after fertilization by a male reproductive cell (sperm). The plural form is ova.

OXYCODONE: A popular semi-synthetic opioid (narcotic) created from an alkaloid of the poppy plant that has a chemistry similar to codeine; marketed as OxyContin and known by the street name "hillbilly heroin" due to its relative low cost and availability. It is combined with ASA to create the prescription analgesic marketed as Percodan, and with acetaminophen to create the product Percocet.

P

P450: Also known as cytochrome P450, this is a large and diverse group of enzymes that oxidize organic compounds and play an important role in drug metabolism.

PANDEMIC: A world-wide epidemic, which is an outbreak of disease in which many more cases than expected occur over a short time.

PANOPTICON: A structure designed by philosopher Jeremy Betham after the French Revolution, designed to work as a prison in which the prisoners are under the impression of constant surveillance; the guard can view all prisoners, but prisoners cannot view the guard.

PARABEN: Esters (chemical compounds) of para-hydroxybenzoic acid commonly used as preservatives in cosmetics and other products. A mixture of parabens, typically methylparaben, propylparaben, and butylparaben, is often included in products to confer activity against a wide range of bacteria and fungi.

PARTHENOGENESIS: The development of a whole organism from an unfertilized egg, which is the main type of reproduction in some organisms, including several insect species, such as some aphids, and certain reptiles and fish.

PASTEURIZATION: The food preservation process named for French chemist and biologist Louis Pasteur (1822–1895) that is designed to extend the useful (shelf) life of food products through the destruction of pathogens. Most often used to treat milk, the method requires food to be heated to a temperature below its boiling point for a fixed period. Pasteurization also tends to limit all bacterial and enzyme activity in food.

PATENT: A form of intellectual property related to a specific invention. A patent is bestowed by a government upon an invention's creator, conferring the exclusive right to produce, use, and sell said invention.

PATENT INTERFERENCE: A proceeding during which rival claimants to a patent or pending patent seek to establish priority.

PATHOGEN: A disease-causing organism or substance.

PEDOLOGY: The division of soil science concerned with the classification, morphology, and formation of different soils as they occur naturally worldwide.

PEMFC: Polymer electrolyte membrane fuel cells (PEMFC) are a common type of fuel cell employed in many commercial applications. A solid polymer membrane is the electrolyte used to transport the protons generated by the chemical reaction within the cell. The PEMFC will typically have an operating temperature of 140°F (60°C) to 212°F (100°C).

PENICILLIN: An antibiotic that is produced from certain strains of blue molds or, more typically, produced synthetically.

PEPTIDE BOND: The bond formed between the carboxyl group of one amino acid and the amino group of a neighboring amino acid that forms the backbone of a protein molecule.

PHARMACODYNAMICS (PD): The study of what a drug does to the body.

PHARMACOGENOMICS: Also know by the term pharmacogenetics, the study of the interactions observed between organisms that have been created or modified by genetic variation and different drugs.

PHARMACOKINETICS (PK): The study of what the body does to drugs, hormones, nutrients, and toxins.

PHARMING: Genetic modification of plants and animals so that they produce pharmaceuticals.

PHENOTYPE: The biochemical, physiological, and physical characteristics of an organism, determined by both interaction with its environment and genetic makeup.

PHOTOSYNTHESIS: The chemical process in which energy provided by sunlight converts carbon dioxide into organic compounds. In all plants and algae the photosynthesis of carbon dioxide and water releases oxygen as the waste product. For this reason photosynthesis is the essential regenerative process for Earth's atmosphere.

PHYSIOLOGY: A branch of biology that focuses on the functions of living organisms and their parts.

PHYTOREMEDIATION: A bioremediation process that uses plants to remove contaminants from soil or water.

PHYTOTHERAPEUTICS: Therapeutic substances naturally occurring in plants and used to make medications.

PLACEBO: A substance or medical procedure that has no therapeutic benefit, which is often used as a control in new drug trials.

PLANT-MADE PHARMACEUTICALS: Proteins for therapeutic use, including vaccines and antibodies, produced by genetically-modified plants and extracted from them after harvest.

PLASMID: A circular strand of DNA molecules capable of self-replication that operates separately from normal chromosomes.

PLASMOLYSIS: The chemical process in which water present in a microbial cell is removed through osmosis, which is triggered by the higher salt concentration present outside the cell wall. In this way salt tends to inhibit microbial growth in foods.

PLATELETS: The cells found in the bloodstream that contribute to blood clotting. Platelets are formed from megakaryocytes, fragments of bone marrow cells. The protein thrombopoietin stimulates platelets to enter the bloodstream when bleeding is detected.

PLURIPOTENCY: Refers to the ability of a human embryonic stem cell (hESC) to differentiate into any human cell type. Pluripotency is not the same as totipotency, which is a feature of plant cells and means that the cell can develop into a whole organism. To develop into a whole organism, an hESC requires extracellular factors present in placenta.

PLURIPOTENT: Able to differentiate into most cell types of an organism. Later stage mammalian embryonic cells are pluripotent.

POLYACRYLAMIDE: A form of acrylamide that can be either a simple linear chain or a cross-linked structure. Polyacrylamide is very water absorbent and is used to form a soft gel in gel electrophoresis. It is a fairly strong neurotoxin, so it must be handled very carefully, using appropriate protective equipment.

POLYGLACTIN: A polymer used in absorbable sutures and surgical mesh.

POLYLACTIDE (PLA): PLA is the generic acronym used to describe a wide range of polylactides, the monomers produced when cornstarch is fermented. PLAs are the foundation for many bioplastic products, including cornstarch packing materials.

POLYMERASE CHAIN REACTION (PCR): A molecular genetics technique facilitating the analysis of DNA or RNA sequences by amplifying small sample strands to create sufficient quantities for accurate analysis.

POLYMORPHISM: A genetic variation with a frequency of at least 1 percent of the population, differentiating it from a mutation, and a causal factor in many hereditary diseases.

POLYPEPTIDE: A single linear organic polymer made up of a large number of amino acids that are peptide bonded together in a chain.

POLYSACCHARIDES: The polymers formed from simple sugars; amylopectin and amylase are two polysaccharides that combine to form cornstarch.

POMPE DISEASE: A lysosomal storage disease caused by deficiency of the enzyme alpha-glucosidase. The deficiency leads to buildup of the sugar glycogen in the cells, which impairs functioning of muscle and nerve cells throughout the body.

POST-TRANSLATIONAL MODIFICATION: Processing of a protein chain after translation, which includes glycosylation, acylation, phosphorylation, and a limited amount of proteolysis.

POTENCY: In biology, the ability of an undifferentiated cell to differentiate to form different cell types.

PRECAUTIONARY PRINCIPLE: A principle stating that any product, action, or process that might pose a threat to public or environmental health should not be introduced in the absence of scientific consensus regarding its safety.

PRECLINICAL DEVELOPMENT: The phase of drug development that involves testing compounds in animals and in non-animal systems for properties such as safety and efficacy. Clinical development, which follows, involves clinical trials in humans.

PREDNISONE: A corticosteroid used to treat inflammatory diseases that acts to suppress the immune system. First isolated in 1955, prednisone is used to treat conditions such as Crohn's disease, asthma, and the prevention of the further spread of tumors.

PREIMPLANTATION GENETIC DIAGNOSIS (PGD): A procedure carried out in conjunction with in vitro fertilization in which DNA from an embryo is subject to PCR and analyzed for the presence of mutations that could lead to single-gene disorders.

PRIMARY STRUCTURE: The sequence of amino acids in a protein chain, held together by peptide bonds.

PRIMER: A strand of DNA, made by chemical synthesis, that binds to the start or end of a single-stranded DNA target sequence in PCR and therefore starts off the action of DNA polymerase in copying the strand.

PRIVACY INTERNATIONAL: An organization that campaigns against privacy intrusions on individuals by government and business.

PROGENITOR CELL: The progenitor cell occupies a stage between the stem cell and its final differentiated cell type. Unlike a stem cell, it cannot renew itself and is committed to differentiation into a limited range of cell types. The terms stem cell and progenitor cell often are used interchangeably but they are not the same. Many stem cell lines actually consist of progenitor cells.

PROKARYOTE: An organism in which the nuclear material is not enclosed within a specialized membrane, but instead is distributed in the cytoplasm.

PROMOTER: A region in a DNA molecule, occurring just before a coding region, where RNA polymerase binds in order to start transcription. The promoter is therefore the starting point for transcription.

PRONUCLEUS: The nucleus of the egg or sperm within the fertilized egg before the fusion of the two together to form a zygote.

PROPHAGE: The genome of a lysogenic phage when incorporated into the chromosome of the host bacterium.

PROSTHETICS: Prosthetics can be either the process of making replacement body parts or the devices themselves. Manufactured replacements for body parts are sometimes custom-made and fitted—such as leg or arm prostheses. Some can be purchased over the counter, such as post-mastectomy bra inserts.

PROTEIN: Large, complex molecules made up of long chains of subunits called amino acids. All living organisms are composed largely of proteins. Humans synthesize more than 30,000 different types of proteins.

PROTEIN EXPRESSION: The process to convert genetic information through synthesis into a protein.

PROTHROMBIN TIME (PT): The period required by the body, through the action of the thrombopoietin protein, to initiate clotting. If PT extends beyond 4 to 5 seconds, this finding is usually an indicator of severe liver damage.

PROTISTS: Eukaryotic single-celled organisms (organisms in which the genetic material is enclosed within a specialized membrane) that belong to the kingdom Protista in the classification used for living organisms.

PROTOPLASM: The thick, semifluid, semitransparent substance that is the basic living matter in all plant and animal cells.

PROTOXIN: A chemical substance that is a precursor to a toxin.

PUBLIC DOMAIN: Intellectual property that does not carry any legal right of enforcement; generally associated with works or creations for which the intellectual property rights have expired by operation of law (e.g. United States patent protection is generally limited to a 20-year term). Also described as the "creative commons," public domain is related conceptually to open source or publicly accessible knowledge.

PYROLYSIS: Degradation of a material into its chemical components achieved through the application of intense heat. Pyrolysis is the first stage in the three-part thermochemical decomposition process that continues to gasification and the eventual combustion of biomass material. A flaming match is a common pyrolysis example. A kiln is the typical enclosed, airtight structure used in pyrolysis.

R

RANDOMIZED CONTROL TRIAL: A type of clinical trial in which participants are randomly assigned to treatment groups and control groups.

RARE DISEASE: A disease affecting fewer than 200,000 people in the United States. Sometimes the definition is extended to cover a disease that affects more than 200,000 people if it is not possible to cover the costs of developing and distributing a treatment by the potential sales revenue.

REAL-TIME PCR: A method of PCR that allows quantitation of the DNA in a sample, by labeling the target with a measurable fluorescent tag.

RECEPTOR: A protein, usually on the surface of a cell, which is the site of specific binding of another protein that recognizes the three-dimensional shape of the receptor.

RECOMBINANT BOVINE GROWTH HORMONE (rBGH): A genetically engineered hormone given to cattle to increase milk production.

RECOMBINANT DNA (rDNA): A form of artificial DNA that contains two or more genetic sequences that normally do not occur together, which are spliced together.

RECOMBINANT DNA (rDNA) TECHNOLOGY: The entire set of processes utilized in order to combine a specific DNA sequence from one organism with the DNA of a specific vector, usually a bacteria or virus.

RECOMBINANT PROTEINS: Proteins derived from genes inserted into the DNA of organisms that originally did not contain the genetic trait.

REFUGE: In management of Bt crops, a refuge is an area planted with non-genetically-modified crop plants. It is an area where pests that are susceptible to Bt toxin can still survive so the location does not become dominated by pests that have evolved Bt resistance.

REGENERATIVE MEDICINE: The branch of medicine dedicated to employing a wide range of bioengineering or biomedical techniques to repair or replace damaged or diseased organs or tissue.

RESIDUAL LIMB: The portion of the limb remaining after amputation, also called the stump.

RESOLUTION: A formal expression of agreement, intent, or opinion made by a formal organization or legislature after voting on the issue.

RESTRICTION ENDONUCLEASE: An enzyme that separates DNA at specific restriction sites, cutting it into sequences.

RESTRICTION ENZYME: A bacterial enzyme able to identify specific sequences of DNA and cut it to produce fragmented pieces. Such enzymes are a useful tool in molecular biology.

RESTRICTION FRAGMENT LENGTH POLYMORPHISM (RFLP): An early and labor-intensive method of DNA analysis. A restriction enzyme is the chemical used to "cut" the two strands of the DNA double helix structure at a predetermined point. The length of the resulting fragments is used to determine whether a potential match exists between the biological sample found at a crime scene and a known source.

REVASCULARIZE: The restoration of natural blood supply to an organ, tissue, or bone. This condition is consistent with the ability of the body to regenerate damaged bone or tissue.

REVERSE TRANSCRIPTASE PCR: Often abbreviated to RT-PCR, not to be confused with real-time PCR, reverse transcriptase PCR allows amplification of RNA. Reverse transcriptase converts RNA to DNA, which is then subject to the usual PCR method.

RFID: This acronym is generally applied to any radio frequency identification process, the identity tracking systems that are used to transmit identifying serial numbers by radio signal for persons, animals, or objects.

RHEUMATOID ARTHRITIS: A chronic inflammatory condition of the joints, thought to be an autoimmune disease.

RIBONUCLEIC ACID (RNA): RNA is similar to DNA in that it is composed of long chains of nucleotides and encodes genetic information. In its messenger form (mRNA) it transmits genetic information utilized in protein synthesis.

RIBOSOME: A small particle in the cytoplasm built of protein and ribonucleic acid that acts as the site for protein synthesis.

RISK ADJUSTMENT: The weight of each existing risk factor, assessed against a context of similarly known risks previously encountered in like circumstances.

ROUNDUP: The commercial name for glyphosphate, a widely used weed killer that acts by blocking an enzyme the plant needs to make amino acids.

RUMINANT: A cloven-hoofed mammal, such as a cow, that chews cud regurgitated from the rumen.

S

SALINITY: The measure of the amount of salt dissolved in water.

SATURATED FATS: Fats containing saturated fatty acids and found mainly in animal sources.

SCAFFOLD: A framework for seeding or growing synthetic or biosynthetic regenerative materials that is porous, often elastic, and typically composed of biodegradable materials.

SCANNING PROBE MICROSCOPE (SPM): The devices that have played a crucial role in the expansion of nanotechnology applications. SPMs are distinct from conventional microscopes that depend on lens alignment and properties to magnify the objects examined. All SPM technologies involve a cantilevered instrument tip that scans the object surface to determine its physical, chemical, and magnetic properties.

SCRUBBERS: Air pollution control devices used to remove dust, particulates, and some acidic gases such as hydrogen chloride from waste gas. Wet scrubber processes direct a liquid solution at the gas, which neutralizes its acids; wet scrubbers generate wastewater that requires further specialized treatment processes. In a dry scrubber, the acidic waste gases are directed through alkaline material such as lime and are neutralized into particulates that require safe disposal.

SECONDARY METABOLITE: A substance produced by plants or microbes that is not essential to survival, such as a primary metabolite, but still plays an important role. Secondary metabolite penicillin, for example, defends bacteria against competing species.

SEED PIRACY: The process of gathering, reusing, or selling genetically modified, patented seeds from one season to the next.

SEED SAVING: Cleaning and reconditioning seeds gathered from one year's harvest for planting during the following growing season.

SEED TRAITS: Desired characteristics such as insect and pest resistance, imperviousness to pesticides and herbicides, and ability to withstand drought or extreme cold weather conditions. The traits are not naturally occurring; they are intentionally created through the use of biotechnology and molecular biology techniques involving isolating the desired traits in other plants or animals and inserting specific segments of their DNA into the seed to be genetically modified. Seed traits are patented by the companies that create or develop them.

SELF-DETERMINATION: Widely acknowledged as a fundamental human right, self-determination includes the rights asserted by a specific people to land and resources.

SEMICONDUCTOR: A solid material in which electrical conductivity varies between a conductor and insulator, depending on the addition of chemicals or differences in temperature.

SEMISYNTHETIC ANTIBIOTIC: An antibiotic made from natural materials enhanced by use of chemical synthesis; a mixture of natural and synthetic chemical substances.

SEQUENCE DATABASE: In this context, a sequence database refers to a relational computerized database containing an ever-increasing body of information about DNA, RNA, protein, and nucleic acid sequences.

SEQUENCING: The generic term for the processes used to determine the order in which the nucleotide bases are arranged in a DNA molecule. The bases are adenine (A), guanine (G), cytosine (C), and thymine (T). The bases form pairs (e.g. AT, GC) that comprise the individual components of the strands of the double helix DNA molecule. Each base pair carries a specific genetic coding.

SERUM: Serum (plural, sera) is a fluid obtained by separating the solid and liquid components of blood.

SEXUAL REPRODUCTION: The production of offspring from sexually polar parents (i.e., male and female) of the same species. The offspring have unique combinations of genes inherited upon fertilization—the union of male and female gametes. In plants, sexual reproduction may require the intervention of a vector (e.g., bees) to unite male and female gametes. In addition, there are organisms, especially plant species, in which meiotically produced gametes fuse to produce a polyploid zygote (an offspring with a greater number of genes sets than its parents).

SHORT TANDEM REPEATS (STR): The process that employs polymerase chain reaction (PCR) principles in combination with known population variability to determine the probability that there is a match between two compared DNA sources. STR analysis is directed to specific known regions of the nuclear DNA. In this way, DNA samples can be distinguished, enabling scientists to differentiate one DNA sample from another. For example, the likelihood that any two individuals (except identical twins) will have the same 13-loci DNA profile can be as high as 1 in 1 billion or greater.

SIBLINGS: Children of the same generation in a family unit related by biological parentage, parental marriage, adoption, or other cultural or legal definition. Siblings usually refer to each other as sisters or brothers.

SINGLE NUCLEOTIDE POLYMORPHISM (SNP): A type of genetic variant involving a variation at a single nucleotide location in a gene sequence. Millions of SNPs have been cataloged in the human genome, and they form the basis of many genetic tests.

SKIN GRAFT: Transplantation of skin, either from a donor or from another part of the recipient's body, intended to cover a wound such as a burn.

SLUDGE: The solid by-product of industrial wastewater treatment. Sludge usually requires further processing once it is removed from the treated water because it will often contain numerous metals, other compounds that do not readily biodegrade, and organisms that have pathogenic potential. Sludge also tends to be rich in nitrogen, phosphorous, and other soil nutrients that make sludge a useful fertilizer component.

SMALL INTERFERING RNA (siRNA): A naturally-occurring or synthetic small double-stranded RNA molecule that can trigger degradation of mRNA or block its translation into protein.

SOLUBILITY: The maximum amount of a solute that can be dissolved in a solvent.

SOLVENT: Any compound or substance, most often a liquid, that is capable of dissolving a different substance. Organic solvents (containing at least one carbon and one hydrogen atom) are commercially-important liquids and are useful for dissolving organic materials such as fats, oil, and plastics.

SOMATIC: Pertaining to the corporeal; the cells of the body that make up tissues and organs, distinct from those involved in reproduction.

SOMATIC CELL NUCLEAR TRANSFER (SCNT): The process by which animal cloning is achieved, involving transfer of a somatic, or body, cell to an egg from which the nucleus has been removed. The transferred somatic cell begins to divide and develop into an embryo.

SOMATIC CELLS: The term that describes all human cells except egg or sperm cells.

SOMATIC GENE THERAPY: The process of embedding healthy or desired genes in an appropriate carrier, such as a virus, and injecting them into an existing individual with the hope of transmitting the new genetic material to the cells, which then will reproduce the "new" genes and incorporate them into the organism.

SPERM: A motile, male reproductive cell capable of fertilizing an ovum.

SPLICING: The use of specific enzymes to introduce foreign genetic material into a genome.

SPORE: An environmentally-resistant form of certain bacteria that can resume normal growth and division when conditions are more conducive to survival.

SPORULATION: A method of asexual reproduction involving the production and release of spores.

SQUID: An acronym for Superconducting QUantum Interference Device; it uses properties of superconductivity to detect extremely weak magnetic fields.

STEM CELL RESEARCH AND ENHANCEMENT ACT: Controversial United States legislation that proposed to extend federal government funding to research directed at the identification of new embryonic cell lines. The Act was vetoed by President George W. Bush in 2006; 2009 legislation confirmed that federal government funding would not be available for projects involving "research in which a human embryo or embryos are destroyed, discarded, or knowingly subjected to risk of injury or death."

STEM CELLS: Cells that have not differentiated into a specific cell type and can divide over and over in the undifferentiated state, but which have the capability to differentiate into several or many cell types.

STRABISMUS: Known colloquially as "cross eyes," this medical condition occurs when the subject cannot control the direction of the eyes while attempting to focus on a specific object. The condition is caused by weakness in the six muscles that control eye direction and movement.

STRATIFIED MEDICINE: The use of advanced diagnostic tests, including genetic testing, to find out which patients will respond better to certain cancer drugs.

SUBSTANTIAL EQUIVALENCE: Substantial equivalence or bioequivalence means that a modified product, whether by biotechnology or conventional methods, contains the same nutrients, minerals, and vitamins found in the products for which it intends to substitute.

SUI GENERIS: Latin legal maxim meaning of its own kind, or unique.

SUPER WEEDS: The term used to describe the type of invasive plants that might develop in a specific ecosystem if cultivated plants that have been modified with GM organisms have contact with natural weed species.

SUPERCONDUCTIVITY: The phenomenon of zero resistance to the flow of electric current in certain materials that have been chilled to very low temperatures.

SUPERFUND SITE: Superfund is a program of the U.S. federal government, administered by the Environmental Protection Agency (EPA), to clean up sites contaminated with hazardous waste.

SURFACTANT: A shortening of the phrase surface-active-agent, the term refers to a substance that can, when used in very low concentrations, markedly decrease the surface tension of water. As a major component of laundry detergents, surfactants facilitate dissolving of grease and removing dirt from fabric/clothing.

SURGICAL SPECIMENS: Samples of damaged organs, tumors, or other abnormal body tissue removed during surgery.

SURROGATE: A woman who completes pregnancy and gives birth to a baby for another woman. A surrogate typically carries a zygote created with one of her own ovum.

SUSPENDED ANIMATION: An organism is said to be in suspended animation when all of its essential functions are temporarily interrupted.

SUSTAINABILITY: The use of natural resources such that the resources in question are not significantly depleted.

SUSTAINABLE: Definitions vary, but broadly speaking sustainable refers to a process, such as drug manufacture or energy production, that can be continued in the long-term with only minimal impact on the environment and minimal depletion of natural resources.

SVALBARD GLOBAL SEED VAULT: A global seed storage vault located below the permafrost level in a remote region (Svalbard) of Norway. The vault stores duplicate copies of nearly every important food crop seed from around the world. It has been built to withstand natural and human-created disasters and is designed as a resource for repopulating the world's food supply, if necessary. Because the seeds are stored at cryonic temperatures, they are able to survive indefinitely.

SYNTHESIS: The way in which proteins are created. Protein synthesis requires DNA, RNA, and protein-specific enzymes.

SYNTHETIC ANTIBIOTIC: Also called biosynthesized antibiotics, these are laboratory produced antibiotics, often chemically similar to naturally occurring antibiotics, that have precise and targeted effects.

SYNTHETIC BIOLOGY: An emerging science combining engineering and biology. The consortium of synthetic biologists at Harvard and Massachusetts Institute of Technology (MIT) define the field as "the design and construction of new biological parts, devices, and systems and the re-design of existing, natural biological systems for useful purposes."

SYPHILIS: A chronic bacterial disease that may result in damage to the cardiovascular or central nervous systems; it may result in death if not treated. Syphilis is typically transmitted through sexual activity but may also be transmitted to a developing fetus, which is known as congenital syphilis.

SYSTEMIC VASCULITIS: The condition marked by the inflammation of blood vessels.

T

T4 PHAGE: One of the largest and most widely-studied phages, T4 is a lytic phage that infects *Escherichia coli* bacteria.

T-LYMPHOCYTE: A type of white blood cell that plays a central role in cell-mediated immunity. Both cell-mediated and antibody-mediated immunity are important in organ rejection.

TAQ POLYMERASE: A heat stable DNA polymerase that is used to copy DNA in PCR.

TELOMERE: Region at either end of each chromosome that functions to maintain the stability of the chromosome through repeated cycles of replication.

TERATOMA: A tumor that contains cell types derived from all three layers of a developing embryo. It is used to check the pluripotency of human embryonic stem cells because it is produced when these are injected into an animal.

TERMINATOR SEEDS: Refers to seeds that will grow into genetically-modified plants that have been made to be sterile and unable to reproduce.

TERMINATOR TECHNOLOGY: A technology that uses genetic modification to render the second generation seeds of a plant sterile. The technology was developed by the United States Department of Agriculture (USDA) and private industry partners and patented in 1998. Use of terminator technology would protect the intellectual property of companies that develop genetically modified seeds and potentially prevent GM plant transgenes from escaping into wild plants. However, it would restrict farmers from saving seeds for future use, affecting small farmers worldwide. The rights to terminator technology were purchased by Monsanto to be incorporated into food crops such as wheat, rice, and soybeans, but terminator seeds have not been made commercially available for food crops as of 2012 due to widespread protest of the technology.

TERRA PRETA: Very fertile anthropogenic soils first identified in the Amazon basin, dated to as far as 9000 BC. The soils were improved through charcoal content that resulted from the use of cooking fires by the indigenous peoples of the region, because charcoal contains high levels of nitrogen, potassium, zinc, and phosphorous.

TERRITORIALITY: A basic premise of all intellectual property rights (IPR), including patents, that limits IPR to a specific legal jurisdiction.

TETRAVALENT VACCINE: Containing four antigens giving protections against four separate serotypes of a disease.

THERAPEUTIC CLONING: The creation of cells through somatic cell nuclear transfer (SCNT) from embryonic stem cells. In cases in which the stem cells are implanted in the donor that provides the DNA for SCNT, the embryonic stem cells are identical to those in the surrounding tissues and guarantee immunocompatiblity.

TI PLASMID: A tumor-inducing plasmid used by scientists to insert genes into plant cells for the development of GMOs. Ti plasmids were originally discovered in *Agrobacterium tumefaciens*.

TISSUE ENGINEERING: The functional remodeling or regeneration of cells or tissue, either inside the body (in vivo) or in a laboratory (ex vivo).

TOTIPOTENT: Able to differentiate into all cell types of an organism, so a totipotent stem cell could form a complete organism. Some types of plant cells are totipotent.

TRAIT: A feature of an organism such as insect resistance in a plant.

TRANS FAT: A partially hydrogenated fat found mainly in processed foods.

TRANSCRIPTION: The process of copying the genetic information in DNA into a complementary mRNA molecule, which then acts as a molecular message ensuring that the gene sequence is translated into the corresponding protein sequence.

TRANSCRIPTION FACTOR: A protein that binds to regulatory DNA sequences and plays a role in gene expression by controlling the transcription (movement) of genetic information from DNA to messenger RNA (mRNA).

TRANSDIFFERENTIATION: The differentiation of a tissue-specific stem cell into another type of cell.

TRANSDUCER: A device that converts variations in a physical quantity, such as pressure or brightness, into an electrical signal, or vice versa.

TRANSESTERIFICATION: The chemical process that occurs during algae diesel biofuel production, in which an ester catalyst (often sodium hydroxide) is combined with an alcohol such as methanol. The bonded hydrogen and oxygen molecule in the ester is replaced by an oxygen molecule. The result is a fuel that consists of a long-chain alkyl ester; the by-products are oxygen and water.

TRANSFEMORAL AMPUTATION: A leg amputation occurring above the knee.

TRANSFORMATION: Genetic change of a cell due to the uptake of deoxyribonucleic acid (DNA), usually DNA that codes for a functional product.

TRANSFORMED: A cell that has taken up a transgene in a genetic engineering experiment.

TRANSGENE: A foreign gene encoding a desirable trait that is introduced into another organism.

TRANSGENIC: Referring to an organism containing DNA from an unrelated organism. Transgenic organisms contain DNA from two or more unrelated organisms, possibly from different species.

TRANSGENIC ORGANISM: An organism that has been genetically modified by the insertion of genes from another species.

TRANSGENIC PLANTS: Plant species created when the genes of another species are inserted into the existing cellular structure to enhance or add a desired genetic feature.

TRANSLATION: The process by which a protein sequence is assembled using the sequence of mRNA as a template.

TRANSTIBIAL AMPUTATION: A leg amputation occurring below the knee.

TREATY: A formal agreement under international law between at least two countries, nations, or autonomous or semiautonomous groups of people, that sets forth the goals and responsibilities of all parties (participants) to the agreement.

TRIGLYCERIDE: The basic type of molecule in a fat, consisting of a glycerol backbone bonded to three fatty acids that may be saturated or unsaturated, depending upon the nature of its carbon-carbon bonds.

TUBER: A tuber refers to a root or other underground part of a plant from which new plants grow.

TUMOR SUPPRESSOR GENE: Also known as an anti-oncogene, a tumor suppressor gene stops a cancer from developing. If mutated, the protective function is lost.

U

U.S. FOOD AND DRUG ADMINISTRATION (FDA): Part of the United States government, the FDA is the office responsible for regulating approval of drugs, including orphan drugs.

UNITED NETWORK FOR ORGAN SHARING (UNOS): UNOS administers the OPTN for the U.S. Department of Health and Human Services (HHS).

UPSTREAM PROCESSING: All the process steps from selection of the cell strain that will be used for production to harvest of the product from the bioreactor.

UREA: A nitrogen-based compound used in solid fertilizers. Human and animal urine are the most common source of organic urea. Urea was the first organic compound to be synthesized (1828).

UTILITY PATENT: The most common type of patent, a utility patent protects any new or improved chemical compound, device, industrial method, machine, manufactured product, or process.

V

VACCINE: A product that produces an immune reaction by inducing the body to form antibodies against a particular agent. Usually made from dead or weakened bacteria or viruses, vaccines cause an immune system response that makes the person immune to (safe from) a certain disease.

VALUE DRIVERS: Facts and factors that positively affect the price of a product or company under valuation.

VARIABLE NUMBER TANDEM REPEAT: A base pair sequence that repeats throughout a specific locus on a genome.

VARIOLATION: From *variola*, a synonym for smallpox, an obsolete practice of intentional infection with the aim of inoculation by inducing a milder form of the disease.

VASECTOMY: A form of male contraception in which the vas deferens from the testis is cut so sperm no longer enter the semen.

VECTOR: In molecular biology, a vector is a carrier that is used to transmit a new piece of DNA into a host organism. In genetic engineering, common vectors are bacterial plasmids or viral phages, because they can be independently grown in a laboratory setting.

VIRION: A single virus particle, made up of RNA or DNA, and a capsid, which may be covered with a lipid envelope.

VIRUS: A microorganism smaller than a bacteria that cannot grow or reproduce outside a living cell host.

VITALISM: A theory that maintains that organic compounds contain a non-material element that could not be explained by science. It was largely superseded by the rise of organic chemistry.

VITRIFICATION: When the term is used in molecular biology, vitrification refers to the use of biotechnology to cool a cell to a glass-like state, removing water from the cell and preventing the formation of ice crystals that could cause cell damage. The process is reversible. As used in cryonics, vitrification refers to a process in which cooling and solidification occur with minimal tissue damage; this is typically done with a patient's brain. In cryonics, vitrification is not currently reversible through reanimation processes. In whole-body cryonics patients, the head/brain are often the only parts vitrified.

VOLATILE ORGANIC COMPOUNDS (VOCs): VOCs are found in the waste gases released from a wide variety of solid and liquid chemical compounds, such as paints, pesticides, glues, and cleaning supplies. Common VOC characteristics are high vapor pressure and relatively low water solubility. VOCs pose a danger to human health when inhaled in significant concentrations.

VON WILLEBRAND FACTOR: The large glycoprotein that binds to other proteins, especially blood factor VIII, during coagulation. The absence of this protein in the bloodstream is the cause of the hereditary bleeding disorder that bears the same name, a condition marked by excessive bleeding, especially from the nose and in the gastrointestinal system.

W

WEEDS: Plants considered undesirable because they compete for nutrients with the species being cultivated.

WET-SPINNING: Extrusion of polymers dissolved in an aqueous environment.

WHITE BIOTECHNOLOGY: Also known as industrial biotechnology, white biotechnology is the use of biological products in industrial products that are normally made from petroleum, such as fuel, polymers, bulk chemicals, and specialty chemicals.

X

X RAY: A form of electromagnetic radiation with a shorter wavelength than light. It is capable of penetrating solids, hence its use in medical imaging. The term refers to the radiograph (image) produced using x rays.

X-RAY CRYSTALLOGRAPHY: A technique in which a beam of x rays interacts with atoms in a crystal, enabling a detailed image of its three-dimensional structure to be built. X-ray crystallography allows the determination of the structure of protein molecules and DNA, and so it is useful in the detailed study of viruses.

XENOGRAFT: A tissue or an organ from one species donated to a recipient of another species, such as from pig to human.

XENOTRANSPLANT: A process in which tissue from one animal species is transplanted into another species. Nuclear transfer procedures delete the genes that would otherwise prompt the rejection of the transplanted tissue to permit acceptance of this tissue in the recipient.

Z

ZEA MAYS: The botanical name for corn, also known as maize, the largest of the cereal grasses.

ZEBRAFISH: A tiny transparent freshwater fish that is popular in aquaria and also widely used as a model organism in genetics research on vertebrates.

ZOONOSIS: An infectious disease that is passed from animals to humans, sometimes through a vector.

Chronology

A chronology of events related to the history of biotechnology.

BC

c.60 BC

Dioscorides writes the first systematic pharmacopoeia. His *De Materia Medica* in five volumes provides accurate botanical and pharmacological information. It is preserved by the Arabs, and when translated into Latin and printed in 1478, becomes a standard botanical reference for the next 1,500 years.

50 BC Crateuas, Greek physician, writes a work on pharmacology in which he is the first to make drawings of plants. These are the earliest known botanical drawings.

AD

c.500 Chinese Daoist philosopher Tao Hung-ching (451–536) compiles a fundamental text of *materia medica* that is the basis of Chinese pharmacology for centuries.

900 Native Americans use the plant *Lithospermum ruderale* as a contraceptive drug.

1241 Frederick II (1194–1250), Holy Roman Emperor, issues a law that allows dissection of cadavers and regulates surgery and pharmacy.

1345 First apothecary shop or drug store opens in London, England.

1584 Sir Walter Raleigh (1554–1618), English explorer, brings curare from Guiana. Obtained from a plant, this drug is a skeletal-muscle relaxant belonging to the alkaloid family. It can be used to paralyze.

1595 Andreas Libavius (1560–1616), German alchemist, publishes *Alchymia* in Frankfurt, Germany. Most now consider it to be the first chemical textbook. Libavius is convinced totally of the significance of chemistry to medicine.

1648 The writings of Jan Baptista van Helmont (c.1577–1644), Flemish physician and alchemist, are published posthumously (*Ortus Medicinae*). He is the first to use quantitative methods in connection with a biological problem and thus is a founder of the study of biochemistry. He also is the first to recognize that more than one air-like substance exists, and he names this vapor or non-solid "chaos," whose phonetic sound in Flemish is "gas." He classifies several types of "gases." He also describes the first quantitative experiment in plant physiology, demonstrating that a plant's growth is directly proportional to the loss of substance by the soil and the air.

1680 Thomas Sydenham (1624–1689), English physician, prescribes quinine as a specific drug for malaria.

1696   Daniel Le Clerc (1652–1728), Swiss physician, publishes *Histoire de la Médecine*. A founder in the study of the history of medicine, Le Clerc makes the first attempt at a synthetic survey of medical history.

1700   Bernardino Ramazzini (1633–1714), Italian physician, publishes the first systematic treatment on occupational diseases. His book *De Morbis Artificum* opens up an entirely new department of modern medicine—diseases of trade or occupation and industrial hygiene.

1711   The first patented medicine in America is "Tuscora Rice," which is hailed as a cure for tuberculosis.

1745   Christian Gottfried Kratzenstein (1723–1795), German physician, first uses electricity to relieve muscle sprains.

1796   Edward Jenner (1749–1823) uses cowpox virus to develop a smallpox vaccine.

1802   Xavier Bichat (1771–1802), French physician, and his colleague, Karl F. Burdach, are the first to use the term "biology" to describe the science of the general properties of organisms.

1817   German pharmacist Frederick Serturner announces the extraction of morphine from opium.

1827   Jacques Babinet (1794–1872), a French physicist, suggests a new standard of length measurement be adopted. Since all standards to this point are based on uncertain and changeable units, he suggests that something unalterable and true, like the wavelength of some particular type of light ray, be used. It is not until 1960 that technology advances to the point where his suggestion becomes fact.

1828   Friedrich Wöhler (1800–1882), German chemist, first synthesizes urea and lays the foundations (along with Jöns Jacob Berzelius [1779–1848], Justus von Liebig [1803–1873], and Robert Bunsen [1811–1899]) of organic chemistry and then later of biochemistry. His preparation of an organic compound from inorganic materials dispels the notion that organic compounds can only be produced by using living organisms.

1852   The American Pharmaceutical Association is founded.

1866   Gregor Johann Mendel (1822–1884), Austrian botanist, discovers the laws of heredity and writes the first of a series of papers on heredity (1866–1869) that formulate the laws of hybridization. His work is disregarded until rediscovered by Hugo de Vries (1848–1935) in 1900. Unknown to both Charles Darwin and himself, Mendel had discovered the proof that Darwin's theory of evolution required.

1869   Francis Galton (1822–1911), English anthropologist, publishes *Hereditary Genius*, in which he introduces the correlation coefficient. He is recognized as the founder of the statistical school of genetics.

1869   Johann Friedrich Miescher (1844–1895), Swiss biochemist, discovers nucleic acid (by isolating the nucleus, in which deoxyribonucleic acid or DNA will later be found).

1881   Louis Pasteur (1822–1895), French chemist and microbiologist, publicly inoculates sheep with an "attenuated culture" of anthrax.

1884   Karl Wilhelm von Naegeli (1817–1891), Swiss botanist, publishes *Mechanisch-Physiologische Theorie der Abstammungslehre*, in which he introduces the notion of "idioplasm," the part of the protoplasm responsible for its genetic makeup.

1892   August F. Weismann (1834–1914), German biologist, publishes his book *Das Keimplasma: Eine Theorie der Vererbung*, which contributes to the founding of the science of genetics. He offers his germplasm theory, which also leads him to predict correctly the phenomenon of meiosis (the splitting of paired chromosomes).

1895   Wilhelm Konrad Röntgen (1845–1923), German physicist, discovers x rays, initiating the modern age of physics and revolutionizing medicine. While working on cathode ray tubes and experimenting with luminescence, he notices that a nearby sheet of paper that is coated with a luminescent substance glows whenever the tube is turned on. For seven weeks he continues to experiment, and near the end of the year is able to report the basic properties of the unknown rays he names "x rays."

1898 Emil Herman Fischer (1852–1919), German chemist, isolates the purin nucleus of uric acid compounds and elucidates their structure in detail. This turns out to be highly significant, for purines are an important part of a group of substances called nucleic acids, which are found in the next century to be the key molecules of living tissues.

1899 Aspirin is first introduced on the market by the German pharmaceutical firm Farbenfabriken Bayer.

1901 William Bateson (1861–1926), British geneticist, coins the terms "genetics," "F1 and F2 generations," "allelomorph" (later shortened to "allele"), "homozygote," "heterozygote," and "epistasis."

1902 Carl Neuberg (1877–1956), German biochemist, introduces the term biochemistry.

1902 Ronald Ross (1857–1932), English physician, is awarded the Nobel Prize in physiology or medicine for his work on malaria, after discovering the *Plasmodium* parasite that causes malaria is carried and transmitted by the *Anopheles* mosquito.

1903 Archibald Edward Garrod (1857–1936), British physician, suggests that errors in genes lead to hereditary disorders. His 1909 book *The Inborn Errors of Metabolism* is the first study in biochemical genetics demonstrating that a single gene mutation blocks a single metabolic step.

1903 Niels Ryberg Finsen (1860–1904), Danish physician, pioneers light therapy and wins the Nobel Prize in physiology or medicine for his treatment of lupus by concentrated light rays.

1904 Ernest-François-Auguste Fourneau (1872–1949), French chemist and pharmacologist, first synthesizes amylocaine. It is used eventually as a local anesthetic known as stovaine.

1904 Lucien Claude Cuénot's crossbreeding experiments on mice provide proof that Mendel's newly rediscovered laws apply to animals as well as to plants. His research on the crossbreeding of yellow mice also provides the first example of genes that are lethal when homozygous. In 1910 other scientists prove that yellow homozygotes always die in utero.

1906 Frederick Gowland Hopkins (1861–1947), English biochemist, first argues that certain "accessory factors" in food are necessary to sustain life. This theory of trace substances becomes the starting point of further work on vitamin requirements.

1906 Freeze-drying is invented by Jacques Arsène d'Arsonval (1851–1940) of France and his colleague, George Bordas. This food preservation process works on the principle of removing water from food. It is not perfected until after World War II.

1906 The Pure Food and Drug Act becomes law in the United States, as does the Meat Inspection Act.

1907 Ross Granville Harrison (1870–1959), American zoologist, develops the first successful animal-tissue culture and pioneers organ transplant techniques. His cultivation of tadpole tissue is the first culture of animal cells in vitro.

1908 Margaret Lewis (1881–1970), American cytologist, first cultures mammalian cells in vitro.

1908 Godfrey Harold Hardy (1877–1947), English mathematician, and German physician Wilhelm Weinberg (1862–1937) publish similar papers (six months apart) on a mathematical system that describes the stability of gene frequencies in succeeding generations of a population. Their resulting "Hardy-Weiberg Law" links the Mendelian hypothesis with actual population studies.

1909 Phoebus Aaron Theodore Levene (1869–1940), Russian-American chemist, discovers the chemical distinction between DNA (deoxyribonucleic acid) and RNA (ribonucleic acid).

1909 Thomas Hunt Morgan (1866–1945), American geneticist, chooses the fruit fly *Drosophila* as a means of investigating genetics with the aid of mutations. He demonstrates the chromosome theory of heredity and perceives the significance of the tendency of certain genes to be transmitted together. He postulates "crossing over," which is demonstrated by Alfred Sturtevant (1891–1970) in 1913.

1911 R. A. Lambert and F. M. Hanes perform the first successful in vitro cultivation of tumoral cells. This technique proves very useful in the advancement of oncological research.

1912 First chair in Biochemistry in the United Kingdom goes to Frederick Gowland Hopkins (1861–1947) at Cambridge University.

1916 First journal devoted purely to genetics in the United States is founded, *Genetics*.

1922 Archibald Vivian Hill (1886–1977), English physiologist, is awarded the Nobel Prize in physiology or medicine for his discovery relating to the production of heat in the muscle. Otto Fritz Meyerhoff (1884–1951), German-American biochemist, is also awarded the Nobel Prize in physiology or medicine for his discovery of the fixed relationship between the consumption of oxygen and the metabolism of lactic acid in the muscle.

1923 Georg von Hevesy (1885–1966), Hungarian chemist, first uses radioactive tracers to follow the path of a substance in an organism.

1923 Gaston Ramon (1886–1963), French bacteriologist, discovers the anti-diphtheria and anti-tetanus vaccines.

1923 Charles Francis Jenkins (1867–1934), American physicist and inventor, transmits radiophoto pictures of President Warren Gamaliel Harding (1865–1923) from Washington, D.C. to Philadelphia, Pennsylvania.

1924 Albert-Léon Charles Calmette (1863–1933), French bacteriologist, invents the antituberculosis vaccine using attenuated live bacilli and vaccinates children.

1924 Ko Kuei Chen (1898–1988), Chinese-American pharmacologist, and Carl Frederick Schmidt, American physician, first introduce ephedrine. Similar to epinephrine but less powerful, it has the advantage of being taken orally. This marks the introduction to Western medicine of a drug called "ma huang," which has been used in China for 5,000 years.

1926 Johannes Andreas Grib Fibiger (1867–1928), Danish bacteriologist, is awarded the Nobel Prize in physiology or medicine for his discovery of the Spiropter carcinoma. He is the first scientist to induce cancer in laboratory animals.

1927 Hermann Joseph Muller (1890–1967), American biologist, induces or causes the first artificial mutations in the *Drosophila* fruit fly by using x rays. His work shows that mutations are the result of some type of chemical change. It also alerts him to the danger of excessive x rays.

1928 N. K. Kolikov (1872–1970) introduces the concept of chromosome duplication, which later becomes the basis of molecular genetics.

1928 Alexander Fleming (1881–1955), Scottish bacteriologist, discovers penicillin. In a report published in 1929, Fleming observes that the mold *Penicillium notatum* inhibits the growth of some bacteria. This is the first of the antibacterials to be discovered, and it opens a new era of "wonder drugs."

1930 Ronald Aylmer Fisher (1890–1962), English biologist, publishes *The Genetical Theory of Natural Selection*, which, together with Sewall Wright's (1889–1988) *Mendelian Populations* (1931), lays the mathematical foundations, and contributes to the advancement of, statistical methods of population genetics.

1931 Joseph Needham (1900–1995), English biochemist, publishes the landmark work *Chemical Embryology*, which shows the relation of biochemistry and embryology and both founds chemical embryology and lays the foundations for modern molecular biology.

1931 Sewall Wright (1889–1988), American geneticist, publishes *Evolution in Mendelian Populations*. This book, along with Fisher's *The Genetical Theory of Natural Selection* (1930), constitutes the mathematical foundation of population genetics.

1931 Phoebus A. Levene (1869–1940) summarizes his work on the chemical nature of the nucleic acids. His analyses of nucleic acids seem to support the hypothesis known as the tetranucleotide interpretation, which suggests that the four bases are present in equal amounts in DNAs from all sources.

This indicates that DNA is a highly repetitious polymer that is incapable of generating the diversity that would be an essential characteristic of the genetic material.

1931 Sewall Wright (1889–1988), American geneticist, presents the first useful picture of genetics and its role in evolution.

1933 Thomas Hunt Morgan (1866–1945), American geneticist, is awarded the Nobel Prize in physiology or medicine for his discoveries concerning the role played by the chromosome in heredity.

1934 George Wells Beadle (1903–1989), working with Boris Ephrussi (1901–1971), in collaboration with A. Kuhn and A. Butenandt, work out the biochemical genetics of eye-pigment synthesis in *Drosophila* fruit flies and *Ephestia* moths.

1936 First non-organic agent (methylene blue) shown to mimic the embryologic organizer (prior to our modern understanding of molecular biology, a chemical component thought capable of—or essential to—organizing and directing the normal sequence of embryological development, e.g., tissue differentiation and organ formation) is described by English biochemist Joseph Needham (1900–1995) and colleagues.

1936 Frank Macfarlane Burnet (1899–1985), Australian immunologist, isolates the first mutant bacteriophages, which are tiny, bacteriolytic viruses that will play a major part in future genetics theory.

1936 Perrin Hamilton Long (1899–1965), American physician, and Eleanor Albert Bliss, American bacteriologist, first introduce sulfa drugs in the United States. They conduct studies of prontosil in a series of wide-scale applications.

1936 Theodosius Dobzhansky (1900–1975) publishes *Genetics and the Origin of Species,* a text considered a classic in evolutionary genetics.

1937 Albert Szent-Gyorgyi (1893–1986), Hungarian-American biochemist, is awarded the Nobel Prize in physiology or medicine for his discoveries in connection with the biological combustion processes, with special reference to vitamin C and the catalysis of fumaric acid.

1937 Pharmacologist Daniel Bovet (1907–1992) discovers antihistamines—compounds that neutralize some of the symptoms of allergic reactions. In later years, he also first synthesizes curare and develops a method to use it during surgery as a muscle paralytic. For his work on both, he wins the Nobel Prize in 1957.

1937 Richard Benedict Goldschmidt (1878–1958), German-American geneticist, theorizes that the gene is a chemical entity rather than a discrete physical structure. His theories cause a major reevaluation of the concepts governing the science of genetics.

1941 The anticoagulant drug dicumarol is first identified and synthesized by American biochemists Mark Arnold Stahmann, Karl Paul Link (1901–1978), and C. F. Huebner.

1941 George Wells Beadle (1903–1989) and Edward Lawrie Tatum (1909–1975) publish their classic study on biochemical genetics titled *Genetic Control of Biochemical Reactions in Neurospora.* Beadle and Tatum irradiate red bread mold *Neurospora* and prove that genes produce their effects by regulating particular enzymes. This work leads to the one gene–one enzyme theory. The discovery that each gene supervises the production of only one enzyme lays the foundation for the DNA discoveries to come.

1943 Carl Ferdinand Cori (1896–1984) and Gerty Theresa Radnitz Cori (1896–1957), Czech-American husband and wife biochemists, first achieve the test-tube synthesis of glycogen and later (1947) win a Nobel Prize in physiology or medicine for their discovery of the role sugar plays in the metabolism of animals.

1943 Project Whirlwind, one of the most innovative and influential computer projects in computer history, begins at Massachusetts Institute of Technology under U.S. Navy sponsorship. It begins as a feasibility study for a general-purpose flight trainer or simulator. It will evolve into a project to design a digital computer. (See 1945.)

1944 Oswald T. Avery (1877–1955), Colin M. MacLeod (1909–1972), and Maclyn McCarty (1911–2005) publish a landmark paper on the pneumococcus transforming principle. The paper is entitled *Studies on the Chemical Nature of the Substance Inducing Transformation of Pneumococcal Types.* Avery suggests that the transforming principle seems to be deoxyribonucleic acid (DNA), but contemporary ideas about the structure of nucleic acids suggest that DNA does not possess the biological specificity of the hypothetical genetic material.

1944 Archer John Porter Martin (1910–2002) and Richard Laurence Millington Synge (1914–1994), both English biochemists, first develop the technique of paper chromatography, for which they share a Nobel Prize in 1952. Their method of using porous filter paper to separate and identify the nearly-identical but different types of amino acids proves an instant success. With this method of being able to separate very closely related compounds, others are able to determine the number of particular amino acids in protein molecules, and the scheme of photosynthesis is eventually worked out.

1944 New techniques and instruments, such as partition chromatography on paper strips and the photoelectric ultraviolet spectrophotometer, stimulate the development of biochemistry after World War II (1939–1945). New methodologies make it possible to isolate, purify, and identify many important biochemical substances, including the purines, pyrimidines, nucleosides, and nucleotides derived from nucleic acids.

1945 Engineers at the Massachusetts Institute of Technology decide to convert Project Whirlwind from analog electronics for their computer to digital electronics.

1946 Joshua Lederberg (1905–2008), American geneticist, and Edward Lawrie Tatum (1909–1975), American biochemist and geneticist, first demonstrate genetic recombination in *E. coli* bacteria. This intermingling of the genetic material of bacteria greatly expands the type of genetic research that can be done.

1946 Edward Mills Purcell (1912–1997), American physicist, discovers nuclear magnetic resonance. This breakthrough becomes widely used to study the molecular structure of materials. It also leads to the field of nuclear medicine, which allows for a sensitive scan of the body to be taken to detect abnormalities.

1947 Fritz Albert Lipmann (1899–1986), German-American biochemist, discovers what he calls coenzyme A. This breakthrough in body chemistry leads to a better understanding of the process of metabolism and the production of energy. He wins the Nobel Prize for this in 1953.

1948 Katherine Koontz Sanford (1948–) isolates a single mammalian cell in vitro and allows it to propagate to form identical descendants. Her clone of mouse fibroblasts was called L929, because it took 929 attempts before a successful propagation was achieved. The picture of this clone has often been reproduced. It was an important step in establishing pure cell lines for biomedical research.

1948 Norbert Wiener (1894–1964), American mathematician, publishes *Cybernetics,* which has a major influence on research into artificial intelligence. He notes that any machine with pretensions to intelligence must be able to modify its behavior in response to feedback from its environment.

1949 Dorothy Crowfoot Hodgkin (1910–1994), English biochemist, is the first to use an electronic computer in direct application to a biochemical problem. She uses a computer to work out x-ray data in her studies of the structure of penicillin.

1950 Erwin Chargaff (1905–2002), Austrian-American biochemist, establishes his rules about the basic chemistry of DNA. His findings contribute to greater understanding made during the next five years about the chemical structures of nucleic acids.

1950 Embryos are implanted in cattle for the first time.

1950 Engineering Research Associates builds the ERA 1101, the first commercially produced computer in the Unites States to offer drum-storage technology. Originally known as the Atlas, the computer has a capacity of two million bits.

1951 An Wang (1920–1990) forms Wang Laboratories. Wang's first major contribution to information technology was his invention of the magnetic "Pulse Transfer Controlling Device." By 1965 Wang Laboratories had designed desktop computers.

1951 Betty Holberton (1917–2001) manufactures a soft merge generator; future developments of this technology yield the compiler.

1951 First successful oral contraceptive drug is introduced. Gregory Pincus (1903–1967), American biologist, discovers a synthetic hormone that renders a woman infertile without altering her capacity for sexual pleasure. It soon is marketed in pill form and effects a social revolution with its ability to divorce the sex act from the consequences of impregnation.

1952 Rosalind Franklin (1920–1958) gives a seminar on her x-ray crystallography studies of two forms of DNA. Her colleague, Maurice Hugh Frederick Wilkins (1916–2004), gives information about her work to James Watson (1928–).

1952 Charles Hufnagel (1916–1989) of the United States inserts the first artificial heart valve.

1953 Francis Crick (1916–2004), English biochemist, and James Watson (1928–), American biochemist, work out the double-helix or double spiral DNA model. This model explains how it is able to transmit heredity in living organisms.

1953 Jonas Edward Salk (1914–1995) begins testing a polio vaccine comprised of a mixture of killed viruses.

1954 John Franklin Enders (1897–1985), American microbiologist, and Thomas Peebles, American pediatrician, develop the first vaccine for measles. A truly practical and successful vaccine requires more time.

1954 Jonas Edward Salk (1914–1995), American virologist, produces the first successful anti-poliomyelitis vaccine, which prevents paralytic polio. It is soon (1955) followed by the Polish-American virologist Albert Bruce Sabin's (1906–1993) development of the first oral vaccine.

1955 Narinder S. Kapany, Indian physicist, introduces fiber optics. By surrounding glass fiber with cladding, he takes advantage of the phenomenon of total internal reflection, meaning that there is very little loss of intensity over a great distance. Optical fibers prove to be a revolutionary tool in such fields as medicine and telecommunications.

1956 John McCarthy (1929–), American mathematician and computer scientist, coins the term "artificial intelligence."

1957 John McCarthy, American mathematician, and Marvin Lee Minsky (1927–), American electrical engineer, founders of artificial intelligence (AI) research, establish the AI Lab at Massachusetts Institute of Technology. This will be the birthplace of McCarthy's LISP programming language.

1957 Arthur Kornberg (1918–2007), American biochemist, first forms synthetic molecules of DNA, although they are not biologically active.

1957 The World Health Organization advances the oral polio vaccine developed by Albert Sabin as a safer alternative to the Jonas Salk vaccine.

1957 The structure of viruses is determined. Through the work of Heinz Fraenkel-Conrat (1910–1999), German-American biochemist, viruses are seen as consisting of a hollow protein shell with a nucleic acid molecule within.

1958 George Wells Beadle (1903–1989), American geneticist, and Edward Lawrie Tatum (1909–1975), American biochemist, are awarded the Nobel Prize in physiology or medicine for their discovery that genes act by regulating definite chemical events. Joshua Lederberg (1925–2008), American geneticist, also is awarded the Nobel Prize in physiology or medicine for his discoveries concerning genetic recombination and the organization of the genetic material of bacteria.

1958 Frederick Sanger (1918–), English biochemist, is awarded the Nobel Prize in chemistry for his work on the structure of proteins, especially for determining the primary sequence of insulin.

1958   Francis Crick (1916–2004), English biochemist, predicts the discovery of transfer DNA.

1959   Severo Ochoa (1905–1993), Spanish-American biochemist, and Arthur Kornberg (1918–2007), American biochemist, are awarded the Nobel Prize in physiology or medicine for their discovery of the mechanisms in the biological synthesis of ribonucleic acid and deoxyribonucleic acid.

1959   Robert Louis Sinsheimer (1920–) reports that bacteriophage +X174, which infects *E. coli,* contains a single-stranded DNA molecule, rather than the expected double stranded DNA. This provides the first example of a single-stranded DNA genome.

1959   English biochemist Rodney Porter (1917–1985) begins studies that lead to the discovery of the structure of antibodies. Porter receives the 1972 Nobel Prize in physiology or medicine for this research.

1960   The first totally implantable pacemakers are introduced for human hearts.

1961   Francis Crick (1916–2004), English biochemist, Sydney Brenner (1927–), and others propose that a molecule called transfer RNA uses a three base code in the manufacture of proteins.

1961   Marshall Warren Nirenberg (1927–) and Severo Ochoa (1905–1993), both American biochemists, work out the beginning of the RNA messenger code. RNA contains instructions for carrying out cellular processes, and it works with DNA, which contains the organism's genetic information.

1961   Marshall Warren Nirenberg (1927–), American biochemist, synthesizes a protein molecule by using artificial DNA.

1962   Francis Crick (1916–2004), James Watson (1928–), and Maurice Wilkins (1916–2004), New Zealand-English physicist, are awarded the Nobel Prize in physiology or medicine for their discoveries concerning the molecular structure of nucleic acids and its significance for information transfer in living material.

1964   The U.S. National Library of Medicine introduces a computer-based system for the analysis and retrieval of medical literature (MEDLARS).

1965   François Jacob (1920–), French biologist, André-Michael Lwoff (1902–1994), French microbiologist, and Jacques-Lucien Monod (1910–1976), French biochemist, are awarded the Nobel Prize in physiology or medicine for their discoveries concerning genetic control of enzyme and virus synthesis.

1965   James M. Schlatter, American chemist, combines two amino acids and discovers the mixture's sweet taste. This chemical is about 200 times as sweet as sugar and is named aspartame. In 1983 it is approved for use in carbonated beverages. For decades, it remains the most widely-used artificial sweetener.

1966   Michael Ellis DeBakey (1908–), American surgeon, designs an artificial left ventricle that he implants in a patient's heart.

1966   Paul D. Parman and Harry M. Myer, Jr., develop a live-virus rubella vaccine.

1967   The new fertility drug clomiphene is introduced. Although it can result in multiple births, it proves very successful in increasing a woman's chances of getting pregnant.

1967   Arthur Kornberg (1918–2007), American biochemist, and colleagues synthesize biologically active DNA using one strand of natural DNA from a virus.

1967   The first volume of *Annual Review of Genetics* (ARG) is published, adding it to the *Annual Review* series. Volume 1 is dedicated to Hermann Joseph Muller (1890–1967).

1967   Charles T. Caskey (1938–), Richard E. Marshall, and Marshall Warren Nirenberg (1927–) state that because amino acids, bacteria, and guinea pigs have identical forms of messenger RNA, there is a universal genetic code shared by all life forms.

1968   Robert William Holley (1922–1993), American chemist, Har Gobind Khorana (1922–2011), Indian-American chemist, and Marshall Warren Nirenberg (1927–), American biochemist, are awarded the Nobel Prize in physiology or medicine for their interpretation of the genetic code and its function in protein synthesis.

1968   Werner Arber (1929–) of Switzerland discovers that bacteria defend themselves against viruses by producing DNA-cutting enzymes. This understanding later proves valuable to future genetic engineering, and the enzymes quickly become important tools for molecular biologists.

1969 Max Delbrück (1906–1981), German-American microbiologist, Alfred Day Hershey (1908–1997), American microbiologist, and Salvador Edward Luria (1912–1991), Italian-American microbiologist, are awarded the Nobel Prize in physiology or medicine for their discoveries concerning the replication mechanism and the genetic structure of viruses.

1969 Stanford Moore (1913–1982) and William Howard Stein (1911–1980), both of the United States, independently identify the active site of an enzyme (ribonuclease from bovine pancreas).

1969 B. Gutte and Robert Bruce Merrifield (1921–2006) complete the chemical synthesis of ribonuclease.

1969 The first artificial heart is implanted in a human being. Denton A. Cooley (1920–), American surgeon, implants a mechanical heart made of silicon. This temporary device keeps the patient alive for 65 hours until a human heart is implanted to replace it. The patient dies 38 hours after this second operation due to pneumonia and kidney failure.

1970 Har Gobind Khorana (1922–2011), Indian-American chemist, and colleagues announce the first complete synthesis of a gene. In making analine-transfer RNA, they do not use a natural gene for a template, but assemble the gene directly from its component chemicals.

1970 Hamilton Othanel Smith (1931–) and Kent Wilcox isolate the first restriction enzyme, HindII, an enzyme that cuts DNA molecules at specific recognition sites.

1970 Patrick Steptoe (1913–1988) and Robert Geoffrey Edwards (1925–), both English physicians, accomplish in vitro fertilization in humans (test-tube babies).

1970 Howard Martin Temin (1934–1994), American virologist, and David Baltimore (1938–), American biochemist, discover reverse transcriptase in viruses. This is an enzyme that causes RNA to be transcribed onto DNA.

1972 Christian Boehmer Anfinsen (1916–1995), American biochemist, is awarded the Nobel Prize in chemistry for his work on ribonuclease, especially the amino acid sequence and the biologically active confirmation. Stanford Moore (1913–1982) and William Howard Stein (1911–1980), both American biochemists, are awarded the Nobel Prize in chemistry for their contribution to the understanding of the connection between chemical structure and catalytic activity of the active center of the ribonuclease molecule.

1972 Recombinant technology emerges as one of the most powerful techniques of molecular biology. Scientists are able to splice together pieces of DNA to form recombinant genes. As the potential uses, therapeutic and industrial, became increasingly clear, scientists and venture capitalists establish biotechnology companies.

1972 The Prolog logic programming language is developed by Robert A. Kowalski and Alain Colmerauer; it is often used for artificial intelligence programming.

1972 Gerald Maurice Edelman (1928–), American biochemist, and Rodney Robert Porter (1917–1985), English immunologist, are awarded the Nobel Prize in physiology or medicine for their discoveries concerning the chemical structure of antibodies.

1973 Concerns about the possible hazards posed by recombinant DNA technologies, especially work with tumor viruses, leads to the establishment of a meeting at Asilomar, California. The proceedings of this meeting are subsequently published by the Cold Spring Harbor Laboratory as a book entitled *Biohazards in Biological Research*.

1973 Genetic engineering is actually conducted as Herbert Wayne Boyer (1936–), American microbiologist, and Stanley H. Cohen (1922–) cut DNA molecules using restriction enzymes, join them together with other enzymes, and reproduce them in the bacterium *Escherichia coli*.

1975 David Baltimore (1938–), American biochemist, Renato Dulbecco (1914–), Italian-American virologist, and Howard Martin Temin (1934–1994), American oncologist, are awarded the Nobel Prize in physiology or medicine for their discoveries concerning the interaction between tumor viruses and the genetic material of the cell.

1976   Martin F. Gellert (1929–), Czechoslovakian-American biochemist, and colleagues discover the enzyme (gyrase) that causes the DNA double helix to form a larger helix called a supercoil.

1976   Har Gobind Khorana (1922–2011), Indian-American chemist, and colleagues build a functional, synthetic gene that has a complete system of regulatory mechanisms.

1977   Robert B. McGhee, American electrical engineer, invents the OSU Hexapod, a robotic walker whose computer-controlled leg motors allow it to climb over obstacles.

1977   Godfrey Hounsfield of England makes magnetic resonance imaging (MRI) practicable and gives physicians their first detailed pictures of the body's soft tissues. MRI uses a computer to control an intense electromagnetic field and radio waves that produce an image.

1977   Frederick Sanger (1918–) English biochemist, reports the sequencing of the complete genetic information of a microorganism.

1977   Genentech, the first genetic engineering company, is founded in order to use recombinant DNA methods to make medically important drugs.

1977   Philip Allen Sharp (1944–) and Richard John Roberts (1943–) independently discover that the DNA making up a particular gene could be present in the genome as several separate segments. Although both Roberts and Sharp use a common cold-causing virus, called adenovirus, as their model system, researchers later find "split genes" in higher organisms, including humans. Sharp and Roberts are subsequently awarded the Nobel Prize in physiology or medicine in 1993 for the discovery of split genes.

1978   Werner Arber (1929–), Swiss microbiologist, Daniel Nathans (1928–1999), American microbiologist, and Hamilton Othanel Smith (1931–), American microbiologist,

are awarded the Nobel Prize in physiology or medicine for the discovery of restriction enzymes and their application to problems of molecular genetics.

1978   The U.S. drug companies Genentech and Eli Lilly announce the production of human insulin by genetic engineering.

1978   The first test-tube baby is born in England. This is the first human being conceived outside the human body to be born. The fertilized egg was implanted by English physicians Patrick Steptoe (1913–1988) and Robert Geoffrey Edwards (1925–) in the mother's uterus and developed normally.

1978   Scientists from Genentech clone the gene for human insulin.

1978   Somatostatin becomes the first human hormone produced by using recombinant DNA technology.

1978   Walter Gilbert (1932–) and others establish Biogen, a pioneering biotechnology company.

1978   Werner Arber, Daniel Nathans, and Hamilton O. Smith are awarded the Nobel Prize in physiology or medicine for the discovery of restriction enzymes and their application to problems of molecular genetics.

1980   Paul Berg (1926–), American biochemist, is awarded the Nobel Prize in chemistry for his fundamental studies of the biochemistry of nucleic acids, with particular regard to recombinant DNA. Walter Gilbert (1932–), American microbiologist, and Frederick Sanger (1918–), English biochemist, are also awarded the Nobel Prize in chemistry for their contributions concerning the determination of base sequences in nucleic acids.

1980   The Supreme Court of the United States rules in the case *Diamond v. Chakrabarty* that a "live, human made microorganism," a microbe developed by General Electric that helps clean up an oil spill, is patentable.

1981   Karl Oskar Illmensee (1939–) clones baby mice.

1981   Researchers in China successfully clone a fish.

1982 The U.S. Food and Drug Administration approves the first genetically engineered drug. Eli Lilly & Company is permitted to market human insulin that is produced by bacteria.

1982 First Jarvik 7 artificial heart is implanted by William DeVries (1933–), American surgeon. The patient, Barney Clark, lives for 112 days.

1983 First artificial chromosome is created by Andrew W. Murray and Jack William Szostak (1952–), Canadian biochemist, who work with yeast.

1983 Barbara McClintock (1902–1992), American geneticist, is awarded the Nobel Prize for physiology or medicine for her discovery of mobile genetic elements.

1983 Los Alamos National Laboratory, a Department of Energy Laboratory (LANL), and Lawrence Livermore National Laboratory, a Department of Energy Laboratory (LLNL), begin production of DNA clone (cosmid) libraries representing single chromosomes.

1984 Allan Charles Wilson (1934–1991), American biochemist, and Russell Higuchi first clone genes from an extinct species, as they take a gene from the preserved skin of a quagga, a type of zebra.

1984 Steen Willadsen successfully clones sheep via nuclear transfer.

1984 The U.S. Department of Energy (DOE), Office of Health and Environmental Research (OHER, now Office of Biological and Environmental Research), U.S. Department of Energy, and the International Commission for Protection Against Environmental Mutagens and Carcinogens (ICPEMC) cosponsor the Alta, Utah, conference highlighting the growing role of recombinant DNA technologies. The Office of Technology Assessment incorporates the proceedings of the meeting into a report acknowledging the value of deciphering the human genome.

1985 Alec Jeffreys (1959–) first develops a method of "fingerprinting" with DNA. He shows there are certain core sequences of DNA that are unique to each individual.

1985 Kary Mullis (1944–), who was working at Cetus Corporation, develops the polymerase chain reaction (PCR), a new method of amplifying DNA. This technique quickly became one of the most powerful tools of molecular biology. Cetus patented PCR and sold the patent to Hoffman-LaRoche in 1991.

1985 The U.S. Food and Drug Administration approves the marketing of a genetically-engineered growth hormone.

1986 Following a Santa Fe, New Mexico, conference, the Department of Energy (DOE) officially initiates the Human Genome Initiative. Pilot projects at DOE national laboratories were given a budget of $5.3 million to develop critical resources and technologies.

1986 The virus that causes AIDS (acquired immunodeficiency syndrome) is isolated. Called the human immunodeficiency virus (HIV), it is part of a group of viruses known as retroviruses. No cure or vaccine has become available as of January 2012.

1987 A Congress-chartered Department of Energy (DOE) advisory committee, the Health and Environmental Research Advisory Committee (HERAC), recommends a 15-year, multidisciplinary, scientific, and technological undertaking to map and sequence the human genome. DOE designates multidisciplinary human genome centers. The National Institute of General Medical Sciences at the National Institutes of Health (NIH NIGMS) begins funding genome projects.

1987 Maynard Olson creates and names yeast artificial chromosomes (YACs), which provide a technique to clone long segments of DNA.

1987 Genetically engineered plants are first developed.

1987 The first criminal conviction is obtained on the basis of genetic information evidence. Following this landmark use of DNA or "genetic fingerprinting" in England, genetic evidence becomes increasingly admissible and accepted in a court of law.

1988 Scientists announce the synthesis of ginkgolide B, which they believe is the active principle in ginkgo herbal remedies. This chemical relieves allergies by suppressing the immune system.

1988   The U.S. National Institutes of Health approves the injection of genetically-altered cells into humans.

1988   Henry A. Erlich of the United States and colleagues develop a method for identifying an individual from the DNA in a single hair.

1988   Philip Leder (1934–) and Timothy Stewart of Harvard Medical School are granted the first U.S. patent for a genetically-altered vertebrate. This "transgenic non-human mammal"—a genetically-altered mouse—is developed by genetic engineering to be highly susceptible to breast cancer.

1989   Seven cloned calves are born from the same embryo.

1989   DNA sequence tagged sites (STSs) are recommended as a method for correlating diverse types of DNA clones.

1989   James Watson (1928–) is appointed head of the National Center for Human Genome Research. The agency was created to oversee the $3 billion budgeted for the American plan to map and sequence the entire human DNA by 2005.

1990   The Human Genome Project begins in the United States with the selection of six institutions to do the work.

1990   Joseph Edward Murray (1919–), American surgeon, and Edward Donnall Thomas (1920–), American physician, are awarded the Nobel Prize in physiology or medicine for their discoveries concerning organ and cell transplantation in the treatment of human disease.

1991   The U.S. Food and Drug Administration announces it will speed up its process for approving drugs. This change in procedure is due to the protests of AIDS activists.

1991   Richard Robert Ernst, Swiss chemist, is awarded the Nobel Prize in chemistry for his contributions to refining the technology of nuclear magnetic resonance imaging.

1991   The Genome Database, a human chromosome mapping data repository, is established.

1992   American and British scientists describe a technique for testing embryos in vitro for genetic abnormalities such as cystic fibrosis and hemophilia.

1992   J. Craig Venter (1946–) establishes The Institute for Genomic Research (TIGR) in Rockville, Maryland. TIGR later sequenced *Haemophilus influenzae* and many other bacterial genomes.

1992   Guidelines for data release and resource sharing related to the Human Genome Project are announced by the U.S. Department of Energy and National Institutes of Health.

1992   Low-resolution genetic linkage maps of the entire human genome are published.

1992   James Watson (1928–), American biochemist, leader of the Human Genome Project, resigns. Many believe he is leaving the project—whose goal is to locate all the genes that the DNA comprises within a human cell's 46 chromosomes and to determine the precise order of the 3 billion gene subparts or nucleotides that form the genetic code—because of differences on the issue of gene patenting.

1992   Francis S. Collins (1950–) replaces James Watson (1928–) as head of the National Center for Human Genome Research at the National Institutes of Health after Watson clashed with J. Craig Venter (1946–), then at NIH, over the patenting of DNA fragments known as "expressed sequence tags."

1992   The first genetically-altered food for human consumption, the Flavr Savr tomato, is approved by the U.S. Food and Drug Administration (FDA).

1993   French Gépnéthon makes its mega-YACs available to the genome community.

1993   George Washington University researchers clone human embryos and nurture them in a Petri dish for several days. The project provokes protests from ethicists, politicians, and critics of genetic engineering.

1993   Leonard Adleman demonstrates the computing potential of DNA (biological computing) by solving a previously unsolvable mathematical problem using sequences of DNA.

1993   Richard Roberts (1943–) and Phillip Sharp (1944–), both American molecular biologists, are awarded the Nobel Prize in physiology or medicine for their independent discovery that genes are often split (that the genetic instructions contained in DNA and used by the living cell to make proteins can be discontinuous).

1993   The U.S. Food and Drug Administration approves bovine somatropin (BST), a genetically-engineered synthetic hormone that increases the amount of milk given by dairy cows.

1994   The Department of Energy announces the establishment of the Microbial Genome Project as a spin-off of the Human Genome Project.

1994   The Genetic Privacy Act, the first United States Human Genome Project legislative product, proposes regulation of the collection, analysis, storage, and use of DNA samples and genetic information obtained from them. These rules are endorsed by the ELSI Working Group.

1994   The Human Genome Project Information Web site is made available to researchers and the public.

1995   A physical genome map with more than 15,000 STS markers is published.

1995   The genome of the bacterium *Haemophilus influenzae* is sequenced.

1995   Scientists develop a potent cocktail of drugs that help stop the progress of HIV. The treatment approach is called highly active antiretroviral therapy (HAART).

1995   Religious leaders and biotechnology critics protest the patenting of plants, animals, and human body parts.

1995   The National Library of Medicine unveils the "Visible Human," a digital image library of volumetric data representing complete, normal adult male and female anatomy.

1996   International participants in the genome project meet in Bermuda and agree to formalize the conditions of data access. The agreement, known as the "Bermuda Principles," calls for the release of sequence data into public databases within 24 hours.

1996   Combined drugs therapy for AIDS is first introduced at the eleventh World AIDS conference held in Vancouver, British Columbia, Canada.

1996   Composites of nanometer-size gold particles and DNA strands are developed for biological sensing and electronics use.

1996   Scientists report that a new microcoil makes it possible to take nuclear magnetic resonance images of nanoliter-size samples.

1996   The genome of baker's yeast, *Saccharomyces cerevisiae,* is sequenced by an international consortium of scientists. The short generation span and ease of genetic manipulation of the yeast provide a suitable system to study basic biological processes that are relevant for many other higher eukaryotes, including humans.

1996   The sequence of the *Methanococcus jannaschii* genome provides further evidence of the existence of a third major branch of life on earth, the archaea. This genome was the first ever sequenced by the whole genome shotgun method.

1996   The Wellcome Trust sponsors a large-scale sequencing strategy meeting for international coordination of human genome sequencing.

1997   The complete genome sequence of the common *Escherichia coli* bacterium is determined. This accomplishment gives researchers a powerful new tool for understanding fundamental questions of biological evolution and function.

1997   Ian Wilmut (1944–) of the Roslin Institute in Edinburgh, Scotland, announces the birth of a lamb called Dolly, the first mammal cloned from an adult cell (a cell in a pregnant ewe's mammary gland).

1997   Plastic biofilms are developed that prevent bacteria from binding to them, helping to prevent the potentially deadly infections that can develop around medical implants.

1997   Scientists announce that they have cloned rhesus monkeys from early stage embryos, using nuclear transfer methods.

1997   Stanley B. Prusiner (1942–) wins the Nobel Prize in physiology or medicine for his studies of prions, a new infectious agent composed of malformed proteins and implicated in neurological disorders such as the human form of mad cow disease.

1997 The FBI announces its new National DNA Index System (NDIS) on December 8, allowing forensic science laboratories to link serial violent crimes to each other and to known sex offenders through the electronic exchange of DNA profiles.

1997 The National Human Genome Research Institute (NHGRI) is recognized as a collaborative institute; physical maps of chromosomes X and 7 are announced.

1998 A new genetically engineered rabies vaccine costs a fraction of the regular vaccine, making the rabies vaccine more available to people in developing countries, where over 40,000 still die every year from the disease.

1998 A professor of cybernetics at the University of Reading in the United Kingdom becomes the first human to host a microchip. The approximately 23mm-by-3mm glass capsule containing several microprocessors stays in Warwick's left arm for nine days, and it was used to test the implant's interaction with computer-controlled doors and lights in a futuristic office building.

1998 British and American scientists complete the genetic map of the nematode, *Caenorabditis elegans*. The genetic map, showing the 97 million genetic letters, in correct sequence, derived from the worm's 19,900 genes, was the first completed genome of an animal.

1998 J. Craig Venter (1946–) forms a company (later named Celera), and predicts that the company will decode the entire human genome within three years. Celera plans to use a "whole genome shotgun" method, which would assemble the genome without using maps. Venter initially asserts that his company will not follow the Bermuda principles concerning data release.

1998 Dolly, the first cloned sheep, gives birth to a lamb that has been conceived by a natural mating with a Welsh Mountain ram. Researchers state the birth of Bonnie proves that Dolly is a fully normal and healthy animal.

1998 Initial AIDS vaccine trials begin with a large-scale test with 5,000 volunteers in 30 cities in the United States and a smaller group in Thailand.

1998 Scientists in Korea claim to have cloned human cells.

1998 University of Hawaii scientists, improving Ian Wilmut's technique, clone a mouse from cumulus cells. The process is repeated for three generations, yielding over 50 cloned mice by the end of July.

1998 The genome of the *Mycobacterium tuberculosis* bacterium is sequenced.

1998 The Department of Energy and National Institutes of Health announces a new five-year plan for the Human Genome Project, which predicts that the project will be completed by 2003.

1998 Two research teams succeed in growing embryonic stem cells.

1999 A milestone in the human genome project is completed when the first full sequence of an entire chromosome is finished (chromosome 22).

1999 Pharmaceutical research in Japan leads to the discovery of donepezil (Aricept), the first drug intended to help ward off memory loss in Alzheimer's disease and other age-related dementias.

1999 Japanese scientists present the first color artificial retina.

2000 A modified version of the Jarvik heart (the Jarvik 2000, about the size of a "C" battery) is implanted in a patient as a bridge to transplantation, the first completely artificial heart to be installed.

2000 An International research consortium publishes the genome of the smallest human chromosome, chromosome 21.

2000 International collaborators publish the genome of the fruit fly *Drosophila melanogaster*, a creature useful in studies of genetics because of its short life span.

2000 Scientists demonstrate that mouse brain stem cells, when placed in physical contact with muscle cells, can differentiate into muscle cells.

2000 The Cartagena Protocol on Biosafety is adopted in Montreal, Canada. The protocol, negotiated under the United Nations Convention on Biological Diversity, is one of the first legally binding international agreements to govern the trade or sale of genetically modified organisms of agricultural importance.

2000 The first retinal implant to move into clinical trials is a subretinally placed Artificial Silicon Retina (ARS), a very small silicon microchip 2 mm in diameter and 25 micrometers thick, composed of miniature solar cells that respond to light similar to natural photoreceptors.

2000 The first volume of *Annual Review of Genomics and Human Genetics* is published. Genomics is defined as the new science dealing with the identification and characterization of genes and their arrangement in chromosomes and human genetics. It is a science devoted to understanding the origin and expression of human individual uniqueness.

2000 At the White House, the Human Genome Project public consortium and Celera jointly announce that both organizations have sequenced a "working draft" of the human genome.

2001 A company in suburban Boston announces that a human cell has been cloned to provide stem cells for research. While the experiment was carried on for only a few cell divisions, the technology required to develop a cloned human is a significant, though controversial, advance.

2001 European countries, including France and Germany, push for tough European Union rules regulating the sale of genetically modified foods. The U.S. State Department brands the new rules without scientific merit.

2001 President George W. Bush (1946–) announces the United States will allow and support limited forms of stem cell research on existing stem cell lines, but will ban federal funding of research on new embryonic stem cell lines.

2001 Researchers create nano-sized bar codes to label molecules.

2001 In a report published in the international scientific journal *Nature*, researchers report findings that indicate the human genome consists of far fewer genes than previous projected by estimation of phylogenic relationships between humans and other species.

2001 Scientists from the Whitehead Institute announce test results that show patterns of errors in cloned animals that might explain why such animals die early and exhibit a number of developmental problems. The report stimulates new debate on ethical issues related to cloning. In the journal *New Scientist* Ian Wilmut (1944–), the scientist who headed the research team that cloned the sheep Dolly, argues that the findings argue for "a universal moratorium against copying people."

2001 Scientists clone the rare, endangered mouflon sheep.

2001 Scientists create the first genetically engineered primate, a rhesus monkey.

2001 Scientists transform embryonic stem cells into insulin-secreting cells, heart cells, and bone marrow cells.

2001 The complete draft sequence of the human genome is published. The public sequence data is published in the British journal *Nature* and the Celera sequence is published in the American journal *Science*, giving genetic researchers easy access to specific gene sequences.

2001 The International Rice Genome Sequencing Project announces that it will complete the sequence of the *Oryza japonica* rice genome by the end of the year in order to ensure that the sequence data will be accurate and freely available.

2001 Scientists from Advanced Cell Technology clone the first endangered animal, a bull gaur (a wild ox from Asia). The newborn dies after two days due to infection.

2002 A national advisory panel of scientists recommends that cloning aimed at creating a child should be outlawed, but recommends allowing medical experiments with cloned human cells.

2002 A private company, Clonaid, announces the birth of the first cloned human, a seven-pound girl nicknamed Eve. The announcement is discounted by scientists, as Clonaid is funded by a religious sect whose tenants hold that humans were initially cloned from extra-terrestrial visitors to Earth, and no verifiable evidence is presented.

2002 Biochemists discover that starchy foods become contaminated by the animal carcinogen acrylamide when fried, and scientists scurry to find the threshold for human exposure and risk.

Chronology

2002 Biologists sequence the genome of the main malaria-causing parasite *Plasmodium falciparum*, along with the mosquito that normally carries the disease.

2002 Reports surface that scaremongering concerning genetically modified foods causes several African countries fighting starvation to reject genetically modified food supplements that would have reduced starvation and death rates.

2002 The Mouse Genome Sequencing Consortium publishes its draft sequence of the mouse genome in the journal *Nature*.

2002 Published reports of genome analysis provide evidence that early migrant populations of humanoids may have been able to intermix with established or indigenous humanoid populations to a greater degree than previously believed.

2002 Spurred by threats of bioterrorism, researchers sequence the genome for the bacterium responsible for anthrax.

2002 The United Nations holds an Earth Summit in Johannesburg, South Africa, to focus on international regulations that address environmental problems: water and air quality, accessibility of food and water, sanitation, agricultural productivity, and land management, that often accompany the human population's most pressing social issues: poverty, famine, disease, and war.

2002 In the aftermath of the September 11, 2001, terrorist attacks on the United States, the U.S. government dramatically increases funding to stockpile drugs, vaccines, and other agents that could be used to counter a bioterrorist attack.

2002 Evidence shows trace amounts of drugs excreted by humans eventually enter waterways, where they combine to harm native microscopic organisms.

2003 Preliminary trials for a malaria vaccine are scheduled to begin in malaria-endemic African areas, where approximately 3,000 children die from the disease every day.

2003 While work still continues on the Human Genome Project, scientists are also beginning a Genomes to Life research program designed to identify and characterize the

protein complexes important in animal, especially human, and microbial cell reactions and to further identify the specific genes that regulate these processes.

2003 Smart passports, fitted with microchips that will allow immigration officials to identify the facial biometric features of the passport holder, are proposed.

2003 The Human Genome Project of the National Institutes of Health culminates in the completion of the full human genome sequence, published in the journal *Nature*.

2003 U.S. Congress enacts the Twenty-first Century Nanotechnology Research and Development Act, which authorizes funding for and government agency participation in the National Nanotechnology Initiative (NNI).

2003 Plagued with a chronic and progressive lung disease, veterinarians are forced to humanely euthanize Dolly, the first cloned mammal.

2003 Canadian scientists at the British Columbia Cancer Agency in Vancouver announce the sequence of the genome of the coronavirus most likely to be the cause of SARS. Within days, scientists at the Centers for Disease Control (CDC) in Atlanta, Georgia, offer a genomic map that confirms more than 99% of the Canadian findings.

2004 The International Human Genome Sequencing Consortium publishes its scientific description of the finished human genome sequence, reducing the estimated number of human protein-coding genes from 35,000 to only 20,000–25,000, a surprisingly low number for the human species.

2005 H5N1 virus, responsible for avian flu, moves from Asia to Europe, as it is found in wild ducks in Romania. The World Health Organization attempts to coordinate multinational disaster and containment plans. Some nations begin to stockpile antiviral drugs.

2005 Korean scientist Woo Suk Hwang's (1953–) claims to have cloned human stem cells lines are proved fraudulent. Despite this setback, the potential for therapeutic cloning for medical purposes continues unabated.

2005 The Food and Drug Administration Drug Safety Board is founded.

2006 U.S. researchers experiment with a genetically modified, flood-tolerant rice.

2006 The U.S. government approves genetically modified (GMO) rice for human consumption.

2006 Norway announces plans to build a "doomsday vault" in a mountain close to the North Pole that will house a two-million-crop seed bank in the event of catastrophic climate change, nuclear war, or rising sea levels.

2006 Defense Advanced Research Projects Agency (DARPA) embarks on the Revolutionizing Prosthetics Program aiming to create by 2016 a prosthetic arm that responds to neural control and has full sensory and motor capabilities.

2006 In an effort to aid vaccine development, World Health Organization influenza pandemic task force officials ask that all countries share H5N1 (avian flu) virus samples and genetic sequencing results.

2007 A team of researchers and engineers at the Johns Hopkins University Applied Physics Laboratory (APL) develops a prototype of a fully integrated, naturally controlled, prosthetic arm.

2007 The U.S. Food and Drug Administration (FDA) concludes that food products containing meat or products from cloned animals and their offspring are safe for human consumption.

2007 The fourth Intergovernmental Panel on Climate Change (IPCC) report is issued (the first segment in February, and the last in November). The IPCC, composed of scientists from 113 countries, issues a consensus report stating that global warming is caused by man, and predicting that warmer temperatures and rises in sea level will continue for centuries, no matter how much humans control their pollution.

2007 Environmental group Greenpeace launches an attack on genetically modified corn developed by U.S. biotech company Monsanto, saying that rats fed on one variety developed liver and kidney problems.

2007 University of Copenhagen in Denmark professor Henrik Clausen announces in the journal *Nature Biotechnology* the discovery of enzymes (proteins that help control the rates of reaction) and a method to convert any blood type into the universal Type O, a discovery that could lead to reduction in blood shortages.

2007 A group of U.S.-based researchers announce a major step toward the synthetic creation of life when they publish work in the journal *Science* showing that they synthetically engineered the construction of the genome (the entire set of genes) of a common laboratory bacterium from the elemental building blocks of DNA molecules.

2007 Stem cells are undifferentiated cells that can give rise to diverse types of differentiated (specialized) cells. Studies announced in 2007 show the ability to create stem cells from cloned monkey embryos and normal adult skin cells rather than from destroyed human embryos and may offer a potential solution to ethical concerns about the origins of stem cells.

2008 The 1000 Genome Project embarks on its mission to sequence the profiles of a large group of people in order to catalogue and better understand variation in humans.

2008 Agricultural testing demonstrates that a genetically modified, drought-tolerant wheat developed to boost harvests in water-challenged areas yields up to 20 per cent more harvestable wheat than similar non-modified crops used as research controls.

2008 In the United Kingdom, the House of Lords Science and Technology Committee issues a lengthy report on nanotechnology and food, warning its country's food industry not to hide the use of nanotechnology.

2008 Oil and food prices rise sharply on a global scale, increasing dangers of famine and poverty. Critics contend increased prices for petroleum lead to the diversion of food crops to biofuel production.

2008 Researchers using sophisticated genetics tests assert that the HIV virus causing AIDS passed into humans in Africa from chimpanzees being butchered for meat no later than the late nineteenth or early twentieth century.

2008 Trials of an HIV vaccine on non-human primates show that a vaccine based on cell-mediated immunity failed to prevent HIV-1 infection. The vaccine also failed to produce significant reductions in levels of the virus in newly infected specimen. Researchers continue to test whether cell-mediated immune responses might reduce replication of HIV.

2009 The U.S. Food and Drug Administration (FDA) approves clinical trials for human embryonic stem cell therapies.

2009 The U.S. Congress renews the Dickey Amendment (also known as the Dickey-Wicker Amendment, first introduced in 1995 and continuously renewed) as part of a larger appropriations bill prohibiting federally appropriated funding of any research involving the creation or destruction of human embryos.

2009 The U.S. Environmental Protection Agency (EPA) embarks on a research inquiry to better understand the nanomaterials used in consumer goods and explore possible strategies for regulation.

2010 A boy in the United Kingdom receives a transplanted artificial esophagus.

2010 The *Deepwater Horizon* oil spill in the Gulf of Mexico is effectively stopped when a well cap is successfully placed. Cleanup efforts include use of microorganisms selected, modified, and tested using an array of biotechnology.

2011 An international team of researchers constructs the first carbon-nanotube yarns to build artificial muscles.

2011 Doctors in Sweden implant the world's first synthetic trachea. The organ is created using a trachea-shaped scaffold lined with the patient's own stem cells to minimize the possibility of rejection.

Advanced Biomass Cookstoves

■ Introduction

Cookstoves have been a feature of all human societies since the time that humans evolved as a distinct animal species. Humans learned that fire provided them with the means to prepare better food. The ability to control the flame produced from the combustion of wood, charcoal, and other fuels permitted primitive human communities to consume cooked meat and boil water. When built inside a dwelling or other enclosed space, cookstoves could be used in any weather. The development of the cookstove is an important benchmark in the course of human evolution.

The prevalence of cookstoves in modern societies is an important distinction between the nations of the developed and the developing world. In addition to the time and effort expended on a daily basis to find sufficient fuel for cookstove fires, the indoor air pollution produced has been identified as a significant health hazard. Deforestation has resulted in many regions where traditional cookstoves are used, as precious wood is consumed at a rate faster than trees can be replaced. The release of "black carbon," the particulates contained in cookstove smoke, is a phenomenon cited by environmentalists as an important contributor to global climate change.

Since 1990, advanced biomass cookstoves have been successfully introduced where environmental and social impacts associated with traditional cookstove use have been the greatest. These modern cookstoves permit more efficient fuel combustion. The harmful smoke and its toxins are vented away from closed interior living spaces.

■ Historical Background and Scientific Foundations

In the twenty-first century, approximately 3 billion people remain dependent on cookstoves for their subsistence. The basic cookstove technology has remained largely unaltered for many millennia. In its most typical form, the stove is constructed from three stones or other similar supports of roughly even height that permit a pot or other cooking vessel to be positioned securely above the fire. Wood and charcoal are the most common biomass products used as cookstove fuel.

To keep the cookstove fire secure from the effects of rain or wind, cookstoves are placed indoors, where they can be tended throughout the day. In its three-stone form, the stove has no external venting. The smoke produced by its fire is released directly into the indoor living environment.

The traditional cookstove design is inefficient in every aspect of its use, from the effort required to obtain sufficient fuel to the actual cooking in the pots placed above the its fire. The time spent by community members to gather cookstove fuel is significant, and in most cultures where use of cookstoves is widespread, it is women and children who are responsible to gather fuel and cook on the stove. Alternatively, where the fuel is purchased commercially, it represents a significant household expense. The cookstove design does not permit fuel used to be consumed to generate maximal heat energy. The heat transfer from the burning fuel to the cooking pot results in a waste of as much as 75 percent of the available energy. Further, the cookstove smoke has been identified both as a contributor to the greenhouse gases that are regarded as a primary culprit in climate change, and as a profound health hazard for those exposed to the pollutants contained in the smoke.

An advanced biomass cookstove has several design improvements that distinguish it from the traditional model. Fuel is burned in a smaller enclosed receptacle, as opposed to an open flame, to encourage more complete fuel combustion and greater energy production. The smoke produced by the advanced stove is externally vented. The cooking surface is semi-enclosed to maximize the effect of the heat generated within the stove. The advanced stoves are constructed from a variety of materials, including metal, cement, and clay.

WORDS TO KNOW

BIOMASS: Organic material used to produce energy. The most common biomass sources are aquatic or terrestrial vegetation, wood, animal waste, and human waste. Generally regarded as renewable energy sources, biomass is the source of fuels such as ethanol that are increasingly important in the maintenance of world energy supplies.

"BLACK CARBON": Black carbon, or soot, is particulate air pollution produced from biomass fuel combustion. When released into the atmosphere, these particles reflect sunlight otherwise absorbed by the Earth's surface and thereby contribute to increased atmospheric temperatures. For this reason, black carbon is closely linked to global warming and climate change.

DEFORESTATION: The permanent destruction of indigenous forests; the removal of forests that otherwise absorb carbon dioxide (CO_2) from the atmosphere contributes to increased CO_2 levels associated with the "greenhouse effect" prominent in all modern discussions concerning climate change.

HEAT TRANSFER EFFICIENCY: This expression is used to describe the efficiency of any process in which heat is generated to achieve a desired result. Expressed as a ratio, efficiency is defined as the amount of energy available for use versus the amount of energy expended in the process.

These materials are fabricated to reduce heat loss. The advanced cookstoves are commonly used with modern pots that are also designed to encourage greater energy efficiency. If the open area of the cooking pot top is reduced by 75 percent, the evaporation and resultant heat loss during cooking is reduced by the same amount.

■ Impacts and Issues

The threat posed to global environmental quality and human health by cookstove use has received prominent scientific attention in recent years. The precise relationship between cookstove smoke and climate change is subsumed in the larger debate concerning global warming, in which increased CO_2 levels are generally accepted as a contributing factor. On this basis, any cookstove that consumes less fuel, or otherwise burns fuel more efficiently, will make a positive contribution to global environmental quality.

The time spent in developing countries by persons engaged in the daily quest for cookstove fuel is time that could otherwise be directed to more profitable human pursuits. In India alone, an estimated 840 million people depend on cookstoves in their daily lives. The National Biomass Cookstoves Initiative (NBCI) sponsored by the Indian government is the most comprehensive project ever undertaken to convert traditional cookstove users to cooking on the advanced biomass models. A more energy efficient cookstove produces a three-pronged result apart from the improved cooking process: Time otherwise spent gathering fuel is saved, because available supplies will last longer; more fuel will be available in a given community; and the need to convert forests into biomass fuel is reduced.

An estimated 1.6 million people worldwide die every year as a result of prolonged exposure to cookstove pollution. The smoke and its particulates are directly associated with conditions such as pneumonia, tuberculosis, and cataracts. Further, the open fires used in traditional cookstoves are a leading cause of scalding, burns, and other domestic accidents, particularly among children, because the stoves are located within dwellings where living space is often confined.

The advanced biomass cookstove does not represent a quantum leap in the energy technologies available to developing countries. The traditional cookstoves and the more advanced modern designs each require combustible biomass in their operation. The modern cookstove confirms the proposition that a series of incremental improvements to a machine or process can often achieve benefits that are exponential. The seemingly modest advances of improved fuel combustion, external vents, greater heat available for cooking, and pots that retain heat more effectively combine to improve the living standards of the people who depend on cookstoves for their survival.

It is anticipated that there will be further advances in cookstove technology in the next decade that will promote further energy efficiency and better human health. However, the benefits of advanced biomass cookstoves are modest when viewed in the broader context of the prevailing living conditions in developing nations versus those found in the developed world. Further, concerns remain that the lessened environmental impacts generated through the use of advanced biomass cookstoves are at best a partial solution to the ongoing environmental impact associated with biomass fuels, no matter how efficiently the modern stoves may function relative to the traditional cooking technology.

SEE ALSO *Biofuels, Gas; Biofuels, Liquid; Biofuels, Solid; Biotechnology; Environmental Biotechnology*

BIBLIOGRAPHY

Books

Hood, Elizabeth E., Peter Nelson, and Randall Worth Powell. *Plant Biomass Conversion*. Ames, Iowa: Wiley-Blackwell, 2011.

U.S. Secretary of State Hillary Clinton tours an exhibition of cookstoves alongside Dr. Kalpana Balakrishnan, a cookstove researcher, during a visit to Chennai, India, July 20, 2011. Clinton is a long-time advocate of using clean-burning cookstoves, such as those using biomass, instead of more traditional stoves that burn wood or solid fuel releasing toxic fumes into poorly ventilated cooking areas in many poor regions around the world, a problem mainly affecting women and small children. © *Saul Loeb/AFP/Getty Images.*

Wrangham, Richard W. *Catching Fire: How Cooking Made Us Human.* New York: Basic Books, 2009.

Periodicals

Dunn, David. "Utility Turns Biomass into Renewable Energy." *Biocycle* 45, no. 9 (September 2004): 34–36.

Rosner, Hillary. "Cooking Up More Uses for the Left-overs of Biofuel Production." *The New York Times,* August 8, 2007.

Web Sites

The Global Alliance for Clean Cookstoves. http://cleancookstoves.org/ (accessed October 5, 2011).

"Report: Household Cookstoves, Environment, Health, and Climate Change: A New Look at an Old Problem." *The World Bank.* http://climatechange.worldbank.org/content/cookstoves-report (accessed October 5, 2011).

Bryan Thomas Davies

Agricultural Biotechnology

■ Introduction

The late twentieth century emergence of agriculture as a key biotechnology subset has generated a range of provocative scientific, legal, and ethical issues that are likely to remain unresolved for the foreseeable future. Early crop plant science was largely composed of trial-and-error experimentation with cross-pollination and other basic forms of genetic modification. Modern agricultural biotechnology is driven by a diverse range of sophisticated DNA-based applications that include the ability to introduce specific genes into existing plant species to achieve a desired agricultural result, such as greater drought or pest resistance. Livestock feed commonly includes material derived from biotech sources. Bovine somatotropin (bST) is a hormone synthesized from natural sources that is often administered to milk cows as a stimulant to increase milk production.

These otherwise laudable scientific developments that have contributed to greater human knowledge and overall improved global food security have generated controversy among various agricultural policy makers and environmentalists. The concerns over the future environmental impacts attributable to the introduction of genetically-modified (GM) organisms to existing crop species are driven by fears that these biotech processes will contaminate native plant species to create invasive weeds harmful to adjacent ecosystems. Also, the long-term effects of GM food consumption on humans and animals have not been definitively determined.

The legal and ethical questions generated by these agricultural applications are equally challenging. Multinational companies that have promoted vigorously the sale of GM seed stocks have been accused of seeking disproportionate profits from subsistence agricultural societies. The legal implications of patent protection being extended to food also are engaged. Further, GM seed use eliminates the ancient practice of seed harvesting, a development opposed by many critics on cultural and biological grounds.

■ Historical Background and Scientific Foundations

The twentieth century marked the development of numerous biotech-based agricultural applications. The prior history of this science sector was notable for the devotion of practical farmers and researchers to methods that were largely trial and error. Crop rotation as it was practiced throughout Europe and in various indigenous societies is a prominent example of how farmers adhered to methods that yielded desired results without truly understanding why.

In many subsistence societies, ancient practices such as seed harvesting formed the essential tools used by farmers to maximize crop yields in a given growing season. Successful animal husbandry was achieved without any specific scientific knowledge of why certain animal species could be improved through breeding; intuition and observation were the basis of an often impressive body of collected knowledge possessed by agricultural societies.

Biologists became important contributors to a number of agricultural sectors in the late nineteenth and early twentieth centuries as a deeper understanding of the genetic structures of crop plants and livestock species began to emerge. Hybrid cereals were first developed in the 1870s from strains of wheat and rye to enhance crop hardiness and resistance to disease. In 1876, British naturalist Charles Darwin (1809–1882) published the results of his experiments with cross-pollinated tomato and tobacco species that confirmed the improved growth and seed fertility of these varieties. The discovery of the DNA double helix in 1953 has been widely hailed as the beginning of modern human genetics; the agricultural science implications associated with the power to modify specific organisms genetically have been equally profound.

The objectives of the geneticists and plant biologists who sought to harness DNA technologies to improve agricultural products and practices varied. The GM food plants conceived through biotechnology were designed to afford better resistance to agricultural threats that have existed since humans began to cultivate crops: droughts, frosts, insects, and various fungal diseases. Additional research efforts were directed to improving the nutritional qualities of staple foods such as rice and wheat through genetic modification.

When it became apparent that agricultural biotechnology would soon possess the tools to promote widespread use of GM processes to alter the fundamental genetic structures of crops and livestock, concerns were expressed that the scientific ability to achieve genetic modification had exceeded the scientific understanding of the true nature and extent of the risks associated with human and animal GM food consumption, as well as the impacts on natural species through contact with GM organisms.

In the 1980s government policy makers took the first tentative steps to regulate how GM organisms of all kinds could be used in commercial agriculture. In the United States, GM regulation is administered by a variety of federal agencies, with the primary role assumed by the U.S. Department of Agriculture (USDA). In the European Union (EU), where the opposition to the sale of GM foods has been the most pronounced, the EU has devised extensive regulatory structures under the supervision of the European Food Safety Authority (EFSA). The EFSA has established itself as a very cautious regulator, and GM products are subjected to extensive testing prior to any commercial sales. The ability to trace precisely any GM component of a food product is a central objective of the EU regulations.

■ Impacts and Issues

There is an intrinsic appeal to the use of biological approaches to improve agricultural production in preference to chemical means. The effort to develop insect-resistant crop strains from *Bacillus thuringiensis* (Bt) is an example. Bt is a naturally-occurring bacterium with the ability to stimulate the production of a specific toxin that inhibits insect development. The genes extracted from the bacterium have been inserted successfully into varieties of American corn and cotton. The genetic modification of these varieties eliminates the need to spray these crops with chemicals that might disturb the existing environmental balance.

However, every biotech success in agriculture raises questions that also tend to arise in other modern scientific endeavors that have commercial potential: To what extent ought scientific knowledge be subject to privately controlled agricultural patents? The need to ensure that innovation is rewarded must be balanced against the need to ensure that essential knowledge is not unduly controlled by private enterprises. The achievement of

---

WORDS TO KNOW

AGRICULTURAL PATENTS: The legal protection sought by developers of biotechnological applications such as GM seeds.

***BACILLUS THURINGIENSIS* (Bt):** A naturally occurring bacterium used as an insecticide; also introduced into transgenic crop species by plant geneticists to increase defenses against insects.

BOVINE SOMATOTROPIN (bST): A natural hormone, the synthesized version of which is used regularly by commercial North American dairy operations to stimulate greater milk production in cows.

SUPER WEEDS: The term used to describe the type of invasive plants that might develop in a specific ecosystem if cultivated plants that have been modified with GM organisms have contact with natural weed species.

TRANSGENIC PLANTS: Plant species created when the genes of another species are inserted into the existing cellular structure to enhance or add a desired genetic feature.

---

such balance is a profound challenge for legal policy makers in all global jurisdictions.

This issue has been brought into its clearest focus in the context of the commercial demands made on various subsistence farming societies to purchase GM seeds, especially farmers located across sub-Saharan Africa. GM seeds have exhibited the potential to improve crop yields. However, some of the large agribusinesses that control the GM seeds require that new seeds are bought each year. For millennia, farming societies have practiced seed harvesting to ensure that a sufficient number of seeds are retained for the next season's planting; many subsistence farmers cannot afford to purchase new seeds each year. This creates an ethical dilemma, especially as the global population is faced with a pending food security crisis. The means must be found to grow more and better food on a planet where the current population of 7 billion is projected to reach upwards of 9 billion by 2050. Finite agricultural land resources have heightened the importance of scientific research directed to improved agricultural yields.

The long-term effects of GM foods on human and animal health are important research and regulatory issues that may only be resolved as longitudinal studies are conducted over periods of years in different human and animal populations. In the United States, it is estimated that most consumers ingest foods with ingredients derived from biotech crops such as corn and soybeans. Recent studies confirm that approximately 70 percent of commercially available processed foods contain at least one ingredient derived from biotechnology.

Farmers hold signs as Greenpeace activists, dressed as Bt brinjal eggplants, look on at a protest during a 2010 public hearing on the genetically modified eggplant crop at Central Research Institute for Dryland Agriculture in Hyderabad, India. © *AP Images/Mahesh Kumar A.*

It is likely that the controversies generated by agricultural biotechnology will not be resolved in the near future. The combined effect of the pressures exerted on agricultural science by population growth and the private pursuit of profit by modern agribusinesses will spawn further conflicts between ethics, science, and legal regulation.

■ Primary Source Connection

Expansive developments in biotechnology have enabled farmers to utilize innovative tools to reap benefits in time, cost, and labor in their farming efforts. The adoption of genetically engineered varieties of major crops is widely accepted by farmers in the United States; however, there is still limited acceptance of these crops throughout Europe. In 2006, when the following article was written, genetically engineered crops had been commercially available for ten years, allowing farmers to save time and use fewer pesticides, while yielding higher crop production. This report examines issues and questions held by three major stakeholders in agricultural technology—seed suppliers and technology providers, farmers, and consumers—to

identify and address concerns that may be determinants in hindering global acceptance of genetically engineered crops.

The First Decade of Genetically Engineered Crops in the United States

Summary

Over the past decade, developments in modern biotechnology have expanded the scope of biological innovations by providing new tools for increasing crop yields and agricultural productivity. The role that biotechnology will play in agriculture in the United States and globally will depend on a number of factors and uncertainties. What seems certain, however, is that the ultimate contribution of agricultural biotechnology will depend on our ability to identify and measure its potential benefits and risks.

What Is the Issue?

Ten years after the first generation of genetically engineered (GE) varieties of major crops became commercially available, adoption of these varieties by U.S. farmers has become widespread. United States consumers eat many products derived from these crops—including some cornmeal, oils, sugars, and other food products—largely unaware of their GE content. Despite the rapid increase in the adoption of GE corn, soybean, and cotton varieties by U.S. farmers, questions remain regarding the impact of agricultural biotechnology. These issues range from the economic and environmental impacts to consumer acceptance.

What Did the Study Find?

This study examined the three major stakeholders in agricultural biotechnology: seed suppliers and technology providers, farmers, and consumers.

Seed suppliers/technology providers. Strengthening of intellectual property rights protection in the 1970s and 1980s increased returns to research and offered greater incentives for private companies to invest in seed development and crop biotechnology. Since 1987, seed producers have submitted nearly 11,600 applications to USDA's Animal and Plant Health Inspection Service for field testing of GE varieties. More than 10,700 (92 percent) have been approved. Approvals peaked in 2002 with 1,190. Most approved applications involved major crops, with nearly 5,000 for corn alone, followed by soybeans, potatoes, and cotton. More than 6,600 of the approved applications included GE varieties with herbicide tolerance or insect resistance. Significant numbers of applications were approved for varieties with improved product quality, viral resistance, and enhanced agronomic properties such as drought and fungal resistance.

Farmers. Adoption of GE soybeans, corn, and cotton by U.S. farmers has increased most years since these varieties became commercially available in 1996. By 2005, herbicide-tolerant soybeans accounted for 87 percent of total U.S. soybean acreage, while herbicide-tolerant cotton accounted for about 60 percent of total cotton acreage. Adoption of insect-resistant crops is concentrated in areas with high levels of pest infestation and varies across States. Insect-resistant cotton was planted on 52 percent of cotton acreage in 2005—ranging from 13 percent in California to 85 percent in Louisiana. Insect-resistant corn accounted for 35 percent of the total acreage in 2005, following the introduction of a new variety to control the corn rootworm.

The economic impact of GE crops on producers varies by crop and technology. Herbicide-tolerant cotton and corn were associated with increased returns, as were insect-resistant cotton and corn when pest infestations were more prevalent. Despite the rapid adoption of herbicide-tolerant soybeans, there was little impact on net farm returns in 1997 and 1998. However, the adoption of herbicide-tolerant soybeans is associated with increased off-farm household income, suggesting that farmers adopt this technology because the simplicity and flexibility of the technology permit them to save management time, allowing them to benefit from additional income from off-farm activities.

Genetically engineered crops also seem to have environmental benefits. Overall pesticide use is lower for adopters of GE crops, and the adoption of herbicide-tolerant soybeans may indirectly benefit the environment by encouraging the adoption of soil conservation practices.

Consumers. Most surveys and consumer studies indicate consumers have at least some concerns about foods containing GE ingredients, but these concerns have not had a large impact on the market for these foods in the United States. Despite the concerns of U.S. consumers, "GE-free" labels on foods are not widely used in the United States. Manufacturers have been active in creating a market for GE-free foods. Between 2000 and 2004, manufacturers introduced more than 3,500 products that had explicit non-GE labeling, most of them food products.

In the European Union and some other countries, however, consumer concerns have spurred a movement away from foods with GE ingredients. Despite the fact that some European consumers are willing to consume foods containing GE ingredients, very few of these foods are found on European grocery shelves.

Jorge Fernandez-Cornejo
Margriet Caswell

FERNANDEZ-CORNEJO, JORGE, AND MARGRIET CASWELL. "THE FIRST DECADE OF GENETICALLY ENGINEERED CROPS IN THE UNITED STATES." *ECONOMIC INFORMATION BULLETIN* 11 (APRIL 2006): III–IV.

SEE ALSO *Agrobacterium; Biofuels, Liquid; Biological Pesticides; Bt Insect Resistant Crops; Corn, Genetically Engineered; Cornstarch Packing Materials; Cotton, Genetically Engineered; Dairy and Cheese Biotechnology; Disease-Resistant Crops; Drought-Resistant Crops; Environmental Biotechnology; Frost-Resistant Crops; GE-Free Food Rights; Genetic Use Restriction Technology; Genetically Modified Crops; Genetically Modified Food; Golden Rice; Patents and Other Intellectual Property Rights; Phytoremediation; Plant Patent Act of 1930; Plant Variety Protection Act of 1970; Plant-Made Pharmaceuticals (Biopharming); Recombinant DNA Technology; Salinity-Tolerant Plants; Terminator Technology; Transgenic Plants; Wheat, Genetically Engineered*

BIBLIOGRAPHY

Books

Entine, Jon, ed. *Let Them Eat Precaution: How Politics Is Undermining the Genetic Revolution in Agriculture.* Washington, DC: AEI Press, 2006.

Heldman, Dennis R., Matthew B. Wheeler, and Dallas G. Hoover, eds. *Encyclopedia of Biotechnology in Agriculture and Food.* Boca Raton, FL: CRC Press, 2011.

McManis, Charles R., ed. *Biodiversity and the Law: Intellectual Property, Biotechnology and Traditional Knowledge.* Sterling, VA: Earthscan, 2007.

Murphy, Denis J. *Plant Breeding and Biotechnology: Societal Context and the Future of Agriculture.* New York: Cambridge University Press, 2007.

Paarlberg, Robert L. *Starved for Science: How Biotechnology Is Being Kept out of Africa.* Cambridge, MA: Harvard University Press, 2008.

Shmaefsky, Brian. *Biotechnology on the Farm and in the Factory: Agricultural and Industrial Applications.* Philadelphia: Chelsea House Publishers, 2006.

Stewart, C. Neal, ed. *Plant Biotechnology and Genetics: Principles, Techniques, and Applications.* Hoboken, NJ: Wiley, 2008.

Tripp, Robert Burnet, ed. *Biotechnology and Agricultural Development: Transgenic Cotton, Rural Institutions and Resource-Poor Farmers.* New York: Routledge, 2009.

Zaikov, Gennadii Efremovich, ed. *Biotechnology, Agriculture and the Food Industry.* New York: Nova Science Publishers, 2006.

Zimmer, Marc. *Glowing Genes: A Revolution in Biotechnology.* Amherst, NY: Prometheus Books, 2005.

Web Sites

"Biotechnology." *U.S. Department of Agriculture (USDA).* http://www.usda.gov/wps/portal/usda/usdahome?navid=BIOTECH (accessed November 12, 2011).

"Food, Genetically Modified." *World Health Organization (WHO).* http://www.who.int/topics/food_genetically_modified/en (accessed November 12, 2011).

"Transgenic Plants: Science, Policy, Politics." *U.S. Department of Agriculture (USDA).* http://www.ars.usda.gov/research/publications/publications.htm?seq_no_115=148389 (accessed November 12, 2011).

Bryan Thomas Davies

Agrobacterium

■ Introduction

Agrobacterium tumefaciens is a common soil bacterium that is closely related to the nitrogen-fixing rhizobium bacteria found on the roots of many plants. It causes crown gall disease in many plants, including grapes, rice, and sugar beet, and is a significant agricultural problem worldwide.

A crown gall is a type of tumor that appears on the trunk, branches, or roots of a plant. Erwin Smith and Charles Townsend of the United States Department of Agriculture were the first to note, in 1907, that *Agrobacterium* causes crown gall disease. Research has shown that the disease-causing potential of *Agrobacterium* can be exploited in genetic engineering. *Agrobacterium* contains a plasmid, the Ti plasmid, which is capable of transferring tumor-inducing genes to the plant genome where they are expressed to trigger the formation of the tumor. Thus, *Agrobacterium* acts as a natural genetic engineer. Genes of scientific or commercial interest may be transferred to the Ti plasmid, and *Agrobacterium* can then be used as a vector, carrying the gene into a host plant. Monocotyledons, which include the cereals such as rice and wheat, are naturally less susceptible to infection with *Agrobacterium* than dicotyledons.

■ Historical Background and Scientific Foundations

The advent of molecular biology techniques in the 1960s meant that the gene transfer process from the *Agrobacterium* Ti plasmid to the plant genome could be studied in detail. Jeff Schell (1935–2003) and Marc Von Montagu (1933–) in Ghent, Belgium, and Eugene Nester's (1931–) team at the University of Washington, Seattle, began to explore the potential of *Agrobacterium* as a vector in plant genetic engineering, the technique in which a gene, or genes, is transferred from one species to another, creating a so-called transgenic species that does not occur naturally.

In 1985 a ground-breaking paper, describing a general method for using *Agrobacterium* to transfer the gene for kanamycin resistance to petunia, tobacco, and tomato plants, was published in the journal *Science*. The kanamycin resistance gene is often used in model genetic engineering experiments because it acts as a marker. Plants that have received this gene can readily be identified because they are resistant to the antibiotic kanamycin, which they were not before the experiment. Another important landmark was the application of *Agrobacterium* gene transfer to rice by scientists at the Japan Tobacco Company in 1994. More recently, *Agrobacterium* has been applied to the genetic modification of all the major cereal plants, including barley and wheat. In 2001 the *Agrobacterium* genome was sequenced and published by both Nester's team and scientists in Cambridge, Massachusetts, giving researchers a powerful tool for better understanding the gene transfer capabilities of the bacterium.

A typical experiment using *Agrobacterium* as a vector would first use specialized enzymes to cut, trim, and splice the Ti plasmid. Some of its natural genes would be removed and a gene of interest, such as one that would confer insect resistance, would then be inserted. Cells of the target plant, often in the form of tiny disks of leaf tissue, are then exposed to the *Agrobacterium* vector so that the genes of interest can be transferred to the plant genome. Plant cells containing the new genes are said to be transformed. Not all cells will take up a new gene; those that do are selected using the antibiotic resistance technique described above. Transformed cells are then placed into a series of special solutions containing growth hormones, where they form first a mass of plant tissue, known as a callus, and then plantlets that will grow into mature plants. The process from transformed plant cell to transgenic plant is known as regeneration.

Development work has resulted in the adaptation of *Agrobacterium*-based genetic engineering techniques to all the major commercial crops. Therefore, plants are being engineered for desirable properties such as pest resistance, which increases yields and aids global food supply.

■ Impacts and Issues

Agrobacterium is currently the basis for one of the main techniques of plant genetic engineering. For those plants that are not readily infected by *Agrobacterium*, other approaches are available such as electroporation, in which foreign genes are introduced through tiny holes in the plant cell wall made by application of an electric field. Another technique involves shooting plant cells with tiny gold particles coated with the DNA of the gene of interest. *Agrobacterium*-mediated gene transfer has been applied on a large scale to a number of commercially significant plant crops including soy, cotton, corn, beet, alfalfa, wheat, canola, and golden rice.

Plants bearing foreign genes that have been transferred by one of the techniques described above are known as genetically modified (GM) or transgenic. Many GM plants have genes for pest resistance, which allows them to resist insects or fungi. Given that a major proportion of food crops are lost to pest attack, planting GM species increases yields and therefore improves food production, which is vital in the context of an increasing global population where too many people already go hungry. Herbicide-resistant plants have also been created with genetic engineering. Weeds are also major enemies of crop plants, competing with them for food, water, and space. Most herbicides are not sufficiently specific to kill the weeds and leave the crops alone. But an herbicide resistant crop is unaffected by application of herbicide to kill surrounding weeds. Genetic engineering can also improve the quality of food. One example is golden rice, which contains two new genes that allow the formation of vitamin A in the edible part of the grain. This development could help prevent blindness and death associated with vitamin A deficiency, which affects between 100 and 140 million children worldwide.

However, plant genetic engineering has attracted some criticism. First, there are safety and environmental issues. Foreign genes could spread from GM crops to other plants, potentially creating weeds that are resistant to herbicides, for example. Most GM plants contain antibiotic resistance genes, which are used to select out the plants that have taken up a foreign gene, and there is concern that if these are consumed by humans, they could lead to the spread of antibiotic resistance, which is already a major public health issue. Finally, GM crops threaten to intensify the dominance of just a few major agrochemical companies in global agriculture at the expense of small farmers using traditional methods and seed varieties. These major companies have the resources to develop the GM seeds and create monopolies on selling them to farmers.

SEE ALSO *Agricultural Biotechnology; Genetically Modified Crops; Genetically Modified Food*

Crown gall caused by the bacteria *Agrobacterium tumefasciens* on a chrysanthemum plant. The gene transfer mechanism used by *Agrobacterium* to infect plants has been used in biotechnology as a way to insert foreign genes into plants. © *Nigel Cattlin/Alamy.*

BIBLIOGRAPHY

Books

Hackworth, Cheryl. *Understanding Mechanisms of Agrobacterium tumefaciens Transformation Genetic Screens, Protein Interactions, and Protein Localization Studies.* Saarbrücken, Germany: VDM Verlag, 2009.

Nelson, Erin Tace. *A Better World Is Possible: Agroecology as a Response to Socio-Economic and Political Conditions in Cuba*. Waterloo, Ontario, Canada: University of Waterloo, 2006.

Nester, Eugene W., Milton P. Gordon, and Allen Kerr. *Agrobacterium Tumefaciens: From Plant Pathology to Biotechnology*. St. Paul, MN: APS Press, 2005.

Thompson, R. Paul. *Agro-Technology: A Philosophical Introduction*. Cambridge: Cambridge University Press, 2011.

Tzfira, Tzvi, and Vitaly Citovsky. *Agrobacterium: From Biology to Biotechnology*. New York: Springer, 2008.

Vandermeer, John H. *The Ecology of Agroecosystems*. Sudbury, MA: Jones and Bartlett Publishers, 2011.

Warner, Keith. *Agroecology in Action: Extending Alternative Agriculture through Social Networks*. Cambridge, MA: MIT, 2007.

Web Sites

"Ecosystems: Agroecosystems." *United States Environmental Protection Agency (EPA)*. http://www.epa.gov/ebtpages/ecosagroecosystems.html (accessed August 23, 2011).

Susan Aldridge

Algae Bioreactor

Introduction

Algae is the generic term for the many species of phytoplankton that are present in all bodies of fresh and salt water. The history of algae runs parallel to the changes to Earth's environment in the past 3 billion years that permitted aerobic life forms to flourish. The emergence of various cyanobacteria strains approximately 3 billion years ago was the crucial evolutionary development. When these bacteria began to occupy the eukaryote cells that are present in all plants, the food produced by the cyanobacteria through photosynthesis could be used by the host cell to multiply and flourish.

Algae is an essentially limitless energy resource. Unsightly greenish algae slicks are capable of forming on most calm bodies of warm water anywhere on Earth's surface. The theoretical understanding of how to unlock the energy contained in algae for commercial purposes has existed since the 1970s. Transesterification is the chemical process that is the essence of all biofuel production. There is a general scientific consensus that algae-based fuel is a clean, sustainable resource that over time will become less expensive to produce as market competition encourages efficiency. The barrier to cost-effective algae biofuels production is the extraction cost to obtain algae lipids necessary for biodiesel manufacture. Lipids are water-insoluble fatty acids whose molecules release energy on combustion.

Historical Background and Scientific Foundations

Algae is a common and easily harvested form of plant-like aquatic life. Able to thrive in virtually any water that is warm, rich in nitrogen and phosphates, and relatively stagnant, algae can be grown without significant disturbance to any other existing life forms. The photosynthesis that occurs within the algae that converts carbon dioxide into oxygen is essential to all terrestrial life. The research and development of algae bioreactors is intended to achieve a crucial triple twenty-first-century objective: the production of readily accessible, environmentally sustainable, and cost-efficient fuels.

The lipids that are used to produce biofuel diesel are contained within the algae cell walls. The ponds and lagoons where algae flourish are natural bioreactors. Modern technology has built on what is observed in nature to enhance the ability to extract algae lipids. In an open-pond reactor, the algae is removed from the water and placed in an oil press, where the cell walls are physically crushed. The lipids are released in this process and collected. Closed pond systems were devised later to provide a controlled environment in which algae grows at an optimal rate, and algae yields are maximized.

The most sophisticated closed systems are the closed-tank bioreactors that have been developed by researchers in the United States, Australia, and New Zealand. In these systems, the bioreactor is constructed close to a large-scale carbon dioxide source such as the flue gas that is emitted from a fossil fuel powered manufacturing plant or generator. The carbon dioxide from the captured flue gas is directed into the bioreactor, where the algae are suspended in nutrient-rich water that increases its growth rate. A blower is used to drive the flue gases through the bioreactor to ensure maximum enrichment of the water with carbon dioxide. Water is withdrawn from the reactor continuously (the process is known as dewatering) to permit the ongoing harvest of algae. The result is highly concentrated algae that is formed into cakes that can be readily processed to extract the lipids that yield biofuels.

Impacts and Issues

In theory, algae bioreactors are the perfect marriage of sustainable energy technology and environmental stewardship. Algae bioreactors have few waste products, and algae are arguably the easiest of all available biomass energy sources to cultivate and harvest. The potential consequences for human and animal health associated with algae growth in open or closed bioreactors are easily safeguarded.

There are two significant issues that have dogged the advance of algae bioreactors into the mainstream of global energy supply. The most important is the cost of production. The algae bioreactor projects that were initiated with significant fanfare in the 1990s were, for the most part, discontinued early in the following decade. At that time, the major players in the world energy markets, primarily the multinational oil companies and their subsidiaries, determined that so long as oil remained approximately one-half the cost of extracted algae biofuels, the incentive to continue algae biofuels development was reduced.

The price rise for petroleum products experienced worldwide after 2005 provided a new incentive for researchers. Since 2008 a number of motor vehicles have been successfully tested using algae biofuels in their engines. However, the relationship between prevailing world oil prices and the cost of algae biodiesel production likely will remain the determining factor in the rate of the full commercialization of this nascent industry.

The secondary issue that has been discussed in this context is the hidden cost of algae bioreactor processes. In the closed systems where flue gases are employed to enrich the algae with carbon dioxide, the fossil fuels that are consumed to produce flue gases represent an energy cost.

It is clear that in the history of algae biofuel development, economic questions have played a greater role in the assessment of bioreactor utility than the obvious environmental benefits associated with these fuel production methods.

■ Primary Source Connection

Individuals at NASA's Ames Research Center have invented an algae photo-bioreactor that grows algae in municipal wastewater, the result of which is the production of renewable biofuel and other products for energy supply. Utilizing their Offshore Membrane Enclosure for Growing Algae (OMEGA) system, the bioreactor enables conversion technology and offers a means for the development of other renewable energy supply sources, such as diesel and jet fuel, as well. Future biorefineries likely will create job opportunities in technical, production, and science-related fields pertaining to the biofuel and energy conversion industry.

NASA Develops Algae Bioreactor as a Sustainable Energy Source

MOFFETT FIELD, Calif.—As a clean energy alternative, NASA invented an algae photo-bioreactor that grows algae in municipal wastewater to produce biofuel and a variety of other products. The NASA bioreactor

WORDS TO KNOW

BIOMASS: The generic term used to describe the living matter found within a given environmental area. In energy generation, biomass describes the plant material, algae, or agricultural waste used as fuel sources.

CYANOBACTERIA: The phylum of single cell aquatic and photosynthetic bacteria commonly known as blue-green algae. Cyanobacteria tend to thrive in warm, stagnant water that is rich in nitrogen and phosphates.

PHOTOSYNTHESIS: The chemical process in which energy provided by sunlight converts carbon dioxide into organic compounds. In all plants and algae the photosynthesis of carbon dioxide and water releases oxygen as the waste product. For this reason photosynthesis is the essential regenerative process for Earth's atmosphere.

TRANSESTERIFICATION: The chemical process that occurs during algae diesel biofuel production, in which an ester catalyst (often sodium hydroxide) is combined with an alcohol such as methanol. The bonded hydrogen and oxygen molecule in the ester is replaced by an oxygen molecule. The result is a fuel that consists of a long-chain alkyl ester; the byproducts are oxygen and water.

is an Offshore Membrane Enclosure for Growing Algae (OMEGA), which won't compete with agriculture for land, fertilizer, or freshwater.

NASA's Ames Research Center, Moffett Field, Calif., licensed the patent pending algae photo-bioreactor to Algae Systems, LLC, Carson City, Nev., which plans to develop and pilot the technology in Tampa Bay, Florida. The company plans to refine and integrate the NASA technology into biorefineries to produce renewable energy products, including diesel and jet fuel.

"NASA has a long history of developing very successful energy conversion devices and novel life support systems," said Lisa Lockyer, deputy director of the New Ventures and Communication Directorate at NASA Ames. "NASA is excited to support the commercialization of an algae bioreactor with potential for providing renewable energy here on Earth."

The OMEGA system consists of large plastic bags with inserts of forward-osmosis membranes that grow freshwater algae in processed wastewater by photosynthesis. Using energy from the sun, the algae absorb carbon dioxide from the atmosphere and nutrients from the wastewater to produce biomass and oxygen. As the algae grow, the nutrients are contained in the enclosures, while the cleansed freshwater is released into the surrounding ocean through the forward-osmosis membranes.

Chief operating officer of AlgaeLink, Peter van den Dorpel, explains a bioreactor in a warehouse in Roosendaal, Netherlands. Van den Dorpel says the bioreactor produces algae in pressure-cooker fashion that he believes will make aviation fuel from algae possible. Experts say it will be years, maybe a decade, before algae, the simplest of all plants, can be efficiently processed for fuel. © AP Images/Arthur Max.

"The OMEGA technology has transformational powers. It can convert sewage and carbon dioxide into abundant and inexpensive fuels," said Matthew Atwood, president and founder of Algae Systems. "The technology is simple and scalable enough to create an inexpensive, local energy supply that also creates jobs to sustain it."

When deployed in contaminated and "dead zone" coastal areas, this system may help remediate these zones by removing and utilizing the nutrients that cause them. The forward-osmosis membranes use relatively small amounts of external energy compared to the conventional methods of harvesting algae, which have an energy intensive de-watering process.

Potential benefits include oil production from the harvested algae, and conversion of municipal wastewater into clean water before it is released into the ocean. After the oil is extracted from the algae, the algal remains can be used to make fertilizer, animal feed, cosmetics, or other valuable products.

This successful spinoff of NASA-derived technology will help support the commercial development of a new algae-based biofuels industry and wastewater treatment.

Ruth Dasso Marlaire

MARLAIRE, RUTH DASSO. "NASA DEVELOPS ALGAE BIOREACTOR AS A SUSTAINABLE ENERGY SOURCE." *NASA.GOV*, NOVEMBER 18, 2009.

SEE ALSO *Biofuels, Gas; Biofuels, Liquid; Biofuels, Solid; Bioreactor; Wastewater Treatment*

BIBLIOGRAPHY

Books

Al Baz, Ismail, Ralf Otterpohl, and Claudia Wendland, eds. *Efficient Management of Wastewater: Its Treatment and Reuse in Water-Scarce Countries.* Berlin: Springer, 2008.

Armstrong, Richard, and Tim Tompkins. *A Look inside Renewed World Energies' Algae Bioreactor.* New York: Knovel, 2010.

Web Sites

DiJusto, Patrick. "Blue-Green Acres." *Scientific American*, August 29, 2005. http://www.scientificamerican.com/article.cfm?id=blue-green-acres (accessed October 8, 2011).

Marlaire, Ruth Dasso. "NASA Develops Algae Bioreactor as a Sustainable Energy Source." *NASA*, November 18, 2009. http://www.nasa.gov/centers/ames/news/releases/2009/09-147AR.html (accessed October 4, 2011).

Bryan Thomas Davies

Amino Acids, Commercial Use

■ Introduction

All amino acids are molecules formed from combinations of amine and carboxyl groups of atoms. There are many different types of amino acids, but 22 amino acids have been identified as proteinogenic (protein creating) amino acids that occur in nature. Proteins, the large molecules that are essential to all human muscle and organ development, are constructed from different formations derived from the 20 amino acids found in the human body. Each amino acid performs a specific function within the body. In addition to the construction of muscle and organ cells, amino acids also play an active role in nervous system performance, because these acids are neurotransmitters that carry messages between cells. Peptides are the combinations of different amino acids that are formed into small molecular chains that are assembled in proteins.

Amino acids are classed as either essential or nonessential to human function. The nonessential amino acids are so called because the body has the inherent ability to manufacture these substances from its own sources. The nine essential amino acids must be ingested from food sources that contain the specific proteins that are parts of the essential acids. It is these acids that have attracted the greatest attention from researchers and the various commercial enterprises that market food supplements.

Commercial amino acids are used in a number of applications. The three most prominent areas are animal feed supplements; food flavors and nutritional additives; and specialty uses, such as aids to physical therapy and rehabilitation.

■ Historical Background and Scientific Foundations

The amino acids that form human proteins constitute approximately 20 percent of body weight and rank second to water as the largest bodily component. Since prehistoric times, all human societies have understood that certain foods promote improved human health and physical performance. The primitive urge in some warrior societies to consume meats rich in blood has a reasonable physiological basis, because these foods contain proteins essential to greater muscle development and endurance.

The first amino acid was discovered through experiments made with asparagus in France in 1806. Key discovery milestones in amino acid research include the discoveries of glutamate (1866), found in the wheat protein gluten; and cysteine (1884), a non-essential amino acid present in many plant and animal proteins. These discoveries provided the foundation for all human supplementation research undertaken in the later twentieth century. American biochemist William C. Rose (1887–1985) is recognized as the most important twentieth century contributor to amino acid science. Among his discoveries was threonine, the 20th human amino acid identified through extensive research conducted during the 1930s and 1940s. Rose identified eight amino acids as essential to human protein formation; it is now widely accepted in nutritional science that there are nine essential amino acids to meet the metabolic needs (the dietary requirements) of a healthy adult. Depending upon their age or their degree of physical development, children may also require the otherwise non-essential amino acids arginine, cysteine, histidine, taurine, and tyrosine to attain optimal physical health.

As twentieth-century science formed its collective understanding that amino acids performed functions within the human body that were essential to overall human health, it is not surprising that the desire to commercialize this knowledge into specific applications was strong. The first serious commercial applications that involved amino acids were food flavorings and preservatives. Monosodium glutamate (MSG) became a favorite in Japan; MSG was exported to the United States and other Western markets in the 1960s as a cost-effective way to enhance the flavor of a wide variety of foods.

As scientific knowledge grew concerning the pivotal role played by nutritional proteins in human health, food supplementation for human and animal feed was the next amino acid research frontier. An important secondary application was the development of amino acid products that assisted in the sustenance of ill persons who required intravenous solutions.

An important element in the study of amino acids is the distinction between the two different types of proteins consumed by humans in their foods. "Whole" proteins (often described in the earlier literature as "complete" proteins) are those derived from animal sources such as poultry products, meat, fish, and dairy products, as well as soybeans and soy products, a rare vegetable source that provides the body with whole protein. These proteins contain all of the essential amino acids necessary for adult metabolic needs. Incomplete protein sources include nuts, grains, fruits, and vegetables. While it is often more difficult for persons who do not consume whole proteins in their food to ensure optimal essential amino acid levels, healthy diets that include a variety of proteins such as beans, seeds, nuts, and grains, as well as other plant foods, can provide all of the necessary amino acids without the need to resort to supplements.

Dietary protein or amino acid supplements tend to promote one of two physical objectives. On one side are the supplements whose amino acid composition is claimed to encourage overall health, muscle development, strength, and endurance. These supplements tend to be popular with athletes. On the other side are the supplements that promote weight loss. Each commercial market is valued in the billions of dollars annually across the western world.

Significant controversy endures over the use of amino acids supplements versus the consumption of these essential foods through natural sources. The controversy has two parts. In the first part, the question is posed concerning whether proteins that are ingested naturally (i.e., through the foods that contain these proteins) are more useful to the human body than those provided through supplements. The second aspect of the controversy remains unsettled in the scientific research conducted to date. It involves whether consuming more than the generally accepted amount of proteins necessary to achieve optimal health could have undesirable consequences.

■ Impacts and Issues

It is a well-accepted principle of nutritional science that a balanced diet that includes proteins, carbohydrates, fats, vitamins, and minerals will provide the foundation of good human health. There is a general consensus among nutritional experts that best way for the human body to receive the necessary proteins for its amino acid needs is through appropriate food consumption. However, supplementation, through nutraceuticals or athletic supplements, is regarded by some consumers as the "quick fix"

WORDS TO KNOW

AMINE GROUP: A chemical formation composed of one nitrogen atom to which two hydrogen atoms are attached. All amino acids have an amine group.

CARBOXYL GROUP: A chemical formation composed of one carbon, one hydrogen, and two oxygen atoms. All amino acids have a carboxyl group.

FERMENTATION: Any process in which a carbohydrate is converted into either an acid or alcohol. Beer, wine, pickles, and yogurts are examples of products that are the result of a specific fermentation process.

MONOSODIUM GLUTAMATE: Known by the acronym MSG, this chemical additive is a popular food supplement that is promoted as a flavor enhancer and preservative. MSG mimics the flavor of sodium chloride (salt).

NUTRACEUTICALS: The generic term that describes any food or food component that is proven to contribute to disease prevention or treatment.

---

to remedy any deficiencies in their diet. If a person has an amino acid deficiency, a specific supplement may remedy the problem.

There are two primary physical risks associated with excessive protein consumption. The breakdown of proteins into amino acids through digestion generates excess nitrogen that must be excreted through urine as urea. This process places stress on the kidneys, and it may contribute to reduced kidney function over time. Excessive protein consumption can also contribute to dehydration as urine is discharged and additional metabolic stress is exerted on the liver. Calcium, the mineral essential to bone formation, may also be excreted as a consequence of a high protein diet. In addition to the heightened risk of osteoporosis, calcium can form painful kidney stones in the urinary tract.

MSG flavoring contains the amino acid glutamate; MSG has been cited in numerous medical studies as a contributor to excessive sodium consumption levels observed in many Western societies. Sodium has been linked conclusively to increased incidence of high blood pressure and cardiovascular disease.

Weight loss amino acid supplements are also problematic. The non-essential amino acid carnitine aids the body in the transfer of fatty acids to working cells for fuel. As metabolic activity is increased (as happens during exercise) greater amounts of fatty acids are necessary to sustain energy output. Various theories have been advanced that if more carnitine is consumed, there will be a greater amount of fatty acids metabolized and therefore greater weight loss achieved. There is limited scientific support for this proposition.

University of California, Santa Cruz, biochemist David Deamer looks at crystalization of amino acids as part of an experiment in his lab in Santa Cruz, California. What researchers are doing is "making the first pieces of life get together," said Deamer. "The thing isn't alive yet, but it's a first step." Far beyond their use as supplements to human health, it is hoped that the ability to build synthetic life forms, starting with building blocks like amino acids, may eventually lead to cures for diseases and alternate means of bioremediation.
© *AP Images/Paul Sakuma.*

Animal feeds, especially those used in broiler chicken and hog operations, have been supplemented with amino acids since the 1960s, as scientists were able to manufacture L-Lysine, L-Threonine, and L-Tryptophan through fermentation processes. Animals that consume supplemented feed tend to grow more quickly and yield meat with higher protein value. The significant risk associated with animal feed supplements is the impact of excess nitrogen production on immediate farm environment. Nitrogen pollution leads to the formation of ammonia and nitrates that poison the soil and water supply.

The global market for commercial amino acids continues to expand in the three main areas identified above. In the United States, market projections suggest that by 2013 the value of the therapeutic supplement market will exceed $1.3 billion. The sale of amino acids as flavoring is expected to reach nearly $600 million, and the animal feeds segment will surpass $520 million in the same period.

SEE ALSO *Agricultural Biotechnology; Biopharmaceuticals; Food Preservation; Protein Therapies; Proteomics*

BIBLIOGRAPHY

Books

Alterman, Michail A., and Hunziker, Peter. *Amino Acid Analysis Methods and Protocols.* Totowa, NJ: Humana, 2012.

Bruückner, Hans, and Noriko Fujii. *D-Amino Acids in Chemistry, Life Sciences, and Biotechnology.* Zurich, Switzerland: Verlag, 2011.

Budisa, Nediljko. *Engineering the Genetic Code: Expanding the Amino Acid Repertoire for the Design of Novel Proteins.* Weinheim, Germany: Wiley-VCH, 2006.

Di Pasquale, Mauro G. *Amino Acids and Proteins for the Athlete: The Anabolic Edge.* Boca Raton, FL: CRC Press, 2008.

Hettiarachchy, Navam S., Kenji Sato, and M. R. Marshall. *Bioactive Food Proteins and Peptides: Application in Human Health.* Boca Raton, FL: CRC Press/ Taylor & Francis Group, 2012.

Janson, Lee W., and Marc Tischler. *The Big Picture: Medical Biochemistry.* New York: McGraw-Hill, 2012.

Web Sites

Alt, Addison. "The Monosodium Glutamate Story: The Commercial Production of MSG and Other Amino Acids." *Journal of Chemical Education,* March 2004. http://www.cornellcollege.edu/chemistry/cstrong/512/MSG.pdf (accessed October 10, 2011).

"Protein in Diet." *MedlinePlus.com.* http://www.nlm.nih.gov/medlineplus/ency/article/002467.htm (accessed October 10, 2011).

Bryan Thomas Davies

Antibiotics, Biosynthesized

■ Introduction

Many of the diseases affecting humans and higher animals arise from bacteria. Bacterial infections can often be treated with antibiotics, which can sometimes also be used prophylactically (as a preventive measure). Natural antibiotics are compounds originating from bacterial microorganisms that are able to kill or markedly reduce the growth rates of other bacteria. Semi-synthetic and biosynthesized (synthetic) antibiotics are generally chemically similar to natural antibiotics and can achieve equivalent results.

Antibiotics typically are classified as broad or narrow spectrum. Broad-spectrum antibiotics are effective in killing a wide range of different types of bacteria, whereas narrow-spectrum antibiotics are successful at combating small, specific families of microorganisms.

Bactericidal antibiotics kill bacteria by interfering with the bacterium's cell contents or with the formation of its cell wall. Penicillin is a bactericidal antibiotic. Bacteriostatic antibiotics impede the multiplication of the bacteria either by interfering with the production of bacterial protein or by inhibiting DNA replication. The tetracyclines, chloramphenicol, and the sulfonamides are bacteriostatic antibiotics.

Bacterial resistance to antibiotics has existed for as long as there have been antibiotics. Some diseases have become resistant to a whole class of antibiotics, some to multiple antibiotic classes. Semi-synthetic and completely biosynthesized (synthetic) antibiotics have been developed in an effort to combat resistance as well as to treat emerging diseases.

■ Historical Background and Scientific Foundations

For more than 5,000 years, humans have been trying to create disease-curing substances. Ancient Sumerian physicians mixed snake skins and turtle shells with soup made from beer to administer to sick patents. Ancient Greeks used many types of herbs and molds to treat infections and to cover wounds. Early Babylonians used ointment made from sour milk and frog bile to treat eye conditions. All of these mixtures contained natural antibiotics.

The germ theory of disease gained general acceptance during the second half of the nineteenth century. As a result, scientists began to research ways of developing pharmacologic agents that could be used to treat or cure disease. In 1871 in the United Kingdom, surgeon Joseph Lister (1827–1912) discovered that urine specimens with mold contamination did not grow bacteria. He began experimenting with human tissue samples. He applied a mold he termed *Penicillium glaucium* to human tissue in the laboratory to test its antibacterial properties.

French bacteriologist Louis Pasteur (1822–1895) hypothesized in 1877 that some types of bacteria would be able to kill other strains of bacteria. He was studying anthrax bacilli at that time. During the 1890s, German physicians Rudolf Emmerich (1852–1914) and Oscar Loew (1844–1941) created a pharmacologic agent from microbes; they called it pyocyanase. It was the first antibiotic used in hospitals, but it was often ineffective. Scottish bacteriologist Alexander Fleming (1881–1955) left an open Petri dish containing staphylococcus bacteria in his lab in 1928. He noted it was growing mold in some areas; there was no bacterial growth near the mold. The mold was named *Penicillium* in 1929, and the drug it produced was called Penicillin, the first formally recognized antibiotic.

Gerhard Domagk (1895–1964), a German bacteriologist and chemist, discovered the first sulfonamide, a synthetic antimicrobial effective against streptococcus, in 1935. This was distributed under the name Prontosil; it was the first commercially marketed antibiotic. Domagk received the Nobel Prize in Medicine in 1939 for his research, which resulted in the first drug effective against a variety of bacterial infections. His research on sulfonamides led to his discovery of thiosemicarbazone and isoniazid, effective treatments for tuberculosis.

Selman Waksman (1888–1973), an American biochemist and microbiologist, discovered numerous antibiotics during the 1940s and 1950s, the most well-known of which are actinomycin, streptomycin, clavacin, and neomycin. Streptomycin was one of the first drugs to be effective against tuberculosis, though it is no longer used because the rate of resistance is high. Waksman is credited with coining the term antibiotic. He was awarded the Nobel Prize in Physiology or Medicine in 1952 for the discovery of streptomycin, as well as the techniques he developed to create streptomycin and many other antibiotics.

Penicillin was not isolated and tested with animals until World War II (1939–1945), when the lack of available antibiotics reached critical proportions. Penicillin was proven both to be extremely effective at combating infections and to have low incidence of side effects during animal research. Though it was discovered in 1928, Howard Florey (1898–1968) and Ernest Chain (1906–1979) were the first to develop a process for large-scale manufacturing of Penicillin G Procaine in 1942. As a result, the use of penicillin with humans became widespread, and this led to increased efforts to develop other similar drugs.

Between the end of World War II and the middle of the 1950s, the antibiotics streptomycin, chloramphenicol, and tetracycline were discovered, tested, and marketed. Lloyd Conover (1923–) obtained the original patent for tetracycline in 1955, which quickly became the most widely prescribed broad spectrum antibiotic in America. Nystatin, a powerful anti-fungal drug, was patented in 1954 by Elizabeth Lee Hazen (1885–1975) and Rachel Fuller Brown (1898–1980).

Semi-synthetic and biosynthesized (synthetic) antibiotics were created in large part to address growing concerns over antibiotic resistant infections, both within hospitals and clinics and in outpatient settings. Semi-synthetic antibiotics combine natural antibiotics with biotechnology. They are naturally occurring antibiotics that have been modified biochemically to enhance efficacy, diminish negative side effects, expand the spectrum of bacteria they treat, and to eliminate or overcome bacterial resistance to their natural counterparts. Synthetic antibiotics are created entirely via laboratory methods, although they are often quite similar to natural or semi-synthetic antibiotics in their action. They are at least as efficacious as natural or semi-synthetic antibiotics.

■ Impacts and Issues

It is likely that antibiotic resistance has been occurring since the first antibiotics were developed, as most bacteria used to create antibiotics are resistant to the antibiotics they produce. In 1946, shortly after the implementation of widespread penicillin use, it was noted that some forms of *Staphylococcus* were resistant to its

WORDS TO KNOW

ANTIBIOTIC RESISTANCE: This occurs when bacteria and other microbes become immune to the effects of antibiotics that previously killed them or halted their proliferation.

GERM THEORY OF DISEASE: This is the theory that bacteria and microorganisms cause disease and that the spread of disease can be at least partially mitigated by hand washing and maintaining basic hygiene and environmental cleanliness.

MOLD: Molds are a form of fungus that grow on organic matter. They aid in decomposition of organic substances and often secrete mycotoxins that inhibit the growth of bacteria. Penicillin is derived from the *Penicillium* mold.

SEMI-SYNTHETIC ANTIBIOTIC: An antibiotic made from natural materials enhanced by use of chemical synthesis; a mixture of natural and synthetic chemical substances.

SYNTHETIC ANTIBIOTIC: Also called biosynthesized antibiotics, these are laboratory produced antibiotics, often chemically similar to naturally occurring antibiotics, that have precise and targeted effects.

effects. Fleming noted that the use of too low a dose, or for an insufficient duration, caused the development of antibiotic resistance to penicillin.

By the late 1950s, it had been discerned that some forms of dysentery-causing bacillus were resistant to the available spectrum of antibiotics: the sulfanilamides, streptomycin, chloramphenicol, and tetracycline. This is called multiple drug resistance. It was also discovered that the bacteria responsible for causing tuberculosis was becoming progressively more resistant to streptomycin, which was previously effective in combating it.

Some infections do not respond to traditional treatments. They are generally termed antibiotic resistant infections and may respond to semi-synthetic or synthetic antibiotics. Some of the most common multiple drug resistant organisms are MRSA (methicillin-resistant *Staphlococcus aureus*), VRE (vancomycin-resistant *Enterococcus*), ESBLs (extended-spectrum beta-lactamases, bacterial enzymes resistant to monobactams and cephalosporins) and PRSP (penicillin-resistant *Streptococcus pneumonia*). MRSA, VRE, and ESBLs are becoming progressively more common in hospital settings as well as in extended-care and skilled nursing facilities. PRSPs are typically found in pediatric and other outpatient settings.

Amoxicillin is a semisynthetic moderate spectrum bacteriolytic antibiotic patented by Smith-Kline Beecham in 1981. It was marketed in 1998 under the names Amoxil, Amoxicillin, and Trimox. Amoxicillin is prescribed very frequently for common infections of

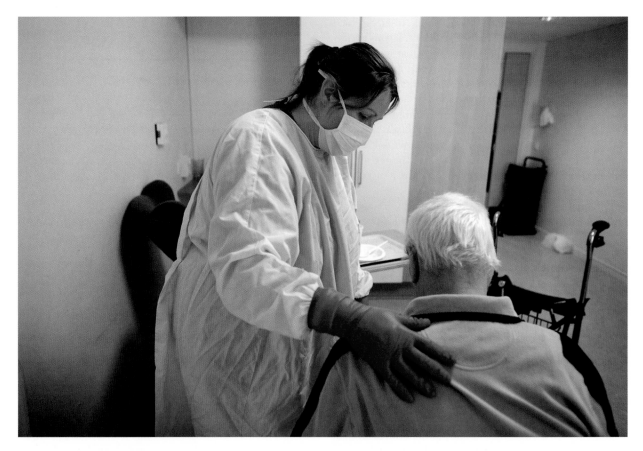

A nurse assists a patient at a nursing home in Oslo, Norway, where patients with MRSA infection reside separately from other hospitalized patients. This cohorting, along with an increase in the use of antibiotics, frequent hand washing, and strict barrier precautions, has led to a reduction in cases of infection with MRSA (methicillin-resistant *Staphylococcus aureus*), a virulent antibiotic-resistant infection. © AP Images/Torbjorn Gronning.

childhood such as acute ear infections, strep throat, urinary tract infections, skin infections, and pneumonia. It also is used as a prophylactic antibiotic to prevent bacterial endocarditis in individuals without spleens or those with internal prosthetics. It is also used for prevention and treatment of anthrax. It acts by preventing the synthesis of bacterial cell walls, effectively killing them.

Synthetic antibiotics are created in the laboratory—they do not originate from naturally-occurring substances such as molds or fungi. Because of the way they are produced and manufactured, they are less likely to cause allergic reactions than traditional antibiotics. Side effects are similar to those of traditional antibiotics: headache, nausea, vomiting, diarrhea, cramping, and abdominal pain. Their mechanism of action is somewhat different than natural or semi-synthetic antibiotics: They terminate the process of protein synthesis before it has an opportunity to occur, entirely preventing replication of the bacterium. They are created to have very precise, targeted effects.

There are two issues simultaneously occurring, driving the need for advances in semi-synthesized and fully biosynthesized antibiotics: Old diseases become progressively more resistant to traditional antibiotic treatments, and new diseases are emerging continually.

As progressively more bacteria became resistant to naturally occurring antibiotics, the pharmaceutical and healthcare industries continue to develop semi-synthetic and biosynthesized antibiotics in order to overcome or eradicate resistance in bacteria and other microorganisms. New antibiotics must be developed and tested continuously in order to assure an adequate supply of available, efficacious antibiotics.

SEE ALSO *Agrobacterium; Antimicrobial Soaps; Bacteria, Drug Resistant; Biopharmaceuticals; Bioreactor; Pharmacogenomics; Wastewater Treatment*

BIBLIOGRAPHY

Books

Armen, M. Boldi, ed. *Combinatorial Synthesis of Natural Product-Based Libraries.* Boca Raton, FL: CRC Press, 2006.

Glazer, Alexander N., and Hiroshi Nikaido. *Microbial Biotechnology: Fundamentals of Applied Microbiology,* 2nd ed. New York: Cambridge University Press, 2007.

Grace, Eric S. *Biotechnology Unzipped: Promises and Realities*, 2nd ed. Washington, DC: Joseph Henry Press, 2006.

Hager, Thomas. *The Demon under the Microscope: From Battlefield Hospitals to Nazi Labs, One Doctor's Heroic Search for the World's First Miracle Drug*. New York: Harmony Books, 2006.

John Wiley & Sons. *Wiley Handbook of Current and Emerging Drug Therapies*. Hoboken, NJ: Wiley-Interscience, 2007.

Okafor, Nduka. *Modern Industrial Microbiology and Biotechnology*. Enfield, NH: Science Publishers, 2007.

Shiva, Vandana. *Monocultures of the Mind: Perspectives on Biodiversity and Biotechnology*. Atlantic Highlands, NJ: Zed Books, 1993.

Strohl, William R. *Biotechnology of Antibiotics*, 2nd ed. New York: Marcel Dekker, 1997.

Vardanyan, R. S., and Victor J. Hruby. *Synthesis of Essential Drugs*. Boston: Elsevier, 2006.

Web Sites

"Get Smart: Know When Antibiotics Work." *Centers for Disease Control (CDC)*. http://cdc.gov/Features/GetSmart/ (accessed November 5, 2011).

"Manufacturing Antibiotics." *Excellence Gateway (UK)*. http://www.excellencegateway.org.uk/VLSP26/030/005/010/015/m_030_005_010_015.htm (accessed November 5, 2011).

Trafton, Anne. "Bacterial Battle for Survival Leads to New Antibiotic." *MIT News*, February 28, 2008. http://web.mit.edu/newsoffice/2008/antibiotics-0226.html (accessed November 7, 2011).

Pamela V. Michaels

Antimicrobial Soaps

■ Introduction

Soaps have a crucial role in both public and personal health because they remove and kill bacteria and other pathogens. More than a century ago, frequent bathing and hand washing gradually was introduced as a means of preventing infections and the transmission of parasites among the population. Soaps contain a number of substances that either directly kill bacteria or facilitate their mechanical removal by friction when rubbing the hands together and then rinsing. The addition of antimicrobial compounds to soaps was intended initially for hospitals, clinics, and medical offices in order to enhance hygiene and to prevent iatrogenic transmission of diseases.

■ Historical Background and Scientific Foundations

As the habit of hand washing and regular bathing became a widespread behavior in many societies, the incidence of infections decreased significantly. Several antimicrobial soaps have been developed since the mid-1950s for use by health professionals in hospitals and clinics in order to prevent the spread of infections among patients via physicians and nurses. Concerned pediatricians also began advising mothers to use antimicrobial soaps at home to wash children's hands and for bathing, and some of these products rapidly were adapted to the commercialized market for the general public.

Because of the introduction and large success of antibiotics and vaccine campaigns, the 1950s and 1960s were a period of great optimism among physicians and researchers as to the power of science and technology to vanquish infectious diseases. The understanding of how bacteria and other microorganisms survive and adapt in hostile environments was in its infancy in the 1950s and was limited to a small circle of microbiologists who performed seminal research on bacterial genetics. As a result, the overuse of the first antibiotics (penicillin, sulfa drugs, tetracycline) as well as the addition of antibiotics to soaps led to the first recognized cases of microbial resistance to antibiotics in the 1960s.

Bacteria are present almost everywhere in the environment. Most bacteria do not cause disease and are, in fact, beneficial for plants, animals, and humans. Pathogenic bacteria, fungi, parasites, and viruses use several different means to be transported from the environment to a host, or from one host to another. Some microbial pathogens can cause skin infections, whereas others access the interior of the body through the eyes, oral or nasal mucosa, or via skin injuries or sexual intercourse. Examples of airborne pathogens include influenza viruses and tuberculosis mycobacteria. Examples of waterborne or foodborne pathogens include *Vibrio cholera*, *Giardia*, *Acanthamoeba*, *Entamoeba*, *Escherichia coli*, *Staphylococcus aureus*, and *Pseudomonas aeruginosa*. Pathogens that are present in the soil or in animal fecal matter may find their way to hosts through food, as in contaminated dairy products, eggs, vegetables, or fruits, usually via skin contact.

■ Impacts and Issues

All soaps, household cleaning products, personal hygiene products, and cosmetics have antimicrobial action when they are used, especially when followed by rinsing with clean water, the universal diluent. Specialized antibacterial soaps are not necessary to make hands clean, according to the U.S. Centers for Disease Control and Prevention (CDC). Washing the hands, however, is the single most effective method of preventing the transmission of infections.

Residues of chemicals found in antibacterial cleaners have been found in groundwater and soil, even miles away from urban centers. Besides the concerns by microbiologists regarding the increasing number of pathogens acquiring resistance to existing antibiotics, there is also a growing concern about the impact that these antimicrobial residues may have upon the soil's microenvironment. Specific concerns center around how antibacterial residues could affect the beneficial soil populations of bacteria, fungi, protozoans, the groundwater biome, and any potential negative effects on human and animal health that could arise from food crops grown in soil containing these residues.

Triclosan is the most widely used antimicrobial agent in the vast range of antibacterial products. Initially used in surgical scrubs, triclosan is a synthetic broad-spectrum agent commonly found in soaps, deodorants, toothpastes, cosmetics, fabrics, household products, and plastics. The addition of triclosan to toothpaste and mouthwash has contributed to a significant improvement in oral hygiene and a decrease in periodontal diseases such as gingivitis, as well as a diminished incidence of dental cavities among the population.

Despite triclosan's four decades of use, no consistent evidence exists that it has induced microbial resistance. Nevertheless, triclosan was found by researchers in 2000 throughout the environment, including surface waters, soil, fish tissue, human breast milk, and urine.

In 2009 the American Public Health Association stated that it would endorse banning triclosan from all non-medical products. The American Medical

WORDS TO KNOW

BIOME: An ecosystem containing indigenous fauna and flora integrated into a local food chain, such as lakes, marshes, or deserts.

DILUENT: A diluting agent.

IATROGENIC: From the Greek *iatros*, meaning physician; and genic, meaning origin. Iatrogenic diseases are transmitted by physicians and other health professionals.

PATHOGEN: A disease-causing organism.

Association (AMA) also expressed concerns about the use of antibacterial chemicals on such a large scale and has asked the U.S. Food and Drug Administration (FDA) to study the issue. In addition, the AMA recommended continued research on the use of common antimicrobial compounds as ingredients in consumer products and their potential impact on public health, as well as upon the environment.

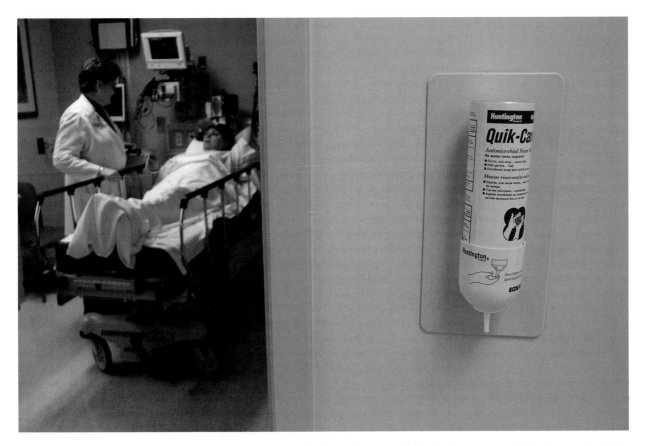

A 2003 study by the Centers for Disease Control (CDC) found if a doctor does not wash her hands, nurses, medical students, and technicians are then five times less likely to wash their hands as well, thus the reasoning for antimicrobial foam hand wash, at right, positioned at the entrance of most patient rooms in the hospital. © *AP Images/John Amis.*

In January 2011 the Alliance for the Prudent Use of Antibiotics (APUA) published a report on triclosan's mode of action and current usage. The report was based upon review of the current literature; it focused on the potential impact upon human and animal health and a possible association between triclosan use and antibiotic resistance. According to the report, triclosan is effective against many (but not all) types of Gram-positive and Gram-negative bacteria, some fungi, and some parasites. Triclosan also stops the growth of microorganisms when present at low concentrations in products; in high concentrations triclosan actually is able to kill these microorganisms. However, recent evidence suggests that some bacteria have the potential to display increased resistance to triclosan.

Antimicrobial soaps containing triclosan can be effective against Methicillin-resistant *Staphylococcus aureus* (MRSA). MRSA is sometimes found in public restrooms, clubs, restaurants, and gymnastic and fitness centers, mostly on wet surfaces and on cloth. MRSA can cause serious illness if infection spreads beyond the skin and into the bloodstream. On skin surfaces, MRSA can cause rashes and infection. Showering or bathing with liquid soaps containing 2 percent triclosan has been shown to be effective for skin decolonization in people who carry MRSA.

Conflicting results among surveys and other studies on bacterial resistance to antimicrobial agents, including triclosan, only add to the ongoing concern and debate. One study on the risk of bacterial-resistance development in association with the wide use of triclosan and other antimicrobial ingredients in consumer products found no evidence that MRSA or *Pseudomonas aeruginosa* have developed resistance to triclosan over a 10-year-period of exposure. Two other studies found evidence that there were no significant differences between the amount of bacteria or the frequencies of antibiotic resistance in homes where antimicrobial products were or were not being used. The latter studies imply that either bacteria were already resistant to those products, or that antimicrobial soaps and cleaning products had the same degree of effectiveness against bacteria as their conventional counterparts.

The FDA is reviewing the safety of triclosan and other antimicrobial ingredients of consumer products. The Environmental Protection Agency (EPA) and the CDC are also conducting surveys and assessments of antibacterial personal hygiene products because of the repetitive daily human exposure to these antiseptic ingredients. Triclosan for instance, is detectable in the urine of 75 percent of Americans over the age of five, which leads to concerns about development of microbial resistance and possible long-term effects on human health.

SEE ALSO   *Antibiotics, Biosynthesized; Bacteria, Drug Resistant; Biosurfactants; Cosmetics; Laundry Detergents*

BIBLIOGRAPHY

Periodicals

Aiello, Allison E., Elaine L. Larson, and Stuart B. Levy. "Consumer Antibacterial Soaps: Effective or Just Risky?" *Clinical Infectious Diseases* 45, suppl. 2 (September 2007): S137–S147.

"Antibacterial Debate: Can Antibacterial Soaps Harm the Environment?" *ICIS Chemical Business* (June 27, 2008): 24–27.

Ballantyne, Coco. "Do Antibacterial Soaps Do More Harm Than Good?" *Scientific American* 298, no. 2 (February 2008): 96.

"Ciba Defends Use of Antibacterial Triclosan in Soaps." *Focus on Surfactants* 2008, no. 8 (August 2008): 4–5.

"The Handiwork of Good Health: Alcohol-Based Hand Sanitizers Are More Effective Than Antibacterial Soaps, but Don't Give Up on Plain Soap and Water." *Harvard Health Letter* 32, no. 3 (January 2007): 1–3.

Riaz, Saba, Adeel Ahmad, and Shahida Hasnain. "Antibacterial Activity of Soaps against Daily Encountered Bacteria." *African Journal of Biotechnology* 8, no. 8 (April 2009): 1431–1436.

Web Sites

Martin, Andrew. "Antibacterial Chemical Raises Safety Issues." *The New York Times*, August 19, 2011. http://www.nytimes.com/2011/08/20/business/triclosan-an-antibacterial-chemical-in-consumer-products-raises-safety-issues.html?_r=1&pagewanted=all (accessed November 11, 2011).

"New Study Demonstrates Simple Handwashing with Soap Can Save Children's Lives." *Centers for Disease Control and Prevention (CDC)*, July 14, 2005. http://www.cdc.gov/media/pressrel/r050714a.htm (accessed November 11, 2011).

"Triclosan: What Consumers Should Know." *U.S. Food and Drug Administration (FDA)*. http://www.fda.gov/forconsumers/consumerupdates/ucm205999.htm (accessed November 11, 2011).

Sandra Galeotti

Antirejection Drugs

Introduction

Antirejection drugs have helped to make transplantation of vital organs, including kidneys, hearts, lungs, livers and pancreases, a clinical reality. In 2008 there were 23,288 solid organ transplants in the United States. Survival rates depend upon the organ transplanted and the previous health of the patient, but a significant number can expect to live full, active lives for five years or more. The successful use of antirejection drugs has increased the demand for organ transplantation and the pressure upon the supply of organs. The recipient's immune system treats a transplanted organ from another individual, the donor, as foreign. The presence of the organ triggers several immune mechanisms, which will destroy the organ either in the short- or long-term. Antirejection drugs, such as cyclosporine and mycophenolate mofetil, work by suppressing these immune responses and allowing the organ to survive. The drugs must be taken for the lifetime of the graft, which carries some risks for the patient. Antirejection drugs produce side effects by themselves, and long-term immunosuppression increases the risk of infection and cancer. Therefore the selection and dose of antirejection drugs involves careful consideration of the balance between benefit and risk to the patient. New scientific approaches to immunosuppression may reduce, or even eliminate, the need for conventional antirejection drugs.

Historical Background and Scientific Foundations

Early animal experiments in organ transplantation during the nineteenth century demonstrated that whereas an autograft was accepted by the body, an allograft would always be rejected. In 1943 British scientist Peter Medawar (1915–1987) discovered that immune responses were responsible for the organ rejection process, research that earned him the Nobel Prize for Physiology or Medicine in 1960. Later it was shown that the antileukemia drug 6-mercaptopurine induced immunological tolerance in dogs given kidney allografts. A related drug called azathioprine, combined with steroids, was the first antirejection therapy to be given to transplant patients. Introduced in the 1960s, it was to be the mainstay of such treatment for the next 20 years.

In 1974 Jean-François Borel (1933–), a researcher at the Swiss pharmaceutical company Sandoz, discovered that cyclosporine, a drug derived from a fungus, had powerful immunosuppressive qualities. Since then, techniques in transplantation have improved as surgeons gained more experience. Thanks to these two developments, survival rates in transplantation have increased dramatically, and the operations, hitherto considered very risky, moved into the mainstream of medicine. Meanwhile, an understanding of the complexity of the rejection process led to the development of additional antirejection drugs.

In the early twenty-first century it is usual to give the transplant recipient a combination of different antirejection drugs. Each type of drug acts at a different point in the immune system to help inhibit the rejection process. The benefit of giving smaller doses of several drugs is that side effects of an individual drug are minimized. Attacking the immune response by more than one mechanism is also more effective in protecting the new organ from rejection. Steroids have long been a mainstay of antirejection therapy, and they inhibit production of an immune molecule called interleukin 2 (IL-2). The macrolides, including cyclosporine and also the more recently developed drugs tacrolimus and sirolimus, block IL-2 by a different mechanism. The antimetabolites, of which mycophenolate mofetil (MMF) is the most important example, inhibit multiplication of immune cells. MMF has replaced the older azathioprine in antirejection therapy in most hospitals. Antibodies, such as rituximab, are also being introduced for their ability to bind onto T-lymphocytes and inhibit their role in the rejection process.

WORDS TO KNOW

ALLOGRAFT: The transplant of an organ, or tissue, from one individual to another of the same species. Organ transplants are allografts, unless they involve identical twins, in which case they are autografts.

AUTOGRAFT: Tissue transferred from one part of the body to another. Common examples include skin grafts to treat burns and blood vessel grafts used in cardiac bypass surgery. Also, transplants involving organs or tissue between identical twins.

IMMUNOLOGICAL TOLERANCE: The extent to which the immune system can ignore a threat such as an invading pathogen, cancer cells, or organ transplant. Prior to the full development of the immune system, a fetus has immunological tolerance. Antirejection drugs induce a state of immunological tolerance in organ transplant recipients.

IMMUNOSUPPRESSION: Reduced or absent normal immune response to disease. Immune deficiency disease and treatment with anticancer drugs or antirejection drugs are major causes of immunosuppression.

T-LYMPHOCYTE: A type of white blood cell that plays a central role in cell-mediated immunity. Both cell-mediated and antibody-mediated immunity are important in organ rejection.

■ Impacts and Issues

All antirejection drugs have side effects. Steroids can cause bone disease; the macrolides are toxic to the kidneys; and MMF affects white and red blood cell counts. Antibodies have fewer side effects, but these treatments are costly. The patient needs to take antirejection drugs for the rest of his or her life, or at least for the lifetime of the graft. Therefore, persistent side effects could be a serious issue for patients and may cause them to skip or stop their antirejection drugs. This may seriously compromise their health, because organ rejection has been found to occur as long as 20 years after a transplant operation, if medication is stopped. Also concerning is the increasingly recognized long-term health risk of immunosuppression. The cause of death in around 27 percent of patients who die with their graft still functioning is related to either infection or cancer arising from chronic impaired immunity.

Despite these concerns, the benefits of organ transplantation should outweigh the risks of antirejection drugs, if these are carefully prescribed and monitored. Indeed, the success of immunosuppressant therapy has been an important element in increasing the demand for organ transplantation, which has never been higher. Increasingly, transplantation is seen as an option for patients who previously might not have been considered, and as the population ages, the number of

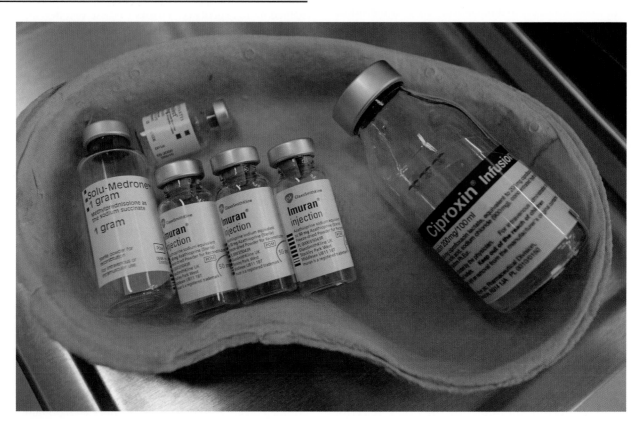

A kidney dish of used anti-rejection drugs sits on a hospital tray in Birmingham, England. © *Christopher Furlong/Getty Images.*

patients needing new organs increases accordingly. In the United States, more than 84,000 adults and children are currently waiting for a suitable organ donor. This has led to developments such as the increasing use of living donors and research into the potential of genetically-modified pig organs, also known as xenotransplants, as an alternative organ supply. The growth of regenerative medicine, in which stem cells are used as a source of replacement tissue, may also eventually reduce the need for conventional organ transplants.

There is also continuing research into the organ rejection process, with the aim of inducing a state of immunological tolerance in the patient without the use of antirejection drugs. Ideally, the tolerance would apply only to that part of the immune system involved in organ rejection, leaving the parts that fight infection and cancer fully functioning. To this end, the Immune Tolerance Network, which brings together some of the world's leading experts in this field, was founded in 1999 by the U.S. National Institute of Allergy and Infectious Diseases. The Network has a number of clinical trials of new drugs that induce immunological tolerance underway. These drugs are far more specific than conventional antirejection drugs because they target just those immune pathways involved in organ rejection. They are also gaining new insights into the biology of organ rejection by studying those rare patients who undergo transplants and do not experience organ rejection.

SEE ALSO *Artificial Organs; Biopharmaceuticals; Organ Transplants; Orphan Drugs; Pharmacogenomics; Stem Cells*

BIBLIOGRAPHY

Books

Bucknell, Duncan Geoffrey. *Pharmaceutical, Biotechnology, and Chemical Inventions: World Protection and Exploitation.* New York: Oxford University Press, 2011.

Cairns, Donald. *Essentials of Pharmaceutical Chemistry,* 3rd ed. London: Pharmaceutical Press, 2008.

Petechuk, David. *Organ Transplantation.* Westport, CT: Greenwood Press, 2006.

Web Sites

"Organ Transplantation." *National Institutes of Health (NIH).* http://health.nih.gov/topic/OrganTransplantation (accessed September 10, 2011).

Susan Aldridge

Ariad Pharmaceuticals, Inc. v. Eli Lilly & Co.

■ Introduction

Ariad Pharmaceuticals, Inc. v. Eli Lilly & Co. is a case heard before the United States Court of Appeals for the Federal Circuit involving an alleged infringement by pharmaceutical company Eli Lilly & Co. of a patent held by Ariad Pharmaceuticals in the development of two drugs. The case had enormous implications for the biotechnology community and other inventors because the case dealt with the breadth and enforceability of certain patent claims. Ultimately, the Federal Circuit ruled that U.S. patent law contains a written description requirement that is separate from the enablement requirement. The court, however, did not set limits for the scope of the written description requirement, stating that "the level of detail required to satisfy the written description requirement varies depending on the nature and scope of the claims and on the complexity and predictability of the relevant technology." The court then decided that Eli Lilly had not infringed on Ariad's patent, because the Ariad patent did not satisfy the written description requirement by failing to describe the invention adequately and in such a way that others could replicate the work.

■ Historical Background and Scientific Foundations

In 2002 the U.S. Patent and Trademark Office (PTO) issued U.S. Patent No. 6,410,516 (the '516 patent) to the Massachusetts Institute of Technology (MIT), the President and Fellows of Harvard College, and the Whitehead Institute for Biomedical Research based on a patent application filed on April 21, 1989. The '516 patent relates to the role of nuclear factor kappa-light-chain-enhancer of activated B cells (NF–κB) in the regulation of gene expression.

NF–κB is a transcription factor—a protein that binds to DNA sequences and regulates the transcription, or copying, of genetic codes from DNA to messenger RNA (mRNA). NF–κB proteins control numerous cellular and organism responses, including immune responses, cellular growth, and inflammatory responses. It is also involved in disease states for arthritis, asthma, cancer, chronic inflammation, heart disease, and neurodegenerative diseases.

The '516 patent specifically addressed the methods for regulating cellular responses to external stimuli by reducing NF–κB activity, which could reduce the harmful symptoms of certain diseases. MIT, Harvard College, and the Whitehead Institute licensed the technology and methods underlying the '516 patent to Ariad. On June 25, 2002—the date that the PTO issued the '516 patent—Araid sued Eli Lilly for infringing the '516 patent. Ariad alleged that two of Eli Lilly's pharmaceutical products—Evista® and Xigris®—infringed on the '516 patent, because they reduce the binding of NF–κB to NF–κB recognition sites. The Federal Circuit's decision noted that claims contained in Eli Lilly's patents on Evista and Xigris "use language that corresponds to language present in the ['516] patent."

On May 4, 2006, a jury in the U.S. District Court of Massachusetts found that the '516 patent was valid and that Eli Lilly had infringed the patent. The jury ordered Eli Lilly to pay Ariad $65 million in back royalties plus 2.3 percent of Eli Lilly's profits on U.S. sales of Evista and Xigris until the patents expire in 2019. In 2007 a separate trial in the District Court of Massachusetts resulted in the judge rejecting additional defenses asserted by Eli Lilly, including that the subject matter of the '516 patent was not patentable under U.S. patent laws because it involved natural processes. Eli Lilly appealed both decisions to the United States Court of Appeals for the Federal Circuit. A panel of judges reversed the jury's verdict and decided that the claims contained in the '516 patent application were invalid because they lacked an adequate written description as required by U.S. patent law. Ariad requested that the Federal Circuit rehear the issue en banc, which involves a hearing by a panel composed of all judges of the court.

■ Impacts and Issues

Upon rehearing *Ariad Phamaceutical, Inc. v. Eli Lilly & Co.*, the Federal Circuit addressed whether the '516 patent application satisfied the first paragraph of 35 U.S.C. §112, part of the Patent Act. The first paragraph of §112 states:

> The specification shall contain a written description of the invention, and of the manner and process of making and using it, in such full, clear, concise, and exact terms as to enable any person skilled in the art to which it pertains, or with which it is most nearly connected, to make and use the same, and shall set forth the best mode contemplated by the inventor of carrying out his invention.

The statute contains the enabling requirement for patentability. In order for an inventor to obtain a patent from the PTO, the inventor must describe the invention in such a way that others will know how to make and use the invention. The enabling requirement encourages the dissemination of knowledge by making the details of the invention public. In exchange for this disclosure, the inventor receives the exclusive right to manufacture and sell the invention under the patent.

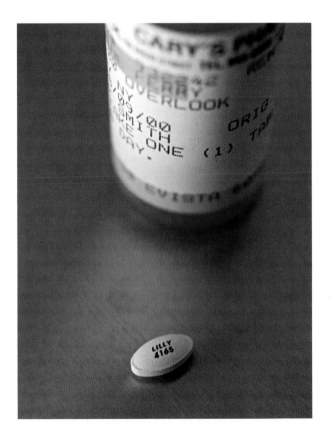

Evista (raloxifene hydrochloride) by Eli Lilly, is used in the prevention treatment of osteoporosis; it reduces the amount of calcium lost from bones. The drug was one of the medications named in the court case *Ariad v. Lilly.* © *Michelle Del Guercio/Peter Arnold, Inc./Alamy.*

In addressing whether the '516 patent satisfied the requirements of the first paragraph of §112, the court also considered whether §112 contains a written description requirement separate from the enabling requirement. The Federal Circuit noted that in 2002 the Supreme Court of the United States held in *Festo Corp. v. Shoketsu Kinzoku Kogyo Kabushiki Co.* that the first paragraph of §112 sets forth three requirements. The Court stated, "[A] patent application must describe, enable, and set forth the best mode of carrying out the invention." From the text of the Court's decision in *Festo*, the Federal Circuit determined that the first paragraph of §112 requires that a patent application must separately and distinctly both "describe" and "enable" the invention. Therefore, the Federal Circuit decided that §112 contains a written description requirement that is separate from the enablement requirement of §112.

The court then noted that Ariad's patent did not adequately describe the invention. "The '516 patent discloses no working or even prophetic examples of methods that reduce NF–κB activity.... Here the specification at best describes decoy molecule structures and hypothesizes with no accompanying description that they could be used to reduce NF–κB activity. Yet the asserted claims are far broader." Essentially, the court asserted that the '516 patent was drafted too broadly and therefore covered processes beyond the scope of work conducted by the researchers involved in the '516 patent.

Critics of the Federal Circuit's decision note that the court did not provide any guidance as to what satisfies the written description requirement and makes it distinct from the enablement requirement. Judge Randall Rader, in his dissenting opinion in *Ariad* asserted that "the court's inadequate description of its written description requirement acts as a wildcard on which the court may rely when it faces a patent that it feels is unworthy of protection." The lack of guidance on the extent and nature of the written description requirement threatens increased litigation and disparate application of the written description requirement by lower courts.

The court's decision in *Ariad* will have a significant impact on biotechnology researchers. The court's failure to include any distinct rules as to the manner in which or how broadly the written description requirement should be drafted will create uncertainty in the industry for years. In light of the court's decision in *Ariad*, researchers will have to be more careful in drafting written description for patents and avoid making the written descriptions overly broad.

■ Primary Source Connection

In a United States Federal Circuit Court of Appeals, Ariad Pharmaceuticals sued Eli Lilly for infringement of Ariad Pharmaceuticals Patent 6,410,516. This patent was licensed from Massachusetts Institute of Technology, Harvard, and the Whitehead Institute. On May 4, 2006, Eli Lilly was ordered to pay both back royalties, and royalties on future sales of Evista and Xigris, two drugs that reduce NF-κB activity (responsible for regulating genes and some specific gene pathways in the body). However, subsequent questions arose about the claim, centering around the scope of the patent, most notably the fact that numerous other marketed drugs can and do affect the NF-κB pathway. In further proceedings, the claims of the patent were found to be invalid for lack of written description. The Court ruled on April 3, 2009, that the Ariad patent was invalid and overturned the previous verdict against Eli Lilly.

ARIAD PHARMACEUTICALS, INC., Massachusetts Institute of Technology, The Whitehead Institute for Biomedical Research, and the President and Fellows of Harvard College, Plaintiffs-Appellees, v. ELI LILLY AND COMPANY, Defendant-Appellant

No. 2008–1248. United States Court of Appeals, Federal Circuit. March 22, 2010. LOURIE, Circuit Judge.

Ariad petitioned for rehearing *en banc*, challenging this court's interpretation of 35 U.S.C. § 112, first paragraph, as containing a separate written description requirement. Because of the importance of the issue, we granted Ariad's petition and directed the parties to address whether § 112, first paragraph, contains a written description requirement separate from the enablement requirement and, if so, the scope and purpose of that

requirement. We now reaffirm that § 112, first paragraph, contains a written description requirement separate from enablement, and we again reverse the district court's denial of JMOL and hold the asserted claims of the '516 patent invalid for failure to meet the statutory written description requirement.

BACKGROUND

In light of the controversy concerning the distinctness and proper role of the written description requirement, we granted Ariad's petition, vacating the prior panel opinion and directing the parties to brief two questions: (1) Whether 35 U.S.C. § 112, paragraph 1, contains a written description requirement separate from an enablement requirement? (2) If a separate written description requirement is set forth in the statute, what is the scope and purpose of that requirement?

DISCUSSION

I.

Although the parties differ in their answers to the court's questions, their positions converge more than they first appear. Ariad, in answering the court's first question, argues that § 112, first paragraph, does not contain a written description requirement separate from enablement. Yet, in response to this court's second question on the scope and purpose of a written description requirement, Ariad argues that the statute contains two description requirements: "Properly interpreted, the statute requires the specification to describe (i) what the invention is, and (ii) how to make and use it."

Ariad reconciles this apparent contradiction by arguing that the legal sufficiency of its two-prong description requirement is judged by whether it enables one of skill in the art to make and use the claimed invention. Thus, according to Ariad, in order to enable the invention, the specification must first identify "what the invention is, for otherwise it fails to inform a person of skill in the art what to make and use." Yet Ariad argues that this first step of "identifying" the invention applies only in the context of priority (i.e., claims amended during prosecution; priority under 35 U.S.C. §§ 119, 120; and interferences) because original claims "constitute their own description."

Lilly, in contrast, answers the court's first question in the affirmative, arguing that two hundred years of precedent support the existence of a statutory written description requirement separate from enablement. Thus, Lilly argues that the statute requires, first, a written description of the invention and, second, a written description of how to make and use the invention so as to enable one of skill in the art to make and use it. Finally, Lilly asserts that this separate written description requirement applies to all claims—both original and amended—to ensure that inventors have actually invented the subject matter claimed.

Thus, although the parties take diametrically opposed positions on the existence of a written description requirement separate from enablement, both agree that the specification must contain a written description of the invention to establish what the invention is. The dispute, therefore, centers on the standard to be applied and whether it applies to original claim language.

A.

As in any case involving statutory interpretation, we begin with the language of the statute itself. Section 112, first paragraph, reads as follows: The specification shall contain a written description of the invention, and of the manner and process of making and using it, in such full, clear, concise, and exact terms as to enable any person skilled in the art to which it pertains, or with which it is most nearly connected, to make and use the same, and shall set forth the best mode contemplated by the inventor of carrying out his invention.

We read the statute to give effect to its language that the specification "shall contain a written description of the invention" and hold that § 112, first paragraph, contains two separate description requirements: a "written description [i] of the invention, and [ii] of the manner *and* process of making and using [the invention"]. 35 U.S.C. § 112, ¶ 1 (emphasis added). On this point, we do not read Ariad's position to be in disagreement as Ariad concedes the existence of a written description requirement. See Appellee Br. 2 ("Under a plain reading of the statute, a patent specification ... must contain a description (i) of the invention, and (ii) of the manner and process of making and using it."). Instead Ariad contends that the written description requirement exists, not for its own sake as an independent statutory requirement, but only to identify the invention that must comply with the enablement requirement.

We see nothing in the statute's language or grammar that unambiguously dictates that the adequacy of the "written description of the invention" must be determined solely by whether that description identifies the invention so as to enable one of skill in the art to make and use it. The prepositional phrase "in such full, clear, concise, and exact terms as to enable any person skilled in the art . . . to make and use the same" modifies only "the written description . . . of the manner and process of making and using [the invention]," as Lilly argues, without violating the rules of grammar. That the adequacy of the description of the manner and process of *making and using* the invention is judged by whether that description enables one skilled in the art to make and use the same follows from the parallelism of the language.

Since 1793, the Patent Act has expressly stated that an applicant must provide a written description of the invention, and after the 1836 Act added the requirement for claims, the Supreme Court applied this description requirement separate from enablement. See infra Section I.B. Congress recodified this language in the 1952 Act, and nothing in the legislative history indicates that Congress intended to rid the Act of this requirement. On the contrary, "Congress is presumed to be aware of a [. . .] judicial interpretation of a statute and to adopt that interpretation when it reenacts a statute without change." Forest Grove Sch. Dist. v. T.A., ___ U.S. ___, 129 S.Ct. 2484, 2492, 174 L.Ed.2d 168 (2009) (quoting Lorillard v. Pons, 434 U.S. 575, 580, 98 S.Ct. 866, 55 L.Ed.2d 40 (1978)).

Finally, a separate requirement to describe one's invention is basic to patent law. Every patent must describe an invention. It is part of the *quid pro quo* of a patent; one describes an invention, and, if the law's other requirements are met, one obtains a patent. The specification must then, of course, describe how to make and use the invention (i.e., enable it), but that is a different task. A description of the claimed invention allows the United States Patent and Trademark Office ("PTO") to examine applications effectively; courts to understand the invention, determine compliance with the statute, and to construe the claims; and the public to understand and improve upon the invention and to avoid the claimed boundaries of the patentee's exclusive rights.

Alan Lourie

LOURIE, ALAN. "ARIAD PHARMACEUTICALS, INC. V. ELI LILLY & CO." *FEDERAL REPORTER* THIRD SERIES (F.3D). VOL. 598 (2010): 1337–1345.

SEE ALSO *Diamond v. Chakrabarty; Eli Lilly & Co. v. Medtronic, Inc.; Labcorp v. Metabolite, Inc.; Patents and Other Intellectual Property Rights; Roche Products v. Bolar Pharmaceutical*

BIBLIOGRAPHY

Books

Gordon, Thomas T., and Arthur S. Cookfair. *Patent Fundamentals for Scientists and Engineers.* Boca Raton, FL: Lewis Publishers, 2000.

Gross, Marc S., S. Peter Ludwig, and Robert C. Sullivan. *Biotechnology and Pharmaceutical Patents: Law and Practice.* New York: Aspen Publishers, 2008.

Mills, Oliver. *Biotechnological Inventions, Moral Restraints and Patent Law.* Rev. ed. Farnham: Ashgate Pub, 2010.

Morgan, Dr. Gareth E. *Patent Litigation in the Pharmaceutical and Biotechnology Industries.* Witney, Oxford, UK: Biohealthcare Publishing, 2011.

Joseph P. Hyder

Arthritis Drugs

■ Introduction

The word arthritis means inflammation and refers to a group of painful conditions affecting the joints, including osteoarthritis, gout, and rheumatoid arthritis. The symptoms of arthritis can be treated using either analgesic drugs or non-steroidal anti-inflammatory drugs (NSAIDS). There are also medicines that modify the disease process itself by tackling the underlying inflammation. These include immunosuppressant drugs, including steroids, that work by suppressing the body's defenses, which otherwise cause inflammation. In recent years a new group of disease modifying drugs called anti-TNF drugs have been introduced for arthritis. The best known of these is etanercept (Enbrel), which is given by injection to patients with rheumatoid arthritis who have failed to respond to other, more conventional, treatments. Tumor necrosis factor (TNF) is an important player in the immune system, which is overproduced in rheumatoid arthritis and some other inflammatory diseases. Blocking the action of TNF with etanercept reduces pain and stiffness in the joints in two-thirds of those with moderate to severe rheumatoid arthritis, according to clinical trials. Moreover, just over half of patients taking etanercept had no progression in their disease after taking it for five years. However, like all biologic drugs, etanercept and related therapies are expensive. They also have some potentially serious side effects.

■ Historical Background and Scientific Foundations

In the nineteenth century high doses of salicylic acid, the active ingredient of willow bark, were widely used in the treatment of rheumatic fever, gout, and arthritis. However salicylic acid is very bitter and causes severe stomach irritation. Felix Hoffman (1868–1949), a chemist at Bayer, the company that manufactured salicylic acid, wanted to find a better drug to treat his own father's arthritis and came up with aspirin, otherwise known as acetylsalicylic acid. Aspirin first went on sale in 1899 and is still used to treat arthritis pain. Aspirin is an NSAID, along with drugs such as diclofenac and indomethacin, which are also used for arthritis. Acetaminophen is another commonly used analgesic for arthritis, although it is not an NSAID. However, NSAIDs can cause stomach pain and bleeding, whereas some of the newer NSAIDS, known as COX-2 inhibitors, have been associated with an increased risk of heart attack. Therefore, there has been active research into alternative treatments for arthritis, preferably one that would do more than just treat the symptoms of the disease.

In 1998 etanercept became the first biologic drug to be introduced for the treatment of rheumatoid arthritis. A biologic drug is one made in living cells. It is often, but not always, a protein molecule; the starting point for its manufacture is the gene that codes for the protein. Biologic drugs often work by targeting a molecule known to be involved in a disease process, thereby blocking its action. Etanercept binds to TNF, which is known to be involved in rheumatoid arthritis. It is given by weekly injection, which patients can administer themselves. There are a number of biologic drugs for the treatment of rheumatoid arthritis, including adalimunab (Humira) and infliximab (Remicade). They are known as anti-TNF drugs or TNF blockers. Adalimubnab and infliximab are monoclonal antibodies, whereas etanercept is a fusion protein. The newer drugs have the advantage of needing less frequent administration. The anti-TNF drugs are also used in the treatment of psoriatic arthritis, ankylosing spondylitis, and in children with juvenile idiopathic arthritis. However, they are not suitable for the treatment of osteoarthritis. Treatment with an anti-TNF drug can often produce dramatic improvement in pain and other symptoms in people with arthritis, as well as slowing down joint damage. However, these medicines cannot reverse joint damage already present.

■ Impacts and Issues

Compared to drugs such as aspirin and penicillin, biologics tend to be very expensive because of the way they are produced. Prices may come down when cheaper copies, known as generic or follow-on biologics, start to be produced, but it will probably be many years before these become widely available. There are around 1.3 million people with rheumatoid arthritis in the United States but not all of those who could benefit from etanercept have access to it. The anti-TNF drugs tend to be restricted to those patients who have failed to respond to other treatments. Partly this is due to high cost, but also because these drugs are still relatively new and their long-term side effects are unknown. When there is more experience with these drugs, and generic versions become available, it is possible they will become first-line treatment, and patients with rheumatoid arthritis will have far less disabling damage to their joints.

Rheumatoid arthritis and related conditions are thought to be autoimmune in nature. That is, an immune response overreaction produces inflammation and related symptoms. What produces the abnormal immune response is not well understood. Because TNF is an important molecule in this response, blocking it with etanercept or a related biologic drug seems rational. However, interfering with the immune response can have unexpected consequences. Blood counts may be lowered by anti-TNF treatment, leaving the patient at increased risk of infection. Patients need to take precautions such as avoiding certain foods, like raw eggs, soft cheeses, and unpasteurized milk, which carry a risk of food poisoning, and they also need to take extra care over food hygiene. Some patients may find such precautions bothersome.

There has long been concern that anti-TNF drugs may also increase the risk of cancer. A review of clinical studies covering more than 40,000 patients suggested that anti-TNF therapy may increase the risk of skin cancer, including melanoma, the most deadly form of the disease. However, there was no increased risk of any other type of cancer. Since a healthy immune system is essential in protecting the body from the development of cancer, it is perhaps not surprising that therapies that block one part of the immune system should increase cancer risk. However, the overall cancer risk with anti-TNFs seems to be relatively small, and most cases of skin cancer can be treated successfully. But these findings will drive research to find even more specific biologic drugs to treat autoimmune diseases such as rheumatoid arthritis.

WORDS TO KNOW

FUSION PROTEIN: A type of biologic drug created from genes coding for two different proteins. This is usually done to improve the property of the protein produced. In etanercept, for instance, one protein binds to TNF while the other makes the drug easier to deliver in the body.

MONOCLONAL ANTIBODY: An immune protein produced in response to the introduction of an antigen protein into the body, usually on the surface of a pathogen or cancer cell. The antibody binds to the antigen as part of the immune response. A monoclonal antibody is a type of biologic drug consisting of a population of identical antibodies that can bind to a target antigen involved in a disease process.

OSTEOARTHRITIS: An inflammatory condition of the joints caused by aging and wear and tear, rather than autoimmunity.

NON-STEROIDAL ANTI-INFLAMMATORY AGENT (NSAID): Any of a group of drugs that act locally, at the site of pain and inflammation, to reduce symptoms. They do not contain steroids, which can also reduce inflammation.

RHEUMATOID ARTHRITIS: A chronic inflammatory condition of the joints, thought to be an autoimmune disease.

SEE ALSO *Biopharmaceuticals; Bone Substitutes; Orphan Drugs; Pain Killers; Pharmacogenomics; Prosthetics; Stem Cells*

BIBLIOGRAPHY

Books

Bucknell, Duncan Geoffrey. *Pharmaceutical, Biotechnology, and Chemical Inventions: World Protection and Exploitation.* New York: Oxford University Press, 2011.

Cairns, Donald. *Essentials of Pharmaceutical Chemistry,* 3rd ed. London: Pharmaceutical Press, 2008.

Lorig, Kate, and James F. Fries. *The Arthritis Handbook: A Tested Self-Management Program for Coping with Arthritis and Fibromyalgia,* 6th ed. Cambridge, MA: Da Capo Press, 2006.

Web Sites

"Arthritis." *Centers for Disease Control and Prevention (CDC).* http://www.cdc.gov/arthritis/ (accessed September 10, 2011).

A woman with rheumatoid arthritis weeds her garden.   *© Sue McDonald/Shutterstock.com.*

"Arthritis." *National Institutes of Health (NIH)*. http://health.nih.gov/topic/Arthritis (accessed September 10, 2011).

"Juvenile Rheumatoid Arthritis." *National Institutes of Health (NIH)*. http://health.nih.gov/topic/JuvenileRheumatoidArthritis (accessed September 10, 2011).

Mayo Clinic Staff. "Arthritis." *Mayo Clinic*. http://www.mayoclinic.com/health/arthritis/DS01122 (accessed September 10, 2011).

"Osteoarthritis." *National Institutes of Health (NIH)*. http://health.nih.gov/topic/Osteoarthritis (accessed September 10, 2011).

Susan Aldridge

Artificial Organs

◼ Introduction

The determined efforts of the medical research community to develop viable artificial organs has particular urgency in most western societies in the twenty-first century. The prominence of cardiovascular disease in its various forms, with its debilitating effects on physical ability, is a significant concern. Kidney failure requires expensive and time-consuming dialysis machines to provide life-giving support for those who await a possible organ transplant. The prevalence of respiratory illnesses and diseases such as lung cancer prompted artificial lung research that led to the development of external membrane oxygenators that tended to extend the lives of persons afflicted with these conditions.

The distinction between artificial organs and life-supporting machines is that organs are inserted in the body as a permanent physical component. The heart, lungs, and liver are the three organs that have received the greatest research attention from regenerative medicine scientists.

In many instances, the condition of persons with diseased or injured organs can be remedied by a transplant. Declining organ donation rates observed throughout the western world have resulted in a profound shortage of donor organs that approaches crisis levels in many nations.

Artificial organ research engages the efforts of scientists from the allied disciplines of regenerative medicine and tissue engineering. The combined research outcomes from work undertaken in cell transplantation, material sciences, and a variety of bioengineering applications have produced a number of prototype biological substitutes that work to restore and maintain normal function in diseased and injured tissues. Biotechnical research advances have taken artificial organs from the realm of futurists and Hollywood filmmakers to scientific reality, with mainstream medical use seemingly near at hand for many devices.

◼ Historical Background and Scientific Foundations

Prior to the late twentieth century, artificial organs were regarded as a concept more suited to the realm of science fiction than as an attainable and practical medical scientific application. As researchers began to understand how a fully functioning organ could be implanted in the human body without being rejected by the surrounding host structures, artificial organ research advanced as part of a wide ranging multidisciplinary regenerative medical science, in which biology, chemistry, stem cell research, material sciences, and tissue engineering are all important constituencies.

Artificial organ research has been spurred by two separate social forces. The first is the increased life expectancy generally enjoyed by people in the western world. Longer-lived persons are more likely to require new organs to counter the effects of aging or disease. This demographic reality is made more acute by the low rate of organ donations observed in these countries. As populations increase, and the number of available organs for transplant relative to population declines, waiting lists for needed organs lengthen, and patients often die while waiting for a suitable transplant.

The second force driving successful artificial organ research and development is that it will provide a complete cure for a particular organ disease or debilitating condition, as opposed to a treatment option that must be continually administered at significant expense. Kidney dialysis machines are a prime example of this distinction: The machines preserve a patient's life, but an artificial kidney is an ultimate solution.

◼ Impacts and Issues

Artificial organ research does not generate as many ethical implications as the other branches of regenerative medicine. One example of the intense controversies that

WORDS TO KNOW

ALBUMIN: A protein produced by the liver that assists in the movement of small molecules through the blood stream. Albumin is essential in the prevention of blood fluid leaking from its vessels into adjacent tissues.

DIALYSIS: The separation of particles in a liquid on the basis of differences in their ability to pass through a membrane. This clinical purification of blood thus is a substitute for the normal function of the kidney.

MEMBRANE OXYGENATION: The artificial process by which carbon dioxide is removed and oxygen simultaneously added to blood. Oxygenation treatments are used if the patient's lungs are so damaged by disease or injury that they do not function. Whereas the current techniques involve external devices that are connected to the body, ongoing artificial lung research seeks to build on membrane oxygenation principles to create an implantable artificial organ.

REGENERATIVE MEDICINE: The emerging medical field that includes three disciplines: artificial organ research, biomaterials and tissue regeneration, and stem cell therapies.

flow through a network of sensors that detect changes in physical activity levels and permit the heart rhythms to increase or decrease accordingly.

The Carpentier heart represents a significant advance from the earlier American devices, such as AbioCor, or the Jarvik unit that was first successfully implanted in a human subject in 1982. The Jarvik design included wiring that protruded through the patient's skin and a large external compressor to power heart function. Clinical trials with AbioCor confirmed a significantly higher risk of blood clots associated with strokes. The Carpentier heart is constructed with a pseudo-skin made from biosynthetic materials created from treated animal skin. The microporous nature of the skin permits more even blood flow, and it reduces the risk of blood clot formation.

Artificial liver research also shows considerable promise, though an implantable artificial liver does not

Program Director Patrick Coulombier holds the prototype of a fully implantable artificial heart at Carmat, a French biomedical firm. French professor and leading heart transplant specialist Alain Carpentier presented the prototype in 2008 and said the prosthetic heart, successfully tested on animals, was ready to be manufactured and tested on humans by 2011. He hoped that it would be ready as a transplant alternative by 2013. *© AP Images/Jacques Brinon.*

continue to stimulate public debate in the United States and Europe is the use of stem cells extracted from discarded human embryos. This cellular material is a by-product of the work carried out in fertility clinics. There are strong views advanced on both sides of the question of whether such cells should be used for any other purpose. Stem cells are an example of how ethical questions often influence the course of scientific research.

In contrast, artificial organs represent the laudable human pursuit of a cure for diseases in which the patient is provided with the opportunity to be independent. Persons who suffer from debilitating conditions that impair vital organ function essentially are tethered to medications or life-supporting machines as they await an organ transplant that may never come. Artificial organs hold the promise of freedom and enhanced quality of life for these persons. The most significant ethical question that arises in the context of artificial organ research, given the prevalence of acute transplant organ shortages worldwide, is whether all citizens ought to be mandated to donate their organs for transplant or research purposes after death.

The research that has been directed at artificial heart development is the most advanced of the three primary organs considered, and it is a body of research work that has generated the greatest optimism among regenerative scientists. In 2008 a fully implantable artificial heart was constructed and made ready for clinical trials. French heart surgeon Alain Carpentier (1933–) devised the unit, which is able to regulate heart rate and blood yet exist. A treatment device that takes science closer to a functioning artificial liver is the Extracorporeal Liver Assist Device (ELAD). The ELAD is placed outside of the patient's body, where it performs the essential liver functions of blood cleansing and the production of albumin. ELAD technology combines artificial machinery and live human liver cells through which blood flows for processing. Liver stem cells have also been employed in a bio-artificial liver, where external bioreactors separate blood into its constituent plasma and cells, and the reactor removes toxic ammonia.

Artificial lung research is the least advanced of the work that has been directed to three primary organs noted above. The principles associated with artificial lung theory have been understood since 1885, when Max von Frey (1852–1932) and Max Gruber (1853–1927) developed a crude processing device that successfully mimicked human lung function. The development of fully functioning external lung machines was crucial to the ability of cardiac surgeons to develop the bypass surgeries that have become staple heart surgery procedures. The development of hollow fiber microporous polypropylene membranes was an important advance that permitted the circulating blood to flow smoothly through the device and allowed the efficient exchange of gases essential to human function. Implantable artificial lungs remain an elusive medical science objective, as the nature of the lung engages complex questions of air volume, durability, and the biocompatibility of the artificial tissue with the surrounding cell structure.

Issues with biocompatibility in all implanted artificial organs may be ultimately overcome if science is able to master the process by which new organs are constructed from human cells. In 2010 American biotech researchers successfully "printed" a blood vessel using human cells. Once assembled, the vessel was repeatedly flushed with nutrients that simulated human blood flow, and the vessel lived for more than one month. In the short term, this technology is likely to permit greater success in heart bypass surgeries. Over the longer term, researchers seek to build on the ability to print numerous vessels to construct a complex organ such as the human kidney.

The cost of artificial organ research and construction remains significantly high, which is a determining factor in the achievement of commercially viable devices. The materials alone used to construct the 2008 Carpentier heart cost more than $100,000. If such artificial devices became widely available, the costs will likely be imposed on the public. This might be either in direct costs or in the form of increased health insurance premiums for all, presenting ethical questions that must be addressed.

SEE ALSO *Antirejection Drugs; Biomimetic Systems; Biorobotics; Nanotechnology, Molecular; Prosthetics; Stem Cell Cloning; Skin Substitutes; Synthetic Organs*

BIBLIOGRAPHY

Books

Bronzino, Joseph D. *Tissue Engineering and Artificial Organs.* Boca Raton, FL: CRC/Taylor & Francis, 2006.

Hakim, Nadey S. *Artificial Organs.* London: Springer-Verlag, 2009.

Hench, Larry L., and Julian R. Jones, eds. *Biomaterials, Artificial Organs and Tissue Engineering.* Boca Raton, FL: CRC Press, 2005.

Lysaght, Michael J., and Thomas J. Webster, eds. *Biomaterials for Artificial Organs.* Oxford: Woodhead Publishing, 2011.

Najjariyan, Siyamik, et al. *Mechatronics in Medicine: A Biomedical Engineering Approach.* New York: McGraw-Hill, 2012.

Web Sites

American Society for Artificial Internal Organs. http://www.asaio.com/ (accessed October 18, 2011).

"Artificial Organs from Discovery to Clinical Use." *Project Bionics: An ASAIO History Project at The Smithsonian.* http://echo.gmu.edu/bionics/ (accessed October 18, 2011).

Glausiusz, Josie. "The Big Idea: Organ Regeneration." *National Geographic*, March 2011. http://ngm.nationalgeographic.com/2011/03/big-idea/organ-regeneration-text (accessed October 18, 2011).

"Medical Devices and Artificial Organs." *McGowan Institute for Regenerative Medicine.* http://www.mirm.pitt.edu/programs/medical_devices/ (accessed October 18, 2011).

"Replacing Body Parts." *NOVA*, March 26, 2011. http://www.pbs.org/wgbh/nova/body/replacing-body-parts.html (accessed October 18, 2011).

Bryan Thomas Davies

Asilomar Conference on Recombinant DNA

■ Introduction

The Asilomar Conference on Recombinant DNA Molecules was a seminal biotechnology conference held in the United States in 1975 to address concerns about the safety of recombinant DNA (rDNA) technologies. Recombinant DNA is a form of DNA that contains two or more genetic sequences that do not occur together naturally. Biologists splice together DNA from multiple sources to create rDNA molecules that may be introduced into a host in order to allow that host to express desired genetic traits.

The Asilomar Conference participants drafted a set of voluntary guidelines for scientists conducting rDNA research. The guidelines addressed the use of appropriate biological and physical controls during rDNA research in order to minimize any threat to humans, plants, and animals posed by potentially harmful rDNA molecules. These voluntary guidelines influenced the development of the biotechnology sector in the United States and elsewhere, as well as the adoption of national laws governing rDNA research.

■ Historical Background and Scientific Foundations

In 1971 Stanford University biochemist Paul Berg (1926–) and other biologists announced plans to introduce genes from the SV40 virus into an *E. coli* bacterium. SV40 is a monkey virus capable of transforming primate cell lines, including those of humans, into cancerous lines. In 1972 Berg introduced DNA from SV40 into the lambda virus, thereby creating the first rDNA molecule. Because of concerns over potential biohazards to laboratory workers, Berg did not complete the final step of the experiment, which involved actually introducing the modified SV40 virus into an *E. coli* bacterium.

Berg's experiment, although incomplete, was the first experiment to use rDNA technology to create a new molecule. rDNA technology involves the use of molecular cloning, polymerase chain reactions, or other laboratory techniques to manipulate genetic material from one or more sources to create a new genetic sequence that is then introduced to an organism. Scientists may manipulate genetic material from any species, because all DNA consists of combinations of nucleobases: adenine (A), cytosine (C), guanine (G), and thymine (T). Scientists may also utilize chemical synthesis of DNA to create unique genetic sequences of nucleobases that do not exist in nature.

Berg's experiment demonstrated the potential for recombinant DNA technology and raised concerns among biologists, government health officials, and medical ethicists about the safety of rDNA experimentation. A group of scientists, including Herbert Boyer (1936–) and Stanley N. Cohen (1935–), inventors of the DNA cloning technique, and Berg, successfully lobbied the president of the National Academy of Sciences (NAS) to appoint a committee to investigate the safety of rDNA technology. In 1974, the NAS committee, the Committee on Recombinant DNA Molecules, proposed an international conference at which leading biologists, attorneys, and other interested parties could debate rDNA biosafety and potential regulations. The NAS also announced a voluntary moratorium on rDNA research until rDNA biosafety issues could be addressed.

In February 1975, 140 participants attended the Asilomar Conference on Recombinant DNA Molecules at the Asilomar State Beach conference center in Pacific Grove, California. Berg organized the conference and served as chairman. The committee adopted a set of recommendations for rDNA research, which appeared in the Proceedings of the National Academy of Sciences in June 1975. The committee also recommended a complete prohibition on certain types of rDNA experiments that posed a high risk to workers or the public at large.

Impacts and Issues

The Asilomar Conference on Recombinant DNA Molecules addressed numerous biosafety issues presented by rDNA technology as reflected in the conference summary statement. First, conference participants addressed whether to end the research moratorium in the field of rDNA technology. Conference participants agreed that the burgeoning field of rDNA posed serious, and often unknown, safety issues. Nevertheless, the conference participants endorsed a continuation of research in the field but with appropriate safeguards in place to prevent a biosafety risk to researchers and the public.

After establishing that rDNA research should continue, conference participants focused on the nature and extent of appropriate safeguards for rDNA research. The conference summary statement includes two primary principles for dealing with the risks involved in rDNA research. First, containment of rDNA molecules should be "made an essential consideration in the experimental design." Researchers should design proper biological and physical barriers before beginning an rDNA project. Biological barriers are the more effective means of containing potentially harmful rDNA molecules. The use of bacterial hosts that are unable to survive in the natural environment, that is, outside of laboratory conditions, is a common biological barrier. Another biological control method relies on using vectors that can grow only in certain hosts. For example, researchers could use a virus that can grow within a certain bacterial host but would not pose a threat to other organisms that are not suitable hosts. Physical containment relies on the implementation of and adherence to strict laboratory procedures designed to reduce the possibility of a potentially harmful rDNA molecule escaping the laboratory. Unlike biological barriers, however, physical barriers cannot eliminate the possibility of spreading an rDNA molecule.

The second principle endorsed by conference participants states that "the effectiveness of the containment should match, as closely as possible, the estimated risk" posed by an experiment. The recommendations of the conference left the level of biological and physical barriers employed in a certain project up to researchers, who must determine whether a project poses a minimal, low, moderate, or high risk. The conference summary statement states that certain experiments that pose an especially high risk of harm to humans, plants, or animals should not be undertaken with even the most stringent biological and physical barriers in place. Examples of experiments that present serious dangers include the

American chemist Paul Berg and his wife Mildred attending the Nobel Prize banquet in the town hall of Stockholm, Sweden. Berg was awarded the chemistry prize for his work with nucleic acid, December 10, 1980. © *Hulton Archive/Getty Images*

cloning of DNA from highly pathogenic organisms, using DNA containing genes from toxins, and the use of large quantities of rDNA molecules that "make products potentially harmful to man, animals, or plants." The voluntary guidelines adopted at the Asilomar Conference remained in place until the U.S. National Institutes of Health (NIH) adopted guidelines for U.S. researchers proposed by the NIH Recombinant DNA Advisory Committee in 1987.

The Asilomar Conference on Recombinant DNA Molecules has had a profound effect on the development of products using rDNA technology. The conference laid the foundation for responsible development of the rDNA biotechnology sector, including the commercial rDNA biotechnology sector that has dominated biotechnology since the early 1990s. Currenty, rDNA technologies are used to produce synthetic insulin, vaccines, herbicide- and insect-resistant crops, high yield agricultural products, and other medical and agricultural uses.

■ Primary Source Connection

To gain a better understanding of biochemical processes at the cellular level, molecular biologists took on the challenge of researching and studying recombinant DNA molecules. The Asilomar Conference took place to review the scientific progress in research on this highly complex subject. Participants discussed ways to deal with potential biohazards of researching recombinant DNA molecules and individuals voiced concerns regarding biohazards, appropriate safeguards, and standards of protection in scientific methodology and experimentation surrounding this type of research.

Summary Statement of the Asilomar Conference on Recombinant DNA Molecules

This meeting was organized to review scientific progress in research on recombinant DNA molecules and to discuss appropriate ways to deal with the potential biohazards of this work. Impressive scientific achievements have already been made in this field and these techniques have a remarkable potential for furthering our understanding of fundamental biochemical processes in pro- and eukaryotic cells. The use of recombinant DNA methodology promises to revolutionize the practice of molecular biology. Although there has as yet been no practical application of the new techniques, there is every reason to believe that they will have significant practical utility in the future.

Of particular concern to the participants at the meeting was the issue of whether the pause in certain aspects of research in this area, called for by the Committee on Recombinant DNA Molecules of the National Academy of Sciences, U.S.A. in the letter published in July, 1974 should end; and, if so, how the scientific work could be undertaken with minimal risks to workers in laboratories, to the public at large, and to the animal and plant species sharing our ecosystems.

The new techniques, which permit combination of genetic information from very different organisms, place us in an area of biology with many unknowns. Even in the present, more limited conduct of research in this field, the evaluation of potential biohazards has proved to be extremely difficult. It is this ignorance that has compelled us to conclude that it would be wise to exercise considerable caution in performing this research. Nevertheless, the participants at the Conference agreed that most of the work on construction of recombinant DNA molecules should proceed provided that appropriate safeguards, principally biological and physical barriers adequate to contain the newly created organisms, are employed. Moreover, the standards of protection should be greater at the beginning and modified as improvements in the methodology occur and assessments of the risks change. Furthermore, it was agreed that there are certain experiments in which the potential risks are of such a serious nature that they ought not to be done with presently available containment facilities. In the longer term, serious problems may arise in the large scale application of this methodology in industry, medicine, and agriculture. But it was also recognized that future research and experience may show that many of the potential biohazards are less serious and/or less probable than we now suspect.

Paul Berg

BERG, PAUL., ET AL. "SUMMARY STATEMENT OF THE ASILOMAR CONFERENCE ON RECOMBINANT DNA MOLECULES." *PROCEEDINGS OF THE NATIONAL ACADEMY OF SCIENCES OF THE UNITED STATES OF AMERICA* VOL. 72, NO. 6. (JUNE 1975): 1981–1984.

SEE ALSO *Bacteriophage; Biosafety Level Laboratories; Genetic Engineering; Recombinant DNA Technology*

BIBLIOGRAPHY

Books

Glick, Bernard R., Jack J. Pasternak, and Cheryl L. Patten. *Molecular Biotechnology: Principles and Applications of Recombinant DNA*, 4th ed. Washington, DC: ASM Press, 2010.

Greif, Karen F., and Jon F. Merz. *Current Controversies in the Biological Sciences Case Studies of Policy Challenges from New Technologies.* Cambridge, MA: MIT, 2007.

Periodicals

Berg, Paul. "Meetings That Changed the World: Asilomar 1975: DNA Modification Secured." *Nature* 455, no. 7211 (September 18, 2008): 290–291.

Hindmarsh, Richard, and Herbert Gottweis. "Recombinant Regulation: The Asilomar Legacy 30 Years On." *Science as Culture* 14, no. 4 (December 2005): 299–307.

Krimsky, Sheldon. "From Asilomar to Industrial Biotechnology: Risks, Reductionism and Regulation." *Science as Culture* 14, no. 4 (December 2005): 309–323.

Web Sites

Berg, Paul. "Asilomar and Recombinant DNA." *Nobelprize.org*, August 26, 2004. http://www.nobelprize.org/nobel_prizes/chemistry/laureates/1980/berg-article.html (accessed September 30, 2011).

Joseph P. Hyder

Bacteria, Drug Resistant

■ Introduction

Bacteria are adaptable to their environment. These adaptations vary depending on the situation and can involve a variety of structural and chemical changes inside a bacterium, along with structural rearrangement of the cell wall or cell envelope layers that enclose the bacterium. Some alterations are reversed when the survival challenge is gone. Other alterations are maintained and, when they have changed because of alterations in the genome, they can even be passed on to succeeding generations of bacteria.

The latter changes that are inherited are the basis of bacterial drug resistance. In some cases, the resistance can be in response to antibiotics. Several types of bacteria have become resistant to most or all known antibiotics. This multiple resistance is very beneficial for the bacteria in environments such as hospitals, where antibiotics are in common use. Indeed, the overuse of antibiotics to encourage weight gain in cattle and to treat illnesses caused by viruses (antibiotics are not effective against viruses) has provided a selective pressure favoring the appearance and spread of drug-resistant bacteria.

Bacteria can also acquire resistance to drugs other than antibiotics by virtue of their mode of growth. Bacteria that are slow-growing or that reside in biofilms can be highly resistant to drugs including ethanol.

■ Historical Background and Scientific Foundations

The first antibiotic, penicillin, was discovered in 1929. Since then, many naturally occurring and chemically synthesized antibiotics have been used to control bacteria. The introduction of an antibiotic is almost always followed by the acquisition of bacterial resistance. Antibiotic research is now literally a race against time. Unfortunately, in 2011 few new antibiotics are under development.

Antibiotic resistance can develop swiftly. Penicillin provides a good example. When first introduced, penicillin was so effective that it was assumed by many scientists that the battle against infectious bacterial diseases was over. Yet, only a few years after the widespread use of penicillin, penicillin-resistant bacteria of various types began to be isolated. Their numbers were small at first, but within a decade, penicillin resistance was global. By the mid-1990s, almost 80 percent of all strains of *Staphylococcus aureus* were resistant to penicillin. Other bacteria, such as *Streptococcus pyogenes*, remain susceptible to penicillin.

Drug resistance can occur in two ways. The first route is inherent (or natural) resistance. Penicillin again provides an example. Gram-negative bacteria such as *Escherichia coli* are often naturally resistant to penicillin because they have two membranes. The outermost membrane is constructed so that the passage of a penicillin molecule across the membrane is very difficult. Because the target of penicillin, like many antibiotics, lies inside the bacterium, an antibiotic that cannot get inside the cell is ineffective. Alterations in the membrane structure can also occur. This process is called adaptation.

Resistance can also be acquired, and it is almost always due to a change in the genetic makeup of the bacterial genome. Acquired resistance occurs when there is change in the genetic material due to a mutation or as a response by the bacteria to the selective pressure imposed by the antibacterial agent. Once the genetic alteration that confers the resistance is present, it can be inherited by subsequent generations.

One of the adaptations in the surface chemistry of gram-negative bacteria is the alteration of a molecule called lipopolysaccharide. Depending on the growth conditions or whether the bacteria are growing on an artificial growth medium or inside a human, the lipopolysaccharide chemistry can become more or less water-repellent. The change to an outer membrane can profoundly affect the ability of antibacterial agents to kill the bacteria.

Another adaptation exhibited by *Vibrio parahaemolyticus*, and a great many other bacteria as well, is the formation of adherent populations on solid surfaces. This mode of growth is called a biofilm. Even in the 1980s, bacterial biofilms were thought to be unimportant and even an artificial by-product of laboratory growth conditions. However, it has since been discovered that bacterial biofilms are the preferred mode of growth for bacteria in nature, in industrial settings, and in the body. In the body, biofilms cause dental decay and some types of heart disease, for example.

Adoption of a biofilm mode of growth induces many changes involving the expression of previously unexpressed genes. As well, deactivation of actively expressing genes can occur. Furthermore, the pattern of gene expression may not be uniform throughout the biofilm. Evidence from studies using techniques such as confocal laser microscopy, where the activity of living bacteria can be measured without disturbing the biofilm, has demonstrated that the bacteria closer to the top of the biofilm, and so closer to the outside environment, are very different than the bacteria lower down in the

biofilm. Deep within a biofilm the bacteria can be almost dormant. Because many antibiotics depend on bacterial growth for their effect, the deep biofilm bacteria can be very resistant to a variety of antibiotics.

Bacteria within a biofilm and bacteria found in other niches, such as in a wound where oxygen is limited, grow and divide at a far slower speed than the bacteria found in the test tube in the laboratory. Such bacteria are able to adapt to the slower growth rate,

Patients sit in the open at the multi-drug resistant tuberculosis hospital in Maseru, Lesotho. Lesotho has the world's fourth-highest prevalence of tuberculosis. Tuberculosis is caused by strains of mycobateria, usually *Mycobacterium tuberculosis*, which is becoming more resistant to existing tuberculosis drugs. *© AP Images/Denis Farrell.*

once again by changing their chemistry and gene expression pattern. When presented with more nutrients, the bacteria can often resume the rapid growth and division rate of their test tube counterparts, often very quickly. Thus, even though they have adapted to a slower growth rate, the bacteria remained "primed" for the rapid another adaptation to a faster growth rate.

■ Impacts and Issues

As of 2011, drug resistant bacteria that pose a serious health threat throughout the world include methicillin resistant *Staphylococcus aureus* (MRSA) extensively drug resistant tuberculosis (XDR-TB), vancomycin resistant *Enterococcus*, and multiple drug resistant *Neiserria gonorrhoeae*. As one example, in 2011, 440,000 new cases of XDR-TB were appearing each year, causing at least 150,000 deaths, according to the World Health Organization.

Molecular biology researchers are trying to identify the genes in disease-causing (pathogenic) bacteria that are crucial for drug resistance. Hopefully, by changing the sequence of the genes or by blocking the activity of the molecules coded for and expressed by the genes, the drug resistance of the target pathogens might be overcome. Still, it is a fact of life that bacterial adaptation will continue. As a result, an era when drug resistant bacteria are no longer a health problem seems very unlikely.

SEE ALSO *Antibiotics, Biosynthesized; Antimicrobial Soaps; Biopharmaceuticals; Penicillins; Pharmacogenomics*

BIBLIOGRAPHY

Books

Ahmad, Iqbal, and Farrukh Aqil. *New Strategies Combating Bacterial Infection.* Weinheim, Germany: Wiley-Blackwell, 2009.

Amabile-Cuevas, Carlos F. *Antimicrobial Resistance in Bacteria.* Norfolk, UK: Horizon Bioscience, 2007.

Bud, Robert. *Penicillin: Triumph and Tragedy.* New York: Oxford University Press, 2007.

Choffnes, Eileen R., David A. Relman, and Alison Mack. *Antibiotic Resistance: Implications for Global Health and Novel Intervention Strategies: Workshop Summary.* Washington, DC: National Academies Press, 2010.

Shlaes, David M. *Antibiotics: The Perfect Storm.* Dordrecht, The Netherlands: Springer, 2010.

Spellberg, Brad. *Rising Plague: The Global Threat from Deadly Bacteria and Our Dwindling Arsenal to Fight Them.* Amherst, NY: Prometheus Books, 2009.

Web Sites

"Antimicrobial Resistance." *World Health Organization.* http://www.who.int/mediacentre/factsheets/fs194/en/ (accessed September 25, 2011).

"Methicillin-Resistant *Staphylococcus aureus* (MRSA) Infections." *Centers for Disease Control and Prevention (CDC).* http://www.cdc.gov/mrsa (accessed September 25, 2011).

"Multidrug-Resistant TB (MDR TB)." *Centers for Disease Control and Prevention (CDC).* http://www.cdc.gov/tb/topic/drtb/default.htm (accessed September 25, 2011).

Brian Douglas Hoyle

Bacteriophage

■ Introduction

A bacteriophage, or phage, is a virus that infects a bacterial cell, taking over the host cell's genetic material, reproducing itself, and eventually destroying the bacterium.

Bacteriophages consist of a protein coat and a nucleic acid core of DNA or RNA. Most DNA phages have double-stranded DNA, whereas phage RNA may be double- or single-stranded. The electron microscope shows that phages vary in size and shape. Filamentous or threadlike phages, discovered in 1963, are among the smallest viruses known. Scientists have studied extensively the phages that infect *Escherichia coli (E. coli)*, bacteria that are abundant in the human intestine. Some of these phages, such as the T4 phage, consist of a capsid or head, often polyhedral in shape, that contains DNA, and an elongated tail consisting of a hollow core, a sheath around it, and six distal fibers attached to a base plate. When T4 attacks a bacterial cell, proteins at the end of the tail fibers and base plate attach to proteins located on the bacterial wall. Once the phage grabs hold, its DNA enters the bacterium while its protein coat is left outside.

■ Historical Background and Scientific Foundations

The word phage comes from the Greek word *phagein*, which means to eat. Bacteriophages were discovered by English bacteriologist Frederick Twort (1877–1950) in 1915. While attempting to grow *Staphylococcus aureus*, the bacteria that most often cause boils in humans, he observed that some bacteria in his laboratory plates became transparent and died. Twort isolated the substance that was killing the bacteria and hypothesized that the agent was a virus. In 1917 French-Canadian scientist Felix H. d'Hérelle (1873–1949) discovered bacteriophages independently.

The significance of phages was not appreciated until years later. One prominent scientist in the field was Salvador E. Luria (1912–1991), an Italian-American biologist especially interested in how x rays cause mutations in bacteriophages. Luria was also the first to obtain clear images of a bacteriophage using an electron microscope. Luria emigrated to the United States from Italy where he met Max Delbruck (1906–1981), a German-American molecular biologist. In the 1940s Delbruck worked out the lytic mechanism by which some bacteriophages replicate. Together, Luria, Delbruck, and others studied the genetic changes that occur when viruses infect bacteria.

Until 1952, scientists did not know which part of the virus, the protein or the DNA, carried the information regarding viral replication. The answer, that DNA transmits genetic information, came from experiments with phages. For their discoveries concerning the structure and replication of viruses, Luria, Delbruck, and Alfred Hershey (1908–1977) shared the Nobel Prize for physiology or medicine in 1969. In 1952 two American biologists at the University of Wisconsin, Norton Zinder (1928–) and Joshua Lederman (1925–2008), discovered that a phage can incorporate its genes into the bacterial chromosome. The phage genes then are transmitted from one generation to the next when the bacterium reproduces. Then in 1980, English biochemist Frederick Sanger (1918–) won a Nobel Prize for determining the nucleotide sequence in DNA using bacteriophages.

Double-stranded DNA phages reproduce in their host cells in two different ways: the lytic cycle and the lysogenic cycle. The lytic cycle kills the host bacterial cell. During the lytic cycle in *E. coli*, for example, the phage infects the bacterial cell, and the host cell starts to transcribe and translate the viral genes. One of the first genes that it translates encodes an enzyme that chops up the *E. coli* DNA. The host now follows instructions solely from phage DNA, which commands the host to synthesize phages.

WORDS TO KNOW

PROPHAGE: The genome of a lysogenic phage when incorporated into the chromosome of the host bacterium.

RESTRICTION ENZYMES: Enzymes that cut DNA at specific points in its sequence and are a useful tool in molecular biology.

T4 PHAGE: One of the largest and most widely-studied phages, T4 is a lytic phage that infects *Escherichia coli* bacteria.

VIRUS: A microorganism smaller than a bacteria that cannot grow or reproduce outside a living cell host.

At the end of the lytic cycle, the phage directs the host cell to produce the enzyme lysozyme, which digests the bacterial cell wall. As a result, water enters the cell by osmosis, and the cell swells and bursts. The destroyed or lysed cell releases up to 200 phage particles ready to infect nearby cells. However, the lysogenic cycle does not kill the bacterial host cell. Instead, the phage DNA is incorporated into the host cell's chromosome, where it is then called a prophage. Every time the host cell divides, it replicates the prophage DNA along with its own. As a result, the two daughter cells each contain a copy of the prophage, and the virus has reproduced

without harming the host cell. Under certain conditions, however, the prophage can give rise to active phages that bring about the lytic cycle.

■ Impacts and Issues

Bacteriophages are one of the most powerful tools in molecular biology. One use of bacteriophages is in genetic engineering, upon which the biotechnology industry is built. During genetic engineering, scientists combine genes from different sources and transfer the recombinant DNA into cells where it is expressed and replicated. Researchers often use *E. coli* as a host because they can grow it easily, and its properties are well known. One method of transferring the recombinant DNA to cells uses phages. Employing restriction enzymes to break into the phage's DNA, scientists splice foreign DNA into the viral DNA. The recombinant phage then infects the bacterial host. Scientists use this technique to create new medical products such as vaccines.

In addition, bacteriophages provide information about genetic defects, human development, and disease, and they have been used extensively in gene mapping. One geneticist has developed a technique using bacteriophages to manipulate genes in mice, while others are using phages to infect and kill disease-causing

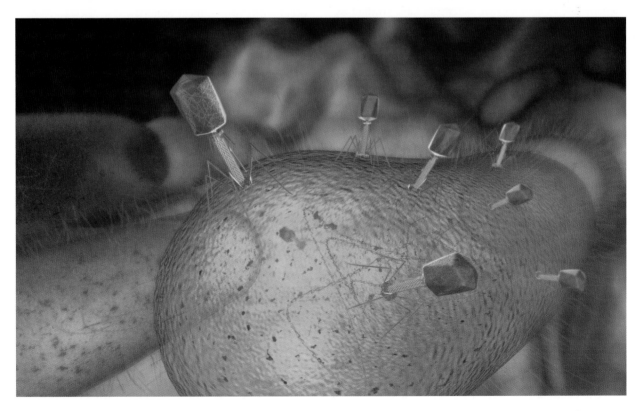

A three-dimensional illustration shows a bacteriophage tacking to the bacterium *Escherichia coli* (adsorption). A bacteriophage (from "bacteria" and Greek *phagein*, "to eat") is a virus that infects bacteria. *© DocCheck Medical Services GmbH/Alamy.*

bacteria in mice. In addition, microbiologists found a filamentous bacteriophage that transmits the gene that encodes the toxin for cholera, a severe intestinal disease that kills tens of thousands worldwide each year. Although this has led to the development of two different vaccines for cholera, they are only available in limited areas (mostly in Asia) and have an effectiveness rate of up to 50 percent. Scientists are continuing to study the role of bacteriophages in developing a highly effective vaccine for cholera that could be manufactured and stored easily, and delivered in a cost-effective manner.

SEE ALSO *Biotechnology; DNA Sequencing; Genbank; Gene Delivery; Genetic Engineering; Genome; Genomics; Molecular Biology; Recombinant DNA Technology; Virology*

BIBLIOGRAPHY

Web Sites

Dale, Jeremy W., and Simon F. Park. *Molecular Genetics of Bacteria*, 5th ed. Chichester, UK: Wiley-Blackwell, 2010.

Kutter, Elizabeth, and Alexander Sulakvelidze, eds. *Bacteriophages: Biology and Applications.* Boca Raton, FL: CRC Press, 2005.

Primrose, Sandy B., and Richard Twyman. *Principles of Gene Manipulation and Genomics*, 8th ed. Malden, MA: Blackwell, 2008.

Srivastava, Sudhir. *Molecular Genetics and Biotechnology.* Delhi: Swastik Publishers & Distributors, 2008.

Stephenson, Frank Harold. *DNA: How the Biotech Revolution Is Changing the Way We Fight Disease.* Amherst, NY: Prometheus Books, 2007.

Wegrzyn, Grzegorz, ed. *Modern Bacteriophage Biology and Biotechnology.* Kerala, India: Research Signpost, 2006.

Web Sites

Beecher, Cookson. "Bacteriophages Eyed as Antibiotic Alternatives." *Food Safety News*, November 16, 2010. http://www.foodsafetynews.com/2010/11/natural-disease-fighters-eyed-as-antibiotic-alternatives/ (accessed November 13, 2011).

Flores, Alfredo. "Phages Eyed as New Way to Control *Salmonella*." *The Journal of Environmental Quality*, April 28, 2006. http://www.ars.usda.gov/is/pr/2006/060428.htm (accessed November 13, 2011).

"General Information about Bacteriophages." *The Phage Phorum.* http://www.phages.org/PhageInfo.html (accessed November 13, 2011).

Bayh-Dole Act of 1980

■ Introduction

The Patent and Trademark Act Amendments of 1980, better known as the Bayh-Dole Act after its original sponsors, former Senators Birch Bayh (D-IN; 1928–) and Bob Dole (R-KS; 1923–), created a uniform patent transfer system for patents emanating from research projects funded by U.S. government agencies. The Bayh-Dole Act permitted universities, non-profits, and small businesses that receive federal research funds to maintain control over patents generated from government-funded projects. Prior to passage of the Bayh-Dole Act, the federal government retained the rights to all intellectual property produced by government-funded research and development projects unless waived by the responsible government agency. This system discouraged innovation and the commercialization of innovation—issues that the Bayh-Dole Act sought to remedy.

■ Historical Background and Scientific Foundations

Prior to World War II (1939–1945), dedicated government laboratories conducted most government-funded research in the United States. During the war, government-funded research projects, conducted in cooperation with university research programs or with the assistance of university professors, resulted in the development of military radar systems and nuclear weapons. These programs also highlighted the need for U.S. technological superiority and the ability to achieve this goal through public-private partnerships and programs.

Following World War II, the United States established several agencies for promoting private and university research through government funding, including the National Science Foundation (NSF), the National Institutes of Health (NIH), and the Office of Naval Research (ONR). Federally-financed university research projects increased greatly. In 1935, the federal government provided $575 million for university research programs, which accounted for less than 25 percent of all university research spending. By the 1960s, the federal government spent over $2 billion annually on university research programs, which accounted for over 60 percent of university spending on research.

The post-war increase in funding for private and university research, however, did not translate into many commercially available products as the result of muddled government policies related to patent transfer. In 1947, the U.S. Attorney General announced that the government should retain title to patents emanating from government-funded research projects unless the head of the agency responsible for the project granted an exception. The Department of Defense (DoD), a major funder of post-war scientific research projects, announced that it would allow private and university contractors to retain all patents resulting from DoD-funded projects. The DoD's stance resulted in unclear title to patents on projects jointly funded by the DoD and other government agencies. For all non-DoD projects, research institutions became bogged down in efforts to license government patents for commercial use or obtain the patent royalty-free.

With unclear patent ownership issues surrounding the majority of government-funded research projects—ultimately, 26 government agencies issued separate patent transfer rules—private investment and commercial innovation dried up. Government agencies also exhibited a reluctance to grant patents to universities outright. In 1980, the U.S. Patent and Trademark Office (PTO) only granted about 250 patents to universities, and only a handful of these inventions made their way to market. The U.S. government, meanwhile, accumulated 30,000 patents before 1980, but only approximately 5 percent of these patents were commercially licensed.

■ Impacts and Issues

By the 1970s, policymakers and universities began to explore ways in which to create a clearly defined uniform patent transfer policy that would promote the commercialization of government-funded research initiatives. The mid-20th century dominance of the U.S. economy had eroded as technological innovation and competitiveness decreased. Asian markets, especially Japan, and Europe had recovered from the economic aftermath of World War II and asserted themselves on the global stage. With input from university research coalitions, the U.S. Congress explored ways in which to increase the return on their investment in technological research.

In September 1978, Senators Bayh and Dole introduced the University Small Business and Patent Act, known as the Bayh-Dole Act. Congress eventually passed the Bayh-Dole Act, and President Jimmy Carter (1924–) signed the act into law on December 12, 1980. The Bayh-Dole Act sought to "promote the utilization of inventions arising from federally supported research or development" through the participation and joint collaboration of the government, small businesses, and non-profit organizations, including universities. The act also aimed to promote the commercialization of these inventions in order to make them available to the public.

The Bayh-Dole Act achieved these goals by enforcing uniform patent transfer provisions across all government agencies. The Bayh-Dole Act adopted the DoD's policy of granting the patent title to the contractor instead of the

WORDS TO KNOW

ACT: A statute, rule, or formal lawmaking document enacted by a legislative body or issued by a government.

PATENT: A patent is government grant of the exclusive right to manufacture, sell, license, or use an invention or process for a limited period.

government. This rule—the defining feature of the Bayh-Dole Act—allowed universities, small businesses, and non-profits to retain patent titles in inventions that they produced through government-funded research projects. The act also specifically stated that the contractors had the right to license the patented inventions.

The Bayh-Dole Act, however, retained one important right for the U.S. government. The law allows the government to "march-in" and reclaim the patent on an invention financed by the government if the contractor-patentholder does not make the product available to the public on reasonable terms. Under this march-in right, the government may then grant the patent to another entity that will make the invention available to the public.

Senators Birch Bayh (D-IN), left, and Bob Dole (R-KS) meet at the Capitol Building in Washington, DC.   © *AP Images/John Duricka.*

The Bayh-Dole Act has had the desired effect of stimulating commercialization of inventions, increasing private sector involvement in research, and stimulating economic growth. Since the passage of the Bayh-Dole Act, over 5,000 new companies have formed around university research. Prior to the Bayh-Dole Act, only a handful of patented innovations produced by joint government-university research were commercially licensed. Today, over 30 percent of all such innovations are commercially licensed. The new patent scheme created by the Bayh-Dole Act has also attracted additional investment in technological research. Between 1980 to 2008, total expenditures on academic research and development, which includes funding from all public and private sources, increased from $6 billion to over $51 billion. Furthermore, the number of U.S. patents granted to universities increased from 390 patents in 1980 to 3,042 patents in 2008.

SEE ALSO *Government Regulations; Patents and Other Intellectual Property Rights; Research Exemption*

BIBLIOGRAPHY

Books

Gordon, Thomas T., and Arthur S. Cookfair. *Patent Fundamentals for Scientists and Engineers.* Boca Raton, FL: Lewis Publishers, 2000.

Gross, Marc S., S. Peter Ludwig, and Robert C. Sullivan. *Biotechnology and Pharmaceutical Patents: Law and Practice.* New York: Aspen Publishers, 2008.

Klemens, Ben. *Math You Can't Use: Patents, Copyright, and Software.* Washington, DC: Brookings Institution Press, 2006.

Mills, Oliver. *Biotechnological Inventions: Moral Restraints and Patent Law.* Rev. ed. Burlington, VT: Ashgate Pub, 2010.

Morgan, Dr. Gareth E. *Patent Litigation in the Pharmaceutical and Biotechnology Industries.* Witney, Oxford, UK: Biohealthcare Publishing, 2011.

Joseph P. Hyder

Belmont Report

■ Introduction

In September 1978 the National Commission for the Protection of Human Subjects of Biomedical and Behavioral Research issued the "Belmont Report: Ethical Principles and Guidelines for the Protection of Human Subjects of Research, Report of the National Commission for the Protection of Human Subjects of Biomedical and Behavioral Research." The Belmont Report promulgates ethical principles for researchers conducting clinical trials or other medical research involving human subjects. The report establishes three main principles for researchers engaged in human experimentation: respect for persons, beneficence, and justice. In order to achieve the goals of these ethical principles, researchers should seek the informed consent of participants by providing information about potential risks and benefits, ensuring that the subject understands that information, and obtaining voluntary consent to proceed.

■ Historical Background and Scientific Foundations

In 1932 the U.S. Public Health Service (PHS) began a study of the effects of syphilis. The study enrolled 600 poor, African-American farmers and sharecroppers in and around Tuskegee, Alabama. The study included 399 participants who had syphilis and 201 participants without syphilis. The PHS promised to provide study participants with free health care. Instead, the PHS merely observed the progress of the disease in the patients and never informed study participants that they had syphilis. The PHS never provided treatment for the disease, even after penicillin became standard treatment for the disease in the 1940s.

In 1966 Peter Buxton (1938–), a PHS public health investigator, learned of the Tuskegee syphilis experiment and sent a letter to the director of the PHS Division of Venereal Diseases to express his concerns about the ethical implications of the study. The U.S. Center for Disease Control (CDC), the government agency responsible for the Tuskegee syphilis experiment at the time, stated that the experiment must continue until all study participants had died and undergone an autopsy, as per the original goal of the study.

In 1972 Buxton leaked information about the Tuskegee syphilis experiment to the media. The U.S. Congress conducted an investigation into the Tuskegee syphilis experiment, and, following a review of the study by the CDC and the U.S. Department of Health, Education, and Welfare (HEW), the CDC terminated the study. At the end of the Tuskegee syphilis experiment in 1972, 28 study participants had died of syphilis and 100 had died of complications related to syphilis. Forty men had infected their wives with syphilis, resulting in 19 children born with congenital syphilis.

Public outcry over the Tuskegee syphilis experiment led to the 1974 passage of the National Research Act, which required HEW to convene the National Commission for the Protection of Human Subjects of Biomedical and Behavioral Research. The National Research Act

---

WORDS TO KNOW

INFORMED CONSENT: The permission granted by the subject of a medical procedure or clinical trial after the individual has been apprised of all potential benefits and risks.

PENICILLIN: An antibiotic that is produced from certain strains of blue molds or, more typically, produced synthetically.

SYPHILIS: A chronic bacterial disease that may result in damage to the cardiovascular or central nervous systems; it may result in death if not treated. Syphilis is typically transmitted through sexual activity but may also be transmitted to a developing fetus, which is known as congenital syphilis.

tasked the National Commission for the Protection of Human Subjects of Biomedical and Behavioral Research with investigating the ethical considerations involved in human experimentation and recommending a code of bioethics for human experimentation to HEW. In September 1978 the commission released the Belmont Report, which was incorporated into the portion of U.S. law that regulates HEW. In 1979 the U.S. Congress reorganized HEW as the U.S. Department of Health and Human Services (HHS) and established the Department of Education as a separate cabinet-level department.

■ Impacts and Issues

The Belmont Report contains ethical principles for the protection of human subjects during human experimentation conducted during medical research. The ethical principles contained in the Belmont Report do not apply to the conduct of medical professionals practicing therapeutic medicine. Medical research carries inherent risks because the therapeutic benefits of a proposed treatment are not known. Without adhering to high ethical standards that emphasize the autonomy and well-being of human subjects, medical professionals conducting human experimentation could cause harm to human subjects.

The Belmont Report contains three ethical principles for researchers conducing human experimentation: respect for persons, beneficence, and justice. The Belmont Report calls on medical professionals to respect test subjects as autonomous agents. An autonomous individual is one with the capacity to deliberate and act in his or her best interests. The ability for an autonomous person to act in his or her self-interest, however, requires informed consent, that is, the patient should enter into the experiment voluntarily and with adequate information about the risks and benefits. In cases where the human subject lacks autonomy through age, mental condition, or any other reason, the researchers should take steps to protect the individual.

Herman Shaw, 94, a Tuskegee Syphilis Study victim, smiles after receiving an official national apology from President Clinton on May 16, 1997. Making amends for the shameful U.S. experiment, Clinton apologized to black men whose syphilis deliberately went untreated by government doctors. © *AP Images/Greg Gibson.*

Unlike other ethical codes related to human experimentation, such as the Nuremberg Code, however, the Belmont Report does not state that informed consent is absolutely necessary. The Belmont Report allows researchers to balance competing interests in cases where the researcher is uncertain if the human subject has given informed consent, such as when prisoners are used in study.

IN CONTEXT: CLINICAL TRIAL STANDARDS MORE LAX IN SOME COUNTRIES

Whereas ethical standards such as those outlined in the Belmont Report dictate research behaviors regarding human subjects, these policies apply only to those studies conducted in the United States. In recent years some multinational pharmaceutical and biotechnology firms have moved research and clinical trials to countries with less stringent standards.

Global pharmaceutical firm Astra Zeneca received news coverage in 2011 for its alleged involvement in unauthorized

clinical trials in Indian hospitals. An October 11, 2011, article by Sujoy Ghar for Inter Press Service notes that "India's powerful industry association Associated Chambers of Commerce and Industry (ASSOCHAM) says the country is set to grab clinical trials business valued at approximately 1 billion dollars by the end of 2010, up from 200 million dollars the previous year, making the subcontinent one of the world's preferred destinations for clinical trials."

The Belmont Report's beneficence requirement calls on researchers to respect the decisions of human subjects, to protect human subjects from harm, and to secure the wellbeing of subjects. In addition to preventing harm to subjects, researchers should seek to maximize the possible benefits to human subjects. This principle was in response to the fallout from the Tuskegee syphilis experiment, in which researchers deliberately denied human subjects a known and effective treatment.

The concept of justice in human experimentation relates to the distribution of benefits and burdens in an equitable and fair manner. Researchers should examine the manner and reasons for selecting certain classes of people to participate in human experimentation. Researchers should evaluate whether the poor, racial minorities, prisoners, or other classes of persons are selected to participate in a study to further the goals of that study or merely because of ease of availability or manipulability.

The Belmont Report, when adopted, only applied to studies conducted under the auspices or with support from HEW. In 1991, 14 additional U.S. government department and agencies adopted the Federal Policy for the Protection of Human Subjects, also known as the Common Rule, a set of rules and regulations for the protection of human research subjects. The Common Rule is identical to the principles of the Belmont Report codified in U.S. law in 1979 that was made applicable to the HEW. The U.S. government also established the Office for Human Research Protections (OHRP) under HHS to provide guidance and regulatory oversight on issues related to human research studies.

SEE ALSO *Bioethics; Government Regulations; Human Subject Protections; Human Subjects of Biotechnological Research; Individual Privacy Rights; Informed Consent; Nuremberg Code (1948)*

BIBLIOGRAPHY

Books

Childress, James F., Eric Mark Meslin, and Harold T. Shapiro. *Belmont Revisited: Ethical Principles for Research with Human Subjects.* Washington, DC: Georgetown University Press, 2005.

Green, Ronald Michael, Aine Donovan, and Steven A. Jauss. *Global Bioethics: Issues of Conscience for the Twenty-First Century.* New York: Clarendon Press, 2008.

Harris, Dean M. *Ethics in Health Services and Policy: A Global Approach.* San Francisco: Jossey-Bass, 2011.

Steinbock, Bonnie, John Arras, and Alex John London. *Ethical Issues in Modern Medicine: Contemporary Readings in Bioethics,* 7th ed. Boston: McGraw-Hill, 2009.

Veatch, Robert M., Amy Marie Haddad, and Dan C. English. *Case Studies in Biomedical Ethics: Decision-Making, Principles, and Cases.* New York: Oxford University Press, 2010.

Web Sites

"Regulations and Ethical Guidelines: The Belmont Report." *National Institutes of Health (NIH).* http://ohsr.od.nih.gov/guidelines/belmont.html (accessed September 25, 2011).

Joseph P. Hyder

Biochemical Engineering

■ Introduction

Biochemical engineering is concerned with the design, construction, and operation of processes that make biotechnology products, including vaccines, enzymes, and antibodies. It is an important component of food, detergent, beverage, pharmaceutical, and other biotechnology industries.

■ Historical Background and Scientific Foundations

The first known application of biochemical engineering was in the brewing of beer in Mesopotamia around 4,000 years ago. Written instructions for the process were discovered on a cuneiform tablet found in Northern Syria in 2001. Chemical engineering became a discipline in its own right in the early twentieth century. Around the same time, before petroleum became the main feedstock of the chemical industry, fermentation was used to produce not just ethanol but other important industrial chemicals such as acetone. Through the study of such processes, the principles of biochemical engineering were developed. In 1964 Shuichi Aiba (1923–) of Tokyo University and colleagues published the first major textbook on biochemical engineering. This multidisciplinary topic can now be studied at many leading universities.

Many of the principles of biochemical engineering are based on chemical engineering. A process is typically first developed on a laboratory scale. Conditions such as temperature, pressure, rate of stirring, and nature of solvent for making a product from raw materials are optimized at this stage. The process is then transferred to a larger scale reaction vessel so that commercial quantities of high purity can be manufactured. Biochemical engineering differs from chemical engineering as it involves growing cells in a reaction vessel called a bioreactor and harvesting the products they make. Often the cell will

have been genetically modified in order to ensure it makes the desired product.

In some cases, the cells themselves are the product; for example, they may be applied in medicine as tissue replacement therapies. Bacterial, fungal, insect, and mammalian cells can all be used as production units in bioreactors. Biochemical engineering involves more challenges than chemical engineering because living cells are more delicate and difficult to work with than chemicals. Issues such as cleanliness, sterility, and safety are therefore very important in designing and operating a biotechnology process.

Biochemical engineering can be divided into two main areas, upstream processing and downstream processing, each of which has a different disciplinary focus.

Upstream processing, which involves the investigation of potential host cells for production, is a microbiology related process. The cell may naturally manufacture the product of interest: For example, yeast turns sugar into ethanol. Increasingly, however, genes that code for a product are inserted into the host cell to create a genetically-modified bacterium, yeast, insect, or mammalian cell. Another important aspect of upstream processing is to find the best medium to grow the host cell so it manufactures the maximum amount of product. Upstream processing includes the production process itself, in which the optimized host cell strain is allowed to multiply in a bioreactor under the conditions most conducive to the manufacture of the product, such as temperature, rate of stirring, and other factors. This stage is known as fermentation or cell culture. The design of the bioreactor and monitoring of conditions during production have a strong engineering focus.

Downstream processing covers all those operations from recovering the product from the host cell to its final purification ready for use. How product is recovered from the cell depends on whether the product is retained inside the host cell or secreted into the growth medium. Following harvest, there are several purification stages in the downstream processing to make the product ready for formulation, packaging, and distribution. These steps usually involve filtration and chromatographic purification, important processes in analytical chemistry.

■ Impacts and Issues

Biochemical engineering plays a part in several industries. It has a profound impact on health care. Several of the top 100 best-selling pharmaceutical drugs are monoclonal antibodies, vaccines, or recombinant proteins. Examples include the antibody infliximab, used in the treatment of rheumatoid arthritis, and erythropoietin, for anemia. These products required many years in development because processes involving cells are particularly challenging to engineer successfully. For example, production must always be carried out under strictly sterile conditions. Failures, through leaks in the bioreactor or technicians not wearing correct protective clothing, usually lead to the introduction of unwanted microbes into the process. Making changes to the process can also have very serious consequences. For instance, in 1990 people being treated with tryptophan, a drug produced

Jonathan Wolfson, CEO and co-founder of Solazyme, speaks to a reporter while standing in front of a 159-gallon (600-liter) fermenter in the Solazyme pilot plant. Many of the U.S. Navy's ships and jets will soon be running on prairie grass, sugar beet pulp, and sawdust that has been eaten by microscopic algae and turned into fuel. In September 2010, the Navy ordered more than 150,000 gallons (567,812 liters) of algae fuel from Solazyme. *© AP Images/Jeff Chiu.*

by microbial fermentation, began to fall ill. They had disabling fatigue, muscle weakness, and inflammation of major organs. Some victims died. The problem was traced back to a biotechnology firm in Japan that had switched production strains to one that, unknown to them, was producing a contaminant that was responsible for the adverse symptoms.

For these reasons, the regulations for production of a biotechnology drug are very strict. Biochemical engineers must be prepared to justify any changes they introduce into a process. This is also why biopharmaceutical drugs tend to be very expensive. However, many of the earlier biopharmaceutical drugs, such as insulin and erythropoietin, have reached the end of their patent life. As patents expire, opportunities open for companies to make cheaper biosimilar versions of these drugs. The debate at present centers around how far a company wanting to make a biosimilar version can benefit from the biochemical engineering knowledge of those scientists who first developed the process.

There have been some technical developments in biochemical engineering that may make biotechnology products more readily available. First, miniaturization technology means that upstream processing is faster and easier. For example, microtiter plates, which contain many tiny wells, each of which resembles a miniature test tube, enable many different experiments to be carried out in parallel over a short time frame. This approach allows rapid identification of the best conditions for growing the cells. As production time decreases, patients may ultimately benefit from the less expensive drugs. In addition, single-use or disposable technology is becoming more popular. This means that instead of using an expensive stainless steel bioreactor, the biochemical engineer can use a giant plastic bag to grow the cells, harvest the product, and then throw it away. There are also disposable filters, tubes, and other components. Some companies are packing single-use components into mobile production units, which are cheaper and more flexible than the traditional production plant. This opens up the possibility of less developed countries being able to locally produce biopharmaceuticals, including vaccines, thus reducing dependency on upon relatively more expensive imports.

SEE ALSO *Bioreactor; Biotechnology Products; Drug Price Competition and Patent Term Restoration Act of 1984; Industrial Genetics; Microarrays; Microinjection*

BIBLIOGRAPHY

Books

Annadurai, B. *A Handbook of Biochemical Engineering and Fermentation*. Cambridge: Woodhead Pub Ltd., 2011.

Beard, Daniel A., and Hong Qian. *Chemical Biophysics: Quantitative Analysis of Cellular Systems*. Cambridge: Cambridge University Press, 2008.

Haracoglou, Irina. *Competition Law and Patents: A Follow-on Innovation Perspective in the Biopharmaceutical Industry*. Cheltenham, UK: Edward Elgar, 2008.

Mills, Oliver. *Biotechnological Inventions: Moral Restraints and Patent Law*. Rev. ed. Burlington, VT: Ashgate Publishing, 2010.

Susan Aldridge

Biochip

Introduction

The term biochip refers to the use of a support (typically composed of glass or nylon) that houses an arrangement of different sequences of DNA; different proteins (including antibodies) or portions of proteins (peptides); or even compounds that can be chemically altered when a suitable enzyme is present. The arrangement is known as a microarray. Microarrays are used as sensors to determine the genetic activity, protein production, or chemical activity of sample cells at any one time or over time.

In a DNA biochip, hundreds to thousands of single strands of DNA are attached. The huge number of DNA sequences present enables the screening of a sample for numerous differences in DNA expression, which can be compared to help clarify the effects of a treatment, environmental condition, or other factor on the activity of the cells from which the DNA was obtained.

More specifically, DNA biochips can be used to detect gene alterations (mutations), as the altered gene will present a different target to which the test probe does not bind. Altered genes in bacteria that have become antibiotic resistant can be detected in this way, which can be a vital step in designing compounds that will kill the resistant bacteria. Other uses for biochips include detection of noxious compounds.

Historical Background and Scientific Foundations

A biochip can be wafer-like in appearance or can appear similar to a glass slide used in light microscopy. Appearances are deceiving, however, because the biochip is much more complex than it appears to the naked eye. A biochip contains row upon row of an orderly and densely packed array of DNA fragments (generated by breaking

DNA using specialized enzymes called restriction endonucleases), proteins or peptide fragments, or chemicals. Whereas the arrays were originally made by spotting the test molecules on the surface by hand, automated techniques have made it possible to build microarrays that contain thousands of different molecules. The pattern of the molecules is known, which allows the results to be meaningfully analyzed.

Each spot in a DNA array is a single-stranded piece of DNA. The sequence of the attached DNA will only allow the binding of another strand of DNA whose sequence is complementary. The term complementary refers to the fact that a particular nucleotide base on one strand of DNA will be able to establish a link with only one of the other three nucleotides. For example, if a bound fragment of DNA has the sequence AATTCCGG (A denotes adenine, C denotes cytosine, G denotes guanine, and T denotes thymine), then only a fragment with the sequence TTAAGGCC will be able to bind.

Binding of sample molecules to molecules making up the microarray on a biochip is usually apparent as the appearance of a fluorescent spot, because the applied sample also includes a probe that will fluoresce under a certain wavelength of light. The pattern of fluorescence can be correlated with the known arrangement of the microarray molecules to determine, for example, which proteins are being manufactured at any one time.

A microarray can also be used to determine the level of expression of a gene. For example, an array can be constructed such that the messenger RNA of a particular gene will bind to the target. Thus, the bound RNAs represent genes that were being actively transcribed, or at least had been transcribed shortly before the sample was collected (messenger RNA is degraded quickly, relative to other RNA species). By monitoring genetic expression, the response of microorganisms to a treatment or condition can be examined. As an example, DNA from a bacterial species growing in suspension can be compared with the same species growing as surface-adherent biofilm in order to probe the genetic nature of the alterations that occur in the bacteria upon association with a surface. Since the method detects DNA, the survey can be all-encompassing, assaying for genetic changes to proteins, carbohydrates, lipids, and other constituents in the same experiment.

Vast amounts of information are obtained from a single biochip experiment. Hundreds of thousands of genes can be probed on a single chip. The analysis of this information has spawned a new science called bioinformatics, that uses both biology and computing to clarify cell behavior at the genetic level.

Biochips are having a profound impact on research. Pharmaceutical companies are able to screen for gene-based drugs much faster than before. In the future, DNA chip technology will extend to the office of the family physician. For example, a patient with a sore throat could be tested with a single-use, disposable, inexpensive chip in order to identify the source of the infection and its antibiotic susceptibility profile. Therapy could commence sooner and would be precisely targeted to the infectious agent.

SEE ALSO *Bioinformatics; Genetic Technology; Microarrays*

BIBLIOGRAPHY

Books

Korenberg, Michael J. *Microarray Data Analysis: Methods and Applications.* Totowa, NJ: Humana Press, 2007.

Rampal, Jang B. *Microarrays Volume 2: Applications and Data Analysis.* Totowa, NJ: Humana Press, 2007.

Voda, Alina. *Micro, Nanosystems, and Systems on Chips: Modeling, Control, and Estimation.* Hoboken, NJ: Wiley, 2010.

Xing, Wan-Li, and Cheng, Jing. *Frontiers in Biochip Technology.* New York: Springer Science & Business Media, 2010.

Web Sites

Sagoff, Jared. "Biochips Can Detect Cancers Before Symptoms Develop." *Argonne National Library,* May 9, 2008. http://www.anl.gov/Media_Center/News/2008/ES080509.html (accessed September 25, 2011).

Brian Douglas Hoyle

Professor Jonathan Dordick holds a biochip at Rensselaer Polytechnic Institute in Troy, New York. The chip looks like a standard microscope slide, but it holds hundreds of tiny white dots loaded with human cell cultures and enzymes. It is designed to mimic human reactions to potentially toxic chemical compounds, meaning animals like rats and mice may no longer need to be on the front line of tests for new blockbuster drugs or wrinkle creams. *© AP Images/Mike Groll.*

Biocomputers

■ Introduction

Biocomputers are devices using biological components that perform traditional computing functions, including calculations, data storage, and data retrieval. Biological materials currently used range from deoxyribonucleic acid (DNA) and proteins to portions of organisms (an example is a computer based on neurons obtained from leeches) to entire organisms (genetically-engineered yeast that can communicate with each other similar to the operation of an electronic circuit). The moniker biocomputer is also synonymous with biological computer, DNA computer, and peptide computer.

First envisioned in the late 1950s, as of 2011 biocomputers remained in the research stage, although several rudimentary versions have been successfully created. However, fully functional biocomputers that can accurately and rapidly execute pre-written programs are still devices of the future.

■ Historical Background and Scientific Foundations

The concept of biocomputers dates back to 1958, with the establishment of the Biological Computer Laboratory, a research facility at the University of Illinois at Urbana-Champaign. Whereas the original intent of the laboratory was the use of conventional computers to study biological reactions and systems, the research evolved to include the use of biological systems and components as the computers themselves. At the time, the intent was to design a biologically-based system that would be capable of producing two responses, corresponding to the numbers 0 and 1. The sequence of these numbers would convey different information. At the time, the idea was not widely embraced. Indeed, the laboratory shut down in 1974 due to lack of funding.

However, the idea persisted, and interest was rekindled two decades later. By 1999 a research team from the United States had created a functioning system that was composed of leech neurons equipped with microelectrodes that were linked together. In the system, each neuron had its own electrical activity and responded independently to electrical stimulus. Each neuron also represented a particular number, so the sequence in which the neurons responded corresponded to an overall number. Different patterns of response generated different numbers. This design mimicked the operation of the human brain, although to a much more rudimentary extent.

By 2003 a computer composed of enzymes and deoxyribonucleic acid (DNA) had been developed and was capable of 330 trillion (330,000,000,000,000) calculations per second. Put into perspective, this was more than 100,000 times faster than the fastest conventional personal computer at that time. Then as now, the latter computers operate by the flow of electrons as the basis of the calculations that are programmed in the software. In a DNA computer, the information contained in the genetic material is the software, and the enzymes that cause reactions to occur are the hardware. The reaction products are the execution of the programs. By controlling the DNA sequences that are present or that are active, the overall operation of the DNA-based biological computer is controlled. For all its lightning fast speed, the DNA biocomputer is still in its infancy. Its use to run sophisticated programs is still a future goal.

Ribonucleic acid (RNA) offers an alternative base for biocomputers. In 2010 scientists reported success with the manufacture of a rudimentary biocomputer that used yeast cells. By harnessing the ability of yeast cells to release and bind specific proteins that act as communication molecules in the real-world environment of the microbes, the yeast cells were turned into living circuits. Genetically-altered yeast cells were capable of sensing their surroundings and excreting

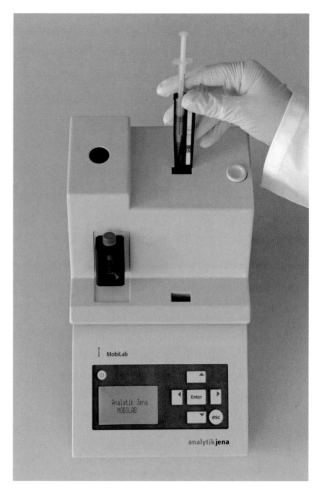

A lab assistant works at a new bioanalytical rapid test system for detecting the influenza virus H1N1. Based on the speed cycler technology, the procedure makes it possible to detect an H1N1 infection in less than two hours. *© AP Images/Jens Meyer.*

compounds. Other yeast cells in the vicinity could detect the released compound and respond by secreting an appropriate compound. By altering the environment, different response patterns could be generated, creating a flow of information along the circuit formed by the linked cells.

The different designs of biocomputers reflect a common philosophy—the use of biological materials to convey information. Instead of the flow of electrons through circuits, the information is passed on by a series of biological reactions (i.e., substrates are used to generate products, which in turn become substrates of other reactions).

■ Impacts and Issues

Although still in their infancy, DNA-based computers have great potential, given their speed, capacity of information (e.g., more than a trillion music CDs) and exceedingly small size (they can fit inside the nucleus of an eukaryotic cell). In a realistic scenario that is being researched as of 2011, DNA biocomputers injected into the body could act as sentinels of cell, tissue, or organ damage and guide the manufacture and release of therapeutic drugs. It is conceivable that the raw materials for the operation of the internalized biocomputer—nucleic acid and protein—could be derived from the person in whom the biological computer has been introduced.

As well, DNA computers are capable of performing a myriad of calculations simultaneously, in contrast to conventional personal computers that are designed for speed, but only for one or a few tasks at a time. Another difference between conventional and biocomputers is in the basis of the calculations. Whereas conventional computers require a complete set of instructions (the program code) to complete a task, biocomputers, at least in theory, can complete a task with only partial information, because, similar to the human intellect, they may be capable of "filling in the blanks." In a real sense, biocomputers may be able to think.

This scenario of a thinking computer is not fantasy: It is real and has occurred. In 1997 a computer built by IBM and dubbed Deep Blue defeated world chess champion Garry Kasparov (1963–) in a six-game match. Fourteen years later, in 2011, another and faster version of the computer, Watson, bested several past champions of the popular television show Jeopardy. The victory by Watson required the computer to perceive, process, and respond quickly to speech, before the humans competitors could. This human-like capability was possible because of the computer's information processing power—80 trillion operations (80 teraflops) per second. The precedent exists for similar accomplishments by a biocomputer.

The promise of biocomputers is not unchallenged. Given their biological basis, some have cautioned that evolution of the machines could occur, perhaps to the extent that the machines become fully autonomous. This concern is not new. In a 1950 publication entitled

"Computing Machinery and Intelligence," British mathematician and computer scientist Alan Turing (1912–1954) proposed a test of a computer's capacity for intelligent behavior. According to the Turing test, if a human carries on a conversation with what she/he perceives to be another human, but which is actually a machine, then the machine has passed the test and can be considered to have intellect, including the ability to think.

SEE ALSO *Biochip; Biodetectors; Biorobotics*

BIBLIOGRAPHY
Books

Altman, Russ B., A. Keith Dunker, and Lawrence Hunter. *Biocomputing 2011 Proceedings of the Pacific Symposium.* Hackensack, NJ: World Scientific Publishing Co. Inc., 2010.

Krishnakumar, M. S. *Bioinformatics: Biocomputing and the Internet.* New Delhi: Campus Books, 2007.

Brian Douglas Hoyle

Biodefense and Pandemic Vaccine and Drug Development Act of 2005

■ Introduction

The Biodefense and Pandemic Vaccine and Drug Development Act of 2005 was a bill designed to advance the research and development of vaccines and drugs needed to respond to a bioterrorism attack or a natural disease outbreak. Although the bill was never enacted into law, one of the bill's key provisions—the establishment of the Biomedical Advanced Research and Development Agency (BARDA)—was enacted into law under another bill. BARDA coordinates the development of medical countermeasures to deal with a bioterrorism attack or outbreak of disease, including the procurement of vaccines and other drugs. The Biodefense and Pandemic Vaccine and Drug Development Act also included provisions that were not enacted into law, including financial incentives for pharmaceutical companies and the extension of some drug patents.

■ Historical Background and Scientific Foundations

Following the terrorist attacks of September 11, 2001, and the subsequent anthrax attacks, the United States government sought to strengthen U.S. defenses against potential bioterrorism actions. During the following years the U.S. Congress passed several bills that reorganized and granted new powers to government agencies responsible for terrorist and public health threats.

In 2002 President George W. Bush (1946–) signed the Public Health Security and Bioterrorism Preparedness and Response Act of 2002, also known as the Bioterrorism Act of 2002, into law. The Bioterrorism Act of 2002 authorizes the U.S. Food and Drug Administration (FDA) to issue regulations to protect U.S. food and drug supplies from a bioterrorism attack or foodborne illnesses. New FDA regulations enacted under the act enable the agency to track the source of a deliberate or unintentional food contamination more easily through information provided by registered food facilities.

The Project BioShield Act of 2004 authorizes the U.S. Department of Health and Human Services (HHS) to expedite procedures for bioterrorism-related spending. The Project BioShield Act allows the U.S. Secretary of HHS to expedite the peer-review procedures for identifying, assessing, and responding to a bioterrorism action. The law also authorized HHS to purchase vaccines to stockpile in the event of a bioterrorist attack.

■ Impacts and Issues

U.S. Senator Richard Burr (1955–) introduced the Biodefense and Pandemic Vaccine and Drug Development Act on October 17, 2005. The Biodefense and Pandemic Vaccine and Drug Development Act was intended to strengthen the biodefenses of the United States in the event of a bioterrorism attack or the natural outbreak of a pandemic illness. The bill called for the establishment of the Biomedical Advanced Research and Development Authority (BARDA), an office within HHS, to promote the research and development

---

WORDS TO KNOW

BIOTERRORISM: A form of terrorism that uses bacteria, viruses, or other germs to attack people or disrupt societies.

VACCINE: A product that produces an immune reaction by inducing the body to form antibodies against a particular agent. Usually made from dead or weakened bacteria or viruses, vaccines cause an immune system response that makes the person immune to (safe from) a certain disease.

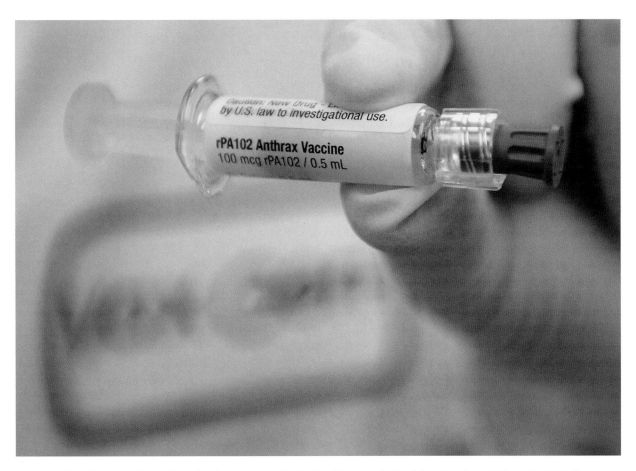

A VaxGen lab technician holds a syringe of anthrax vaccine at the VaxGen laboratory in South San Francisco, California, in 2006. One year after the Biodefense and Pandemic Vaccine and Drug Development Act of 2005 was passed by the House of Representatives but failed in the Senate, and five years after anthrax attacks left five dead, sickened 17, and terrified America, millions of vaccine shots developed through cutting-edge genetic engineering were supposed to be filling a new national stockpile of biodefense drugs. © AP Images/Benjamin Sklar.

of vaccines and other drugs needed to respond to a bioterrorism attack or natural outbreak of illness.

The Biodefense and Pandemic Vaccine and Drug Development Act of 2005 contained a provision that would have exempted BARDA from requests under the Freedom of Information Act (FOIA), a federal law that allows the public or media to obtain information held by government agencies. The Biodefense and Pandemic Vaccine and Drug Development Act also called for financial incentives for vaccine manufacturers in the event of a bioterrorism attack or natural disease outbreak. In addition, the bill specified liability protections for manufacturers, distributors, health care workers, and government and health care administrators for injuries caused by vaccines or other drugs used during a bioterrorism attack or natural disease outbreak.

Although the U.S. House of Representatives passed the Biodefense and Pandemic Vaccine and Drug Development Act in September 2006, the U.S. Senate did not pass the bill, thereby preventing it from becoming law. Some provisions of the bill were proposed for other bills during the 109th Congress. The Senate version of the Pandemic and All-Hazards Preparedness Act contained a provision for the creation of BARDA under the HHS Office of the Assistant Secretary for Preparedness and Response. In the final hours of the 109th Congress, the House and Senate reconciled the bills to establish BARDA.

The Biodefense and Pandemic Vaccine and Drug Development Act's provision to shield pharmaceutical companies from liability for injuries caused by drugs or vaccines used during a public health emergency was also included in the Public Readiness and Emergency Preparedness Act (PREPA), which was enacted into law in December 2005. The liability provision of PREPA allows the Secretary of HHS to declare a public health emergency and trigger PREPA's liability immunity.

SEE ALSO *Bioethics; Biopharmaceuticals; Bioterrorism; Government Regulations; Human Subjects of Biotechnological Research; Individual Privacy Rights; Patents and Other Intellectual Property Rights; Vaccination; Vaccines*

BIBLIOGRAPHY

Books

Bucknell, Duncan Geoffrey, ed. *Pharmaceutical, Biotechnology, and Chemical Inventions: World Protection and Exploitation.* New York: Oxford University Press, 2011.

Centers for Disease Control and Prevention (CDC). *Interim Pre-pandemic Planning Guidance: Community Strategy for Pandemic Influenza Mitigation in the United States.* Washington, DC: U.S. Department of Health and Human Services (HHS), 2007.

Fong, I. W., and Kenneth Alibek, eds. *Bioterrorism and Infectious Agents: A New Dilemma for the 21st Century.* New York: Springer, 2009.

Gross, Marc S., S. Peter Ludwig, and Robert C. Sullivan. *Biotechnology and Pharmaceutical Patents: Law and Practice.* New York: Aspen Publishers, 2008.

Homeland Security Council (U.S.). *National Strategy for Pandemic Influenza: Implementation Plan.* Washington, DC: Homeland Security Council, 2006.

Mullner, Ross M. *Pharmaceutical Marketing.* Bradford, UK: Emerald Group Publishing, 2005.

Torrence, Paul F., ed. *Antiviral Drug Discovery for Emerging Diseases and Bioterrorism Threats.* Hoboken, NJ: Wiley-Interscience, 2005.

Periodicals

Cohen, Hillel W., Robert M. Gould, and Victor W. Sidel. "The Pitfalls of Bioterrorism Preparedness: the Anthrax and Smallpox Experiences." *American Journal of Public Health* 94, no. 10 (October 2004): 1667–1671.

Monto, Arnold S. "Vaccines and Antiviral Drugs in Pandemic Preparedness." *Emerging Infectious Diseases* 12, no. 1 (January 2006): 55–60.

Web Sites

"Bioterrorism Agents/Diseases." *Centers for Disease Control and Prevention (CDC).* http://emergency.cdc.gov/agent/agentlist.asp (accessed October 21, 2011).

"Flu Pandemics." *Centers for Disease Control and Prevention (CDC).* http://www.flu.gov/individualfamily/about/pandemic/ (accessed October 21, 2011).

"WHO Programs and Projects: Global Alert and Response (GAR)." *World Health Organization (WHO).* http://www.who.int/entity/csr/en (accessed October 21, 2011).

Joseph P. Hyder

Biodegradable Products and Packaging

■ Introduction

Biodegradable products and packaging (biodegradables) are items engineered to degrade with minimal environmental impact using only biological processes or natural weathering. Biodegradable products may be made from plant or animal materials or from conventional industrial materials such as petroleum to which degrading agents are added.

Biodegradables aim to reduce solid municipal waste streams. Greater waste streams require more space for landfills and produce more greenhouse gas emissions. Common non-engineered biodegradables include food and paper wastes. Engineered biodegradables include biodegradable plastics, fabrics, detergents, and inks.

■ Historical Background and Scientific Foundations

Consumer desire for convenience products led to the development of disposable containers and packaging. Consumer consumption of disposable products grew rapidly in the second half of the twentieth century. Processed convenience foods such as frozen dinners gained popularity in the 1950s. They were attractively packaged for simple cooking and easy cleanup, often in tinfoil containers inside of paper or cardboard outer packages. In the following decades, household cleaning products packaged in plastic bottles became popular. Consumers flocked to disposable versions of everyday items, such as pens and diapers. The advent of fast food and carryout restaurants further increased the use of disposable materials. All of these contributed to an increase in municipal waste streams.

In 2010 plastics took up an estimated 25 percent of landfill volume in the United States. Bioplastics accounted for as much as 10 percent of the total plastic used by consumers. Whereas paper and cardboard, being wood products, have the ability to degrade fairly rapidly under the right conditions, traditional petroleum-based plastics degrade slowly over thousands of years. Biodegradable plastics were invented to reduce the environmental impacts of plastics and reduce landfill burdens.

Biodegradable plastics are engineered to break down into base materials and chemicals with no or minimal harm to the environment. They rely on different natural processes, from sunlight to bacteria, to fully degrade. Oxo-biodegradable (OBD) plastics degrade when exposed to sunlight and oxygen. Many shopping bags are made with OBD. Bioplastics—plastics made from corn, potatoes, or tapioca food starches—use heat and moisture to degrade. Bio-active bioplastics combine bioplastic technology with the addition of swelling or digesting agents to speed degradation. Even petroleum-based plastics can be engineered to degrade faster and better with the addition of bio-active compounds or digesting bacteria that activate to help breakdown the plastics into tiny pieces.

Corn-based bioplastic (PLA) packaging is the most common bioplastic currently used by consumers. The plastic was designed to mimic traditional petroleum-based PET (polyethylene terephthalate) plastics. PLA has several advantages over traditional plastics. According to bioplastic producer NatureWorks (owned by agribusiness giant Cargill), depending on manufacturing processes, PLA production may use as much as 70 percent less energy and generate 68 percent fewer greenhouse gases than traditional plastic production. Neither the manufacture nor the decomposition of PLA produces toxins. PLA plastic containers do not leach potentially harmful chemicals (such as vinyl chloride) into stored foods. The first PLA products were only for use with cold foods and beverages and could not be exposed to high temperatures, but more recent research has produced PLA that is heat resistant up to about 230°F (110°C). PLA is made from renewable plant resources, helping to reduce the more than 200,000 barrels of oil used daily in plastic production in the United States alone.

WORDS TO KNOW

AEROBIC: Any process or organism that requires the presence of oxygen.

ANEROBIC: A process or organism not dependent on the presence of oxygen.

BIODEGRADABLE: A material that uses biological processes or natural weathering to degrade rapidly with minimal environmental impact.

COMPOSTABLE: Organic and engineered wastes able to be composted; that is, able to be broken down by various aerobic microorganisms.

MUNICIPAL SOLID WASTES: Solid refuse, including household and business trash, that is disposed of by local sanitation agencies.

A man holds a tomato plant in a biodegradable peat pot. The peat pot can be planted in the garden along with the plant rather than using a petroleum-based plastic pot that will be thrown out.
© *forestpath/Shutterstock.com.*

In 2005 Wal-Mart, one of the largest North American retailers, introduced PLA containers into its stores. The company used 114 million PLA containers in the first year, saving an estimated 800,000 barrels of oil.

■ Impacts and Issues

In 2010 two-thirds of the 250 million tons of trash thrown out in United States ended up in landfills. Biodegradable products and packaging rely on certain environmental conditions to degrade properly and rapidly as designed. Photodegradable items need adequate light to degrade; oxy-degradable plastics require exposure to humid air. Common municipal solid waste disposal practices undermine or even completely negate the degradability of some products.

As many as 1 trillion plastic shopping bags are used worldwide each year. Many of these bags are manufactured to be compostable, biodegradable, or photodegradable. However, anaerobic landfills do not offer the light, air, and moisture necessary for effective breakdown of most bioplastic bags. PLA and PET containers buried in the same level of a landfill may take equal time to degrade. Similarly, many recycling centers that accept petroleum-based plastics do not have the facilities to process plant-based bioplastics, meaning more bioplastics end up in landfills.

Current waste disposal technologies in most areas are insufficient to promote full degradation of most biodegradable materials. Composting facilities and large-scale anaerobic digesters capable of handling biodegradable wastes could help communities move from capturing as much as 30 percent of their solid municipal wastes through recycling to capturing as much as 90 percent of their waste before it reaches a landfill.

Products labeled as biodegradable appeal to environmentally-conscious consumers. Without commonly accepted definitions for the term biodegradable and regulations on use of the word in product labeling, however, the term is susceptible to "greenwashing," the specious promotion of a product as environmentally sound without sufficient scientific basis.

In the United States, use of the terms degradable or biodegradable in product marketing is governed by the Federal Trade Commission's (FTC) Green Guide (Guide for the Use of Environmental Marketing Claims). For example, plastic products advertised as biodegradable must disclose the environmental requirements and time-frame necessary for the claimed biodegradation to take place. Manufacturers must also note the circumstances under which a product does not degrade completely or degrades into smaller components. In 2007 California passed more stringent criteria for biodegradable labeling of products sold in that state. The first product lawsuits under the act targeting two bottled water sellers that claimed to have biodegradable bottles were

A hand holds a pair of bottle caps made from Mirel, a microbe-produced, biodegradable plastic, one new, and one broken-down by 60 days in seawater. © *AP Images/Josh Reynolds.*

Although biodegradable products may offer environmental benefits over non-biodegradable products, some environmental advocates worry that biodegradable containers impart to consumers a false sense of diminished environmental impact. Biodegradable plastics in particular encourage consumer "take-out" behavior that promotes over-packaged, single-serving, disposable products. Instead of focusing on proliferating biodegradable products, advocates argue that emphasis should be shifted to using fewer containers and disposable products overall.

SEE ALSO *Bioplastics (Bacterial Polymers); Biotechnology Products; Cornstarch Packing Materials; Laundry Detergents; Solid Waste Treatment; Waste Gas Treatment*

BIBLIOGRAPHY

Books

1001 Easy Ways for Earth-Wise Living: Natural and Eco-Friendly Ideas That Can Make a Real Difference to Your Life. Ultimo, New South Wales, Australia: Reader's Digest, 2006.

Web Sites

"Pollution Prevention: Recycling." *United States Environmental Protection Agency (EPA).* http://www.epa.gov/ebtpages/pollrecycling.html (accessed November 3, 2011).

"Pollution Prevention: Recycling: Plastics." *United States Environmental Protection Agency (EPA).* http://www.epa.gov/ebtpages/pollrecyclingplastics.html (accessed November 3, 2011).

Royte, Elizabeth. "Corn Plastics to the Rescue." *Smithsonian.com,* August 2006. http://www.smithsonianmag.com/science-nature/plastic.html (accessed November 3, 2011).

"Wastes: Solid Waste—Nonhazardous: Municipal Solid Waste." *United States Environmental Protection Agency (EPA).* http://www.epa.gov/ebtpages/wastsolidwastemunicipalsolidwaste.html (accessed November 3, 2011).

"Wastes: Waste Disposal: Landfills." *United States Environmental Protection Agency (EPA).* http://www.epa.gov/ebtpages/wastwastedisposallandfills.html (accessed November 3, 2011).

Adrienne Wilmoth Lerner

brought in 2009. The United Kingdom, Australia, and the European Union also have labeling laws that govern use of the term biodegradable.

Biodegradable products manufactured from food plants raise socioeconomic ethical concerns. Like biofuels such as corn ethanol, bioplastics made from corn products divert possible food stores to industrial production. High global petroleum prices make corn-based materials more economically attractive—encouraging the diversion of more corn to increased packaging production—at the same time that scarcer food and rising fuel costs raise the market price of food corn. Higher food prices disproportionately affect the world's poorer citizens, but corn plastic biodegradable products are disproportionately consumed in wealthy industrialized countries.

A desire for biodegradable products also led to revival of older technologies. Use of petroleum-based plastics increased as they became cheaper to produce. A campaign begun by U.S. school children in the 1980s prompted international fast-food chain McDonalds to change from Styrofoam containers to paper wrappers and containers. Seeking a more biodegradable alternative to plastic drinking straws and cups, some consumers chose to return to older style waxed paper products.

Biodetectors

Introduction

Biodetectors, or biological recognition systems, are analytical devices with microelectronic processing capabilities. Bodiotectors are used in forensics, medicine, food analysis, quality control in laboratory and food or beverage production settings, agriculture, in-home medical monitoring and diagnostics, pharmaceutical development, environmental diagnostics, military operations, biological warfare, and anti-terrorism efforts.

Typically, biodetectors consist of a biological or biotechnologically mediated detector used to recognize discrete, specific analytes and a transducer that is either closely connected to or part of the detector element. It receives the signal recognized by the detector component and transforms it into an identifiable, analyzable, and measurable output. The signal component receives the information from the transducer and produces an output signal based on the significance of the element being measured. For a blood glucose detector, the signal is generally digital; for a home pregnancy test, the signal is often a simple color change. In the case of detectors designed to sense pathogens in food, water, or the air, the signal is usually a loud tone. If detecting an act of bioterrorism or impending natural disaster such as an earthquake or tsunami, the signal would be a loud and persistent siren.

Scientists and engineers currently are developing and implementing extremely sensitive, accurate, and portable biodetectors. Input signal identification and analysis times have been significantly decreased, and rapid or instantaneous readouts are becoming progressively more common. Detectors capable of recognizing multiple biological inputs reliably are being developed and tested.

DNA-based detectors capable of breaking open cells to extract DNA for amplification through the use of PCR technology and then performing rapid, accurate, and reliable DNA sequencing for organism identification are particularly useful for discernment and identification of pathogens and toxins. Research and development are ongoing for deployment of portable, rapid, and accurate response DNA-based biodetection devices.

Historical Background and Scientific Foundations

The American biochemist Leland C. Clark (1918–2005) developed enzyme electrodes for the detection of glucose in 1962, setting the stage for the development of biodetector technology. He invented the oxygen electrode in 1956.

There are many types of biodetectors currently available for qualitative and quantitative analysis, the most common being thermal, mass, optical, and electrochemical. Thermal biodetector signal output is measured in absorption or production of heat, resulting in systemic temperature changes. In a mass-change sensitive biodetector, when the molecule to be analyzed attaches to the biosensing element, the atomic weight of the substrate changes. The transducers of optical biodetectors create an output signal measured by changes in luminescence or illumination. Electrochemical biodetectors are used most frequently to measure glucose concentration in the blood and to detect the presence of DNA and DNA-binding drugs, causing a measurable change in the electrical properties or overall electrical charge of the biorecognition system.

The sensor mechanism consists of a highly sensitive and specific biosensor layer that interacts with the molecule to be analyzed, causing a reaction that is recognized by a transducer. The recognition layer typically contains enzymes, proteins, antibodies, nucleic acids, bacteria, single-celled organisms, or specific types of tissue reactive to the target molecule. The reaction between the molecule to be analyzed and the detector produces physical or chemical changes in the biorecognition layer. These are detectable by the transducer, which then converts the output to an electrical signal. The signal is amplified, analyzed, and displayed numerically by a visual display or by means of an audible signal commensurate with the significance of the reaction.

A biodetector measuring blood glucose levels for a person with diabetes usually has a numerical output if the result is within the normal range; it might add an audible signal for a result outside of normal. A detector used with a home pregnancy test generally has a visible display change to signify a positive or negative result. However, a biodetector designed to indicate the presence of an environmental disaster such as an act of bioterrorism in the form of an aerosolized biochemical weapon would be likely to have a loud, persistent alarm.

■ Impacts and Issues

Two of the most common uses for biodetectors are home pregnancy tests and blood glucose monitoring systems. Research and development of smaller, lighter, and progressively more efficient and effective biodetectors is ongoing. Instantaneous and very rapid, accurate detection systems currently are being utilized, as are hand-held, ultra-sensitive devices. Historically, most commercial (rather than individual medical or diagnostic) biodetectors have been quite specific, targeted to one type of analyte. Research, particularly that designed for the food, military, and antiterrorism sectors, is underway to design multi-target biodetectors able to recognize and identify multiple analytes very precisely.

In the United Kingdom, biodetector systems also are being developed for widespread use in medical diagnostics. These biodetectors use antibodies capable of detecting, rapidly and accurately, a variety of biomarkers indicative of different serious disease processes. Currently they are able to diagnose several types of cancer, multiple sclerosis, cerebrovascular events, fungal infections, and cardiac disease. Coordinators of this biosensor project (called ELISHA) anticipate that the biodetection-developing technology will become the next generation in diagnostic testing, have increasing versatility, and be able to diagnose a widening disease set accurately, including the HIV spectrum and tuberculosis.

Biodetectors used for antiterrorism are generally one of three types: tissue-based systems using living cells from mammals for recognition of toxic chemicals or bioagents; chemical mass spectrometry systems that fragment analytes into their elemental components and then compare their weights against known toxins or bioagents; and biochemical systems capable of recognizing and identifying proteins or DNA sequences of identified toxins. Because bioagents may be aerosolized or airborne, it is essential that detection be extremely rapid, occur in real time, be highly accurate, and be able to analyze minute or trace amounts of biotoxic substances.

Much work currently is focused on the application of DNA-based, rapid response, reliable biodetectors. They utilize cell disruption technology in which the analyte organism's DNA is extracted and rapidly sequenced after PCR amplification. This is especially important for the swift identification of biological weapons in suspected acts of bioterrorism.

SEE ALSO *Agricultural Biotechnology; Biochip; Biocomputers; Biorobotics; Bioterrorism; Digital-Biological Computer; Environmental Biotechnology; Food Preservation; Forensic Biotechnology; Microarrays; Nanotechnology; Molecular*

BIBLIOGRAPHY

Books

Gorton, L. *Biosensors and Modern Biospecific Analytical Techniques.* Boston: Elsevier, 2005.

Morrison, Dennis R., ed. *Defense against Bioterror Detection Technologies, Implementation Strategies and Commercial Opportunities.* Dordrecht, The Netherlands: Springer, 2005.

Rasooly, Avraham, and Keith E. Herold, eds. *Biosensors and Biodetection: Methods and Protocols.* New York: Humana Press, 2009.

WORDS TO KNOW

ANALYTE: A substance to be measured or analyzed.

BIOLOGICAL RECOGNITION SYSTEM: Also called a biodetector or biosensor, this is an instrument that recognizes, records, analyzes, and transmits data in response to specific biochemical, chemical, or physical changes recorded by the system.

BIOTERRORISM: The use of microorganisms, toxic agents, or biological diseases as an act of terrorism. Bioterrorism acts fall into three categories: Category A acts pose a large threat to national security due to rapid transmission or dissemination, severity of human/animal symptoms and high mortality rate; Category B acts pose a moderate security risk due to relative ease of dissemination and transmission, leading to moderate symptoms and low mortality rates; and Category C acts involve developing or emerging pathogens capable of causing extremely severe mass effects due to their ease of dissemination and gravity of effects.

MICROELECTRONICS: A subset of electronic technology specializing in the creation, design, development, and utilization of compact, very small, and energy efficient components.

POLYMERASE CHAIN REACTION (PCR): A molecular genetics technique facilitating the analysis of DNA or RNA sequences by amplifying small sample strands to create sufficient quantities for accurate analysis.

The cartridge for NRL's (Naval Research Laboratory) Bead ARray Counter (BARC) uses technologies such as microfluidics and giant magnetoresistance (GMR) sensors to detect antibodies and DNA in a biological sample. A biosensor is a type of biodetector. *© Getty Images.*

Zhang, Xueji, Huangxian Ju, and Joseph Wang, eds. *Electrochemical Sensors, Biosensors, and Their Biomedical Applications.* Boston: Academic Press, 2008.

Zourob, Mohammed, Souna Elwary, and Anthony Turner, eds. *Principles of Bacterial Detection Biosensors, Recognition Receptors, and Microsystems.* New York: Springer, 2008.

Web Sites

"Biodetection." *Los Alamos National Laboratory.* http://www.lanl.gov/natlsecurity/threat/biothreat/biodetection.shtml (accessed October 19, 2011).

"Biosensors and Biochips for Environmental and Biomedical Applications." *Oak Ridge National Laboratory.* http://www.ornl.gov/sci/biosensors/ (accessed October 19, 2011).

Wampler, Stephen P. "Lab's Work Provides BASIS for Biodetection." *Lawrence Livermore National Laboratory*, September 2011. http://www.lanl.gov/natlsecurity/threat/biothreat/biodetection.shtml (accessed October 19, 2011).

Pamela V. Michaels

Biodiversity

■ Introduction

Biodiversity is a term that has distinct biological and genetic connotations. In biology, biodiversity describes the total number of ecosystems and plant and animal species in a given terrestrial or aquatic region. For geneticists, biodiversity refers to the variety of genes and the proliferation of the organisms that have provided the impetus for the forces of evolution that have shaped the history of Earth.

This perspective on biodiversity is challenged when the relationship of organisms to the environment is considered from anthropocentric standards. These may be distilled into a single question—what boundaries ought to govern the extent to which humans are permitted to alter the natural environment and reduce biodiversity? This difficult question simultaneously engages ethical, moral, economic, cultural, and spiritual issues if a correct answer is one gauged by personal viewpoint as much as it is influenced by science. Current environmental concerns that are centered on observed climate change, its relationship to anthropogenic sources such as greenhouse gas emissions, and the impact on human, plant, and animal life in all global regions have given biodiversity greater prominence in the twenty-first century.

■ Historical Background and Scientific Foundations

Charles Darwin (1809–1882) formulated his theories of evolution through his mid-nineteenth century studies of the ecosystems found on the remote Galapagos Islands in the Pacific. Darwin is a useful point of commencement in the modern understanding of biodiversity. In this geographically isolated environment, Darwin determined that many previously unknown species to thrive as a result of their ability to adapt to their environment. "Survival of the fittest" is the expression often associated with Darwinian theory; the modern understanding of biodiversity includes recognition that in a stable ecosystem, the number of emerging plant or animal species will generally equal the number of extinctions.

There are two essential biodiversity hierarchies. The number of functional organisms or ecosystems is the first. If a specific region has more plants and animals within its boundaries (often stated as a function of its latitudinal gradient), it is deemed to possess greater biodiversity. The Amazon rain forest has millions of different organisms; a stark, arid, high elevation mountainous region has far less. The second hierarchy is the number of species found within a particular genotype that appear to perform similar functions within the region ("species turnover," or the correlation between species emergence and extinction). A region that provides the habitat for a variety of predators that occupy the top of their respective food chains is biodiverse, because the existence of animal co-equals means that each species has sufficient food to thrive.

As the environmental impacts associated with global warming and anthropocentric activities such as the timber cutting of the world's rain forests to create more agricultural land received closer scientific scrutiny in the late twentieth century, biologists began to critically evaluate the relationship between biodiversity and ecosystem health. A general scientific consensus formed to support the position that greater biodiversity tended to ensure that processes fundamental to the existence of all earthly life were maintained. The role of the Earth's forests to promote the exchange of carbon dioxide and oxygen, and the ability of wetlands to purify water are prominent examples. The monocultural crop systems developed to maximize agricultural profits have been identified as ones harmful to biodiversity.

The central issue that drives the importance of biodiversity is whether humans possess an inherent right to exploit the natural world as they determine best, so long as the environment is managed appropriately. Management is a concept that is distinct from biodiversity, in that ecosystems managed by humans will sustain plant,

WORDS TO KNOW

ANTHROPOCENTRIC: Human-centered or from a human source.

ECOSYSTEM: The product of the interactions that occur among all living organisms (including all plants, animals, and microorganisms such as bacteria), and the cumulative effect of these organic interactions with the surrounding physical environment (sunlight, climate, water, and soil).

HOLOCENE EXTINCTIONS: The widespread animal species extinctions observed in the last 11,500 years since the end of the last Ice Age. This period is known as the Holocene epoch. Human activity is the primary cause of many of the extinctions recorded during this period, as forests and range land that supported many species have been converted to human use.

LATITUDINAL GRADIENT: The relationship between biodiversity and the latitudinal position of a geographic region on Earth's surface. As a general rule, greater genetic and species diversity is observed the closer an ecosystem is located to the equator.

MONOCULTURE: The low diversity agricultural practice of cultivating a single crop species over a large area to promote efficient farming. A prominent feature of industrial agriculture, monoculture tends to expose crops to disease because pathogens are able to spread with greater speed through a single crop.

animal, and microbial species extinction. For many scientists, it is the maintenance of appropriate levels of biodiversity in a given ecosystem that is the primary objective.

■ Impacts and Issues

Species diversity is a particularly pressing contemporary environmental question. The rate of plant and animal extinction is greater in the Holocene era than at any other time in the history of the Earth, including the widespread destruction of life that occurred during the Cretaceous-Tertiary event approximately 66.5 million years ago. Alpha-diversity is the measure of the number of species found in a specific environment. The proliferation of a given species is a function of how organisms interact and compete for the same finite resources in their environment. In contrast, beta-diversity is a spatial measure of how different species occupy a given region. The geographical diversity observed in a species is termed its gamma-diversity. Biodiversity is reduced where any one of these three measures is reduced.

Habitat preservation is one of the most profound contributions made by biotechnology to species diversity. The use of genetically modified organisms to degrade otherwise toxic substances in the treatment of industrial wastewater, chemical by-products, and atmospheric emissions is one example. There is a direct relationship observed between the preservation of natural habitats and species diversity because habitat quality is essential to the maintenance of the evolutionary potential of all species. Other, more direct biotechnology interventions such as assisted reproduction (IVF) techniques and gamete banking have helped maintain populations of some endangered species, including giant pandas and big cats. Polar bears and endangered frogs are some of the species that have their eggs and sperm stored in cryobanks. Tissue cultures of several critically endangered species are also maintained at some laboratories. Additionally, the Svalbard Global Seed Vault in northern Norway safeguards frozen seeds from more than 500,000 plant species.

It is a generally accepted scientific proposition that greater biodiversity is the hallmark of a healthy and sustainable environment. A larger number of plant species will generally mean that a greater variety of available food crops. Research biologists have established a correlation between natural life form sustainability and greater species diversity. The ability of ecosystems to withstand disasters, both those of anthropocentric origins such as an oil spill or natural phenomena like drought is a direct function of their health as measured in part by its diversity.

Humans have long recognized that biodiversity had an intrinsic value to human health and the vitality of their world. In the intrinsic value model, the argument is that biodiversity should be saved not because it is of value to humans, but because it is intrinsically valuable as it is, and all species should be preserved. From this perspective, biodiversity must be protected notwithstanding that an ecosystem may include many equivalent species. In his famous treatise on evolution, Darwin identified 14 different variations of the finch that inhabited the Galapagos Islands; all are equal in the intrinsic value model. Such recognition is mirrored in the moral duty articulated in major world religions that humans are the stewards of nature.

The strength of the intrinsic position is buttressed by empirical data, and to some extent can align with an anthropocentric view of biodiversity. When biodiversity is examined from an anthropocentric value perspective, the potential of the distinct economic, aesthetic, and recreational value of the world's biodiversity to humans is gauged. The central anthropocentric premise is that humans have a right to exploit the natural environment for their own purposes that is superior to any other consideration, but biodiversity is valued because it is helpful to humans. Supporters of this biodiversity viewpoint rely on facts such as how the exploitation of the natural environment has led to many discoveries that have improved the quality of foods and the ability to manufacture medicines that promote greater human health. For example,

Orangutans such as this mother orangutan and her baby are an endangered species. Habitat loss due to human activities and poaching are two of the greatest threats to such primates and to species diversity in general.   © *Eric Gevaert/Shutterstock.com.*

biodiversity promotes opportunities to pursue further plant-based medical research. The United Nations estimates that up to 40 per cent of the global economy is directly attributable to the human ability to exploit available biological resources.

The decline in the North American honeybee population observed in the past decade is partially attributed to reduced biodiversity by many environmental scientists. An estimated 130,000 plants depend on the pollination achieved by honeybees for their continued survival; a sustained decline in this prolific pollinator may compromise the sustainability of many plant species on the continent. The ancient pact between flowers and pollinators was described by American entomologists Diana Cox-Foster and Dennis van Engelsdorp as an essential food supply safeguard.

The positive impacts associated with biodiversity are confirmed when the myriad of ongoing biological interactions that occur in a typical agricultural environment are uninterrupted by development or pollution. For example, soil-based microbacteria that feed on cellulose present in straw fibers are the primary food source for amoebas that produce lignite fibers, which are converted into plant food. Algae that grows in

watercourses contributes to the absorption of carbon dioxide from the atmosphere through photosynthesis. Rodents and similar small mammals create underground habitats that permit better soil aeration and its ability to retain moisture. The insects that inhabit the soil process available existing organic matter, and their waste is correspondingly enriched with nutrients that contribute to soil richness. The actions of earthworms within the soil also make a vital contribution to soil fertility, as they promote aeration, soil drainage, and the maintenance of soil structures conducive to human agriculture.

There is common ground between the intrinsic and anthropocentric approaches to biodiversity. The different degrees of importance attached to the promotion of ecosystem diversity in each concept do not displace the scientific fact that such environments are more sustainable and more likely to promote environmental stability.

SEE ALSO *Agricultural Biotechnology; Biochemical Engineering; Biodiversity Agreements; Bioethics; Biological Confinement; Biomimetic Systems; Biopreservation; Biotechnology;*

Wild populations of the Panamanian golden frog are thought to have been extinct in the wild since 2007 due to habitat loss, illegal collection for the pet trade, and a fungal outbreak of chytridiomycosis that is destroying many amphibian populations globally. The decline of amphibian species has many scientists concerned about decreasing biodiversity. Panama has great biodiversity within its borders, with unique species of birds, mammals, amphibians, and insects found in its dense mountain forests and the mangrove forests in the coastal areas, but is subject to many of the pressures felt around the world in terms of deforestation and climate change.   *© Darren Green/Shutterstock.com.*

Cartegena Protocol on Biosafety; Cloning; DNA Sequencing; Environmental Biotechnology; Genbank; Gene Banks; Genetically Modified Crops; Nanotechnology, Molecular; Plant Variety Protection Act of 1970; Recombinant DNA Technology; Synthetic Biology; Tissue Banks; Transgenic Plants

BIBLIOGRAPHY

Books

Ammann, Klaus. *Biodiversity and Biotechnology.* Basel, Switzerland: Birkhauser, 2003.

Boreém, Aluizio, Fabricio R. Santos, and David E. Bowen. *Understanding Biotechnology.* Upper Saddle, NJ: Prentice Hall, 2003.

Cooper, Joseph, Leslie Lipper, and David Zilberman, eds. *Agricultural Biodiversity and Biotechnology in Economic Development.* New York: Springer, 2005.

Dronamraju, Krishna R. *Emerging Consequences of Biotechnology: Biodiversity Loss and IPR Issues.* Singapore and Hackensack, NJ: World Scientific, 2008.

Dwivedi, Padmanabh, S. K. Dwivedi, and M. C. Kalita. *Biodiversity and Environmental Biotechnology.* Jodhpur, India: Scientific Publishers, 2007.

Kumar, Aravind, and Govind Das. *Biodiversity, Biotechnology and Traditional Knowledge: Understanding Intellectual Property Rights.* New Delhi: Narosa, 2010.

Lévêque, C., and Jean-Claude Mounolou. *Biodiversity.* Chichester, UK: John Wiley & Sons, 2004.

McManis, Charles R, ed. *Biodiversity and the Law: Intellectual Property, Biotechnology and Traditional Knowledge.* Sterling, VA: Earthscan, 2007.

Melendez-Ortiz, Ricardo, and Vicente Sanchez, eds. *Trading in Genes: Development Perspectives on Biotechnology, Trade, and Sustainability.* Sterling, VA: Earthscan, 2005.

Participants of a demonstration by Greenpeace Kids, the youth organization of the environmental protection organization Greenpeace, hold banners and wear colorful costumes as they stage a protest in Bonn, Germany, on May 19, 2008. The protection of flora, fauna, and even food sources was on the agenda of the 191 governments attending the ninth conference of the UN Convention on Biological Diversity in Bonn held in May 2008.   *© AP Images/Hermann J. Knippertz.*

Shiva, Vandana. *Monocultures of the Mind: Perspectives on Biodiversity and Biotechnology.* Atlantic Highlands, NJ: Zed Books, 2003.

Web Sites

"Biodiversity." *The Guardian UK.* http://www.guardian.co.uk/environment/biodiversity (accessed November 7, 2011).

"Biodiversity." *Natural History Museum (UK).* http://www.nhm.ac.uk/nature-online/biodiversity/ (accessed November 7, 2011).

"Biodiversity." *United Nations Environment Programme (UNEP).* http://www.unep.org/themes/biodiversity/ (accessed November 7, 2011).

"Biodiversity." *USAID Environment.* http://www.usaid.gov/our_work/environment/biodiversity/ (accessed November 7, 2011).

Carrington, Damien. "Encyclopedia of Life Catalogues more than One-third of Earth's Species." *The Guardian UK*, September 5, 2011. http://www.usaid.gov/our_work/environment/biodiversity/ (accessed November 7, 2011).

Nellemann, Christian, and Emily Corcoran, eds. "Dead Planet, Living Planet—Biodiversity and Ecosystem Restoration for Sustainable Development." *United Nations Environment Programme*, 2010. http://www.unep.org/publications/contents/pub_details_search.asp?ID=8196 (accessed November 7, 2011).

United Nations Decade on Biodiversity. http://www.cbd.int/2011-2020/ (accessed November 7, 2011).

Bryan Thomas Davies

Biodiversity Agreements

■ Introduction

Biodiversity refers to the variety of life forms present in an area, whether within a given ecosystem or on Earth as a whole. Earth's biodiversity is often understood in terms of the wide variety of its plants, animals, and microorganisms. So far, about 1.75 million species of plants, animals, and microorganisms have been identified. Most are small, rapidly breeding varieties of life such as insects. Scientists estimate that there are likely 13 million to 100 million different species on Earth.

Biodiversity is crucial to environmental health because individual species of plants and animals often play complex and interrelated roles within an ecosystem. An imbalance between predator and prey animals, for example, may have effects as diverse as species overpopulation, food scarcity, resource depletion, species migration, or even species extinction.

Conservation of biodiversity ensures that humans will have access to desired resources. Some of the world's most vulnerable ecosystems—including the Amazon rainforests, Arctic tundra, and tropical coral reefs—produce crucial medicines, construction materials, and food sources. These ecosystems also drive local economies and help filter out pollution, mitigating environmental degradation.

Localized biodiversity problems can be caused by natural phenomena such as extreme weather events. However, humans pose a more significant threat to global biodiversity. Human actions and human-created pollution that degrade the environment negatively affect biodiversity. Climate change and deforestation threaten animal and plant habitats worldwide, altering the biodiversity of threatened regions. Overhunting and overfishing deplete species populations and cause ecosystem imbalance, often damaging the economies and health of human populations that depend on such resources.

Evaluation of global biodiversity is typically divided into three categories. Ecosystem diversity looks at habitats and ecosystems as a whole by studying a region's environmental integrity, the flora and fauna present, the ecological processes that take place within the region, and the overall contribution of the region to the global environment. Genetic diversity is concerned with the genetic variations within one species and differences among diverse species. Species diversity encompasses the variety of species and their populations within a region.

Because ecosystems, habitats, and waterways cross national borders, international cooperation is essential to protect global biodiversity. Beginning in the mid-twentieth century, nations implemented a range of cooperative biodiversity initiatives governed by treaties or agreements. Individual governments, international environmental agencies, and non-governmental environmental advocacy groups monitor biodiversity conservation efforts made to meet treaty obligations.

■ Historical Background and Scientific Foundations

The first international agreements to impact biodiversity focused on fishing, hunting, and the capture and trade of certain animals. In 1946 several nations gathered to form the International Convention for the Regulation of Whaling (ICRW). The ICRW did not ban international whaling around the globe, but established an international body to limit and regulate whaling in an effort to conserve dwindling whale populations. Whereas some participating nations had ceased commercial whaling decades before as demand for whale oil and bone products disappeared, others agreed to limiting the number of whales taken by fishermen, indigenous populations, and scientific researchers. In the decades following the initial treaty, international bans on commercial whaling have tightened, and in 1982 the ICRW adopted a global moratorium on commercial whaling. However, critics of the ICRW assert that loopholes in the treaty allow traditional whaling nations such as Japan to kill an unsustainable number of whales each year.

In 1963 the international community turned its attention to the protection of endangered species threatened by overhunting, habitat destruction, overharvesting, and commercial appetite for rare goods such a ivory and exotic skins and furs. The Convention on the International Trade in Endangered Species of Flora and Fauna (CITES) treaty was first proposed at the 1963 meeting of the International Union for Conservation of Nature (IUCN) after the passage of a resolution calling for the regulation of the international trade in animals and plants. Eighty member nations agreed to adopt the CITES treaty in March 1973; the treaty entered into force in 1975. As of 2011, 175 countries had adopted CITES. The United Nations Environment Programme (UNEP) credits CITES with playing a significant role in the conservation of threatened African big game animals and the preservation of biodiversity in protected African savanna ecosystems.

In 1979 the Bonn Convention on Migratory Species (CMS or Bonn Convention) was adopted as the first global agreement to protect migratory species and their cross-border migration routes. The convention lists threatened migratory species and calls on participating nations to adopt laws protecting migratory habitats. The Bonn Convention covers terrestrial, marine, and aerial migration routes and species, requiring concerted cooperation of international environmental bodies such as UNEP to coordinate conservation measures among governments. Participating nations are encouraged to enact binding agreements with other nations in their region, but can also issue nonbinding memoranda of understanding that establish voluntary preservation guidelines. The CMS further encourages member nations to protect rangelands and migration avenues within their own national borders. As of July 2011 there were 116 parties to the CMS. Regional agreements enacted under CMS have extended international protection to—among others—European bat populations, seabirds such as petrels and albatross, seals, and gorillas.

The United Nations Conference on Environment and Development (UNCED), also known as the Earth Summit, was held from June 3–14, 1992, and marked a significant turning point for global biodiversity agreements. More than 2,500 representatives of national governments and nongovernmental organizations, as well as 108 heads of state, attended the conference. A further 17,000 environmental advocates, scientists, environmental policy advisors, and others participated in the advisory Global Forum that ran concurrent with the Earth Summit. Various global environmental issues—from conservation to pollution, environmental justice to global warming—were discussed, but the convention marked the first large-scale international meeting that addressed biodiversity globally and holistically. Participants sought to find solutions to global biodiversity challenges that balanced concerns for threatened species, human desire for environmental resources and its impact on ecosystems, human population ties to local environments and

WORDS TO KNOW

AGREEMENT: An arrangement among parties to do something or not do something in order to achieve a certain goal. It also refers to the document in which the agreement is recorded. It also may be called a contract, protocol, or treaty.

BIODIVERSITY: The variety of life forms present in an area, whether a within a given ecosystem or on Earth as a whole.

ENVIRONMENTAL DEGRADATION: The process by which the health and integrity of an environment declines as a result of human action.

RESOLUTION: A formal expression of agreement, intent, or opinion made by a formal organization or legislature after voting on the issue.

TREATY: A formal agreement under international law between at least two countries, nations, or autonomous or semi-autonomous groups of people, that sets forth the goals and responsibilities of all parties (participants) to the agreement.

traditional uses, environmental degradation, ecosystem health (including human health), water scarcity, pollution, and economic and scientific development.

Participants drafted Agenda 21, the Rio Declaration on Environment and Development, the Statement of Principles for the Sustainable Management of Forests, and the Convention on Biological Diversity (CBD or Biodiversity Convention), all of which addressed biodiversity issues in a global environmental context. In December 1992 the Commission on Sustainable Development (CSD) was established to help nations implement the agreements reached at the Earth Summit and to monitor their future progress. Adoption of the Biodiversity Convention was also rapid. The legally binding treaty aims to conserve biodiversity, promote sustainable use of biological and environmental resources, and ensure a fair distribution of resource benefits. It entered into force on December 29, 1993.

Developments in agricultural biotechnology and the growth of global land dedicated to agriculture prompted debate about the growth and export of genetically modified (GM) foods and genetically modified organisms (GMOs), especially crop seeds. Agricultural biotechnologists assert that GM foods are safe; the most frequently manipulated food crops were created to boost crop yields by making plants resistant to common diseases or drought. Critics assert that the foods have not been proven safe. Most GM food critics advocate adherence to the precautionary principle, asserting that because GM foods may contain a risk of harm, they should be regulated as harmful until scientifically proven safe.

Hymenocallis coronaria, also known as the Cahaba Lily or the Shoals spider-lily, grows only in a few areas in Alabama, Georgia, and South Carolina. It is under consideration for federal protection because much of its habitat has been destroyed to make way for dams. Human population growth, unsustainable demands for land and water resources, climate change, and pollution have contributed to an exponential increase in the extinction of species worldwide, creating a major loss of genetic resources and impacting Earth's ecosystems. The United States signed but never ratified the Convention on Biological Diversity.   © *Bobby Johnson/Shutterstock.com.*

In 2000 the Cartagena Protocol on Biosafety to the Convention on Biological Diversity (Cartagena Protocol) addressed CBD member concerns about the identification and transboundary movement of GMOs and GM foods. The Cartagena Protocol only regulates GM foods that contain living modified organisms (LMOs) capable of reproducing or transferring genetic material. Under the protocol, which entered into force in 2003, producers cannot export LMOs into another country without disclosing that the foods are modified and obtaining permission from the country of import to permit the goods to enter. Although the protocol is intended to assuage concerns over safety to humans, nations have used it to stop the import of organisms they suspect as harmful to local plant or animal biodiversity. Routine export laws, trade agreements, and treaties between nations have also addressed biodiversity concerns by prohibiting the import of plant and animal species that could adversely affect vulnerable native flora and fauna.

The Codex Alimentarius Commission (Codex), part of the United Nations Food and Agriculture Organization (FAO) and World Health Organization (WHO), keeps the Codex Alimentarius, a database of international laws, standards, regulations, and categorizations governing food. The Codex is also tasked with developing principles and processes for assessing any human health risks possibly posed by GM foods. GM foods are evaluated for both potential direct effects and potential unwanted effects (unintended possible consequences) to human health. However, there is no Codex protocol governing mandatory assessment of risk to local or global biodiversity.

Participants in the International Youth Conference on Biodiversity in Aichi, Japan, study plants in Mie Prefecture on August 24, 2010. The conference is associated with the 10th meeting of the Conference of the Parties to the Convention on Biological Diversity, the so-called CBD COP10, that was held in Aichi Prefecture in October 2010 and at which 191 countries participated and the Nagoya Protocol on Biodiversity Preservation was adopted.   © *AP Images/Kyodo.*

■ Impacts and Issues

The United Nations declared 2010 the Year of Biodiversity. To highlight the importance of the global issues, on December 22, 2010, the United Nations further declared the decade spanning from 2011 to 2020 its Decade on Biodiversity.

The development and increased global implementation of agricultural biotechnology and environmental biotechnology continues to influence global debates on biodiversity. Many environmental advocates, especially in industrialized nations, are opposed to the increased global introduction of genetically modified food crops into sensitive ecosystems, noting for example studies that have charted possible negative effects on beneficial insect and bee populations in some regions. At the same time, genetic biotechnology has helped researchers better understand biodiversity by aiding in the genetic identification and comparison of species and charting the effects of environmental degradation on species within various ecosystems. Biotechnology has aided with cross breeding and population restoration efforts that have conserved threatened species or helped to reintroduce species to restore biodiversity and ecological balance.

Participants at the October 2010 Convention on Biological Diversity Conference of Parties (CBD COP-10) in Nagoya, Japan, adopted the Nagoya Protocol on Access to Genetic Resources and the Fair and Equitable Sharing of Benefits Arising from their Utilization. The protocol established an international legal framework to help achieve the CBD's stated goal of fair sharing of benefits from biological and environmental resources, but specifically refers to genetic resources that aid in species and habitat conservation.

In order to monitor global biodiversity efforts and facilitate the sharing of scientific resources, the international community established the Intergovernmental Science-Policy Platform on Biodiversity and Ecosystem Services (IPBES). The organization is not a research organization in own right, but rather a convention and clearinghouse to help researchers share, assess, compile, and report on their findings. It aims to function like the Intergovernmental Panel on Climate Change (IPCC) and is planning to issue a comprehensive report on the state of global biodiversity similar to the IPCC's climate change reports. IPBES's first scheduled plenary meeting was in October 2011, in Nairobi, Kenya.

Technology has also helped researchers, environmental monitoring groups, and policymakers to share information on biodiversity more easily. The Global Biodiversity Information Facility (GBIF) facilitates information sharing via online research databases and Internet-accessible libraries of research reports, academic papers, and peer-reviewed articles.

The United Nations Environment Programme's World Conservation Monitoring Centre reported in *Global Biodiversity Outlook 3* that worldwide efforts had failed to significantly reduce the rate of biodiversity loss, and warned that certain regions, such as unprotected areas in the Amazon basin and tropical coral reefs, may be close to critical tipping points in biodiversity from which they cannot recover. One indicator of biodiversity, the Living Planet Index, charted a 30 percent decline in global vertebrate populations since 1970. Additionally, the Index noted that the world likely has less than half of the coral reef cover it once had.

■ Primary Source Connection

The Convention on Biological Diversity, a legally binding treaty, promotes living well around and within nature. Entered into force in 1993, it serves as a framework for the conservation and sustainable use of biological diversity and in equitable sharing of benefits brought about by the utilization of genetic resources. Functioning at both the national and international level, the Convention promotes the protection and preservation of such things as nature's landscapes and ecosystems, while advancing environmental awareness of endangered species throughout the world. Measures to enhance and preserve nature are top priorities, as is education and public awareness on biological diversity on both scientific and technical fronts. The Convention strives to build or improve policies, legislation, and fiscal programs and practices that impact the environment. In 1992 world leaders attended the United Nations Conference on Environment and Development in Rio de Janeiro, Brazil. From that meeting, the Convention on Climate Change and the Convention on Biological Diversity agreements were signed. More than 150 governments signed the biodiversity treaty while at the conference and more than 187 countries have since ratified the agreement.

Convention on Biological Diversity

Preamble

The Contracting Parties,

Conscious of the intrinsic value of biological diversity and of the ecological, genetic, social, economic, scientific, educational, cultural, recreational and aesthetic values of biological diversity and its components,

Conscious also of the importance of biological diversity for evolution and for maintaining life sustaining systems of the biosphere,

Affirming that the conservation of biological diversity is a common concern of humankind,

Reaffirming that States have sovereign rights over their own biological resources,

Reaffirming also that States are responsible for conserving their biological diversity and for using their biological resources in a sustainable manner,

Concerned that biological diversity is being significantly reduced by certain human activities,

Aware of the general lack of information and knowledge regarding biological diversity and of the urgent need to develop scientific, technical and institutional capacities to provide the basic understanding upon which to plan and implement appropriate measures,

Noting that it is vital to anticipate, prevent and attack the causes of significant reduction or loss of biological diversity at source,

Noting also that where there is a threat of significant reduction or loss of biological diversity, lack of full scientific certainty should not be used as a reason for postponing measures to avoid or minimize such a threat,

Noting further that the fundamental requirement for the conservation of biological diversity is the in-situ conservation of ecosystems and natural habitats and the maintenance and recovery of viable populations of species in their natural surroundings,

Noting further that ex-situ measures, preferably in the country of origin, also have an important role to play,

Recognizing the close and traditional dependence of many indigenous and local communities embodying traditional lifestyles on biological resources, and the desirability of sharing equitably benefits arising from the use of traditional knowledge, innovations and practices relevant to the conservation of biological diversity and the sustainable use of its components,

Recognizing also the vital role that women play in the conservation and sustainable use of biological diversity and affirming the need for the full participation of women at all levels of policy-making and implementation for biological diversity conservation,

Stressing the importance of, and the need to promote, international, regional and global cooperation among States and intergovernmental organizations and the non-governmental sector for the conservation of biological diversity and the sustainable use of its components,

Acknowledging that the provision of new and additional financial resources and appropriate access to relevant technologies can be expected to make a substantial difference in the world's ability to address the loss of biological diversity,

Acknowledging further that special provision is required to meet the needs of developing countries, including the provision of new and additional financial resources and appropriate access to relevant technologies,

Noting in this regard the special conditions of the least developed countries and small island States,

Acknowledging that substantial investments are required to conserve biological diversity and that there is the expectation of a broad range of environmental, economic and social benefits from those investments,

Recognizing that economic and social development and poverty eradication are the first and overriding priorities of developing countries,

Aware that conservation and sustainable use of biological diversity is of critical importance for meeting the food, health and other needs of the growing world population, for which purpose access to and sharing of both genetic resources and technologies are essential,

Noting that, ultimately, the conservation and sustainable use of biological diversity will strengthen friendly relations among States and contribute to peace for humankind,

Desiring to enhance and complement existing international arrangements for the conservation of biological diversity and sustainable use of its components, and

Determined to conserve and sustainably use biological diversity for the benefit of present and future generations.

Have agreed as follows:

Article 1 Objectives

The objectives of this Convention, to be pursued in accordance with its relevant provisions, are the conservation of biological diversity, the sustainable use of its components and the fair and equitable sharing of the benefits arising out of the utilization of genetic resources, including by appropriate access to genetic resources and by appropriate transfer of relevant technologies, taking into account all rights over those resources and to technologies, and by appropriate funding.

Article 3 Principle

States have, in accordance with the Charter of the United Nations and the principles of international law, the sovereign right to exploit their own resources pursuant to their own environmental policies, and the responsibility to ensure that activities within their jurisdiction or control do not cause damage to the environment of other States or of areas beyond the limits of national jurisdiction.

UNITED NATIONS. "CONVENTION ON BIOLOGICAL DIVERSITY." 1992.

SEE ALSO   *Agricultural Biotechnology; Bt Insect Resistant Crops; Cartegena Protocol on Biosafety; Corn, Genetically Engineered; Environmental Biotechnology; GE-Free Food Rights; Genetic Engineering; Genetically Modified Crops; Genetically Modified Food*

BIBLIOGRAPHY
Books

Chivian, Eric, and Aaron Bernstein. *Sustaining Life: How Human Health Depends on Biodiversity.* New York: Oxford University Press, 2008.

Hilty, Jodi A., William Zander Lidicker, and Adina Maya Merenlender. *Corridor Ecology: The Science and Practice of Linking Landscapes for Biodiversity Conservation.* Washington, DC: Island Press, 2006.

Hulot, Nicholas. *One Planet: A Celebration of Biodiversity.* New York: Abrams, 2006.

Kumar, Aravind, and Govind Das. *Biodiversity to Biotechnology: Intellectual Property Rights.* New Delhi, India: Narosa Publishing House, 2010.

Periodicals

Araúo, Miguel B., and Carsten Rahbek. "How Does Climate Change Affect Biodiversity?" *Science* 313, no. 5792 (September 8, 2006): 1396–1397.

Willis, Katherine J., and H. John B. Birks. "What Is Natural? The Need for a Long-Term Perspective in Biodiversity Conservation." *Science* 314, no. 5083 (November 24, 2006): 1261–1265.

Web Sites

"Biodiversity." *United Nations System-Wide EarthWatch.* http://www.un.org/earthwatch/biodiversity/index.html (accessed October 1, 2011).

"Biodiversity." *U.S. Department of the Interior, U.S. Geological Survey (USGS).* http://www.usgs.gov/science/science.php?term=92 (accessed October 1, 2011).

Global Biodiversity Information Facility (GBIF). http://www.gbif.org (accessed October 1, 2011).

"World Conservation Monitoring Centre." *United Nations Environment Programme (UNEP).* http://www.unep-wcmc.org/ (accessed October 1, 2011).

Adrienne Wilmoth Lerner

Bioeconomy

■ Introduction

Human economics relies on a plentiful supply of raw materials to create the energy and chemicals to drive industrial activities. Much of the world's economy is based on gaining these supplies from fossil fuels, such as coal and petroleum, which are nonrenewable resources. In the bioeconomy, these supplies come from renewable plant resources. Biotechnology is central to the realization of a bioeconomy because it will enable the transformation of renewable plant sources into substitutes for petroleum-based chemicals in many products and the use of renewable biomass rather than energy sources such as oil and gas, which are nonrenewable fossil fuels.

■ Historical Background and Scientific Foundations

In the nineteenth century, coal was the source of energy that enabled the growth of industrial nations. Coal was then superseded in the twentieth century by petroleum, which drove the growth of the chemical industry. The products of the latter transformed human lives in many ways—from the widespread adoption of plastics to the manufacture of modern pharmaceuticals. Petroleum was also, of course, responsible for a revolution in transportation. However, beginning in the 1960s the harm that oil and gas-based industry does to the environment became increasingly apparent. Dependence on oil and increasing oil prices have also led to political conflict. The publication in 1962 of Rachel Carson's *Silent Spring*, a classic text for the environmental movement, was an important turning point in the shift towards a more sustainable bio-based economy. Another milestone was legislation such as the Clean Air Act coming into force in the 1970s. The United States Department of Agriculture and the Organization for Economic Cooperation and Development published position papers in 2008 and 2010 respectively, which crystallize some of the basic principles of bioeconomy and outline the challenges lying ahead.

Worldwide, more than $400 billion worth of industrial and consumer products are already being manufactured from biomass each year. These include chemicals, pharmaceuticals, soaps, detergents, pulp and paper, lumber, biofuels, paints, and lubricants. In the United States, 12 billion pounds of biomass are used annually in the production of industrial bio-based products, including biofuels, chemicals, and other materials traditionally produced from petroleum feedstock. Success in commercializing new bio-based products relies on advances in industrial biotechnology and "green" chemistry, which focus upon developing environmentally-friendly processes with biomass as the raw material. Central to these efforts is the discovery and development of enzymes, which are the renewable tools that enable transformations to take place under mild conditions of temperature and pressure.

■ Impacts and Issues

The term knowledge-based is often applied to bioeconomy, referring to the application of knowledge gained from molecular biology and genomics in biotechnology processes—knowledge that is needed to develop biofuels and other bio-based products. Bioeconomy has many advantages and may even be the only way in which a growing human population can survive. Bioeconomy is more environmentally friendly than a petroleum-based economy because it is associated with fewer carbon emissions and may alleviate global warming and other pollution, as well as reducing waste and reducing production costs. Its resources are renewable because a plant crop can be grown continuously. A shift towards biomass also reduces dependence upon oil. Therefore, most governments around the world have committed to developing a knowledge-based bioeconomy for their country. What is needed to make bioeconomy work is extensive policy and infrastructure development, investment in research and development, and consumer education to accept bio-based products such as biofuels.

According to the United States Department of Agriculture, the bioeconomy is set for substantial growth in the next decade. Biobased chemicals could increase from 2 percent of the global chemical market in 2006 to 22 percent by 2025. The chemical market itself is likely to grow at 3 to 6 percent per year until 2025 and was worth a total of $1.2 trillion in 2005. Currently, the emphasis in the bioeconomy is upon biofuels. In the United States the main biofuels are ethanol, manufactured from corn, and biodiesel, from soybeans.

If the bioeconomy does grow, the benefits could be substantial. Replacing fossil fuels with renewable biofuels and other products from biomass could reduce air pollution, such as sulfur dioxide and nitrogen oxides, and also greenhouse gas emissions. A cleaner environment would improve human health and also counter the threat of climate change. Bioeconomy could also improve economic performance and increase political security by reducing dependence on oil and oil-rich states. Farmers and agricultural suppliers can expect to benefit from increased demand for industrial biomass crops such as corn and soybeans. Consumers are taking a greater interest in the environmental and ethical aspects of what they buy, and the bioeconomy can start to satisfy their aspirations toward a more sustainable way of living.

However, there are many challenges involved in making bioeconomy into a global reality. Thus far, corn and soybean have been the main industrial biomass crops. These are also potential food crops and there has been some criticism that their widespread cultivation could compromise and destabilize worldwide food supplies and further harm those who already go hungry. There is an urgent need to develop a wider range of plants for biomass. Researchers are looking, in particular, for a way of utilizing cellulosic biomass for biofuels by developing enzymes that can break down previously intractable cellulose polymers into sugar for fermenting into alcohol. There is also a requirement for more effective pretreatment processes for cellulosic biomass to make its conversion a more viable commercial proposition. Grass and other cellulose-based crops can be grown on land that is not suitable for cultivation of cereals, which would mean that food crops need not be diverted for industrial use.

There are many other scientific and technical challenges involved in developing the bioeconomy. Processes that work well in the laboratory often cannot be scaled up to make commercial quantities of a bio-based product. Often the limitation is that the enzyme tool needed for transforming the biomass is too unstable, or not available in industrial scale quantities. For instance, in 2004 researchers at the Idaho National Laboratory discovered an enzyme that could be used

---

WORDS TO KNOW

BIOFUEL: A biofuel is manufactured from either biomass, such as corn, or from treated municipal or industrial waste. The most common biofuels are ethanol, biodiesel, and methane.

CELLULOSIC BIOMASS: Biomass, including grass and trees, whose main carbohydrate component is cellulose rather than starch.

ENZYME: An enzyme, or biocatalyst, is a protein produced by a living cell that is capable of catalyzing, or speeding up, a specific biochemical reaction such as the breakdown of starch to glucose. Enzymes generally work at room temperature and in aqueous solution.

INDUSTRIAL BIOTECHNOLOGY: The use of bio-products (such as enzymes) and biomass feedstock in industrial products that are normally made from petroleum-based products such as fuel, polymers, petrochemicals, bulk chemicals, and specialty chemicals.

SUSTAINABLE: There are many definitions of sustainable but the one most commonly used originates from the 1987 declaration of former Chair of the World Commission on Environment and Development, Gro Harlem Brundtland, that sustainable development meets the needs of people today without compromising the needs of others in the future by having minimal impact on resources and the environment.

---

instead of bleach in the paper and textile industries, potentially cutting wastewater treatment costs and environmental impact. The enzyme, Ultrastable Catalase, comes from a bacterium found in Yellowstone Park that is capable of withstanding very high temperatures.

IN CONTEXT: BIOECONOMY AND BIOPHARMACEUTICALS

Bioeconomy also encompasses health care because biopharmaceutical drugs, such as monoclonal antibodies, tend to be manufactured in a way that is more friendly to the environment than traditional pharmaceutical methods, which use large amounts of energy and toxic solvents. Biopharmaceutical drugs are generally made by cells, which are a renewable resource, in aqueous solution. On the patient-centered level, biotechnology drugs promise more efficient use of health care resources. Use of genetics in manipulation of special properties of biopharmaceutical drugs facilitates a more personalized approach to the prescribing drugs. A patient is more likely to respond positively and less likely to experience side effects if their medication is matched to their genetic profile.

A biodiesel production plant in Italy uses transesterification to turn vegetable oils extracted in the plant into biodiesel. The plant then refines the fuel. © *Moreno Soppelsa/Shutterstock.com.*

Normal catalase would decompose under industrial conditions, but this new one does not. However, there is a need for research and development investment from an enzyme company to make Ultrastable Catalase in industrial quantities.

Although many governments have signaled an interest in developing bioeconomy, they need to put policy and investment in place to make sure it happens. In the United States there are tax credits available for the producers of biofuels, but these could usefully be extended to those developing other bio-based products. There are also many programs to ensure the development of biofuels. For instance, the U.S. Department of Energy announced in 2010 that nearly $600 million would be used to fund building and operating new biorefinery facilities producing fuels and products from wood chips and other forms of biomass. A biorefinery can be regarded as an integral unit that can accept different biological feedstocks, which are fractionated and then converted into a range of useful products including chemicals, energy, and materials. Wood biorefineries are often

adaptations or extensions of traditional mills where wood pulp is produced to make paper, adding value to the operation. Industry and government will need to make more funding available to build an infrastructure of biorefineries. Countries such as Norway, Sweden, Austria, and Japan are already operating large-scale biorefineries.

SEE ALSO   *Biofuels, Liquid; Biofuels, Solid*

BIBLIOGRAPHY

Books

Cooke, Philip. *Growth Cultures: The Global Bioeconomy and Its Bioregions.* London: Routledge, 2007.

Copeman, Jacob. *Special Issue on Blood Donation, Bioeconomy, Culture.* London: SAGE, 2009.

Organisation for Economic Co-operation and Development and OECD International Futures Programme. *The Bioeconomy to 2030: Designing a Policy Agenda.* Paris: Organization for Economic Co-operation and Development, 2009.

Safferman, Steven. *Engineering the Bioeconomy.* Reston, VA: ASCE, 2009.

Sawaya, David. *Agricultural and Health Biotechnologies: Building Blocks of the Bioeconomy.* Paris: OECD, 2009.

Styhre, Alexander. *Venturing into the Bioeconomy: Professions, Innovation, Identity.* New York, NY: Palgrave Macmillan, 2011.

Washington State University, College of Agricultural, Human and Natural Resource Sciences and the Agricultural Research Center. *On the Road to a Bioeconomy.* Pullman, WA: College of Agricultural, Human and Natural Resource Sciences and the Agricultural Research Center, 2008.

Susan Aldridge

Bioethics

■ Introduction

As advances in technology and life sciences move forward, it is important not only to study those advances, but further to consider how those advances are undertaken and developed as well as their social, legal, and philosophical impacts. Bioethics has a wide scope, covering in various definitions the span of human existence by touching on topics as diverse as stem cell research and genetic engineering, end-of-life decisions, organ transplants, food manufacture, space research, and clean energy.

Whereas in the past concerns about the need for a bioethical standpoint were few, in the modern world, and with increasing frequency, the need for a collaborative, fluid, and progressive framework underlines many scientific, cultural, and philosophical affairs. Ethics has come to be offered as a degree in many universities; major corporations have established ethics boards; and many legislations worldwide are tailored to maintain ethical integrity even as they promote progress and innovation.

■ Historical Background and Scientific Foundations

Fritz Jahr (1895–1953), a German scientist and philosopher, coined the word bioethics in 1927, but the foundations of its principles were laid in ancient history with, for example, the writing of the Hippocratic Oath, an oath still taken by physicians. When the newly formed American Medical Association (AMA) met in Philadelphia in May 1847, one of the first orders of business was to formulate a comprehensive code of ethics, a living document that would evolve as medicine and the world did. The atrocities of World War II (1939–1945) led to the creation of the Nuremberg Code during the war crimes trials that followed. The Code directly addressed the topic of human experimentation, setting into place rules concerning the necessity of informed consent.

The rise of biotechnology complicated the field of bioethics with new, unprecedented and complex technologies and medical treatments. The 1970s saw one system of ethics developed by Tom Beauchamp and James Childress, "The Four Principles Approach." This was so named because of the four elements the authors deemed essential to ethical decision making: respecting individual autonomy and beliefs, helping society as a primary goal, avoiding harm or nonmalificence, and lastly, upholding the concept of justice as it pertained to the fair distribution of both benefits and burdens. In 1990 the Human Genome Project established its Ethical, Legal and Social Implications (ELSI) program, at that point the largest bioethics research body globally.

■ Impacts and Issues

In May 2010 Craig Venter (1946–), a pioneering American geneticist, and a team of researchers created a new semi-synthetic organism with a computer as one of its "parents," by inserting synthetic DNA into a enucleated bacterial cell. Venter envisions this as a step toward building synthetic organisms that can be controlled and directed to produce vaccines or absorb carbon dioxide. Later that year, U.S. President Barack Obama (1961–) directed the Presidential Commission for Bioethics to review the field as a response to Venter's experiment. The commission undertook open dialogue with a wide range of representatives of engineering, medical, secular, and faith-based groups along with representatives of society at large through organized public meetings. It finally advised a stance of prudent vigilance. This approach would allow for progress while maintaining the forward momentum of scientific progress, maximizing benefits while minimizing risks.

The critics who accused Venter of playing God however, reveal an insight into how genetic engineering can touch the nerves of deeply held spiritual beliefs. When the European Court of Justice banned patents based on embryonic research in October 2011, religious groups applauded the move. In the words of Peter Liese (1965–) of the EPP Christian Democrat group, "We are in favor of research and development in biotechnology, but human beings must not be destroyed, not even in the early stages of their development." Wherever the blueprint of life is being adapted or the fundamental ideas of what constitutes life are challenged, such as in the areas of cloning, stem cell research, transgenics, or genetic engineering, there has been and should continue to be serious, and respectful debate and understanding of all points of view.

Another charge, levied at biopharmaceutical companies in particular, has been that of using developing world test subjects for new drugs, a violation of the Nuremberg Code. For desperate patients in disease-ridden or war-torn areas, the appearance of a doctor and the promise of any medication at all, married with language and information deficits regarding risk, effectively destroys the notion of informed consent. In some cases there have been accusations of a kind of medical abandonment, as the drug supply chain to the patients has dried up after the test period, with focus turning to marketing in the United States and elsewhere.

WORDS TO KNOW

ETHICS: Moral principles that determine how individuals and society make decisions and behave.

GENETICALLY-MODIFIED ORGANISM (GMO): An organism in which the genetic material has been manipulated, usually in order to produce a more disease-resistant, higher yield version.

TRANSGENIC: An organism or cell into which the genes of another species have been combined.

In the case of the antibiotic Trovan, developed by the Pfizer company, the drug was tested on children in Nigeria during a meningitis epidemic in 1996 and resulted in the deaths of 11 children. An article in the *Washington Post* in 2000 maintained that Pfizer had bypassed the usual protocols for human testing of drugs, not having received signed permission from the

Visitors stand next to a poster which reads, "Save the girl child, finding out the gender of a fetus is a sin" at a hospital in New Delhi, India. Internet giants Google and Microsoft have pulled advertisements for sex selection products and services considered illegal in India after being threatened with legal action, activists said September 18, 2008. In 2008 India's Supreme Court asked the two companies, as well as Yahoo, to respond to a complaint that they were illegally advertising do-it-yourself home kits and expensive genetic techniques to find out an unborn baby's gender. The preference in parts of India is for male children, leading some parents to terminate pregnancies if there is a female fetus. © Pedro Ugarte/AFP/Getty Images.

Genetic Savings and Clone displays one of their cloned cats (R) and the original (L) at NextFest 2005 on June 24, 2005, in Chicago, Illinois. The company was in business from 2004 to 2006, cloning cats for $32,000 and working on technology to clone dogs. The company was criticized by animal rights activists for cloning animals for high fees when thousands of animals are euthanized daily, as well as by those who believe the cloning of animals is a slippery slope towards the cloning of humans. © *Scott Olson/Getty Images.*

children's guardians nor the Nigerian government before testing began. In fact, they were accused of having falsified a letter of permission from the Nigerian ethics board. Some reports indicated that earlier animal studies on dogs and rats by Pfizer had shown liver failure and joint damage. The civil case was settled in 2010 with a $75 million payout by Pfizer to settle the civil and criminal cases, although the company continues to deny any liability. A federal case for $6 billion in damages is ongoing. In the United States the drug was never licensed for use in children, and the European Union banned it completely in 1999.

In 2006 Ventria Biosciences, a U.S.-based biotechnology company, became embroiled in a controversial case involving a transgenic rice serum it had developed. The rice was genetically modified to produce two human antimicrobial proteins, lactoferrin and lysozyme, aimed at reducing the recovery period from diarrhea. It was tested on 140 children suffering from the illness at the Institute for Child Health and the Nutrition Research Institute, Lima, Peru. When the case came to light, there was a huge public outcry. On investigation it was claimed that there were no published records of prior animal testing, and some of the parents stated they had not been informed that their children were taking a

genetically-modified product nor the potential risks that it entailed. Some children on the trial serum subsequently suffered from allergic reactions, but more worrying is the idea of children in a developing country being used as "guinea pigs" for a treatment unapproved for testing by the U.S. Food and Drug Administration (FDA).

In the developed world there are strict regulations and restrictions on the use of GM crops, and ongoing discussion regarding their safety and the necessity of keeping the public informed, illustrated by the case in U.S. federal courts battling to have GM foods labeled. In the developing world, the situation is different; there is a lack of biosafety protocols to assess risks to health and environment. As stated by the U.S. think-tank International Food and Policy Research Institute (IFPRI), "… as many as 100 developing countries lack the technical and management capacity needed to review tests and monitor compliance [of GM crops]." In India the introduction of Bt brinjal, a genetically modified eggplant, has been stalled since 2009 amid controversies regarding safety, allegations of the sidestepping of local biodiversity laws by the biotechnology firms involved, and conflicts of interest within the Indian safety council, some of the members of which are said to have received payoffs from the GM company. Apart from the troubling mire of confusion within official channels, it also questions the ethics of continuing to grow and market a product if the necessary safeguards and education for farmers are not in place.

The scope of bioethics is very diverse, depending on the viewpoint of those using the term. However, the general principle of applying ethical and moral concerns to science will become increasingly important for all as emerging technologies continue to push the boundaries of medical and technological knowledge and understanding. It is not only the job of professional bioethicists, but also for governments, educational establishments, and the general public to engage in open discussion and decision making, taking responsibility for understanding the impacts new technology will have not only in the current time but in tomorrow's world. As the National Academy of Sciences points out, "Science and technology are essential social enterprises, but alone they only can indicate what can happen, not what should happen."

■ Primary Source Connection

In the April 25, 1953, issue of *Nature*, James Watson (1928–) and Francis Crick (1916–2004) published the first part of their seminal paper identifying the double helix structure of the DNA molecule. The work of Maurice Wilkins (1916–2004) and Rosalind Franklin (1920–1958) assisted Watson and Crick in their discovery. Separate articles by Wilkins and Franklin, which supported the discovery of Watson and Crick, also appeared in the same issue of *Nature*. Watson, Crick, and Wilkins were awarded the Nobel Prize in Physiology or Medicine in 1962. (Franklin, who had died in 1958, was ineligible for nomination for the Nobel Prize.) In 1990 the U.S. National Institutes of Health (NIH) appointed Watson as the head of the Human Genome Project. Watson left his position at the Human Genome Project in 1992 after disagreeing with the desire of NIH Director Bernadine Healy (1944–2011) to patent gene sequences discovered during the project.

Ethical, Legal, and Social Issues Related to the Human Genome Project

When James Watson and Francis Crick's experiments fifty years ago led to the discovery of the structure of deoxyribose nucleic acid (DNA), the science of genetics took a major step toward deciphering the genetic code. Watson and Crick's discoveries set in motion the late twentieth century interest in the biological underpinnings of heredity. Since then, the focus of scientific inquiry has moved from genetics—that is, the study of individual genes and the way traits pass between generations—to genomics—that is, the study of an organism's entire complement of DNA. Today, the scientific landscape is dominated by the Human Genome Project (HGP), an international research consortium that completed the first draft of the human genome in June 2000. The end product, that is, the complete sequence of all 3.1 billion base pairs of human DNA, reveals the biological blueprint for human life in the form of our genetic code.

The uses of information from the HGP are numerous and will continue to be so. Individuals, institutions (such as schools, the workplace, and government), and society will have to deal with situations in which some interests are advanced and some are impaired. When the interests of everyone cannot be advanced, or when some interests are advanced at the expense of others, whose interests ought to receive priority? Questions about "ought" or "should" are properly addressed by the disciplines of ethics and public policy. Because such questions will continue to arise from the HGP, it is essential to consider its ethical and public policy dimensions.

Since soon after its inception in the late 1980s, the HGP pledged three to five percent of its annual budget to support research about the societal implications of findings from the HGP. This research program is called ELSI, the acronym for the Ethical, Legal, and Social Implications of mapping and sequencing the human genome, and has explored numerous questions over the last dozen years.

In the earlier years of the HGP, questions concerning the management and use of genetic data dominated discussions. The concerns here regard ownership and the privacy of information, how to secure free and informed consent from donors, who should have access to such information, and how ought we to regulate such information. The importance of distinguishing between non-anonymous or personal genetic information and anonymous or aggregate genetic information has become clear. Where heightened concerns regarding the violations of individuals' rights arise in the former case, the justification of "big science" projects (in an era when there are clearly other needs) arise in the latter case. Related to these concerns is the profit motive: Who ought to benefit financially from the storage and use of genetic tissue, data, and information?

As the HGP produced more and more information that could be put to clinical use, discussions arose concerning the diagnosis and prediction of disorders, genetic testing and screening, gene therapy, and reproductive decision-making. When is it beneficial to patients to diagnose a disease for which there is no treatment, as in the case of Huntington disease? How can one ensure that health care professionals and patients understand how to interpret predictive genetic information, with its highly probabilistic nature? When ought genetic testing (for individuals) and screening (for populations) to be offered, and how ought one balance the benefits and risks of such information? Obligations to future generations become a topic when one asks whether couples who have a family history of a genetic disorder have an obligation to undergo genetic testing before conceiving offspring. The permissible boundaries of gene therapy received attention when people considered when it is justified to use somatic-cell gene therapy given that the therapy carries the possibility of unpredictable risks and whether it is ever permissible to use germ line therapy; which inevitably affects future generations. These and other issues focused on the hopes invested in what some refer to as the "genetic revolution" in science and medicine.

Some of these hopes for enhanced diagnostic and therapeutic means have not yet materialized. We still do not have widespread use of genetic tests and genetic therapies in the clinical setting. Those with mental illness continue to wait for genomics-based diagnostic tests for psychiatric disorders, such as schizophrenia and bipolar disorder, particularly for their off-spring. Those with rare diseases and those who have not responded well to a standard drug hope for the benefits of pharmacogenomics and the development of specialized drugs based on single nucleotide polymorphisms (SNP) research, which investigates the locations along the DNA chain where spelling differs from one person to the next. We continue to be faced with the diagnosis-treatment gap, as in the case of amyotropic lateral sclerosis (ALS). Perhaps initial promises were inflated. Perhaps we have become wiser about the complexity of the human genome and the central role of variation in genetic expression and the difficulty in developing a single test for a specific disease type. Increasingly; then, discussions have focused on how one ought to respond to the uncertainty of genetic information and the highly probabilistic nature of the information. Tied to these challenges are those concerning genetic illiteracy. One is well-advised to watch out for promises and interpret genetic information very carefully and spread such messages to others.

Then again, some promises and predictions have materialized. We have tests for cystic fibrosis, neurofibromatosis type 2 and certain breast cancers (BRCA 1), and many other diseases. There has been some success in somatic-cell gene therapy for individuals with severe combined immune deficiency or cystic fibrosis. On a more troublesome note, one continues to hear concerns about individuals losing health insurance or jobs because of genetic diagnoses, a phenomenon that has come to be known as "genetic discrimination"; that is, making choices on the basis of the irrelevant use of genetic information. Despite attempts in the 1990s by numerous local and state governments to prohibit such actions, concerns about violations of what has come to be known as "genetic privacy"—that is, the rights to control access to genetic information about oneself continue to surface. The public has begun to recognize that "genetic information is family information" and thus is calling for greater protection of samples stored in tissue banks and of information stored in clinical charts. The widespread use of computers in science and medicine, especially in "bioinformatics"—that is, for instance, the study of the collecting, sorting, and analyzing DNA and protein sequence information using computers and statistical techniques—accelerates concerns about violations of genetic privacy. Concerns about the management and use of genetic information remain center stage and focus our attention on the rights and welfare of individuals.

Perhaps we should not be surprised about the advances and challenges of the last fifty years. One of the strengths of science and medicine is its openness to new ideas and its use of checks and balances. When evidence supports an explanation (hypothesis), the explanation gains validity. When evidence does not support the explanation, the explanation must be abandoned and a new one sought. Science and medicine are durable but dynamic. For example, genetics could explain the general inheritance of Huntington disease for many years before it could explain the variability in the age of onset and the severity of its symptoms. With the use of molecular techniques for cloning and sequencing DNA, scientists discovered unstable trinucleotide repeats, thus providing the foundation of an explanation for some of the variability in the age of onset and severity seen in Huntington disease. This is such an example of the durable but dynamic character of science and medicine, which may also affect ELSI considerations.

As we move into the twenty-first century, we continue to worry about the management and use of genetic information as well as the boundaries and violations of genetic privacy in clinical and research settings. We are increasingly faced with concerns regarding gene therapy's permissible boundaries and its potential benefits and harms. As more tests and therapies are made available, the issue of fair or just access in the selection of recipients will receive greater attention. Outside clinical medicine, those in criminal justice seek clarification in how to use genetic information in the courts, especially in cases where it may assist in deciding the guilt or innocence of a defendant. These and other ELSI issues regarding who gets to know and how we should use genetic information remind us about the need to continue our reflections on privacy; confidentiality; rights, benefits and risks, and justice in the context of the management and use of genetic information. The last fifty years has provided us much to think about, particularly with regard to whether we should engage in that which we could.

Mary Ann G. Cutter

CUTTER, MARY ANN G. "ETHICAL, LEGAL, AND SOCIAL ISSUES RELATED TO THE HUMAN GENOME PROJECT." THE NATURAL SELECTION. *BIOETHICS*, SPRING 2003.

SEE ALSO *Asilomar Conference on Recombinant DNA;
Biodiversity; Biodiversity Agreements; Biotechnology
Organizations; Cell Therapy (Somatic Cell Therapy);
Cloning; Designer Babies; Designer Genes; Egg (Ovum)
Donors; GE-Free Food Rights; Gene Therapy; Genetic
Bill of Rights; Genetic Engineering; Genetic Privacy;
Genewatch; Germ Line Gene Therapy; Government
Regulations; HeLa Cell Line Case; Human Cloning;
Human Embryonic Stem Cell Debate; Human
Subject Protections; In Vitro Fertilization (IVF);
Indigenous Peoples and Traditional Resources Rights;
Individual Privacy Rights; Informed Consent; Life
Patents; Nuremberg Code (1948); Orphan Drugs;
Pharmacogenomics; Recombinant DNA Technology;
Synthetic Biology; Therapeutic Cloning; Transgenic
Animals; Transgenic Plants*

BIBLIOGRAPHY

Books

Bedau, Mark, and Emily C. Parke, eds. *The Ethics of Protocells: Moral and Social Implications of Creating Life in the Laboratory.* Cambridge, MA: MIT Press, 2009.

Bowring, Finn. *Science, Seeds and Cyborgs: Biotechnology and the Appropriation of Life.* New York: Verso, 2003.

Finegold, David. *Bioindustry Ethics.* Boston: Elsevier Academic Press, 2005.

Grace, Eric S. *Biotechnology Unzipped: Promises & Realities,* 2nd ed. Washington, DC: Joseph Henry Press, 2006.

Mitchell, C. Ben. *Biotechnology and the Human Good.* Washington, DC: Georgetown University Press, 2007.

Morris, Jonathan. *The Ethics of Biotechnology.* Philadelphia: Chelsea House Publishers, 2006.

Thompson, Paul B. *Food Biotechnology in Ethical Perspective,* 2nd ed. Dordrecht, The Netherlands: Springer, 2007.

Twine, Richard. *Animals as Biotechnology Ethics: Sustainability and Critical Animal Studies.* Washington, DC: Earthscan, 2010.

Web Sites

"The Ethical, Legal, and Social Implications (ELSI) Research Program." *National Human Genome Research Institute.* http://www.genome.gov/10001618 (accessed November 5, 2011).

McLean, Margaret R. "A Framework for Thinking Ethically about Human Biotechnology." *Santa Clara University.* http://www.scu.edu/ethics/publications/submitted/mclean/biotechframework.html (accessed November 5, 2011).

Kenneth Travis LaPensee

Biofuels, Gas

■ Introduction

Biofuels have emerged as one of the most successful biotechnology-based innovations. This family of fuel products includes solid, liquid, and gas fuels that are derived from processes that recover energy from sources that were previously regarded as waste and carried no commercial value. Biofuel gases include biogas, biosyngas, and biohydrogen.

Animal manure, food industry organic wastes, household wastes, sludge extracted from wastewater, and all forms of vegetable biomass comprise the major biomass substrates that are used in all biofuel gas production.

The common chemical thread that connects these fuels is that each utilizes carbon found in various forms in the source waste materials. Petroleum, coal, and natural gas are excluded from the biofuels definition, notwithstanding their carbon composition. It is generally accepted that these fossil fuels represent carbon sources that have been removed from the regular carbon cycle that is the basis for the carbon–oxygen exchange fundamental to all plant and animal life. The fundamental appeal of all biofuel processes is the contribution made to the restoration of the natural carbon cycle.

To be labeled as biofuel, 80 percent or more of all biofuel content must be derived from living organisms (such as algae) or metabolic by-products. Gas biofuels are used widely in industry, as commercial and residential heating products, and in chemical manufacture. There are few environmental drawbacks to the various technologies that transform waste products in biofuel gases.

■ Historical Background and Scientific Foundations

The existence and basic chemical composition of the gases that have become designated as biofuels have been known to science for several hundred years, albeit in the rudimentary understanding that various types of human and industrial waste are capable of producing gaseous by-products. The swamp gas generated in marshy areas composed primarily of methane, carbon dioxide, and sulfur dioxide is a prominent example of naturally occurring gases formed through the operation of the carbon cycle.

The commercial development of biofuel gases began with the use of the gases generated in the first London sewage treatment plants in the 1890s. The gas was captured and used as fuel for the city's streetlights. The next technological advance was the introduction of anaerobic digestion in sewage treatment that accelerated the biodegradation of wastes. After 1960 both China and India promoted micro-systems on small farms that used gases from collected animal manure as the fuel source for cooking and residential heating.

Modern biotechnology has elevated biofuel production from its small-scale, inefficient origins to a sophisticated suite of low environmental risk processes that capture and utilize what was previously wasted energy stored in forms that posed significant human health hazards. The 1973 global energy crisis that was driven by significant price increases for oil from the member nations of the Organization of the Petroleum Exporting Countries (OPEC) was a key stimulus to biofuel gas research. Alternatives to petroleum-based energy sources were perceived as an important way to contribute to greater energy security in all western economies.

The collection of gas from landfill sites for commercial application has been a worldwide energy initiative. Treatment of agricultural wastes to produce commercially viable biogas has been a particular success in Europe. In Germany, more than 400 biogas facilities have been constructed. By the year 2030 the current primary biogas sources, namely municipal sludge treatment, industrial wastewater purification by-products, and treatment of agricultural waste treatment, likely will be expanded as anaerobic digestion systems become more sophisticated. Greenhouse gas emissions, the source of significant environmental concerns associated with climate change and global warming, are reduced demonstrably where biofuel gases are manufactured.

The production of biofuel gases from any form of biodegradable waste involves some fundamental steps no matter what specific manufacturing process is employed. The first stage is pretreatment, in which pathogens and other undesirable components found in raw waste are removed. In the second stage, anaerobic digesters commence the breakdown of the waste into its constituent parts. Separation of waste into solids and liquids occurs at the third stage, in which the gases released from digestion are captured for further processing to render them suitable for commercial use. It is at this stage that hydrogen sulfide is removed to ensure the gases are suitable for efficient combustion in a wide range of engines, including boilers, gas turbines, and fuel cells. All of these biofuel gases can be burned with fewer pollutants than conventional gasoline and diesel fuels.

■ Impacts and Issues

As with most biofuels, the theoretical downside risks associated with biofuel gas production are very low, and the corresponding environmental benefits are high. It is the cost associated with the implementation of the necessary biotechnologies that has proved to be the rate-determining step in their universal application.

Each of the following serious environmental challenges reduces Earth's natural ability to cleanse itself thorough the fullest possible carbon cycle function: greenhouse gas emissions, water pollution through unsatisfactory waste management processes, increased global water consumption rates that are driven by population increases, reduced soil fertility, and deforestation. When combined, the cumulative effect of these impacts remains a monumental threat to global environmental quality. Biofuel gas and its associated technologies are crucial elements in the environmental countermeasures that likely will be pursued in all societies, particularly in the face of a global population that has surpassed 7 billion persons.

It is important to distinguish between the theoretical and practical aspects of biofuel gas production. The growth in the world's population has been accompanied by a significant change in how specific societies seek to produce food. The years since the early 1980s have seen a significant growth in the production of animal meats for human consumption. Industrial-scale hog, cattle, and chicken production facilities generate large amounts of animal waste; the grains that are cultivated as feed for these animals require large amounts of fertilizer to facilitate maximum crop yields. The nitrogen-based fertilizer runoff from these facilities tends to contribute to greater water and soil pollution. Aggressive waste treatment that promotes biofuel gas production is an essential element of the modern strategies necessary to combat these specific environmental impacts.

WORDS TO KNOW

ANAEROBIC DIGESTION: The process by which microorganisms convert biodegradable materials such as human waste to biofuel forms. Methanogens are the specific organisms used in the conversion of waste to methane gas and carbon dioxide.

CARBON CYCLE: The biogeochemical exchanges that ultimately result in the rotation of all existing carbon atoms (such as the carbon dioxide / oxygen exchange that occurs during photosynthesis in terrestrial plants and marine biota). The cycle includes carbon movement between the biosphere, atmosphere, oceans, and geosphere.

GASIFICATION: The conversion of a carbon-based solid (e.g., wood) into carbon monoxide, hydrogen, carbon dioxide, and methane gases. These products generally are known as syngas, used as a fuel or in the manufacture of chemicals such as the propellant dimethyl ether.

LIGNOCELLULOSIC MATERIALS: Any materials made from wood; when gasified, the components are used to manufacture various types of biosynthetic natural gas (syngas).

PATHOGEN: A disease-causing organism.

Biofuel gases displace the use of fossil fuels that otherwise contribute to greater levels of air pollution and the depletion of non-renewable energy resources. As importantly, biofuel gases contribute to greater carbon cycle efficiency, as carbon-rich wastes are restored to the environment. An additional residual benefit of biofuel gas production is the removal of coliform bacteria, as well as the anaerobic destruction of various other pathogens and internal parasites transmitted through contact with raw waste products.

The few appreciable risks associated with biofuel gas production are largely beyond the control of the manufacturer. Waste solids often are required to be stockpiled for future processing; the stockpiles or operating sewage lagoons are sources of unpleasant odors and insect pests. Excessive rainfall may contribute to lagoon or stockpile runoff that adversely impacts adjoining soil and water quality.

SEE ALSO *Advanced Biomass Cookstoves; Agricultural Biotechnology; Biochemical Engineering; Biofuels, Liquid; Biofuels, Solid; Environmental Biotechnology; Solid Waste Treatment; Waste Gas Treatment; Wastewater Treatment*

BIBLIOGRAPHY

Books

Abbasi, Tasneem, S. M. Tauseef, and S. A. Abbasi. *Biogas Energy.* New York: Springer, 2011.

Deublein, Dieter, and Angelika Steinhauser. *Biogas from Waste and Renewable Resources: An Introduction*, 2nd ed. Weinheim, Germany: Wiley-VCH, 2011.

Hoogendoorn, Alwin, and Han van Kasteren, eds. *Transportation Biofuels: Novel Pathways for the Production of Ethanol, Biogas and Biodiesel.* Cambridge, UK: RSC Publishing, 2011.

Web Sites

"Anaerobic Digestion: What Is Biogas?" *National Non-Food Crop Center (UK).* http://www.biogas-info.co.uk/index.php/biogas-qa.html (accessed October 27, 2011).

"Biogas." *Ashden Awards for Sustainable Energy.* http://www.ashdenawards.org/biogas (accessed October 27, 2011).

Bryan Thomas Davies

Biofuels, Liquid

Introduction

Liquid biofuels have achieved the greatest degree of prominence among the three environmentally friendly fuel sources derived from biomass. This prominence has proven to be a two-edged sword, because the reduced environmental impacts associated with clean burning ethanol, biodiesel, and bioethers are often contrasted with the cumulative impacts of the crop cultivation and manufacturing costs associated with the various components of these fuels. Ethanol produced from corn fermentation emerged as an attractive alternative to petroleum products in the aftermath of the energy crisis precipitated by the rise in global oil prices in the early 1970s. As biotechnological processes have become more efficient, the easy transportability and relatively high energy yields from liquid biofuels remain important incentives to further research.

The advanced biofuels that have been developed from sources such as cellulose and algae since the beginning of the twenty-first century are the tantalizing new frontier for liquid biofuel applications. Myco-diesel and bioether formulations are examples of conventional chemical compounds that carry the promise of future commercial viability while producing minimal adverse environmental impacts. These technological advances face a crucial present day challenge—the cost of research and development that takes these environmentally sound alternatives past the accepted price points for commercial and industrial uses. In many respects, it is the contest between reduced environmental impact and increased costs associated with their implementation that constitutes the barrier to mainstream acceptance of liquid biofuels.

Historical Background and Scientific Foundations

Of the three biofuel types, liquid varieties have the greatest commercial advantages. Liquids are easier to transport than gases or solids, and they have a higher energy density than the gaseous forms. Liquid fuels are safer to handle than gases, which require specialized equipment to maintain pressure, and liquids tend to be less prone to explosion or other accidents. They also have a greater range of uses than solid biofuel products.

The first liquid biofuels, known in the industry as first generation or conventional biofuels, were manufactured from readily available sugar, starches as extracted from corn and other grains, and vegetable oils (used and raw). The World Bank began programs to fund biofuel projects in various developing nations, including Brazil, in 1983. The potential importance of biofuels to both global energy security and to the environment through reduced greenhouse gas emissions remains a powerful research incentive in this field.

Brazil, using sugar cane, and the United States, using corn, are the world's leading ethanol producers. Germany utilizes rapeseed oil to occupy the top biodiesel position. Brazil historically has been the most efficient ethanol producer. As with all liquid biofuel production, the global price of the crop used to produce fuel will have the most prominent impact on ultimate fuel prices. When world sugar prices surged by 25 percent in the first decade of the twenty-first century, American ethanol producing costs fell as corn prices declined during the same period.

All liquid biofuels are produced through the fermentation of grains that have been processed into liquid form. Fermentation is an anaerobic (without oxygen) process that breaks down the natural organic compounds in the plant through the action of enzymes. The fermentation by-products are ethanol, carbon dioxide, and water. In most instances the fermentation process is environmentally neutral because the carbon dioxide is captured for further commercial application and not simply released onto the environment.

Impacts and Issues

Ethanol and biodiesel have emerged as the two most important liquid biofuels. The environmental appeal of these products often carries a hidden price tag: The reduced

WORDS TO KNOW

ADVANCED BIOFUELS (SECOND GENERATION BIOFUELS): Liquid fuels manufactured from sustainable sources that do not impact the crops otherwise available for human food or feedstock. These fuels include alcohols such as ethanol and butanol converted from biomass.

BIODIESEL: Liquid fuel produced from refined animal fats and vegetable oils.

BIOETHERS: Ethers produced from biomass sources that are used as additives in conventional fuels such as gasoline to enhance engine performance and reduce carbon monoxide emissions. Methyl tert-butyl ether (MTBE) is a commonly used bioether.

MYCO-DIESEL: Based on the interaction of the fungus species *Gliocladium roseum* and cellulose materials to produce hydrocarbon molecules that mimic those found in conventional diesel fuels.

environmental impacts associated with the end use of these otherwise convenient biofuels tend to disguise the economic fragility of this industry. Ethanol is the culmination of a complex series of agricultural and industrial processes that each consumes vast amounts of energy. The crops must be planted, cultivated, and sprayed with various chemical agents to prevent insect infestations and fungal diseases. The chemicals are largely manufactured from petroleum, a resource whose extraction and processing has its own inherent energy requirements. Once harvested, the crops must be transported to an ethanol processing facility where each step in ethanol production has its own significant energy demand; non-renewable natural gas is a common fuel source in ethanol plants, where industrial fermenters, drying devices, and secondary processors are employed to produce fuel. The commercial product is ultimately shipped to retail outlets along highways and railways, using transportation systems that require vehicles that are primarily powered by conventional fossil fuel engines.

Ethanol has a lower net energy content than either gasoline or conventional diesel; with all of the energy required to produce ethanol added to the equation, the

Raw sugarcane is offered for sale in Kerala, India. India, like Brazil, is increasingly diverting its sugarcane crop towards the production of the biofuel ethanol. Critics of using food crops for biofuels state that such practices raise food costs and promote monocultural crops, endangering the environment and food supplies in a world in which the population is growing.   © *Lucy Baldwin/Shutterstock.com.*

economic value achieved through ethanol production is questionable. Further, the environmental impacts of the petroleum-based products that contribute to commercial ethanol production and distribution appear to cast doubt on claims that ethanol is a "clean" energy alternative. Further, where productive lands are removed from food production to produce raw materials for ethanol, there are price and supply impacts felt by food consumers. Arable land is generally regarded as a finite resource. Where crop lands are converted to use as a source for ethanol, either additional land must be converted to agriculture through the deforestation of existing natural regions (correspondingly reducing the ability of Earth to process carbon dioxide into oxygen) or methods must be found to improve crop yields.

Government policies developed in the United States and other Western nations such as Germany (where rapeseed is the primary source for ethanol) to encourage crop cultivation for ethanol production are also a controversial issue in this field. Policies that mandate biofuel use (usually as a fixed percentage contribution to a conventional fuel such as gasoline), with programs that include producer subsidies and reduced fuel taxes to encourage biofuel production are examples. Empirical studies confirm that liquid biofuels have achieved only marginal economic viability, and only then with the significant government assistance provided through these various subsidies and incentives. However, the environmental appeal and the ever-present Western world concerns regarding energy security are likely to continue to drive future research. The environmental promise of all liquid biofuels remains sufficiently attractive to industry and environmentalists alike that economic viability concerns are regarded more as hurdles than as permanent barriers to mainstream acceptance.

■ Primary Source Connection

This article details how one man from West Virginia converts frying oil into biodiesel for use in various motorized vehicles such as cars, trucks, motorcycles, and farm equipment. "Uncle Dave" gives a detailed explanation about the materials, steps, and equipment necessary to make a product that begins as a tool used for cooking food, and ends as a fuel used for work and travel. Aspects of the chemistry involved in the conversion process are discussed, along with the ingredients used in his successful recipe. The article also reveals the costs incurred and the challenges faced in the process of obtaining the final product.

Lyle Rudensey refines homemade biodiesel from used cooking oil in his garage in Seattle, Washington. Because of the relative ease of making biodiesel from existing oils, people have moved towards creating their own fuel rather than buying petroleum-based gasoline for their vehicles. © *AP Images/Ted S. Warren.*

Home Brewed: How to Turn Grease into Biodiesel (for Cheap!)

Dave Hubbard turns grease from restaurants into fuel for his black Jetta TDI, a BMW motorcycle, bulldozers, tractors for five local farms and a tree nursery, and his neighbors' pickups—all for the low cost of about 50 cents a gallon. The process is basic chemistry and can be done by any DIYer with a steady supply of restaurant oil, a strong winch, and a willingness to get greasy. Here's how he does it.

For Dave Hubbard, the process of making biodiesel begins in the gloaming under an old walnut tree in West Virginia, where behind a rural tavern a week's worth of frying oil sits in a 55-gal drum. I meet my uncle here on a Sunday evening so that I can see how the oil gets from the drum to the tank of his black Jetta—plus those of a motorcycle, bulldozers, and tractors for five local farms and a tree nursery—all for the low cost of 50 cents a gallon.

With the full drum secured to a homemade steel hitch on Uncle Dave's car, and an empty in its place under the tree, we drive the few miles back to his workshop. It's a 100-ft long space filled with equipment—at least four drill presses, an old still for experimenting with ethanol (built from plans purchased on eBay) and, under a sheet, a BMW motorcycle he converted to run on biodiesel. There are makeshift devices as well—practical solutions to an engineer's immediate needs, crafted to last long and cost little. To maneuver the 365-pound drum, for example, Uncle Dave uses a $120 winch rigged to a steel I-beam—which, in a past life, was part of a chairlift for carrying the disabled up stairs.

Uncle Dave pries the lid off the drum to show me what's inside: The oil is amber and smells like chicken (not surprising since the tavern is known for its wings). It's also still hot from the fryer, where it was heated to 350 F. Biodiesel is formed from a reaction between vegetable oil and an alcohol like methanol. "The minimum temperature for the reaction is 120 F, but I can't go above 148 F—the boiling point of methanol," Uncle Dave says. To avoid expending energy bringing cold oil up to that range, he uses the winch to pour the fresh, hot oil into an older drum, then mixes the two with a long metal paddle.

The Chemistry

"Now I need to know how burned it is," Uncle Dave says. "Free fatty acids build up in the oil when you burn stuff over and over again." He pulls a 2-liter pop bottle off the shelf that is filled with fryer oil from a bar near the local college—it looks like motor oil. "I get this once a year because the cook uses the same oil over and over," he says. "I wouldn't want to eat anything they cooked in there."

When making biodiesel, you want to convert these free fatty acids to soap (glycerin) with a base like sodium hydroxide (lye)—but you need to put in a little extra to act as a catalyst for the virgin bean oil, too. The oil from the college bar is burned so far beyond recognition that it's not worth the time and materials it would require to convert it—Uncle Dave burns it in a furnace for heat. A chemistry experiment will tell us exactly how much catalyst we need to add to the good stuff.

We walk over to an old wooden workbench covered with bottles of various shapes and sizes. "If it were water, I'd need to know the pH. But you can't take the pH of oil," Uncle Dave says. Instead, we have to go by color. He takes a bottle of isopropyl alcohol, in the form of Iso Heet, and measures out 10 milliliters. "I don't buy anything fancy," he says. "My other special stuff here is turmeric from Kroger's. It's a hot pepper, so it's acidic." He dips the end of a toothpick into the spice and swirls it in the alcohol, which turns yellow.

Then Uncle Dave pulls a 1-liter bottle of distilled water off the shelf. He adds a milligram of sodium hydroxide, which makes the water slightly basic. "This is my titration fluid," he says. Using an eyedropper, he drips the titration fluid into the alcohol mixture until it just barely turns red. "That's telling me it's back to neutral."

This neutral mixture will help determine the acidity of the oil. "This cook is very precise," Uncle Dave says. "He fries the same amount of stuff all the time and changes it every week, so I can almost guarantee what it's going to be." He takes a milliliter of the cooking oil and mixes it into the alcohol, which turns from red to yellow again—back to acid. "So now I need to know what it takes to balance it back to neutral with my titration fluid," he says. Uncle Dave adds a milliliter of the distilled water and sodium hydroxide. Still yellow. He adds a half—back to red, just as he suspected.

The Recipe

Making biodiesel is like following a recipe, Uncle Dave tells me. The drum contains 45 gal of oil, and the reaction requires 20 percent methanol—or 9 gal. The magic number for the catalyst is 6.5, because it takes 5 milliliters to return 1 liter of pure oil to neutral, plus—as our chemistry experiment indicated—1.5 milliliters to counteract the amount of fatty acids in the burned product. Multiply all that and convert it to grams, and Uncle Dave needs to add 1100 grams of sodium hydroxide to make the reaction work.

Okay, enough with the math—back to the machinery. Uncle Dave pours the sodium hydroxide into the methanol and turns on a converted drill press as a mixer. "If you put your hand on the side of the container, you can feel the heat," he says. "It's an exothermic reaction. It's going to get really hot in there in a second."

While the methoxide is blending, he turns back to the oil. He measures the temperature to ensure it's 120 F, and then uses the winch to lift and pour the oil into another 55-gal drum. A window screen stretched over the top filters out any chunks—I think I spot what used to be a French fry.

Uncle Dave activates a 1-in. electric transfer pump, and the filtered oil flows into a used 300-gal insulated bulk milk tank sitting nearby. He activates another pump and the methoxide is transferred to the milk tank, too. The tank is equipped with its own mixing motor with two stainless-steel propellers attached. Uncle Dave closes the top and sets a timer. For the next 45 minutes the propellers will blend the oil and methoxide, sustaining a reaction.

"When the recipe is right, I'll end up with about 40 gal of biodiesel and 5 gal of glycerin," Uncle Dave says, "but that depends on how much the oil has been burned." He'll let the mixture sit in the tank overnight, and the dense glycerin will fall to the bottom where he can drain it off with a valve. What's left on top is the biodiesel, which he'll wash to remove any residual soap or methanol.

The Cost

All told, the cost of materials, energy for heat and power, and odds and ends like gloves and paper towels, still only adds up to about 50 cents per gallon, Uncle Dave says. He sells it to friends and neighbors for $2 a gallon— a transaction that helps keep some local farmers in business. Steve Paul, who raises beef cattle, says he would otherwise spend $15,000 a year on fuel. Uncle Dave provides him with 1200 gallons at the beginning of each summer and Steve pays him when he sells his calves in October.

While compact, off-the-shelf biodiesel units may be convenient, Uncle Dave says, his system allows a lot more room for experimentation. "I can have a bad batch and set it off to the side, and put another drum in its place and check each thing as I go," he says, "I can probably confine all this into one package to take up less space, but if I've got a problem with it I'm locked up."

The real challenge now, he says, is just getting enough oil. In the eight years that he's been experimenting with biodiesel, people have begun stealing it from behind restaurants—when the owners used to have to pay companies to take it away. Uncle Dave now gets his feedstock from several establishments, which has introduced him to yet another advantage of home-brewed fuel: Scavenging for fryer oil provides excellent insight into where to buy wings.

Jennifer Bogo

BOGO, JENNIFER. "HOME BREWED: HOW TO TURN GREASE INTO BIODIESEL (FOR CHEAP!)." *POPULAR MECHANICS,* FEBRUARY 5, 2009.

SEE ALSO *Agricultural Biotechnology; Agrobacterium; Advanced Biomass Cookstoves; Biofuels, Solid; Environmental Biotechnology; Ethanol; Soybeans*

BIBLIOGRAPHY

Books

Goettemoeller, Jeffrey, and Adrian Goettemoeller. *Sustainable Ethanol: Biofuels, Biorefineries, Cellulosic Biomass, Flex-Fuel Vehicles, and Sustainable Farming for Energy Independence.* Maryville, MO: Prairie Oak, 2007.

Langeveld, Hans, Marieke Meeusen, and Johan Sanders. *The Biobased Economy: Biofuels, Materials and Chemicals in the Post-Oil Era.* Washington, DC: Earthscan, 2010.

Levy, Sarah L., ed. *Biofuels, Biorefinery & Renewable Energy: Issues & Developments.* Hauppauge, NY: Nova Science, 2011.

Sexton, Steven, and David Zilberman. *How Agricultural Biotechnology Boosts Food Supply and Accomodates Biofuels.* Cambridge, MA: National Bureau of Economic Research, 2011.

Periodicals

Cressey, Daniel. "Advanced Biofuels Face an Uncertain Future." *Nature* 452, no. 7188 (April 10, 2008): 670–671.

Raghu, S., et al. "Adding Biofuels to the Invasive Species Fire?" *Science* 313, no. 5794 (September 22, 2006): 1742.

Scharlemann, Jörn P. W., and William F. Laurance. "How Green Are Biofuels?" *Science* 319, no. 5859 (January 4, 2008): 43–44.

Searchinger, Timothy. "Use of U.S. Croplands for Biofuels Increases Greenhouse Gases through Emissions from Land-Use Change." *Science* 319, no. 5867 (February 29, 2008): 1598–1600.

Web Sites

"Biofuels: The Original Car Fuel." *National Geographic Society.* http://environment. nationalgeographic.com/environment/global-warming/biofuel-profile (accessed October 23, 2011).

"Generate Your Own Energy." *Energy Savings Trust.* http://www.energysavingtrust.org.uk/generate_ your_own_energy/how_renewable_energy_works (accessed October 24, 2011).

Bryan Thomas Davies

Biofuels, Solid

■ Introduction

Solid fuels are the oldest human fuel source, and they remain the most common fuel consumed in human societies. Approximately 3 billion of the world's 7 billion people depend to some degree on the direct combustion of biomass solids for food preparation or warmth. The indirect contribution made by solid biofuels such as biochar to electrical energy production and large-scale commercial heating systems indirectly improves the lives of additional hundreds of millions of people across the industrialized world.

Solid biofuels arguably have attracted the least amount of public interest among the three biofuel family members that have emerged as clean energy alternatives since the early 1980s. Conversely, solid biofuels have proven to be the most commercially viable bioenergy applications, because they are far less dependent on government subsidies or other economic incentives that have been essential to the development of liquid biofuels such as ethanol.

Wood, sawdust, and the waste materials collected from the crop plants used to produce liquid biofuels are well known biomass forms. A broad biomass definition is any plant or vegetable matter that is suitable for use as fuel, or any such matter that can be converted into a useable fuel prior to combustion. Animal wastes, sulphates generated in paper manufacturing, and any other solid biomass is capable of conversion into solid fuel. Biochar, the charcoal produced through a combination of biochemical processes, has become a key industrial fuel used primarily in power plants and large-scale operations that require large amounts of on-site energy production. The research conducted with terra preta soils confirms the additional benefits of charcoal as an effective fertilizer and soil nutrient.

The current research and development of new and more efficient solid biofuels parallels the work that has been undertaken with respect to liquid and gaseous biofuel forms. As the global population continues its inexorable rise past the 7-billion-person mark, increased demands for energy will continue to drive the pursuit of more accessible and cost-effective solid biofuel applications.

WORDS TO KNOW

BIOMASS: The biodegradable component found in all agricultural products, their waste and residues, vegetation, forest products, and industrial and municipal waste.

CARBON SEQUESTRATION: The intentional removal of carbon from the atmosphere by various geoengineering methods, including the separation of carbon dioxide from industrial flue gases for permanent underground storage. Carbon sequestration alters the function of the carbon cycle, including the biogeochemical exchanges that involve carbon atoms, such as the carbon dioxide / oxygen exchange in photosynthesis that occurs in terrestrial plants and marine biota.

PYROLYSIS: Degradation of a material into its chemical components achieved through the application of intense heat. Pyrolysis is the first stage in the three-part thermochemical decomposition process that continues to gasification and the eventual combustion of biomass material. A flaming match is a common pyrolysis example. A kiln is the typical enclosed, airtight structure used in pyrolysis.

TERRA PRETA: Very fertile anthropogenic soils first identified in the Amazon basin, dated to as far as 9000 BC. The soils were improved through charcoal content that resulted from the use of cooking fires by the indigenous peoples of the region, because charcoal contains high levels of nitrogen, potassium, zinc, and phosphorous.

A tractor moves past piles of biomass fuel used to power the Wheelabrator generation plant in Anderson, California. At the plant, semitrailers line up three at a time around the clock during busy periods. Giant hydraulic lifts effortlessly tip the trucks on end, spilling out their cargo of wood chips, nut hulls, and other forest and agricultural waste onto giant conveyer belts that eventually feed immense steam-fed generators. © *AP Images/Rich Pedroncelli.*

■ Historical Background and Scientific Foundations

The use of wood, vegetation, and animal wastes for fuel is a biomass application as old as humankind itself. There is an elemental connection between heat produced by the burning materials and sustainable human life. The use of biomass fuels has also tended to enrich the soils where fuel residues have been deposited, such as the terra preta found throughout the Amazon basin.

As the term suggests, biomass can take many forms. The residues generated by all forestry operations (including sawmill wood wastes and those produced in pulp and paper manufacturing); agricultural byproducts such as straw, cattle manure, and poultry wastes; and the discarded materials from construction and furniture manufacture are examples of this diverse and readily identifiable fuel source. It is difficult to quantify the exact amount of biomass solids used to generate heat on a global basis. Wood or other vegetation is used to provide cookstove heat for approximately 2 billion persons worldwide.

Charcoal is generally defined as any natural organic material produced from biomass that has been subjected to pyrolysis. The black carbon byproduct is the generic name for the various charcoal derivatives such as soot produced when biomass is consumed.

Biochar is an complex and stable carbon form that is created by heating biomass under controlled conditions to temperatures between 662° and 1112°F (350° and 600°C) in a virtually anaerobic environment. Pyrolysis is the process in which the relative amounts of biochar, liquids, and gases derived from biomass will vary in relation to process temperature. Lower operating temperatures (from 752° to 932°F or 400° to 500°C) tend to increase biochar yields. Greater amounts of gases and liquids are produced through pyrolysis when the temperatures reach 1292°F (700°C); the resulting combination of hydrogen, carbon dioxide, and carbon monoxide is the syngas frequently used in industry to power turbines used in the generation of electricity.

Biochar is also a proven means to increase soil fertility. In addition, significant environmental research concerning possible ways to counter the accumulation of greenhouse

Dried hay and plant stalks lie covered at a local bioenergy plant in Juehnde, Germany. Juehnde is the first village in Germany to become energy self-sufficient by building its own bioenergy electrical plant, which uses wood chips, cow dung, and plant remains gathered from the community to create electricity and heat. Interest in bioenergy has grown as a means to cut greenhouse gas emissions. © *Sean Gallup/Getty Images.*

gases in the Earth's atmosphere has focused on biochar and its ability to trap carbon dioxide indefinitely. The use of biochar to improve soil quality may provide increased carbon sequestration as a secondary benefit. Research also has revealed that because the charcoal found in Amazonian terra preta has remained stable in the soil for as long as several thousand years, wide-scale charcoal production offers attractive environmental possibilities.

■ Impacts and Issues

Although biofuel solids are multi-purpose and tend to create fewer environmental impacts and safety concerns in handling and storage than either liquid or gaseous biofuels, the greater commercial use of these fuels poses a number of significant logistical difficulties. The most prominent of these questions is that these bulky,

large-volume fuels are more expensive to transport than gases or liquids; in most cases, biofuel solids for commercial and industrial applications must be stored immediately next to the desired power generation facility.

Vigorous humanitarian efforts have been undertaken in the developing world to replace the traditional three-cornered cookstove powered by wood that provides cooking heat for approximately 2 billion people daily with charcoal powered units that would also contribute to soil remediation. The traditional stoves are a major source of bronchial disorders and generally reduced air quality. Cleaner-burning charcoal stoves are regarded as an important and environmentally preferred substitute.

Solid biofuels are unlikely to displace liquid and gas forms in any other residential settings, except where charcoal is used for cooking and heating purposes. The equipment necessary to manufacture biochar on a large scale is expensive and a factor likely to restrict the compound to its current industrial applications. The adaptability of biochar at high temperatures to liquid gas production will make it doubly attractive in coming years as population growth continues to exert pressure on scientists to discover other efficiencies in fuel production. These research directions will also be affected profoundly by factors such as government policies that influence commercial viability (subsidies and tax incentives), the relative cost of other fuel alternatives (conventional and biotechnological), and the state of global energy security.

SEE ALSO *Advanced Biomass Cookstoves; Agricultural Biotechnology; Agrobacterium; Biofuels, Gas; Biofuels, Liquid; Environmental Biotechnology; Ethanol; Soybeans*

BIBLIOGRAPHY

Books

Goettemoeller, Jeffrey, and Adrian Goettemoeller. *Sustainable Ethanol: Biofuels, Biorefineries, Cellulosic Biomass, Flex-Fuel Vehicles, and Sustainable Farming for Energy Independence.* Maryville, MO: Prairie Oak, 2007.

Hood, Elizabeth E., Peter Nelson, and Randall Worth Powell, eds. *Plant Biomass Conversion.* Ames, Iowa: Wiley-Blackwell, 2011.

Langeveld, Hans, Marieke Meeusen, and Johan Sanders, eds. *The Biobased Economy: Biofuels, Materials and Chemicals in the Post-Oil Era.* Washington, DC: Earthscan, 2010.

Levy, Sarah L., ed. *Biofuels, Biorefinery & Renewable Energy: Issues & Developments.* Hauppauge, NY: Nova Science, 2011.

Periodicals

Cressey, Daniel. "Advanced Biofuels Face an Uncertain Future." *Nature* 452, no. 7188 (April 10, 2008): 670–671.

Scharlemann, Jörn P. W., and William F. Laurance. "How Green Are Biofuels?" *Science* 319, no. 5859 (January 4, 2008): 43–44.

Web Sites

"Generate Your Own Energy." *Energy Savings Trust.* http://www.energysavingtrust.org.uk/ generate_your_own_energy/how_renewable_ energy_works (accessed October 23, 2011).

"Survey of Energy Resources 2007: Biomass for Electricity Generation." *World Energy Council.* http:// www.worldenergy.org/publications/survey_of_ energy_resources_2007/bioenergy/713.asp (accessed October 23, 2011).

Bryan Thomas Davies

Bioglass

■ Introduction

Bioglass is the generic name for the family of bioactive glass compositions that evolved from efforts to better understand how to rehabilitate injured Vietnam War (1955–1975) veterans. Originally conceived as a means of repairing bone damaged in combat, Bioglass compositions now are used widely in a variety of reconstructive surgical procedures and soft tissue repair. When metal or plastic prosthetics are implanted into the body, these devices tend to be rejected by the surrounding host tissue. Bioglass is designed to reduce the possibility of rejection in hard tissue regeneration procedures.

Bioglass is a chemical compound assembled from phosphorus, calcium salts, sodium salts, and silicates, the essential elements and compounds found in mineralized human bone. When Bioglass is introduced to the surface of a damaged bone, its components tend to promote higher levels of osteoblast formation on the bone surface. The low rate of bioactivity as Bioglass becomes attached to the bone surface contributes to a greatly reduced rejection risk. The most common Bioglass formulation used to repair human bone is Bioglass 45S5.

Additional Bioglass compositions have been developed to promote soft tissue regeneration when the nature of the wound or other injury carries a high risk of infection for the patient. Nanotechnology research involving compounds such as lithium borate has yielded a new generation of infection resistant Bioglass products that can be directly applied to the damaged tissue to promote quicker and infection-free recovery.

While Bioglass is the generic term used to describe these bioactive compounds, Bioglass is also the registered trademark owned by the University of Florida. The university was the site of the initial Bioglass research studies that began under the leadership of Professor Larry Hench in 1969.

WORDS TO KNOW

BIOACTIVITY: Bioactivity is the term that describes the specific effect observed when living animal or plant tissue is exposed to a particular substance.

CERAVITAL: The registered trademark for the first Bioglass material that was developed for clinical applications in 1973. Ceravital remains the most commonly used Bioglass material in middle ear reconstructive surgery.

NANOFIBERS: Fibers that have a diameter of less than 1,000 nm; a nanometer is the unit of measure defined as 1 billionth of 1 meter.

RFID: This acronym is generally applied to any Radio Frequency Identification process, the identity tracking systems that are used to transmit identifying serial numbers by radio signal for persons, animals, or objects.

■ Historical Background and Scientific Foundations

The high amputation rate documented by American military surgeons in the Vietnam War provided the impetus for the research that ultimately led to the development of Bioglass. Military surgeons could not use the metal or plastic parts then available for reconstructive surgery given the likely rejection of these foreign objects by the human body shortly after surgery. Amputation of the limb was often the only sensible option. The challenge to science provided by this conflict was whether a material could be developed to aid in hard tissue restoration that would form a living bond with the host tissue. The work of the research team led by University of Florida engineering professor Larry Hench (1938–) that commenced in 1969 provided the basis for the various Bioglass applications in use throughout the world today.

The Hench research proceeded on the hypothesis that hydrated calcium phosphate, a compound essential to bone formation, could be induced to form a living layer between the materials used to repair the damaged structure and the bone and eliminate the rejection risk. Hench successfully proved this hypothesis: By 1985, Bioglass 45S5 (marketed under the trademark Ceravital) was used as the compound in prosthetics implanted to replace damaged ossicle bones, the small, three-part middle ear structure essential to the quality of human hearing.

Unlike conventional glasses that are used in windows or cookware, Bioglass generally has a much lower percentage of silicates. It also has a correspondingly higher quantity of sodium, calcium, and phosphorus.

The development of Bioglass 45S5 was premised on the theory that the compound would only facilitate bone repair. In the mid 1980s a series of research developments spurred the development of soft tissue regeneration Bioglass products. Bioglass is now widely used in bone repair, dental surgery that requires implants, and the treatment of serious wounds and soft tissue injuries.

■ Impacts and Issues

The development of Bioglass from concept to commercial application has spanned 40 years. Bioglass continues to be regarded as a fruitful research area, given the advances in nanotechnology that enable different elements and compounds to be introduced to known Bioglass formulations to alter composition and performance. An

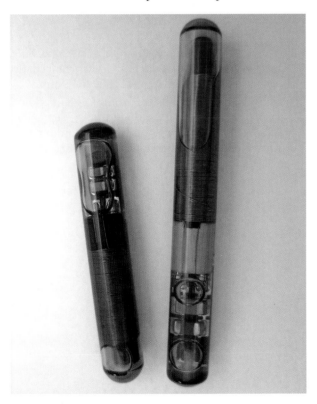

An implantable RFID tag made with bioglass.   © *Albert Lozano/ Shutterstock.com*

example is the use of borate glass nanofibers to promote soft tissue regeneration. Borate glass nanofibers that are between 300 nanometers to 5 micrometers in diameter are used to construct a mesh that is fashioned into a healing scaffold. The scaffold is designed to mimic the clotting process that is the first stage of all soft tissue repair.

An important current Bioglass research focus is the ongoing pursuit of stronger hard tissue regeneration formulations. The most significant drawback observed in most commercial Bioglass compounds is their general mechanical weakness and the relatively low resistance to fracture. Bioglass is generally used to repair non-load bearing bone structures, such as the inner ear, where the relative weakness of the bond created between Bioglass and the bone is of little concern. Research has sought to improve fracture resistance through the alteration of the ratio between calcium oxide and silicon dioxide used in Bioglass 45S5. This adjustment has generated research results that suggest that osteoblast production on the surface of the damaged bone is increased through the use of Bioglass 45S5; the healing process is both faster and more complete.

Bioglass has also made an important contribution to the development of the microchip and RFID transponder technology that is essential to the tracking of objects, animals, and humans. The tracking devices are encased in a shell manufactured from Bioglass 8625, a specialized compound that permits these tiny devices to transmit and receive radio signals without the compound binding to adjacent tissue, skin, or bone. The Bioglass exterior over time becomes coated with calcium, and it remains a stable communications portal with no known adverse physiological effects for the carrier.

SEE ALSO *Antirejection Drugs; Bioplastics (Bacterial Polymers); Bone Substitutes*

BIBLIOGRAPHY

Books

Deng, Hong-wen, and Yao-zhong Liu. *Current Topics in Bone Biology.* Hackensack, NJ: World Scientific, 2005.

Hall, Brian Keith. *Bones and Cartilage: Developmental and Evolutionary Skeletal Biology.* San Diego, CA: Elsevier Academic Press, 2005.

Smith, Douglas G., John W. Michael, and John H. Bowker. *Atlas of Amputations and Limb Deficiencies: Surgical, Prosthetic, and Rehabilitation Principles,* 3rd ed. Rosemont, IL: American Academy of Orthopaedic Surgeons, 2004.

Periodicals

Madan, Natasha. "Tooth-remineralization Using Bioactive Glass – A Novel Approach." *Journal of Academy of Advanced Dental Research* 2, no. 2 (May 2011): 45–50.

John Hassell uses a ZigBeef handheld reader to inventory cattle as they pass through a chute. The cattle have ear tags attached that transmit a unique radio frequency identification (RFID) signal to the receiver with ranges up to 300 feet (91.4 m). Bioglasses are typically used in the production of RFID tags for wildlife and livestock. © *AP Images/Mel Root.*

Rahaman, Mohamed N., et al. "Bioactive Glass in Tissue Engineering." *Acta Biomaterialia 7*, no. 6 (June 2011): 2355–2373. Available online at http://www.ncbi.nlm.nih.gov/pmc/articles/PMC3085647/.

Web Sites

"Amputees." *National Institutes of Health (NIH).* http://health.nih.gov/topic/Amputees (accessed September 10, 2011).

"Bioactive Glass Nanofibers Produced." *Science Daily,* December 18, 2009. http://www.sciencedaily.com/releases/2009/12/091218094646.htm

"Bioactive Glass Scaffolds for Bone Regeneration." *Lawrence Berkeley National Laboratory,* September 28, 2011. http://www.als.lbl.gov/index.php/science-highlights/science-highlights/588

Bryan Thomas Davies

Bioinformatics

Introduction

Contemporary bioinformatics represents an interface between the biological sciences, information technology, and computer science. As the acquisition of scientific data burgeoned with the rapid development of biotechnology and genomics, it became critical to create computerized means of organizing the information and making it publicly usable. Bioinformatics represents the confluence of information and computer technology, genomics, and biotechnology. The evolving multidiscipline is designed to facilitate long-term biological sciences information storage; allow for ready public access and global data-sharing; and support the rapid evolution of biotechnology by making available vast bodies of data readily available for utilization in further analysis and research.

The term bioinformatics was first used by Dutch biologists Paulien Hogeweg (1943–) and Ben Hesper in 1978; they conceptualized it as relating to informatic processes and biotic systems. Other scientists and researchers have broadened the definition to include the organization, storage, categorization, interpretation, and analysis of data acquired through biotechnology, genomics, and information technology processes. Bioinformatics also includes the development of statistical algorithms for manipulation and meta-analysis of the discovered or modeled data.

Historical Background and Scientific Foundations

Austrian monk Gregor Mendel (1822–1884) laid the groundwork for bioinformatics in the 1860s while accomplishing pioneering work in genetics. Mendel kept detailed notes and logs of his work that categorized and illustrated his laws of inheritance.

As originally conceptualized by Hogenweg and Hesper in the 1970s, bioinformatics related to the study of information processes on biotic systems. There has been a significantly accelerated generation of information in the biological sciences since the 1970s, largely as a result of advances in molecular biology and genomics. The scientific information surge created a demand for computerized repositories for the organization and storage of the data, as well as a need to index, catalog, access, and analyze it. Biological databases are central to bioinformatics. Two key components of computerized databases are ready accessibility and ease of mining the desired data.

The evolution of bioinformatics incorporates the history of molecular biology with that of information technology and computer science. Sequencing of proteins, beginning in the mid-1950s, fueled the need for development of informational databases. The Advanced Research Projects Agency (ARPA), begun in 1958, presaged the creation of the Internet and the World Wide Web. In 1965, American chemist Margaret Dayhoff (1925–1983) compiled all of the then-available sequence data and published the first bioinformational data collection, called the *Atlas of Protein Sequences*, which contained data on 65 proteins.

In 1969 ARPANET was created by linking computers from Stanford University in California with those at The University of Utah and the University of California at Los Angeles. Paul Berg (1926–), an American biochemist, constructed the first molecule of recombinant DNA in 1972. Later that year, scientists engineered the first recombinant DNA organism. In 1973 American biochemists Herbert Boyer (1936–) and Stanley N. Cohen (1935–) successfully cloned DNA, and the Brookhaven Protein DataBank was created. That year also heralded the beginning of the Internet, with the inception of Transmission Control Protocol (TCP) technology, designed by Vinton Cerf (1943–) and Robert Khan (1938–), a pair of American computer scientists, as a means of linking remote networks of computers. Working independently, molecular biologists Walter Gilbert (1932–) and Allan Maxam (1942–) in the United States

WORDS TO KNOW

BIOLOGICAL DATABASE: A compilation of data gathered from laboratory or literature research, experimental outcomes, clinical trials, biotechnology, and computer analyses in the biological sciences.

DATA MINING: The process of utilizing statistical methods and relevant algorithms to search for comparisons, correlations, and patterns in large databases.

GENBANK: GenBank is a database containing real-time data on all publicly available nucleotide sequences, their protein translations, and all relevant bibliographic data. It is updated daily.

GENOMICS: The study of the complete DNA sequencing of organisms, aimed at gathering data about the structure and function of genes, as well as the mapping of genomes.

HUMAN GENOME PROJECT: An international research project with the goal of identifying and labeling every gene in the human genome and storing all of this information in a computerized database. The project took 13 years, beginning in 1990 and concluding in 2003.

and Frederic Sanger (1918–) in the United Kingdom published methods for the sequencing of DNA in 1976. Rodger Staden of the United Kingdom pioneered the first DNA sequence analysis software in 1977; processes were created for rapid sequencing of long sections of DNA in the same year. The first Usenet connection was established in 1978 and linked Duke University and the University of North Carolina at Chapel Hill. The first complete genome, for the virus pi-x174, was published in 1980. By 1981, 579 human genes were mapped. Bioinformatics work was significantly enhanced by the development of automated DNA sequencing by two Americans, biologist Leroy Hood (1938–) and biochemist Marvin Carruthers (1940–) in 1981.

In 1982, GenBank, a comprehensive, publicly-available nucleic acid sequence database, was launched by the National Institutes of Health (NIH). GenBank has been in partnership with the European Molecular Biology Laboratory (EMBL) since 1986 and with the DNA Databank of Japan (DDBJ) since 1987. It shares information daily in order to ascertain that all sequence data is comprehensive and up-to-date. The three systems make up the International Nucleotide Sequence Database Collaboration (INSDC). The records in each database must contain the same parameters in order to be searchable easily. Free to users and publicly available through the Internet, GenBank is considered the first stop for biomedical and biological researchers studying potential new organisms. The Collaboration accepts thousands of submissions daily; the database doubles in size roughly every 18 months. In order to assure data consistency, data entries can be updated only by the database that was the original recipient.

The SWISS-PROT database, collaboratively created by the EMBL and the Department of Medical Biochemistry at the University of Geneva, was developed in 1986. SWISS-PROT is part of UniProt, designed to be a global comprehensive resource of protein sequence information. Like GenBank, it is accessible via the internet using BLAST (Basic Local Alignment Search Tool) or FTP (file transfer protocol) and available at no cost to users.

In 1990 the international Human Genome Project (HGP) was launched. Jointly sponsored by the NIH's National Human Genome Research Institute and the U.S. Department of Energy, the HGP took 13 years to complete. The BLAST Program, designed to locate regions of similarity between nucleotide or protein sequences was also launched in 1990.

■ Impacts and Issues

Bioinformatics bridges the biological sciences, information technology, and computer science. During the past several decades the three disciplines have evolved rapidly. Advances in genetics and molecular biology in particular have fueled the need for comprehensive, real-time information storage and retrieval systems capable of processing and organizing vast bodies of data in ways that make it possible to mine discrete pieces of information as well as to compare large arrays and sequences. The National Center for Biotechnology Information (NCBI) was created in 1988 as a comprehensive computerized literature resource and data warehouse for the disciplines of bioinformatics and biomedicine.

The scope of bioinformatics data includes very large relational, real-time databases such as GenBank, DDBJ, EMBL, and UniProt, critical for the storage and utilization of the ever-growing body of information resulting from genomics and molecular biology research and data collection.

The purposes of bioinformatics extend far beyond the computerized collection and categorization of data. Researchers utilize biological and information technologies to deepen understanding of the roles of proteins and DNA in health and disease processes, particularly as they affect humans. Bioinformatics affords ready and reliable access to vast amounts of data, making it possible to analyze and interpret ever growing bodies of biological information. As a result, researchers have access to tools necessary to understand biological processes in healthy living organisms. Acquiring an understanding of the healthy functioning of organisms provides a framework for understanding the processes that lead to malfunctions in biological systems. This enables researchers

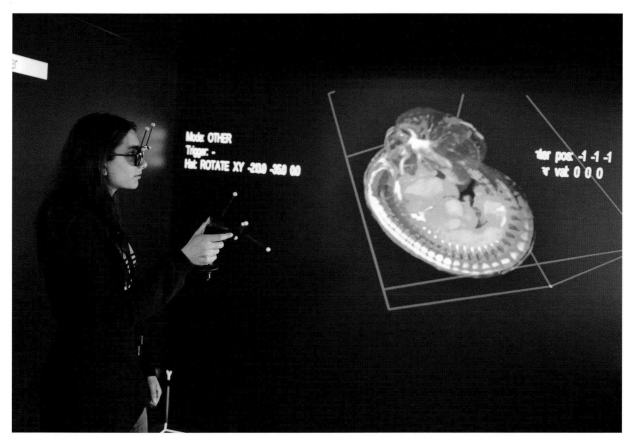

A woman works with advanced medical imaging, an example of 3-D bioinformatics.   © *Hans Verleur Photo & Film/Getty Images.*

to create disease models, furthering the development of pharmaceutical and therapeutic protocols.

SEE ALSO   *Biochemical Engineering; Biocomputers; Bioeconomy; Biotechnology; Biotechnology Valuation; Combinatorial Biology; DNA Sequencing; Forensic DNA Testing; Gene Banks; Genetic Engineering; Genetic Technology; Genome; Genomics; HapMap Project; Microarrays; Molecular Biology; Polymerase Chain Reaction (PCR); Pyrosequencing; Recombinant DNA Technology; Proteomics*

BIBLIOGRAPHY

Books

Buehler, Lukas K., and Hooman H. Rashidi. *Bioinformatics Basics: Applications in Biological Science and Medicine*, 2nd ed. Boca Raton, FL: CRC Press, 2005.

Gopal, Shuba, et al. *Bioinformatics: A Computing Perspective*. Boston: McGraw-Hill Higher Education, 2009.

Hodgman, T. Charlie, Andrew French, and David R. Westhead. *Bioinformatics*. New York: Taylor & Francis, 2010.

Krishnakumar, M. S. *Bioinformatics: Biocomputing and the Internet*. New Delhi: Campus Books, 2007.

Lesk, Arthur M. *Introduction to Bioinformatics*, 3rd ed. New York: Oxford University Press, 2008.

Zvelebil, Marketa J., and Jeremy O. Baum. *Understanding Bioinformatics*. New York: Garland Science, 2008.

Web Sites

"Bioinformatics Factsheet." *National Center for Biotechnology Information.* http://www.ncbi.nlm.nih.gov/About/primer/bioinformatics.html (accessed October 19, 2011).

"Bioinformatics." *NIH Library.* http://nihlibrary.nih.gov/ResearchTools/Pages/Bioinformatics.aspx (accessed October 19, 2011).

Pamela V. Michaels

Biological Confinement

■ Introduction

Biological confinement (also known as biological containment) refers to strategies and procedures that are adopted to prevent the environmental spread of organisms that have been genetically modified.

Biological confinement is a precaution to minimize the possibility that the genes introduced into a genetically modified organism (GMO) will spread to native species. Biological confinement also refers to the use of specialized research facilities to prevent the movement of organisms from the laboratory in which research is taking place. Depending on the threat posed by the organism, biological confinement can be increasingly stringent.

■ Historical Background and Scientific Foundations

Biological containment, whether conducted in the field or within the confines of a building, is concerned with keeping a target organism at its present site. The goal is to prevent the spread from the organism to other sites.

The biological containment conducted in the real-world environment is almost always done in response to the use of GMOs, with the aim of preventing the spread of the organisms from their point of use to the wider world. This can include measures to contain bacteria that are being used to clean up spills of compounds such as oil or radioactive compounds (this approach is known as bacterial remediation), or farmed fish that have been genetically modified for faster growth. Most often, however, the approach involves genetically modified plants including canola and cotton that have been engineered to be resistant to certain pesticides and/or to enhance certain commercially attractive qualities of the plant.

In this real-world environment, the containment can be physical, through use of barriers. Most often, especially with bacteria and plants, the containment is

genetic. It involves additional genetic modifications that are designed to absolutely prevent the organism from growing should it escape the confines of the crop field. A typical example is the genetic modification that renders the target crop absolutely dependent on a nutrient that is supplied and that is rare or entirely absent in the natural environment.

The type of biological containment conducted in buildings uses four levels of containment stringency, from biological safety level BSL-1 to BSL-4. Each level has its own specific guidelines for the type of laboratory that must be used, the equipment housed in the laboratory, and the way the tests and procedures are done. Laboratories that are used for the most dangerous of disease-causing (pathogenic) microbes have to undergo a rigorous approval process before they can begin operation and are evaluated frequently to ensure that the standards are being maintained. These evaluations extend to the testing of the lab personnel to make sure that they are qualified and skilled at the tasks they perform.

The lowest level, which applies to any biological laboratory, is BSL-1. This type of laboratory handles organisms that pose little risk to people (importantly, the laboratory personnel), animals, plants, and the environment. There is no need for special containment facilities in the laboratory, and all procedures are done on a regular lab bench. Of course, the regular procedures to ensure cleanliness and sterile conditions are maintained (e.g., wearing lab coats and, as needed, sterile gloves, use of disinfectants on work surfaces, autoclave/gas sterilization of media and lab ware), as they should be in any laboratory. A BSL-1 facility is supervised by an experienced microbiologist. Examples of some organisms handled in a BSL-1 facility include *Escherichia coli* K12, *Agrobacterium radiobacter*, *Bacillus thuringiensis*, and *Pseudomonas aeruginosa*.

The next level is BSL-2. This laboratory works with organisms that are of higher risk, but still do not pose a major danger. However, the microbes can cause infection when spread by casual contact. Examples include

Mycobacterium species, *Streptococcus* pneumonia, and *Salmonella choleraesuis*. Whereas access to a BSL-1 laboratory is unrestricted, access to a BSL-2 facility is restricted to laboratory personnel and approved visitors. Procedures and equipment must be in place, and personnel must be trained to deal with accidents. Lab coats, gloves, and face masks are mandatory. The laboratory supervisor must have experience in containment procedures.

A BSL-3 facility is used for microorganisms that are dangerous to humans, animals, or plants. These microbes can cause serious infections that can be lethal, particularly when inhaled. Such microbes are usually not spread by causal contact; examples include *Mycobacterium tuberculosis* (a cause of tuberculosis), *Bacillus anthracis* (the cause of anthrax), West Nile virus, Eastern equine encephalitis virus, the coronavirus that causes severe acute respiratory syndrome (SARS), *Coxiella burnetii* (the cause of Q fever), and yellow fever virus. A BSL-3 laboratory must be separated from other

WORDS TO KNOW

CLONING: The procedure used to create an organism that is an exact copy of the original organism.

EBOLA: An extremely infectious virus with a high death rate. Research on Ebola must be carried out in the most stringent biological containment facilities (BSL-4).

GENE: A sequence of deoxyribonucleic acid (DNA) that harbors the information for a functional product.

laboratories in the facility; this can be done by housing the BSL-3 laboratory in a separate building or by utilizing dedicated air and water systems. The air exiting the laboratory is filtered to ensure that microbes cannot escape. The windows (if there are any) cannot be opened. The laboratory is maintained under negative pressure, which means that air can flow in but cannot flow out. The facility is entered through a double-door system, with personnel having to change into protective clothing and undergo disinfection in the space between the doors; the clothing worn while in the laboratory must be removed and placed in a special container before leaving. Work is never done on the open lab bench; all work is done in enclosed hoods to prevent air from moving from the vicinity of the organisms.

The most stringent biological containment laboratory, which is used for highly dangerous microbes that are still not greatly understood (including a full understanding of their danger and capability to spread) and that cause diseases that are often untreatable, is the BSL-4 laboratory. Examples of microbes for which a BSL-4 facility is required are Ebola virus, Marburg virus, smallpox virus, and Lass fever virus. Indeed, many of the microbes handled by a BSL-4 facility are viral and cause diseases that progress swiftly to death. All the requirements of a BSL-3 laboratory apply to a BSL-4 facility, but the requirements concerning aspects such as air movement and use of enclosed work spaces are even more stringent. The laboratory personnel are highly-trained experts. All procedures must be conducted with great care, as any mistake could literally prove lethal.

■ Impacts and Issues

In general, biotechnology experts assert that biological containment in the context of physical facilities has been very successful in both containing the danger posed by the target microbes and allowing research on the organisms that is vital in finding cures for the diseases they cause and/or treatment strategies. Still, perfection is impossible to achieve, and accidents happen. This has occurred famously with Ebola virus, documented in

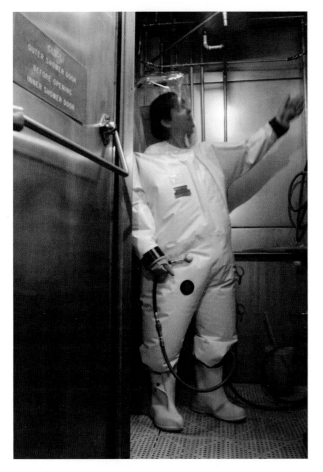

Dr. Catherine Wilhemson, biosecurity manager for the United States Army Medical Research Institute of Infectious Diseases (USAMRIID), steps into a decontamination room during a tour of USAMRIID's Patient Containment Lab at Fort Detrick in Frederick, Maryland. Wilhelmson is wearing a Biosafety Level 4 positive pressure suit. *© AP Images/Timothy Jacobsen.*

Biologist Ralph Bock looks into a glass holding a genetically transformed tomato plant at the University of Freiburg, southern Germany. Together with his team, the scientist created a transgenetic tomato plant with a safety lock. This means that the release of transformed genes into the environment, for example through pollen, is not possible, creating biological confinement of the genes. Bock has a worldwide patent on the plant. © *AP Images/Winfried Rothermel.*

riveting and frightening fashion in a 1994 book entitled *The Hot Zone*, in which author Richard Preston (1954–) described an Ebola release at a BSL-4 facility in Reston, Virginia, which was located less than 15 miles from downtown Washington, DC. Indeed, because

of the sophisticated nature of a BSL-4 facility, both in terms of equipment and necessary staff, such buildings tend not to be isolated geographically, but rather located in the heart of major centers, including in the midst of university campuses.

IN CONTEXT: GMOS IN THE WILD

Genetically modified organisms (GMOs) have escaped into the wild. Agricultural practices and transport accidents have allowed genetically engineered canola plants to grow along roadsides in North Dakota. Experts claim that the discovery of the wild-growing genetically engineered canola offers proof that genetically modified crops can become established in the wild. During a recent study of wild-growing canola plants along North Dakota highways, up to 80 percent of samples showed evidence of genetic engineering or of being derived from genetically engineered plants resistant

to the herbicides glyphosate (commonly known by its trade name, Roundup) or glufosinate. An herbicide is a plant-killing chemical and a form of pesticide. Genetically altered canola has also been found growing in the wild in Oregon and Japan.

Some GMO-use proponents contend that it would be prudent if experimental organisms carried genes (often termed "suicide genes") that render genetically altered organisms unable to survive openly without specific support (e.g., special nutrient mixtures).

At the heart of this issue is the concern by some that the presence of introduced microbes or genes could confer detrimental properties on the affected organisms, such as viruses or resistance to pesticides. In terms of GMOs, some believe that genetic modification of foods is undesirable because it risks introducing undesirable traits in non-modified organisms.

The rule of imperfection applies to biological containment of genetically modified crops. Numerous examples of windborne spread of seeds and pollen from the modified plants to adjacent fields have been documented. In a famous case from Canada, Monsanto sued farmer Percy Schmeiser (1931–) for patent infringement after finding their genetically modified plants in his field. Though Schmeiser had not deliberately planted the seeds, and the crops had appeared due to natural spread, he did save some of the seeds for the following years. After a protracted court battle, the decision went in Monsanto's favor.

SEE ALSO *Agricultural Biotechnology; Biologics Control Act; Biosafety Level Laboratories; Genetically Modified Crops;* Monsanto v. Schmeiser

BIBLIOGRAPHY

Books

Australian Farm Institute. *Can Agriculture Manage a Genetically Modified Future?* Surrey Hills, New South Wales: Australian Farm Institute, 2011.

Bodiguel, Luc, and Michael Cardwell. *The Regulation of Genetically Modified Organisms: Comparative Approaches.* Oxford: Oxford University Press, 2009.

Ferry, Natalie, and A. M. R. Gatehouse. *Environmental Impact of Genetically Modified Crops.* Wallingford, UK: CABI, 2009.

Gupta, Kavita, J. L. Karihaloo, and R. K. Khetarpal. *Biosafety Regulations of Asia-Pacific Countries.* Bangkok: Asia-Pacific Association of Agricultural Research Institutions, 2008.

Halford, Nigel. *Plant Biotechnology: Current and Future Applications of Genetically Modified Crops.* Chichester, UK: John Wiley, 2006.

The Regulation of Genetically Modified Organisms: Comparative Approaches. Oxford: Oxford University Press, 2009.

Thomson, Jennifer A. *Seeds for the Future: The Impact of Genetically Modified Crops on the Environment.* Ithaca, NY: Comstock Publishing Associates, 2007.

Web Sites

"Biosafety." *Centers for Disease Control and Prevention (CDC).* http://www.cdc.gov/biosafety (accessed August 23, 2011).

"Biosafety." *World Health Organization (WHO).* http://www.who.int/topics/biosafety/en (accessed August 23, 2011).

Brian Douglas Hoyle

Biological Pesticides

■ Introduction

Pesticides have a long history of being employed in the raising of crops as humans turned to agriculture and the domestication of crops in order to support a non-nomadic life. Eventually knowledge of chemistry led to chemical pesticides, but increasing resistance among pests, weeds, and pathogens to these synthetic products over the years has led agriculturists and scientists to turn to biological pesticides, also known as biopesticides. These are natural pesticides in the form of plants, animals, bacteria, and minerals. While some have been employed for thousands of years, for example neem (from the *Azadirachta indica* tree on the Indian subcontinent), canola oil, and nicotine, advancements have been made and a sophisticated science has grown up around their development, improvement, and modes of application.

■ Historical Background and Scientific Foundations

For as long as people have employed agriculture as a way of providing sustenance as opposed to foraging, there have been means of managing the pests and weeds that feed or compete with them. Records from the Zhou dynasty in China show a realization of the need for pest control around 1000 BC, when ants were used to protect citrus groves. Over the centuries, ancient apothecaries recommended various plant extracts to deter insects, some of which are still in use, for example lemon oil. They also recommended animal waste, such as silkworm droppings and clam ash, which were used topically, as well as minerals such as sulfur and mercury.

The chemical revolution saw the rise of synthetic pesticides in the 1880s. Synthetic pesticides are artificial, chemically-based pesticides with wide-ranging impact. One well-known example is dichlorodiphenyltrichloroethane or DDT, massively popular on an international

level during and after World War II (1939–1945) as a way of combating malaria and typhus. However, mounting concern over the effects such a widely used pesticide could have on the environment culminated in the book *Silent Spring* by environmentalist Rachel Carson (1907–1964) in 1962. She claimed synthetic pesticides such as DDT were having a detrimental effect not only on the insects they were aimed at controlling, but also on birds and wildlife as the poison worked its way up the food chain, accumulating in concentration as it did so. The book led to a public outcry, investigation, debate, and finally a ban in 1970. The events acted as an impetus in the search for effective, non-toxic ways of controlling pests.

Currently the most widely used biopesticide, *Bacillus thuringiensis* (Bt), is a bacteria named after the German city where it was rediscovered in 1911 by Ernst Berliner (1880–1957). It had been first discovered ten years earlier by Shigetane Ishiwata, a Japanese biologist. In 1938 it was marketed as Sporeine in France. In 1996 transgenic plants that could produce the Bt bacteria entered the market, and presently there are multiple international companies involved in the development of GM crops using Bt technology.

■ Impacts and Issues

Biopesticides can be divided into three types. First, biochemical pesticides that are defined as non-synthetic and non-toxic might include pheromones to disrupt mating behavior or plant scent extracts used in combination with insect traps. The second type is microbial pesticides, which consist of microorganisms that are more specific to a particular target pest. Microorganisms can include fungi, bacteria, and viruses. The aforementioned *Bacillus thuringiensis*, or Bt, is one of the most widely used microbial biopesticides on the market in the early twenty-first century. It is extremely specific, each strain producing different proteins that kill a certain species of insect larvae. Each of these related species has a particular enzyme in their gut that binds to a protein, paralyzing the gut and preventing the larva from absorbing food, leading to starvation. Mammals and other vertebrates, including humans, do not possess this enzyme, meaning the Bt bacteria is of no documented threat.

Bt, when used as a topical biopesticide, is sprayed directly onto the plant, where it is ingested by the insects feeding on that crop. Most commercial products include dried spores (which can lead to new bacteria) and active Bt crystals containing the protoxin. An interesting fact is the observed quality that even dead bacteria when ingested can cause death in the targeted insect. There is a danger that, over time, resistance within an insect population can build up. In order to avoid this situation, farmers are required by the Environmental Protection Agency (EPA) to use crop rotation, whereby a different crop is grown in the field intermittently, interrupting the possibility the insect will mutate over time and become immune to the Bt CRY proteins. Bt breaks down under UV light, meaning both that it has a limited span of effectiveness, but also that it degrades to leave no trace in the environment.

The third group is termed plant incorporated protectants or PIPs. PIPs involve the use of transgenics to

WORDS TO KNOW

CRY PROTEIN: An abbreviation of crystal proteins, such as those produced by the *Bacillus thuringiensis* bacteria, that are activated by an insect's digestive enzymes.

PROTOXIN: A chemical substance that is a precursor to a toxin.

TRANSGENIC: An organism or cell into which the genes of another species have been combined.

change the plant's genetics so that it can produce the substance to destroy the pest directly. The gene for the protein is inserted into the plant's genetic code, and that gives it the ability to produce the protein itself on a continuous basis, doing away with the need for spraying throughout the year. For farmers, it cuts down on the practical costs such pest management might entail. Bt is also being used in this way, and Bt crops of many kinds (e.g. corn, cotton, potatoes) are planted across millions of hectares worldwide. Besides pest control, one positive side effect in the case of Bt corn has been the reduction in toxic fungi infection in insect damaged

A ladybug picks up an aphid. Ladybugs are used as a natural, biological way to eliminate pests like aphids without the use of pesticides. © *Joseph Calev/Shutterstock.com.*

Pyrethrum-based insecticidal products are shown with flowering chrysanthemums in Australia. Pyrethrum is a natural, biochemical insecticide created from dried flower heads of several members of the *Chrysanthemum* genus. © *Bill Bachman/Alamy.*

crops. In the United States, there are regulations in place from the EPA for these transgenic crops in order to address the fears of pest resistance. Fields with transgenic Bt crops are required to have refuge areas that must cover at least 20 percent of the whole crop coverage. This allows for non-mutant insects to mate with possible resistant insects in the Bt area, diluting the resistant genotypes.

Genetically-modified organisms (GMOs) are highly controversial, and there is ongoing and mounting opposition to their use. In April 2000, a petition headed by Greenpeace maintained that Bt products should be banned, given their "unreasonable adverse effect on the environment," but the petition was rejected by the EPA. Another study has shown that in the case of Bt corn manufactured and marketed by Monsanto, the pollen from the genetically modified corn was contaminating plants adjacent to the crop fields and reducing the survival rates of the Monarch butterfly larva. A later study found no such link. For farmers, the case is a simple one, as the Monarch does not directly impact their crops. However, taking a wider view, some scientists are concerned that the possibility of contamination and the effects of a pesticide previously declared safe may have larger ramifications. Still another concern is the idea of cross-pollination with wild versions

of the crop, in which case the genetic makeup of an entire species could be affected with unknown repercussions. In some cases, other pests are invading crop plantations in the absence of the original pest, requiring pesticides to be sprayed in order to contain the new crop threat.

Due to its specificity, Bt biopesticides have long been used by farmers as an alternative to chemical pesticides. One fear expressed is that although they might spray at certain times of the year when the insect threat is high, transgenic crops produce the Bt toxin all year round, increasing the likelihood of "superpests" resistant to the toxin. In that case, farmers would be left with less recourse to fight off the pests in an environmentally safe way. Research undertaken regarding pyramided plants incorporating two dissimilar Bt genes has shown that pest resistance is significantly reduced even where there is a smaller refuge area.

As costs of synthetic pesticides, worries about toxicity, and a demand for more environmentally sound practices all rise, there is increasing demand for biopesticides and other alternatives. The benefits must be weighed against concerns about GM plants in the case of PIPs and resistance to biopesticides across the board. Achieving the right balance between these benefits and concerns is critical to the future of agriculture.

SEE ALSO *Agricultural Biotechnology; Agrobacterium; Biological Confinement; Bt Insect Resistant Crops; Genetically Modified Crops; Genetically Modified Food; Plant Variety Protection Act of 1970; Terminator Technology; Transgenic Plants*

BIBLIOGRAPHY

Books

Bailey, Alastair. *Biopesticides: Pest Management and Regulation.* Cambridge, MA: CABI, 2010.

Maczulak, Anne E. *Sustainability Building Eco-Friendly Communities.* New York: Facts on File, 2010.

Ohkawa Hideo, Hisashi Miyagawa, and Philip W. Lee, eds. *Pesticide Chemistry: Crop Protection, Public Health, Environmental Safety.* Weinheim, Germany: Wiley-VCH, 2007.

Van Driesche, Roy, Mark Hoddle, and Ted D. Center. *Control of Pests and Weeds by Natural Enemies: An Introduction to Biological Control.* Malden, MA: Blackwell, 2008.

Vincent, Charles, Mark S. Goettel, and George Lazarovits, eds. *Biological Control: A Global Perspective.* Cambridge, MA: CABI, 2007.

Web Sites

"Pesticides: Pesticide Types: Biopesticides." *U.S. Environmental Protection Agency (EPA).* http://www. epa.gov/ebtpages/pestpesticidetypesbiopesticides. html (accessed October 27, 2011).

"What Are Biopesticides?" *U.S. Environmental Protection Agency (EPA).* http://www.epa.gov/ oppbppd1/biopesticides/whatarebiopesticides.htm (accessed October 27, 2011).

Kenneth Travis LaPensee

Biological Weapons: Genetic Identification

■ Introduction

The advent of molecular technologies and the application of genetic identification in clinical and forensic microbiology have greatly improved the capability of laboratories to detect and identify organisms used in biological weapons. Not only does this ability enhance national defense capabilities, but also the development and administration of countermeasures, including vaccines.

■ Historical Background and Scientific Foundations

Groundbreaking research by American geneticist James D. Watson (1928–) and British biophysicists Rosalind Franklin (1920–1958), Francis Crick (1916—2004), and Maurice Wilkins (1916–2004) in the 1950s at Cambridge University led to the discovery of the role and structure of deoxyribonucleic acid (DNA), the molecule that contains the genetic codes of all living organisms. American biochemists Marshall Nirenberg (1927–1910), Robert Holley (1922–1993), and Har Gobind Khorana (1922–2011) further deciphered the genetic code and protein synthesis in the following decade. Their research and discoveries unlocked the key to understanding and identifying the 22 protein-building amino acids found in all life forms, which act as the building blocks of cells, tissues, and organ systems of all living organisms.

The genome of *Caenorhabditis elegans*, a flatworm, was assembled in 1998, setting the stage for the technology and scientific procedures required to map the human genome. The Human Genome Project, completed in 2003, successfully mapped the 25,000 genes in human DNA and handed scientists and physicians the tools to determine not only each person's genetic strengths and vulnerabilities to potential illnesses, but also enabled the concept of DNA fingerprinting and instigated the field of bioinformatics. Bioinformatics is not restricted to humans, but rather is focused on analyzing, categorizing, and storing biological and genetic information for all living organisms including viruses, bacteria, and protozoa, which may be used in biological warfare.

Biological weapons have played a part in warfare for thousands of years, and up to the 1960s played a prominent part in military research and development programs. The loss of state programs and support for biological weapons in the early 1970s left the field open to fanatical organizations, small factions, and cults, which increasingly began to utilize bioweapons for terrorism.

■ Impacts and Issues

Biological weapons are weapons with payloads containing microorganisms that can cause infection, or the toxic components or outputs of the microorganisms. Examples of infectious microorganisms include viruses (e.g., the viruses that cause smallpox, Ebola, influenza), bacteria (e.g., *Bacillus anthracis* and *Clostridium botulinum*), and protozoa. The most prominent example of a toxic component is the variety of toxins that are produced and released from bacteria, such as neurotoxins produced by strains of *Clostridium*.

The genetic identification of microorganisms utilizes molecular technologies to evaluate specific regions of the genome and to determine the genus, species, or strain of a microorganism. This work grew out of the similar, highly successful applications in human identification using the same basic techniques. Thus, the genetic identification of microorganisms also has been referred to as microbial fingerprinting, and it is a key way in which bioinformatics can assist in the identification of pathogens.

Genetic technologies are especially useful in the detection of biological weapons. Of particular note is the polymerase chain reaction, or PCR, which uses selected enzymes to make copies of genetic material.

If the genetic material is unique to the microorganism (e.g., a gene encoding a toxin), then investigators can use PCR to detect a specific microorganism from among the other organisms present in the sample. Traditional PCR detects RNA at the end point of the process (the plateau stage), however advances in the technology led to real-time PCR detection. This gave scientists the ability to collect data in the exponential growth phase, making DNA and RNA quantitation more efficient and accurate, and facilitated the development of hand-held detectors. Hand-held PCR detectors used by United Nations inspectors in Iraq during their weapons inspections efforts of 2002/2003 were sensitive enough to detect a single living *Bacillus anthracis* bacterium (the agent of anthrax) in an average kitchen-sized room.

The production of biological weapons can be accomplished with relatively unsophisticated technology by a typically trained microbiologist. Furthermore, the equipment necessary to accomplish weaponization (i.e., incubators, autoclaves, fermenters, centrifuges, refrigerators, and lyophilizers) can be housed in less than a few thousand square feet. Thus, biological weapons manufacture is not difficult to conceal. Furthermore, although biological weapons can be deployed in

> ## WORDS TO KNOW
>
> **BIOINFORMATICS:** The application of models and statistical analysis from the fields of computer science, molecular biology, and mathematics to define biological systems.
>
> **NEUROTOXIN:** A biological or chemical complex that damages or destroys tissues in the nervous system.
>
> **TRANSGENIC:** An organism or cell into which the genes of another species have been combined.

traditional weaponry (i.e., rockets), the weapons also can be carried literally in someone's pocket to the target site. This can make the deployment of biological weapons virtually impossible to stop, unless the carrier passes near an instrument designed to detect the biological agent.

The microbial agents used as biological weapons typically are highly infectious. The direct exposure of even a small number of people to the weapon can lead quickly to a large number of illnesses or casualties. Bacteria such as *Clostridium botulinum* and various species

Health Ministry official Kazuhiko Kawauchi, second from right, talks with another ministry official, February 21, 2011, as Japan starts excavation at the site of the former medical school in Tokyo linked to Unit 731, a germ and biological warfare outfit during World War II (1939–1945). Experiments conducted by the unit on war prisoners have never been officially acknowledged by the government but have been documented by historians and participants. © *AP Images/Koji Sasahara.*

IN CONTEXT: BIOSAFETY LABORATORIES

In the United States, the principle laboratory for research into the medical aspects of biological warfare is the United States Army Medical Research Institute of Infectious Diseases (USAMRIID). The facility is operated by the Department of Defense and serves mainly to develop vaccines for infectious diseases and tests to detect and identify disease-causing microorganisms. While some of the research conducted at USAMRIID is classified, other research findings of the resident civilian and military scientists are used to benefit the larger public community.

Laboratories have a rating system with respect to the types of microbes that can safety be studied. There are four levels possible. A typical university research lab with no specialized safety features (i.e., fume hood, biological safety cabinet, filtering of exhausted air) is a Biosafety Level 1 lab. Progression to a higher level requires more stringent safety and biological controls. A Biosafety Level 4 laboratory is the only laboratory that can safely handle microbes such as the Ebola virus, *Bacillus anthracis* (the cause of anthrax), the Marburg virus, and hantavirus.

USAMRIID has Biosafety Level 4 laboratories. Other Biosafety Level 4 laboratories exist in the United States, including laboratories operated by the Centers for Disease Control and Prevention (CDC).

Entry to the Biosafety Level 4 area requires passage through several checkpoints and passage through various security checks. When existing vaccines are available, researchers are typically required to be immunized against microorganisms under study. All work in the Biosafety Level 4 lab is conducted in a pressurized and ventilated suit, with filtered air supplied by both portable and centralized systems.

of *Salmonella* readily cause contamination, either by their growth in food or by the production of potent toxins. Such foodborne microbial threats also are considered biological weapons. Indeed, in the aftermath of the U.S. anthrax attacks in 2001, the vulnerability to sabotage of the food production and supply systems in many countries has become evident.

The anthrax outbreak of 2001 illustrated the significance of diagnostic genetic technologies. Because an anthrax infection can mimic cold or flu symptoms, the earliest victims did not realize they were harboring a deadly bacterium. After confirmation that anthrax was the causative agent in the first death, genetic technologies were utilized to confirm the presence of anthrax in other locations and for other potential victims. Results were available more rapidly than would have been possible using standard microbiological methodology, and appropriate treatment regimens could be established immediately. Furthermore, unaffected individuals were informed quickly of their status, alleviating unnecessary anxiety.

The second stage of the investigation included locating the origin of the anthrax cells. The evidence indicated that this event was not a random, natural phenomenon, and that an individual or individuals had most likely dispersed the cells as an act of bioterrorism. In response to this threat, government agencies collected samples from all sites for analysis. A key element in the search was the genetic identification of the cells found in patients and mail from Florida, New York, and Washington, DC. The PCR studies clearly showed that all samples were derived from the same strain of anthrax, known as the Ames strain, as the cell line was established in Ames, Iowa. Although this strain has been distributed to many different research laboratories around the world, careful analysis revealed minor changes in the genome that enabled investigators to narrow the search to about 15 U.S. laboratories. With time, total genome sequencing of these 15 strains and a one-to-one base comparison with the lethal anthrax genome may detect further variation that will allow a unique identification to be made.

A further defense against the spread of infectious agents within a population was established with the use of infrared thermal imaging technology such as that used in airport body scanners. During the 2003 outbreak of Severe Acute Respiratory Syndrome (SARS) and the 2009 H1N1 swine flu epidemic, some countries employed the scanners at airports to assess body temperature in passengers attempting to board flights. The scanners were shown to be 90 percent effective in identifying high body temperatures. They had the added advantage of being noninvasive, given they were able to assess temperatures from a distance of between three and six feet. An Nguyen of the U.S. Centers for Disease Control and Prevention (CDC), who with a team of researchers studied the use of these scanners in various hospital emergency rooms, pointed out that because fever is a common indicator of infectious disease, identifying people with fever is a major component of screening efforts. Studies also showed that people often lied about having a fever for fear of being detained, or having their travel disrupted. Scanners bypass the need to rely on information requested from the passengers. In this case, it was a successful attempt to help curb the rapid spread of disease.

In the context of biological warfare, infectious agents have the express purpose of causing illness and even death. This is all the more disturbing given both the difficulties with spontaneously detecting those agents once they are in the body and also the propensity most agents have for spreading, whether that be through body contact or from inhalation of airborne pathogens. The ability to sense an elevated temperature could help to better identify and quarantine those passengers who may be unwittingly or otherwise carrying dangerous and infectious pathogens in their bodies, halting the spread of disease in both national and international settings.

Detection of known pathogens is an involved process as illustrated by the previous examples, but the field of genetic identification of biological weapons becomes more complicated when considered in the context of genetic engineering. Prior to the 1972 Biological Weapons Conventions (an agreement for the cessation of biological weapons programs, the storage of said weapons, and destruction of any existing stockpiles), international governmental science departments were working on various bioweapons and utilizing genetic engineering to manipulate the codes for viruses and bacteria. The results were transgenic pathogens, composed of, for example, disease-causing bacterium housing a pathogenic virus. Other experiments manipulated the anthrax spore so that it was antibiotic resistant. The purpose of genetically modifying pathogens was twofold: to create a more potent or deadly agent and to make identification and subsequent treatment more difficult. Any detection system in use must take the possibility of genetically-modified pathogens into account, and defense programs must run concurrently with research into genetic identification of hybrids.

The advent of biotechnology has brought great benefits across many fields, from medicine to agriculture, manufacturing, and engineering. These benefits spring from the fact that biosciences offer innovative, advanced, and forward-looking solutions. It cannot be ignored, however, that these solutions and benefits can be a double-edged sword: Knowledge and skills are not contained easily and could be adapted, along with equipment and resources, for nefarious reasons, ending up as biological weapons. As biotechnology gains an ever more secure foothold within science and medicine, it is paramount that global governments draw up comprehensive strategies to protect this technology, and indeed to utilize it in the realms of national and global security. The identification of bioweapons represents an important defense against insidious invasion and attack on both military personnel and civilians.

■ Primary Source Connection

On April 10, 1972, government officials in Washington, London, and Moscow signed a document pertaining to the prohibition of the development, production, and stockpiling of bacteriological (biological) and toxin weapons and on their destruction. The document was ratified in 1975 and entered into force on March 26, 1975.

Convention on the Prohibition of the Development, Production and Stockpiling of Bacteriological (Biological) and Toxin Weapons and on their Destruction

Signed at Washington, London, and Moscow April 10, 1972

Ratification advised by U.S. Senate December 16, 1974

Ratified by U.S. President January 22, 1975

U.S. ratification deposited at Washington, London, and Moscow March 26, 1975

Proclaimed by U.S. President March 26, 1975

Entered into force March 26, 1975

The States Parties to this Convention,

Determined to act with a view to achieving effective progress towards general and complete disarmament, including the prohibition and elimination of all types of weapons of mass destruction, and convinced that the prohibition of the development, production and stockpiling of chemical and bacteriological (biological) weapons and their elimination, through effective measures, will facilitate the achievement of general and complete disarmament under strict and effective international control,

Recognizing the important significance of the Protocol for the Prohibition of the Use in War of Asphyxiating, Poisonous or Other Gases, and of Bacteriological Methods of Warfare, signed at Geneva on June 17, 1925, and conscious also of the contribution which the said Protocol has already made, and continues to make, to mitigating the horrors of war,

Reaffirming their adherence to the principles and objectives of that Protocol and calling upon all States to comply strictly with them,

Recalling that the General Assembly of the United Nations has repeatedly condemned all actions contrary to the principles and objectives of the Geneva Protocol of June 17, 1925,

Desiring to contribute to the strengthening of confidence between peoples and the general improvement of the international atmosphere,

Desiring also to contribute to the realization of the purposes and principles of the Charter of the United Nations,

Convinced of the importance and urgency of eliminating from the arsenals of States, through effective measures, such dangerous weapons of mass destruction as those using chemical or bacteriological (biological) agents,

Recognizing that an agreement on the prohibition of bacteriological (biological) and toxin weapons represents a first possible step towards the achievement of agreement on effective measures also for the prohibition of the development, production and stockpiling of chemical weapons, and determined to continue negotiations to that end,

Determined, for the sake of all mankind, to exclude completely the possibility of bacteriological (biological) agents and toxins being used as weapons,

Convinced that such use would be repugnant to the conscience of mankind and that no effort should be spared to minimize this risk,

Have agreed as follows:

Article I

Each State Party to this Convention undertakes never in any circumstances to develop, produce, stockpile or otherwise acquire or retain:

(1) Microbial or other biological agents, or toxins whatever their origin or method of production, of types and in quantities that have no justification for prophylactic, protective or other peaceful purposes;

(2) Weapons, equipment or means of delivery designed to use such agents or toxins for hostile purposes or in armed conflict.

Article II

Each State Party to this Convention undertakes to destroy, or to divert to peaceful purposes, as soon as possible but not later than nine months after the entry into force of the Convention, all agents, toxins, weapons, equipment and means of delivery specified in article I of the Convention, which are in it's possession or under its jurisdiction or control. In implementing the provisions of this article all necessary safety precautions shall be observed to protect populations and the environment.

Article III

Each State Party to this Convention undertakes not to transfer to any recipient whatsoever, directly or indirectly, and not in any way to assist, encourage, or induce any State, group of States or international organizations to manufacture or otherwise acquire any of the agents, toxins, weapons, equipment or means of delivery specified in article I of the Convention.

Article IV

Each State Party to this Convention shall, in accordance with its constitutional processes, take any necessary measures to prohibit and prevent the development, production, stockpiling, acquisition, or retention of the agents, toxins, weapons, equipment and means of delivery specified in article I of the Convention, within the territory of such State, under its jurisdiction or under its control anywhere.

Article V

The States Parties to this Convention undertake to consult one another and to cooperate in solving any problems which may arise in relation to the objective of, or in the application of the provisions of, the Convention. Consultation and cooperation pursuant to this article may also be undertaken through appropriate international procedures within the framework of the United Nations and in accordance with its Charter.

Article VI

(1) Any State Party to this Convention which finds that any other State Party is acting in breach of obligations deriving from the provisions of the Convention may lodge a complaint with the Security Council of the United Nations. Such a complaint should include all possible evidence confirming its validity, as well as a request for its consideration by the Security Council.

(2) Each State Party to this Convention undertakes to cooperate in carrying out any investigation which the Security Council may initiate, in accordance with the provisions of the Charter of the United Nations, on the basis of the complaint received by the Council. The Security Council shall inform the States Parties to the Convention of the results of the investigation.

Article VII

Each State Party to this Convention undertakes to provide or support assistance, in accordance with the United Nations Charter, to any Party to the Convention which so requests, if the Security Council decides that such Party has been exposed to danger as a result of violation of the Convention.

Article VIII

Nothing in this Convention shall be interpreted as in any way limiting or detracting from the obligations assumed by any State under the Protocol for the Prohibition of the Use in War of Asphyxiating, Poisonous or Other Gases, and of Bacteriological Methods of Warfare, signed at Geneva on June 17, 1925.

Article IX

Each State Party to this Convention affirms the recognized objective of effective prohibition of chemical weapons and, to this end, undertakes to continue negotiations in good faith with a view to reaching early agreement on effective measures for the prohibition of their development, production and stockpiling and for their destruction, and on appropriate measures concerning equipment and means of delivery specifically designed for the production or use of chemical agents for weapons purposes.

Article X

(1) The States Parties to this Convention undertake to facilitate, and have the right to participate in, the fullest possible exchange of equipment, materials and scientific and technological information for the use of bacteriological (biological) agents and toxins for peaceful purposes. Parties to the Convention in a position to do so shall also cooperate in contributing individually or together with other States or international organizations to the further development and application of scientific discoveries in the field of bacteriology (biology) for prevention of disease, or for other peaceful purposes.

(2) This Convention shall be implemented in a manner designed to avoid hampering the economic or technological development of States Parties to the Convention or international cooperation in the field of peaceful bacteriological (biological) activities, including the international exchange of bacteriological (biological) agents and toxins and equipment for the processing, use or production of bacteriological (biological) agents and toxins for peaceful purposes in accordance with the provisions of the Convention.

Article XI

Any State Party may propose amendments to this Convention. Amendments shall enter into force for each State Party accepting the amendments upon their acceptance by a majority of the States Parties to the Convention and thereafter for each remaining State Party on the date of acceptance by it.

Article XII

Five years after the entry into force of this Convention, or earlier if it is requested by a majority of Parties to the Convention by submitting a proposal to this effect to the Depositary Governments, a conference of States Parties to the Convention shall be held at Geneva, Switzerland, to review the operation of the Convention, with a view to assuring that the purposes of the preamble and the provisions of the Convention, including the provisions concerning negotiations on chemical weapons, are being realized. Such review shall take into account any new scientific and technological developments relevant to the Convention.

Article XIII

(1) This Convention shall be of unlimited duration.

(2) Each State Party to this Convention shall in exercising its national sovereignty have the right to withdraw from the Convention if it decides that extraordinary events, related to the subject matter of the Convention, have jeopardized the supreme interests of its country. It shall give notice of such withdrawal to all other States Parties to the Convention and to the United Nations Security Council three months in advance. Such notice shall include a statement of the extraordinary events it regards as having jeopardized its supreme interests.

Article XIV

(1) This Convention shall be open to all States for signature. Any State which does not sign the Convention before its entry into force in accordance with paragraph (3) of this Article may accede to it at any time.

(2) This Convention shall be subject to ratification by signatory States. Instruments of ratification and instruments of accession shall be deposited with the Governments of the United States of America, the United Kingdom of Great Britain and Northern Ireland and the Union of Soviet Socialist Republics, which are hereby designated the Depositary Governments.

(3) This Convention shall enter into force after the deposit of instruments of ratification by twenty-two Governments, including the Governments designated as Depositaries of the Convention.

(4) For States whose instruments of ratification or accession are deposited subsequent to the entry into force of this Convention, it shall enter into force on the date of the deposit of their instruments of ratification or accession.

(5) The Depositary Governments shall promptly inform all signatory and acceding States of the date of each signature, the date of deposit of each instrument of ratification or of accession and the date of the entry into force of this Convention, and of the receipt of other notices.

(6) This Convention shall be registered by the Depositary Governments pursuant to Article 102 of the Charter of the United Nations.

Article XV

This Convention, the English, Russian, French, Spanish and Chinese texts of which are equally authentic, shall be deposited in the archives of the Depositary Governments. Duly certified copies of the Convention shall be transmitted by the Depositary Governments to the Governments of the signatory and acceding states.

IN WITNESS WHEREOF the undersigned, duly authorized, have signed this Convention.

DONE in triplicate, at the cities of Washington, London and Moscow, this tenth day of April, one thousand nine hundred and seventy-two.

UNITED NATIONS. "THE CONVENTION ON THE PROHIBITION OF THE DEVELOPMENT, PRODUCTION AND STOCKPILING OF BACTERIOLOGICAL (BIOLOGICAL) AND TOXIN WEAPONS AND ON THEIR DESTRUCTION." APRIL 10, 1972.

SEE ALSO *Biodefense and Pandemic Vaccine and Drug Development Act of 2005; Biodetectors; Bioterrorism; Genetic Technology; Government Regulations; Polymerase Chain Reaction (PCR); Security-Related Biotechnology; Weaponization*

BIBLIOGRAPHY

Books

Ainscough, Michael J. *Next Generation Bioweapons: The Technology of Genetic Engineering Applied to Biowarfare and Bioterrorism.* Maxwell Air Force Base, AL: USAF Counterproliferation Center, Air War College, Air University, 2002.

Anderson, Burt, Herman Friedman, and Mauro Bendinelli, eds. *Microorganisms and Bioterrorism.* New York: Springer, 2006.

Brannigan, Michael C., ed. *Cross-Cultural Biotechnology.* Lanham, MD: Rowman & Littlefield, 2004.

Edelstein, Sari. *Food and Nutrition at Risk in America: Food Insecurity, Biotechnology, Food Safety, and Bioterrorism.* Sudbury, MA: Jones and Bartlett Publishers, 2009.

National Research Council (U.S.), and Institute of Medicine (U.S.). *Globalization, Biosecurity, and the Future of the Life Sciences.* Washington, DC: National Academies Press, 2006.

National Research Council (U.S.), and National Academies Press (U.S.). *Biotechnology Research in an Age of Terrorism.* Washington, DC: National Academies Press, 2004.

Nestle, Marion. *Safe Food: Bacteria, Biotechnology, and Bioterrorism.* Berkeley: University of California Press, 2003.

Rappert, Brian, and Caitriona McLeish, eds. *A Web of Prevention: Biological Weapons, Life Sciences and the Governance of Research.* Sterling, VA: Earthscan, 2007.

Sherman, Robert. *Avoiding the Plague: An Assessment of US Plans and Funding for Countering Bioterrorism.* Washington, DC: Center for Strategic and Budgetary Assessments, 2007.

Sussman, Joel, and Paola Spadon. *From Molecules to Medicines: Structure of Biological Macromolecules and Its Relevance in Combating New Diseases and Bioterrorism.* Dordrecht, The Netherlands: Springer, 2009.

Web Sites

"Biological Weapons." *Gene Watch.* http://www.genewatch.org/sub-396425 (accessed November 13, 2011).

van Aiken, Jan, and Edward Hammond. "Genetic Engineering and Biological Weapons." *EMBO Reports,* June 2003. http://www.ncbi.nlm.nih.gov/pmc/articles/PMC1326447/ (accessed November 13, 2011).

Kenneth Travis LaPensee

Biologics Control Act

■ Introduction

The Biologics Control Act, also known as the Virus-Toxin Law, is a United States federal law passed in 1902 that regulated the handling, labeling, testing, and distribution of antitoxins, sera, toxins, and vaccines. The law was passed after a series of deaths resulted from failures to identify and recall adulterated batches of newly developed diphtheria antitoxin and smallpox vaccines.

■ Historical Background and Scientific Foundations

At the close of the nineteenth century, researchers in the emerging field of immunology identified the mechanisms of several common diseases, allowing also for the development of preventative inoculations and cures. German physician Robert Koch (1843–1910) isolated the bacteria responsible for tuberculosis, anthrax, and cholera. Researchers in Koch's lab, German physiologist Emil von Behring (1854–1917) and Japanese bacteriologist Shibasaburo Kitasato (1853–1931), discovered that animals injected with toxins produced by diphtheria and tetanus developed antitoxins to the diseases. Unlike the antibiotics discovered in the twentieth century, the antitoxin sera did not kill the disease-causing bacteria itself; the sera worked to counteract the poisons that the disease bacteria released. Thus, these sera produced immune reactions in other animals that could help cure active diphtheria and tetanus infections in sick animals and prevent future infection in healthy animals. Injections developed from the sera proved effective in human subjects during clinical trials in Berlin in 1891. Commercial production of antitoxin sera followed. Both Koch and von Behring won Nobel Prizes in medicine for their work.

The earliest record of independently produced diphtheria antitoxin in the United States dates to 1895. The first U.S. mass production of serum was undertaken by government health services, including the Hygienic Laboratory in Washington, DC, and the New York City Health Department's Bacteriological Laboratory.

At the turn of the twentieth century, diphtheria was one of the top 10 causes of death in the United States. The development of diphtheria antitoxin serum had a significant positive impact on public health. In the United States, the serum was made by injecting horses with a regimen of increasingly greater doses of diphtheria-causing bacteria. The horse's immune systems responded to the inoculations with increased diphtheria immunity and high levels of antitoxin production. The antitoxin-rich blood was then taken from the horses to produce a serum for injecting into humans.

In the earliest years of antitoxin serum production, there were few controls on quality, stability, purity, storage, and efficacy. Different batches of serum could have different levels of antitoxins. There was no standardized testing protocol to check for adulterated batches. There existed no tests for many disease-causing agents. Reliable electrical refrigeration was expensive and rare, and there were no universal regulations for serum storage. Despite a lack of government regulation in the industry, most serum production labs did have individual safety protocols that promulgated standards for lab cleanliness, animal wellness evaluation, batch labeling, and sporadic batch testing for known possible contaminants.

In early October 1901, lab workers noticed that a long-serving serum-production horse named Jim was ill. The horse showed signs of having contracted tetanus and was euthanized after his condition deteriorated. On September 30, Jim had donated blood for serum, but his illness was either not noted before his final batch of serum was distributed or producers failed to identify the batch and keep it from distribution. An investigation into the death of a young girl in St. Louis, Missouri, showed she had received an injection of antitoxin serum shortly before her death.

The origin of the serum was traced back its production facility, and through lab records to one donor animal: Jim. Tests indicated that the September 30 batch of serum was contaminated with tetanus. Lab records indicated that some of the serum in the September 30 batch could have been misplaced or mislabeled into vials with a September 24 batch date, but the potential donor animals were not indicated. However, the lab had tested the September 24 batch before distribution and found it free of contaminants. Investigators concluded that the September 30 batch of serum had not been tested before distribution. Failure to catch the contaminated injections resulted in the deaths of 13 children.

The St. Louis incident was followed by nine deaths in Camden, New Jersey, after people had received contaminated smallpox vaccines. Public health workers worried that the vaccine deaths would undermine inoculation programs by making the public afraid of vaccines. While assuring the public that injections and vaccinations produced better health outcomes than unprevented or untreated diseases, health workers also pushed for greater government regulation of vaccine, serum, and drug production. Public health officials and the Medical Society of the District of Columbia proposed regulation through legislation.

■ Impacts and Issues

Congress passed the Biologics Control Act on July 1, 1902. The act established the Center for Biologics Evaluation and Research (CBER) and tasked a specially created board of health officials from the military and public health agencies to set forth rules controlling the sale and preparation of biologics, including "viruses, serums, toxins, and antitoxins and analogous products in international or foreign commerce." The act also specified that biologics must be produced in licensed facilities, be recorded and tested before distribution, and be marked with expiration dates.

Capped vials of medication on an autosampler. Sampling is one of the many quality control measures mandated by the federal government before medicines reach the market. © *Ioana Davies (Drutu)/Shutterstock.com.*

From 1903 to 1906, the federal government added more regulations governing standards for lab cleanliness, inspections, recordkeeping, quality control, and safety testing. The Pure Foods and Drug Act, passed in 1906 after public outcry over adulterated meat products, required that some medicines list their ingredients and forbade the sale of many dangerous patent medicines (commercial remedies often of little or no therapeutic value).

In 1912 the Public Health Service was granted expanded research authority. In addition to the research spurred by the Biologics Control Act in the emergent field of immunology, federal researchers began to study noncontagious diseases. In 1913 the Virus-Serum-Toxin Act aided ranchers and farmers by addressing the safety, purity, and quality of animal biologics.

The 1938 Federal Food, Drug, and Cosmetic Act (FDCA or FFDCA) shifted U.S. regulation of most medicines and cosmetics to the newly created Food and Drug Administration (FDA). However, the FDA was not responsible for regulating vaccines until 1972. The Public Health Service, and later the National Institutes of Health, regulated vaccines until then.

■ Primary Source Connection

An article published November 16, 1901, in *The New York Times*, offers information about individuals in Camden, New Jersey, who fell critically ill after injection of the tetanus vaccine. It also gives an account of the number of victims who died following injection of the vaccine and provides information regarding the origin of the vaccine and speculation as to why five of the seven victims died. In 1902, the Biologics Control Act was passed to regulate the handling, labeling, testing, and distribution of antitoxins, sera, toxins, and vaccines. Today, in the twenty-first century, questions still arise regarding the risks versus benefits of receiving this, and other, vaccines.

Five Victims of Lockjaw

Special to The New York Times

CAMDEN, N.J., Nov. 16.—Out of seven victims of lockjaw following vaccination in Camden, five have died. The fifth victim, Lillian Carty, sixteen years old, of 742 Berkley Street, died early this morning.

William Brower, the boy at 217 North Front Street, who has been ill from the disease for nearly two weeks, appears to be getting better, but is still in a critical condition.

The seventh victim was found today in the person of Mamie Winters, a girl of seven, of 746 Mount Vernon Street. Her condition is serious.

It is alleged that in every case where death has ensued the virus used in vaccination was the sort that is sold in hermetically sealed tubes, and it is said further that no evil results have followed the use of the dry vaccine points. Many physicians argue that the deaths were due to carelessness, possibly exposure of the vaccination wounds to cold.

"FIVE VICTIMS OF LOCKJAW." *NEW YORK TIMES,*
NOVEMBER 16, 1901.

SEE ALSO *Agricultural Biotechnology; Biotechnology; Government Regulations; Vaccination; Vaccines*

BIBLIOGRAPHY
Books

Ekins, Sean, and Jinghai J. Xu. *Drug Efficacy, Safety, and Biologics Discovery: Emerging Technologies and Tools.* Hoboken, NJ: John Wiley & Sons, 2009.

Plotkin, Stanley A., Walter A. Orenstein, and Paul A. Offit. *Vaccines*, 5th ed. Philadelphia: Saunders/ Elsevier, 2008.

Termini, Roseann B. *Life Sciences Law: Federal Regulation of Drugs, Biologics, Medical Devices, Foods, and Dietary Supplements*, 4th ed. Wynnewood, PA: FORTI Publications, 2010.

Web Sites

"Vaccine Safety." *Centers for Disease Control and Prevention (CDC).* http://www.cdc.gov/ vaccinesafety/Vaccine_Monitoring/Index.html (accessed September 28, 2011).

"Vaccines." *World Health Organization (WHO).* http://www.who.int/topics/vaccines/en (accessed September 28, 2011).

"Vaccines & Immunizations." *Centers for Disease Control and Prevention (CDC).* http://www.cdc .gov/vaccines/ (accessed September 28, 2011).

"Vaccines & Preventable Diseases." *Centers for Disease Control and Prevention (CDC).* http://www.cdc .gov/vaccines/vpd-vac/ (accessed September 28, 2011).

Adrienne Wilmoth Lerner

Bioluminescence

Introduction

Bioluminescence is the production of light by living organisms, including bacteria and protista, as well as by multicellular organisms such as insects, marine life, and fungi. A well-known bioluminescent organism is the firefly, whose glow is a common nighttime sight during the summer months in some areas. In nature, bioluminescence serves functions such as communication and attracting a breeding partner. Bioluminescence is a common feature of some marine environments, which support a number of bioluminescent organisms including dinoflagellates, jellyfish, coral, shrimp, and fish. This natural process has been harnessed in biotechnology to produce probes that detect target molecules on the basis of emission of light.

Historical Background and Scientific Foundations

The production of light in organisms capable of bioluminescence occurs when the chemical luciferin reacts with oxygen in a reaction catalyzed by an enzyme called luciferase to produce light, along with another compound called oxyluciferin. The structure of the luciferases and luciferins found in different organisms varies. In addition to luciferin, oxygen, and luciferase, other molecules that are termed cofactors are needed. In their absence, bioluminescence will not occur. Cofactors such as calcium and adenosine triphosphate (ATP) aid luciferase in catalyzing the reaction of luciferin with oxygen.

The terms luciferin and luciferase were coined in 1885 by German scientist Emil du Bois-Reymond (1818–1895), who was researching the nature of the bioluminescence of extracts from clams and beetles. When mixed, the extracts produced light. Du Bois-Reymond heated each of the extracts separately, and then combined them in various ways. The heating of one of the extracts prevented the luminescence, but heat had no effect on the other of the extracts. This led to the suggestion that bioluminescence involved at least two compounds, not just a single compound. The fact that bioluminescence was destroyed by heating one compound was interpreted as evidence of a changed structure. Surmising that the compound destroyed by heat must be an enzyme, du Bois-Reymond named it luciferase. He called the heat-resistant compound luciferin.

Isolated luciferin and luciferase can be used to determine the concentrations of important biological molecules such as the cofactors ATP and calcium. To do this, a known amount of luciferin and luciferase is added to the blood or tissue sample; the cofactor concentrations can then be determined from the intensity of the light emitted (which is measured in a machine called a spectrometer).

Bioluminescence has been very useful in detecting and even determining the quantity of target molecules, even if the molecules do not take part in the actual bioluminescence reaction. Most typically, this is done by attaching luciferase to an antibody to the target protein. The attachment does not destroy the enzymatic activity of the luciferase. So, when the antibody-luciferase complex binds to the target protein, the appropriate chemicals are added and luciferase-catalyzed bioluminescence occurs.

Luciferase has been used to study viral and bacterial infections in living animals, based on the use of luciferase-antibody complexes, and to detect bacterial contaminants in food. These sorts of assays are accurate and can be done quickly and relatively inexpensively, so they have become widely used. Because such applications can also rely on a complex created between luciferase and specific sequences of deoxyribonucleic acid (DNA), bioluminescence can be used to determine DNA sequences, based on the binding of the DNA sequence in

Crystal jellies (*Aequorea victoria*) are bioluminescent hydrozoans that have a fluorescent protein that appears bright green when exposed to blue light. The protein, which has become an important tool as a reporter gene, was isolated by scientist Osamu Shimomura, who won the Nobel Prize in Chemistry with Martin Chalfie and Roger Y. Tsien for their work on the protein.
© Dwight Smith/Shutterstock.com.

WORDS TO KNOW

ANTIBODY: A protein that is produced by the immune system in response to the presence of another protein termed an antigen.

ENZYME: A protein that allows a chemical reaction to proceed more easily than it otherwise would and is not altered by the reaction.

PROTISTS: Eukaryotic single-celled organisms (organisms whose genetic material is enclosed within a specialized membrane) that belong to the kingdom Protista in the classification used for living organisms.

IN CONTEXT: LIGHT IN THE OCEAN DEPTHS

In deep ocean waters, starting at 4 miles (6.5 km) below the surface, there is no light except that emitted by marine organisms. Over 90 percent of the organisms in the deep ocean are bioluminescent, using this attribute as a means of communication, a defense from predators, or as a lure for prey. Thousands of species, including bacteria, crustaceans, jellyfish, squid, and fish, emit flashes of light or glows produced by bioluminescence. Though a few ocean organisms display red or yellow bioluminescence, most bioluminescent ocean creatures exhibit green or blue-green light. Scientists believe the trait evolved separately in different species because bioluminescent jellyfish, shrimp, and some fish have utilize a type of luciferin called coelenterazine, while the luciferin found in organisms like krill, dinoflagellates, and other types of fish is different.

the complex to a complementary DNA sequence in the sample genetic material.

Luciferase often is utilized to monitor the activity of genes. This purpose (to monitor when genes are active and how their activity can be repressed) is a part of what is termed "reporter gene" technology, because the probe in effect reports on the activity of the target gene. The approach relies on the binding of a complex containing luciferase to a region preceding the gene that is called the promoter. This region is where activity of the gene begins and is controlled. A specific gene promoter can be attached to the DNA that codes for firefly luciferase and introduced into an organism. The activity of the gene promoter can then be studied by measuring the bioluminescence produced in the luciferase reaction.

Another form of bioluminescence involves proteins that, as a consequence of their shape, are able to absorb light of different wavelengths (i.e., colors). A well-known example of a biofluorescent protein is the green fluorescent protein (GFP).

■ Impacts and Issues

Bioluminescence has revolutionized biotechnology by enabling the detection of target molecules and gene activity. At the heart of many of the applications is the GFP molecule. The gene encoding GFP was originally isolated from *Aequorea victoria*, a type of jellyfish. The role of the protein in jellyfish is still unclear. The gene can be incorporated into an organism, and the appearance of the fluorescent protein can indicate gene activity. Alternately, the protein itself can be supplied as part of a probe to determine the

Scientist Tomoko Nakanishi points to three 48-hour-old glowing mice at a laboratory in the Osaka University where the world's first fluorescent mammals were created in 1997. The geneticists injected mouse embryos with the DNA of a bioluminescent North American jellyfish. *© AP Images/KK.*

cellular location of target molecules, most often using the technique of fluorescence microscopy. Research that established such biosensor functions for the GFP gene and protein garnered the 2008 Nobel prize in Chemistry for Martin Chalfie, Osamu Shimomura, and Roger Tsien.

The scientific uses of bioluminescence have included the detection of specific cells in living organisms, such as in the brain, where fluorescent probes have been used to reveal the mechanisms of neuron communication and the activity of surface receptors on brain cells. As well, bioluminescence has been important in revealing the mechanisms of the infection process of several types of viruses, particularly how virus enters the host cell.

Less scientific applications of GFP include the breeding of transgenic glowing rabbits, zebrafish marketed as GloFish (which were originally bred to be indicators of polluted water), and fluorescent mice (NeonMice).

SEE ALSO *Cell Imaging; Security Related Biotechnology*

BIBLIOGRAPHY

Books

Rodgerson, David J. *Bioluminescence: Characteristics, Adaptations and Biotechnology.* Nova Science Pub Inc., 2011.

Shimomura, Osamu. *Bioluminescence: Chemical Principles and Methods.* Singapore: World Scientific Publishing Company, 2006.

Wink, Michael, ed. *An Introduction to Molecular Biotechnology: Fundamentals, Methods and Applications.* Weinheim, Germany: Wiley-VCH, 2011.

Brian Douglas Hoyle

Biomagnetics

■ Introduction

Magnetism is a fundamental feature of the natural world. Magnetism that is produced by a living creature is referred to as biomagnetism. By measuring the changing magnetic fields within a person, the corresponding electrical currents that produce those magnetic fields can be deduced. Because magnetism and electricity are intertwined, biomagnetism is considered a part of bioelectromagnetism, which entails both the electric currents and magnetism produced by a living organism. Sometimes the word biomagnetics is used synonymously with biomagnetism.

The study of magnetism produced within the human body did not become a serious field of study until the latter half of the twentieth century. This is because the magnetic fields produced by humans have a very low intensity (are very weak), and certain technologies and techniques first had to appear to enable researchers to precisely measure such weak fields. The first detailed measurements of human-produced magnetic fields were made at the close of the 1960s; continued improvements in both the sensitivity and cost of such diagnostic equipment has made the study of biomagnetism a promising avenue of research, including for the investigation of various diseases.

■ Historical Background and Scientific Foundations

In 1791 Luigi Galvani (1737–1798) observed that an electric current caused a dead frog's leg to twitch. Galvani's discovery indicated that electricity plays an essential role in organisms. The subsequent discovery by Hans Christian Oersted (1777–1851) that an electric current passing through a wire produces a circular magnetic field demonstrated the intimate relationship between electricity and magnetism. Because all living things produce electric currents within their bodies (known as bioelectricity), then it follows that all living things also produce magnetism.

The magnetic fields produced in the human body arise from the electrical currents produced there. The heart, for instance, relies on rhythmic electric impulses to pump blood. Those electric impulses produce pulsating magnetic fields. By measuring the varying magnetic fields produced in the heart, the currents that produced those fields can be determined. Although electric currents can be measured by placing electrodes on the skin (as in electrocardiography, or ECG), such measurements can be distorted by the varying conductivity of bone, tissue, and other body parts. Because magnetic fields pass through the body almost unaltered, measuring the magnetic fields produced by the body can sometimes provide more accurate information about the body's electrical activity.

Starting in the early 1960s, researchers attempted to image the magnetic fields produced in the human body. Two problems that the researchers encountered were background interference from external magnetic fields, and the low precision, or sensitivity, of the magnetic detection equipment available at the time. External magnetic fields (fields produced outside the human body) include the magnetic field of Earth, as well as magnetic fields produced by electrical wiring and equipment. Such interference is sometimes referred to as signal noise—in this case the signal is the magnetism produced by the human body, and the outside interference is the noise.

The problem of extraneous magnetic fields, and of insufficient magnetic detection equipment, was overcome towards the end of the 1960s. Interference from outside magnetic fields was greatly reduced by improved shielding, which absorbed or reflected away the vast majority of extraneous magnetism in the area where the subject's magnetic field was being measured. And the detection equipment was greatly improved by the introduction of superconducting quantum interference device (SQUID) technology, which allows for the precise detection of extremely small magnetic fields, such as those produced within the human body.

WORDS TO KNOW

ELECTROMAGNETISM: The branch of physics that deals with electricity and magnetism, and the interrelation between the two.

MAGNETIC FIELD: A region that arises due to the presence of magnet(s) or electric currents, and in which a measurable magnetic force exists at each point within the region.

SQUID: An acronym for Superconducting QUantum Interference Device; it uses properties of superconductivity to detect extremely weak magnetic fields.

SUPERCONDUCTIVITY: The phenomenon of zero resistance to the flow of electric current in certain materials that have been chilled to very low temperatures.

■ Impacts and Issues

The introduction of SQUID technology enabled researchers to probe the very weak magnetic fields produced by the human body. Continued development of the technology has reduced the size and cost of the magnetic detection equipment so that there are currently hundreds of such machines being used for diagnostics worldwide. Magnetic imaging is mostly performed on the brain and heart. Imaging the magnetic fields produced within the brain is called magnetoencephalography (MEG). The use of MEG is helpful in a variety of medical conditions, such as pinpointing the source of epileptic seizures in the brain, which are associated with abnormal electrical activities. Chronic depression is also thought to be associated with abnormal electrical activity in the brain, so MEG could also potentially be useful in the investigation of that brain disorder. MEG can precisely "map out" various regions of a person's brain, such as the area responsible for speech, which is of use to surgeons when performing brain surgery, as when removing a tumor.

Magnetic imaging of the heart is known as magnetocardiography (MCG). Imaging the magnetic field produced by the heart is noninvasive and benign for the patient, and so such tests can be performed as often as needed without fear of side effects. MCG is useful for detecting an array of cardiac problems, and its sensitivity means that it is especially useful in the early detection of disease, including arrhythmias (irregular heartbeats) and ischemia (inadequate blood flow).

Several challenges confront the wider use of MEG and MCG for diagnosis, research, and disease prevention. The superconducting magnets used in MEG and MCG devices are expensive and require extremely low temperatures. The use of so-called high temperature superconductors could translate into lower costs for purchasing and operating magnetic detection equipment, making MEG and MCG more affordable to hospitals and clinics for routine use.

Another challenge is to expand the use of magnetic imaging to improve the understanding and treatment of a variety of conditions, especially those associated with the brain. For instance, researchers still do not understand the basic underlying causes of depression, schizophrenia, obsessive-compulsive disorder, and other serious mental illnesses. It has been shown that some of the neuronal activity (and therefore the electrical activity) of certain regions of the brain is abnormal in those suffering from depression and other mental disorders. An expanded use of MEG to study the magnetic and electrical activity of the brain would seem to be appropriate, given the multitude of people suffering from brain disorders and the efficacy of MEG imaging techniques to learn more about how the brain functions.

SEE ALSO *Biopharmaceuticals; Cell Imaging; Magnet-Assisted Transfection; Neuroimaging*

BIBLIOGRAPHY

Periodicals

Ueno, Shoogo, and Masaki Sekino. "Biomagnetics and Bioimaging for Medical Applications." *Journal of Magnetism and Magnetic Materials* 304, no. 1 (September 2006): 122–127.

Web Sites

"Gastrointestinal SQUID Technology Laboratory." *Vanderbilt University.* http://www.vanderbilt.edu/biomag/research.htm (accessed September 21, 2011).

Philip Edward Koth

Biometrics

Introduction

Biometrics or biometry is a field of studies that identifies and measures unique physical individual characteristics including hand geometry, palm veins, iris features, facial features, or behavioral traits such as voice, signature, or gait recognition. Biometric technologies are applied primarily in personal identification and authentication situations in secure environments.

Historical Background and Scientific Foundations

Prior to the digital era, biometrical methods were much simpler and less accurate. Fingerprints often were analyzed in order to identify an individual, polygraphs helped indicate whether or not a suspect was telling the truth, and blood type analysis could identify a suspect as belonging to a vast number of people sharing the same blood type. Although these techniques remain in use, newer automated technologies such as optical scanners with advanced software and stored databases of individual biometric traits quickly help to identify unique individual anthropometric characteristics. These scanners evaluate segmentation patterns of the iris, retina, palm veins, fingerprints, hand geometry, and facial traits.

Behavioral biometric technologies such as voice, signature, and gait recognition are used for a variety of purposes, from identification at passport controls to helping police identify persons of interest in solving a crime. This type of biometric technology is also in the initial stages of use for identifying customers and allowing them access to their account information when making purchases on the Internet.

Impacts and Issues

With the rapid development of computer technologies and telecommunications, and the growing number of Internet users worldwide, protecting individual data as well as safeguarding the sensitive information of banks, businesses, and governments has become a high priority. Espionage, identity theft, fraud, and network sabotage with viruses or worms are the main concerns facing e-commerce and Internet banking. Biometric systems are employed in increasing numbers when responding to these security challenges.

Biometric technologies have high accuracy and reliability when used as authentication systems, because unique individual physical traits are not transferable or easily copied. Access to highly sensitive data and classified information can be better protected by multimodal biometric systems combining, for instance, iris identification with voice recognition and palm vein spectral scanning. Many ATM machines provide a hand recognition device along with the other more familiar security requirements such as password, personal identification number (PIN), and smartcard.

The iris (the part of the eye that shows color) recognition system is considered one of the most reliable biometric methods because of the iris's complexity and the uniqueness of its texture in each individual. Therefore, iris recognition is used when maximum security is required, usually in combination with other modalities. Retina biometrics, which analyzes the unique patterns of veins and arteries at the back of the eye, is also considered a highly reliable system, with a margin of error of about one in 10 million scans. In comparison, biometric fingerprint identification systems misclassify a fingerprint in about one in 500 attempts.

Hand biometrics analyzes a set of measurements taken from fingers, joints, and the general structure of the palm, as well as palm skin surface, ridges, and creases. The most accurate hand biometry devices measure about 90 different parameters of hand geometry and detect the temperature of the hand. They also require the user to move the fingers, in order to avert fraud by the use of artificial hands. Hand geometry biometrics, however, is less accurate than palm vein spectrometry in spite of being one of the most commonly used technologies to verify

135

an authorized person's entrance into apartment buildings, banks, hospitals, airports, offices, and governmental buildings. Palm vein spectrometry uses near-infrared light to capture the singular pattern formed by the blood vessels inside the hand of a person, and is also highly accurate. Moreover, because it is not an invasive process and does not offer any risk of trauma or damage, palm vein spectral biometrics is becoming a common system of authentication in banking, especially at ATM machines.

Face biometry is used for identification purposes and analyzes shapes and proportions of facial features, such as the distance between the eyes; the shape, dimensions, and locations of nose and lips; bone structures of the forehead and jaws; shape and outlines of eye sockets; different facial expressions of the same individual; facial images taken from different angles; and facial thermography, thus yielding a reliable identification tool.

Voice recognition is a biometric tool based on two vocal traits: voice tract (a physiological trait) and accent (a behavioral trait). The level of accuracy depends on the audio signal quality and the number of variables covered during voice recording. This system is commonly used for voice verification in e-commerce, customer authentication, and the authentication of persons under house arrest or on parole. Voice recognition systems have relatively high rates of false matches or non-matches compared to other authentication systems.

Voice recognition and face biometrics are also targets for civil liberties controversies because they can be recorded without the individual's consent or knowledge. With video cameras virtually everywhere in public places, including factories, offices, banks, shopping centers, stadiums, elevators, streets, and buildings, being filmed and photographed has become an inevitable annoyance, half-consciously endured by the majority of urban populations.

As biometric technologies become the preferred tools for identification and authentication in banking operations, the inventiveness and creativity of individuals searching to bypass the system also increases. Published research reports have demonstrated that several optical and silicon sensors for fingerprints also accepted artificial fingers at least in 60% of the attempts. High-resolution

photos were successfully used to fool face recognition systems, and dummy hands were used with some success to bypass hand geometry devices. Even some iris biometric scanners were shown not to be as reliable as initially assumed. Therefore, experts recommended the introduction of multifactor authentication systems, in which more than one biometric modality is combined with PIN and smartcards, to reduce the risk of authentication forgery.

Adding spectral biometry to a system also reduces the risk of forgery. Spectral biometry uses spectroscopy to examine the properties of matter (for example, skin, blood, fat, melanin pigment in the eye) through the analysis of absorbed and scattered light, particles, or sound. Whereas most counterfeiting techniques use artificial materials such as contact lenses, plastic fingertips, or high-resolution photos, spectroscopic readings require an authentic, living person to generate their spectral bio-signature with their unique and multiple physiological characteristics.

A parent looks in an iris-recoginition scanner to gain entrance to an elementary school. If they are registered, parents, teachers, and other adults can use the scanner and the doors will unlock automatically. Iris recognition systems use a video camera to record the colored ring around the eye's pupil. More accurate than fingerprints and other biometric markers, iris technology is considered a nearly foolproof way of identifying people because markings in the iris are unique to each person and do not change as people age. Others who have reason to visit the area schools can get in the old fashioned way, by pressing a buzzer. © *AP Photo/Daniel Hulshizer.*

SEE ALSO *Bioinformatics; DNA Fingerprinting; Forensic Biotechnology; Government Regulations; Security-Related Biotechnology*

BIBLIOGRAPHY

Books

Gates, Kelly. *Our Biometric Future: Facial Recognition Technology and the Culture of Surveillance.* New York: New York University Press, 2011.

Nelson, Lisa S. *America Identified: Biometric Technology and Society.* Cambridge, MA: MIT Press, 2011.

Pato, Joseph N., and Lynette I. Millett, eds. *Biometric Recognition: Challenges and Opportunities.* Washington, DC: National Academies Press, 2010.

Pugliese, Joseph. *Biometrics: Bodies, Technologies, Biopolitics.* New York: Routledge, 2010.

Vacca, John R. *Biometric Technologies and Verification Systems.* Boston: Butterworth-Heinemann/Elsevier, 2007.

Wayman, James, et al., eds. *Biometric Systems Technology, Design, and Performance Evaluation.* London: Springer, 2005.

Periodicals

Jarjes, Ann A., Kuanquan Wang, and Ghassan J. Mohammed. "A New Iris Segmentation Method Based on Improved Snake Model and Angular Integral Projection." *Research Journal of Applied Sciences, Engineering and Technology* 3, no. 6 (June 2011): 558–568.

Web Sites

Biometrics.gov. http://www.biometrics.gov/ (accessed November 1, 2011).

"Fingerprints & Other Biometrics." *U.S. Federal Bureau of Investigation (FBI).* http://www.fbi.gov/about-us/cjis/fingerprints_biometrics (accessed November 1, 2011).

"US-VISIT Biometric Identification Services." *U.S. Department of Homeland Security (DHS).* http://www.dhs.gov/files/programs/gc_1208531081211.shtm (accessed November 1, 2011).

Sandra Galeotti

Biomimetic Systems

Introduction

Artists and poets have long drawn inspiration from the beauty and complexity of living things. In an analogous way, engineers and biochemists have drawn inspiration for human-made designs and processes from those exhibited by organisms. Biomimetics means adapting the designs and mechanisms found in living creatures to the solution of practical problems in disciplines such as engineering and architecture. The word bionics is often used interchangeably with biomimetics.

Since the concept was first precisely formulated in the mid-twentieth century, biomimetics has grown immensely and has become a widely-used and accepted method for meeting numerous modern challenges. Biomimetic systems exploit the features found in living systems, whether it is the structure of a single organism, or the group behavior exhibited by a collection of organisms.

Biomimetics is analogous to the process of reverse engineering. Engineers utilize reverse engineering by analyzing the function, structure, operation, and purpose of things—anything from computer software to mechanical devices—in order to discern the basic principles that make the thing function effectively at a given task. But instead of reverse engineering human-made devices, biomimetics seeks to "reverse-engineer" from living things in order to make better devices and structures for human use.

Historical Background and Scientific Foundations

Long ago people were applying the forms and techniques they saw in nature to solve various problems. The Renaissance artist and inventor Leonardo da Vinci (1452–1519) considered the idea of human flight and based many designs for flying contraptions on the physiology of birds. In the mid-twentieth century, the current terms that are used to describe the process of mimicking biological function and form were first defined. The American professor of physics and electrical engineering Otto Schmitt (1913–1998) officially introduced the term biomimetics in a paper in 1969, though he reportedly created the term in the 1950s. Dr. Schmitt was a pioneer in several fields, especially electronics, and much of his work was inspired by the behavior and structures of organisms. In the 1930s, for instance, Schmitt invented the "thermionic trigger," which is better known as the Schmitt trigger. This essential electronic device was inspired by Dr. Schmitt's study of signal propagation in the nerves of squid. The term bionics—often used synonymously with biomimetics—was first introduced publicly in 1960 by Jack Steele (1924–2009), a U.S. Air Force officer and medical doctor.

One does not necessarily have to be a scientist or researcher to utilize biomimetics. Joseph Cox (1905–2002), an American logger and part-time inventor, drew inspiration from how the jaws of the timber beetle's larva moved—in a side-to-side motion—to create improved cutting surfaces for chainsaws. Swiss inventor George de Mestral (1907–1990) noticed that the husks of cocklebur seeds stuck tenaciously to clothing and fur. Upon examining their structure under a microscope, de Mestral noticed tiny hooks on each seed husk. From this inspiration, de Mestral produced the first hook-and-loop fastening system, which he named Velcro.

The features of living things that can be exploited for humanity's benefit can be separated into general categories: structures that function to perform a given task (such as the hooks present on cockleburs used to hitch a ride on fur); the functioning of a given structure (such as the side-to-side movement of a larva's jaws); and the social organization and behavior of groups of organisms, such as exhibited by the social insects (e.g., bees and ants) or flight behavior of flocks of birds.

Even the principles that guide the development of traits in different species through evolution can be exploited. The principles of natural selection and mutation can be applied within software programs to "evolve" so-called artificial life. The same scheme can be applied to the field of robotics, as well as to creation of synthetic life, which seeks to produce new (generally simple) living systems from a starting set of biochemicals.

■ Impacts and Issues

Biomimetics has resulted in many novel applications of biological systems to artificial ones. Products abound that capitalize on the unique attributes of different organisms. The hook-and-loop fastener has become ubiquitous in the modern world—it is used in automotive products, sports gear, and clothing by regular consumers, but astronauts and soldiers have used it as well.

Increasingly, advanced imaging and analysis techniques are used to understand the structure and functioning of organisms so that their capabilities can be mimicked artificially. For example, researchers wanted to mimic the climbing ability of gecko lizards. Using high magnification, they were able to decipher the extremely fine hairs (ten times finer than human hair) on

> ### WORDS TO KNOW
>
> **COMPOSITE MATERIAL:** A material composed of two or more substances that remain distinct from one another. A low-tech example is straw combined with mud to create bricks (adobe). The resulting composite material exhibits properties different from that of its individual components.
>
> **ENGINEERING:** The discipline of applying principles found in mathematics and physics to the design, production, and operation of various devices, machines, systems, and processes.
>
> **MIMIC:** To simulate, or resemble closely.
>
> **ORGANISM:** An individual living thing.

the gecko's toes that enable them to climb, even up smooth surfaces such as glass. Researchers then created climbing robots based on the hair-like structures found on geckos.

Former NBA basketball player Mike Williams catches a ball during a December 2010 physical therapy session. After a near-fatal shooting in 2009 that left him mostly paralyzed from the waist down, Williams is teaming up with a Chicago doctor in hopes of being outfitted with robot-like legs that used to be the stuff of science fiction, but now are being made by a handful of biotech companies using biomimetics to mimic human functions.   © *AP Images/Nam Y. Huh.*

Biomimetics has influenced a great number of fields, only some of which are listed below (along with a few examples):

1. improved aircraft and boats (aerodynamics and hydrodynamics)—for instance, mimicking the structures found in the skins of dolphins and whales to reduce the drag on ship's hulls

2. electronics—the Schmitt trigger, a comparator that measures electrical input to see if it is above or below a certain threshold

3. computer and software engineering

4. materials engineering—developing strong, lightweight materials based on those produced by organisms

5. sustainable architecture—passive ventilation of buildings based on observations of termite mounds

6. improved drug delivery

7. robotics—locomotion of robots; foraging behavior based on that of social insects, such as ants

One of the main challenges for the field of biomimetics is to accelerate the process of identifying and adapting useful attributes found in life forms to aid humanity. The solutions for problems such as climate change, increased demand for food and energy from a growing world population, and the health needs of an aging population potentially could be found by utilizing solutions that are embodied in the structures and behaviors of animals and plants.

Another challenge is to master particular features of certain organisms that researchers have been working on for years. Just two examples are the strong and durable composite materials of which marine shells are made and the strong, yet lightweight and elastic, properties of spider silk. Researchers continue to uncover the basic chemical and physical processes that animals use to synthesize these key materials.

SEE ALSO *Biochip; Biocomputers; Biorobotics*

BIBLIOGRAPHY

Books

Li, Junbai, Qiang He, and Xuehai Yan. *Molecular Assembly of Biomimetic Systems.* Weinheim, Germany: Wiley-VCH, 2011.

Narayan, Roger, and Prashant N. Kumta. *Advances in Biomedical and Biomimetic Materials: A Collection of Papers Presented at the 2008 Materials Science and Technology Conference (MS & T08), October 5–9, 2008, Pittsburgh, Pennsylvania.* Hoboken, NJ: J. Wiley & Sons, 2009.

Vepa, Ranjan. *Biomimetic Robotics: Mechanisms and Control.* New York: Cambridge University Press, 2009.

Web Sites

Mueller, Tom. "Biomimetics." *National Geographic,* April 2008. http://ngm.nationalgeographic.com/2008/04/biomimetics/tom-mueller-text (accessed September 23, 2011).

Philip Edward Koth

Biopharmaceuticals

■ Introduction

Pharmaceuticals are compounds that have some medicinal benefit. Traditionally, such compounds are discovered, extracted, and purified from their natural origin, or are chemically made (synthesized). Biopharmaceuticals refer to pharmaceutical drugs that are made by living organisms. They are an example of biotechnology—the use of organisms to achieve a desired outcome—in this case the creation of medically-useful drugs. Many biopharmaceuticals are entire protein molecules, but protein fragments, deoxyribonucleic acid (DNA), and ribonucleic acid (RNA) can also be used.

A large number of biopharmaceuticals are produced by the genetic alteration of bacteria and fungi. Essentially, these organisms are converted to living factories that produce the desired product. Ideally, the manufactured product is secreted from the cells, making it easy to obtain the product in highly-purified form. Biopharmaceuticals can also be obtained from blood and blood plasma and are designated as biologics.

■ Historical Background and Scientific Foundations

The first commercially available biopharmaceutical was human insulin produced by the common intestinal bacterium *Escherichia coli* that had been genetically engineered with the insertion of the human gene encoding insulin (this approach is known as recombinant DNA technology). Developed by Genentech in 1978, the manufacturing license was acquired by Eli Lilly & Company, who began to manufacture the product (dubbed Humulin) in 1982.

The burgeoning research and development of biopharmaceuticals in the intervening decades attests to the huge therapeutic benefits of these compounds, because the development and approval process is lengthy (a decade or more) and very expensive (in the tens of millions of dollars). In the United States, approval for sale and post-approval safety monitoring of products in the marketplace is the responsibility of the Food and Drug Administration (FDA). The many criteria associated with product approval include the guarantee that the conditions used in creating the biological agents are sterile and minimize the chance of alterations of the organisms that would affect production of the target compound and/or production of other compounds that could be toxic.

Typically, the growth of the organisms involved, and the manufacture and purification of the target compounds takes place in a "clean room"—a facility especially designed to minimize the presence of contamination. Clean room features can include the wearing of special clothing (e.g., masks, gloves, and gowns) by personnel and filtering of incoming air to exclude airborne contaminants.

Aside from bacteria and fungi, plants are another important source of biopharmaceuticals. The facilities required for plant rearing are often less complex and less expensive than those required for microorganisms. However, the extraction and purification of the target compounds from plants can be more laborious, because plants do not secrete the manufactured compound into the growth medium the way that microorganisms can. Purification of biopharmaceuticals involves a variety of techniques including electrophoresis (various kinds), filtration, and chromatography (various kinds).

It is essential that the molecules produced have the same three-dimensional shape (conformation) as the parent molecule. The production of the target compound per se can be ineffective if the compound's shape precludes the desired outcome. This requirement adds to the challenge of biopharmaceutical design. The organism selected as the factory for production of the particular biopharmaceutical must be capable of producing the compound in large quantities and in the conformation that is exactly the same as in the normal host (for

WORDS TO KNOW

BIOGENERICS: Compounds that have the same structure and activity as the originally-developed compounds and can be licensed for sale when the period of patent protection of the original compound has expired.

HORMONE: A protein that produces a certain physiological response.

MOLECULAR FARMING: The use of plants to produce a target compound.

of antibodies), nucleic acids, and vaccines. Plants can be especially valuable with the latter, because the consumption of the plant can bestow the protection related to the vaccine. Such plants have been dubbed edible vaccines.

The rationale behind an edible vaccine is that when the plant is eaten, the target molecule (typically a protein or peptide) will circulate in the bloodstream following digestion of the plant. In the blood, the protein acts as an antigen that is recognized as foreign by the host, stimulating an immune response that results in the production of antibodies, such as antibodies to a disease-causing (pathogenic) microbe. By consuming the plant, protection against subsequent infection by the pathogen can be conferred.

Edible vaccines are hampered by the need for digestion of the plant. Digestion is a very destructive process for molecules such as proteins. One way researchers have tried to lessen this barrier is to use a target molecule that does not need to be fully intact to elicit an immune response. Peptides are especially useful.

The industrial-scale use of microorganisms in biopharmaceutical production has created the need for

example, human insulin produced by *E. coli* must be exactly the same as that normally produced in people).

A variety of molecules can function as biopharmaceuticals. These include proteins, peptides (fragments of the entire proteins that do not have the intended biological function of the particular protein, but which can elicit a beneficial reaction, such as the production

Marine organism samples in the PharmaMar laboratory in Madrid. Founded in 1986 by Jos-Maria Fernandez-Sousa Ph.d., PharmaMar is Spain's principal biopharmaceutical company and one of the world's leaders in research and development of antitumor compounds from marine organisms. The company claims a strong commitment to good corporate citizenship and protecting the resources of the sea. © Pedro Armestre/AFP/Getty Images.

A wintergreen plant grown by Phytomedics, Inc., a small biotech company founded by researchers at Rutgers University, is seen at a greenhouse on the campus of Cook College in New Brunswick, New Jersey. Phytomedics produces plants with exact concentrations of specific compounds with medicinal properties. The compounds are being used to develop everything from prescription medicines to dietary supplements, cosmetic ingredients, and foods with distinct health benefits. *© AP Images/Mike Derer.*

large-scale techniques of organism growth and product extraction/isolation. In contrast to the beakers containing a few cupfuls of growth medium commonly used in a laboratory, the production on an industrial scale involves bioreactors that hold thousands or tens of thousands of gallons of culture. Ensuring that the growth conditions in such a large volume of fluid remain constant in the short-term (i.e., the days or weeks that particular batch is being grown) and in the long-term (i.e., from batch-to-batch in the months and years that a facility is operating) is a daunting challenge. The dynamics of fluid movement can be very different in a large volume of fluid compared to the far smaller volume that is typically used in the laboratory. This means that the delivery of nutrients and removal of noxious waste products can be different in the industrial setting, sometimes in ways that can be detrimental to growth and production of the biopharmaceutical. Designing an industrial system that is reliable over time and permits the efficient and abundant production of the target compound is, thus, very important.

A specialized form of biopharmaceutical is the monoclonal antibody. A monoclonal antibody is an antibody that is produced exclusively (no other antibody types are made) by an engineered cell that is a fusion between a cell type that undergoes limitless cycles of growth and division without altering the cell characteristics and the genetic information for the production of the particular antibody. When such cells are cultured in the large volumes described above, huge amounts of the particular monoclonal antibody are produced. These can then be introduced by injection or other routes to bestow protection.

■ Impacts and Issues

The interest in biopharmaceuticals, particularly in their for-profit use, has grown enormously since the development of techniques of molecular biology in the 1970s and the subsequently ongoing refinement of these techniques. By 2010, for example, the budget for research and development of biopharmaceuticals by companies in the United States was approximately $65 billion annually.

Another indicator of the explosive growth of biopharmaceuticals is the number of patents granted for the exclusive manufacturing rights. In the United States and

Europe, the number of patents issued for biopharmaceuticals in 1978 was 30. By 1995, 15,600 patents were issued. By 2001 the number had more than doubled to 34,527. The trend continues.

The market for biopharmaceuticals continues to grow. For companies, the prospect of huge revenue is enticing. The result can be beneficial for the treatment of a variety of ailments. However, the approach is not without its critics, who decry the engineering of life-forms, especially plants, as vehicles of therapy. A central part of this debate is the patent process. Whereas a patent is designed to provide a company with a period of guaranteed marketplace exclusivity to the product, it results in the access to a food source being under the control of one or a few private firms. This market-driven food monopoly is of great concern to some.

SEE ALSO *Biotechnology Products; Drug Price Competition and Patent Term Restoration Act of 1984; Patents and Other Intellectual Property Rights*

BIBLIOGRAPHY

Books

Bucknell, Duncan Geoffrey. *Pharmaceutical, Biotechnology, and Chemical Inventions: World Protection and Exploitation.* Oxford: Oxford University Press, 2011.

Gross, Marc S., S. Peter Ludwig, and Robert C. Sullivan. *Biotechnology and Pharmaceutical Patents: Law and Practice.* New York: Aspen Publishers, 2008.

Prasad, S. K. *Pharmaceutical Research and Development.* New Delhi: Discovery Pub. House, 2011.

Rathore, Anurag S., and G. K. Sofer. *Process Validation in Manufacturing of Biopharmaceuticals: Guidelines, Current Practices, and Industrial Case Studies.* Boca Raton, FL: Taylor & Francis, 2005.

Ukwu, Henrietta. *Global Regulatory Systems: A Strategic Primer for Biopharmaceutical Product Development and Registration.* Boston: CenterWatch, 2011.

Brian Douglas Hoyle

Bioplastics (Bacterial Polymers)

■ Introduction

The term plastic usually refers to a class of synthetic materials first introduced in the early 1900s. Subsequently a host of other plastic materials were invented, with familiar-sounding names such as nylon, rayon, and PVC (short for polyvinylchloride). The properties of plastics range from the thin, soft, and flexible plastic wrap used in kitchens, to strong, stiff, heat-resistant parts used in vehicles and aircraft.

Most plastics manufactured in the early 2010s are made from chemicals derived from fossil fuels such as petroleum (oil). However, some plastics are composed of chemicals derived from biomass—such plastics are referred to as bioplastics. Biomass refers to materials from plants, microorganisms, and other living things.

Bioplastics represent a small proportion of all plastics produced, in part due to their typically higher costs when compared to plastics made from fossil fuels. Nevertheless, bioplastics are gaining a bigger share of the plastics market. One reason for the market share increase is the rising cost of oil, which makes the plastics derived from oil more expensive, and hence bioplastics more cost competitive. Another reason for the increased use of bioplastics is that biologically-derived materials are generally considered to be more environmentally-friendly than traditional plastics derived from fossil fuels.

The terms biodegradable and bioplastic should not be confused. A material is biodegradable if it decomposes in the soil or other natural environment fairly rapidly. Bioplastics may or may not be biodegradable; the same is true for conventional plastics made from fossil fuels. Indeed, some companies have created plastic bags and other items from conventional plastics that decompose in the soil in a matter of weeks.

■ Historical Background and Scientific Foundations

All plastics are made from molecules called polymers. Most polymers are organic (contain the element carbon), but there are some types that are inorganic, such as silicone. Polymers are composed of smaller subunits called monomers. In different polymers, the type of monomer, as well as the way in which they are linked together, can vary.

Rubber originally was made from latex, a naturally-occurring polymer that forms the sap of certain plants. Rubber is thought to have been introduced to France in the early 1700s. In Europe, natural rubber was used to waterproof cloth, but it had a tendency to crack and split. American inventor Charles Goodyear (1800–1860) discovered the vulcanization of rubber around 1839, from which a semi-synthetic rubber could be formed with qualities much superior to those of natural rubber.

The first purely synthetic plastic, called Bakelite after its inventor Leo Baekeland (1863–1944), was introduced in the early twentieth century. After the introduction of Bakelite, many other plastics followed with names such as polyethylene, polystyrene, polyvinylchloride (PVC), and polyamide (nylon).

The production of plastic begins with basic chemicals, which can vary depending upon the type of plastic. For instance, Bakelite is formed from the chemicals phenol and formaldehyde. The production of polystyrene begins with the chemicals benzene and ethylene. Both these chemicals typically are derived from petroleum. Reaction of the two chemicals creates ethylbenzene. This synthesized chemical is exposed to heat and catalysts (compounds that speed up chemical reactions), resulting in liquid styrene. Styrene is the monomer (subunit) that forms polystyrene. Styrene molecules are linked together, or polymerized, using heat, pressure, and

WORDS TO KNOW

BIOMASS: Any sort of material recently derived from living things such as plants, animals, or microorganisms. The qualifier "recently derived" is added to distinguish biomass from fossil fuels and similar substances that were produced millions of years ago from the remains of plants and animals.

FOSSIL FUEL: Hydrocarbons (molecules containing only hydrogen and carbon atoms) derived from ancient organisms that typically lived millions of years ago. Petroleum, coal, and natural gas are all examples of fossil fuels.

GENETIC ENGINEERING: The human alteration of the genetic material of an organism. Recombinant DNA is the technique used to perform genetic engineering.

GREENHOUSE GAS: Any gas in Earth's atmosphere that contributes to the greenhouse effect, wherein solar radiation (such as visible light) largely passes through such gases, which block infrared radiation from Earth or the atmosphere from escaping into space.

MICROORGANISM: Any organism that is too small to be seen unless magnified, as with the use of a microscope.

SUSTAINABILITY: The use of natural resources such that the resources in question are not significantly depleted.

An NEC Corp. researcher unveils bioplastic made by bonding cardanol, a primary component of cashew nut shells, and cellulose, a main component of plant stems, in Tokyo, Japan, on August 25, 2010. *© AP Images/Kyodo.*

chemical reactions to produce the polymer called polystyrene. Generally speaking, polystyrene, along with most other modern plastics, is produced using chemicals derived from petroleum or natural gas.

Bioplastics are a specific type of organic or semi-organic plastic—the basic chemicals from which bioplastics are made are derived from some sort of biomass. For instance, in the United States and many other countries large quantities of ethanol are produced from corn and other grains. Also, sugars are derived from sugar cane and sugar beets. The basic compounds of ethanol and various sugars can be processed into chemicals, which in turn can be made into bioplastics.

Besides plants, microorganisms such as bacteria, as well as certain types of algae, show great potential for producing the chemical constituents of plastics. Like plants, algae use photosynthesis to grow. In the early 2010s, algae are being researched intensively to produce large quantities of biofuels to power cars and trucks. Such fuels, including biodiesel, are chemically similar to petroleum-based fuels. Because chemicals used to make plastics are derived from petroleum, those same chemicals also could be derived from the biofuels that researchers one day hope will be produced in large quantities by algae.

Bacteria are another group of promising microorganisms for producing chemicals for plastics. Bacteria have long been cultured in large vats on an industrial scale to yield beneficial substances such as penicillin and insulin. More recently, bacteria have been genetically engineered to produce precursor (starting) chemicals for plastics.

A more direct approach has also been demonstrated: Instead of culturing bacteria that produce chemicals, which in turn are further processed to make the actual plastic, some researchers have developed strains of bacteria that produce the actual polymer of which the plastic is composed. The polymers are harvested directly from the bacteria to produce bioplastic.

■ Impacts and Issues

There are a number of reasons to replace traditional plastics—almost all of which are derived from fossil fuels—by bioplastics derived from microorganisms and other biomass. Petroleum is a limited resource in the world. In the early years of the petroleum industry, oil deposits that were relatively close to the surface and easy to drill into were tapped first. Although there still exist large reservoirs of oil in the world that are fairly easy to extract, in other regions oil is being recovered from sources that were once considered not economical to utilize. Besides being very costly as compared to more traditional oil deposits, these more recently developed techniques raise significant environmental concerns. For instance, oil wells are being drilled into petroleum deposits that are far beneath the ocean floor. For example, in 2009, the *Deepwater Horizon* rig was able to drill some of the deepest undersea oil wells up to that time, to a depth of about 6 miles (9.7 km). In 2010, the Gulf of Mexico oil spill occurred in the wake of an explosion that destroyed the *Deepwater Horizon* oil platform and killed 11 crewmen. The Gulf spill released nearly 5 million barrels of crude oil before the leak at the bottom of the Gulf was sealed, becoming the largest marine oil spill in history. Tremendous damage to both the environment and regional economies resulted from the nearly three-month leak.

Also, the extraction of oil embedded in the Canadian tar sands is increasing each year. A large quantity of oil is trapped within the Canadian tar sands, mostly located in the province of Alberta. The United States now imports more oil from Canada than any other country, and much of that oil is derived from Canada's tar sand deposits. Unfortunately, the oil is mixed with sand, clay, and soil, and forms a thick, viscous paste that requires heat (usually steam) to extract. In the process to recover the oil, topsoil is destroyed; also, the energy value of the oil is lessened because of the energy needed to heat it; and greenhouse gases such as carbon dioxide are released by the extraction process. Increased worldwide demand for oil is driving these measures to increased production, in spite of the environmental impacts.

Natural gas is another fossil fuel that is used to form the basic chemical units used to make plastics. The supply of natural gas in the United States has been increasing. However, much of this increase in supply is due to a method of liberating the gas from shale deposits (a kind of rock) called hydraulic fracturing, or fracking. Some people are concerned that this method of injecting liquid under high pressure to fracture the shale may contaminate ground water supplies (although fracking is typically carried out thousands of feet below most ground water reservoirs) and potentially trigger small earthquakes.

By deriving the materials needed to make plastics from biological sources, the use of oil and natural gas can be reduced. The production of chemicals needed to produce plastics from plants and microorganisms is referred to as a sustainable process. This means that the basic feedstock used to make bioplastic, whether it come from plants or microbes, can be renewed continuously, as opposed to the Earth's oil and natural gas reserves, which are limited.

One factor limiting the greater utilization of bioplastics is their higher cost compared to conventional plastics derived from fossil fuels. This situation may change if the price of fossil fuels—especially that of petroleum—increases dramatically. Additionally, the use of microorganisms holds out great promise for producing bioplastics that are more affordable. Bacteria especially are the focus of intensive research, either to make the polymer materials directly, or to produce chemicals that in turn are processed into plastics. Genetically-engineered bacteria have been used to produce butadiene, a chemical used in the production of carpeting, latex, and the rubber used in tires. Another line of bacteria has been used to produce limited quantities of a chemical called butanediol (BDO), which is used to create the plastics in products such as running shoes, sports clothes, and computers.

Scientists have also developed particular types of bacteria that will produce the polymers of plastics directly. For instance, researchers have demonstrated that a particular strain of the soil bacterium *Pseudomonas putida* can convert the chemical styrene into the polymer polyhydroxyalkanoate (PHA). The PHA produced by the bacteria is a polymer that can be used directly to make various plastic products. Styrene is used industrially to make the plastic polystyrene, but the chemical also ends up in industrial wastes of various sorts, and disposal of it is difficult. The *Pseudomonas putida* bacteria identified by the researchers could turn toxic wastes into a useful polymer for making plastic products.

Many companies produce bioplastic products, mostly from plant byproducts such as sugars. Their market is currently limited, but growing. In many workplaces and universities, for example, bioplastic products have replaced food service items, janitorial supplies, and additional items in everyday use. Several companies are also working on producing chemicals for plastics, or the plastics themselves in the form of polymers, from microorganisms such as bacteria. Ultimately, the marketplace success of bioplastics most likely will depend upon a combination of factors, including the cost of fossil fuels, government incentives to manufacturers for producing bioplastics, and the attitude of consumers towards purchasing plastics made from renewable resources.

SEE ALSO *Agricultural Biotechnology; Agrobacterium; Algae Bioreactor; Biochemical Engineering; Biodegradable Products and Packaging; Bioglass; Cornstarch Packing Materials; Environmental Biotechnology; Green Chemistry*

BIBLIOGRAPHY
Books

Clark, James H., and Fabien E. I. Deswarte. *Introduction to Chemicals from Biomass.* Hoboken, NJ: John Wiley & Sons, 2008.

National Institute of Industrial Research (India) Board of Technologists. *Hand Book on Biodegradable Plastics: Eco-Friendly Plastics.* Delhi, India: Publication Division, National Institute of Industrial Research, 2007.

Smith, Ray, ed. *Biodegradable Polymers for Industrial Applications.* Boca Raton, FL: CRC Press, 2005.

Tolinski, Michael. *Plastics and Sustainability: Towards a Peaceful Coexistence between Bio-Based and Fossil Fuel-Based Plastics.* Hoboken, NJ: John Wiley & Sons, 2012.

Web Sites

Dell, Kristina. "The Promise and Pitfalls of Bioplastic." *Time*, May 3, 2010. http://www.time.com/time/magazine/article/0,9171,1983894,00.html (accessed November 2, 2011).

Green Plastics. http://green-plastics.net/ (accessed November 2, 2011).

Royte, Elizabeth. "Corn Plastics to the Rescue." *Smithsonian .com*, August 2006. http://www.smithsonianmag.com/science-nature/plastic.html (accessed November 3, 2011).

Philip E. Koth

Biopreservation

■ Introduction

The term biopreservation refers to a variety of techniques used for the preservation of biological materials, such as food products, pharmaceutical bio-compounds, life sciences research materials such as human animal or plant tissues, seeds, sex cells (sperm and eggs), surgically removed specimens, DNA, blood or other body fluids, bacteria, and viruses. Therefore, biopreservation sciences are comprised of three main segments: food preservation, medical research, and species biobanking. Biopreservation techniques are used in a wide range of industrial, commercial, and scientific activities including the food processing industry, cosmetic and medication manufacturing, clinical assays, forensic sciences, and disease research.

■ Historical Background and Scientific Foundations

Food biopreservation is a modern concern in view of the need for transporting large amounts of perishable products long distances, and to store and distribute them in urban centers far away from the region where the food was produced. Whether *in natura* or precooked, the risk of food contamination and food poisoning remains high, in spite of the widespread use of refrigeration, chemical preservatives, and pasteurization. Therefore new technologies, involving addition of beneficial microorganisms or their byproducts to food, are being developed to improve food safety and decrease the use of chemical additives.

Biopreservation is also important in medical research and drug development. The collection and study of large numbers of samples extracted from patients with infectious diseases enables scientists to compare variations and features of each disease and record its natural history, including the pattern of disease onset and progression. Better understanding of what causes a disease and how it starts and develops leads not only to

improved diagnostic methods, but also to the development of new therapeutic protocols, vaccines, and drugs.

Human exploitation of natural resources (mining, deforestation, excessive hunting and fishing, agricultural malpractices, and environmental pollution) has resulted in the extinction of innumerable species and pushed others to the brink of extinction. The growing understanding that all life forms are interdependent within their ecology, including humans, has led scientists to develop projects for genetic sequencing and DNA preservation of endangered plant and animal species.

■ Impacts and Issues

Improvements in food preservation technologies, as well as measures to prevent foodborne illnesses or the onset of food-related epidemics, are important public health issues. According to the World Health Organization (WHO), foodborne diseases are the second cause of morbidity (illness) in Europe, with up to 300,000 yearly cases of gastroenteritis per each million population. In the United States, between 250 and 350 million people have at least one acute gastroenteritis infection per year. According to the U.S. Centers for Disease Control and Prevention (CDC), approximately 25 percent of Americans experience at least one food-related illness per year. The CDC states that the most common bacterial pathogens found in food are *Escherichia coli, Staphylococcus aureus, Salmonella, Clostridium botulinum*, and *Campylobacter jejuni*. Depending on the age and general health status of the affected victim, these food-related diseases sometimes can be fatal.

Since the early 1990s, a new approach to food preservation has been taken with the development of biotechnologies based on the understanding of how some bacterial properties and by-products can be used as natural and safer protectants of perishable food products. These new biological methods for food preservation take advantage of the natural ability of some microorganisms

WORDS TO KNOW

CRYOPRESERVATION: A method for storage of living cells or tissues at ultra-low-temperatures, to be later revived for use (e.g., in vitro fertilization) or study (e.g., stem cell research).

DNA: Deoxyribonucleic acid, or DNA, is a complex molecule containing the genetic code of a plant, animal, or bacteria.

GASTROENTERITIS: Stomach and intestinal infection caused by bacteria or virus present in contaminated food, soil, or water.

IN NATURA: In its natural state, not processed by cooking, smoking, or other methods of alteration.

MORBIDITY: Disease incidence.

PASTEURIZATION: Method of sterilizing wine, milk, and other beverages developed by French physician Louis Pasteur (1822–1895) and Claude Bernard (1813–1878), in 1862.

PATHOGENS: Microorganisms or substances that cause diseases.

SURGICAL SPECIMENS: Samples of damaged organs, tumors, or other abnormal body tissue removed during surgery.

to kill pathogens. One such case includes lactic acid bacteria (LAB) and their antimicrobial compounds (bacteriocins), which are used for preserving perishable foods for longer periods of time. Lactic acid kills most harmful foodborne bacteria. LAB bacteriocins are small proteins (peptides) present in lactic acid, which are nontoxic to humans or animals and are usually non-allergenic. In some cases, this biological preservation method is used in combination with other physical or chemical preservation technologies. The combination of several preservation methods is known as hurdle technology.

Because bacteriocins are resistant to temperatures used in sterilization and pasteurization processes, and are destroyed only during digestion in the stomach, they can be combined with other methods to increase food quality and duration. LAB-derived bacteriocins and other types of bacteriocins have been tested in dairy products, vegetable products, meat, and meat products. Examples of hurdle technologies applied to improve food safety are the combined use of high or low temperature sterilization, mild chemical additives, vacuum packaging, pH control, LAB or other bacteriocins, and a modified atmosphere.

In addition to the usefulness of biopreservation for food, medical research and drug experimentation also benefit from research into various methods of biopreservation. Preserved body tissues that were extracted during

Fishermen dry fish in the traditional way for preservation at Lotfoten in the north of Norway.   © *Andreas Gradin/Shutterstock.com.*

Lactic acid bacteria added to packaged beef has been shown to reduce the growth of harmful bacteria.   © *Losevsky Pavel/Shutterstock.com.*

surgery led to a better comprehension of the characteristic features associated with many diseases and also allowed scientists to develop new diagnostic methods and tests. This methodology is known as histopathology or the study of tissues. Many surgical specimens are preserved in paraffin blocks, an old, inexpensive, and effective way of preventing decay in the specimen.

Studies of other diseases, however, are better performed with preserved live specimens, such as live cancer cells cultures in Petri dishes or the cryopreservation of body fluids or DNA for further studies. One example of the importance of conserving large numbers of specimens or samples extracted from different individuals with the same disease is shown by a study conducted in the early 1980s using more than 150,000 blood samples from women who died of breast cancer. Scientists discovered that all 150,000 breast cancer patients had DDT in their blood, which raised a red flag concerning the use of DDT in agriculture. Prompted by those results, new investigations were conducted worldwide, which finally proved that small amounts of DDT, when present in food, accumulate in the blood and other tissues throughout the years. This accumulation is associated with developing breast and prostate cancers. As a result, the use of DDT was banned throughout most of the world. Since that time, however, DDT has been used

again in limited amounts to help reduce insect populations in areas at high risk for malaria and other diseases.

Other preserved live cells and tissues that increasingly are stored in biobanks include umbilical cord blood, which is rich in stem cells that can be used to treat some blood disorders and genetic disorders, and reproductive cells (ova and sperm). However, the American Academy of Pediatrics (APA) currently discourages physicians from recommending private cord blood banking. This recommendation is based on a lack of evidence that an autologous cord blood transplant is effective in treating cancer; the likelihood that an individual child will need their own stored cells has not been determined; and private collection and storage is expensive. The APA does recommend cord blood donations to public banks that are part of the National Marrow Donor Program network.

The large collections of biological samples stored at academic institutions or commercial specialized facilities are known as biobanks. The CDC and many university research departments have their own biobanks, whereas biotechnology industries and agribusinesses sometimes outsource sample storage to commercial biobanks. The CDC collects and preserves a large variety of contagious viruses and bacteria involved in past epidemics in order to compare them with newly mutated lineages and to be able to develop more effective vaccines and antibiotics.

The World Health Organization (WHO) recommends that countries keep tumor banks of several types of cancer. WHO itself maintains a World Tumor Bank as do several American cancer research centers, including the National Cancer Institute. Under the leadership of Brazil's National Institute of Cancer, a network of tumor and DNA banks was formed throughout the Latin American countries in the early 2000s. All tumor and DNA banks share information and samples, thus allowing investigators throughout the world to study different molecular and genetic aspects of a given tumor type. New therapies and cancer vaccines have been developed as a result of this joint effort and shared information among the banks.

Species biobanking or ecobanking refers to efforts to preserve animal DNA, semen, and eggs, as well as plant seeds through preservation technologies in the hope of being able to preserve and increase the gene pool of threatened species. Reduced populations of a given species lead to a loss of genetic variability and the likelihood of increased abnormalities in the offspring, such as genetic defects or degenerative diseases associated with inbreeding. Genetic variability is crucial for species survival because it prevents the prevalence of unfavorable traits in the general population. Confined small populations are at risk of passing degenerative traits to an ever-increasing number of descendants, condemning the species to extinction.

Ecobanking policies have met strong opposition from some environmentalist groups and biology researchers, however, who argue that better than preserving DNA, semen, and eggs of endangered species in highly expensive facilities, more resources and stronger efforts should be allocated to prevent near-extinction situations. Many critics instead call for laws and regulations to limit deforestation, to manage land, and to avert nonsustainable use of land and water near natural ecosystems.

SEE ALSO *Biotechnology; Biotechnology Products; Blood Transfusions; Blood-Clotting Factors; Cell Therapy (Somatic Cell Therapy); Cryonics; Egg (Ovum) Donors; Food Preservation; Forensic Biotechnology; Frozen Egg Technology; Genbank; Gene Banks; In Vitro Fertilization (IVF); Stem Cell Lines; Tissue Banks; Tissue Engineering*

BIBLIOGRAPHY

Books

Baust, John G., and John M. Baust, eds. *Advances in Biopreservation.* Boca Raton, FL: CRC/Taylor & Francis, 2007.

Lacroix, Christophe, ed. *Protective Cultures, Antimicrobial Metabolites and Bacteriophages for Food and Beverage Biopreservation.* Philadelphia: Woodhead, 2011.

Theron, Maria M., and J. F. Rykers Lues. *Organic Acids and Food Preservation.* Boca Raton, FL: Taylor & Francis, 2011.

Toldra, Fidel, ed. *Meat Biotechnology.* New York: Springer, 2008.

Vittal, Ravishankar Rai, and Rajeev Bhat. *Biotechnology: Concepts and Applications.* Oxford: Alpha Science International, 2009.

Periodicals

Cambon-Thomsen, Anne, Gudmundur A. Thorisson, and Laurence Mabile. "The Role of a Bioresource Research Impact Factor as an Incentive to Share Human Bioresources." *Nature Genetics* 43, no. 6 (June 2011): 503–504

Clark, Brian J., John M. Baust, and Glyn Stacey. "How much Will the Biobanking Industry Come to Rely on Private Companies?" *Biopreservation and Biobanking* 8, no. 4 (2010): 179–180.

Galvez, Antonio, et al. " Bacteriocin-based Strategies for Food Biopreservation." *International Journal of Food Microbiology* 120, nos. 1–2 (November 30, 2007): 51–70.

McDonald, Sandra A. "Principles of Research Tissue Banking and Specimen Evaluation from the Pathologist's Perspective." *Biopreservation and Biobanking* 8, no. 4 (November 2010): 197–201.

Perrera, Frederica P., and I. Bernard Weinstein. "Molecular Epidemiology and Carcinogen-DNA Adduct Detection: New Approaches to Studies of Human Cancer Causation." *Journal of Chronic Diseases* 35, no. 7 (1982): 581–600.

Web Sites

"Donate Cord Blood." *National Marrow Donor Program.* http://marrow.org/Get_Involved/Donate_Cord_Blood/Donate_Cord_Blood.aspx (accessed October 31, 2011).

Office of Biorepositories and Biospecimens Research, U.S. National Cancer Institute. http://biospecimens.cancer.gov/default.asp (accessed October 31, 2011).

Sandra Galeotti

Bioreactor

Introduction

A bioreactor is a vessel in which cells use biochemical processes to transform raw materials into useful products. The biotechnology industry relies upon well-designed bioreactors for many different types of operation, from the manufacture of biopharmaceutical drugs to processing municipal waste.

Cell function relies upon the conditions inside the bioreactor, so factors such as oxygen concentration, pH, temperature, and the composition of nutrient medium used to feed the cells must be very carefully controlled, usually by automated systems. The manufacturing bioreactor must also be designed so that it can be thoroughly cleaned at the end of the production process, as contamination can adversely affect the quality of the product. Cells that are commonly cultured in bioreactors to produce pharmaceuticals include bacteria, yeast, mammalian, and insect cells. Conventional bioreactors are usually made of stainless steel, but there is an increasing trend towards using disposable bioreactors made of plastic. Bioreactors come in various sizes, depending on their application.

There are miniature bioreactors, used in the research laboratory, and bioreactors with volumes of thousands of liters producing commercial quantities of drugs such as monoclonal antibodies. Living organisms can also act as bioreactors, when transgenic plants and animals manufacture the proteins coded for by their transgenes and secrete them in their leaves, or milk, respectively.

Historical Background and Scientific Foundations

The first bioreactors were probably the containers used in the fermentation of grains and fruits to make beer and wine, a process that dates back thousands of years. The earliest documented vessels used in brewing were wooden casks, but these were superseded from the nineteenth century by stainless steel vessels, which are easier to clean and sterilize. There are two basic types of bioreactor, and several variants of these two types. In the batch bioreactor, all the raw materials, including the cells, are added at once, and the vessel then sealed until the process is complete. By contrast, the continuous flow bioreactor has materials flow through it continually as the process takes place. Continuous flow is used to process waste, rendering it harmless by microbial action within the bioreactor. It also may produce a form of energy, such as biogas.

There are a number of factors that need to be carefully and continually controlled within a bioreactor. It will usually be fitted with a stirrer, to ensure that cells have access to nutrients and oxygen, and with some form of oxygen supply. The pH of the medium will also be measured continually, because cells will die if this falls outside a neutral range. All of this process control must be done without breaching the sterile conditions within the bioreactor. Once the process is complete and product harvested, the bioreactor must be cleaned and sterilized, generally using steam, before further processes can be started.

Impacts and Issues

Advances in biotechnology have greatly expanded the scope of the bioreactor beyond the conventional stainless steel vessel. There has been a trend towards the use of the single-use bioreactor, made from a special polymer. After one process, the bioreactor is discarded. This saves on cleaning and sterilization costs, as well as reducing the capital investment needed to set up a biotechnology process. Single-use technology has led to the development of mobile, lower cost production units. With these, vaccine and other biopharmaceutical production can be set up in less developed countries, saving the cost of importing these drugs from the West.

Biopharmaceutical drugs are expensive, and there has been an increasing focus on alternative production systems which could bring down the cost per dose.

WORDS TO KNOW

AEROBIC: An organism requiring oxygen to function is described as aerobic. The term may also be applied to any process taking place in a bioreactor that requires oxygen.

ANAEROBIC: An organism that does not require oxygen to function is described as anaerobic. Oxygen may be toxic to some anaerobes. The term may also be applied to any process taking place in a bioreactor that does not require oxygen and, in some cases, it may be important to exclude oxygen.

FERMENTATION: The process by which yeast turns the sugar in fruits and grains into alcohol and carbon. The term is often used more generally to describe any microbial biotransformation taking place in a bioreactor.

A few biotechnology companies have developed technology whereby therapeutic human proteins can now be produced, in commercial quantities, in the milk of transgenic rabbits or cows. There have also been a number of clinical trials of edible vaccines, in which vaccines against hepatitis B and cholera have been manufactured in the leaves of transgenic plants such as tobacco or tomato. In these instances, the animal or plant is the bioreactor, with herds or crops replacing the traditional biopharmaceutical production unit with stainless steel bioreactors. This kind of transgenic technology is known as pharming or molecular farming, and it will probably be many years before it reaches its true potential as a low cost and efficient way of manufacturing biopharmaceuticals.

Meanwhile, the application of bioreactor technology to the treatment of municipal waste offers many environmental benefits and is being actively researched and promoted, particularly by the U.S. Environmental Protection Agency. In a landfill bioreactor, waste water and solid waste are broken down by either aerobic or anaerobic bacteria producing methane gas, which can be used on site or sold. Bioreactor treatment uses less land than conventional landfills and produces less hazardous byproducts. It is also capable of breaking down waste in years, rather than decades.

SEE ALSO *Agrobacterium; Algae Bioreactor; Tissue Engineering; Wastewater Treatment*

Process engineers Ranti Ogunsola, left, and Justine Niamke watching over the medium exchange on a bioreactor growing cells during the first stage of production. © *Washington Post/Getty Images.*

BIBLIOGRAPHY

Books

Chaudhuri, Julian, and Mohamed Al-Rubeai. *Bioreactors for Tissue Engineering: Principles, Design and Operation.* Dordrecht: Springer, 2005.

Judd, Simon. *The MBR Book: Principles and Applications of Membrane Bioreactors for Water and Waste Water Treatment.* Amsterdam: Elsevier, 2011.

Wells, George Fraser, Craig Criddle, Christopher Francis, and Perry L. McCarty. *Reexamining the Engineered Nitrogen Cycle Microbial Diversity, Community Dynamics, and Immigration in Nitrifying Activated Sludge Bioreactors* (Ph.D. Thesis). Stanford University: 2011. Available online at http://purl.stanford.edu/zn813dy5100

Susan Aldridge

Bioremediation

■ Introduction

Bioremediation refers to the use of natural processes to remove contaminants from the soil or water. Naturally occurring or deliberately introduced microorganisms, plants, or enzymes break down environmental contaminants into water or other harmless substances. Whereas bioremediation occurs naturally, biologists, environmental engineers, and other scientists apply bioremediation technologies to expedite these natural processes by increasing the metabolic processes of organisms.

Bioremediation technologies may employ either in situ technologies, which remove toxic substances at the site of contamination, or ex situ technologies, which require the removal of the contaminated soil or water for treatment at a remote location. In situ bioremediation technologies are often cheaper, less likely to expose workers or others to the contaminants, and less likely to spread the toxic material through the air or water. When the contaminated soil or water is easily accessible near the surface, ex situ bioremediation technologies may remove contaminants more quickly that in situ technologies.

■ Historical Background and Scientific Foundations

Bioremediation utilizes the metabolic processes of organisms to remove toxins. The process occurs naturally, but humans often desire to accelerate the removal of elements or compounds that are toxic or harmful to humans or the environment. Bioremediation generally relies on promoting the growth of naturally occurring organisms, such as microbes, by making the environment hospitable to growth.

The introduction of nutrients, oxygen, or other substances often is used to promote the metabolic processes of the desired organisms in a process known as biostimulation. Occasionally, specific naturally-occurring or genetically-engineered organisms are introduced to the environment

to expedite the removal of toxins. The use of oil-eating bacteria to clean up oil spills is a well-known example of biostimulation. Bioventing, a process in which oxygen and nutrients are introduced to groundwater to encourage the growth of beneficial microbes, is a common form of biostimulation. Bioventing typically is used to remove petroleum compounds or volatile organic compounds.

Phytoremediation is a commonly employed bioremediation technique in which specific plants are introduced to clean up the environment. Phytoremediation involves utilizing certain plants to remove heavy metals, petroleum compounds, pesticides, solvents, or other contaminates from soil, water, or the air. The type of plant used in a phytoremediation project must be suited for the particular climate and target contaminant. Phytoremediation typically involves long-term projects aimed at removing chronic contamination through the bioaccumulation of the contaminant in plants. Plants must be removed and disposed of properly once they have absorbed the contaminant. Phytoremediation has several advantages, including low cost and no excavation, a process that can spread contaminants through airborne particulates or by leaching into waterways.

Mycoremediation is another form of bioremediation in which particular organisms, in this case, fungi, are introduced to the environment to aid the removal of contaminants. Mycelia, or the vegetative portion of fungi, secrete enzymes that decompose hydrocarbons, chlorinated compounds, and other compounds. Mycoremediation is useful for in situ treatment of soil containing petroleum, pesticides, or other contaminants. Mycoremediation may also be used to treat contaminated water moving through soil in a process known as mycofiltration.

■ Impacts and Issues

Bioremediation is a safe process that relies on naturally-occurring microbes that are not harmful to humans. People should take precautions at bioremediation sites, however, and should avoid contact with soil or water that contains contaminants targeted by the bioremediation process. Most bioremediation processes break down the components of contaminants into water, carbon dioxide, or other harmless compounds. Some bioremediation processes, however, may increase the risk of human or animal exposure to contaminants if not performed properly. Phytoremediation, for example, traps heavy metals and other contaminants within plants. If these plants are not removed and disposed of properly and are allowed to decompose, then the plants will deposit bioaccumulated contaminants on the surface.

Targeted bioremediation processes allow for in situ remediation in some instances when ex situ remediation was required previously. Bioremediation processes, such as bioventing, phytoremediation, and mycoremediation, remove contaminants without excavation, thereby reducing remediation costs and reducing the likelihood of spreading contaminants to other areas. Some forms of bioremediation also remove contaminants from areas where excavation and treatment would be impractical or impossible. Bioventing removes contaminants from groundwater trapped in soil or located in fractures in rock formations below the surface.

The duration of bioremediation projects may vary from a few months to several years, based on several environmental and site-specific factors. The type and amount

of contaminants will affect the duration of a bioremediation project. Larger sites or sites where contaminants are located at greater depths will lengthen the time required for cleanup. The type of soil and whether weather conditions are favorable to microbial, fungal, or plant growth will also affect the duration of a bioremediation project.

A bioremediation landfill contains soil contaminated with crude oil in Ecuadorian Amazon.   © *Morley Read/Alamy.*

Ultimately, however, the effectiveness and expediency of a specific bioremediation project requires knowledge about biological pathways and the implementation of the best methods and processes for that project.

SEE ALSO Diamond v. Chakrabarty; *Bioremediation: Oil Spills; Patents and Other Intellectual Property Rights; Phytoremediation; Soil Modifying Bacteria; Wastewater Treatment*

BIBLIOGRAPHY

Books

Ahmad, Wan Azlina, Zainoha Zakaria, and Zainul Akmar Zakariab, eds. *Bacteria in Environmental Biotechnology: The Malaysian Case Study-Analysis, Waste Utilization and Wastewater Remediation.* New York: Nova Science Publishers, 2011.

Atlas, Ronald M. and Jim Philp, eds. *Bioremediation: Applied Microbial Solutions for Real-World Environmental Cleanup.* Washington, DC: ASM Press, 2005.

Fingerman, Milton, and Rachakonda Nagabhushanam, eds. *Bioremediation of Aquatic and Terrestrial Ecosystems.* Enfield, NH: Science Publishers, 2005.

Singh, Ajay, and Owen P. Ward, eds. *Biodegradation and Bioremediation.* New York: Springer, 2004.

Trivedi, Pravin Chandra. *Bioremediation of Wastes and Environmental Laws.* Jaipur, India: Aavishkar Publishers, Distributors, 2010.

Web Sites

"Emergencies: Oil Spills." *United States Environmental Protection Agency (EPA).* http://www.epa.gov/ebtpages/emeroilspills.html (accessed October 27, 2011).

"Treatment/Control: Treatment Technologies: Bioremediation." *United States Environmental Protection Agency (EPA).* http://www.epa.gov/ebtpages/treatreatmenttechnbioremediation.html (accessed October 27, 2011).

Joseph P. Hyder

Bioremediation: Oil Spills

Introduction

Oil spill bioremediation is the process of using microorganisms and their enzymatic action to return a body of water to its natural state after it has been fouled with oil. The ever-increasing demands of global industry and the distances over which petroleum is transported have made the world's ocean shipping routes prominent examples of how ineffective controls over vessel hull design, crew training, and proper navigation techniques can place entire ecosystems at risk in the event of an oil spill. Oil spills have been synonymous with environmental disasters since the 1967 wreck of the supertanker *Torrey Canyon* in the English Channel.

Modern oil exploration techniques have also heightened the risk that the accidental release of crude oil or natural gas into the environment can pose long-term threats to all immediate human, animal, and plant life. The *Deepwater Horizon* drilling rig explosion of April 2010 is the most recent incident to underscore the gap between the technical ability to extract natural resources from beneath the earth and environmentally safe resource transportation.

These issues have driven the modern pursuit of bioremediation techniques that are designed not only to counter the immediate effects of an oil spill on land or water, but also to limit irreversible damage to human and natural environments. The controversies prompted by the use of fire and chemical dispersal techniques in cleanup initiatives underscore the fundamental importance of bioremediation of affected sites that do not trigger additional adverse environmental consequences.

Historical Background and Scientific Foundations

The sophistication of modern oil transportation systems that connect wells to distant refineries have their origins in the Pennsylvania oil industry that emerged in the mid-nineteenth century. The world's first successful commercial oil well was drilled at Titusville, Pennsylvania, in 1859. The earliest oil shipments were made by horse-drawn wagon trains maneuvering along rough and dangerous roads. The advent of barge transport in the 1860s was the first stage in the evolution of efficient oil transportation. The development of specialty tanker railway cars was followed by pipeline construction projects; the first Pennsylvania line was completed in 1879.

The 1870s discovery of oil in the Russian territory of Azerbaijan prompted the construction of the first oil tanker ships. Unlike the primitive barges used in Pennsylvania, these steel-hulled vessels had reservoirs to secure cargo more effectively than the often leaky individual wooden barrels.

Various oil discoveries in the Arab Gulf region led to the rapid growth of its oil production capabilities and a concurrent rise in global oil tanker traffic. By the 1960s, supertankers up to 930 feet (300 m) long carried in excess of 100,000 tons of crude oil or refined petroleum products over routes that spanned the globe. The 1967 *Torrey Canyon* spill, which released 120,000 tons of oil into the English Channel and destroyed marine habitats along the Cornwall and Guernsey coastlines, was the first significant oil-related environmental disaster. The environmental effects of the *Torrey Canyon* persist into the early 2010s.

Conventional clean up technologies used to combat the *Torrey Canyon* spill included highly toxic detergents to counter the hydrophobic crude oil; the chemical napalm was also applied to burn the spreading slick. The 1989 *Exxon Valdez* wreck in environmentally sensitive Prince William Sound, Alaska, confirmed that the technological ability to safely transport oil was exceeded by the risk of spills. Chemical dispersants and containment booms were not effective due to adverse wave conditions. As oil reached nearby shores, high-pressure hot water was applied to remove it. The high temperature water destroyed many shoreline microbial populations essential to marine food chains.

Dispersants played a prominent role in the efforts to control the oil released when the *Deepwater Horizon* oil

WORDS TO KNOW

2-BUTOXYLETHANOL: A chemical dispersant used in the 2010 Deepwater Horizon cleanup efforts in the Gulf of Mexico, subsequently determined to have carcinogenic properties.

ALCANIVORAX BORKUMENSIS: The oil-consuming microorganism identified in bioremediation research conducted during the 2010 Gulf oil spill. The microbe has a set of genes that apparently allow it to reduce oil alkanes to a usable food source.

BIOACCUMULATION: The toxic buildup of chemicals within the body. Bioaccumulation can occur through the inhalation or ingestion of toxins; it may also result from direct contact between tissues and contaminated water, soil, or materials.

CONTAINMENT BOOMS: The physical barriers placed around oil spills that occur on a body of water. At the height of the Gulf of Mexico cleanup operations undertaken in response to the *Deepwater Horizon* disaster, more than 345 miles (550 km) of containment booms were deployed by emergency crews. Containment booms are distinct from the sorbent booms designed to absorb oil directly from the water surface.

DISPERSANTS: The general term used to describe all chemical agents utilized to remove spilled oil from a water body surface. As the name suggests, dispersants are one of the "first response" options available to combat oil spill effects, because these agents act to transfer accumulated spilled surface oil into the water subsurface. In the subsurface, the oil is diluted, and the speed of natural degradation is increased.

HYDROPHOBIC: Water insoluble; oil is an example of a substance with a chemical composition that repels water.

Six months after the March 1989 *Exxon Valdez* oil spill, Otto Harrison, head of Exxon's oil spill cleanup operations in Alaska, shows how the bioremediation treatment cleans under the rocks as well as on top during an inspection tour of Eleanor Island. The island was one of the heaviest hit islands during the oil spill in Prince William Sound. Bioremediation in this case included use of the fertilizer Inipol, which was sprayed on the oiled beaches after they were steam cleaned. The fertilizer enhances the oil eating bacteria already present in the area. *© AP Images/Al Grillo.*

drilling platform exploded in April 2010. At the height of the disaster, oil flowed into the Gulf of Mexico from 5,000 ft (1,500 m) below sea level at a rate of 200,000 gallons per day. The urgency of the looming environmental catastrophe somewhat masked the risks to human health and the environment posed by the use of dispersant 2-butoxylethanol.

Natural oil slick dispersion occurs when wave action (generally observed at speeds greater than 11 mph [17.7 kmh]) breaks up a slick. Larger droplets form, return to the surface, and coalesce into a thinner slick; smaller droplets disperse into the water below. Chemical dispersants accelerate this natural process by countering the hydrophobic character of oil droplets that otherwise create interfacial tension between the oil and water surfaces.

Intense biotechnological research was undertaken during the *Deepwater Horizon* clean up. The oil slick spread to cover more than 3,850 square miles (10,000 square kilometers) of the Gulf. Bioremediation processes typically involve microbial agents, micronutrients, and biostimulants that reduce oil and associated toxins to water, carbon dioxide, and other harmless compounds. Biostimulation is the process that combines the aeration of the affected area with specific micronutrients that speed the degradation of oil. If the surrounding oil spill environment is not rich in microbial life, fertilizer is added to increase the biodegradation achieved by native microorganisms that naturally promote the elimination of hydrocarbon molecules.

■ Impacts and Issues

Bioremediation techniques offer a number of attractive options for oil spill cleanup efforts on any scale. The effectiveness of conventional technologies such as dispersants, physical containment, and removal of surface oil with booms and skimmers depends to a large degree on wave and climatic conditions. Whereas climatic conditions inherently will affect the speed with which fertilizers will work to boost microorganism activity, there are no significant environmental drawbacks to processes that contribute to oil-eating microbial function.

A variety of new bioremediation products are poised to enter the commercial market as a means to counter the environmental impact of an oil spill. One is Aerogel, the class of super-light structures that also act as a highly efficient absorbent material. Manufactured from a mixture of clay, plastic, and air, when Aerogel is added to water polluted by oil, it extracts the oil, and clean water remains. The captured oil is available for recycling. Nanobiotechnologies are an emerging important aspect of Aerogel applications. The addition of nanoscale-sized Aerogel molecules to the material being used to clean up a spill increases the available surface area of the absorbent compound.

The living laboratory that was inadvertently created by the *Deepwater Horizon* explosion confirmed the importance of scientific research conducted in an evolving real world environment. The Gulf slick and its profound threat to all immediate ecosystems was the product of the plumes of oil that rose from the fractured drilling structure through the water to the surface. As efforts to combat the spreading slick intensified, researchers identified specific microbes such as *Alcanivorax borkumensis* that depend solely on oil for nutrition. It is these microorganisms that are likely to provide the surest long-term remediation of large-scale environmental problems. The impacts on the Gulf of Mexico observed from not only the massive 2010 oil spill, but also those attributed to the suspected carcinogenic chemicals used to disperse the slick, suggest that most available future research resources likely will be directed to how safer bioremediation technologies will contribute to the ever-present threat of such events.

■ Primary Source Connection

In this article by David Biello, the discussion centers on how and why bacteria and other microbes are the only solutions that ultimately will clean up the oil spill that began in the Gulf of Mexico on April 20, 2010. Some biochemists maintain that hydrocarbon-chewing microbes are the sole means of defense against the Deepwater Horizon oil spill. They theorize that utilizing

Oil from the *Deepwater Horizon* oil spill is shown on a beach on June 12, 2010, in Gulf Shores, Alabama. Bioremediation was one of the options for trying to rid the ocean of some of the oil from the spill. © *Danny E Hooks/Shutterstock.com.*

hundreds of gallons of chemical dispersants above and below the oil-slick water surface will enable the oil to break down into smaller droplets that the bacteria then can consume easily. Pros and cons of utilizing this technique for use in oil cleanup in the sea are explored.

Slick Solution: How Microbes Will Clean Up the Deepwater Horizon Oil Spill

The last (and only) defense against the ongoing Deepwater Horizon oil spill in the Gulf of Mexico is tiny—billions of hydrocarbon-chewing microbes, such as *Alcanivorax borkumensis*. In fact, the primary motive for using the more than 830,000 gallons of chemical dispersants on the oil slick both above and below the surface of the sea is to break the oil into smaller droplets that bacteria can more easily consume.

"If the oil is in very small droplets, microbial degradation is much quicker," says microbial ecologist Kenneth Lee, director of the Center for Offshore Oil, Gas and Energy Research with Fisheries and Oceans Canada, who has been measuring the oil droplets in the Gulf of Mexico to determine the effectiveness of the dispersant use. "The dispersants can also stimulate microbial growth. Bacteria will chew on the dispersants as well as the oil."

For decades scientists have pursued genetic modifications that might enhance these microbes' ability to chew up oil spills, whether on land or sea. Even geneticist Craig Venter forecast such an application last week during the unveiling of the world's first synthetic cell, and one of the first patents on a genetically engineered organism was a hydrocarbon-eating microbe, notes microbiologist Ronald Atlas of the University of Louisville. But there are no signs of such organisms put to work outside the lab.

"Microbes are available now but they are not effective for the most part," says marine microbiologist Jay Grimes of the University of Southern Mississippi. At this point, there are no man-made microbes that are more effective than naturally occurring ones at utilizing hydrocarbons.

The natural world is replete with a host of organisms that combine as a community to decompose oil—and no single microbe, no matter how genetically enhanced, has proved better than this natural defense. "Every ocean we look at, from the Antarctic to the Arctic, there are oil-degrading bacteria," says Atlas, who evaluated genetically engineered microbes and other cleanup ideas in the wake of the *Exxon-Valdez* oil spill in Alaska. "Petroleum has thousands of compounds. It's complex and the communities that feed on it are complex. A superbug fails because it competes with this community that is adapted to the environment."

Nor is it easy to help the existing communities of thousands of microbes, such as various species of *Vibrio* and *Pseudomonads*, to eat the oil faster—seeding experiments have generally failed. "Microbes are a lot like teenagers, they are hard to control," says marine chemist Chris Reddy of the Woods Hole Oceanographic Institution. "The concept that nature will eat it all up is not accurate, at least not on the time scale we're worried about."

Just like your automobile, these marine-dwelling bacteria and fungi use the hydrocarbons as fuel—and emit the greenhouse gas carbon dioxide (CO_2) as a result. In essence, the microbes break down the ring structures of the hydrocarbons in seaborne oil using enzymes and oxygen in the seawater. The end result is ancient oil turned into modern-day bacterial biomass—populations can grow exponentially in days. "Down in the Gulf of Mexico there is an indigenous population [of microbes] adapted to oil from so much marine traffic and daily spills. Oil is not new," says Lee, who has also been monitoring the plumes of oil beneath the surface. "There are so many natural seeps around the world that if it wasn't for microbes we would have a lot of oil in the oceans."

Already, measurements of oxygen depletion of as much as 30 percent in the Gulf of Mexico seawater suggest that the microbes are hard at work eating oil. "I take the 30 percent depletion of oxygen in water near the oil as indicating bacterial degradation," Atlas says.

That happens best near the surface, whether at land or sea, where warm-water bacteria such as *Thalassolituus oleivorans* can thrive; colder, deeper waters inhibit microbial growth. "Metabolism slows by about a factor of two or three for every 10 degree[s] Celsius you drop in temperature," notes biogeochemist David Valentine of the University of California, Santa Barbara, who just received funding from the National Science Foundation to characterize the microbial response to the ongoing oil spill. "The deeper stuff, that's going to happen very slowly because the temperature is so low."

Unfortunately, that's exactly where some of the Deepwater Horizon oil seems to be ending up. "They saw the oil at 800 to 1,400 meters depth," says microbial ecologist Andreas Teske of the University of North Carolina at Chapel Hill, whose graduate student Luke McKay was on the research vessel *Pelican* that first reported such subsurface plumes—as predicted by small-scale experiments, such as the U.S. Minerals Management Services Project "Deep Spill". "It is either at the surface or hanging in the water column and possibly sinking down to the sediment."

Yet, microbes are the only process to break down the oil deeper in the water, far away from physical processes on the surface such as evaporation or waves. "The deep waters are dominantly microbial" when it comes to oil degradation, although these communities are not as well studied as those at the surface, notes microbial geochemist Samantha Joye of the University of Georgia. "As long as there is oxygen around, it will get chewed up."

To understand how the microbes will work and how quickly, however, will require a better understanding of exactly how much oil is out there. "It's a function of size, and we don't know size," Joye says. "We need to know how much oil is leaking out. Without that information we can't begin to make any kind of calculation of potential oxygen demand or anything else." BP now admits that its original estimate of roughly 200,000 gallons per day was far too low without providing an alternative; independent experts have offered estimates as high as four million gallons per day.

It is possible to add fertilizers, such as iron, nitrogen and phosphorus, to stimulate the growth of such bacteria, an approach used to speed up microbial activity in the sediment along the Alaska coast after the *Exxon-Valdez* spill. "We saw a three to five times increase in rate of biodegradation," Atlas says, suggesting the technique might prove effective along the oil-inundated Louisiana coast as well. "It was hundreds of miles of shoreline, the largest bioremediation project ever."

But that's strictly onshore. "In the ocean, how do you keep the nutrients with the oil?" Lee asks. "It's much easier to add to soil. That's why you don't see bioremediation in the open ocean." And aerating soils in wetlands can have its own problems; Lee tried tilling oil-soaked wetlands in Nova Scotia where there was limited oxygen to increase microbial activity. "That didn't work. We had large erosion as a result," he says. "If the oil reaches shore, our recommendation was to leave the oil alone and let nature do it."

But sediment, whether the muck of Louisiana marshland or the deep ocean seafloor, suffers from a dearth of oxygen. That means it's up to anaerobic microbes—ancient organisms that live via sulfate rather than oxygen—to do the dirty work of consuming the spill. "What occurred in 10 days aerobically, took 365 days to occur anaerobically," says Atlas of the breakdown of oil in the wake of the *Amoco Cadiz* spill off the coast of France in 1978. Adds Teske: "The heavy components are sinking to the sediment and forming an oily or tarry carpet or getting buried. Then they are much harder to degrade."

Such anaerobic environments can develop locally in the seawater itself, thanks to a ready supply of oil and blooming microbes eager to devour it. In deepwater, where there's less mixing with the surface waters to provide fresh supplies of oxygen, a dead zone may result. "It's not exchanging with the atmosphere," Joye notes. "Once the oxygen is gone, how are you going to replace it? It's not going to get mixed up by winter storms."

That's bad news for the speedy breakdown of oil as well as for the *Lophelia coral* and other sessile deepwater life.

At the same time, the addition of 130,000 gallons of dispersants deep beneath the surface is having uncertain effects; it may even end up killing the microbes it is meant to help thanks to the fact that Corexit 9527A contains the solvent 2- butoxyethanol, which is a known human carcinogen and toxic to animals and other life. But the U.S. Environmental Protection Agency, National Oceanic and Atmospheric Administration and others are monitoring whether adding such dispersants ends up boosting microbe-growth and hence dangerously depletes oxygen levels, among other potential environmental ill effects.

Nor is it clear how fast the microbial community will respond. "Which microbial communities are the fastest responders?" Teske asks. "That would be interesting to know" and this oil spill may provide the real-world answer. Some research suggests that oil spills may actually feed themselves nitrogen by stimulating the growth of various bacteria that fix the vital nutrient, Joye notes. At the same time, microbial predators such as protozoa tend to dampen the efficiency of would-be oil-eating microbes.

Scientists are still working to deploy known oil-eaters, such as *Alcanivorax*, in the form of booms laced with slow-release fertilizer and the microbes. In experiments such microbial booms ate heavy fuel oil in two months and "the experimental waste water was clean enough to be released back to the sea," says environmental geneticist Peter Golyshin of Bangor University in Wales. But "in the Gulf of Mexico, the amount of oil is simply too big. The oil gets dispersed but there is not enough [nitrogen] and [phosphorus] to feed bacterial growth."

Ultimately, it is only microbes that can remove the oil from the ocean. "In the long run, it's biodegradation that removes most of the oil from the environment in these situations," Lee says. Or, as Joye puts it, "They're clever, they're tough, they can basically eat nails. ... The microbes have to save us again."

Regardless, the oil will linger in the environment for a long time. The microbes break down hydrocarbons in "weeks to months to years, depending on the compounds and concentrations—not hours or days," Atlas notes. "Much of the real tar or asphalt compounds are not readily subject to microbial attack. ... Tar tends to persist. Asphalt tends to persist."

Adds Valentine: "We wouldn't make roads out of them if the bacteria ate them."

David Biello

BIELLO, DAVID. "SLICK SOLUTION: HOW MICROBES WILL CLEAN UP THE DEEPWATER HORIZON OIL SPILL." *SCIENTIFIC AMERICAN* (MAY 25, 2010): 1–4.

SEE ALSO   *Biochemical Engineering; Bioremediation; Biosurfactants; Biotechnology Products; Environmental Biotechnology; Metabolic Engineering; Microbial Biodegradation; Phytoremediation*

BIBLIOGRAPHY

Books

Atlas, Ronald M., and Jim Philp, eds. *Bioremediation: Applied Microbial Solutions for Real-World Environmental Cleanup.* Washington, DC: ASM Press, 2005.

Birkin, Danielle M., and Jonathan J. Asher, eds. *Deepwater Horizon Oil Spill and Related Issues.* New York: Nova Science Publishers, 2011.

Fingerman, Milton, and Rachakonda Nagabhushanam, eds. *Bioremediation of Aquatic and Terrestrial Ecosystems.* Enfield, NH: Science Publishers, 2005.

International Maritime Organization. *Bioremediation in Marine Oil Spills: Guidance Document for Decision Making and Implementaion of Bioremediation in Marine Oil Spills.* London: IMO, 2004.

International Maritime Organization, and United Nations Environment Programme. *IMO/UNEP Guidance Manual on the Assessment and Restoration of Environmental Damage Following Marine Oil Spills.* London: International Maritime Organization, 2009.

Kriipsalu, Mait. *Biotreatment of Oily Sludge and Sediments: Biopile Technology.* Saarbruücken, Germany: VDM Verlag, 2009.

Ollivier, Bernard, and Michel Magot, eds. *Petroleum Microbiology.* Washington, DC: ASM Press, 2005.

Safina, Carl. *A Sea in Flames: The Deepwater Horizon Oil Blowout.* New York: Crown Publishers, 2011.

Vazquez-Duhalt, Rafael, and Rodolfo Quintero Ramirez. *Petroleum Biotechnology: Developments and Perspectives.* Boston: Elsevier, 2004.

Web Sites

Biello, David. "Slick Solution: How Microbes Will Clean Up the Deepwater Horizon Oil Spill." *Scientific American*, May 25, 2010. http://www.scientificamerican.com/article.cfm?id=how-microbes-clean-up-oil-spills (accessed November 8, 2011).

"Oil Spill Bioremediation Research." *U.S. Environmental Protection Agency (EPA).* http://www.epa.gov/ORD/NRMRL/lrpcd/esm/oil_spill_bioremediation_researc.htm (accessed November 8, 2011).

Bryan Thomas Davies

Biorobotics

■ Introduction

Biorobotics is a multidisciplinary subject that contains elements of biomechanics, genetic engineering, nanotechnology, and synthetic and systems biology, among others. It can be divided broadly into two overlapping subsets.

First, it is the study of biological systems from an engineering standpoint. Understanding the principles of how these systems function aids in the development of innovative technology with wide-ranging applications and also provides a platform for investigation and research in areas such as neuroscience, with particular focus on the healthcare field.

Secondly, biorobotics is considered a direct branch of robotics, creating artificial life using biological inspiration or incorporating biological systems into robotic forms. An in-depth understanding of mechanical and chemical functions in the biological realm has led to the development of biomedical instruments and neuroprosthetics. It has also aided the advancement of nanotechnology in the diagnosis and treatment of disease and its future use in surgery and rehabilitation.

■ Historical Background and Scientific Foundations

Although the idea of mechanical beings was mentioned in literature as far back as the Homer's *Iliad*, it was Leonardo da Vinci's designs for a humanoid robot, based on his studies and drawings of the Vetruvian Man, that mark the first documented link between detailed investigation of anatomical structure and robotics. The onset of the Industrial Revolution in the late eighteenth century saw automation on a wide scale in manufacturing, but it was not until the twentieth century and significant progress in computer technology, along with the evolution of the concept of cybernetics, that the practical application of robotics and its link to bio-systems made significant progress.

In 1963 a robotic arm, "Rancho Arm," was designed to assist handicapped patients in a hospital in California. It had six joints, which enabled flexibility approaching that of a human arm. Over the next decade, variations of this idea were created in the fields of engineering, research, and prosthetics. One notable design was David Silver's "Silver Arm" in 1974. Designed to replicate the fine movement of human fingers, along with the use of pressure sensors that provided feedback to a control computer, Silver Arm was capable of small-part assembly.

In the healthcare field, research led to the creation of the first robotic radiosurgery system, "Cyberknife," in 1994; a non-surgical way of treating inoperable tumors. This robotic system utilizes biofeedback, which enables it to compensate for movement of the heart, lungs, and other body essentials while operating with the precision of a surgeon.

In the 1980s the rise of nanotechnology coupled with advances in computer programming, artificial intelligence, and robotics led to the beginnings of research into its possible use in the medical field in areas such as drug delivery systems, tumor therapy, and physiological investigation. However, concerns about the impact of nanoscale materials on the environment and health has made this area of biorobotics controversial.

■ Impacts and Issues

The word robot is said to come from the Czech word *robota*, meaning servitude. What are the ways in which biorobotics can serve human interests, and what are the impacts of advances in this area?

Prosthetics

Since the early 1990s considerable advancements have been made in the field of prosthetics. Where once the options were limited to crude wooden structures, prosthetics are constructed with carbon fiber, titanium, and

WORDS TO KNOW

BIOMECHATRONIC: Incorporating biology, electronics, and mechanics.

BIOMIMETIC: The study and application of functions and structures found in nature to engineering and technology.

MYOELECTRIC: Using electrical signals from muscles to operate devices, especially prosthetics.

NANOTECHNOLOGY: Also referred to as molecular manufacturing, nanotechnology covers engineering, research, and technology at an atomic and molecular level of under 100 nanometers.

NEURAL INTERFACE: Also known as a brain-computer interface, a neural interface is a path of communication between the brain and an external computer system.

VISUAL SERVOING: Coming from the term "servomechanism," visual servoing is the use of visual feedback, e.g. from a mounted camera, to control a robot.

thermoplastic sockets. But the real advancements have come in the incorporation of computer and robotic systems.

Myoelectric prostheses pick up muscle-generated electrical signals in the amputee's stump, which are then interpreted by a processor in the prosthetic limb. Touch Bionics' i-Limb series illustrates this evolution of prosthetic design. The technology enables the user to perform fine motor movement, as is needed in many everyday activities such as picking up and manipulating small objects. The next advancement involves interfacing directly with the nervous system, a technique known as "thought-controlled prosthetics," as seen in designs such as the Otto Bock arm, which makes use of targeted muscle reinnervation (TMR) to "rewire" nerves from the damaged limb to a large working muscle group in the chest, which in turn controls the prosthesis.

Recent innovations include the use of Bluetooth technology that enables the wearer to access and program features of the prosthetic so that it adapts to his or her individual needs. Bluetooth is also being employed in double leg amputees, letting the two artificial robotic prosthesis "talk" to each other, optimizing load bearing and compensating for variations in gait, speed, and pressure.

Intelligent, powerful exoskeletons at the prototype stage, such as HULC (Human Universal Load Carrier) and eLEGS of Berkeley Bionics, are addressing physical restrictions of the human body. The first enables humans to carry greater weights than would normally

be possible, which would be of tremendous benefit in many fields (e.g., the military—enabling soldiers to carry heavy packs up to 200 pounds in weight), or in emergency situations (e.g., fire fighters carrying injured victims). The second prototype example enables wheelchair users to stand upright and even walk.

A major issue with all of these uses of the technology is the high cost. With specific reference to prosthetics, there is a growing debate about whether funds should be redirected into the design of lower-cost, simpler, and less lifelike (and therefore less cosmetically attractive) technology that would be accessible to a wider patient base. Another ethical concern of some less able-bodied people is that these advances are being marketed as "solutions" for their "sub-normal" lives.

Biorobotics in Surgery

Cardiac surgeries and procedures have serious potential complications ranging from bleeding following the surgery to iatrogenic stroke and myocardial infarction during the surgery. In robotic heart surgery, many elements of the surgery are controlled simultaneously from a centralized processing unit. Robert D. Howe, of the Harvard Biorobotics Laboratory, and Dr. Pedro del Nido, of the Children's Hospital of Boston, designed such a robot; "Cathbot" combines a 3-D ultrasound imaging device with vision-based robot control (or visual servoing) that uses feedback from the visual images to control the movement of the robot and catheter.

The catheter itself is able to both compensate for the movement of the still-beating heart, by bracing itself against tissue walls, and also undertake delicate repairs. The 3-D imaging allows for constant feedback to both the processing unit and the surgical team without the need for a large incision in the chest, and enables the surgeon to assess the effectiveness of the repair operation immediately. In this case, the biorobot works with the body's structures to affect an efficient, precise procedure. This process could be adapted for other operations, and given its less invasive nature, could open up these operations to a wider patient base, for example to older patients who would not be medically fit enough for traditional open heart surgery.

Neural Interfaces

Project Cyborg is an ambitious biorobotics experiment conceived of by Professor David Warwick of the University of Reading in the United Kingdom. In 2002 Warwick implanted an electrode array into his arm that would interface directly with his own nervous system. The chip was connected to supporting electronics housed in an external "gauntlet." The implant enabled Warwick to control the movements of a robotic arm and also to receive feedback from fingertip sensors in the artificial limb.

Chris Borer demonstrates a snake robot climbing up the inside of a transparent plastic tube in the biorobotics lab at the Robotics Institute at Carnegie Mellon University in Pittsburgh, Pennsylvania, in 2006. The labs researchers have been developing the snake-like robots to slither through collapsed buildings in search of victims trapped after natural disasters or other emergencies.   © *AP Images/Keith Srakocic.*

The success of the experiment has implications for the future of prosthetics, but also for the development of more lifelike robots incorporating biological neural networks into their build. For Warwick, whose research continues, this was one of the first steps to becoming a cyborg. In his own words, "There is no way I want to stay a mere human." The question for scientists and philosophers alike, is how much biorobotics can be incorporated into the human body before it ceases to be human?

Nanotechnology

Nanoscale robots are already in use in the electronics industry. When this technology is taken into consideration alongside other developing areas such as the creation of new biomaterials, nanoscale biorobotics promises sophisticated future drug delivery systems via the human vascular system. It may also enable drug delivery across cellular membranes to less accessible regions such as the brain, and facilitate simplified and earlier ways of diagnosing disease and novel surgical approaches, such as for nerve repair.

Treating brain injuries and disease is particularly difficult because of the blood-brain barrier (BBB), a protective membrane encasing the brain, which does not allow large molecules to pass across it. Smaller molecules, such as caffeine and nicotine, pass across unimpeded. So too, perhaps, it would be possible for nanorobots to cross the BBB and deliver drugs to treat the brain, to make repairs on biosensors implanted there, or to operate on tumors or blood clots.

Biomimetics and Engineering

Engineers have in the past taken inspiration from nature in their designs, one example being the cat's eye reflectors found on roads, but for the most part they have ignored natural blueprints. In the exploration and research of inhospitable environments such as space, the deep sea, deserts, and the poles, and in the design of robots, which can assist humans to survive in those challenging environments, it has become increasingly necessary to turn to the natural world to solve practical problems. Nature has already, over millions of years, developed strategies to conquer and thrive in these unpredictable environments. As Mark Cutkosky, professor of mechanical engineering at Stanford University, points out, "One of the things that natural organisms do wonderfully well is coping with unexpected variations in the environment." Biorobotics, in this sense, can be seen as engineering and technology taking their cue from nature.

Cutkosky's own research led to the design of a gecko-inspired robot, Stickybot. A gecko's feet have tiny hairs, which allow for easy movement over practically any surface. Stickybot was built with a similar system and can cope with a myriad of terrains, including scaling vertical surfaces of concrete and plaster. Research into the movement of fish is also leading to new propulsion systems for underwater exploration based on the design of fins instead of traditional propellers.

Taking all of these varied and wide-ranging applications of biorobotics into consideration, and keeping in mind the possible future development of the field, it is apparent that biorobotics is not only bringing biology and technology closer together, but also testing the boundaries of what being human means, both in terms of physical and neurological capacities, but also with regard to ethics and social responsibility.

SEE ALSO *Biochip; Biomimetic Systems; Nanotechnology, Molecular; Prosthetics; Silk Making; Space Research*

BIBLIOGRAPHY

Books

Akay, Metkin. *Neural Nanotechnology: Biorobotics, Artificial Implants & Neural Prosthesis, Vol. 3.* Hoboken, NJ: Wiley, 2005.

Danion, Frédéric. *Motor Control: Theories, Experiments, and Applications.* New York: Oxford University Press, 2011.

Floreano, Dario, and Claudio Mattiussi. *Bio-Inspired Artificial Intelligence: Theories, Methods, and Technologies.* Cambridge, MA: MIT Press, 2008.

Hamdi, Mustapha, and Antoine Ferreira. *Design, Modeling and Characterization of Bio-Nanorobotic Systems.* New York: Springer, 2011.

Periodicals

Dario, Paolo. "Biorobotics." *Journal Robotics Society of Japan* 23, no. 5 (2005): 552–554.

Web Sites

"Bio-Nano Robotics" *Northeastern University.* http://www.bionano.neu.edu/ (accessed October 4, 2011).

Kenneth Travis LaPensee

Biosafety Level Laboratories

Introduction

In a wide range of laboratories, biosafety protocols work to identify the level of hazard and risk posed by the biological substances used therein. Diagnostic laboratories for disease or research; public health and medical laboratories used for screening, treatment decision-making, and surveillance; and also laboratories involving animal or environmental testing, all must be subject to an organized system of bio-containment precautions. The levels range from Biosafety Level 1 (BSL-1), posing a low biohazard threat, to Biosafety Level 4 (BSL-4), which requires the strictest forms of isolation of materials and codes of practice for laboratory procedures.

Historical Background and Scientific Foundations

American biologist Arnold G. Wedum (1903–1976) is considered to be one of the principal founders of the field of biosafety. His interest in the subject grew from his work as Director of Safety at the U.S. Army Biological Warfare Laboratories at Fort Detrick in Frederick, Maryland, during the 1950s and 1960s. He focused on the need to keep research staff safe from the biological agents with which they were experimenting. On April 18, 1955, Wedum brought together 14 experts with interests in not only biosafety, but also chemical, radiological, and industrial safety, to take part in a conference at Fort Detrick. Given the sensitive nature of the information and the consequences of the work being undertaken at the biological warfare laboratories these representatives worked in, this meeting was restricted to those with high security clearance.

In 1957 the conferences began to include non-classified sessions. Over the next two decades, they increasingly included representatives from various government research facilities, universities, and hospitals, as well as private laboratories and industry. In 1984

the American Biological Safety Association (ABSA) was established as an extension of Wedum's Safety Conferences. As of 2011, ABSA has more than 1,600 professional members and aims to continue to pursue biosafety as a scientific discipline, providing an ongoing forum for the exchange and discussion of biosafety information and practices.

In 1983 the World Health Organization (WHO) published the Laboratory Biosafety Manual, which laid out basic models for biological safety, supporting countries in formulating national codes of practice. This manual and subsequent editions have assisted many countries since 1983 in creating and enforcing such codes.

Impacts and Issues

Biosafety levels encompass a combination of lab facilities, safety equipment, and lab practices. The level required is set after a comprehensive microbiological risk assessment. Each biosafety level is dependent on the biomaterials present, their individual make-up, suspected routes of transmission and the risk they pose to lab personnel, the environment, and society at large. Biohazardous materials can be categorized into four risk groups, and although they do not correlate directly with the biosafety levels, they are useful in informing decisions about the latter. Risk groups rank biohazards depending on the danger they represent to individuals and communities. The first risk group (Risk Group 1) poses no or a low-level risk. Risk Group 2 poses a moderate risk, with treatment available and a limited risk of infection spread. Risk Group 3 is rated as having a high risk to the individual but is a substance that usually does not spread between hosts, giving a low community risk. Effective treatment is available. Risk Group 4 are the highest-rated pathogens and have both a high risk to individuals and the community because they can cause serious disease, are transmitted easily, and there are no effective treatments available. Where Risk Group 2 and above agents are

WORDS TO KNOW

BIOHAZARD: In the field of biological materials and research, any agent or situation that constitutes a threat to health or environment.

BIOSAFETY: Applicable to all facilities in which biological hazards are present, biosafety covers the protocols and ongoing process of hazard and risk identification, assessment, and prevention.

ZOONOSIS: An infectious disease that is passed from animals to humans, sometimes through a vector.

children, or immunosuppressed individuals. These laboratories are part of the general traffic pattern of a research facility, with work being performed in the open, on bench tops without the need for special containment equipment.

At Biosafety Level 2, access to the laboratory is restricted when work is taking place, recommendations are in place for the use of safety cabinets and/or protective equipment (especially if there is a risk of generating airborne particles). Laboratory personnel will have training in pathogenic materials and are familiar with microbiological practices. Agents in these laboratories can cause disease in even the healthy and are of moderate risk to the environment.

Building on the procedures at BSL-2, Biosafety Level 3 deals with agents that cause serious or even lethal results, necessitating special containment facilities. The Level 3 laboratory is restricted to those directly undertaking work with these agents, and the laboratory itself is designed specifically to adhere to special safety protocols. Examples would be an airlock at the entrance, directional airflow (negative pressure to the surrounding areas), and a sealed floor. Laboratory personnel would have specialized training and must observe stringent protocols.

present, a biohazard sign must be displayed at the entrance to the laboratory.

Biosafety Level 1 laboratories are used if work involves agents in Risk Group 1. It is important to note that considerations for this risk group apply to healthy individuals and due diligence should be applied where there is a possibility of transmission to the elderly, young

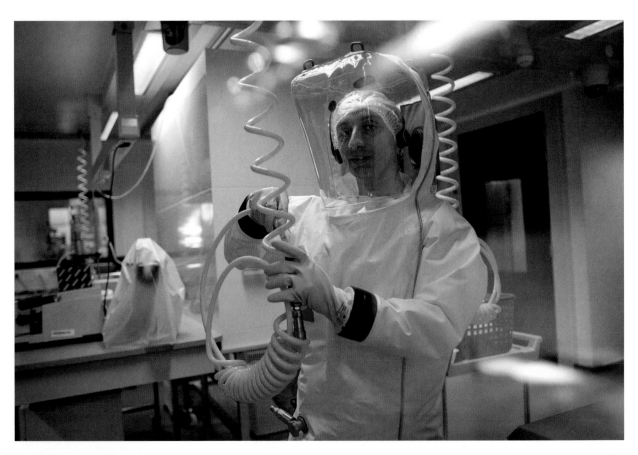

A researcher wears protective gear in the P4 European High Level Security Laboratory (EHLS) in Lyon, France. The lab, which normally handles only the most deadly viruses, such as Ebola, is also preparing to receive the new swine flu influenza virus H1N1. P4 stands for pathogen, or protection, level 4. *© AP Images/Laurent Cipriani.*

Biosafety Level 4 reflects the highest level of risk to individuals, and further, the agents involved are dangerous aerosol-transmitted pathogens with a likelihood of fatal consequences upon infection. BSL-4 is deemed a maximum containment unit; it is either a separate building within a complex or an isolated area, both options incorporating specially designed waste and ventilation systems. In addition to following the procedures from the previous biosafety levels, BSL-4 personnel are expected to use complete laboratory clothing (changing from street clothes in an isolated room before entry into the main lab area), to shower before leaving the containment area, and to be completely isolated from the materials being used. Many BSL-4 facilities also include a "buffer" corridor around the laboratory area to protect it in case of a bomb attack or earthquake.

One additional biosafety level is BSL-3-Agriculture (BSL-ag), where the facility itself works as a primary barrier between the laboratory and the outside environment. BSL-ag is used if research involves High Consequence Pathogens or necessitates the accommodation of large or loose-housed animals.

The International Standards Organization (ISO), along with the Clinical and Laboratory Standards Institute (CLSI), also lays out protocols that cover many aspects of laboratory facilities and the path of workflow under a quality management system. Although ISO standards are often considered the domain of manufacturing and business establishments, adherence to the ISO standards, especially as they refer to safety and security, is extremely important in laboratories concerned with biosafety, as well.

Every year a significant number of accidents involving bio-hazardous materials are reported. Taking as an example *Mycobacterium tuberculosis*, there were 223 reported instances of laboratory-based infections in the United States from 1979 to 1999. *Brucella* bacteria were indicated in a further 81 accidents over the same period. *Brucella* causes brucellosis, or undulant fever, a highly contagious zoonosis, and in one incidence was spread to 94 people from the basement of a research facility as far as the third floor due to centrifugation of the bacteria leading to airborne contamination. These instances underscore the importance of rigorously adhering to safety measures within the laboratory, but also having an effective emergency response provision in the event of an accident.

There is an ongoing need for research pertaining to hazardous biological materials from the perspectives of disease diagnosis, treatment and vaccine production, the threat of biological weapons, and defense against such weapons. Biosafety level laboratories function with the aim of providing safe, efficient, specialized facilities for laboratory personnel, but also indirectly for the benefit of the public and the wider environment.

SEE ALSO *Biodefense and Pandemic Vaccine and Drug Development Act of 2005; Biological Confinement; Biological Weapons; Biologics Control Act; Biotechnologists; Bioterrorism; Security-Related Biotechnology; Vaccines; Virology*

BIBLIOGRAPHY

Books

Delany, Judy R., et al. *Guidelines for Biosafety Laboratory Competency.* Atlanta: U.S. Department of Health and Human Services, Centers for Disease Control and Prevention, 2011.

Kostic, Tanja, Patrick Butaye, and Jacques Schrenzel. *Detection of Highly Dangerous Pathogens: Microarray Methods for the Detection of BSL 3 and BSL 4 Agents.* Weinheim, Germany: Wiley-VCH, 2009.

World Health Organization. *Laboratory Biosafety Manual*, 3rd ed. Geneva: World Health Organization, 2004.

Web Sites

"International Health Regulations: Biosafety and Laboratory Biosecurity." *World Health Organization (WHO).* http://www.who.int/ihr/biosafety/en/ (accessed November 11, 2011).

"Laboratory Biosafety Level Criteria." *U.S. Centers for Disease Control and Prevention (CDC).* http://www.cdc.gov/biosafety/publications/bmbl5/BMBL5_sect_IV.pdf (accessed November11, 2011).

"Reconstruction of the 1918 Influenza Pandemic Virus." *U.S. Centers for Disease Control and Prevention (CDC).* http://www.cdc.gov/flu/about/qa/1918flupandemic.htm#biosafety (accessed November 11, 2011).

Kenneth Travis LaPensee

Biosurfactants

■ Introduction

Biosurfactants are compounds manufactured by living organisms that change the chemistry of either the region where a liquid (typically water) meets air (an interface) or a region deeper in the liquid. Typically, the chemical change makes it easier for hydrophobic compounds such as constituents of petroleum that would otherwise tend not to disperse in water to do so. Biosurfactants also reduce surface tension, which makes it easier for liquids that would normally remain separate on a fluid surface (such as water and oil) to mix together. This can be useful in making contaminants more available to be degraded by chemical means or by organisms.

■ Historical Background and Scientific Foundations

A variety of compounds can act as biosurfactants. These include peptides (portions of proteins), lipopeptides, fatty acids, phospholipids (lipid molecules with a phosphate group at one end), glycolipids, and even antibiotics. Glycolipids are the most common type of biosurfactant.

Biosurfactants are used in the decontamination of soil, groundwater, and surface water, and in enhanced oil recovery (a process in which underground oil that is hard to extract is obtained). Each process involves the exploitation of compounds that are produced by organisms (mainly species of bacteria and yeast). In natural settings, where nutrients may be less available due to their relatively hydrophobic nature or because they are bound avidly to surfaces, the production of compounds by bacteria and yeast increases the availability of the nutrients.

The production of biosurfactants is usually coincident with the manufacture of compounds that function in the degradation of proteins and lipids. As the compounds selectively degrade the proteins and lipids, the presence of the biosurfactant drives the formation of a solution containing the degraded compounds, which can be taken up by the biosurfactant-producing bacteria or yeast.

The bacterium *Pseudomonas aeruginosa* is a well-known and extensively-studied producer of biosurfactant. The production of biosurfactant is one reason why *P. aeruginosa* is successful at growing in a wide range of habitats, which range from natural waters to the lungs of people with cystic fibrosis. Other biosurfactant producing bacteria include species of *Rhodococcus*, *Mycobacterium*, *Arthrobacter*, and *Corynebacterium*. Yeasts that produce biosurfactants include species of *Candida*.

Biosurfactants are able to accomplish the association of water with compounds that naturally tend not to associate with water by virtue of their chemical structure. A typical biosurfactant is amphiphilic, which means it possesses regions that have a net positive or negative charge and associate with water (hydrophilic regions) and regions that tend not to associate with water (hydrophobic regions). This intermediate structure makes an amphiphilic compound capable of acting as a chemical bridge between hydrophobic and hydrophilic (water-loving) compounds.

■ Impacts and Issues

Biosurfactants have proven to be useful in natural settings. Two important reasons for this are their tendency to be degraded for use as a nutrient by other living organisms (they are described as being biodegradable) and their non-toxic nature. Thus, after their role in decontamination is done, biosurfactants will tend not to persist in the environment, with remnant biosurfactant not posing a threat to the soil or water, or the resident organisms.

Biosurfactants promote emulsification (the rapid mixing of polar and nonpolar liquids that normally repel each other). This is useful when water, which is polar, is contaminated with nonpolar hydrocarbons during oil or

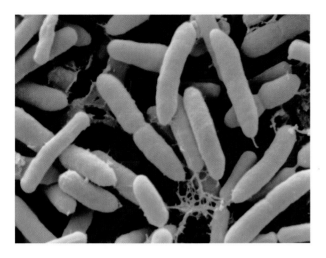

A colored scanning electron micrograph of *Pseudomonas putida*, a gram-negative, aerobic, fluorescent bacterium with multitrichous flagella. It is a unique soil bacterium that can resist adverse effects of organic solvents and is capable of decontaminating soil contaminated with such solvents including toluene and paints. Some strains secrete rhamnolipids, which are biosurfactants. © *Medical-on-Line/Alamy.*

chemical spills. Biosurfactants act to create emulsified solutions of water, and the hydrocarbons are more readily degraded by microorganisms naturally present in the water. Microbes, including specially engineered microbes, are also used to degrade herbicides and pesticides.

Microbial enhanced oil recovery, which depends on biosurfactants to loosen tightly-bound oil from the surface of rocks, is being used in locations such as Alberta, Canada, to extract oil that would otherwise remain unrecovered; this source of oil has been estimated to comprise about one-quarter of all the underground oil in North America.

The ability to add select genes to organisms to drive the production of the desired compound is also being exploited in biosurfactant production. This is especially the case in enhanced oil recovery, where the bacteria used must withstand harsh conditions such as elevated temperature and nutrient-poor conditions. By selecting bacteria that thrive in such challenging environments and instilling in them the genetic capability to produce the desired biosurfactant, the process of enhanced oil recovery can be optimized.

WORDS TO KNOW

BIOREMEDIATION: Use of living organisms or their constituents to reduce or remove contaminating compounds from soil or water.

ENZYMES: Proteins that enable reactions to proceed more easily or faster than in their absence and are not changed by the reactions.

HYDROPHOBIC: The tendency to exclude water.

SEE ALSO *Bioremediation; Bioremediation: Oil Spills; Biotechnology Products; Laundry Detergents; Wastewater Treatment*

BIBLIOGRAPHY

Books

Atlas, Ronald M. and Jim Philp, eds. *Bioremediation: Applied Microbial Solutions for Real-world Environmental Cleanup.* Washington, DC: ASM Press, 2005.

Fingerman, Milton, and Rachakonda Nagabhushanam, eds. *Bioremediation of Aquatic and Terrestrial Ecosystems.* Enfield, NH: Science Publishers, 2005.

Web Sites

"Bioremediation." *United States Environmental Protection Agency (EPA).* http://www.epa.gov/ebtpages/treatreatmenttechnbioremediation.html (accessed September 7, 2011).

Knox, Richard. "Morning Edition: How Will the Gulf Oil Spill Affect Human Health?" *National Public Radio,* June 23, 2010. http://www.npr.org/templates/story/story.php?storyId=128008826 (accessed September 7, 2011).

Tutton, Mark. "Lessons Learned from the Largest Oil Spill in History." *CNN.com,* June 4, 2010. http://articles.cnn.com/2010-06-04/world/kuwait.oil.spill_1_slicks-oil-spill-rigs-and-pipelines?_s=PM:WORLD (accessed September 9, 2011).

Brian Douglas Hoyle

Biotechnologists

■ Introduction

The rapidly developing field of biotechnology spans many scientific disciplines. For the people working in the industry this can pose challenges and offer tremendous opportunities for research. A biotechnologist can work in multiple sectors, depending on his or her area of specialty and interest. The career options can be diverse, such developing medicines; bioengineering crops and livestock; working alongside police agencies in DNA fingerprinting and forensics; or conducting research to advance the technology of the field itself. Combining elements of human-made technology, biological systems, and chemistry, and applying those to problems of medicine, engineering, environmental issues, and space research to name but a few, requires a creative yet scientific approach.

■ Historical Background and Scientific Foundations

Biotechnology is the coming together of various disciplines of science in a rapidly developing scientific field. As such, the story of biotechnology can be seen to have many strands running through the history of chemistry, biology, engineering, and information technology.

Using organisms to work for the benefit of human beings through biochemical reactions has a long history. Some of the oldest recipes in the world include traditional ways of preserving food through fermentation to produce products such as Korean kimchi and Sumerian beer. The production of yogurt and cheese and the use of yeast in bread have also been documented for thousands of years. The historical selective breeding of animals to create strains with desirable traits, for example in goats, dogs, and horses, can also be viewed as a type of biotechnology. This aim of "enhancing" organisms is a major part of biotechnology, although it has become possible through genetic engineering in the lab to determine the specific characteristics required, or even create new phenotypes.

The publication of English naturalist Charles Darwin's (1809–1882) *On the Origin of the Species* in 1859 began a process in which biotechnology ultimately became an established scientific discipline. Darwin laid the groundwork for a revolutionary way of conceptualizing adaptation in biological organisms. Shortly afterwards, the Austrian monk Gregor Mendel (1822–1884) discovered the laws of heredity, the beginning of genetic research. The French microbiologist Louis Pasteur's (1822–1895) work in the nineteenth century also fed into the growing body of thought about how the processes in nature worked at molecular levels and what that meant for scientific discovery and research.

The discovery of the double helix structure of DNA was a fundamental step forward, enabling scientists to envision genetic manipulation, although the latter development had to wait until 1970, when the first restriction enzyme was identified. A restriction enzyme can cut through DNA and became the first tool of biotechnology, enabling the cutting and splicing of genetic code and unlocking the future of genetic engineering.

■ Impacts and Issues

An interdisciplinary approach gives biotechnologists a superior advantage in tackling many problems worldwide. Because biotechnologists are involved in so many areas, they can have great economic value both to companies and on wider national scales. It is estimated that the biotechnology industry in the United States will be worth $90 billion annually by 2013, and other countries globally are investing in their emerging markets with both grants and tax incentives. In particular, Germany is working with its own global biotech companies such as Pfizer to back up the home market; German universities and government agencies also have their own funds to invest in biotechnological research. The Swiss are active investors in emerging markets in Eastern Europe, Africa and Asia, while Taiwan and Australia are investing much capital in their own research and products in order to secure a place in the field's future.

In 2004 the Mexican government made the decision to cut grants for overseas study in order to keep more biotechnology students in the country. "We lack a critical mass of experts in many advanced disciplines such as genome sciences and nanotechnology," stated Octavio Paredes-Lopez (1943–), then president of the Academia Mexicana de Ciencias (Mexican Academy of Sciences.) The value to the future economic and scientific development of the country was clear.

Biotechnologists, while looking to the future, must also be in touch with contemporary society, in regard to both cultural standards and popular technology. A recent innovation has been the adaptation of the iPhone as a medical device for quality imaging and chemical detection. For the price of a regular app, the smart phones can be fitted with a small magnifying lens that enables doctors and researchers without access to proper laboratories to diagnose blood and skin diseases simply by taking a photo of the sample on a slide and sending it to colleagues internationally. Other options include installing

special software onto the iPhone and a spectrometer attachment that would work with the iPhone's imaging capabilities to enable more detailed chemical analysis of specimens, including estimating oxygen levels in blood. The research was presented in October 2011 at the Optical Society's Annual Meeting in California.

Whereas ethics committees and official regulation agencies must exist to inform and direct progress

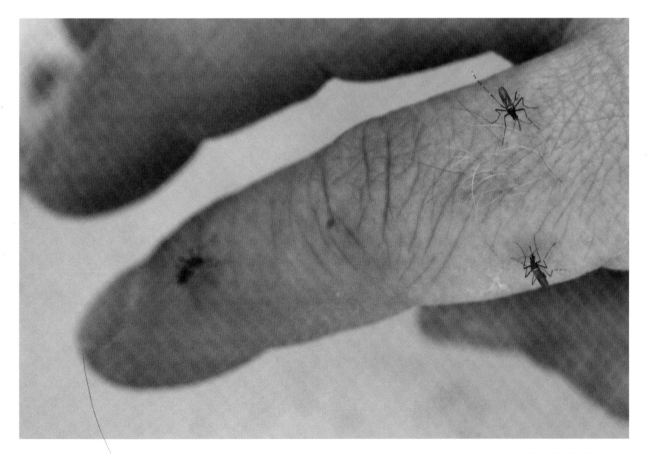

Mosquitoes alight on a scientist at the University of Maryland Biotechnology Institute. Biotechnologists there carefully beheaded hundreds of the mosquitoes, taking care to preserve their salivary glands, in which malaria parasites reside, in order to ship them to a research facility working on developing a vaccine against malaria. © *AP Images/Jacquelyn Martin.*

in the field, on a more informal level, Web sites such as AmericanBiotechnologist.com and EuropeanBiotechnologist.com provide open forums for biotechnologists of various experiences and levels. A biotechnologist, who may be working in the isolation of a laboratory, cannot work disconnected from the rest of the scientific community or society. Developments and inventions in other fields have a big part to play across the scientific board, but are especially important when it comes to biotechnology because so many sciences feed into the concepts of this particular area.

SEE ALSO *Bioeconomy; Bioethics; Biopharmaceuticals; Biotechnology; Combinatorial Biology; Genomics; Molecular Biology; Pharmacogenomics*

BIBLIOGRAPHY

Books

Brown, Sheldon S. *Opportunities in Biotechnology Careers*, Rev. ed. New York: McGraw Hill, 2007.

Echaore-McDavid, Susan. *Career Opportunities in Science*, 2nd ed. New York: Ferguson, 2008.

Madhavan, Guruprasad, Barbara A. Oakley, and Luis G. Kun, eds. *Career Development in Bioengineering and Biotechnology*. New York: Springer, 2008.

Web Sites

"How the U.S. Can Develop Its Biotech Industry." *Biotechnology Industry Organization (BIO)*. http://www.bio.org/content/how-us-can-develop-its-biotech-industry (accessed October 19, 2011).

United States Regulatory Agencies Unified Biotechnology Website. http://usbiotechreg.nbii.gov/ (accessed October 19, 2011).

"Welcome to NCBI." *National Center for Biotechnology Information (NCBI)*. http://www.ncbi.nlm.nih.gov/ (accessed October 19, 2011).

Kenneth Travis LaPensee

Biotechnology

Introduction

Biotechnology is the use of laboratory and engineering techniques to create, alter, or shape biological processes. The goals and outcomes of the application of biotechnology range from the fundamental manipulation of genetic and cellular processes to create life (i.e., new chimera species) to the production of consumer and manufacturing goods that meet specific market needs.

Although modern biotechnology relies on the science of molecular biology, traditional biotechnology utilizes enzymes within cells and microbes to carry out reactions that transform substances. Typical examples include the biotransformation of milk and grain to make cheese and beer. Advances in molecular biology, particularly those concerning the structure of DNA and developments in genomics, have greatly expanded the scope of biotechnology and turned it into a global industry.

Biotechnology makes possible the creation of new drugs and vaccines through genetic engineering techniques. It also shapes modern agriculture by creating plants with superior agronomic characteristics that could increase the global food supply. Microbes and enzymes are now used to clean oil spills and other forms of environmental pollution. Biotechnology may also play a part in providing sustainable energy alternatives to traditional fossil fuels.

Some applications of biotechnology also raise controversial issues, with issues ranging from the proper collection and use of stem cells to the safety of foods derived from genetically-modified plants and animals.

Historical Background and Scientific Foundations

Biotechnology has a history that stretches back for thousands of years. The ancient Egyptians used to apply moldy bread to infected wounds for antibiotic effect. The modern version of this practice is to put the fungus *Penicillium chrysogenum* to work in a giant fermenter to produce penicillin, a drug that transformed medicine in the twentieth century. Meanwhile, the fermentation of fruits and grains to make wines, beers, and spirits has a long tradition all around the world. The ethanol produced by fermentation was also an important feedstock for the chemical industry before it was replaced by petroleum in the 1920s. Ethanol and other fermentation products, such as acetone, have been used as the starting point for the manufacture of textiles, rubber, explosives, and soap. The discovery of the structure of DNA in 1953, by James Watson (1928–) and Francis Crick (1916–2004), pioneered the modern generation of biotechnology, mainly through the discovery of genetic engineering by researchers at Stanford University in the early 1970s. Genetic engineering is a broad-based technology that can be applied in medicine, agriculture, and the food and beverage industry. It can be used either to make an end product, such as human insulin, or a tool used in manufacturing, such as genetically-modified chymosin, an enzyme used to make vegetarian cheese.

Biotechnology is sometimes described according to a color spectrum with each color corresponding to a different application sector. Thus, red biotechnology describes the medical or healthcare sector. In this sector, genetic engineering has allowed the introduction of a new generation of drugs, generally known as biopharmaceuticals. These differ from the traditional small molecule drugs such as aspirin and penicillin. Most biopharmaceuticals are proteins, and they fall into two main categories. Recombinant proteins, such as insulin and erythropoietin, are usually enzymes or hormones produced naturally in the body but which may be deficient or lacking in certain disorders, such as hemophilia or anemia. Then there are the monoclonal antibodies, such as bevacizumab (Avastin), which are used increasingly in the treatment of cancer and inflammatory disease. Meanwhile, a number of new vaccines, such as one against the chronic liver infection hepatitis B, are also made by genetic engineering. Genomics has also led to

WORDS TO KNOW

BIOTRANSFORMATION: Chemical conversion of one substance into another by the action of enzymes, either isolated or within a cell.

DEOXYRIBONUCLEIC ACID (DNA): In DNA-utilizing organisms, genes are segments of DNA. DNA is a polymer; its repeating units contain the chemical code for a gene.

GENETIC ENGINEERING: A technology involving the transfer of a gene of interest from one organism to another. The host organism, which receives the gene, is said to be genetically modified and can be used to manufacture the protein the foreign gene codes for. It can also be a product in its own right, such as a genetically-modified plant.

GENOMICS: The study of the structure and function of genes, particularly in the context of a whole genome, which is the sum total of an organism's genes.

MOLECULAR BIOLOGY: A branch of biology concerned with the structure and function of key biological molecules, including nucleic acids and proteins.

the development of a number of new diagnostic tests based upon the detection of disease genes, or genes from infectious organisms. Therefore, women carrying the breast cancer genes BRCA1 or BRCA2 can be tested and, if positive, opt for close monitoring or preventive treatment. Genetic tests also allow for detection of the human papilloma virus as an additional test in cervical cancer screening, because certain strains of this virus are known to be the cause of the disease.

Green biotechnology refers to the application of biotechnology in agriculture. Here again, genetic engineering has been the key, for it enables the creation of important crop plants, such as wheat, corn, and cotton, with improved characteristics. Genes for pest resistance have been transferred to plants to increase yields. Crop loss to weeds has also been alleviated by transferring herbicide resistance to crop plants, allowing them to be sprayed with herbicide that kills the weeds but not the crop. In horticulture, new types of flowering plant, such as blue roses, can be created by transferring the genes that control petal color. Plant biotechnology does not always involve genetic engineering, however. Tissue culture has also led to the production of cloned plantations of plants such as oil palm with higher yields.

A scanning electron microscope (JSM-6510) is on display at the international exhibition of analytical and laboratory equipment in Russia and CIS on April 28, 2011, in Moscow, Russia. Such equipment is part of the medical, or red, sector of biotechnology. © *Dikiiy/Shutterstock.com.*

Blue biotechnology is also known as marine biotechnology and refers to the exploitation of marine natural resources in medicine and other industries. The development of genomics has led to an increased interest in biodiversity, with a focus upon the oceans as possibly the least explored regions in this respect.

Finally, white biotechnology or industrial biotechnology is a broad definition that covers many other applications of biotechnology. It has a focus upon using enzymes to make industrial processes more environmentally friendly and to provide more sustainable energy supplies. A well known example is the use of enzymes in detergents to allow washing at lower temperatures to save energy. Also growing in significance is the use of enzymes to produce biofuels from plant-based raw materials.

■ Impacts and Issues

Biotechnology products and services serve a huge, and growing, market worldwide. By 2014, experts predict, the top five drugs will be biopharmaceuticals, rather than small molecules. Moreover, there are around 14 million

farmers in 25 countries planting genetically-modified crops. In the next decade or so, it is likely that biofuels, such as ethanol, will have made a significant impact in replacing at least some of the gasoline volumes used in transportation. Biotechnology fits the demands of both government and the public for a cleaner, more sustainable approach to energy generation and use, and industrial processes. It has also opened up new horizons in medicine. Not only can people with genetic disorders be treated with replacement enzymes and proteins and, potentially, gene therapy, but there is a much larger market for regenerative medicine, which includes stem cell therapies. An increase in the number of genomes being sequenced, and a deeper understanding of the human genome, are sure to drive biotechnology forward in years to come.

However, the benefits of biotechnology can be fully realized only with further scientific research, which will require commitment of substantial amounts of public funding. Understanding genes, and how they work, is in its infancy, despite the vast amount of genomic information that has been amassed in recent years. On a more practical level, cells and enzymes, which are at the heart of biotechnology processes, are fragile and require

Signs for biotech companies line the Torrey Pines Science Park in San Diego, California. Biotechnology industries can bring jobs and revenues into state economies. Many medical research firms are looking for cheaper land, million-dollar incentives, and opportunity for new scientific fronts in other states. In 2006, Florida successfully lured three facilities, the Torrey Pines Institute for Molecular Studies, the Burnham Institute for Medical Research, and The Scripps Research Institute, with dreams of creating a biotechnology hub along the Treasure Coast. *© AP Images/Denis Poroy.*

careful handling. Contamination is an ever present threat in a fermentation process, so safety and sterility are more crucial than they would be in a conventional chemical process.

Biotechnology also raises some ethical and safety issues. In the medical arena, there is concern over the use of stem cells to repair and regenerate damaged tissue in diseases such as Parkinson's disease, stroke, and diabetes. Although the benefits may be significant, these cells may be derived from human embryos or fetuses. There is also concern that stem cells transplanted to the human body could perhaps grow into a tumor, instead of into the desired tissue. In agriculture, fears have been expressed over the potential environmental harm in introducing plants bearing foreign genes into ecosystems. A foreign gene is one from another organism, such as a bacterium that confers a desirable property, such as pest resistance, on the plant. If the gene spreads to other plants nearby, it may result in unforeseen consequences. In addition, there has been criticism over the widespread use of cereals as a feedstock for making biofuels, as these crops are a potential food source.

Concerns and objections about the use of genetically-modified foods continue. Although the vast majority of evidence finds no causative links between the consumption of genetically modified foods and various ill effects, there are studies that find some correlation between the consumption of genetically-modified foods and undesirable side effects. For example, some studies suggest that antibiotic resistance genes, often used as a selection marker in genetic engineering experiments, might mix with native bacterial cells in the human gut to produce bacteria with higher levels of antibiotic resistance. There is, however, no evidence establishing a causative mechanism for this transfer.

SEE ALSO *Agricultural Biotechnology; Biotechnologists; Genetic Technology; Industrial Genetics; Molecular Biology; Nanotechnology, Molecular*

BIBLIOGRAPHY

Books

Achrekar, Jayanto. *Concepts in Biotechnology.* New Delhi, India: Dominant Publishers and Distributors, 2005.

Morris, Jonathan. *The Ethics of Biotechnology.* Philadelphia: Chelsea House Publishers, 2005.

Nigam, Seemanta Charit. *Dictionary of Biotechnology.* New Delhi, India: Dominant Publishers and Distributors, 2005.

Web Sites

"Welcome to NCBI." *National Center for Biotechnology Information (NCBI).* http://www.ncbi.nlm.nih.gov/ (accessed August 7, 2011).

Susan Aldridge

Biotechnology Organizations

■ Introduction

The Biotechnology Industry Organization, or BIO, is an international organization with membership from biotechnology, fuel, health, and other corporations with ties to the biotech field. Forged as an umbrella organization for industrial needs, BIO includes more than 1,100 companies, academic institutions, and non-profit centers in its membership rolls. That a professional industry organization could grow to this size and incorporate such diverse corporations as Johnson & Johnson, Merck, Canadian firm Ag-West Bio, accounting and financial services firm Ernst & Young, and the Juvenile Diabetes Research group speaks to the reach of biotechnology in the corporate, academic, and non-profit realms.

Organizations such as BIO are corporate-heavy, with a focus on biotech promotion and industry conventions to display new products and services from members, whereas other biotechnology organizations focus on ethical issues, industrial concerns, and professional standards. In addition to international biotechnology organizations, many states in the United States have statewide organizations, and countries such as Japan, the United Kingdom, and Germany have their own, country-specific groups.

■ Historical Background and Scientific Foundations

The first professional organizations devoted to biotechnology emerged in the 1980s as the field grew quickly. The Industrial Biotechnology Association (IBA), after joining with the Association of Biotechnology Companies (ABC) in 1993, became the organization now known as BIO. It was the first such organization created to bring together a variety of actors in biotech, examine market forces, and research funding, policy issues, and more. As advances in medicine, agriculture, fuel technology, and pure science crossed different fields, biotechnology and

its unique needs became a growth area with a need for some cohesion and professional support.

Medical discoveries spur medical biotech growth, for example, the discovery of and research on insulin's role in the body started a race to create an working insulin pump. The first to be manufactured was introduced in the late 1970s, and the pumps have undergone constant improvements in dosing accuracy and size reduction since its introduction.

In agriculture, research on genes lead to the ability to genetically engineer plants, which allowed for hardier plants, as well as greater yields. Seemingly pure science discoveries had cross-field applications. As the mapping of various species' genomes developed throughout the late 1980s and 1990s, biotechnology deepened and broadened, with pharmaceutical companies such as Merck and AstraZeneca and agribusiness firms such as Monsanto and Dow taking the lead.

Meanwhile, work by researchers and faculty members at universities provided fertile ground for new developments, especially in the genetic engineering field. As increasingly-complex products, devices, and drugs entered laboratories, the marketplace, and homes,

---

WORDS TO KNOW

BIOTECHNOLOGY INDUSTRY ORGANIZATION: Also known as BIO, this leading international trade organization contains more than 1,110 biotechnology firms, academic institutions, and non-profit organizations.

GENETIC ENGINEERING: The field in which specialists manipulate genes to elicit a desired outcome.

GENETICALLY MODIFIED ORGANISMS: Also known as GMOs, genetically-modified organisms are food products that involve genetic engineering to produce a desired outcome in the crop. United States agribusinesses provide the vast majority of GMO-based foods in the world.

non-governmental organizations such as the United Nations began to research and monitor the impact of these changes on the population, devoting the 53rd World Health Assembly to the topic of genetically modified foods. The UN's World Health Organization also focuses on any form of biotechnology involving health, ranging from vaccine development and use to gene therapy and cloning.

Advocacy and regulatory policy monitoring are two major issues for industry organizations in biotechnology. Others include group purchasing, promotion, evaluating state and regional development issues that may affect the biotechnology and pharmaceutical industries, and standards compliance.

■ Impacts and Issues

Smaller biotechnology organizations such as those at the state level, or niche organizations that work to provide services to a small subsector in the field, fill a gap left by international organizations that tend to be dominated by multinational corporations, government agencies, and non-governmental organizations (NGOs). State funding

Michael J. Fox, founder of the Michael J. Fox Foundation for Parkinson's Research speaks at the Biotechnology Industry Organization's 2007 biotechnology conference, at Boston `Convention & Exhibition Center in Boston, Massachusetts. Fox, a keynote speaker at the conference, appealed to scientists and investors to aggressively translate scientific research into creative treatments for debilitating diseases, including the Parkinson's disease he has fought for more than a decade. © AP Photo/Chitose Suzuki.

Actress Brooke Shields speaks to a luncheon on women's health issues at the 2004 Biotechnology Industry Organization convention in San Francisco. Shields credited biotechnology with helping her to conceive a child. © AP Images/Eric Risberg.

for biotechnology institutes at state-funded public universities is crucial for drawing new biotech facilities and jobs to regions. For legislators determined to make economic development a top priority, smaller, local biotechnology organizations and the more local arms of larger groups such as BIO can help draw desired development.

Universities can act as incubating organizations for smaller, local companies trying to launch new products and services. One example is Ohio University in Athens, Ohio, which is the site of the Edison Biotechnology Institute, a state-funded institute that not only generates new research in the field, but helps biotech entrepreneurs create business plans, apply for federal funding, and manage intellectual property issues.

In the 1990s industry groups such as BIO began negotiating with the U.S. Food and Drug Administration (FDA) to determine fee levels for administrative review of new creations. Over time, the fees paid by biotechnology entities have increased, with the firms able to use the financial influence to push which drugs, devices, and products are reviewed in which order. Advocating for priority review status with the FDA involves claims that a therapeutic advance is inherent in the drug or device.

The FDA's Center for Drug Evaluation and Research works to evaluate all prescription and over-the-counter drugs and also contains the Office for Biotechnology Products, the mission of which is "to protect public health by assuring the quality, safety, efficacy, availability and security of therapeutic protein and monoclonal antibody products." However, critics charge that biotechnology and pharmaceutical companies bias the process through fees, whereas industry officials claim that the process is a neutral one, with federal employees at government agencies at the helm. By comparison, in other countries, including Canada, neither biotechnology organizations nor member corporations have direct financial input into the new-product evaluation process.

Biotechnology organizations and their members point to the fee structure when comparing drug approval times with Europe and Canada, with the FDA claiming that in most cases the FDA approves new drugs long before other countries. For instance, new drug approvals in Japan can take two to three times as long to make their way through federal channels than drugs in the United States. Influencing such policies to promote members' interests is a critical part of biotechnology organizations' missions. As the field continues to grow, transnational policies and procedures will make up a larger percentage of these organizations' activities.

SEE ALSO *Bioeconomy; Bioethics; Biotechnology Products; Biotechnology Valuation; Environmental Biotechnology; Forensic Biotechnology; Genetic Engineering; Government Regulations; Industrial Genetics; Patents and Other Intellectual Property Rights; Pharmaceutical Advances in Quantitative Development; Recombinant DNA Technology*

BIBLIOGRAPHY

Books

Kleinman, Daniel Lee. *Science and Technology in Society: From Biotechnology to the Internet.* Malden, MA: Blackwell, 2005.

Schmidt, Markus, ed. *Synthetic Biology: The Technoscience and Its Societal Consequences.* New York: Springer, 2009.

Terziovski, Milé. *Building Innovation Capability in Organizations: An International Cross-Case Perspective.* London: Imperial College Press, 2007.

West, Darrell M. *Biotechnology Policy across National Boundaries: The Science-Industrial Complex.* New York: Palgrave Macmillan, 2007.

Web Sites

Biotechnology Industry Organization. http://www.bio.org/ (accessed November 13, 2011).

UK Bioindustry Association. http://www.bioindustry.org/home/ (accessed November 13, 2011).

Melanie Barton Zoltan

Biotechnology Products

■ Introduction

Biotechnology refers to a wide variety of technologies utilized by a broad range of industrial, commercial, and service activities in the biological sciences. As a result, biotechnology products are as diverse as the technologies that created them. As the term indicates, biotechnology products are based on the study and utilization of living organisms (such as viruses, fungi, and bacteria, as well as animals and plants) or their products (including enzymes, toxins, hormones, living cells, seeds, and DNA). Products derived from biotechnology are present in medications, vaccines, bioinformatics (biochips), chemical industrial products, agricultural products, laboratory reagents, and the food and cosmetic industries.

■ Historical Background and Scientific Foundations

Biotechnology research and products have affected every aspect of modern society, from the global economy to social behavior. They are present in virtually every aspect of modern human activity, including food processing, environmental technologies, human and veterinary health sciences, animal breeding, and farming.

Since the 1960s, concerns about how to feed a rapidly growing world population started to be taken seriously in academic circles, especially among researchers at departments of agricultural and life sciences. Genetic manipulation of livestock and crops is not a novelty; it has been around for centuries through conventional breeding and selection of livestock. Plant grafting to produce better hybrids of plants and crops has also been known for centuries. When the world's population dramatically increased, and along with it the need for more food, the introduction of more pesticides, new herbicides, and fertilizers seemed for a while to be the solution to help increase food production. In the early twenty-first century, herbicides and pesticides have become a growing public health concern.

In addition, several insects, bacteria, and fungi that plague crops have developed resistance to the existing products. New biotechnologies are creating crops that are heartier in their environment and require fewer pesticides.

Advances in molecular biology research since the early 1970s and the development of new techniques for the study of genes in the mid-1980s led to an entirely new field of possibilities for agriculture and animal husbandry, including genetic engineering and agricultural biotechnologies. The same is true of human life sciences, a field in which biotechnologies based on molecular genetics have become a source of important new therapies, early diagnoses, and disease prevention protocols.

The rationale behind biotechnologies is the development of products based on naturally occurring organic compounds, using as production sources either microorganisms or plants. It is interesting to note that most biological molecules (hormones, enzymes, non-catalyst proteins), and even gene sequences involved in mammalian physiology and embryogenesis, were first developed and perfected by bacteria and other microorganisms or by plants that primarily used them for their own purposes. Only later along the evolutionary path were those already established genetic resources incorporated by new and more complex emerging organisms, a phenomenon known as evolutionary conservation.

Because microorganisms have a very fast reproduction rate, it makes sense to use them as a source for producing large quantities of molecules needed for research or products. To do so, genetic engineering techniques are applied to transfer the gene encoding the desired molecule to the microorganism that will be cultivated in larger scale.

■ Impacts and Issues

Genetic engineering technologies are being used to change undesirable traits, to add new traits, to enhance existing properties, or to improve resistance to disease in several crops and animal products. Some genetically-modified (GM) crops include soy, cotton, canola, corn, Hawaiian papaya, eggplant, alfalfa, tobacco, zucchini, tomato, yellow squash, and rice. Chicken eggs are also genetically modified to produce more beneficial Omega 3 fatty acids along with less cholesterol. GM by-products, such as flavorings, processing agents, and sweeteners (e.g., aspartame and sucralose), are used in processed foods and medications.

Medical science is another field that has benefited from biotechnologies directly, especially with the development of new molecules and new ways of delivering them to specific organs and cells. Some of these new drugs were designed to become active only after their incorporation into their biological target (a specific organ or tissue), thus averting major systemic side effects. As a result of these new biotechnologies, a novel class of treatments for several diseases has emerged, known as molecular-targeted therapies. Targeted therapies allow the transport and delivery of drugs or biological molecules directly and specifically into biological features present on the surface of cell membranes or into their nuclei. As a result, new and more effective vaccines, anticancer drugs, and psychiatric medications have been developed. Included among the medical treatments in

use that are recent products of biotechnology are interleukin, plasminogen activators, growth hormones, biosynthetic insulin, erythropoietin, blood clotting factors, growth factors, monoclonal antibodies, interferon alpha and beta, and skin tumor growth-factor inhibitors. Also, new diagnostic tools such as molecular markers and PET-scans are being improved continually through innovations in biotechnology.

New production of some drugs used daily by patients with chronic diseases such as diabetes has reduced costs and/or improved efficacy of the drugs. In the case of insulin production for diabetes, the cost of the hormone was reduced by the introduction of

Delegates stroll among the exhibits presented by various biotechnology companies at a large exhibition of medical technologies in Moscow, Russia. © *Losevsky Pavel/Shutterstock.com.*

insulin-producing genes into corn DNA. Insulin traditionally has been extracted from the pig pancreas and then purified for human use, a costly and complex procedure. Genetically-modified corn crops that produce insulin can be cultivated easily to increase insulin production at lower costs than pig insulin. Moreover, GM-corn insulin requires simpler purification methods.

Biotechnology products, such as GM grains, potatoes, and fruits bearing a resistance trait against insects or other pests, led to a significant reduction in the use of chemical pesticides. As a result, both soil and water resources are less exposed to chemical residues that pollute ground water and are harmful to the environment. Some GM fruits such as tomatoes also are designed to slow down the decay process and prolong their shelf life. This represents an economical advantage to both businesses and consumers.

Examples of products derived from GM bacteria that are used in food processing include the enzymes alpha-acetolactate decarboxylase (this removes bitter substances from beer), alpha-amylase (converts starch into sugars), and chymosin (induces milk coagulation to make cheese). GM fungi are also sources of enzymes for the food industry, such as the enzymes catalase and glucose oxidase that slow down food deterioration, especially in those foods containing cholesterol and other fatty acids. In contrast with non-organic chemical additives used for food preservation, GM enzymes do not become an integral part of the food and are broken down rapidly in the digestive tract. As catalysts, enzymes do not react with food compounds, but act as protective agents or as stabilizers of the food microenvironment. Moreover, the above-mentioned enzymes and others present in GM microorganisms are natural proteins, and also are produced by plants, animals, and human organisms, and cause many less allergies than chemical additives.

As in any other scientific field, biotechnologies and the basic science that develops them are subjected to trial and error, and therefore regulatory legislation for the resulting products is also evolving. The concept of gene transfer and transgenic food products initially has caused significant public controversy. Besides the concerns with toxicity and immunogenicity of GM food, regulatory agencies also had to deal with questions raised by both scientists and public opinion regarding the unknown long-term effects that genetically-modified food could have on humans and animals. Could a transgene present in food travel to human or animal DNA? Would the consumption of GM foods cause new forms of autoimmune disorders? What could be the environmental impact of a gene flow from GM crops to non-GM crops? What if, after such environmental contamination,

Models walk the runway wearing clothing made from genetically modified corn fibers during a fashion show at the World Congress on Industrial Biotechnology in Toronto, Canada, in 2006. © *AP Images/Canadian Press/Frank Gunn.*

it would be impossible to revert to the wild-type species? In response to such understandable worries, strict regulation was created in the 1980s for the application of transgenic technologies in agriculture and GM-based product development.

Since the early 1990s, some of those concerns were proven to be relevant. Gene flow among GM and non-GM crops through natural pollination was observed, for example. Controversies about the immunogenicity or allergenic nature of GM foods continues, despite mounting evidence that the foods are safe and have passed the rigorous allergenic and immunogenic tests required for food and pharmaceutical product approval. Notwithstanding the ongoing debate, advances in the understanding of molecular genetics have provided evidence that agricultural biotechnologies are reliable and GM products are safe for both human and animal consumption. Therefore, several researchers and scientific governmental panels are discussing the need for changes in the regulatory requirements in order to adjust them to be based on product analysis instead of process analysis. The key regulatory principle is that genetically-engineered products should be regulated according to their characteristics, safety, environmental impact, and unique features, rather than whether or not they were created through biotechnology.

SEE ALSO *Agricultural Biotechnology; Biochemical Engineering; Biodegradable Products and Packaging; Biodetectors; Biofuels, Liquid; Biopharmaceuticals; Bioremediation; Biorobotics; Forensic Biotechnology; Forensic DNA Testing; Genetic Engineering; Genetically Modified Crops; Genetically Modified Food; Industrial Genetics; Metabolic Engineering; Nanotechnology, Molecular; Pharmacogenomics; Recombinant DNA Technology; Security-Related Biotechnology; Solid Waste Treatment; Synthetic Biology; Tissue Engineering; Transgenic Animals; Transgenic Plants; Vaccines*

BIBLIOGRAPHY

Books

Egorov, Alexei M., and Gennady E. Zaikov, eds. *New Research on Biotechnology and Medicine.* New York: Nova Biomedical Books, 2006.

Guebitz, Georg M., Artur Cavaco-Paulo, and R. Kozlowski, eds. *Biotechnology in Textile Processing.* New York: Food Products Press, 2006.

Guzman, Carlos A., and Giora Z. Feuerstein, eds. *Pharmaceutical Biotechnology.* New York: Springer Science&Business Media, 2009.

Havkin-Frenkel, D., and Faith C. Belanger, eds. *Biotechnology in Flavor Production.* Ames, Iowa: Blackwell, 2008.

Mascia, Peter N., Jurgen Scheffran, and Jack Milton Widholm, eds. *Plant Biotechnology for Sustainable Production of Energy and Co-Products.* New York: Springer, 2010.

Mine, Yoshinori, ed. *Egg Bioscience and Biotechnology.* Hoboken, NJ: Wiley-Interscience, 2008.

Mousdale, David M. *Biofuels: Biotechnology, Chemistry, and Sustainable Development.* Boca Raton, FL: CRC Press, 2008.

Okafor, Nduka. *Modern Industrial Microbiology and Biotechnology.* Enfield, NH: Science Publishers, 2007.

Proksch, Peter, and Werner E. G. Muüller, eds. *Frontiers in Marine Biotechnology.* Norfolk, UK: Horizon Bioscience, 2006.

Schacter, Bernice Zeldin. *Biotechnology and Your Health: Pharmaceutical Applications.* Philadelphia: Chelsea House Publishers, 2006.

Shmaefsky, Brian. *Biotechnology on the Farm and in the Factory: Agricultural and Industrial Applications.* Philadelphia: Chelsea House Publishers, 2006.

Trigiano, Robert N., and Dennis J. Gray. *Plant Development and Biotechnology.* Boca Raton, FL: CRC Press, 2005.

Web Sites

"Biotechnology and Its Applications." *North Carolina State University Cooperative Extension.* http://www.ces.ncsu.edu/depts/foodsci/ext/pubs/bioapp.html (accessed November 4, 2011).

"Medical Uses of Biotechnology." *New Zealand Organisation for Rare Disorders.* http://www.nzord.org.nz/research/modern_biotechnology/medical_uses_of_biotechnology (accessed November , 2011).

"What Is Biotechnology?" *Biotechnology Industry Organization (BIO).* http://www.bio.org/node/517 (accessed November 4, 2011).

Sandra Galeotti

Biotechnology Valuation

Introduction

Biotechnology valuation consists of the use of financial-assessment methodologies to attribute value to a new biotechnology product in order to attract investors to finance its research and development (R&D). Before the product reaches the market, biotechnology valuation is an estimate of the potential financial and commercial value of a future biotechnology product. Biotechnology valuation is also used to assess a biotechnology company dedicated to research and development of new products.

Historical Background and Scientific Foundations

Biotechnology valuation is an important factor in raising capital for companies that research and develop new drugs and food plants. Among the many industries using and developing new biotechnologies, pharmaceutical companies have been the most profitable. Drug and diagnostic biotechnologies have brought highly significant improvements to medicine, such as revolutionary new medications, vaccines, and early-detection diagnosis for life-threatening diseases, along with great profits to all those involved, from bench research to final products sales.

However, many new molecules with a potential for new therapies initially are investigated and developed by small R&D ventures, founded by individuals with a background in life sciences research. These small R&D companies depend heavily on venture capital and/or research grants until a very promising molecule or an outright breakthrough attracts the attention of large corporations.

Despite the risks involved for both small venture companies and outside investors, they have often been a frequent story of financial success. Since the early 1990s, big biotechnology corporations have either acquired many small companies, signed joint-venture partnerships with them, or outsourced preliminary basic research of new products to them.

The ability of a small biotech venture to raise money depends on the potential for big financial returns associated with the product being developed. The R&D of a new product can be especially attractive to investors when other similar products, already in the market, have proven to be extremely profitable. Their valuation can then factor in comparative-worth values (comparables) in order to calculate a probable value for the new product. Other sources of capital for small biotechnology ventures are governmental agencies and R&D incentive grants provided by either regular legislation or special R&D programs promoted in many countries, especially the United States, Canada, Brazil, Germany, France, and Italy.

Impacts and Issues

Valuating biologic-derived molecules (or biologics) in their early development stages is a challenging task, because many years are required for basic research and testing and selecting the most promising molecules among many candidates. The next step consists of a series of trials with animal models, which, if successful, prompts the design of three different phases of clinical trials with humans. Uncertainties are present in each step of the entire process. Examples of uncertainties associated with R&D of a new drug include unclear or negative preliminary laboratory results, the degree of efficacy in animal models, and unintended side effects in human trials. The entire lengthy process of approval by regulatory agencies for new drugs or treatments averages 10 to 15 years from bench to bedside, and costs many millions of dollars.

In spite of being a risky endeavor, advances in knowledge and know-how have overcome many technical hurdles initially involved in manufacturing bio-derived products, also known as biologics. As a consequence, biologics have become manufactured on a large scale and commercialized with good returns for investors, which makes it possible to quantify more accurately the risks involved in the development of new ones. In the case of pharmaceutical products, the success rates of clinical trials and new drug approval for marketing have provided pivotal value drivers for sound estimates of new drugs.

Other essential value drivers for both drugs and other R&D biotech products are the cost of capital, the existence of fiscal incentives for research, the duration of R&D, the success rates of previous trials, and estimated marketing costs and sales projections. One biotechnology valuation method is the risk-adjusted net present value (rNPV). Predominantly used for new drug valuation, the rNPV is calculated by taking into consideration four parameters: clinical previous success rates, projected costs, sales projections, and the discount rate.

Comparative worth value is another analysis tool that compares the financial success of an upcoming biotechnology product or company with the current value of several similar biotech products already in the market. In addition, this valuation includes traditional financial criteria used in conventional corporate valuations, such

as the value of real assets, the company's current earnings power, and potential for growth.

Biotechnology valuation projections and estimates can turn out to be wrong, especially when valuing an R&D venture company or a product in the early stages of development. Therefore, experts advise the

Visitors look at posters and exhibition during the 3rd Annual International Congress of Antibodies on March 24, 2011, in Beijing, China. © *testing/Shutterstock.com.*

An attorney holds a sample packet of Vioxx during 2006 court proceedings, where he argued that the drug increased the likelihood of heart attacks in persons taking the drug. Vioxx was a beneficiary of the FDA Modernization Act of 1997, which helped streamline the drug approval process and bring new innovations in biotechnology to the market sooner. The drug was later voluntarily recalled by the manufacturer, and the fast-track drug approval process is again under review by the FDA. The possibility for such lawsuits may affect the valuation of pharmaceutical products. © *AP Images/Mel Evans.*

total profile analysis of a biotech company as the best decision-making tool for private investors willing to take a chance in the biotechnology investment market.

SEE ALSO *Biopharmaceuticals; Biotechnology Products; Drug Price Competition and Patent Term Restoration Act of 1984; Government Regulations; Human Subject Protections; Human Subjects of Biotechnological Research; Pharmaceutical Advances in Quantitative Development*

BIBLIOGRAPHY

Books

Ahuja, Satinder, and M. W. Dong, eds. *Handbook of Pharmaceutical Analysis by HPLC.* Boston: Elsevier Academic Press, 2005.

MacKinnon, George E., ed. *Understanding Health Outcomes and Pharmacoeconomics.* Sudbury, MA: Jones & Bartlett Learning, 2012.

Turner, J. Rick. *New Drug Development: Design, Methodology, and Analysis.* Hoboken, NJ: Wiley-Interscience, 2007.

Walley, Tom, Alan Haycox, and Angela Boland, eds. *Pharmacoeconomics.* New York: Churchill Livingstone, 2004.

Yuryev, Anton, ed. *Pathway Analysis for Drug Discovery: Computational Infrastructure and Applications.* Hoboken, NJ: John Wiley & Sons, 2008.

Web Sites

Johnson, Gary. "The Trouble with Expert Judgment in Forecasts." *Pharmaphorum,* January 19, 2011. http://www.pharmaphorum.com/2011/01/19/the-trouble-with-expert-judgment-in-forecasts-2/ (accessed November 6, 2011).

Ranade, Vinay. "Early-Stage Valuation in the Biotechnology Industry." *The Walter H. Shorenstein Asia-Pacific Research Center,* February 2008. http://aparc.stanford.edu/publications/earlystage_valuation_in_the_biotechnology_industry (accessed November 6, 2011).

Sandra Galeotti

Bioterrorism

Introduction

Terrorism as a means of coercion for political or ideological reasons is targeted sometimes at civilians with the specific aim of instilling terror. Biological weapons are employed to this end in a bid to cause death, illness, social disruption, and panic. Viruses, bacteria, and toxins all can be used as biological weapons disseminated through air, water, and food. The resultant illnesses in some cases can be further spread by physical contact or infection by inhalation, and symptoms may not occur until hours or even days after the actual attack. Whereas many biological agents are naturally occurring, the use of genetic engineering and modifications in the laboratory complicates the process of protection, detection, and treatment, which will challenge the capabilities of scientists working in defense against such attacks in the future.

Historical Background and Scientific Foundations

Bioweapons have long been a battlefield partner to more traditional methods of war. Tales of poisoned arrows had their place in the ancient Greek myths. From as far back as the sixth century BC, bioterrorism has had a place in history. In China, the military work *The Art of War*, attributed to Chinese general Sun Tzu (c.544–c.496 BC), stated: "Subjugating the enemy's army without fighting is the true pinnacle of excellence." The Chinese were known to have used poisons in water, wine, and grain at this time. In Mesopotamia (an area overlapping parts of modern day Iraq, Turkey, Syria, and Iran) in the same era, the Assyrians used rye ergot, a toxic fungus, to poison the water supply of their enemies. The fungus caused convulsions and circulatory problems that could lead to gangrene and subsequent loss of limbs. In AD 1346 defenders of the besieged Crimean city of Kaffa (modern Feodosija, Ukraine) were subject to plague-ridden corpses catapulted over the city battlements by the Tartar army that brought the bubonic plague into the city.

Over the centuries, further examples of biological weapons appeared in the records, but it was not until the twentieth century that organized and systematic programs for their research, development, and manufacture were carried out. The 1930s and 1940s in the period of World War II (1939–1945) and beyond saw the Japanese, Americans, and Soviets stockpiling anthrax (*Bacillus anthracis*). The United States alone produced 5,000 bombs filled with the spores, but in 1972 it, along with an original 21 other countries worldwide, signed the Biological Weapons Convention that prohibits the development and accumulation of such weapons and promised the destruction of any existing stockpiles. Although this agreement has been reviewed at intervals over the decades that followed, there is still no agreed system of monitoring to ensure compliance.

The loss of state programs and support for biological weapons left the field to small factions and cults, which increasingly began to utilize bioweapons for terrorism. In 1994 and 1995 in Japan, the nerve agent sarin was released with a combined death toll of 20, but it affecting thousands. The first attack took place in Matsumoto and served as a trial run for an attack the following year on the Tokyo subway. These acts were perpetrated by the Aum Shrinriyko, a religious sect.

Impacts and Issues

Considering the severity of resulting illnesses or death and the ease of dissemination, bioterrorism agents fall into three categories. These range from high- to low-priority agents. The high-priority agents in category A are easily spread, have high death rates, and require special preparedness protocols. Category B poses a moderate dissemination threat and low mortality rates (for example, *Salmonella*). Category C generally refers to agents that could also have the potential for high

WORDS TO KNOW

BIOHAZARD: In the field of biological materials and research, any agent or situation that constitutes a threat to health or environment.

STOCKPILE: Accumulation of materials or weapons held in reserve for future emergency or shortage.

VACCINE: Attenuated or live pathogen, or part thereof, used to confer immunity by inducing the production of antibodies.

death rates (for example, SARS—severe acute respiratory syndrome), but generally would only pose a threat in the future given further engineering. According to the U.S. Centers for Disease Control and Prevention (CDC), the top threats from bioterrorism are posed by anthrax, botulism, plague, and smallpox, all Category A agents.

In 2001 anthrax spores were sent by mail to the offices of two U.S. Democratic senators and several news agencies with a death toll of five. Because of these attacks the U.S. Postal Service instigated biohazard detection systems at various distribution centers to scan for anthrax being sent in the mail. On June 12, 2002, the Bioterrorism Act became law in the United States, mainly in response to the events of September 11, 2001. The Bioterrorism Act covers possession, use, and transfer of certain toxins and agents, and also sets strict criteria for risk assessment of individuals with access to these materials.

Members of the British and American armed forces are inoculated against anthrax, and members of public agencies are educated about its symptoms, treatments, and the emergency response required should an attack be suspected. Although the French bacteriologist Louis Pasteur (1822–1895) created the first vaccine for anthrax back in 1881, new improved vaccines have been developed since then alongside topical cleaners to kill and remove spores. More recently, new innovative technology has come into play. Researchers from the University of Wisconsin Center for Space Automation and Robotics, working alongside horticulture researchers on the International Space Station (ISS), developed an anthrax-killing system based on an air-purifying device used to screen out bacteria in the ISS greenhouses. The

A hazardous materials unit worker is hosed down on Capitol Hill after inspecting buildings and offices for anthrax contamination after anthrax was mailed to several people in the aftermath of the September 11, 2001, terrorist attacks. © *AP Images/Ron Thomas.*

device could have practical application in public buildings and anywhere a large scale release of anthrax might be anticipated.

Advances in molecular biology and genetic engineering have meant that even as vaccines are being employed to deal with known threats, new, more ominous threats are emerging. An interview with the Russian defector Sergei Popov for the NOVA television program on PBS provided insights into the now-defunct Soviet bioweapons program. Among the research mentioned was development of a strain of anthrax created to be resistant to 10 types of antibiotics. Also discussed was the Hunter program, which created hybrid viruses through genetic engineering, and transgenic bacteria that house viruses. In this last example, the bacterial infection would cause a disease such as plague, which would clear upon treatment with antibiotics, but those same antibiotics would release a virus (such as encephalomyelitis) that would cause a new, unexpected illness days later. The Soviet program wound down in the 1990s, but the technology and research continues to be significant because the knowledge is untraceable. Combined diseases would complicate the treatment regime, confuse the tracking of those responsible, and slow down a national response system if these types of bioweapons ever fell into the wrong hands.

Bioterrorism has as its roots ideologies and beliefs that drive individuals, sects, and larger organized groups to use terror and widespread destruction in order to achieve their aims. Apart from the ethics of using such stealth tactics, especially where they are used almost indiscriminately and intentionally against civilians, bioterrorism is ultimately the employment of lethal and uncontrollable biological agents that wreak havoc with health, economies, and the normal running of society. One of the consequences of the mere threat of bioterrorism is that governments and corporations must spend large sums to protect against the potential use of bioweapons. Even though their actual use has long been minimal, the threat of their existence diverts resources away from what many public health workers may consider to be more pressing public health priorities.

■ Primary Source Connection

A February 2011 report, prepared for Members of Committees of Congress, discusses federal efforts to address the threat of bioterrorism. Questions directed at congressional policy makers center around existing efforts to address the threat of bioterrorism, potential changes in activities and existing programs surrounding bioterrorism, and the federal government's role in response to the threat of bioterrorism. Congressional policymakers, through Congress, may need to make choices relating to the size and role of existing programs, influencing the federal response to threats of bioterrorism.

Federal Efforts to Address the Threat of Bioterrorism: Selected Issues and Options for Congress

Summary

Reports by congressional commissions, the mention of bioterrorism in President Obama's 2010 State of the Union address, and issuance of executive orders have increased congressional attention to the threat of bioterrorism. Federal efforts to combat the threat of bioterrorism predate the anthrax attacks of 2001 but have significantly increased since then. The U.S. government has developed these efforts as part of and in parallel with other defenses against conventional terrorism. Continued attempts by terrorist groups to launch attacks targeted at U.S. citizens have increased concerns that federal counterterrorism activities insufficiently address the threat.

Key questions face congressional policymakers: How adequately do the efforts already under way address the threat of bioterrorism? Have the federal investments to date met the expectations of Congress and other stakeholders? Should Congress alter, augment, or terminate these existing programs in the current environment of fiscal challenge? What is the appropriate federal role in response to the threat of bioterrorism, and what mechanisms are most appropriate for involving other stakeholders, including state and local jurisdictions, industry, and others?

Several strategy and planning documents direct the federal government's biodefense efforts. Many different agencies have a role. These agencies have implemented numerous disparate actions and programs in their statutory areas to address the threat.

Despite these efforts, congressional commissions, nongovernmental organizations, industry representatives, and other experts have highlighted weaknesses or flaws in the federal government's biodefense activities. Reports by congressional commissions have stated that the federal government could significantly improve its efforts to address the bioterrorism threat.

Congressional oversight of bioterrorism crosses the jurisdiction of many congressional committees. As a result, congressional oversight is often issue-based. Because of the diversity of federal biodefense efforts, this report does not provide a complete view of the federal bioterrorism effort. Instead, this report focuses on four areas under congressional consideration deemed critical to the success of the biodefense enterprise: strategic planning; risk assessment; surveillance; and the development, procurement, and distribution of medical countermeasures.

Congress, through authorizing and appropriations legislation and oversight activities, continues to influence the federal response to the bioterrorism threat. Congressional policymakers may face many difficult choices about the priority of maintaining, shrinking, or expanding existing programs or creating new programs to address identified deficiencies. Augmenting or creating programs may result in additional costs in a time of fiscal challenges. Maintaining or shrinking programs may pose unacceptable risks, given the potential for significant casualties and economic effects from a large-scale bioterror attack....

Introduction

Reports by congressional commissions, the mention of bioterrorism in President Obama's 2010 State of the Union address, and the issuance of executive orders have increased congressional attention to the threat of bioterrorism. Federal efforts to combat the threat of bioterrorism predate the anthrax attacks of 2001 but have significantly increased since then. The U.S. government has developed these efforts as part of and in parallel with other defenses against conventional terrorism. Continued attempts by terrorist groups to launch attacks targeted at U.S. citizens, including those in transit to U.S. soil, have increased concerns that federal counterterrorism activities, and the investments that underlie them, insufficiently address the threat.

Experts differ in their assessments of the threat posed by bioterrorism. Some claim the threat is dire and imminent. The congressionally mandated Commission on the Prevention of WMD Proliferation and Terrorism concluded that

> unless the world community acts decisively and with great urgency, it is more likely than not that a weapon of mass destruction will be used in a terrorist attack somewhere in the world by the end of 2013.

> The Commission further believes that terrorists are more likely to be able to obtain and use a biological weapon than a nuclear weapon.

In contrast, other experts assert that the bioterrorism threat is less severe or pressing than that posed by more conventional terrorism or other issues facing the United States. The Scientists Working Group on Biological and Chemical Weapons concluded that

> public health in the United States faces many challenges; bioterrorism is just one. Policies need to be crafted to respond to the full range of infectious disease threats and critical public health challenges rather than be disproportionately weighted in favor of defense against an exaggerated threat of bioterrorism.

Stakeholders often measure federal efforts against the perceived magnitude of the threat. Thus, those who believe that bioterrorism poses a relatively low threat tend to conclude that the government has done too much. In contrast, those who perceive a greater threat conclude that the federal government needs to do more, whether under existing programs or new ones. Many experts come to mixed conclusions: they regard some programs as effective but identify others as insufficient.

The federal government's biodefense efforts span many agencies and vary widely in their resources, scope, and approach. For example, the Departments of State and Defense have cooperated with foreign governments and nongovernmental organizations to engage in nonproliferation, counterproliferation, and foreign disease outbreak detection efforts. The Departments of State and Commerce have strengthened export controls of materials that could be used for bioterrorism. The Department of Health and Human Services (HHS) has made investments in public health preparedness; response planning; foreign disease outbreak detection; and research, development, and procurement of medical countermeasures against biological terrorism agents. The intelligence community has engaged in intelligence gathering and sharing regarding bioterrorism. The Department of Justice performs background checks on people who want to possess certain dangerous pathogens. The Department of Homeland Security (DHS) has engaged in preparedness, response, and recovery-related activities, developed increased capabilities in environmental biosurveillance, and invested in expanding domestic bioforensics capabilities. The Environmental Protection Agency (EPA) has explored post-event infrastructure decontamination. Many agencies, jointly or separately, have invested in expanded biodefense infrastructure, including public and private high-containment laboratories for research, diagnostic, and forensics purposes. Lastly, the Executive Office of the President and other executive branch coordinating groups have engaged in risk assessment and strategic planning exercises to coordinate and optimize federal investment against bioterrorism and response capabilities.

Conflicting views of the bioterrorism threat and the breadth of the federal biodefense effort, which crosses congressional committee jurisdictions, complicate congressional oversight of the overall biodefense enterprise. Providing oversight and direction for individual biodefense agencies or programs is easier than addressing the entirety of the biodefense enterprise at once, but such an approach may focus too narrowly to improve the overall effort. An alternative approach identifies key areas or activities that shape federal agency efforts. The Bush Administration identified four such "pillars" as organizing principles for the federal biodefense efforts: threat awareness; prevention and protection; surveillance and detection; and response and recovery. Each of these pillars may have several agencies performing critical parts of the activity. Congressional oversight and direction of biodefense efforts has followed a similar but not identical path. Congress has provided oversight and direction on the basis of both individual agency biodefense activity and on those cross-agency themes and policies deemed most important by congressional policymakers....

Conclusion

While no mass-casualty bioterrorism event has yet occurred in the United States, some experts and policymakers assert that terrorist organizations are attempting to develop such a capability. The federal government has been preparing for a bioterrorism event for many years. Multiple programs in many agencies attempt to prepare for and respond to a bioterrorism event. Whether these programs are sufficient, redundant, excessive, or need improvement has been a topic of much debate. Congress, through oversight activities as well as authorizing and appropriations legislation, continues to influence the federal response to the bioterrorism threat. Congressional policymakers may be faced with many difficult choices about the priority of maintaining, shrinking, or expanding existing programs versus creating new programs to address identified deficiencies. Augmenting such programs may incur additional costs in a time of fiscal challenges while maintaining or shrinking such programs may be deemed as incurring unacceptable risks, given the potential for significant casualties and economic effects from a large-scale bioterror attack.

Frank Gottron
Dana A. Shea

GOTTRON, FRANK, AND DANA A. SHEA. "FEDERAL EFFORTS TO ADDRESS THE THREAT OF BIOTERRORISM: SELECTED ISSUES AND OPTIONS FOR CONGRESS." *CRS REPORT FOR CONGRESS,* FEBRUARY 2, 2011, I–II, 1–3, 14–15.

SEE ALSO *Agricultural Biotechnology; Biodefense and Pandemic Vaccine and Drug Development Act of 2005; Biodetectors; Biological Confinement; Biological Weapons; Biologics Control Act; Biosafety Level Laboratories; Forensic Biotechnology; Security-Related Biotechnology; Weaponization*

BIBLIOGRAPHY

Books

Brannigan, Michael C., ed. *Cross-Cultural Biotechnology.* Lanham, MD: Rowman & Littlefield, 2004.

Fong, I. W., and Ken Alibek. *Bioterrorism and Infectious Agents: A New Dilemma for the 21st Century.* New York: Springer, 2005.

Grey, Michael R., and Kenneth R. Spaeth. *The Bioterrorism Sourcebook.* New York: McGraw-Hill Medical Publishing Division, 2006.

National Research Council (U.S.), and Institute of Medicine (U.S.). *Globalization, Biosecurity, and the Future of the Life Sciences.* Washington, DC: National Academies Press, 2006.

National Research Council (U.S.), and National Academies Press (U.S.). *Biotechnology Research in an Age of Terrorism.* Washington, DC: National Academies Press, 2004.

Rasco, Barbara A., and Gleyn E. Bledsoe. *Bioterrorism and Food Safety.* Boca Raton, FL: CRC Press, 2005.

Wenger, Andreas, and Reto Wollenmann, eds. *Bioterrorism: Confronting a Complex Threat.* Boulder, CO: Lynne Rienner Publishers, Inc., 2007.

Zubay, Geoffrey L. *Agents of Bioterrorism: Pathogens and Their Weaponization.* New York: Columbia University Press, 2005.

Web Sites

"Bioterrorism." *U.S. Centers for Disease Control and Prevention (CDC).* http://www.bt.cdc.gov/bioterrorism/ (accessed November 4, 2011).

Brower, Vivki. "Biotechnology to Fight Bioterrorism." *European Molecular Biology Organization (EMBO) Reports,* vol. 4, no. 3, March 2003. http://www.ncbi.nlm.nih.gov/pmc/articles/PMC1315911/ (accessed November 4, 2011).

Lewis, Susan K. "History of Biowarfare." *NOVA scienceNOW,* April 1, 2009. http://www.pbs.org/wgbh/nova/military/history-biowarfare.html (accessed November 4, 2011).

Popov, Sergei. "Interviews with Biowarriors: Sergei Popov." Interview with Kirk Wolfinger. *NOVA Online.* http://www.pbs.org/wgbh/nova/bioterror/biow_popov.html (accessed November 4, 2011).

Kenneth Travis LaPensee

Blood-Clotting Factors

■ Introduction

A blood clot is the natural defensive mechanism provided by the body when it sustains an injury that promotes bleeding beyond the walls of an injured blood vessel. Clot formation is prompted by the action of the blood-clotting factors, the proteins that facilitate blood clotting wherever a blood vessel is breached and bleeding results. Clots assemble at the site of the bleeding, and bleeding will generally cease in a healthy person through clotting action.

If the blood clotting factors function properly, escaping blood will coagulate and the bleeding at the wound site will cease. When the body systems stimulate indiscriminate blood clotting, hyper-coagulation is often the result. If clots form in blood vessels that are not damaged, the result for is thrombophlebitis, a potentially dangerous condition that includes deep vein thrombosis, in which clot pieces may potentially break away from the blood clot mass and enter an artery to cause pulmonary embolism (a lung artery is entered and obstructed), stroke (an artery to the brain is obstructed), or heart attack. Such dislodged blood clots often prove to be fatal.

Blood conditions that occur if clotting factors do not function properly typically are treated through the administration of blood thinning medications such as warfarin and heparin, a naturally occurring anticoagulant. Heparin is manufactured from tissues extracted from pig intestines or bovine lung membrane. In more extreme instances, "clot-busting" medications are injected directly into the affected site. Recent experiments in which rats have been injected with specialized artificial platelets constructed from plastic nanoparticles demonstrate that the prior injection of these artificial cells may speed the ability of the body to form livesaving clots at the wound site in advance of medical treatment.

■ Historical Background and Scientific Foundations

The observed coagulation of blood in all mammals has been studied since ancient times. Blood clots are the result of a series of intricate biological responses within the bloodstream that are initiated immediately upon the body recognizing that a blood vessel has been severed and bleeding has resulted. Platelets, the blood cells that are primarily responsible for the body's defensive response to the injury, are released into the bloodstream in response to the stimulus provided by the protein thrombopoietin. Collagen and thrombin are the additional response proteins directed to the wound site that facilitate the binding of the platelets to form blood clots. The platelets slowly change shape at the site of injury as the body's forces respond to the threat posed by the assault on the blood vessel. Their round shapes at their initial arrival at the wound become increasingly elongated. The platelets then release other proteins into the surrounding bloodstream. The rough, irregular platelet surfaces are designed to bind to other available platelets and clot-forming proteins. As the cellular mass enlarges at the wound site, the clots formed become a plug that seals the wound and promotes gradual healing.

The observed coagulation of blood in all mammals has been studied since ancient times. The fibrous protein fibrin, which is essential to blood clotting processes, was discovered in the mid-nineteenth century. Later research confirmed that fibrin was the product of fibrinogen, a soluble fibrin precursor. Fibrinogen then was classified as one of six factors (Factor I) known at the time, of which prothrombin (Factor II) and the tissue platelet factor (Factor III) were the most prominent in the applicable research. All six factors are produced in the liver.

Once the biological mechanisms associated with blood clotting were understood, scientific attention in the mid-twentieth century turned to the physical disorders associated with blood clotting dysfunction. This research lead to the associated identification of more than a dozen additional factors that play a part in coagulation. Liver disease is one of the most common causes of blood clotting dysfunction. Whenever the levels of any of the originally discovered six blood clotting factors decrease in the blood stream, the prothrombin time (PT) becomes abnormally prolonged. The subject is at risk of excessive blood loss, the eventual loss of blood pressure, and death. PT is a primary indication of dysfunction in the ability of the liver to manufacture essential proteins through synthesis. As the platelets assemble at the site of the wound, the additional proteins collagen and thrombin and the Von Willebrand factor accumulate. Collagen and thrombin help platelets stick together. As platelets gather at the site of injury, they change in shape from round to spiny, releasing proteins and other substances that help catch more platelets and clotting proteins. This enlarges the plug that becomes a blood clot.

Hemophilia is the omnibus term that describes a number of congenital medical conditions marked by excessive bleeding. The absence of the von Willebrand factor is a key determinant of whether a particular subject will have hemophilia.

■ Impacts and Issues

As hyper-coagulation conditions are genetic in origin, their onset is often noted early in the subject's life. The diagnosis means that the subject must take special measures throughout his or her life to promote proper platelet and blood clot factor function.

Factor testing is the process often initiated when an inherited blood factor deficiency is suspected. The determination of blood factor levels is also undertaken if the patient is suspected of having an acquired condition that contributes to excessive bleeding, such as vitamin K deficiency or chronic liver disease. The sample is a simple one, as the necessary blood extracted for analysis is taken from the patient's arm by way of a needle.

Blood loss is the leading factor in all traumatic deaths for persons between the ages of 5 and 44 years of age. The inability to stop serious bleeding is a particularly difficult problem in circumstances including accidents or the course of battle. If a soldier sustains a significant wound that cannot be treated quickly, the risk of death for the soldier is exponentially increased. Recent research initiatives have been directed to the development of plastic polymers that are able to bind with platelets in the bloodstream at the wound site to form a nano-structural barrier; these polymers are able to distinguish between these platelets and those found elsewhere in the bloodstream where clot formation might otherwise lead to the onset of thrombosis.

WORDS TO KNOW

DEEP VEIN THROMBOSIS (DVT): The sometimes fatal condition associated with thrombophlebitis, in which blood clots form in a major vein, usually in the lower legs of the patient.

HEMOSTATIS: The process in which blood loss is reduced and ultimately stopped through the combined action of a platelet and fibrin, a fibrous protein. Fibrin creates a nanoparticle-sized mesh that forms over the wound where blood is lost.

PLATELETS: The cells found in the bloodstream that contribute to blood clotting. Platelets are formed from megakaryocytes, fragments of bone marrow cells. The protein thrombopoietin stimulates platelets to enter the bloodstream when bleeding is detected.

PROTHROMBIN TIME (PT): The period required by the body, through the action of the thrombopoietin protein, to initiate clotting. If PT extends beyond 4 to 5 seconds, this finding is usually an indicator of severe liver damage.

VON WILLEBRAND FACTOR: The large glycoprotein that binds to other proteins, especially blood factor VIII, during coagulation. The absence of this protein in the bloodstream is the cause of the hereditary bleeding disorder that bears the same name, a condition marked by excessive bleeding, especially from the nose and in the gastrointestinal system.

These developments also hold special attraction for all first responders at accident scenes that cannot necessarily treat an internal wound during the period when the accident or wound victim is being transported to hospital for medical assistance. The current research suggests that if first responders were able to inject these synthetic platelets into the victim's bloodstream immediately upon their arrival at the scene, the victim would acquire a greater ability to produce blood clots at the internal wound site that, in turn, could better sustain the victim until they could be treated at a formal medical facility. Prior injection of these platelets is also a possible option for future medical treatment of combat soldiers.

SEE ALSO *Antibiotics, Biosynthesized; Biomimetic Systems; Blood Transfusions; Metabolic Engineering; Nanotechnology, Molecular; Synthetic Biology*

BIBLIOGRAPHY

Books

Beaulieu, R. A. *Engineered Nanomaterials, Sexy New Technology and Potential Hazards.* Washington, DC: United States Department of Energy, 2009.

U.S. Army Master Sgt. Michael Brochu, a medic who is the senior enlisted medical advisor to the Special Operations Command, poses with a bandage designed to stop soldiers from bleeding to death in the battlefield. Soaked in a blood-clotting agent, the 4-by-4 inch (10.2-by-10.2 cm) cloth bandage could save soldiers wounded in such places as the neck, groin, or armpit, where bleeding is particularly hard to stop. *© AP Images/Chris O'Meara.*

Gatti, Antonietta M., and Stefano Montanari. *Nanopathology: The Health Impact of Nanoparticles.* Singapore: Pan Stanford Publishers; Hackensack, NJ: Distributed by World Scientific Publishing, 2008.

Maniatis, Alice, Phillipe van der Linden, and Jean-François Hardy. *Alternatives to Blood Transfusion in Transfusion Medicine.* Hoboken, NJ: John Wiley & Sons, 2010.

Winslow, Robert M. *Blood Substitutes.* Boston: Elsevier Academic Press, 2006.

Periodicals

Chang Thomas Ming Swi. "Nanobiotechnology for Hemoglobin-Based Blood Substitutes." *Critical Care Clinics* 25, no. 2 (April 2009): 373–382.

Web Sites

"Biotechnology: Super Sticky Barnacle Glue Cures like Blood Clots." *Scientist Live.* http://www.scientistlive.com/European-Science-News/Biotechnology/Super_sticky_barnacle_glue_cures_like_blood_clots/23534/ (accessed October 20, 2011).

"Blood Clots." *MedicineNet.com.* http://www.medicinenet.com/blood_clots/article.htm (accessed October 20, 2011).

"Your Guide to Preventing and Treating Blood Clots." *U.S. Department of Health and Human Services.* http://www.ahrq.gov/consumer/bloodclots.htm (accessed October 20, 2011).

Bryan Thomas Davies

Blood Transfusions

■ Introduction

Blood is the essential means of transport of essential nutrients, gases, and biochemical compounds necessary for cells and physiological body systems. Blood also transports oxygen, which is vital to the body's main metabolic processes and energy production. Red blood cells (or erythrocytes) transport oxygen bound to a molecule called hemoglobin to the body's cells and tissues.

Blood transfusions were first used in wartime hospitals, when critically wounded soldiers needed rapid blood replenishment during life-saving surgical procedures. In the United States, it was not until 1947 that blood transfusions became available to the civilian population. Whole blood was then the only available form of transfusion and sometimes caused serious adverse effects in some recipients, such as infections, allergic reactions, and even fatal anaphylactic shock. Since then, safety protocols have been implemented worldwide to prevent such reactions, as well as the transmission of diseases such as viral hepatitis, HIV, and syphilis through blood or plasma transfusions.

■ Historical Background and Scientific Foundations

Several tests for blood typing were introduced in the first half of the twentieth century, such as ABO blood typing and Rh antigen typing, which solved many immunogenic complications associated with whole blood transfusions. However, blood is also a vehicle for viruses, fungi, and bacteria, as well as antibodies against infectious microorganisms, which could cause not only infections but also serious allergies in blood recipients.

The first immunoassay introduced for blood typing was the ABO, which permitted the identification of four hereditary blood types: A, B, O, and AB. ABO typing uses the presence of antigens on red blood cells that are encoded by the ABO gene on human chromosome 9. The importance of ABO typing for blood transfusions is that in this blood group system, antibodies are detected consistently and predictably in the serum of individuals whose red blood cells do not contain the antigen. ABO compatibility between donor and recipient is the foundation upon which all other transfusion tests were based.

In the ABO system, the A gene and the B gene are co-dominants and the O gene is recessive. Individuals with type O blood are considered universal donors, because all the other blood types can receive type O blood transfusions. However, universal donors are not universal recipients and can receive only type O blood transfusions. The AB types are universal recipients, because they can receive blood transfusions from types A, B, AB, and O. Type A and type B can receive only blood of their own type or blood type O.

Further understanding of blood composition and genetic variations among individuals led to the development of new immunoassays for other factors, such as Rh(D) antigens, which are present in blood red cells. There are several different Rh antigens, but the most important is the Rh antigen known as the D protein, or Rh(D).

The D protein is a highly immunogenic antigen and elicits strong immune responses. Individuals with the Rh(D) antigen in their blood are called Rh positive, whereas those with absence of this antigen in their erythrocytes (red blood cells) are called Rh negative. Rh(D) immunoassays expose blood samples from patients to antibodies that bind to Rh antigens. This can lead to three different results: a strong reaction, if person is Rh(D) positive; no reaction, which indicates the blood is Rh(D) negative; and a weak reaction, if the person's blood contains D proteins that are structurally incomplete or mutated. This type of blood is known

WORDS TO KNOW

ALBUMIN: A protein involved in blood coagulation.

ANAPHYLACTIC SHOCK: Violent reaction by the immune system against an allergenic substance, leading to serious systemic complications and even death.

ANEMIA: Deficient levels of red blood cells (erythrocytes), causing poor transport of oxygen to cells in tissues and organs.

ANTIGEN: Any protein, sugar, or toxin recognized by the immune system as alien, therefore eliciting an immune response.

HEMOPHILIA: A recessive hereditary disorder affecting only males, which causes a bleeding disorder associated with deficiencies of clotting factors.

IMMUNOASSAY: Laboratory tests using antigens to identify molecular properties of proteins, DNA sequences, and tissues.

PLATELETS: The term platelets means "small dishes" and refers to circulating fragments of a cell known as megakaryocyte. Plasma platelets help in blood coagulation and also contain growth factors used in tissue repair.

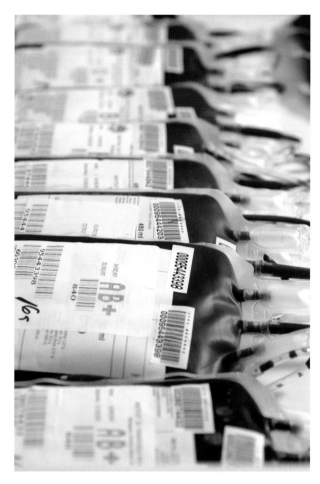

Human blood, with the types showing on labels, sits in storage in readiness for blood transfusions. © *vladm/Shutterstock.com.*

as Weak D, which is rare. Individuals who are Rh(D) negative are likely to produce an allergic response when exposed to blood transfusions containing Rh(D) antigens.

The Rh(D) antigen is a hereditary dominant trait. The predominance of Rh(D) positive individuals in different ethnic groups is as follows: Asians and Africans: more than 99 percent are Rh positive; Native Americans: 99 percent are Rh positive; Europeans: 84 percent are Rh positive; African Americans: 93 percent are Rh positive; Indigenous Australians: more than 99 percent are Rh positive; Australian Non-Indigenous population: 83 percent are Rh positive.

If a woman is Rh negative, but her husband is Rh(D) positive, their children will be Rh positive. Rh negative pregnant women may develop a strong immune reaction against Rh(D) positive fetuses, especially after the first pregnancy, when the maternal immune system has built a specific memory of Rh(D) antigens. The maternal immune response will produce antibodies that will cross the placenta and destroy the fetus' erythrocytes (blood red cells), causing a disorder known as hemolytic disease of the newborn (HDN), which if left untreated can be fatal. Moreover, an Rh(D) negative person who receives a blood transfusion containing even a small number of Rh(D) positive red cells will mount an immune response to destroy the Rh positive erythrocytes and, in some cases, transfusion-related complications may lead to anaphylactic shock or death.

■ Impacts and Issues

Despite the advances of hematology (the medical field that studies and treats blood-related diseases and blood transfusion issues), it was not until the 1980s that a consistent set of safety protocols for blood collection and screening was established, along with a rigorous criteria for selecting blood donors. New biotechnologies have brought great advantages to hematology, such as blood fractionation techniques for therapeutic purposes. Blood fractionation is the separation of blood into its component parts. Simple centrifugation can separate blood into plasma and some of its component cells.

Blood is composed of different types of cells that are suspended in a fluid called plasma. Blood composition includes red blood cells (erythrocytes), white blood cells (leukocytes), antibodies, antigens, cholesterol, minerals (calcium, potassium, iron), sugar (glucose), and factors (e.g., clotting factors, growth factors, and other proteins and hormones that regulate different physiological functions). Blood also transports a wide range of vitamins, the role of which is to protect cells and DNA against several toxins and oxidative processes.

Blood plasma contains thousands of different proteins, but only 20 are extracted and used for therapeutic purposes. Plasma can be separated from other blood contents through a process called plasmapheresis. Once separated, plasma can be frozen and stocked as fresh frozen plasma (FFP). Plasma can also be used to produce other products derived from plasma, generically known as blood-products or hemo-derivates. FFP is used as plasma transfusions to treat hemorrhage or low-levels of blood clotting factors in hospitalized patients.

Plasma-derived products are used in medications and include immunoglobulins (antibodies used in injections to enhance the immune response to infections and immune disorders); coagulation factors (to treat hemophilia and other bleeding disorders); and albumin (used

against dehydration and albumin-low plasma levels). Some genetically-engineered (recombinant) clotting factors are available, and saline or starch solutions also are used as alternatives to albumin for treating fluid loss (dehydration).

Presently, transfusion with blood red cells is done only in emergency rooms and hospitals for cases of severe blood loss due to disease or accidents, surgical procedures, or severe anemia. Advances in molecular genetics and biotechnology have introduced recombinant erythropoietin, which can serve as an alternative to red blood cell transfusion to treat moderate to severe or persistent anemia. Erythropoietin is an erythrocyte growth factor that prompts the body to produce more red blood red cells (erythrocytes). Recombinant erythropoietin is also used to treat cancer patients with chemotherapy-induced anemia. Recombinant erythropoietin, like other recombinant therapeutic biomolecules, is less allergenic than the conventional counterparts, which are purified through conventional techniques. Nevertheless, it is not completely free of side effects.

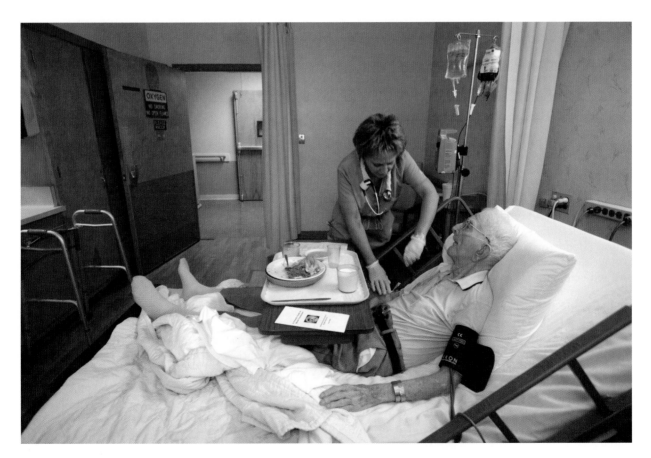

A nurse examines a patient as he is being given a blood transfusion.   © *AP Images/Rich Pedroncelli.*

IN CONTEXT: OXYGEN CARRIERS

Blood transfusions are a common medical procedure. The American Red Cross estimates that every two seconds a patient in the United States needs blood and that a unit of blood is transfused in the United States every 3–4 seconds. Critical and periodic shortages in the supply of donated blood have increased research interest in alternative blood products. Research on alternative blood products, or artificial blood products, began during World War II (1939–1945). The discovery of transfusion-transmissible illnesses such as human immunodeficiency virus (HIV) and hepatitis in the 1980s drew attention to blood supply safety (all donated blood is now screened to exclude blood with such diseases) and accelerated interest in human blood alternatives. Using human blood in transfusions also presents medical challenges: donated blood must be antigen typed and matched to the recipient to avoid transfusion reactions and life-threatening complications. Researchers, therefore, seek to develop artificial blood products that are inexpensive, easy to manufacture, simple to store, have a long shelf-life, can be pasteurized or otherwise free of pathogens, are freely transfusable to any recipient, and function in the body as closely as possible to real human blood.

The term artificial blood is a scientific misnomer. No products have been created in a lab that can perform all of the many functions of human blood—from oxygen transport to immune support—nor can any currently available products permanently replace human blood within the body. Researchers and physicians prefer the term oxygen carriers to describe products created to aid oxygen transport within the body (a typical function of red blood cells). Researchers are also working on platelet products that promote hemostasis, but as of 2011, these were less well-developed than oxygen carriers.

Almost all oxygen carriers are red blood cell substitutes. The two main types are hemoglobin-based and perfluorocarbon (PFC) based. Hemoglobin-based substitutes (typically known as hemoglobin-based oxygen carriers or HBOCs) are derived from hemoglobin from human, animal, or recombinant blood sources. Hemoglobin is obtained from donated blood and red blood cells collected during plasma donation. HBOCs do not have to be typed and matched; they are universally transfusable. Usually, hemoglobin-based substitutes are created from donated blood products that have reached their expiration date and can no longer be used for whole-blood transfusion. However, it takes two units of donated human blood to make one unit of hemoglobin solution.

Perfluorocarbon-based substitutes (PFCs or PFC-based solutions) are completely synthetic hydrocarbon-based compounds with halide substitutions. They are 1/100th the size of healthy human red blood cells and can dissolve nearly 50 times more oxygen than plasma. PFC-based solutions contain PFC compounds mixed with fats or oils so to better mix with blood. PFCs are limited by their short effective duration (only a few days) and the necessity for patients receiving them in transfusion to be breathing 100 percent pure oxygen.

Compared to healthy human red blood cells that can last approximately 100 days, first and second generations of both types of red blood cell substitute oxygen carriers are short lasting—from 24 hours to a few days. With estimated manufacturing costs of $500 per unit, they are also very expensive to produce.

Once collected, whole blood is centrifuged and separated into red blood cells, white blood cells, platelets, and plasma, so recipients can receive the most appropriate treatment they specifically need. All red blood cells and platelets are filtered to remove white blood cells. Platelets then are screened for bacterial contamination. Mandatory testing for ABO and Rh(D) blood typing, along with viral disease and syphilis screening and red cell antibody screening are performed. Individuals who have been previously infected with malaria, viral hepatitis, typhus, or HIV cannot donate blood.

Once adequately tested for safety and purity, hemo-derivates are properly labeled and stored. Different blood components will be used by different patients who specifically need one of these hemo-derivates. For example, cancer patients undergoing systemic chemotherapy often need platelet replacement because some anticancer drugs destroy platelets. Another patient may need erythrocytes (red blood cells), due to a severe anemia. Patients suffering from blood clotting disorders may require clotting factors, and so on. Therefore, plasma and blood fractionation enable many patients to benefit from a single blood donation.

SEE ALSO *Blood-Clotting Factors; Cancer Drugs; Genetic Technology; Genome; Organ Transplants; Recombinant Antibody Therapeutics; Recombinant DNA Technology; SNP Genotyping; Synthetic Biology*

BIBLIOGRAPHY

Books

Hemning, Denise M., ed. *Modern Blood Banking and Transfusion Practices*, 6th ed. Philadelphia: F A Davis, 2012.

Waite, Lee, and Jerry Michael Fine. *Applied Biofluid Mechanics*. New York: McGraw-Hill, 2007.

Web Sites

"Artificial Blood Could Soon Be on the Way." *Zeit News.* http://www.zeitnews.org/biotechnology/

artificial-blood-could-soon-be-on-the-way.html (accessed November 9, 2011).

"Keeping Blood Donations Safe." *U.S. Food and Drug Administration (FDA).* http://www.fda.gov/BiologicsBloodVaccines/SafetyAvailability/BloodSafety/ucm095522.htm (accessed November 9, 2011).

"What is a Blood Transfusion?" *National Heart, Lung, and Blood Institute, U.S. Department of Health and Human Services.* http://www.nhlbi.nih.gov/health/health-topics/topics/bt/ (accessed November 9, 2011).

Sandra Galeotti

Bone Substitutes

Introduction

Medical science has long appreciated that in many circumstances, the natural healing processes of the body do not always adequately deal with many debilitating bone conditions. Bone substitutes are employed in a variety of these cases, either to speed the healing process or as supplements to prosthetics or surgical repair. Some of these conditions include cases of multiple fractures; bone damage caused by the onset of diseases such as arthritis and osteoporosis; or bone cancer, if the affected bones required amputation.

Bone substitutes are defined as any natural or synthetic materials that are used to replace bone or bone tissue. The range of possible bone replacement products has expanded in the twenty-first century through the combined efforts of modern medical science, materials engineers, and biologists. These products include specialized polymers employed in hard tissue replacement procedures, natural coral derivatives, and the newly developed biomaterials DMX, hydroxylpatite, and beta-tricalcium phosphate. These bone substitutes are inert materials; for this reason they can be readily incorporated into adjacent human tissues. Alternatively, bone substitutes can be replaced over time by the growth processes of natural human tissue.

The primary difficulties associated with bone substitutes encountered by surgeons and rehabilitative specialists have been the risk of rejection posed by the host body and the inability of the substitutes to bear the physical stresses that are absorbed by natural bone structures. Contemporary research initiatives have been directed to the development of substitute products that are able to adapt to these forces.

Irrespective of the source, ideal modern bone grafts combine biocompatibility, ease of surgical use, chemical properties similar to natural bone, and the ability to be physically integrated into the original skeletal structure in a cost-effective manner.

Historical Background and Scientific Foundations

Bone substitute research represents the intersection between modern orthopedic research and ancient understandings of human skeletal function. The mainstream medical commentaries refer to the use of autogenuous bone substitutes as the gold standard. There is powerful scientific support for this proposition, because these substitutes provide the patient with the inherent ability to regenerate the injured site. In many instances, autogenuous bone can be revascularized to the extent that the newly formed bone will grow with the patient.

Early attempts to insert foreign bone substitute sources into damaged human bodies met with failure.

WORDS TO KNOW

AUTOGENUOUS BONE GRAFT: A bone sample used for substitution that is taken from the patient's own bones.

AUTOGRAFT: Human tissue that is transplanted from one part of a patient's body to another; also known as autologous grafts or autotransplants. Common autograft forms include skin transplants to treat burn injuries.

DEMATERIALIZED BONE MATRIX (DBM): Glycoprotein that comprises the transforming growth factor family (TGF-β) aids in the development of human tissue and organs. DBM is a commercially available compound that is used to heal problematic fractures. DBM does not contribute to increased structural support in the damaged bone.

OSSEOINTEGRATION: A direct interface formed between an implant (artificial or by autograft) and an original bone structure.

REVASCULARIZE: The restoration of natural blood supply to an organ, tissue, or bone. This condition is consistent with the ability of the body to regenerate damaged bone or tissue.

The concepts associated with the development of new mineral-based structures that mimicked other known natural sources such as ivory or animal bone had a powerful inherent logic. Bone matched to damaged bone ought to provide a reasonable treatment alternative, but these earlier bone substitute pioneers did not possess the knowledge of human immunology to appreciate why the body naturally rejected these products when they were surgically implanted.

Inert substances such as calcium sulfate (commonly known as plaster of Paris) were also explored as bone substitute alternatives to fill existing bone cavities, with little success. It was the advent of biocompounds such as hydroxylpatite, known as bone mineral, that spurred the development of a new class of bone substitutes. Widely used as a substitute for damaged tooth enamel, hydroxylpatite is a naturally occurring compound found in all human teeth and bones. Most modern artificial joints, such as those used in hip and knee replacements, are coated with hydroxylpatite prior to their insertion into the body to reduce the risk of rejection by the host, and to promote osseointegration. If the process is successful, there will be no scar tissue formed between the bone and the surface of the implanted substitute.

Dr. Dong Kim, director of the Mischer Neuroscience Institute at Memorial Hermann, discusses the cranial implant he inserted in U.S. Representative Gabrielle Giffords (D-AZ) during a briefing May 19, 2011, in Houston, Texas. Giffords, who was shot in the head January 8, 2011, while holding a meeting with constituents outside a Tucson, Arizona, grocery store, underwent a cranioplasty procedure to repair damage to her skull May 18, 2011. The implant, a piece of molded hard plastic, was attached to her skull using tiny screws; doctors expect that the bone will eventually fuse with the plastic's porous material.   © *Dave Einsel/Getty Images.*

■ Impacts and Issues

Because all global societies tend to live longer now than at any other time in human history, the ability to replace bones damaged through injury or disease is highly valued. The harvest of bone or tissue suitable for autografts in appropriate cases necessitates additional and often painful surgery for the transplant candidate. Any additional surgical procedure also includes the risk of inflammation and infection at the point of entry, as well as the onset of chronic pain that may persist well past the limits of the original surgery. The amount of bone that can be harvested safely in an autograph procedure is limited. The corresponding option to introduce safe and rejection-free bone substitutes into a damaged area necessarily spares this surgical step.

Allographs have emerged as a viable alternative to autographs when bone substitutes are required. An allograph is a bone substitute removed either from living donors or cadavers. These procedures eliminate issues associated with donor-site infection and inflammation; they reduce the problems associated with limited bone substitute supply because deceased donors represent ready availability in a variety of shapes and sizes. Allographs are often performed through the utilization of tissue freezing and gamma irradiation techniques. The obvious risks associated with the transmission of diseases or other infectious agents from the donor to the recipient have not been entirely eliminated. It is the quality of the tissue bank that provides the desired bone substitute sources, especially the attention paid by the bank to the preservation of the integrity of the biomechanical and biochemical properties of the tissues intended for bone substitution, that will generally dictate surgical success in this area.

See Also *Artificial Organs; Biomimetic Systems; Biorobotics; Prosthetics; Skin Substitutes; Synthetic Biology; Therapeutic Cloning; Tissue Banks; Tissue Engineering; Translational Medicine*

BIBLIOGRAPHY

Books

Calandrelli, Luigi, Paola Laurienzo, and Adriana Oliva. *Biodegradable Composites for Bone Regeneration.* New York: Nova Science Publishers, 2010.

Lieberman, Jay R., and Gary E. Friedlaender, eds. *Bone Regeneration and Repair Biology and Clinical Applications.* Totowa, NJ: Humana Press, 2005.

Nather, Aziz, ed. *Bone Grafts and Bone Substitutes: Basic Science and Clinical Applications.* Hackensack, NJ: World Scientific, 2005.

Petite, Hervé, and Rodolfo Quarto, eds. *Engineered Bone.* Georgetown, TX: Landes Bioscience, 2005.

Vallet-Regi, Maria, and Daniel Arcos. *Biomimetic Nanoceramics in Clinical Use: From Materials to Applications.* Cambridge, UK: RSC Publishing, 2008.

Web Sites

Laurencin, Cato T., et al. "Bone Graft Substitute Materials." *MedScape Reference.* http://emedicine. medscape.com/article/1230616-overview (accessed October 31, 2011).

Bryan Thomas Davies

Botox

Introduction

Botox is the well-known trade name for the bacteria-based toxin that causes botulism, the paralytic illness that is triggered when the bacteria enter the body through an unprotected wound or by the consumption of improperly prepared canned foods. Most consumers associate wrinkle reduction as the only physiological objective desired when Botox is injected into the human body, but it has several other medical uses. Botox has also been used to treat severe underarm sweating and cervical dystonia, which causes muscle contractions in the shoulders and neck. In 2010 Botox was approved by the United States Food and Drug Administration (FDA) as a possible treatment for chronic migraine headaches, and in 2011 it was approved as a treatment for some forms of urinary incontinence.

Botox originally was developed in the 1970s as a medication to treat strabismus (crossed eyes) in children. Blepharospasm, the condition where eyelashes twitch uncontrollably, was also treated through Botox injections. These early treatments established that in small doses injected into the site of the affliction, the toxin acted on the local neurons to prevent their signals from being transmitted to the brain. This neurological intervention was observed to cause the adjacent muscles to relax. As a consequence, skin wrinkles in the vicinity are reduced and the skin surface appears smoother.

The reduction or temporary elimination of wrinkled skin is the primary objective of most Botox treatments. Botox treatments provide only temporary relief from wrinkles and other changes to the skin precipitated by the natural aging process. To retain the effect of the toxin on the skin surface, the treatments must generally be administered every three to six months. The repeated injection of Botox carries risks that include muscle weakness. Also, according to the FDA, those using Botox may experience mild side effects including "pain at the injection site, flu-like symptoms, headache, and upset stomach." In addition, "injections in the face may also cause temporary drooping eyelids."

Historical Background and Scientific Foundations

During the 1970s, scientists first determined that the destructive power of botulinum toxin type A found in *Clostridium botulinum* bacteria could be harnessed for cosmetic purposes. Botulism generally occurs when the bacteria enter the bloodstream through a wound, through the ingestion of tainted preserved foods, or when infants are exposed to contaminated dust or honey. When the naturally occurring toxin affects a human or animal subject, it causes paralysis as the toxin enters the axon terminals, the extension of the neurons that transmit electrical impulses to the central nervous system and brain. If untreated, botulism can be fatal.

The use of the botulinum toxin to treat cosmetic problems grew out of a series of research initiatives undertaken in the early 1970s. This work confirmed the connection between the administration of minute amounts of the toxin and relaxation of the surrounding muscle and skin tissues for a period of months. Botox was first administered in American cosmetic commercial applications to reduce wrinkles in 1987. It has become the most common cosmetic procedure performed in the United States; regulations of the procedure and the persons permitted to administer the Botox injections vary widely from state to state and country to country.

Impacts and Issues

The primary purpose of Botox is the enhancement of a person's physical appearance. There are undoubted psychological benefits for the patient that will flow from a successful cosmetic procedure, in that the desired youthful appearance will contribute to a greater sense of personal well-being. In some instances, particularly in people with deep wrinkles, the treatment includes a combination of Botox injections and the application of collagen.
In many respects, the decision to undergo a Botox procedure is a classic example of a simple cost to benefit

WORDS TO KNOW

BOTOX: The trade name used for botulinum toxin type A formulations, one of seven forms of the toxin isolated. Allergan, Inc. is one of several pharmaceutical companies that use this trade name; in all commercial preparations, Botox is a purified form of the toxic bacterium *Clostridium botulinum* commonly found in various soils.

COLLAGEN: Naturally occurring proteins that are found in animal and human connective tissues.

MIGRAINE HEADACHES: Serious and often debilitating headaches that afflict up to one in 10 North American adults. The causes of migraines are not entirely understood; research has established likely connections between migraine onset and hormonal changes (in women), as well as chemical imbalances in the brain, most notably the level of the pain regulator serotonin.

STRABISMUS: Known colloquially as "cross eyes," this medical condition occurs when the subject cannot control the direction of the eyes while attempting to focus on a specific object. The condition is caused by weakness in the six muscles that control eye direction and movement.

analysis. The physical risks associated with the procedure are relatively small: Aside from the localized bruising that often occurs at the injection site, most patients are able to resume their everyday activities, including sports, within a few hours of the toxin being administered. The number of reported incidents in which the botulinum toxin spread from the desired area to infect adjacent tissues or the body as a whole are few; no deaths directly attributable to the FDA-approved version of the toxin or botulism used for cosmetic purposes have been reported in the United States or the United Kingdom. The cost of the injection varies between $200 and $500 per site injected. The consumer must weigh the expenses associated with a temporary cosmetic remedy and the personal benefits that they derive as a result.

The emergence of Botox injections as a treatment for chronic migraines is a development that is not entirely understood in the current medical research. The possible linkage between the toxin and migraines reduced in both frequency and intensity was first based on the anecdotal evidence of those who had Botox injections for cosmetic purposes. It is uncertain whether the toxin provides relief from migraine symptoms, or whether it is the interaction between the toxin and the axon terminals. Given the debilitating nature of migraine headache attacks, this particular research avenue may extend the benefits of Botox from cosmetic applications to actual physiological relief.

IN CONTEXT: BOTOX FRAUD

In 2004 four patients at a clinic in Florida were hospitalized with temporary paralysis after receiving injections of what they believed to be Botox Cosmetic, a drug approved by the U.S. Food and Drug Administration (FDA) for minimizing facial lines and wrinkles. The FDA's Office of Criminal Investigations launched an investigation into the Florida incident and discovered that the physician at the Florida clinic had used an unapproved Botox Cosmetic substitute. The FDA broadened its investigation and uncovered widespread use of unapproved Botox Cosmetic substitutes by doctors and other approved clinicians. The FDA worked with the U.S. Department of Justice and other law enforcement agencies to bring criminal charges against dozens of unscrupulous medical professionals.

The FDA ultimately traced the source of the substitute Botox Cosmetic used in the Florida case and several other cases to Toxin Research International, Inc. (TRI), an Arizona-based business that sold a cheaper, unapproved Botox Cosmetic substitute to clinics. TRI obtained its Botox Cosmetic substitute from a California laboratory that produced the toxin *Clostridium botulinum* for research purposes. The vials that TRI obtained from the California laboratory were labeled "For Research Purposes Only, Not For Human Use." TRI, however, sold them to physicians as a cheaper

alternative to Botox Cosmetic. The FDA found that more than 200 medical professionals across the country had purchased counterfeit Botox Cosmetic from TRI. The president of TRI was sentenced to nine years in prison and other company employees received lesser sentences.

The use of counterfeit Botox Cosmetic by unlicensed individuals and the illegal sale of Botox Cosmetic of unknown origin directly to consumers may pose even more serious health threats. Law enforcement agencies in the United States and other countries have uncovered numerous examples of unlicensed persons performing Botox Cosmetic injections in beauty salons, apartments, and elsewhere. Many of these unlicensed persons have used counterfeit Botox Cosmetic. The direct sale of do-it-yourself kits to consumers also poses a health risk. In 2008 and 2009, consumer advocacy groups in the United Kingdom discovered advertisements on the online auction website eBay for do-it-yourself Botox Cosmetic injection kits without a prescription. eBay removed the advertisements following complaints by consumer advocacy groups and the UK Medicines and Healthcare Products Regulatory Agency. The FDA and health care professionals recommend that consumers verify that the person performing Botox Cosmetic injections is licensed and that the product is legitimate.

A plastic and reconstructive surgeon injects Botox into the forehead of a patient in her early 30s. Botox, a muscle-paralyzing neurotoxin, is commonly used in cosmetic procedures to reduce facial lines and prevent underarm sweating.   © *AP Images/Damian Dovarganes.*

S<small>EE</small> A<small>LSO</small>   *Collagen Replacement; Cosmetics; Tissue Engineering*

BIBLIOGRAPHY

Books

Benedetto, Anthony V., ed. *Botulinum Toxins in Clinical Aesthetic Practice.* New York: Informa Healthcare, 2011.

Cooper, Grant. *Therapeutic Uses of Botox.* Totowa, NJ: Humana Press, 2007.

May, Suellen, and D. J. Triggle. *Botox and Other Cosmetic Drugs.* New York: Chelsea House, 2008.

U.S. Food and Drug Administration (FDA). *Botox.* Rockville, MD: FDA Office of Women's Health, 2007.

Periodicals

Beck, Susanne. "Enhancement as a Legal Challenge." *Journal of International Biotechnology Law* 4, no. 2 (2007): 75–81.

"Patient Awarded $212 Million for Adverse Reaction to Botox." *Biotechnology Law Report* 30, no. 4 (August 2011): 482.

Shukla, H., and S. Sharma. "Clostridium Botulinum: A Bug with Beauty and Weapon." *Critical Reviews in Microbiology* 31, no. 1 (2005): 11–18.

Web Sites

"Botox." *MedLine Plus.* http://www.nlm.nih.gov/medlineplus/botox.html (accessed October 24, 2011).

"FDA Approves Botox to Treat Chronic Migraine." *U.S. Food and Drug Administration,* October 15, 2010. http://www.fda.gov/NewsEvents/Newsroom/PressAnnouncements/ucm229782.htm (accessed October 24, 2011).

Bryan Thomas Davies

Bovine Growth Hormone

■ Introduction

A variety of methods and technologies have been devised to increase food production worldwide. The injection of a synthetically derived hormone into dairy cows called bovine growth hormone (BGH) constitutes one such food production technology. Dairy farmers can purchase this synthetic hormone commercially. Giving cows synthetic growth hormone causes their daily milk production to increase. The term bovine somatotropin (BST) is synonymous with BGH.

Technological advances introduced to the market by agribusinesses have led to greatly increased plant and animal production over the last half century. Some of these innovations, including the use of BGH, have met with resistance from some consumers, scientists, and government officials. The use of BGH by dairy farmers is legal in the United States, but many countries have opted to ban its use altogether in their agricultural sectors, as well as to ban products containing milk from cows given the hormone.

■ Historical Background and Scientific Foundations

All plants and animals produce and use molecules called hormones. In mammals and other animals, hormones are secreted by cells and by particular glandular organs, and circulate throughout the body (often through the bloodstream). Hormones elicit particular responses from various cells and organs in the body, typically signaling other cells or organs to perform some action or set of actions. For example, the thyroid gland releases two different hormones that help regulate the body's metabolism. Only small amounts of hormones are needed in order to perform their necessary functions inside the body.

In the 1930s, it was discovered that milk production in cows is regulated by the BGH that the animals naturally produce. Therefore, milk production in dairy cows would be expected to increase if the animals were exposed to higher levels of BGH. With the techniques of the day, it was impractical to give cows additional BGH, because the only way it could be obtained was by distilling the BGH from the glands of slaughtered cattle—a cost-prohibitive method.

The 1980s witnessed the first artificially produced hormone, insulin. The hormone insulin is produced by the pancreas and is essential for proper metabolism. Some people with diabetes mellitus (diabetes) require regular injections of insulin to manage their disease, which can be life-threatening if untreated. Before the development of synthetic insulin, the hormone was obtained from the pancreases of pigs (called porcine insulin) or cattle (bovine insulin). Although a life-saving medication for many diabetics, a minority of those who used the animal-derived form of insulin had a severe reaction to it. In the early 1980s, the U.S. Food and Drug Administration (FDA) gave its approval for production of artificial insulin using recombinant DNA technology, thus becoming the first drug derived from recombinant DNA techniques to be approved by the FDA.

All living creatures on Earth use DNA (deoxyribonucleic acid) within their cells for storing and transmitting hereditary information (some hereditary information is also contained in RNA—ribonucleic acid). The DNA molecule is very large, biologically speaking, and forms long strands of repeating subunits consisting of smaller molecules. Different stretches of DNA code for the various traits of an organism. These stretches of DNA are called genes. In recombinant DNA, a selected gene, or genes, is extracted from one organism and then inserted into the genetic sequence of another organism. In the case of insulin, the gene responsible for its production in humans was isolated and cloned—meaning that many identical copies of the gene were produced. These cloned sections of human DNA were then inserted into the DNA of the *Escherichia coli* (*E. coli*) bacterium. The genetically engineered bacteria were cultured (supplied with the proper nutrients and maintained at the correct temperature for multiplying); the bacteria then produced an insulin molecule identical to that made by humans.

Around the mid-1990s, the same basic recombinant DNA technology used to make synthetic insulin was applied to the production of synthetic BGH. The synthetic form of the hormone is often denoted as rBGH, meaning recombinant bovine growth hormone. The synthetic BGH was sold under the brand name Posilac.

■ Impacts and Issues

When properly administered as per the manufacturers' guidelines, dairy cows injected with rBGH demonstrate a marked increase in milk production over time. Increases in milk production of five to 15 percent are to be expected with use of synthetic BGH. In a world of increasing population and finite resources, the availability of products such as rBGH means that a greater food yield can be obtained from the same number of cows.

In spite of increased milk production, the United States is alone among the developed nations of the world in allowing the use of rBGH by dairy farmers. Other

countries, including Canada, Japan, Australia, and the members of the European Union, all ban the use of rBGH in agriculture. Various governmental organizations in those countries contend that the use of synthetic BGH is potentially harmful to human health under certain circumstances. For instance, Canadian researchers conducted animal studies that they asserted showed negative side effects from excessive amounts of the synthetic hormone. However, the FDA later issued its own report that refuted the claims made by the Canadian researchers.

Studies have shown that milk from dairy cows treated with rBGH has a higher concentration of a hormone called insulin-like Growth Factor-1 (IGF-1). The IGF-1 hormone is produced in the human body, and it is important in the physiological development of children. IGF-1 levels affect adult health as well. Because of its essential role in normal human development, higher levels of IGF-1 in food have been a source of concern for both consumers and health officials (although the FDA stated in 1991 and maintains today that there is no significant risk to human health from rBGH). Cows treated with BGH have shown higher incidence of mastitis (a bacterial infection of the udder) and therefore require antibiotics more often than cows not given BGH.

Cattle eat a hay mix on the Reece farm in Knob Lick, Kentucky. Farmer Darrell Reece used Posilac on his cattle, but in 2004 Monsanto Co. began severely rationing the amount of genetically engineered growth hormone designed to increase dairy cow milk production after federal regulators found quality control problems at the factory where it was made. *© AP Images/Patti Longmire.*

Foreign countries are not alone in their suspicions about the possible deleterious effects of using rBGH in dairy cows. Many dairy operations in the United States refuse to inject the synthetic hormone into their cows. Many consumers of milk and milk products in the United States, whether commercial or the buying public, also choose milk that is not derived from cows treated with BGH. Consumer demand has led several mainstream retailers, including Wal-Mart, Safeway, Kroger, and Starbucks, to stock rBGH-free products. Many U.S. companies advertise that their milk or milk products, such as cheese, are derived from rBGH-free dairy cows.

SEE ALSO *Agricultural Biotechnology; Dairy and Cheese Biotechnology; Genetic Engineering; Genetically Modified Food; Insulin, Recombinant Human; Metabolic Engineering; Recombinant DNA Technology*

BIBLIOGRAPHY

Books

Barbee, Michael. *Politically Incorrect Nutrition: Finding Reality in the Mire of Food Industry Propaganda.* Ridgefield, CT: Vital Health Pub, 2004.

Holstead, Joseph R., and David K. Leff. *Recombinant Bovine Growth Hormone.* Hartford, CT: Connecticut General Assembly, Office of Legislative Research, 2007.

Khaliq, T., Zia-ur-rahman, and Javed H. Ijaz. *Bovine Growth Hormone.* Tulsa, OK: Gardners Books, 2010.

Larsen, Laura. *Environmental Health Sourcebook: Basic Consumer Health Information About the Environment and Its Effects on Human Health.* Detroit, MI: Omnigraphics, 2010.

Mills, Lisa Nicole. *Science and Social Context: The Regulation of Recombinant Bovine Growth Hormone in North America.* Montreal: McGill-Queen's University Press, 2002.

Web Sites

"Food Safety: From the Farm to the Fork: Report on Public Health Aspects of the Use of Bovine Somatotrophin." *European Commission*, March 15–16, 1999. http://ec.europa.eu/food/fs/sc/scv/out19_en.html (accessed November 10, 2011).

"Report on the Food and Drug Administration's Review of the Safety of Recombinant Bovine Somatotropin." *U.S. Food and Drug Administration (FDA)*. http://www.fda.gov/AnimalVeterinary/SafetyHealth/ProductSafetyInformation/ucm130321.htm (accessed November 10, 2011).

Breeding, Selective

■ Introduction

Selective breeding has been practiced for thousands of years, beginning with the domestication of wild plants to create food or medicinal crops. Farm and domestic animals have been selectively bred to reinforce necessary or desired traits since the advent of farming.

Charles Darwin's (1809–1882) observations and the work of Gregor Mendel (1822–1884) brought the science of genetics to the practice of selective breeding. The discovery of the structure of DNA and the development of gene mapping, coupled with recombinant DNA technology, fostered scientific precision in genetic engineering. Desired genes could be isolated, marked, and transferred from one member of a species to another, reinforcing the desired characteristics in successive generations and creating heritable, permanently changed DNA. Transgenic organisms can be created by insertion of DNA/genes from one species to another. Repeated selective breeding of homologous individuals leads to the creation of pure breeds in animals and specific strains or varieties in plants.

Traditional selective breeding involves successive pairings of individuals, whether plant or animal, showing desired traits. The gene pools of two individuals are combined, which results in offspring evincing a combination of characteristics. The characteristics could include the desired traits, but it is also possible that the desired traits may not be expressed for many generations, if ever. The process is imprecise and may require repeated breeding over many generations in order to develop strains of the organism reliably expressing the desired traits.

Contemporary genetic engineering techniques involve the precise implantation of specific genes carrying the desired traits into the host organism. Often the characteristic is expressed with the next reproductive cycle. By successive pairings of organisms bearing the engineered genes, the traits can be reinforced. Humans can utilize scientific selective breeding through the use of sperm or egg donors or genetic screening of embryos fertilized in vitro.

■ Historical Background and Scientific Foundations

Unlike natural selection, selective breeding is a planned process whereby individuals with desired traits are systematically bred in order to strengthen specific characteristics. Natural selection occurs when plants pollinate by having seeds blown by wind or spread by birds or insects, creating chance genetic combinations. In animals, natural selection can be somewhat more purposeful and is related to species-specific mating rituals, pheromones, and attraction between individuals.

Selective breeding has been occurring for thousands of years. Crop plants with the best taste, most appealing appearance, hardiness, and/or resistance to pests were bred together or replanted over successive growing seasons. Similarly, animals with desired traits such as size, coat density or texture, strength, hardiness, longevity, specific appearance, and meat quality were bred over many generations to produce the desired traits successfully and reliably. Selective breeding involves reproduction of the entire organism in order to reinforce or strengthen specific characteristics and occurs with different members of the same species.

Horses were among the earliest domesticated animals to be selectively bred, with some researchers citing 5000–5500 BC as the approximate time period of domestication. Selective breeding was determined by prevailing culture and need. In warm nomadic cultures, it was preferable to have horses bred for quickness, agility, and endurance for long travel. In cold northern and agrarian areas, horses were selected for size and heaviness, as well as strength for work in farming communities or areas where horses were used to move goods and equipment from one location to another.

Animals with similar appearance and traits are known as specific or pure breeds; they have been selectively inbred to bring out specific characteristics and to eliminate others. The goal is to reinforce the desired traits so as to cause them to be reliably reproduced across generations

WORDS TO KNOW

GENOTYPE: The physical expression of the genetic makeup of an individual. An individual's genetic identity is physically and biochemically expressed through his or her entire set of genes, called the genotype.

HYBRIDIZATION: The reproductive crossing of two organisms with different genotypes. Hybridization can involve two different varieties of a species, or it can result from the crossing of two different species.

INBREEDING: When two closely related individuals, whether plant or animal, reproduce. As part of a selective breeding program, this is often continued over many generations.

NATURAL SELECTION: Charles Darwin described natural selection as the process that leads to evolution of a species. When those with the traits most suited to survival in a particular set of circumstances reproduce with similar, though unrelated, members of the same species, the desirable traits are fostered. Undesired, or less suited, traits or characteristics are gradually eliminated.

American botanist Luther Burbank (1849–1926, left) with inventor Thomas Edison (1847–1931), circa 1890. Burbank used selective breeding to create many new varieties of plants from vegetables to flowers. © *FPG/Archive Photos/Getty Images.*

and throughout the totality of the breed. This is exemplified in dog breeding. Most breeds were created over dozens, even hundreds, of generations to have specific size, coloration, conformation, temperament, and abilities. Purebred animals have a single, recognizable, and identifiable breed; mixed breed animals are the result of random (sometimes intentional) interbreeding of two or more breeds. They often retain some characteristics of each breed, and they may express some of the distinguishing physical features associated with them.

In agriculture, selective breeding in both crop plants and animals is common. Plant strains (similar to breeds in animals) are progressively bred in order to reinforce characteristics having to do with yield, hardiness, rapid and hardy reproduction, taste, nutritional content, disease and pest resistance, and marketability. Animals are bred for similar traits: speed of maturity, amount and taste of meat, ability to successfully and prolifically reproduce (i.e., chickens that produce significantly greater-than-average numbers of eggs), appearance, strength, and longevity.

Within the human race, selective reproduction occurs in several ways. Some cultures place a high value on homogeneity or have prohibitions against marriage or reproduction with non-group members. Sperm banks have been in existence since the mid-1950s; egg banks are a more recent phenomenon. Individuals can select donor sperm or eggs based on a multitude of desired characteristics such as ethnicity, height, weight, career choice, academic record, hair color and type, eye color, and skin tone. Through the use of assisted reproductive technology, fertilized human embryos can be screened for many lethal or debilitating genetically-transmitted

diseases. The healthy embryos can then be implanted in the female host. Selective reproduction utilizing genetic screening also allows for sex selection of offspring.

■ Impacts and Issues

Selective breeding in agrarian communities began with the domestication and intentional reinforcement of desired traits in wild plants. With the advent of biotechnology, selective breeding has advanced to involve the ongoing reproduction of traits created by the insertion of specific and marked genes from the same or different species. In the case of transgenic plants or animals (animals carrying inserted genes from a different species), the goal is heritability, having the desired characteristics reliably appear throughout successive generations. In order to assure expression of the transgene, organisms carrying the inserted gene are often mated.

One of the primary challenges of "natural" (not involving genetic engineering) selective breeding has to do with the reproduction of entire organisms in the service of one or a few selected traits. Sometimes, promotion of a positive characteristic reinforces negative traits as well. It may take many generations to eliminate the undesired

Charlene, a plains zebra (L) and Storm (R) graze happily at Faure, outside of Cape Town, South Africa. The two zebras are part of the "quagga project." The quagga became extinct in 1883, and the project is an attempt to re-breed it through a selective breeding program. © *Anna Zieminski/AFP/Getty Images.*

traits. Because traditional selective breeding results in a mix of the DNA of both parent organisms, desired traits may not be immediately, or ever, expressed.

In the mid-1800s, Gregor Mendel conducted experiments on pea plants, fostering the inheritance of specific traits in successive generations. Mendel's methods for selective breeding began to be used early in the twentieth century to systematically alter, enhance, and create new plant species.

In 1944 Colin McLeod (1909–1972) and Maclyn McCarty (1911–2005) determined that DNA was responsible for genetic heritability. In 1953 James Watson (1928–) and Francis Crick (1916–2004) described the double helix structure of DNA, leading to the development of recombinant DNA technology by Herbert Boyer (1936–) and Stanley N. Cohen (1935–) in the early 1970s. This biotechnology advanced scientific selective breeding by engineering the transfer of a section of DNA from one cell to another. Recombinant DNA technology provided the foundation for modern genetic engineering techniques.

The use of biotechnology and genetic engineering has made the science of selective breeding efficient, precise, and rapid. Specific, mapped sections of DNA known to carry the gene for a desired trait or traits can be marked

and inserted into the species to be altered. The engineered change is often immediately expressed and becomes a permanent part of the organism's DNA, transmitted to offspring in successive reproductive cycles. Because only a single gene (or a few genes, if more than one trait is to be altered) and marker is inserted into the host, only the desired characteristics are reproduced. It is also possible to transfer a gene with a desired characteristic from one species to another, resulting in a transgenic organism.

Currently, selective breeding often consists of a combination of natural and engineered techniques. Individuals showing groups of sought after characteristics are selectively bred in order to create a foundation breeding stock, which can then be genetically altered to enhance or introduce other desired traits.

SEE ALSO *Agricultural Biotechnology; Biodiversity Agreements; Dairy and Cheese Biotechnology; Designer Babies; Designer Genes; Disease-Resistant Crops; Drought-Resistant Crops; Environmental Biotechnology; Genetically Modified Crops; Genetically Modified Food; Human Subjects of Biotechnological Research; Lysenkoism; Transgenic Animals; Transgenic Plants*

BIBLIOGRAPHY

Books

Fullick, Ann. *Inheritance and Selection*. Chicago: Heinemann Library, 2006.

Hood, Kathryn E. *Handbook of Developmental Science, Behavior, and Genetics*. Malden, MA: Wiley-Blackwell, 2010.

Keller, Greg G. *The Use of Health Databases and Selective Breeding: A Guide for Dog and Cat Breeders and Owners*, 5th ed. Columbia, MO: Orthopedic Foundation for Animals, 2006.

Kelley, Richard S. *You Did Something Right: The Case for Selective Breeding of Beef Cattle 1950–2000*. Indiantown, FL: 5K Ranch Publishing, 2005.

Kingsbury, Noel. *Hybrid: The History and Science of Plant Breeding*. Chicago: University of Chicago Press, 2009.

Rechi, Leopold J. *Animal Genetics*. New York: Nova Science Publishers, 2009.

Pamela V. Michaels

Bt Insect Resistant Crops

Bacillus thuringiensis (Bt) is a naturally occurring soil bacterium that produces a number of insecticidal toxins. Bt is also found on the surfaces of some plants as well as in the digestive tract of some species of Lepidoptera (butterflies and moths). More than 200 variations of *B. thuringiensis* toxins have been isolated and identified in various strains of the bacterium.

Bt has agricultural biotechnology applications. The Cry toxin of Bt is often extracted for use as a common agricultural pesticide. Bt genes inserted into common crop plants have produced plants with increased pest resistance. Bt modified plants, also called Bt crops, are modified to self-produce and release one or a combination of several of Bt's insecticidal toxins. Bt modified corn, for example, initially was created to limit damage from corn borers, an insect that destroys corn stems before they can produce harvestable ears (the edible part of the corn plant). A Bt corn variety resistant to root-worm was introduced in 2003.

■ Historical Background and Scientific Foundations

Bacillus thuringiensis is closely related to other toxin-producing bacteria, including *B. anthracis*, the bacterium that causes anthrax. Like anthrax, *B. thuringiensis* forms gene-encoded proteinaceous insecticidal delta-endotoxin crystals (Cry proteins or crystal proteins) during sporulation. Bt Cry genes used in genetic engineering typically are located in the bacteria's plasmids.

Biotechnologists use *Agrobacterium tumefaciens*, which induces the growth of tumors on plants, to facilitate the transfer of Bt genes to recipient plants. Scientists replace portions of the ti-plasmids that coordinate the insertion of certain subsets of DNA (t-DNA) into the plant chromosomes with Bt Cry genes.

Japanese biologist Shigetane Ishiwatari first identified Bt in 1901 when studying the worms used in Japanese silk production industry. German scientist Ernst Berliner (1880–1957) rediscovered and named the bacterium *Bacillus thuringiensis* in 1911. By the 1920s, Bt was commercially produced and used as an agricultural pesticide in Europe.

Starting in 1976, the identification of the plasmids in *B. thuringiensis*—and later, the identification of its endotoxin crystal formation—made genetic engineering using Bt possible. The first genetically modified Bt crop, a tobacco plant with increased insect resistance, was created in Belgium in 1985.

The development of Bt food crops soon followed, but government regulations necessitated initial study of the new crops for possible negative environmental and human health consequences. Different countries approved the introduction of Bt crops at various times. The U.S. Environmental Protection Agency (EPA), U.S. Department of Agriculture (USDA), and Food and Drug Administration (FDA) approved the first Bt crop—a potato—for planting, harvest, and sale in 1995. By the following year, government regulators had approved several varieties of Bt corn and maize, Bt cotton, Bt potatoes, and Bt tobacco. By 2009 the USDA reported that Bt corn grew on 55 million acres in the United States and represented 63 percent of all of the nation's corn acreage.

Bt-derived pesticides are still used in agriculture. Common Bt pesticides are sold under the trade names Dipel and Thuricide. Bt insecticides are considered a safer alternative to many synthetic pesticides because Bt degrades rapidly, is not toxic to humans and other mammals, and has minimal to no effect on beneficial pollinators such as bees. Some Bt pesticides are approved for use around USDA certified organic crops.

Whether applied as an insecticide or produced by Bt crops, Bt works by disrupting the digestive tract of insects. When insects ingest toxin crystals by eating plants or cleaning their feet, their alkaline digestive tract activates the toxin. When transferred to cells during the digestive process, the Cry toxins cause the formation of a pore that rapidly causes cell death.

■ Impacts and Issues

In the early 2010s, Bt toxins are used worldwide to control pests that threaten forests, kill insect vectors of disease including mosquitoes, and aid crop resistance to harmful insects. Bt-modified crops have several advantages. Crop plants that self-produce Bt toxins reduce the need for applied chemical pesticides that may damage nearby ecosystems or pose a greater risk to human health. Bt crops reduce pesticide accumulation in local soils and watersheds. The crops are safe to touch and handle, thus saving agricultural workers from the skin and lung exposure commonplace with agricultural chemicals.

Bt crops are biotechnology products. As such, the intellectual property rights to most varieties of Bt crops are held by corporations that developed or currently produce Bt crop seeds. Farmers have complained of onerous contract terms with seed makers, including higher seed prices for modified products and limitations on saving seeds from one year's crop to plant the following year. Small-hold farmers in industrialized nations dominated by large farms note that some modified seeds are available domestically only in very large quantities, making such purchases fiscally impossible or impractical,

A farm worker collects cotton in Phetchabun province, north of Bangkok, Thailand. Local activists alleged that cotton grown in this area contained genetically-altered plants that contained a gene to produce *Bacillus thuringiensis* toxin for increased insect resistance. The Thais worry that gene-altered products could contaminate their exports and close the doors to other nations that ban genetically modified (GM) crops. They also fear that banning GM crops will alienate another important partner, the United States. *© AP Images/Thaksina Khaikaew.*

but that the same companies make smaller-volume purchasing more available and more affordable in developing markets.

In 1999 public concern prompted researchers to investigate the possible effects of Bt plants on monarch butterfly populations. Concern rested on the possible effects of Bt corn pollen on the leaves of the ubiquitous milkweed upon which caterpillars fed. Bt corn sheds pollen for about two weeks per year, and pollen was found to be ingested by caterpillars. Bt pollen levels greater than 1,000 grains/cm2 showed some toxic effects on caterpillars. However, a study by the USDA Agricultural Research Service (ARS), peer-reviewed and published by the National Academy of Sciences, found non-toxic pollen levels, ranging from 6 pollen grains/cm2 to 170 pollen grains/cm2, on milkweed leaves near Bt corn crops.

In 2007 Bt crops and Bt pesticides were evaluated for a possible link to colony collapse disorder (CCD) in North American bee populations. The cause of CCD remains unknown, but several studies, including one from the Mid-Atlantic Apiculture Research and Extension Consortium, an American bee research institute, found no evidence linking Bt to the phenomenon.

A 2010 study indicated that Bt corn confers some of its insect resistant benefits to non-modified corn in fields adjacent to Bt corn. Non-modified corn planted adjacent to Bt corn showed a 28 to 73 percent reduction in borer insect populations compared to non-modified corn grown alone. In turn, the non-modified corn helps reduce insect resistance to Bt toxin in the modified crops. Insect resistance to Bt crops has occurred, but is largely limited to early-generation Bt plants that expressed only one type of Bt toxin. In 2009 agribusiness conglomerate Monsanto reported that one of its Bt cottons was no longer effective on pink bollworm pests in the Amreli, Bhavnagar, Junagarh, and Rajkot regions of India. Plants engineered for production of multiple toxins and crop management techniques have stemmed insect resistance.

Bt-modified food crops have no known adverse health effects in humans. Bt foods are nutritionally equivalent (same calorie or vitamin content, for example) to non-modified varieties of the same food. However, critics of GM foods assert that there have not been sufficient long-term studies about Bt toxin accumulation or exposure in humans. They advocate that governments should exercise the Precautionary Principle, banning GM foods until they are proven safe. The EU bans many types of GM crops intended for human consumption, including many Bt crops. The EU has not banned the use of externally-applied Bt pesticides. In the United States, Bt crops are used in many common food products, including corn syrups used to sweeten processed foods and sodas.

IN CONTEXT: MONARCH BUTTERFLIES AND GM CROPS

Monarch butterfly caterpillars (larvae) predominantly feed on milkweed, a North American weed that grows in wild and cultivated areas. Milkweed commonly grows amongst the corn crops found in the butterfly's habitat and migration zone. This prompted researchers to investigate the potential effects of Bt crops on the Monarch population, especially at the larval stage.

In 1999, a study from Cornell University published in the journal *Nature* suggested that Bt corn could be toxic to Monarch butterfly caterpillars. The laboratory study involved feeding three-day-old caterpillars milkweed leaves with corn pollen from both Bt and non-Bt corn plants. Milkweed without any pollen was used as a control. Four days later, researchers measured, weighed, and counted the surviving caterpillars. They found the larvae that lived on milkweed with Bt corn pollen ate less, were smaller, and died in greater numbers. The amount of pollen used was not recorded.

Subsequent studies in the field, including some financed in part by U.S. government regulatory agencies and agro-biotech firms, did not replicate the 1999 lab study. The results of several independent field studies were coordinated and published in the *Proceedings of the National Academy of Sciences* in 2003. These studies found that almost all types of Bt corn pollen had no significant effect on Monarch populations, even in pollen concentrations much greater than typically occur in the agricultural environment. One early Bt corn variety, Bt 176, planted on a maximum estimated 2 percent of U.S. farmland was found to have some negative effects on Monarch caterpillars, but the USDA slated the variety for a complete phase-out by 2003.

Fears of GM crop effects on Monarch butterflies resurfaced in 2011 when the *New York Times* published a story on July 11, 2011 that cited findings in a study published in the *Journal of Insect Conservation and Diversity* that attributed a decline in Monarch butterflies at least in part to a milkweed habitat loss associated with planting genetically modified Round-Up Ready corn. The study also mentioned that severe weather events and habitat loss from logging and development likely contributed to diminished Monarch populations.

However, there are significant differences between Bt crops and Roundup Ready crops; the two should not be conflated. The former is engineered to secrete Bt (an organic insecticide) itself, reducing the need for applied insecticides. The latter is engineered to be resistant to the trademarked commercial herbicide Roundup or similar herbicides so that farmers can control weeds with the chemical without harming crops. The former has no effect on the growth of milkweed, especially weeds Monarchs prefer located in and around crop fields. However, the use of herbicides in combination with Roundup Ready crops would destroy the milkweed on which Monarch larvae feed and is itself selectively toxic to many caterpillars.

Greenpeace activists dressed as a sheep and cow seek an audience with the Agriculture Minister, as policemen laugh, at the Ministry of Agriculture in New Delhi, India. The activists protested against genetic engineering of crops. Opponents of the GM crop saying it is a threat to indigenous varieties and creates environmental concerns. © *AP Images/Mustafa Quraishi.*

■ Primary Source Connection

The health of transgenic crops producing *Bacillus thuringiensis* (Bt) toxins for insect pest control may be threatened due to the evolution of pest resistance to the Bt toxin. Discussion includes the definition of field-evolved resistance, field control problems, strategies for delaying resistance, and resistance monitoring methods. Data from five continents are reported and analyzed pertaining to field populations, target pest populations, and resistance monitoring. The degree of various pest populations' susceptibilities and outcomes is evaluated against field-evolved resistance. Analysis of resistance monitoring data is considered as it pertains to the durability and field-evolved resistance of insecticidal crops, followed by recommendations based on study findings.

Field-Evolved Insect Resistance to *Bt* Crops: Definition, Theory, and Data

Crop plants genetically engineered to produce toxins from *Bacillus thuringiensis (Bt)* for insect control have been planted on >200 million ha since 1996. The first

generation of *Bt* crops was dominated by plants producing single toxins to kill key caterpillar pests: corn producing *Bt* toxin Cry1Ab and cotton producing the closely related toxin Cry1Ac. Although *Bt* corn and *Bt* cotton still dominate, varieties of these crops currently registered in the United States collectively produce 18 different combinations of 11 *Bt* toxins. Each variety produces 1–6 *Bt* toxins that kill caterpillars, beetles, or both.

The primary threat to the continued success of *Bt* crops is evolution of resistance by pests. The literature on this topic has grown rapidly, with hundreds of papers published in the past 5 yr and confusion arising from differences in definitions of resistance, resistance monitoring methods, and interpretation of data. This paper aims to help clarify the current status of field-evolved resistance to *Bt* crops. Here we define field-evolved resistance, describe its relationship to field control problems, and explain how it is measured. We summarize the theory underlying the refuge and pyramid strategies for delaying resistance. We analyze resistance monitoring data for *Bt* crops that produce Cry1 and Cry2 toxins targeting caterpillar pests, including many studies published since we last reviewed this topic. We compare the observed outcomes in the field with the expectations based on theory and conclude by considering the implications of current knowledge and prospects for the future of insect control with transgenic crops.

Field-Evolved Resistance: Definition and Relationship to Field Control

We define field-evolved (or field-selected) resistance as a genetically based decrease in susceptibility of a population to a toxin caused by exposure of the population to the toxin in the field. In contrast, laboratory-selected resistance occurs when exposure to a toxin in the laboratory causes a heritable decrease in susceptibility of a laboratory strain. Because both field-evolved and laboratory-selected resistance entail changes in gene frequency across generations, they exemplify evolution.

Our definition of field-evolved resistance is based on the definition provided by a group of resistance experts convened by the United States National Academy of Sciences that was later paraphrased and applied to *Bt* toxins and *Bt* crops. Like the National Research Council (1986) definition, our definition of field-evolved resistance does not necessarily imply loss of economic efficacy in the field. Nonetheless, field-evolved resistance to toxins in *Bt* crops is expected to confer decreased susceptibility to *Bt* crops in the field, whereas laboratory-selected resistance achieved by feeding insects on toxin-treated diet does not always increase survival on *Bt* crops. Although the terms "field-evolved resistance" and "evolution of resistance" refer to populations, the word "resistance" can be used to indicate heritable, lower susceptibility of an individual relative to conspecific individuals.

Insect populations often have natural genetic variation affecting response to a toxin, with some alleles conferring susceptibility and others conferring resistance. Alleles conferring resistance are typically rare in insect populations before the populations are exposed to a *Bt* toxin, with empirical estimates often close to one in a thousand. Field-evolved resistance occurs when exposure of a field population to a toxin increases the frequency of alleles conferring resistance in subsequent generations. Hence, inherently low susceptibility of a species to a toxin does not signify field-evolved resistance. Likewise, merely detecting resistance-conferring alleles without demonstrating that their frequency has increased does not constitute evidence of field-evolved resistance.

The main goal of monitoring resistance to *Bt* crops is to detect field-evolved resistance early enough to enable proactive management before control failures occur. Therefore, as noted above, documentation of field-evolved resistance does not necessarily imply that control problems have occurred in the field. This means that regulatory decisions about continued use of *Bt* crops should incorporate information about the relationship between field-evolved resistance and field control problems.

The relationship between field-evolved resistance and field-control problems depends on many factors including the frequency of resistance alleles, the magnitude of resistance, the extent to which resistance increases survival in the field, the number and spatial distribution of resistant populations, the insect's population density, the availability of alternative control tactics, and the extent to which the insect is a pest. For example, field-evolved resistance to insecticides in species such as *Drosophila melanogaster (L.)* may have virtually no implications for control, whereas field-evolved resistance in natural enemies can enhance pest control. Even for major target pests, field-evolved resistance to a *Bt* crop may not cause widespread problems if alternative control methods are effective. Furthermore, if a *Bt* crop targets several species of pests, its efficacy may be fully maintained against species that remain susceptible, even though field-evolved resistance reduces its efficacy against other species. Accordingly, regulatory decisions should be based on the net benefits and drawbacks of a particular *Bt* crop in a particular location, so that detection of field-evolved resistance in one or more pest populations does not automatically trigger large-scale removal of valuable varieties from the marketplace.

Conclusions and Implications

Bt cotton and corn have been remarkably successful since their commercial introduction more than 12 yr ago. Despite a few documented cases of field-evolved resistance to the *Bt* toxins in these transgenic crops, most pest populations remain susceptible. The evidence from field monitoring is generally consistent with the principles of resistance management underlying the refuge and pyramid strategies. To more rigorously test the correspondence between evidence and theory, scientists must thoroughly document and systematically analyze current and future examples of field-evolved resistance to transgenic crops, as well as cases in which pest susceptibility is sustained. For example, if additional data confirm trends in China, it would be useful to know why *H. armigera* resistance to Cry1Ac in *Bt* cotton is evolving faster in Qiuxian County than in neighboring areas. The results of such analyses can bolster the scientific basis for improving resistance management strategies.

We favor continued use of the long-standing definition of field-evolved resistance cited here because it promotes early detection and proactive management of resistance. In contrast, definitions that incorporate criteria about field control problems are likely to delay recognition of resistance and postpone management actions that can limit the negative consequences of resistance. Rather than debating the definition of resistance, we think it will be more productive to focus discussions on which regulatory actions, if any, are appropriate in response to specific data on the magnitude, distribution, and impact of field-evolved resistance.

As use of transgenic insecticidal crops increases, resistance management will be increasingly important. Expanded use of transgenic crops for insect control will likely include more varieties with combinations of two or more *Bt* toxins, novel *Bt* toxins such as vegetative insecticidal proteins, and modified *Bt* toxins that have been genetically engineered to kill insects resistant to standard *Bt* toxins. Transgenic plants that control insects via RNA interference are also under development. Increasing use of transgenic crops in developing nations is likely, with a broadening range of genetically modified crops and target insect pests. Incorporating enhanced understanding of observed patterns of field-evolved resistance into future resistance management strategies can help to minimize the drawbacks and maximize the benefits of current and future generations of transgenic crops.

Bruce E. Tabashnik
J. B. J. Van Rensburg
Yves Carriere

TABASHNIK, BRUCE, J. B. J. VAN RENSBURG, AND YVES CARRIERE. "FIELD-EVOLVED INSECT RESISTANCE TO *BT* CROPS: DEFINITION, THEORY, AND DATA." *JOURNAL OF ECONOMIC ENTOMOLOGY* 102, NO.6 (2009): 2011–2022.

SEE ALSO *Agricultural Biotechnology; Agrobacterium; Breeding, Selective; Corn, Genetically Engineered; GE-Free Food Rights; Genetic Engineering; Genetically Modified Crops; Genetically Modified Food; Patents and Other Intellectual Property Rights; Plant Patent Act of 1930; Plant Variety Protection Act of 1970*

BIBLIOGRAPHY

Books

Australian Farm Institute. *Can Agriculture Manage a Genetically Modified Future?* Surrey Hills, New South Wales: Australian Farm Institute, 2011.

Ferry, Natalie, and A. M. R. Gatehouse, eds. *Environmental Impact of Genetically Modified Crops.* Cambridge, MA: CABI, 2009.

Gordon, Susan, ed. *Critical Perspectives on Genetically Modified Crops and Food.* New York: Rosen Publishing Group, 2006.

Wolf, Timm, and Jonas Koch, eds. *Genetically Modified Plants.* New York: Nova Science Publishers, 2008.

Web Sites

Cranshaw, W. S. "*Bacillus Thuringiensis.*" *Colorado State University Extension.* http://www.ext.colostate.edu/pubs/insect/05556.html (accessed November 6, 2011).

"Pesticides." *National Institutes of Health (NIH).* http://health.nih.gov/topic/Pesticides (accessed October 31, 2011).

"Pesticides." *U.S. Environmental Protection Agency (EPA).* http://www.epa.gov/ebtpages/pesticides.html (accessed October 31, 2011).

Adrienne Wilmoth Lerner

Cabilly Patents

■ Introduction

The Cabilly patents are two U.S. patents (U.S. Patent 4,816,567, known as the Cabilly I patent and U.S. Patent 6,331,415, known as the Cabilly II patent) granted to Genentech, a U.S. biotechnology company. The Cabilly patents cover the technology necessary for the artificial synthesis of antibody (immunoglobulin) molecules. The patents are named after Shmuel Cabilly, one of the inventors of the technology.

The Cabilly patents have been the subject of legal proceedings since the Cabilly I patent was granted in 1989. Genentech challenged a patent held by Celltech, a British biotechnology company, that covered similar antibody synthesis technology. Later, MedImmune, a U.S. biotechnology company that licensed the antibody synthesis technology from Genentech, sued after the issuance of the Cabilly II patent, which considerably lengthened Genetech's patent term on the underlying technology. The legal disputes over the Cabilly patents have had a major effect on the biotechnology industry and have even led for calls to modify the U.S. patent system.

■ Historical Background and Scientific Foundations

The Cabilly patent issue began in 1983 when Genentech and Celltech filed separate patent applications that covered the technology necessary for the artificial synthesis of antibody molecules. Celltech filed the initial application on March 25, 1983, in the United Kingdom, which named Michael Boss and others as the inventors of the technology. Celltech then filed an international patent application. On April 8, 1983, Genentech filed its patent for the antibody technology, naming Cabilly and others as inventors. The U.S. Patent and Trademark Office (PTO) issued both the Celltech patent ("Boss patent") and Genentech (Cabilly I patent) on March 28, 1989.

In March 1990 Genentech filed an amendment to its patent that copied the language contained in Celltech's patent. Genentech then asked the PTO to declare an interference between the competing patents. An interference in which two patent holders claim the same patent results in a process by which the PTO determines which patent holder has priority and therefore the ability to enforce its patent.

After years of interference proceedings, the PTO ultimately decided that Celltech's Boss patent had priority. Genentech sued in U.S. courts, and in 2001 the two sides reached a settlement. Celltech acknowledged that Genentech's Cabilly patent had priority over their Boss patent. The PTO, therefore, revoked the Boss patent and issued a new patent covering artificial antibody synthesis to Genentech, known as the Cabilly II patent. Whereas both the Cabilly I and Boss patents were set to expire in 2006, the PTO granted Genentech's Cabilly II patent a term that extended until 2018. The Cabilly II patent issued by the PTO, therefore, granted Genentech a patent term of 29 years on their antibody synthesis technology—far in excess of the original 17-year term granted to the Cabilly I patent.

The Cabilly patents describe the fundamental technology required for making recombinant cells expressing both an immunoglobulin light chain and heavy chain. The Cabilly patents also cover the synthetic production of immunologically functional immunoglobulin fragments. These methods are useful for making any of several genetically-engineered antibody molecules that are utilized in numerous biotechnology drugs with billions of dollars of revenue per year.

■ Impacts and Issues

The legal proceedings involving the Cabilly patents have had a profound effect on the biotechnology community and have led to calls for reforming the U.S.

and vaccines manufactured using artificial antibody synthesis account for a multi-billion dollar per year segment of the biotechnology industry.

Many of the biotechnology companies that licensed the artificial antibody synthesis technology developed their products under the belief that the Cabilly I and Boss patents would expire in 2006, after which they no longer would have owed money for licensing the products. The issuance of the Cabilly II patent obligated companies that had licensed Genentech's technology to an 12 additional years of paying license fees.

In 2003, MedImmune, a biotechnology company that had licensed Genentech's technology for the production of Synagis, a monoclonal antibody used to prevent respiratory syncytial virus (RSV) infections, sued Genentech and Celltech. MedImmune asserted that Genentech and Celltech had conspired to secure an extended patent term for the artificial antibody synthesis technology covered under the Cabilly I and Boss patents with favorable benefits to both companies. Genentech received the patent with an extended term, and, under the legal settlement, Celltech seems to have secured a payment from Genentech and preferential access to the technology covered under the Cabilly II patent.

patent system. Lawsuits related to the Cabilly and Boss patents have had a significant impact on other biotechnology companies that licensed the antibody synthesis technology covered under those patents. The use of the artificial antibody synthesis technology covered by the Cabilly and Boss patents resulted in the creation of recombinant-antibody drugs and vaccines by numerous biotechnology companies that have licensed the technology from Genentech and, formerly, Celltech. The drugs

A sign is displayed in front of the biotech firm Genentech's headquarters in South San Francisco, California.　© *Justin Sullivan/Getty Images.*

The Supreme Court of the United States heard *Med-Immune, Inc. v. Genentech, Inc.* in 2006 and issued a decision in 2007. In an 8-1 decision, the Supreme Court declared that MedImmune, as a licensee, had standing to challenge the underlying licensed patent. The PTO subsequently reviewed the Cabilly II patent and declared it invalid because the claims contained in Cabilly II were anticipated under the previous patents and other references.

Rather than face additional years of legal action, in May 2008 MedImmune and Genentech reached a settlement over MedImmune's use of the antibody technology in the production of Synagis. In February 2009 the PTO announced that it would issue a re-examination certificate regarding the Cabilly II patent, which ultimately confirms the patentability of the claims in the Cabilly II patent.

The entire legal history of the Cabilly and Boss patents has raised the question of whether the U.S. patent system should move to a first-to-file regime, which is used by most countries. Under a first-to-file patent system, the first person to file a patent covering a particular invention or process is entitled to a patent regardless of whether someone else independently devised the same invention or process at an earlier date. Under the United States' first-to-invent patent system, a party may obtain a patent over another party that filed first by proving that their invention predates the other party's invention. The first-to-invent system often results in litigation regarding invention dates that generally do not exist within a first-to-file system.

The Supreme Court's decision in *MedImmune, Inc. v. Genentech, Inc.* has also altered the balance of power between patent holders and licenses. Before the *MedImmune* decision, a licensee had to breach its license contract in order to challenge a patent that the licensee believed to be invalid. Such a breach typically has drastic consequences for licensees, including additional litigation and the loss of the license. Under *MedImmune*, a licensee may now challenge the validity of an underlying patent without breaching their license agreement.

SEE ALSO Ariad v. Lilly; *Biotechnology;* Eli Lilly & Co. v. Medtronic, Inc.; Medimmune, Inc. v. Genentech, Inc.; *Patents and Other Intellectual Property Rights;* Roche Products v. Bolar Pharmaceutical

BIBLIOGRAPHY

Books

Dressler, Marc. *Biotechnology under the European Patent Regime with Regard to Science and Industry.* Rotterdam: Erasmus Universiteit, 2010.

Einsiedel, Edna F., and Julie A. Smith. *Canadian Views on Patenting Biotechnology.* Ottawa, Ontario, Canada: Canadian Biotechnology Advisory Committee, 2005.

Gross, Marc S., S. Peter Ludwig, and Robert C. Sullivan. *Biotechnology and Pharmaceutical Patents: Law and Practice.* New York: Aspen Publishers, 2008.

Haracoglou, Irina. *Competition Law and Patents: A Follow-on Innovation Perspective in the Biopharmaceutical Industry.* Cheltenham, UK: Edward Elgar, 2008.

Mills, Oliver. *Biotechnological Inventions Moral Restraints and Patent Law,* Rev. ed. Farnham, Surrey, UK: Ashgate Publishing, 2010.

Morgan, Dr. Gareth E. *Patent Litigation in the Pharmaceutical and Biotechnology Industries.* Witney, Oxford, UK: Biohealthcare Publishing, 2011.

Periodicals

Górski, Andrzej. "The Ethics of Intellectual Property Rights in Biomedicine and Biotechnology: Papers from the 5th International Conference on Bioethics: The Ethics of Intellectual Property Rights and Patents, Warsaw, Poland, 23–24 April 2004: Special Issue." *Science and Engineering Ethics* 11 (2005).

Joseph P. Hyder

Cancer Drugs

■ Introduction

Around one American in two will develop cancer at some point in their lives, according to the National Cancer Institute. Cancer drugs will be an important part of the treatment plan for many of these people. There are more than 100 cancer drugs in common use. They act to either kill cancer cells outright or to at least slow down their rate of growth. Cancer cells differ from most healthy cells in that they divide rapidly; it is this property that cancer drugs exploit. Unfortunately, many cancer drugs also target normal cells that divide rapidly, causing various side effects. Cancer cells also tend to become resistant to many cancer drugs. Treatment with cancer drugs is often administered before or after surgery. It may be combined with radiotherapy, and it is increasingly common to prescribe combinations of cancer drugs, each of which attacks the cancer cell in a different way, which helps overcome the problem of resistance. Drug treatment may be given with the aim of curing the patient or, in the case of advanced cancer in which cure is unlikely, as a palliative therapy to prolong life and relieve symptoms. Biotechnology has begun to transform cancer treatment, with the advent of more effective drugs, particularly the monoclonal antibodies.

■ Historical Background and Scientific Foundations

The notion that drugs might cure cancer came from the observation of the effects of mustard gas on soldiers during World War I (1914–1918). A chemical warfare agent, mustard gas belongs to a group of compounds known as alkylating agents, which can severely damage DNA. It was noted that men who died from mustard gas exposure had severely lowered white blood cell counts. In some forms of leukemia, the white blood cell count increases dramatically, so there was a rationale for applying compounds such as mustard gas, known as nitrogen mustards, in

these types of cancers. However, cancer drugs did not enter the mainstream of cancer treatment until the 1940s.

The alkylating agents such as mustine and cyclophosphamide add methyl groups at locations along the cancer cell's DNA, which is enough to induce apoptosis. Two other commonly used cancer drugs, doxorubicin and actinomycin, act by getting between the building blocks of the cancer cell's DNA so its genes cannot make essential proteins. One of the most widely used cancer drugs, methotrexate, blocks an enzyme called dihydrofolate reductase, which the cell must have to synthesize its DNA. Taxol, which is derived from the yew tree, works differently; it freezes the cytoskeleton of the cancer cell so it cannot divide.

Breast and prostate tumors are often responsive to sex hormones, and cancer drugs that block these hormones, such as tamoxifen or flutamide, may also be effective. Cancer drugs are usually given in tablet form, by injection, or infusion through a catheter or device known as a port that is permanently inserted into a blood vessel or body cavity. Each dose kills just a proportion of cells in a tumor, so it is usual to give a course of treatment, such as six cycles of an infusion every three weeks. Cancer drugs can also be used as adjuvant therapy, to kill any cells remaining after surgery, thereby reducing the risk of recurrence should these invisible cells multiply to form a new tumor. Drugs that kill cancer cells are generally known as chemotherapy. There are also several drugs known as biologics that work differently, by encouraging the immune system to deal with the cancer.

■ Impacts and Issues

The two main problems with conventional cancer drugs are side effects and resistance. Several cancer drugs affect healthy, rapidly-dividing cells in the hair follicles, bone marrow, and lining of the stomach, leading to hair loss, low blood counts (with lowered resistance to infection), and nausea, respectively. Meanwhile, many drug-resistant cancers have high levels of a molecule called p-glycoprotein that pumps the drug out of the cells. A better understanding of cancer biology has led to the development of newer cancer drugs that can address these issues and target therapy more precisely and effectively. In years to come, increased use of these new drugs offers the promise of improving cancer survival rates.

Several of the new generation of cancer drugs, known as biologics, are monoclonal antibodies. These are protein molecules that target specific molecular targets, called antigens, on the cancer cells, thereby flagging them for destruction by the immune system. They are made by the large-scale culture of mammalian cells containing the gene for a specific antibody. One example is Herceptin (trastuzumab), a monoclonal antibody that targets the human epidermal growth factor receptor 2 (Her 2) on breast cancer cells. Testing classifies breast cancers as either Her 2 positive or as Her 2 negative, depending upon the amount of Her 2 present. Herceptin is effective only with Her 2 tumors, so testing is necessary before it can be prescribed. The test saves women with Her 2 negative tumors from being treated with Herceptin, as the drug will not be effective for them. The use of companion diagnostics, which use the testing of the molecular biology of a tumor

WORDS TO KNOW

ADJUVANT THERAPY: Adjuvant therapy is the administration of cancer drugs after surgery has removed all visible cancer. The aim is to prevent recurrence by killing off any invisible cancer cells that may remain after surgery and otherwise go on to form a new tumor.

ANGIOGENESIS: Growth of new blood vessels. Anti-angiogenesis drugs inhibit the growth of new blood vessels.

APOPTOSIS: A programmed series of events leading to the death of a cell. Often known as cell suicide, apoptosis plays a role in the development of the fetus and is often faulty in cancer when damaged cells continue to survive instead of dying. Apoptosis is distinct from necrosis, a type of cell death caused by oxygen deprivation or other injury.

COMPANION DIAGNOSTIC: A test given before a drug, including a cancer drug, is prescribed to ascertain, by measurement of a marker molecule such as surface antigen, whether the patient will respond.

CYTOTOXIC: Refers to a drug that is directly toxic to a cell, the action of which often leads to the death of the cell.

A close up shot of *Catharanthus roseus*, also known as *Vinca rosea* or the Madagascar periwinkle, a flower from which cancer drugs have been made. The drugs are created from four vinca alkaloids—vinblastine, vincristine, vindesine, and vincrelbine—which are each used for different types of cancer. © *Norman Chan/Shutterstock.com.*

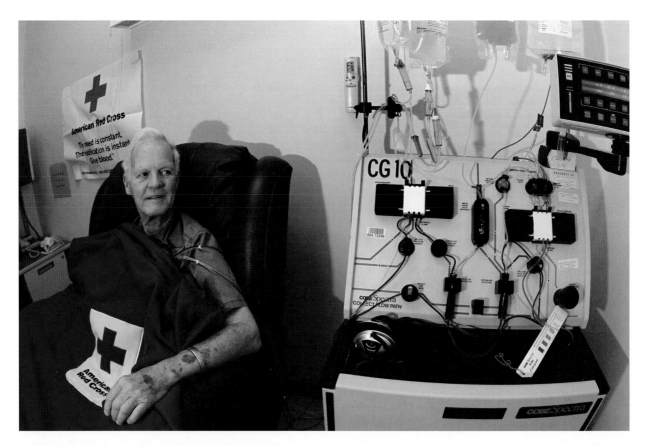

A patient is hooked up to a blood infusion machine at the American Red Cross in Dedham, Massachusetts, as he undergoes a $93,000 prostate cancer treatment. The machine uses apheresis to remove white blood cells, then mixes the cells with a drug that target prostate cancer cells and returns the blood to the patient. The Provenge therapy adds four months' survival, on average, for men with incurable prostate tumors. *© AP Images/Elise Amendola.*

to determine a prescribed treatment, is being used increasingly.

Erbitux (cetuximab) also targets an epidermal growth factor receptor antigen and is used in the treatment of colorectal and head and neck cancers. It is additionally being investigated for potential treatment of a number of other cancers. Erbitux is effective only against tumors with a normal version of a gene called KRAS. Therefore, the tumor must be tested before treatment starts; if the gene is mutated, another treatment must be chosen. Rituxan (rituximab) is a monoclonal antibody that targets an antigen called CD20 on cancer cells in non-Hodgkin's lymphoma and in chronic lymphocytic lymphoma. In the next few years, it is likely that increasing numbers of these monoclonal antibodies will be used in targeted treatments for cancer.

Killing cancer cells directly is no longer the only way of treating a tumor. Avastin (bevacizumab) is a monoclonal antibody that targets angiogenesis by blocking a protein called vascular endothelial growth factor, which helps blood vessels to form. The effect of Avastin is to starve the tumor by inhibiting the

formation of its blood supply. Another drug, Gleevec (imatinib mesylate), is prescribed for a form of leukemia characterized by an abnormal chromosome known as the Philadelphia chromosome. The presence of the Philadelphia chromosome is linked with a protein called bcr-abl, which causes a buildup of abnormal white cells. Gleevec is able to block production of this protein and reduce white cell count. Gleevec can also target an abnormal protein called c-kit in a rare form of stomach cancer called gastrointestinal stromal tumor.

In years to come, cancer drugs will become even more specific and more effective. One goal is to develop drugs that target the interior of a tumor, which is often oxygen-deficient, with molecules that work in a low oxygen environment. There is also research into drugs that could target cancer stem cells, which are thought to be resistant to most cancer drugs and responsible for the recurrence of cancers after treatment.

See Also *Biopharmaceuticals; Chemotherapy Drugs; Monoclonal Antibodies; Oncogenomics; Pharmacogenomics; Stem Cells*

BIBLIOGRAPHY

Books

Enna, S. J., and Williams, Michael. *Contemporary Aspects of Biomedical Research: Drug Discovery.* San Diego, CA: Academic Press, 2009.

Esteller, Manel. *DNA Methylation, Epigenetics and Metastasis.* New York: Springer, 2005.

Fialho, Arseno M. *Emerging Cancer Therapy: Microbial Approaches and Biotechnological Tools.* Hoboken, NJ: Wiley, 2010.

Nathan, David G. *The Cancer Treatment Revolution: How Smart Drugs and Other New Therapies Are Renewing Our Hope and Changing the Face of Medicine.* Hoboken, NJ: John Wiley & Sons, 2007.

Neidle, Stephen. *Cancer Drug Design and Discovery.* New York: Academic Press, 2007.

Schacter, Bernice Zeldin. *Biotechnology and Your Health: Pharmaceutical Applications.* Philadelphia: Chelsea House Publishers, 2006.

Skeel, Roland T. *Handbook of Cancer Chemotherapy,* 7th ed. Philadelphia: Lippincott Williams & Wilkins, 2007.

Web Sites

"Cancer Chemotherapy." *National Institutes of Health (NIH).* http://health.nih.gov/topic/CancerChemotherapy (accessed September 23, 2011).

Susan Aldridge

Cartagena Protocol on Biosafety

Introduction

The Cartagena Protocol on Biosafety is a supplementary agreement to the Convention on Biological Diversity (CBD) that regulates trade of living modified organisms (LMOs) among countries. In 1992 the CBD produced an international treaty that promotes the conservation and sustainable use of biological diversity. Adopted on January 29, 2000, the Cartagena Protocol took effect on September 11, 2003. It addresses the impact that LMOs, a subtype of genetically modified organisms (GMOs), might have on biodiversity, particularly on conventional plants and animals used in agriculture.

The Cartagena Protocol contains reference to and utilizes the precautionary principle in its regulation of LMOs. The precautionary principle states that a new product, action, or process suspected of posing a risk to the public or environment should not be introduced in the absence of scientific consensus that the product, action, or process is safe. It is the responsibility of the party seeking to introduce the new product, action, or process to shoulder the burden of proof of its safety. The Cartagena Protocol authorizes countries to prohibit the importation of LMOs that the country asserts to be a threat to its biodiversity. In addition, the Protocol requires strict labeling of LMOs and establishes a center for sharing scientific and technical information about LMOs.

Historical Background and Scientific Foundations

A living modified organism (LMO), like other genetically modified organisms (GMOs), is created through genetic engineering—a process in which genetic material is altered or added to an organism's genome. Most genetic engineering processes typically involve the use of recombinant DNA (rDNA) technology. Recombinant DNA is an artificial strand of DNA formed by combining two or more genetic sequences into a sequence that normally would not occur.

The genetic sequences used in rDNA technology may come from the same species, a closely related species, or different species. GMOs that use DNA from the same or closely related species are referred to as cisgenic organisms. GMOs created from the genetic material of two or more substantially different species are known as transgenic organisms. Regardless of the source of the genetic material, the use of rDNA technology enables geneticists to modify an organism so that it expresses desired traits that otherwise would not be expressed within that organism.

Genetic engineering differs from other methods used by humans to produce species that express desired traits. For centuries, farmers have produced hybrid plants with greater yields, drought resistance, and pest resistance by cross breeding related species. Cross breeding livestock has produced breeds that produce more milk, yield more meat, and more easily adapt to harsh climates, among other desirable traits.

Following the discovery of the double helix structure of DNA in 1953, geneticists theorized about the ability to identify and splice genes from one organism to another. In 1973 American scientists Herbert Boyer (1936–) and Stanley N. Cohen (1935–) created the first recombinant DNA organism, a transgenic *Escherichia coli* bacterium that expressed an exogenic gene from *Salmonella*. Boyer later co-founded Genentech, a U.S. biotechnology company, which genetically modified *E. coli* to produce human insulin. In 1992 Calgene, a U.S. agricultural biotechnology company, received approval for the Flavr Savr tomato, the first commercially produced, LMO whole food crop.

Impacts and Issues

The Conference of the Parties to the Convention on Biological Diversity adopted the Cartagena Protocol on Biosafety on January 29, 2000. Crafted to addresses the concerns that many parties to the CBD expressed about potential threats posed by LMOs, the Protocol primarily seeks to protect biological diversity from any potential risks posed by GMOs.

Many member nations of the CBD raised concerns about potential cross-pollination of imported transgenic crops with native species. Cross-pollination of native or hybrid species with transgenic species could reduce genetic diversity if dominant genes within the transgenic crop are expressed in the resulting hybrid. Pollen spread over great distances by wind or insects could cross-pollinate conventional crops over a large area because of a single farmer raising transgenic crops.

The introduction of terminator genes posed an even greater threat to farmers of local conventional crops. Terminator genes are genes inserted into a transgenic crop that prevent the crop from producing seeds that may be planted the following year. Terminator technology ensures that farmers growing genetically modified crops will buy seeds from the seed manufacturer each year as required under typical GM seed contracts. If terminator genes cross-pollinate with native conventional crops, the conventional farmer's crops could produce sterile seeds,

WORDS TO KNOW

CISGENIC ORGANISM: A genetically modified organism that contains genetic sequences from the same, or closely related, species that have been spliced together in a different order.

LIVING MODIFIED ORGANISM (LMO): Any living organism that has been genetically modified through the use of biotechnology.

PRECAUTIONARY PRINCIPLE: The directive that any product, action, or process that might pose a threat to public or environmental health should not be introduced in the absence of scientific consensus regarding its safety.

RECOMBINANT DNA: A form of artificial DNA that contains two or more genetic sequences that normally do not occur together, which are spliced together.

TRANSGENIC ORGANISM: A genetically modified organism that contains genetic sequences from multiple species.

The fifth meeting on the Cartagena Protocol on Biosafety, in Nagoya, Japan, 2010. The supplement to the biosafety protocol set new rules for redress from damage caused to ecosystems by the movements of genetically modified crops.   © *AP Images/Kyodo.*

thereby forcing the conventional farmer to purchase new seeds the next year. There are no food crops being grown using this technology as of early 2012.

The Cartagena Protocol grants nations control over protecting conventional species from any potential threats posed by LMOs. The Protocol requires exporters of LMOs to obtain an advanced informed agreement (AIA) before introducing a new LMO into a country. An AIA essentially requires the permission of the target nation before the first shipment of a new LMO is introduced. AIAs enable nations to choose which genetically modified seeds, plants, fish, animals, or microorganisms might pose a threat within their country before the first shipment arrives.

The Cartagena Protocol also mandates specific labeling of shipments of genetically modified commodities, such as corn or soybeans, intended for direct use as food, feed, or processing. Shipments composed in whole or part of LMO commodities must state that the shipment "may contain" LMOs and are "not intended for intentional introduction into the environment."

The Cartagena Protocol established the Biosafety Clearing-House, which enables nations to exchange scientific and technical data on LMOs. Countries may use this data to make informed decisions regarding potential threats to biodiversity or public health posed by a particular LMO.

In 2011 the Nagoya-Kuala Lumpur Supplementary Protocol on Liability and Redress to the Cartagena Protocol on Biosafety was opened for signatures. The Nagoya-Kuala Lumpur Supplementary Protocol seeks to provide international rules and procedures on liability and redress for damages resulting from the introduction of LMOs. The agreement has not entered into force.

The Cartagena Protocol addresses biodiversity issues, but it does not consider issues involving genetically modified foods or food safety. The import and export requirements of the Cartagena Protocol only apply to LMOs, not to food produced using LMOs. Any non-living foods derived from genetically modified plants or animals, including milled grains and processed foods, are not subject to the Cartagena Protocol.

■ Primary Source Connection

The opening sections of the Cartagena Protocol on Biosafety—including the preamble, objectives and general provisions set forth in the accompanying primary source excerpt—are integral and interrelated to a number of international agreements (e.g., regulations on trade among World Trade Organization members). The Protocol "aims to ensure the safe handling, transport and use of living modified organisms (LMOs) resulting from modern biotechnology that may have adverse effects on biological diversity" and human health.

Cartagena Protocol On Biosafety To The Convention On Biological Diversity

The Parties to this Protocol,

Being Parties to the Convention on Biological Diversity, hereinafter referred to as "the Convention",

Recalling Article 19, paragraphs 3 and 4, and Articles 8 (g) and 17 of the Convention,

Recalling also decision II/5 of 17 November 1995 of the Conference of the Parties to the Convention to develop a Protocol on biosafety, specifically focusing on transboundary movement of any living modified organism resulting from modern biotechnology that may have adverse effect on the conservation and sustainable use of biological diversity, setting out for consideration, in particular, appropriate procedures for advance informed agreement,

Reaffirming the precautionary approach contained in Principle 15 of the Rio Declaration on Environment and Development,

Aware of the rapid expansion of modern biotechnology and the growing public concern over its potential adverse effects on biological diversity, taking also into account risks to human health,

Recognizing that modern biotechnology has great potential for human well-being if developed and used with adequate safety measures for the environment and human health,

Recognizing also the crucial importance to humankind of centres of origin and centres of genetic diversity,

Taking into account the limited capabilities of many countries, particularly developing countries, to cope with the nature and scale of known and potential risks associated with living modified organisms,

Recognizing that trade and environment agreements should be mutually supportive with a view to achieving sustainable development,

Emphasizing that this Protocol shall not be interpreted as implying a change in the rights and obligations of a Party under any existing international agreements,

Understanding that the above recital is not intended to subordinate this Protocol to other international agreements,

Have agreed as follows:

Article 1

Objective

In accordance with the precautionary approach contained in Principle 15 of the Rio Declaration on Environment and Development, the objective of this Protocol is to contribute to ensuring an adequate level of protection in the field of the safe transfer, handling and use of living modified organisms resulting from modern biotechnology that may have adverse effects on the conservation and sustainable use of biological diversity, taking also into account risks to human health, and specifically focusing on transboundary movements.

Article 2

General Provisions

1. Each Party shall take necessary and appropriate legal, administrative and other measures to implement its obligations under this Protocol.

2. The Parties shall ensure that the development, handling, transport, use, transfer and release of any living modified organisms are undertaken in a manner that prevents or reduces the risks to biological diversity, taking also into account risks to human health.

3. Nothing in this Protocol shall affect in any way the sovereignty of States over their territorial sea established in accordance with international law, and the sovereign rights and the jurisdiction which States have in their exclusive economic zones and their continental shelves in accordance with international law, and the exercise by ships and aircraft of all States of navigational rights and freedoms as provided for in international law and as reflected in relevant international instruments.

4. Nothing in this Protocol shall be interpreted as restricting the right of a Party to take action that is more protective of the conservation and sustainable use of biological diversity than that called for in this Protocol, provided that such action is consistent with the objective and the provisions of this Protocol and is in accordance with that Party's other obligations under international law.

5. The Parties are encouraged to take into account, as appropriate, available expertise, instruments and work undertaken in international forums with competence in the area of risks to human health.

SECRETARIAT OF THE CONVENTION ON BIOLOGICAL DIVERSITY. *CARTAGENA PROTOCOL ON BIOSAFETY TO THE CONVENTION ON BIOLOGICAL DIVERSITY: TEXT AND ANNEXES.* MONTREAL, QUEBEC, CANADA: SECRETARIAT OF THE CONVENTION ON BIOLOGICAL DIVERSITY, JANUARY 29, 2000. AVAILABLE ONLINE AT WWW .CBD.INT/DOC/LEGAL/CARTAGENA-PROTOCOL-EN.PDF (ACCESSED SEPTEMBER 29, 2011).

■ Primary Source Connection

Individuals met at the United Nations Conference on Environment and Development held in Rio de Janeiro in June 1992 to establish a global partnership of cooperation among States, key sectors of society, and people. Reaffirming the 1972 Declaration of the United Nations Conference on the Human Environment, the conference attendees established Principles 14 and 15, specifying goals to take measures to protect the environment and to prevent environmental degradation.

Rio Declaration on Environment and Development

The United Nations Conference on Environment and Development,

Having met at Rio de Janeiro from 3 to 14 June 1992,

Reaffirming the Declaration of the United Nations Conference on the Human Environment, adopted at Stockholm on 16 June 1972, and seeking to build upon it,

With the goal of establishing a new and equitable global partnership through the creation of new levels of cooperation among States, key sectors of societies and people,

Working towards international agreements which respect the interests of all and protect the integrity of the global environmental and developmental system,

Recognizing the integral and interdependent nature of the Earth, our home,

Proclaims that:

Principle 14

States should effectively cooperate to discourage or prevent the relocation and transfer to other States of any activities and substances that cause severe environmental degradation or are found to be harmful to human health.

Principle 15

In order to protect the environment, the precautionary approach shall be widely applied by States according to their capabilities. Where there are threats of serious or irreversible damage, lack of full scientific certainty shall not be used as a reason for postponing cost-effective measures to prevent environmental degradation.

UNITED NATIONS CONFERENCE ON ENVIRONMENT AND DEVELOPMENT. "RIO DECLARATION ON ENVIRONMENT AND DEVELOPMENT," JUNE 14, 1992.

SEE ALSO *Biodiversity; Biodiversity Agreements; Bioethics; Environmental Biotechnology*

BIBLIOGRAPHY

Books

Convention on Biological Diversity. *Handbook of the Convention on Biological Diversity: Including Its Cartagena Protocol on Biosafety.* Montreal, Canada: Secretariat of the Convention on Biological Diversity, 2005.

Thomson, Jennifer A. *Seeds for the Future: The Impact of Genetically Modified Crops on the Environment.* Ithaca, NY: Comstock Publishing Associates, 2007.

Periodicals

Koester, Veit. "The Compliance Mechanism of the Cartagena Protocol on Biosafety: Development, Adoption, Content and First Years of Life." *Review of European Community & International Environmental Law* 18, no. 1 (2009): 77–90.

Secretariat of the Convention on Biological Diversity. *Cartagena Protocol on Biosafety to the Convention on Biological Diversity: Text and Annexes.* Montreal, Quebec, Canada: Secretariat of the Convention on Biological Diversity, January 29, 2000. Available online at www.cbd.int/doc/legal/cartagena-protocol-en.pdf (accessed September 29, 2011).

Web Sites

"The Cartagena Protocol on Biosafety." *Convention on Biological Diversity.* http://bch.cbd.int/protocol/ (accessed September 29, 2011).

Joseph P. Hyder

Cell Imaging

■ Introduction

Cell imaging enables cells, their internal structures such as nuclei and mitochondria, and distribution of DNA and protein molecules, to be visualized under a microscope. The latest technologies in optics and cell biology are combined to create ever more detailed and complex images of the cell. For instance, the distribution of specific molecules within a cell can be visualized by tagging them with fluorescent antibodies, which is informative about cellular biochemical processes taking place in both health and disease. It is also possible to produce images of living, rather than fixed, cells, as well as three-dimensional cellular images.

Cell imaging has revolutionized the understanding of cellular structure and function. It is widely used in drug discovery and development to identify the effect drug molecules have on cell structures. Cell imaging also plays an important role in the accurate diagnosis of infection, cancer, and other diseases by identifying infectious agents and cellular abnormalities. The latest developments in cell imaging allow long-term studies of live cells, which give new insights into some fundamental biological questions, such as how stem cells differentiate into specific cell types, which could help in the development of stem-cell based therapies to replace damaged tissue in organs like the heart and the eye.

■ Historical Background and Scientific Foundations

Most cells are too small to be seen with the naked eye; their diameters range from a few micrometers to 100 micrometers at most. During the nineteenth century, the microscope became a key item in the scientist's toolkit, allowing detailed studies of cells for the first time. Cell imaging began with the discovery of the nucleus in 1831 by Scottish scientist Thomas Brown. He observed the nucleus as an opaque spot within the cells of an orchid

plant. However, it was the discovery of intensely-colored dyes, such as Perkin's mauve, by the rapidly developing German chemical industry that led to the exploration of the complex inner world of the cell under the microscope. It soon became apparent that there was more to the cell than just the nucleus and the cytoplasm. In 1879, the German biologist Walther Flemming (1843–1905) discovered tiny thread-like structures that absorbed dye within the nucleus. These were later named chromosomes and shown to contain DNA, the genetic code of the cell.

The twentieth century saw exciting developments in cell imaging. Electron microscopy, first discovered in the 1930s, reveals the structure of the cells at much higher resolution and magnification than light microscopy. Meanwhile, developments such as confocal and fluorescence microscopy have allowed the imaging of cells in ever more detail, including studies of living cells and of the location of specific molecules within the cell. In 2007 scientists at Massachusetts Institute of Technology revealed the first three-dimensional images of the living cell. Their technique, applied initially to cervical cancer cells and cells of the tiny worm *Caenorhabditis elegans*, resembles the computed tomography scan used by doctors to image the inside of the human body. X rays are used to produce images of slices through the cell, which are combined, using advanced software, to create a three-dimensional picture.

The technique selected in cell imaging depends upon the amount of detail required. Ordinary light microscopy gives a resolution of 200 micrometers, whereas fluorescence microscopy goes down to around 10 nanometers. The electron microscope can resolve down to around 1 nanometer. The cells need to be prepared carefully to present a good quality image. First, a tissue sample containing the cells is fixed with chemicals to preserve them in a lifelike condition, and they are then dehydrated, because water content can blur the image. The tissue must subsequently be embedded in wax or resin before being cut into very thin sections by an instrument

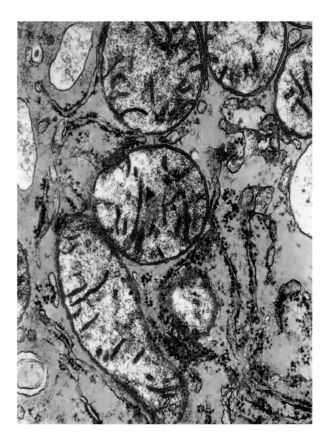

Tinted black and white transmission electron micrograph of a cell, showing rough ER, mitochondria, and ribosomes at 11,300x. Mitocondria function primarily to convert energy from food into a form the cell can use. Ribosomes provide the site for the synthesis of proteins. Rough endoplasmic reticulum (ER) is covered in ribosomes.
© Getty Images/Ron Boardman.

called a microtome. The sections will be a few micrometers thick for light microscopy and only a few nanometers thick for electron microscopy. Because most tissue is transparent, the sample needs to be stained to increase the contrast between different structures within it. Specific molecules like protein and DNA can be tagged with an antibody linked to a fluorescent dye to make their location in the cell show up in fluorescent microscopy.

■ Impacts and Issues

The applications of cell imaging range from routine clinical diagnostics to fundamental research in cell biology. For example, crystal violet, also known as Gram's stain, can be used to distinguish bacteria whose cell wall structures differ. Gram positive bacteria retain the stain and appear purple under the microscope. Gram negative bacteria do not retain the stain and are usually stained pink by another dye for imaging. This can be an important step in the diagnosis of a bacterial infection because Gram positive and Gram negative bacteria respond to different antibiotic drugs.

Cancer cells look different from normal cells under the microscope, therefore cell imaging is an important part of cancer diagnosis. Cells are taken from a tumor sample and exposed to fluorescent tags that can light up molecules found in cancer cells but not in normal cells. Increasingly, cell imaging is used to distinguish subtypes of cancer that would previously have been considered to be one disease. These distinctions help treatments to be tailored more accurately to the patient and the specific disease, with tumors identified by imaging as having a poorer prognosis being treated more aggressively.

Cell imaging is also playing an increasingly important role in drug discovery and development. Assays are laboratory tests used on potential drugs at an early stage to find out how they might act in the body. Using cells, and imaging them in assays, provides more realistic and detailed information about the drug molecule's action by answering questions such as which cell structures are affected by exposure to the drug. Advanced cell-based assays may even be able to replace, or at least complement, the preclinical testing of drugs in animals before clinical trials in humans are carried out.

Fundamental questions in cell biology can be answered by cell imaging, particularly now that long-term in vivo imaging is possible. For instance, researchers at the Clinical Research Institute in Montreal, Canada, have been tracking stem cells using fluorescent marker

An applications specialist works with the ELYRA microscope system behind a monitor where cancer cells are visualized. With the PAL-M and SR-SIM microscopy techniques, the ELYRA microscope system offers superresolution beyond the classic diffraction limit of microscopes. This enables scientists in biomedical research to image cellular structures marked with fluorescent dyes with maximum spatial resolution below the classic diffraction limit of microscopes, i.e. under 200 nanometers. This offers new possibilities for the analysis of the spatial relationship between the smallest cellular components down to the individual molecule. © *AP Images/Jens Meyer.*

molecules to see how they develop into retinal cells. This research could help to produce cells for treating blindness.

However, cell imaging does have some limitations. When a tissue sample goes through the fixing and staining routine, the cells could be damaged by the chemicals used, and the results may not be well related to what happens in a living cell. Being able to image living cells is a great advance, but cells are delicate, and the conditions within the incubator that is used in place of a microscope slide must be monitored with great care. It is also possible to misinterpret results from a cell imaging experiment. For instance, movement that might seem like genuine cell migration may merely be the result of cells following a temperature gradient introduced by inadequate monitoring of the apparatus.

SEE ALSO *Biopharmaceuticals; Biotechnology; Cell Therapy (Somatic Cell Therapy)*

BIBLIOGRAPHY

Books

Alberts, Bruce. *Essential Cell Biology.* New York: Garland Science, 2009.

Alberts, Bruce, et al. *Molecular Biology of the Cell*, 5th ed. New York: Garland Science, 2007.

Bourgaize, David B. *Biotechnology: Demystifying the Concepts.* Sudbury, MA: Jones & Bartlett Learning, 2011.

Lee, Kimberly Fekany. *Cell Scientists: Discovering How Cells Work.* Huntington Beach, CA: Teacher Created Materials, 2008.

Susan Aldridge

Cell Therapy (Somatic Cell Therapy)

■ Introduction

Cell therapy is the treatment of injury or disease using cells rather than drugs. The cells may be combined with a polymer support to create a tissue-like material resembling a bandage. The most important distinction in cell therapy is between stem cells and somatic cells. Stem cells are primitive cells that can be differentiated into various cell types whereas somatic cells are mature cells of one specific type. Cell therapies may also be autologous or allogeneic, depending upon the origin of the cells. Cell therapies may offer cures, rather than just treatment, for diseases for which current treatments are either non-existent or limited, such as stroke or diabetes. They also offer an alternative to organ transplantation. Stem cell therapies appear to have more potential for treating disease, but therapies based on somatic cells may be easier to develop. Currently, somatic cell therapies are available for skin replacement in ulcers and burns and for repairing injuries to knee cartilage. Somatic cell therapies are also being developed for treating hair loss and wrinkles. More ambitious projects include using cells, including somatic cells, to create whole organs such as the bladder and even the heart. However, there are still many technical, safety, and manufacturing challenges in bringing cell therapies to all who might benefit.

■ Historical Background and Scientific Foundations

The foundations of cell culture science were laid down early in the twentieth century, opening up the possibility to industrial manufacture of cell-based therapies. In 1977 there was a ground-breaking publication on Epicel, a somatic-cell product comprising keratinocytes and supporting fibroblasts, which is now available as a permanent skin replacement for life-threatening burns. Two years later, scientists at Massachusetts Institute of Technology reported on another skin substitute, this one made of fibroblasts and collagen. The term tissue engineering, referring to this new branch of science, was first used by Yuan-Cheng Fung (1919–) of the University of California, San Diego, in 1985.

This early work led to the development of a number of somatic cell therapy products now in clinical use. Besides Epicel there is Apligraf, a skin replacement composed of dermis and epidermis, which is used for the treatment of chronic venous ulcers and diabetic foot ulcers. Chronic venous ulcers affect around 2.5 million people in the United States. These ulcers often refuse to heal and can seriously affect mobility. Meanwhile, diabetic foot ulcers are a complication of diabetes that can result in hospital admission and even amputation of the foot. Diabetes is increasingly common, and it is to be expected that the incidence of diabetic foot ulcers will increase accordingly. Dermagraft is a dermis-only skin replacement cell therapy used to treat diabetic foot ulcers. Previously there were no effective treatments for these kinds of chronic ulcers.

Somatic cell therapy has also been applied to treatment of cartilage injury. Carticel is a product consisting of autologous cultured chondrocytes, the cells that form cartilage. Currently it is used to treat injury to the knee cartilage. In a four-year clinical trial, improvement in knee function was noted from six months and persisted up to the four year mark, with 77 percent of patients reporting good to excellent benefit from the cell therapy.

Manufacturing of a somatic cell therapy starts with the isolation of cells from either a donor, in the case of allogeneic treatment, or from a patient biopsy if the treatment is autologous. The cells are separated and purified to obtain a population of the right kind of cells. They are then grown in culture until there are enough cells for the intended treatment. Often some kind of polymer gel will be added at this point to create a tissue-like material, although the cells can also be injected as they are. Storage and shipping of the cells has to be carried out under carefully controlled conditions to be sure they are in good condition when they reach the patient.

■ Impacts and Issues

Tissue engineering, including somatic cell therapy, is challenging because cells are fragile and prone to contamination with viruses and mycoplasma, a type of parasitic bacteria. Therefore, safety issues are at the forefront when the regulatory authorities, such as the U.S. Food and Drug Administration, consider whether to approve a cell therapy. Any treatment involving cells, tissues, and even organs is known as an advanced therapy and is subject to rules that are still being developed. So for biotech companies and their investors there is much to do to satisfy the authorities in regard to a cell-based product's safety and efficacy before it can reach the commercial stage.

However, these issues have not deterred scientists from moving forward with somatic cell therapy. Further work is being carried out on skin replacements, and there is hope that the chondrocyte-based therapies may one day be applicable to joint degeneration in osteoarthritis, which affects far more people than knee injury, particularly as the population ages. Other research involves the transplantation of the dermal papilla cells in the hair follicles to treat hair loss, and, as an alternative to filler treatments with hyaluronic acid or collagen, injection of fibroblasts to treat wrinkles and restore skin elasticity.

WORDS TO KNOW

ALLOGENEIC: In cell therapy, allogeneic refers to cells taken from a donor, cultured in a lab, and then applied to a patient as therapy.

AUTOLOGOUS: In cell therapy, autologous refers to cells taken from the patient's body, cultured in a lab, and then applied to the site where therapy is needed.

CELL CULTURE: The growth of cells outside the body from an initial small sample of cells.

FIBROBLAST: A connective tissue cell, found in the structural parts of the body such as bone, skin, and cartilage, and also in between other tissues, playing a supportive role.

KERATINOCYTE: A type of skin cell found in the upper layer known as the epidermis.

However, the most ambitious plans involve combining somatic cells with polymer scaffolds to create two-dimensional tissue or even organs manufactured from cells. Scientists at Wake Forest University have pioneered

A research associate at Sangamo BioSciences Inc. isolates DNA at the Sangamo BioSciences lab in Point Richmond, California. After a man was cured of AIDS in California when he received a bone marrow transplant (a form of cell therapy) from an individual with a genetic mutation that prevented HIV infection, Sangamo developed a related therapy that renders cells in the human immune system resistant to HIV. © *David Paul Morris/Bloomberg/ Getty Images.*

the creation of a bladder from the urothelial cells that line the organ. The Wake Forest team grew the cells on a biodegradable polymer scaffold and then grew a layer of muscle cells on the outside to produce an organ that was shown to function in dogs. The first clinical trial, starting in 1999, involved several children with spina bifida and was successful, with all the artificial bladders functioning well throughout the following years. The Wake Forest team currently is working on a number of other organs, using the same technique.

Making organs such as the heart or liver, with their complex structures and layers of different cells, is a distant goal but one that the Living Implants from Engineering Initiative, set up in 1998, thought achievable. Their goal was to create a human heart by tissue engineering within 10 years, but the group has long since admitted that this was unrealistic. Further experience with somatic and stem cell therapies may, however, help push the development of these more complex structures forward.

SEE ALSO *Biopharmaceuticals; Bone Substitutes; Cloning; Gene Therapy; Pharmacogenomics; Skin Substitutes; Stem Cell Lines; Stem Cells; Stem Cells, Adult*

BIBLIOGRAPHY

Books

Atala, Anthony. *Principles of Regenerative Medicine.* Boston: Elsevier/Academic Press, 2008.

Ding, Sheng. *Cellular Programming and Reprogramming: Methods and Protocols.* New York: Humana Press, 2010.

Fong, Calvin A. *Stem Cell Research Developments.* New York: Nova Biomedical Books, 2007.

Li, Song, Nicolas L'Heureux, and Jennifer H. Elisseeff. *Stem Cell and Tissue Engineering.* Singapore: World Scientific, 2011.

Sullivan, Patrick J., and Elena K. Mortensen. *Induced Stem Cells.* Hauppauge, NY: Nova Science, 2011.

Web Sites

"Gene and Cell Therapy for Diseases." *American Society of Gene & Cell Therapy.* http://www.asgt.org/general-public/educational-resources/gene-and-celltherapy-for-diseases (accessed September 23, 2011).

Susan Aldridge

Chemotherapy Drugs

■ Introduction

Sometimes the word chemotherapy is used to cover all drugs used to treat cancer but, strictly speaking, the word applies only to drugs that kill cancer cells directly. Therefore, chemotherapy drugs are distinct from the newer biologic drugs, which target cancer cells and flag them for destruction by the immune system.

According to the National Cancer Institute, half of Americans will develop cancer at some point during their lifetimes. For many, chemotherapy will be used to either cure their disease or slow its progression. Cancer cells are distinguished from most healthy cells in that they divide rapidly, having lost the usual controls over the rate of cell division that would normally keep numbers in check. Chemotherapy drugs are compounds that kill rapidly dividing cells by various mechanisms. Unfortunately, they also kill other rapidly dividing cells in the body such as bone marrow cells, which means they have a number of side effects. The success rate of chemotherapy varies with the drug and type of tumor. There have been improvements in the survival rate of those treated with chemotherapy as the treatments have evolved. In the future, even more lives will be saved as developments such as stratified medicine, better drug delivery systems, and increased use of biologic drugs enter the mainstream of medicine.

■ Historical Background and Scientific Foundations

Chemotherapy dates back to observations on the effects of mustard gas on soldiers in World War 1 (1914–1918). Mustard gas is one of a group of compounds called alkylating agents that can damage DNA. Men who died from mustard gas chemical warfare had very low white cell counts. In some forms of leukemia, white cell counts increase dramatically, so compounds such as mustard gas began to be used in the treatment of the disease. One of these, mustine, is still used in the twenty-first century.

The pathologist Sidney Farber (1903–1973), cofounder of the great Dana-Farber Cancer Institute, was the originator of modern chemotherapy and introduced a number of new drugs, including aminopterin and actinomycin D, during the 1940s. He also pioneered the use of chemotherapy in mainstream clinical practice.

There are more than 50 chemotherapy drugs in common use as of 2011, and they are classed according to the way in which they kill cancer cells. The alkylating agents, including mustine and cyclophosphamide, attack the cells' DNA directly. Antimetabolites, including 5-fluorouracil, aminopterin, and methotrexate, interfere with the cells' synthesis of DNA and stop them from growing. The antitumor antibiotics, such as actinomycin D and doxorubicin, are made from natural substances such as soil fungi and interfere with various functions of the cancer cell. Plant alkaloids, such as vincristine and vinblastine from the periwinkle and taxol from the yew, prevent cells from dividing normally. Finally, steroid hormones slow the growth of hormone-dependent cancers. For example, tamoxifen is used to treat breast cancers that grow in the presence of estrogen. The end result of interfering with a cancer cell's functioning is usually its death by apoptosis, or cell suicide.

Cancer treatment normally consists of some combination of surgery, radiotherapy, and chemotherapy, the exact regime depending on the patient and the type of cancer that is present. It is increasingly common to use chemotherapy drugs in combination with one another for specific cancers. For instance, a combination of cyclophosphamide, methotrexate, and 5-fluorouracil is a common choice in breast cancer treatment. Chemotherapy can be delivered in tablet form, by injection, or by infusion through a catheter or device known as a port that is permanently inserted into a blood vessel or body cavity.

■ Impacts and Issues

Chemotherapy drugs have many side effects arising from their impact upon fast-growing cells other than cancer cells. Thus, many bone marrow cells that produce all the

WORDS TO KNOW

APOPTOSIS: A programmed series of events leading to the death of a cell. Often known as cell suicide, it is often faulty in cancer when damaged cells continue to survive instead of dying, but can be triggered by interfering with the cancer cell's functioning. Apoptosis is distinct from necrosis, a type of cell death caused by oxygen deprivation or other injury.

CATHETER: A hollow, flexible tube inserted into a blood vessel or body cavity to allow passage of fluids in and out of the body.

LIPOSOME: A microscopic particle composed of an outer layer of lipid molecules around an aqueous center. Liposomes can encapsulate drug molecules to make them have a longer lifetime in the body.

NEUTROPENIA: A decrease in neutrophil count in the blood. Neutrophils are immune cells that are vital in defending the body against infection.

STRATIFIED MEDICINE: The use of advanced diagnostic tests, including genetic testing, to find out which patients will respond better to certain cancer drugs.

types of blood cells are destroyed, resulting in lower-than-usual blood counts and lowered immunity. If the patient develops severe neutropenia (a loss of immune cells that are vital in defending the body against infection), he or she is very vulnerable to infection. Any infection that is contracted may escalate very rapidly and may even be life-threatening. Lowered platelet counts may increase the risk of bruising and bleeding, whereas lowered hemoglobin leads to anemia and tiredness. The cells that line the digestive tract also divide rapidly, and chemotherapy causes damage to these cells, which results in nausea, vomiting, and diarrhea. Finally, the impact of some chemotherapy drugs on rapidly dividing hair follicle cells often causes hair loss. Some chemotherapy drugs produce more side effects than others, and patients vary in the severity of the side effects they experience. Generally, the side effects of chemotherapy cease once treatment is complete.

Chemotherapy drugs sometimes give disappointing results because tumors may become resistant to them, or a drug may be too toxic to allow delivery of a sufficient dose to kill the tumor. Some cancers, such as brain cancer, are simply not very responsive to most

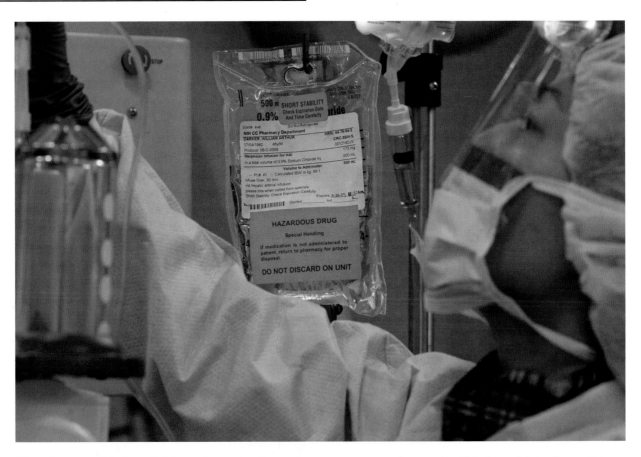

Chemotherapy medicine is readied for a patient undergoing a unique cancer treatment that uses ultra-high doses of chemotherapy that are isolated to the liver, at the National Institutes of Health in Bethesda, Maryland. Surgeons thread balloons up blood vessels to the liver to block off its normal blood supply. Then ultra-high doses of chemotherapy are flooded directly into the liver. Blood exposed to the drugs is drained out of the body, filtered, then tubed back in. The chemo only hits the liver and nowhere else. © AP Images/J. Scott Applewhite.

chemotherapy drugs. Cancer also tends to recur, so the success of chemotherapy may be temporary. These limitations can be overcome by using existing drugs in a better way, as well as developing new and more effective drugs.

For instance, doxorubicin and other chemotherapy drugs are available in a liposome delivery system that both reduces side effects and ensures that more of the drug reaches the cancer cell. Meanwhile, the biologic drugs such as Erbitux (cetuximab) are being used increasingly with, or instead of, chemotherapy drugs. Biologics are monoclonal antibodies that home in upon specific molecules on the surface of the cancer cell, thereby signaling the immune system to destroy the tumor.

Perhaps the best way of improving cancer treatment, however, is to pay careful attention to patient characteristics when prescribing a cancer drug. In a population of patients given the same chemotherapy drug, some will have a very good response while others will not respond at all. The response depends in large part on the molecular and genetic characteristics of the patient's tumor. If responders could be detected by molecular diagnosis of their tumor, then they could be offered a drug with confidence that it will work. Non-responders would be offered another drug. Researchers are building databases of tumor types and the drugs to which they respond. Known as stratified medicine, this more personalized approach to treatment enables cancer drugs to be used more effectively, offering the patient more benefit and less risk.

SEE ALSO *Biopharmaceuticals; Cancer Drugs; Monoclonal Antibodies; Oncogenomics; Pharmacogenomics; Stem Cells*

BIBLIOGRAPHY

Books

Enna, S. J., and Williams, Michael. *Contemporary Aspects of Biomedical Research: Drug Discovery.* San Diego, CA: Academic Press. 2009.

Esteller, Manuel. *DNA Methylation, Epigenetics and Metastasis.* New York: Springer, 2005.

Fialho, Arseno M. *Emerging Cancer Therapy: Microbial Approaches and Biotechnological Tools.* Hoboken, NJ: Wiley, 2010.

Nathan, David G. *The Cancer Treatment Revolution: How Smart Drugs and Other New Therapies Are Renewing Our Hope and Changing the Face of Medicine.* Hoboken, NJ: John Wiley & Sons, 2007.

Neidle, Stephen. *Cancer Drug Design and Discovery.* New York: Academic Press, 2007.

Schacter, Bernice Zeldin. *Biotechnology and Your Health: Pharmaceutical Applications.* Philadelphia: Chelsea House Publishers, 2006.

Skeel, Roland T. *Handbook of Cancer Chemotherapy*, 7th ed. Philadelphia: Lippincott Williams & Wilkins, 2007.

Web Sites

Brice, Philippa. "New UK Stratified Medicine Program for Cancer." *PHG Foundation*, September 6, 2011. http://www.phgfoundation.org/news/9837/ (accessed September 23, 2011).

"Cancer Chemotherapy." *National Institutes of Health (NIH).* http://health.nih.gov/topic/Cancer Chemotherapy (accessed September 23, 2011).

Susan Aldridge

Cloning

■ Introduction

A clone is an identical genetic copy of a biological entity. Cloning occurs in nature, for example if single celled organisms such as bacteria reproduce through asexual reproduction. In mammals, identical twins are created when a fertilized egg splits, giving them almost identical DNA to each other, but different DNA from either parent. Clones do not always look identical however, given different ways that genes can be interpreted and also the role the environment plays in how an organism develops. Artificial cloning can refer to overlapping but separate technologies that are focused on a range of materials such as genes, cells, organs, and entire organisms. These technologies are, firstly, recombinant DNA cloning—reproductive cloning that is the cloning process most detailed in the press—and secondly, therapeutic cloning.

■ Historical Background and Scientific Foundations

Although the term clone was not used until 1963 in a speech "Biological Possibilities for the Human Species of the Next Ten-Thousand Years" by British-born Indian geneticist J. B. S. Haldane (1892–1964), investigation into genetics had begun much earlier with the work of August Weismann (1934–1914) in the late 1880s. Weismann proposed that cell differentiation would reduce the genetic information contained within a cell. This theory pervaded until 1902 when the German embryologist Hans Spelmann (1869–1941) showed how split salamander embryos could still grow to adulthood. In 1938, Spelmann stated that the way forward for this area of research should be by inserting the nucleus of a differentiated cell into an enucleated egg.

In 1952 the tadpole of a northern leopard frog was the first in a long line of animals to be cloned from embryonic cells. It was followed in 1953 by the discovery by molecular biologists James D. Watson (1928–) and Francis Crick (1916-2004) of the structure of DNA at Cambridge's Cavendish Laboratory. Two decades later, in 1979, researchers cloned mice by splitting mouse embryos (in a similar fashion to how twins occur in nature) and implanting the split embryos into surrogate mice. However the first animal to be cloned by somatic (mature) cell nuclear transfer (SCNT) was Dolly the sheep, in 1997, using a cell from the udder of a six-year-old sheep. The cloning was achieved by scientists at Roslin Institute, Scotland, after 276 unsuccessful attempts. By 2006 domestic ferrets were being cloned successfully with the aim of eventually using them to study diseases such as cystic fibrosis.

With regard to human cloning, past experiments have been few, controversial, and limited in success. Advanced Cell Technology (ACT), a Massachusetts biotechnology company, cloned human embryos by inserting a skin cell into enucleated donor eggs in 2001. The chemical ionomycin was used to incite division. However, of eight eggs, only one divided as far as a six cell stage before the process ceased. A year later, Clonaid, part of a religious group called the Raelians, announced the birth of the first cloned human, but has yet to reveal any evidence.

■ Impacts and Issues

Recombinant DNA Technology

Recombinant DNA technology or molecular cloning has been in use since the 1970s and is commonplace in the field of molecular biology. In this type of cloning, a fragment of DNA is inserted into an element capable of self-replication such as a bacterial plasmid, which in turn is inserted into a host cell, such as bacteria, yeast, or mammalian cells, in order to create multiple copies of the original gene. The Human Genome Project, completed in 2003, used molecular cloning to ensure enough identical genetic material for study and research.

Recombinant DNA technology is a fundamental part of many other related technologies. It is used in genome sequencing, and also gene therapy, which can be used to correct faulty genes involved in the manifestation of genetic conditions. In genetic engineering, transgenic crops or genetically-modified organisms (GMOs) can be created to improve taste, yield, or hardiness. Genetically-modified and cloned microbial cells can also be produced to digest crude oil spills.

Reproductive Cloning

When cloning is mentioned in the press, it is often in reference to reproductive cloning in animals. Reproductive cloning is used initially to produce a cell with the same nuclear DNA as its parent cell, a process also referred to as nuclear transfer technology, with the final goal of producing an entire cloned organism. After division has

WORDS TO KNOW

PLASMID: A circular strand of DNA molecules capable of self-replication that operates separately from normal chromosomes.

SOMATIC: Pertaining to the corporeal; the cells of the body that make up tissues and organs, distinct from those involved in reproduction.

SOMATIC CELL NUCLEAR TRANSFER (SCNT): SCNT involves the use of adult cells to provide the nucleus in the cloning process.

TRANSGENIC: An organism or cell into which the genes of another species have been combined.

In Guadalajara, Mexico, a lab worker places a group of agave plants in containers for a cloning reproduction process designed to obtain two main results: making the plant stronger and for faster reproduction than the plants grown in the fields. Agave, a cactus-like plant that provides the main ingredient of Mexico's fiery tequila, became valuable in recent years as tequila demand has begun to far outstrip agave supply, sending tequila prices sky-rocketing. © *Susana Gonzalez/Newsmakers/Getty Images.*

Got, the country's first cloned Spanish fighting bull, is seen hours after being born at a ranch in Melgar de Yuso, northern Spain, on Wednesday, May 19, 2010. A dairy cow has given birth to the calf, which should be an exact genetic replica of a ferocious, hefty beast of the type faced in bullrings by matadors. © *AP Images/I. Lopez.*

been encouraged with either electricity or chemicals, the now multi-cell embryo is implanted in a surrogate host where it can develop through a normal in utero developmental period until birth. The clones have the same nuclear DNA as their parent, but the presence of genetic material in the mitochondria of the donor egg means that the clone is not completely identical.

Embryonic stem cells in an organism divide and take on specialized roles, eventually becoming the organs and tissues of the new animal. Scientists for a long time saw this as a permanent one-way process, with the unused genes turning off, but the experiments with Dolly the sheep at the Roslin Institute changed this view, because Dolly was cloned using a somatic cell from the udder of another, older sheep. Even though the nuclear DNA had originally resided in a very specialized cell, the new organism was able to access the genes for all the requisite differentiation of cells throughout the new lamb's body.

Reproductive cloning in animals could be used in the future in tandem with genetic engineering to mass produce transgenic animals with applications in both medicine and farming. Genetically altered livestock with increased disease resistance or desirable qualities such as leaner meat or higher milk production could become the norm. The FDA stated in 2008 that any cloned animal products are safe for human consumption. Transgenic animals already are being used to produce useful products in their milk, a practice known as 'pharming.' Goats are being engineered to produce spider silk, with a myriad of uses in medicine such as a suture material and artificial tendons. Sheep are producing the emphysema drug alpha-1 antitrypsin; and rabbits are producing milk with human interleukin. Reproductive cloning would allow larger scale production of these animals to be possible.

From an environmental standpoint the technology could address the issue of endangered species by helping to rejuvenate a population, something that has already been tested with the Banteg ox. Whether cloning would ensure long-term viability of a species is yet to be seen, given the inherent lack of genetic variability that cloned animals would involve. Recreating extinct animals, a popular subject of science fiction, raises even more problems. Because the donor egg would come from a related but not identical species, the mitochondrial DNA contained therein would mean that the cloned animals could never be exact reproductions of their extinct ancestors.

The success rates for reproductive cloning are low, with only about 10 percent of cloned animals being viable, and often the statistics are much lower. Primates, for example, are very difficult to clone. One reason for this involves spindle proteins. These two proteins are essential for correct division of cells. Whereas other animals have spindle proteins spread throughout the cell, in primates these cells lay close to the chromosomes, meaning removing the nucleus can damage or destroy them easily. The adverse effects of cloning on animals are well documented and manifest in a multitude of ways, from abnormally high birth size to serious defects in organs and a compromised immune system. At the Whitehead Institute for Biomedical Research in Cambridge, Massachusetts, scientists discovered that in 4 percent of the genes in cloned animals there was an abnormality. One element is the idea of cell age. Telomere sequences are regions of repetitive DNA at the ends of chromosomes. Cell division causes the telomeres to decrease in size, and once they have disappeared, cell senescence occurs. Clones created from adult cells as in SCNT would already have shortened chromosomes. In the case of Dolly the sheep, whose genetic material was taken from a six-year-old sheep, she died at half the expected age for her breed. Dolly also illustrates the effect these aging implications, programming errors, and abnormalities can have on an organism; she was euthanized by lethal injection because of lung cancer and severe arthritis.

Human Reproductive Cloning

It is these aforementioned multivariate problems with the reproductive cloning process that have resulted in an almost worldwide ban against using these techniques in humans. Additionally, alongside the medical considerations, is the idea of cognition. Humans are higher-order organisms who rely on intellect and emotional variances to function socially. There is no substantial body of work addressing how cloning would affect mental functioning. Furthermore, it is considered by many unethical to proceed with human cloning when animal cloning has produced organisms of such ill health, especially considering the problems of extracting intact genetic information from primates.

The concept of producing a genetically identical clone is a controversial one. Although it could allow for sterile couples to have a genetically linked child or could be used with genetic engineering to wipe out genetic defects in family lines, it also would appear to conflict with the ideas of individuality, human dignity, and religious principles. Further, the implication that it would allow for the cloning of a genetically identical lost loved one is a provocative topic.

Therapeutic Cloning

Also referred to as embryonic cloning, this refers to the creation of a cloned embryo for research. Specifically it refers to the production of embryonic stem cells with the same DNA as the donor cell from which they are copied as opposed to the making of a fully formed human being. Stem cells are harvested from the embryo (at this stage of development referred to as a blastocyst), after five days of cell division, at which point there are about 100 cells in the cluster. Stem cells are of interest to scientists and biomedical researchers because of their ability to differentiate into any type of cell within the body. Study of stem cells could therefore open a window to deeper understanding of the development and working systems of organisms and the molecular roots of disease. This could eventually allow for tailored medical treatments for individual patients, for example in cases of degenerative diseases such as Parkinson's and Alzheimer's.

Another use of cloned stem cells could be the growth of replacement tissues in the laboratory. These could be used for tissue repair, for example in skin grafting after extensive burn damage, or for the replacement of damaged organs. The ability to produce complete organs for transplant would save thousands of lives per year and reduce the burden on the organ donation system. Given that the tissue would be cloned from the patient's own cells, the resulting cells or organs would be genetically compatible, reducing the likelihood of rejection by the body. Alternatively, research with pigs, whose tissues and organs are remarkably compatible with humans, led to a breakthrough in the field in 2002 by PPL Therapeutics, the company behind Dolly the sheep. In this case cloning is used to produce genetically engineered pigs from which organs could be harvested for use in humans. One of the major pitfalls in the process would be how to manipulate the organ so that it would not provoke a hyper-acute immune response. PPL scientists managed to switch off both copies of the alpha 1,3 galactosyl transferase (GT) gene, effectively preventing the production of its enzyme and the carbohydrate it produces on cell surfaces. This sugar is what triggers the human immune system. Although PPL is no longer in existence, investigation and research is ongoing in other laboratories, and in 2011 Chinese scientists successfully created their own "knock-out" pigs in a bid to address the staggering 1.5 million patients on China's transplant waiting list.

A significant concern regarding therapeutic cloning is the similarity between stem and cancer cells, both being able to divide rapidly and quickly. In some cases, stem cells have developed mutations leading to cancerous cells. From an ethical standpoint, therapeutic cloning raises some serious objections. Traditionally, the process for collecting embryonic stem cells has involved the creation and subsequent destruction of embryos. Moral and religious objections are raised by some people who consider this process the destruction of a human life. Even leveraged against the huge medical potential for combating disease and dealing with injury, there are many who do not feel humans have the right to value one life above another. In October 2011 the European Union banned patents based on research involving embryos, a move that angered many in the research industry who called it the death knell for stem cell research.

From anesthesia to recombinant DNA, breakthroughs in science and technology have pushed humanity to debate and ultimately decide on the best way to move bioscience forward. The same decisions will continue to affect cloning, balancing innovation and the potential for huge benefits to humankind against ethical and medical concerns to temper the progress of cloning in its various manifestations.

■ Primary Source Connection

Using DNA from skin samples taken from a Pyrenean ibex found dead in northern Spain, scientists have been successful at cloning a female Pyrenean ibex. This is especially important, as the Pyrenean ibex, declared protected in 1973, was officially declared extinct in 2000. Although the newborn ibex died soon after birth due to lung defects, there is hope that continued successful cloning, through the resurrection of frozen tissue, may give way to saving this and other species that are endangered or newly extinct.

Extinct Ibex is Resurrected by Cloning

An extinct animal has been brought back to life for the first time after being cloned from frozen tissue.

The Pyrenean ibex, a form of wild mountain goat, was officially declared extinct in 2000 when the last-known animal of its kind was found dead in northern Spain.

Shortly before its death, scientists preserved skin samples of the goat, a subspecies of the Spanish ibex that live in mountain ranges across the country, in liquid nitrogen.

Using DNA taken from these skin samples, the scientists were able to replace the genetic material in eggs from domestic goats, to clone a female Pyrenean ibex, or bucardo as they are known. It is the first time an extinct animal has been cloned.

Sadly, the newborn ibex kid died shortly after birth due to physical defects in its lungs. Other cloned animals, including sheep, have been born with similar lung defects.

But the breakthrough has raised hopes that it will be possible to save endangered and newly extinct species by resurrecting them from frozen tissue.

It has also increased the possibility that it will one day be possible to reproduce long-dead species such as woolly mammoths and even dinosaurs.

Dr Jose Folch, from the Centre of Food Technology and Research of Aragon, in Zaragoza, northern Spain, led the research along with colleagues from the National Research Institute of Agriculture and Food in Madrid.

He said: "The delivered kid was genetically identical to the bucardo. In species such as bucardo, cloning is the only possibility to avoid its complete disappearance."

Pyrenean ibex, which have distinctive curved horns, were once common in northern Spain and in the French Pyrenees, but extensive hunting during the 19th century reduced their numbers to fewer than 100 individuals.

They were eventually declared protected in 1973, but by 1981 just 30 remained in their last foothold in the Ordesa National Park in the Aragon District of the Pyrenees.

The last bucardo, a 13-year-old female known as Celia, was found dead in January 2000 by park rangers near the French border with her skull crushed.

Dr Folch and his colleagues, who were funded by the Aragon regional government, had, however, captured the bucardo the previous year and had taken a tissue sample from her ear for cryopreservation.

Using techniques similar to those used to clone Dolly the sheep, known as nuclear transfer, the researchers were able to transplant DNA from the tissue into eggs taken from domestic goats to create 439 embryos, of which 57 were implanted into surrogate females.

Just seven of the embryos resulted in pregnancies and only one of the goats finally gave birth to a female bucardo, which died seven minutes later due to breathing difficulties, perhaps due to flaws in the DNA used to create the clone.

Despite the highly inefficient cloning process and death of the cloned bucardo, many scientists believe similar approaches may be the only way to save critically endangered species from disappearing.

Research carried out by Japanese geneticist Teruhiko Wakayama raised hopes that even species that died out long ago could be resurrected after he used cells taken from mice frozen 16 years ago to produce healthy clones.

But attempts to bring back species such as woolly mammoths and even the Dodo are fraught with difficulties. Even when preserved in ice, DNA degrades over time and this leaves gaps in the genetic information required to produce a healthy animal.

Scientists, however, last year published a near-complete genome of the woolly mammoth, which died out around 10,000 years ago, sparking speculation it will be possible to synthesise the mammoth DNA.

Professor Robert Miller, director the Medical Research Council's Reproductive Sciences Unit at Edinburgh University, is working with the Royal Zoological Society of Scotland on a project to use cloning on rare African mammals including the northern white rhino.

They have set up the Institute for Breeding Rare and Endangered African Mammals in the hope of using breeding technologies to conserve species including the Ethiopian wolf, the African wild dog and the pygmy hippo.

Professor Millar said: "I think this is an exciting advance as it does show the potential of being able to regenerate extinct species."

"Clearly there is some way to go before it can be used effectively, but the advances in this field are such that we will see more and more solutions to the problems faced."

A number of projects around the world are now attempting to store tissue and DNA from endangered species. The Zoological Society of London and the Natural History Museum have set up the Frozen Ark project in a bid to preserve DNA from thousands of animals before they disappear entirely.

Richard Gray
Roger Dobson

GRAY, RICHARD, AND ROGER DOBSON. "EXTINCT IBEX IS RESURRECTED BY CLONING." *THE TELEGRAPH*, JANUARY 31, 2009.

SEE ALSO *Bioethics; Bioreactor; Cell Therapy (Somatic Cell Therapy); Combinatorial Biology; Cryonics; Designer Babies; Designer Genes; Egg (Ovum) Donors; Gene Banks; Genetic Bill of Rights; Genetic Discrimination; Genetic Engineering; Genetic Privacy; Genetic Technology; Germ Line Gene Therapy; Human Cloning; Human Embryonic Stem Cell Debate; Human Subjects of Biotechnological Research; In Vitro Fertilization (IVF); Industrial Genetics; Molecular Biology; Nuclear Transfer; Recombinant Antibody Therapeutics; Recombinant DNA Technology; Silk Making; Stem Cell Cloning; Stem Cell Laws; Stem Cell Lines; Stem Cells, Embryonic; Therapeutic Cloning*

BIBLIOGRAPHY

Books

Dale, Jeremy, Malcolm von Schantz, and Nick Plant. *From Genes to Genomes: Concepts and Applications of DNA Technology*, 3rd ed. Chichester, West Sussex, UK: John Wiley & Sons, 2011.

Davis, Dena S. *Genetic Dilemmas: Reproductive Technology, Parental Choices, and Children's Futures*, 2nd ed. Oxford and New York: Oxford University Press, 2010.

Deech, Ruth, and Anna Smajdor. *From IVF to Immortality: Controversy in the Era of Reproductive Technology*. New York: Oxford University Press, 2007.

Herzfeld, Noreen L. *Technology and Religion: Remaining Human in a Co-Created World*. West Conshohocken, PA: Templeton Press, 2009.

Howe, Christopher J. *Gene Cloning and Manipulation*, 2nd ed. New York: Cambridge University Press, 2007.

Korobkin, Russell, and Stephen R. Munzer. *Stem Cell Century: Law and Policy for a Breakthrough Technology*. New Haven, CT: Yale University Press, 2007.

Lim, Hwa A. *Multiplicity Yours: Cloning, Stem Cell Research, and Regenerative Medicine*. Hackensack, NJ: World Scientific, 2006.

Panno, Joseph. *Animal Cloning: The Science of Nuclear Transfer*, Rev. ed. New York: Facts on File, 2011.

Silver, Lee M. *Remaking Eden: How Genetic Engineering and Cloning Will Transform the American Family*. New York: Harper Perennial, 2007.

Smith, Gina. *The Genomics Age: How DNA Technology Is Transforming the Way We Live and Who We Are*. New York: AMACOM—American Management Association, 2005.

Web Sites

"Cloning." *National Human Genome Research Institute*. http://www.genome.gov/25020028 (accessed October 31, 2011).

"Cloning / Embryonic Stem Cells." *National Human Genome Research Institute*. http://www.genome.gov/10004765 (accessed October 31, 2011).

"Cloning Fact Sheet." *Human Genome Project Information*. http://www.ornl.gov/sci/techresources/Human_Genome/elsi/cloning.shtml (accessed October 31, 2011).

Kenneth Travis LaPensee

Collagen Replacement

■ Introduction

Collagen is the most abundant protein in the human body and is found in large quantity in the dermis. It gives skin its resilience, suppleness, elasticity, and contour. The body is constantly producing collagen, which is perpetually sloughing off. As humans age, the rate at which collagen degrades exceeds the speed at which the body can manufacture it. As a result, skin appears to age, by developing lines and wrinkles, along with loss of tone and suppleness. This is particularly noticeable in the face.

Some animal (primarily bovine and porcine) collagen is quite similar in structure and function to the kind that occurs in humans. Because of this, cow or pig collagen has been used in a variety of medical applications for more than a century. In the early 1970s, a group of researchers at Stanford University developed a process for purifying bovine skin collagen to the point that it could be used to replace missing human skin. By the late 1970s, injectable forms of bovine collagen were developed to improve the appearance of aging skin.

Currently one of the most common medical cosmetic procedures involves the injection of human or animal collagen beneath facial skin in order to give a more youthful appearance. Collagen injections stimulate the production of the individual's collagen, but are not a permanent solution. It is necessary to add small amounts of collagen to existing sites up to several times yearly.

■ Historical Background and Scientific Foundations

Skin, the largest organ in the body, contains two layers: epidermis (outer layer) and dermis (inner layer). The dermis contains blood vessels, hair follicles, and nerves, as well as collagen-producing fibroblasts, the most prevalent cells in the dermis. Collagen is a fibrous protein that provides scaffolding for the growth of cells; it is the most common protein in the human body. It is found throughout the organism and is most abundant in blood vessels, bones, and connective tissue such as tendons, ligaments, cartilage, fascia, and skin. Its presence in the dermis creates a support structure for the epidermis and provides firmness, texture, elasticity, and contouring, keeping younger skin smooth and supple in appearance.

The body is constantly producing collagen. Over time, the rate at which collagen fibers deteriorate increases beyond the speed at which the body is able to manufacture them. In the epidermis, accelerated collagen breakdown results in the traditional signs of aging: wrinkles, loss of tone, and looser skin.

Collagen replacement injections are considered most therapeutically effective for smoothing raised scars as well as diminishing the appearance of facial wrinkles, including forehead lines, lines running between the eyebrows, crow's feet (lines at the corner of the eyes), loss of cheek contouring, lines extending from the sides of the nose down and around the corners of the mouth, and lines outlining the edges of the lips.

Collagen is present in animals as well as humans. Animal collagen, particularly that found in the skin of cows and pigs, has been most commonly used for human collagen replacement therapy.

The molecular structure of collagen has been studied since the 1930s. The most accurate description of the collagen molecule is a triple-stranded left-handed helix: It consists of three very long strands of polypeptides tightly woven together and coiled to the left. Roughly thirty different types of collagen have been identified in the human body. Type I is the most abundant, accounting for approximately 90 percent of all human collagen.

Collagen has been used in medical applications since the beginning of the twentieth century, particularly for sutures, surface coverage while preparing for skin grafts, burn coverage, and replacement heart valves. In the early 1970s, a team of physicians and biochemists working at Stanford University sought to develop alternatives to traditional skin grafts. It had already been determined that bovine (cow) collagen closely resembled human collagen on a molecular level; the Stanford group developed a protocol for processing and purifying bovine collagen so that it could be used as a substitute for human skin collagen. It was a short scientific evolutionary step to move from skin grafts to using minute amounts of animal collagen in an injectable form to replace diminished human collagen under the surface of the skin. In 1976 the first human patients were treated with injectable animal collagen.

■ Impacts and Issues

Collagen is injected directly under the skin, in areas where there are depressions, lines, or scars. The injected collagen helps to fill in the depressions and lines or flatten the scars, while stimulating the body to produce more natural collagen. Initially, all injectable collagen was made from purified and processed cow, or occasionally pig, hide. Near the end of the twentieth century, human collagen was successfully isolated and produced in laboratory settings.

Before collagen injections can be administered, it is mandatory for patients to undergo skin sensitivity testing. A small amount of collagen is injected under the skin of the forearm. If a reaction such as excessive or prolonged swelling, redness, itching, or skin ulceration

A Texas A & M molecular biologist uses a gene gun to introduce genes into sugar cane at the university's Agricultural Research Center in Weslaco, Texas. Scientists there have genetically engineered sugar cane with a human gene to produce a human therapeutic protein. Such a protein could replace the need to farm cadavers for face-plumping collagen. © *AP Images/Joe Hermosa.*

occurs within a period of four weeks, the individual is not considered a candidate for collagen use. If there is a slight reaction, the skin test is repeated on the other forearm, with an additional four-week waiting period. Human collagen almost never causes reactions; animal collagen occasionally does (in 3 to 5 percent of patients).

Collagen injections are considered far more effective than topical application of collagen creams or ingestion of collagen supplements. Collagen creams do not penetrate the skin or stimulate the body's own production of collagen, they act simply to temporarily slow the rate of water loss from the skin's surface giving it a fuller, more youthful appearance. Their effect lasts for several hours. Collagen injections can remain effective for a period of several months. Patients typically get "top up" injections two to four times annually in order to maintain the appearance enhancements.

Collagen replacement injections afford immediate results. The injections are administered by a physician in a medical office. There is little recovery time, and the cost is significantly less than surgical procedures designed to restore a more youthful appearance. The risk is considered minimal, as the collagen does not migrate within the body, and pre-injection skin tests are made before the injections in order to avoid sensitivity or allergic reactions.

SEE ALSO *Skin Substitutes; Tissue Engineering*

BIBLIOGRAPHY

Books

Barnes, Steven J., and Lawrence P. Harris. *Tissue Engineering: Roles, Materials, and Applications.* New York: Nova Science Publishers, 2008.

Carruthers, Jean, and Alastair Carruthers. *Soft Tissue Augmentation*, 2nd ed. Philadelphia: Saunders Elsevier, 2008.

Eremia, Sorin. *Office-Based Cosmetic Procedures and Techniques.* New York: Cambridge University Press, 2010.

Fratzl, Peter, ed. *Collagen: Structure and Mechanics.* New York: Springer Science + Business Media, 2008. Available online at http://www.springerlink.com/content/978-0-387-73906-9#section=163861&page=2&locus=67

Goldberg, David J. *Fillers in Cosmetic Dermatology.* Abingdon, Oxon, UK: Informa Healthcare, 2006.

Health Canada. *Injectable Cosmetic Treatments.* Ottawa: Health Canada, 2011.

Mecham, Robert P. *The Extracellular Matrix: An Overview.* New York: Springer Verlag, 2011.

Web Sites

Goodsell, David. "Collagen: Molecule of the Month." *Protein Data Bank*, April 2000. http://www.pdb.org/pdb/101/motm.do?momID=4 (accessed September 23, 2011).

Hadlington, Simon. "Synthetic Self-assembling Collagen for Tissue Engineering." *Royal Society of Chemistry*, August 28, 2011. http://www.rsc.org/chemistryworld/news/2011/august/28081101.asp (accessed September 23, 2011).

Pamela V. Michaels

Combinatorial Biology

Introduction

Combinatorial biology refers to a biotechnology approach in which compounds are deliberately and artificially created (biosynthesis), instead of relying on natural organic chemical reactions. Typically, the molecules that are created are proteins or fragments of proteins, which are termed peptides. Peptides can be especially useful in provoking the production of antibodies. The antibodies can be protective to the intact protein molecule and can be used as the basis of vaccines. The process of combinatorial biology, in which molecules can be tailor-made and screened for their successful function, represents a biotechnology-driven form of evolution, which can be faster than natural evolution.

Historical Background and Scientific Foundations

The process of combinatorial biology was independently developed in the 1980s by two scientists, Richard Houghten, an organic chemist and pioneering scientific entrepreneur at the University of California, Berkeley, and Mario Geysen, still serving in 2011 as a professor emeritus at the University of Virginia.

As the name implies, combinatorial biology uses various techniques to produce combinations of molecules that would otherwise be difficult, if not impossible, to achieve naturally. A widely-used technique is known as phage display. This approach uses bacteriophages—viruses that infect bacteria. By adding gene(s) to the bacteriophage genetic material, the target gene can be transported into the infected bacteria. The gene product can be produced as the inserted genetic material is transcribed and translated along with the host genes. The target gene can be modified so that information for the transport of the manufactured product to the bacterial surface is included. This transport typically involves the movement of the protein, which has water-loving (hydrophilic) regions, across the interior of the one or two bacterial membranes, which are water-excluding (hydrophobic).

As different combinations of target gene-bearing phages are used to infect bacteria, the result is a library of bacteria that express different molecules on their surfaces. Analogous to a conventional library of books, a particular surface molecule can be searched out, in a process that is known as high-throughput screening (essentially, a process in which many samples are screened in a rapid and accurate way to detect a particular target molecule, usually by the binding of the target molecule to another molecule).

High-throughput screening can be done using plastic dishes that are subdivided into wells (originally 96 wells, although plates containing hundreds of wells are now available). Each well houses a separate sample. The binding of the target surface protein can be revealed as a color change or the development of fluorescence, enabling the dish to be scanned by a spectroscopic reader that will register the color change or fluorescence of the particular well. Data on the results of each well can be stored as a computer file for future reference and analysis. More sophisticated versions of the approach exist, in which the positive well is then automatically sampled using a robotic arm equipped with a pipette.

Impacts and Issues

Combinatorial biology has had a huge impact on areas such as drug development and development of industrially-active enzymes and other compounds. The process of high-throughput screening has been refined (including the use of computers capable of storing and processing huge amounts of data); as of 2011, millions of reactions can be screened in a typical laboratory workday.

Advances in gene technology have profoundly influenced combinatorial biology. Whereas in the 1980s the genes used were recovered from existing organisms, in

WORDS TO KNOW

ENZYMES: Proteins that enable reactions to proceed more easily or faster than in their absence, and which are not changed by the reactions.

FLUORESCENCE: Absorption of light of one wavelength and emission of light of a different wavelength.

LIGAND: A molecule, such as a charged atom (ion), that is bound to another atom.

the 2010s genes can be synthetically constructed, allowing the tailor-made production of molecules.

The ability to create molecules and then test for their successful function has created a synthetic form of evolution, which is known as molecular evolution. The discovery of useful molecules through molecular evolution can be faster than relying on the natural process of biological evolution. In industrial research and development, the time savings can equate to millions of dollars in research and development costs.

An important aspect of drug production associated with combinatorial biology is the production of vaccines. Screened molecules can be made with fine detail, enabling the production of antibodies that might otherwise not be produced in the real-world of the body. Nonetheless, these antibodies can still be effective in protecting against the target disease or infection.

Combinatorial biology has also been very useful in revealing interactions between proteins and deoxyribonucleic acid (DNA), which has helped reveal details of the process of transcription.

SEE ALSO *Antibiotics, Biosynthesized; Biotechnology; Biotechnology Products; Genetic Technology; Transcription; Vaccines*

BIBLIOGRAPHY

Books

Achrekar, Jayanto. *Concepts in Biotechnology.* New Delhi, India: Dominant, 2005.

Advances in Biochemistry and Biotechnology. Witney, Oxford, UK: Daya Publishing House, 2005.

Akinloye, Oluyemi. *Biotechnology Trends in Advancement of Life Science Research and Development in Nigeria.* Göttingen: Cuvillier, 2011.

Barun, Susan R. *Biotechnology: An Introduction.* Australia: Thomson Books Cole, 2005.

Bourgaize, David B. *Biotechnology: Demystifying the Concepts.* Sudbury, UK: Jones & Bartlett Learning, 2011.

Brian Douglas Hoyle

Competitions and Prizes in Biotechnology

■ Introduction

Numerous competitions and prizes in biotechnology encourage innovation and recognize achievement within the biotechnology industry. Competitions establish parameters for research and demand certain results in order to win the competition. Teams then conduct research and compete to satisfy the conditions of competitions, which typically involve a financial incentive. Prizes recognize previous achievement within the field of biotechnology. Although prizes often contain a monetary component, the financial incentive bestowed by prizes is not the driving force behind the research recognized by the award. Prizes may be granted for a particular advancement within the field or conferred in recognition of a lifetime of work. The awarding of biotechnology prizes publicizes particular achievements, recognizes the work of researchers within the field, and inspires other researchers.

■ Historical Background and Scientific Foundations

The field of biotechnology saw a proliferation of competitions and prizes for research within the field in the last quarter of the twentieth century. The most prestigious prize awarded for biotechnology research, however, predates the coining of the word "biotechnology" in 1919. The Nobel Prize in Physiology or Medicine was first awarded in 1901 under the conditions set forth in Swedish chemist and industrialist Alfred Nobel's (1833–1896) will. In the second half of the twentieth century, the Karolinska Institute, the Swedish medical university responsible for awarding the Nobel Prize in Physiology or Medicine, routinely awarded the prize for advancements in the field of biotechnology.

Since 1946 the Lasker Foundation, which was established by American advertising executive Albert Lasker (1880–1952) and Mary Woodard Lasker (1900–1994),

has awarded the Lasker Awards for medical research. The Lasker Foundation annually presents awards for basic medical research, clinical research, and public service. In 1994 the Lasker Foundation established a special achievement award in medical science, although the foundation does not award this prize annually. The Lasker Award is often considered predictive of Nobel Prize recipients, as 76 Lasker Award laureates have also received the Nobel Prize in Physiology or Medicine.

Since 1978 the Wolf Foundation, an Israeli non-profit organization founded by Dr. Ricardo Wolf (1887–1981), has awarded prizes to scientists and artists for achievement. The Wolf Foundation routinely recognizes achievements in the field of biotechnology with both the Wolf Prize in Medicine and the Wolf Prize in Agriculture. The Wolf Prize in Agriculture is one of the most prestigious prizes awarded for agricultural research and achievement along with the World Food Prize. The World Food Prize, which recognizes contributions to food and agriculture, is awarded annually and often recognizes biotechnology research, including the development of genetically modified or hybridized crops or pest control.

■ Impacts and Issues

Competitions and prizes in biotechnology encourage development within the field by offering a financial incentive to conduct research, as with competitions, or by providing recognition for a particular discovery or a lifetime of work. Individuals, foundations, and other organizations that award prizes do so with the hope that the awards will inspire achievement and encourage innovation within the field. Additionally, prominent prizes, such as the Nobel Prize, may also call public attention to a certain area of research.

Unlike prizes, which promote general interest in and advancement of biotechnology, competitions typically promote the development of a particular technology or

WORDS TO KNOW

AMYOTROPHIC LATERAL SCLEROSIS (ALS): Also referred to as Lou Gehrig's Disease in the United States, ALS is a progressive neurological diseases that results in the degeneration of nerve cells in the brain and spinal cord, leading to muscle weakness and atrophy, paralysis, and death.

PHYSIOLOGY: A branch of biology that focuses on the functions of living organisms and their parts.

seek a solution to a particular issue. The Archon X Prize for Genomics is one of the most lucrative biotechnology awards in history. In October 2006, the X Prize Foundation and J. Craig Venter Science Foundation announced the establishment of a $10 million prize for the first team that builds a device that can sequence 100 human genomes in ten days or less within specified accuracy and cost guidelines. Prize4Life, a non-profit organization, established two specific biotechnology prizes

in November 2006. The first prize offers $1 million for the discovery of a biomarker for tracking the progression of amyotrophic lateral sclerosis (ALS). The second Prize4Life prize is a $1 million award for the discovery of a therapy that extends the life expectancy of mice with ALS by 25 percent.

Prizes and competitions in biotechnology may be sponsored by educational institutions, non-profit organizations, or companies. Corporate sponsorship of biotechnology prizes raises concerns among consumer advocates and other groups about the impartiality of the prizes or competitions. Critics claim that corporate-sponsored prizes and competitions promote the development of technologies with commercial viability while eschewing purely scientific research or beneficial technological innovation without the potential for commercialization.

SEE ALSO *Bayh-Dole Act of 1980; Biochemical Engineering; Biotechnologists; Biotechnology; Biotechnology Organizations; Biotechnology Products; Patents and Other Intellectual Property Rights; Registry of Standard Biological Parts*

Elizabeth Blackburn was the 2009 Nobel Prize winner in medicine along with Carol Greider and Jack Szostak. The scientists won the prize for discovering how chromosomes protect themselves as cells divide, work that has inspired experimental cancer therapies and may offer insights into aging. © *AP Images/Paul Sakuma.*

Shinya Yamanaka, a Kyoto University professor known for developing technologies to generate induced pluripotent stem cells, or iPS cells, receives a thread ball as a memento at an award ceremony in Kyoto on November 10, 2010, for the 2010 Kyoto Prize, an international prize presented by the Inamori Foundation. Yamanaka was recognized for his contributions in the field of biotechnology and medical technology.
© *AP Images/Kyodo.*

BIBLIOGRAPHY

Books

Achrekar, Jayanto. *Concepts in Biotechnology.* New Delhi: Dominant, 2005.

Betz, Frederick. *Managing Technological Innovation Competitive Advantage from Change,* 3rd ed. Hoboken, NJ: Wiley, 2011.

Block, Fred L., and Matthew R. Keller, eds. *State of Innovation: The U.S. Government's Role in Technology Development.* Boulder, CO: Paradigm Publishers, 2011.

Castle, David, ed. *The Role of Intellectual Property Rights in Biotechnology Innovation.* Cheltenham, UK: Edward Elgar, 2009.

Rhodes, Catherine. *International Governance of Biotechnology.* New York: Bloomsbury Academic, 2010.

Web Sites

Lasker Foundation. http://www.laskerfoundation.org/ (accessed October 24, 2011).

Nobel Foundation. Nobelprize.org (accessed October 24, 2011).

Wolf Foundation. http://www.wolffund.org.il/main.asp (accessed October 24, 2011).

X Prize Foundation. http://www.xprize.org/ (accessed October 24, 2011).

Joseph P. Hyder

Contract Research Organizations

■ Introduction

Contract research organizations (CROs), also called clinical research organizations, are facilities specializing in the organization and oversight of the full spectrum of biotechnology-oriented medical and pharmaceutical research processes, from study design through publication of results and marketing rollout of the final product. Although research laboratories at universities, hospitals, and pharmaceutical companies conduct some clinical trials, progressively more trials are outsourced and handled by CROs.

CROs often have access to multiple sites, are able to attract a wide variety of study participants through myriad recruitment sources, and have the infrastructure to manage data collection and subject tracking. CROs range in size and scope from small, niche companies to large, multispecialty organizations. Since the beginning of the twenty-first century, consulting firms with expertise in providing liaisons between research sponsors and the outsourced clinical trial performing CROs have gained popularity. The task of the consultants is to assure that the CROs produce maximum quality and benefit for the products they are testing.

■ Historical Background and Scientific Foundations

The first report of an interventional, randomized clinical trial was in 1754, when surgeon James Lind (1716–1794) created a design involving several different treatment types for scurvy. His laboratory was a ship named the HMS *Salisbury*. Lind randomly assigned sailors with scurvy to several different treatment groups to ascertain most effective means of combating the disease. Although the outcome was decisive in favor of citrus ingestion, the cost of treatment with oranges and lemons aboard ship was prohibitive. Ultimately, British sailors were given limes to eat during long voyages, effectively preventing scurvy.

Medical experimentation conducted by the Germans and Japanese on humans during World War II (1939–1945) was eventually adjudicated to be unethical and illegal. To prevent future occurrences, to define the parameters of acceptable medical research, and to protect the rights and safety of research program participants, the Nuremberg Code was created. The Code contains 10 points. Those specific to clinical research are, in essence:

- a requirement for informed consent and freedom from coercion;

- a mandate that the ultimate aim be the betterment of humankind and the research be necessary and not frivolous or gratuitous in nature—the potential benefits must outweigh the risks;

- research must be based on a solid scientific foundation, and animal research should precede human clinical trials;

- overall avoidance of discomfort or harm;

- ability to cease participation or to end the research program entirely without fear of reprisal;

- the research program must be supervised and overseen by scientifically trained and qualified individuals.

Drug Development and CROs

Drug development is a multi-stage process. Generally, clinical pharmaceutical trials encompass four or five phases prior to approval by the Federal Drug Administration (FDA) for commercial production and distribution. Preclinical studies are designed to test whether the proposed drug may have clinical efficacy. This is the period of laboratory and animal research. Phase I studies involve small groups of healthy human subjects (typically from 20 to 100) and are meant to discover potential dose ranges, drug safety, tolerability, mechanism of action, and major side effects.

Randomization and control groups typically enter at Phase II. Phases IIA and IIB (if Phase I is deemed a success) involve larger subject pools (generally 100 to 300) composed of healthy volunteers as well as clinical patients. Phase IIA looks at dosing patterns and ranges; IIB examines efficacy at prescribed doses. If a drug is going to fail in clinical trials, it generally occurs during Phase II. Phase III employs large subject groups of 300 to 3,000 or more, involves control groups, utilizes randomization, frequently encompasses multiple study locations, and is designed to fine-tune dosing for maximum effectiveness and to quantify rates and types of side effects.

Phase III compares the treatment outcomes of the investigational drug with current best practice standards. The regulatory application process for successful drugs or medical devices begins in Phase III. In Phase IIIB, broader patient categories may be examined in order to expand the range of possible applications. If regulatory authorities in the country (or countries) in which the drug will be made available give their approval, and there are no reports of significant adverse outcomes, the drug is marketed.

Phase IV takes place after regulatory approval and is used for gathering post-marketing safety and efficacy data on groups not involved in the original clinical trials, such as women who are pregnant or nursing. Phase IV allows for longitudinal data collection, and sometimes results in the recall of previously marketed drugs such as Vioxx. When they occur, Phase V trials are used as part of the process of incorporating a new treatment protocol into public health practice.

The entire clinical trial process generally takes several years, in addition to the years spent on concept and early development (pre-clinical trials). It is a very costly, time-consuming process to bring a drug or device from concept through approval, marketing, and long-term data collection. CROs help streamline this process.

■ Impacts and Issues

The several decades long growth and development curve for clinical/contract research organizations has paralleled that of the rise and expansion of biotechnology. During the 1970s, CROs found an early niche in drug and medical device discovery and pre-clinical research and development. By 2010, CROs had expanded internationally to all phases of clinical trials as well as marketing and manufacturing.

The World Health Organization has created an electronic International Clinical Trials Registry Platform so that individuals with an interest in any aspect of the clinical research process can have public access to detailed information about current clinical trials occurring around the globe.

According to the Tufts Center for the Study of Drug Development, CROs are both time- and

---

WORDS TO KNOW

INFORMED CONSENT: Receiving full information about the intent, process, length of study, principal investigator, study coordinator, and relevant staff contact lists, risks and benefits, and ways in which to withdraw from or to register a complaint or concern about a research study before agreeing to participate in a research study or clinical trial program.

INTERVENTIONAL STUDY: A type of research protocol in which individuals are assigned to groups in which they will receive some sort of treatment or clinical (or other) intervention.

NUREMBERG CODE: In 1947, in Nuremberg, Germany, a panel of American judges presided over the Doctors' Trial, involving 23 individuals charged with murder and torture occurring during medical experimentation performed on non-voluntary human subjects housed in concentration camps during World War II (1939–1945). As a result, a set of 10 criteria for designing ethical human subjects research was created, called the Nuremberg Code.

OUTSOURCE: When research or development organizations opt to contract with a third party to carry out specific functions that they might have otherwise done in-house, such as a clinical trials process.

---

cost-efficient means of conducting clinical trials research, when compared to clinical trials conducted in-house. CROs continue to grow at an annual rate that is significantly outperforming the economy and the current job market. In addition to the individual Web sites maintained by individual CROs, there are increasing numbers of directories aimed at showcasing CROs and providing sponsor, consultant, and CRO matching services.

An ongoing issue for sponsors and CROS has been subject recruitment, enrollment, and retention for clinical trials. With the explosive popularity growth of social networking Web sites, more people around the world are able to access understandable medical information and to find sites specific to their particular needs and interests. The major social networking websites, Twitter, YouTube, and Facebook, are being used increasingly as informational and educational vehicles by medical and research facilities, including CROs. Social media websites are able to deliver information in small segments, are infinitely repeatable (can be viewed over and over) and can be experienced without potential pressure from recruiters or other interested parties. Although little data has yet been collected, it is hypothesized that individuals may be sufficiently motivated by positive exposure to patient education delivered through social media that they will be inclined to take the next step of engaging

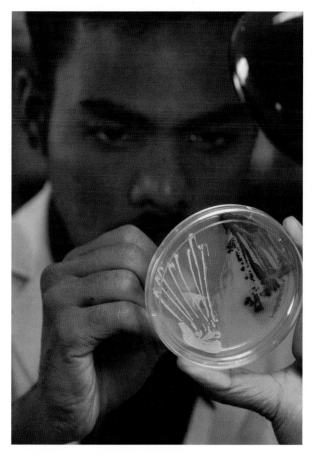

A research scientist checks for cells that are showing the presence of enzymes in bacteria in a petri dish in a laboratory in Bangalore, India. He is working in conjunction with a U.S.-based pharmaceutical company to develop a drug that uses benign viruses to kill the deadly *E. coli* bacteria. © *AP Images/Gautam Singh.*

in direct contact with research organizations. It is hoped that social media may, by educating patients before they begin to interact with research institutions, encourage

positive participation leading to improved patient retention. Electronic media are expected to become an integral part of the research participation process, as portable electronic medical record-keeping starts to incorporate tablet-style, hand-held computers for research subject use.

SEE ALSO *Biodefense and Pandemic Vaccine and Drug Development Act of 2005; Biopharmaceuticals; Biotechnology Organizations; Biotechnology Valuation; Government Regulations; Human Subject Protections; Human Subjects of Biotechnological Research; Informed Consent; Nuremberg Code (1948); Pharmaceutical Advances in Quantitative Development; Pharmacogenomics*

BIBLIOGRAPHY

Books

Chorghade, Mukund S. *Drug Discovery and Development.* Hoboken, NJ: Wiley-Interscience, 2006.

Freedman, Toby. *Career Opportunities in Biotechnology and Drug Development.* Cold Spring Harbor, NY: Cold Spring Harbor Laboratory Press, 2008.

Turner, J. Rick. *New Drug Development: An Introduction to Clinical Trials,* 2nd ed. New York: Springer, 2010.

Web Sites

Akst, Jeff. "Contract Research on the Rise." *The Scientist,* August 5, 2009. http://classic.the-scientist.com/blog/display/55878/ (accessed November 1, 2011).

Association of Clinical Research Organizations (ACRO). http://www.acrohealth.org/ (accessed November 1, 2011).

Pamela V. Michaels

Corn, Genetically Engineered

■ Introduction

Corn, also known as maize, is the third most important human food crop in the world, after wheat and rice. Around 300 million tons of corn are produced annually in the United States, which is around half of total global output. Planting of genetically-engineered corn is on the increase, with such crops being found in 11 countries around the world. Genetically-engineered corn accounts for the majority of corn in the United States.

Corn is an important source of human and animal nutrition. It is a source of a multitude of food products and, increasingly, as feedstock for biofuel. Given its economic importance, it is hardly surprising that corn has long been subject to improvement by plant breeding. In recent years, scientists have sought to improve corn still further by genetic engineering. These efforts have led to development of pest-resistant Bt corn and herbicide-tolerant corn, which increase the yield and quality of the crop and reduce the farmer's overall use of chemicals. Corn can also be genetically engineered to grow medicines and for easier processing into biofuels. Further developments will be driven by the recent availability of the corn genome. However, the planting of genetically-engineered crops remains controversial because of concerns about the environmental impact of transgenic plants.

■ Historical Background and Scientific Foundations

Corn has been cultivated for more than 7,000 years in the Americas, with Mexico being home to thousands of different corn plant varieties. *Zea mays* has long been subject to conventional plant breeding. Currently a few general varieties, including popcorn and sweet corn, are grown. Each variety consists of several different cultivars. Corn was one of the first plants to be subjected to genetic engineering technology. In 1996 pest-resistant Bt corn was planted in the United States, and Roundup herbicide-tolerant corn

followed two years later. According to the U.S. Department of Agriculture, 65 percent of corn currently grown in the United States is Bt corn, while 72 percent is genetically engineered for herbicide tolerance.

The European corn borer is one of the most significant crop pests worldwide. It burrows into the stalks of the corn plant and causes the stalks to fall over, which means the corn cannot be harvested. However, a soil bacterium called *Bacillus thuringiensis* (Bt) produces a toxin called Cry1Ab, which perforates the gut of the corn borer and various other insects. A spray containing the bacteria has been used for some years as a biological pesticide. In Bt corn, the Cry1AB gene has been transferred to the plant, conferring built-in resistance to the corn borer. Meanwhile, Roundup herbicide-tolerant corn has been genetically modified to contain a different version of a plant enzyme, which the herbicide would otherwise destroy.

All genetic engineering involves transferring a gene from one species to another in order to confer some beneficial trait, like insect resistance, on the recipient, thereby creating a transgenic organism. There are two main methods for introducing a transgene into a crop plant like corn. The first involves using the bacterium *Agrobacterium tumifaciens*, which naturally infects plants, as a vector to carry the gene into plant cells. In the second, a gene gun is used to shoot particles coated with the DNA of the transgene into the plant cells. The transformed cells are then grown into mature plants under very carefully controlled conditions. The resulting corn plants are thoroughly tested for their characteristics and cultivated in small-scale outdoor field trials before moving to large scale cultivation.

■ Impacts and Issues

The largest agricultural crop in the Americas, corn is the source of a multitude of products including cornflakes, grits, cornmeal, whiskey, animal feed, corn starch, corn syrup, and bioethanol. Therefore, optimizing the quality and yield of corn is of great economic importance. Losses of corn to insect damage amount to an estimated annual 40 million tons. The cost of crop damage by the corn borer and the cost of controlling the insect in North America are thought to exceed $1 billion per annum. Studies have also shown a health benefit from Bt corn in protecting the plant from infection. The tunnels created by the corn borer provide an entry point for pathogenic fungi, which reduce the quality and yield of the crop.

Weeds can also be major enemies of crop plants, as they compete with them for space, light, water, and nutrients. Chemical herbicides are usually not specific enough to leave plants unharmed when they are applied to a plot, and hand weeding is labor intensive. Engineering a plant for herbicide resistance means a field can be sprayed and the weeds die while the corn remains unscathed. Corn plants that are genetically modified with both Bt toxin and Roundup tolerance are now available. Field studies show that genetically engineered corn does give higher yields than traditional corn, so there are clear economic benefits to the farmer.

However, the introduction of genetically engineered corn into commercial agriculture has raised a number of environmental concerns. In 1999 researchers at Cornell University reported that the Monarch butterfly was harmed by consuming milkweed coated with Bt pollen, at least in laboratory experiments. The Environmental Protection Agency then conducted a number of field studies to ascertain whether the Bt corn did threaten the butterfly and concluded that the risk was in fact negligible. There is also the question of whether the corn borer might become resistant to Bt corn in the way pests often become resistant to chemical pesticides.

For this reason, the regulations governing genetically-modified crops state that non-Bt corn must be grown around Bt corn, and insect populations are regularly monitored for resistance. So far, there has been no sign of widespread resistance leading to crop failures, but this could emerge in the longer-term, particularly if farmers are tempted to ignore the regulation and grow more Bt corn than it permits.

Another concern is whether so-called "superweeds" may be created by the transfer of herbicide tolerance genes from crop plants to common weeds. There is some evidence that such plants are indeed emerging where fields are planted with genetically-modified corn. Farmers deal with the issue by treating the land with ever-increasing doses of herbicide.

Finally, it is not known what the long-term impact of introducing a new species, on a large scale, might be on local biodiversity. This is why Mexico, the birthplace of corn, banned the cultivation of genetically-modified (GM) corn in 1998, though it did allow the sale of GM corn from other countries, including the United States. Despite the ban on cultivation, in the early 2000s, genetically modified corn was found growing in several remote places in southern Mexico due to cross-contamination from transgenic corn sold to Mexico. For many years thereafter, Mexico continued to resist allowing GM corn cultivation. However, some Mexican farmers feel they are losing out on the benefits of being able to grow their own genetically-modified crops, and, after pressure from farmers and large agricultural companies, experimental field trials of GM corn began in Mexico in 2011. Cultivation of GM corn is still banned in some states, such as Oaxaca and Chiapas, to protect native species.

In 2009 a team led by scientists at The Genome Center at Washington University School of Medicine announced the sequencing of the corn genome. This opens up the detailed study of the plant's 32,000 or so genes and is likely to result in the creation of many more different varieties of genetically engineered corn. There is already corn carrying the gene for amylase, which makes it easier to convert corn starch into bioethanol, and corn that can produce a protein for the treatment of cystic fibrosis. If corn could be engineered for traits such as nutritional value or resistance to harsh environmental conditions, such as cold or salinity, then there could be significant benefits to the world's food supply.

SEE ALSO *Agricultural Biotechnology; Biodiversity; Bt Insect Resistant Crops; Cornstarch Packing Materials; Genetically Modified Crops; Genetically Modified Food*

BIBLIOGRAPHY

Books

Ferry, Natalie, and Angharad M. R. Gatehouse. *Environmental Impact of Genetically Modified Crops.* Cambridge, MA: CABI, 2009.

Giorgio Fidenato picks up corn on 350 square meters (almost an acre) of nearly mature corn genetically altered to resist pesticides that just a day before had been trampled by 70 anti-GMO activists, near Pordenone, northern Italy. Fidenato's corn is genetically modified, grown in fields of surreptitiously, and, detractors say, illegally, planted Monsanto seed in northeastern Italy not far from the Austrian and Slovene borders. More activist than farmer, Fidenato's cultivation of GM corn is a rogue act aimed at forcing the authorization of genetically engineered crops in Italy. The Italian battle is shaping up at a critical moment for the future of genetically modified crops in Europe, where the population has generally viewed the technology with suspicion.   *© AP Photo/Paolo Giovannini.*

Fiechter, Armin, and Christof Sautter. *Green Gene Technology: Research in an Area of Social Conflict.* New York: Springer, 2007.

Gordon, Susan, ed. *Critical Perspectives on Genetically Modified Crops and Food.* New York: Rosen Publishing Group, 2006.

Halford, Nigel. *Plant Biotechnology: Current and Future Applications of Genetically Modified Crops.* Hoboken, NJ: John Wiley, 2006.

Lacey, Hugh. *Values and Objectivity in Science and Current Controversy about Transgenic Crops.* Lanham, MD: Lexington Books, 2005.

Web Sites

"Food, Genetically Modified." *World Health Organization.* http://www.who.int/topics/food_genetically_modified/en (accessed September 21, 2011).

"Oxfam International's Position on Transgenic Crops." *Oxfam International.* http://www.oxfam.org/en/campaigns/agriculture/oxfam-position-transgenic-crops (accessed September 21, 2011).

Susan Aldridge

Cornstarch Packing Materials

Introduction

The safe and efficient transport of consumer goods is essential to business profitability and consumer satisfaction. Goods that are damaged in transit represent a non-recoverable loss to their manufacturer or distributors; broken items can pose a safety threat when the package is opened. Materials such as newspaper, straw, cardboard, and foam peanuts have been, and continue to be, used at various times to create a secure and cushioned package for merchandise that reduces the risk of damage caused during shipping and handling.

Petroleum-based polystyrenes were developed to create more secure packing materials for any products shipped in a container. Shaped like peanuts, the polystyrene pieces provided loose fill that could be packed easily between the protected item and the container walls. Polystyrene packing proved to be economical and highly effective, but it was a product that came with a significant environmental cost. Polystyrene is not biodegradable; its manufacturing processes generate a number of byproducts that potentially harm human health and the environment.

The polysaccharide cornstarch was identified as a safe alternative to polystyrene loose fill packing in the early 1990s. Cornstarch packing proved to be as effective as polystyrene products in the ability to protect merchandise from damage during transport, with the additional advantage of being biodegradable. Cornstarch is also used in a variety of packaging applications, such as bioplastic food containers. As with many technological advances that are more environmentally friendly than traditional alternatives, the costs associated with the manufacture of cornstarch packing materials has been a limiting factor in their wider commercial use.

Historical Background and Scientific Foundations

The ability to ship merchandise of all kinds effectively is a hallmark of modern consumer society. When goods are damaged in transit due to rough handling or accidents,

commercial profits are reduced. Overall consumer confidence in the ability of the business enterprise to deliver goods safely is eroded when package contents arrive in a damaged state.

In earlier commercial eras, a variety of materials were used to fill the spaces between transported goods and the walls of their container. Straw, old newspapers, and peanuts are common packing materials that have been utilized in the United States.

The plastic styrene was first distilled in 1831; it provided the basis for the insulating and packaging products popularized as Styrofoam, a material patented by Dow Chemical in 1941. Styrofoam also became a generic term used to describe any extruded polystyrene foam. When formed into peanut shapes, polystyrene proved to be a very effective, lightweight, and reusable packing material.

The manufacture of polystyrene foam generates significant environmental impacts at all points in its life cycle. Styrene production involves the use of hydrochlorofluorcarbons (HCFCs), the gas used to blow the styrene into foam. The release of HCFCs into the atmosphere during polystyrene manufacture is linked to the reduction of the Earth's ozone layer that has been observed since the 1980s. Benzene, another chemical used in styrene production, is a highly toxic substance that poses specific health risks to anyone engaged in styrene manufacture. It is estimated that when styrene products are disposed of in landfill sites, the product does not degrade for several hundred years.

Cornstarch packing materials address most of the environmental concerns generated by polystyrene manufacture. The cornstarch production process is relatively simple. Corn germ and its endosperm are separated, and polysaccharide starch is extracted from each source. When pressed into kernels, peanut or "s" shaped pellets, starch packing has all of the advantages of polystyrene, with minimal negative environmental impacts. Cornstarch is water soluble; the only significant drawback associated with starch-based packing materials is that they degrade when exposed to water.

■ Impacts and Issues

Cornstarch packaging has emerged as a cost-effective and environmentally-friendly alternative to conventional polystyrene packing products. The environmental qualities of cornstarch packaging highlight the distinction between the terms biodegradable, compostable, and recyclable, all commonly used in the assessment of the environmental impact associated with a particular product. The terms are often used interchangeably yet without proper scientific precision. Biodegradable products are those that naturally decompose as a result of microorganism activity. Wood and animal wastes are common examples. A compostable product is one that can be reduced to its constituent biomass, carbon dioxide and water, at the same rate as paper. Recyclable is the term used to describe any used or waste materials that can be refashioned into new products. International standards (ISO standards) have been developed to ensure that where materials are recycled, the applicable processes conform to environmental quality standards.

WORDS TO KNOW

ENDOSPERM: Plant tissue; the endosperm of all grains and flowering plants contains starch.

LOOSE FILL: Any form of packaging material used to cushion items shipped in boxes or larger containers. The fill is placed in the spaces between the shipped item and the container wall.

POLYLACTIDE (PLA): PLA is the generic acronym used to describe a wide range of polylactides, the monomers produced when cornstarch is fermented. PLAs are the foundation for many bioplastic products, including cornstarch packing materials.

POLYSACCHARIDES: The polymers formed from simple sugars; amylopectin and amylase are two polysaccharides that combine to form cornstarch.

A quality assurance analyst pours pellets of corn plastic into a dish at the Cargill-Dow plant in Blair, Nebraska. The United States' abundant corn supply is being converted into environmentally friendly plastics and fibers for use in products ranging from mattress pads to golf shirts to packaging. Products made from corn-based plastic are giving birth to a new industry built on highly biodegradable products. © *AP Images/Nati Harnik.*

In accordance with international standards, cornstarch packaging is deemed biodegradable if the product achieves 60 percent biodegration within 180 days of disposal. Cornstarch products can be disposed of safely into household or commercial waste systems, because the only adverse environmental impact is the additional volume of waste generated. Cornstarch packing is a less attractive compost option, as these materials will generally take over 180 days to degrade unless heat, moisture, and air exposures are ideal. The American Society for Testing and Materials defines compost as the outcome of this three-stage process:

- The materials degrade at the same rate observed in cellulose to produce biomass, carbon dioxide, and water.

- The materials disintegrate to a degree that they are not visible in surrounding compost and they do not require screening.

- No toxic materials are produced and the compost is able to support plant life.

The development of cornstarch packing represents a biotechnology success story. It is anticipated that most materials packaging will be manufactured from similar bioplastics as processing efficiencies increase.

SEE ALSO *Bioplastics (Bacterial Polymers); Corn, Genetically Engineered; Environmental Biotechnology*

BIBLIOGRAPHY

Books

Cava, David, and Alan Campbell. *Biodegradable and Compostable Packaging Materials for Foodstuffs.* Chipping Campden, UK: Campden BRI, 2010.

Janssen, Leon P. B. M., and Leszek Moscicki, eds. *Thermoplastic Starch: A Green Material for Various Industries.* Weinheim, Germany: Wiley-VCH, 2009.

Rudnik, Ewa. *Compostable Polymer Materials.* Oxford and Boston: Elsevier, 2008.

Yam, Kit L. *The Wiley Encyclopedia of Packaging Technology,* 3rd ed. Hoboken, NJ: Wiley, 2009.

Web Site

American Friends of Tel Aviv University. "Sweet and Biodegradable: Sugar and Cornstarch Make Environmentally Safer Plastics." *Science Daily,* December 14, 2010. http://www.sciencedaily. com/releases/2010/12/101214111919.htm (accessed October 10, 2011).

Bryan Thomas Davies

Corticosteroids

■ Introduction

The discovery and subsequent commercial development of the corticosteroid family of medications is one of the most important pharmaceutical advances in the twentieth century. Few medications have proved to be more influential over a wider range of applications than these compounds. Corticosteroids have been developed primarily for use as anti-inflammatories, with particular demonstrated utility in the treatment of conditions as diverse brain injuries; Crohn's disease, a chronic bowel inflammation disorder; asthma; and a variety of muscle injuries that arise through participation in various sports or in the workplace.

All corticosteroids are derived from cholesterol, the steroid produced in the human liver and within intestine walls. Cholesterols that are synthesized within the adrenal cortex are the basis for all corticosteroids. Synthesized cortisone and prednisone are the best known members of this family of medications.

These medications are administered either locally through topical application or injections or systemically by oral means or intravenous/intramuscular injections. When administered systemically, corticosteroids are intended to reduce inflammation at multiple sites within the body.

Corticosteroids are distinct from the anabolic steroids that are notorious in the athletic community for their ability to assist in muscle and strength development. The prescription of anabolic steroids is strictly regulated in most Western nations. However, some corticosteroids are also banned for use in international sports by the Olympic Committee and the World Anti-Doping Agency, which states that corticosteroids such as prednisone and prednisolone can create a sense of "euphoria" that might give an athlete an unfair advantage. Since the early 1990s, COX-2 inhibitors and other types of nonsteroidal anti-inflammatory drugs (NSAID) have been developed. These NSAID medications tend to cause fewer side effects in patients than the corticosteroid varieties.

■ Historical Background and Scientific Foundations

Anti-inflammatory medicines have been the subject of concerted research since the beginnings of formal medical science. Prior to the understanding of the role played by the hormone cortisol within the adrenal glands and the immune response generally, a variety of plants were recognized as possessing anti-inflammatory powers. The discovery of the salicylate compound in the mid-nineteenth century, from the active ingredient found in the willow bark previously known for its anti-inflammatory properties, was an important research milestone. Salicylate is the active chemical in aspirin and other over-the-counter pain relief medications.

There are many degenerative diseases that cause inflammation of joints, tissues, or internal organs. Arthritis in various forms, myositis, systemic vasculitis, and Crohn's disease are among the most prominent and debilitating of these conditions.

Corticosteroids act on the area of inflammation to decrease its effect through the suppression of normal immune system activity. Bacteria and viruses are the most common targets. The hormone cortisol is produced naturally within the adrenal glands; cortisol is released when the body encounters stress. In these circumstances cortisol triggers an increase in blood sugar and suppressed immune system activity. Inflammation results within a joint, tissue, or organ when phagocytes (white blood cells normally released to fight infection and to repel invading foreign substances from the body) do not function as expected. As phagocytes commence their destruction of the foreign invaders in the bloodstream, the signaling hormones known as cytokines communicate the message that a greater than necessary initial immune reaction is required. This phenomenon is known as the immune cascade. When the immune system malfunctions, the immune response is excessive. The response directs excess amounts of blood and phagocytes to the site of an injury, causing inflammation marked by redness on the adjacent skin surface, increased tissue temperatures, and pain.

In normal circumstances, the inflammatory response terminates when the threat to the body posed by invading organisms has ended. The ultimate risk to the patient from an excessive immune response is that the afflicted cells will commit apoptosis, the cellular death that results when an incorrect chemical signal is communicated to the cell during the immune cascade.

As with all medications, corticosteroids carry a risk of side effects. These are largely dependent on the specific steroid administered, its dose, and the treatment length. Typical side effects include weight gain (due to increased appetite), susceptibility to mood swings, and muscle weakness or instability.

■ Impacts and Issues

Corticosteroids are medications that have a wide variety of beneficial uses, and unlike most medications, the known side-effects associated with their use are relatively few. For the treatment of many forms of joint inflammation, including rheumatoid arthritis, corticosteroids are effective and essentially non-toxic. Athletes or persons who suffer from a potentially debilitating inflammation may undergo corticosteroid injections administered directly into the afflicted joint. The active steroidal agent will take effect typically within 24 hours; the relief will continue for a period of weeks due to the slow rate of diffusion of the corticosteroid within the body. The dosage levels are dictated by the combined effect of factors

such as age, other health concerns, and the degree of disability or discomfort prompted by the condition.

As the administration of the injection may produce significant pain (especially if the injection is directed to a joint with little tissue coverage, such as a knee or wrist), the corticosteroid injection may be preceded by the administration of a smaller local injection of an anesthetic such as lidocaine. The excess use of topical corticosteroids can cause skin damage.

If inflammation poses a threat to the function of the kidneys, lungs, or other essential organs, corticosteroid administration may preserve the life of the patient. Kidney infections are especially problematic, because persons with systemic vasculitis or other circulatory disorders may die from a kidney failure precipitated by infection. The medications assist in the preservation of the organ and thus reduce the incidence of organ failure that otherwise requires ongoing dialysis treatments or transplant.

The advances made in early twenty-first-century pharmaceutical research have taken corticosteroid applications in new directions from their accepted use as anti-inflammatory medications. Earlier intensive research directed to the reduction of gastrointestinal upset caused by salicylate-based medications led to the development of new COX-2 inhibitors. These NSAIDs have fewer side effects and excellent anti-inflammatory properties.

As the global population now collectively enjoys greater longevity than at any time in human history, the incidence of physical conditions that involve the excess inflammation of joints is likely to increase. The effective prescription and availability of corticosteroids and NSAIDs will assume even greater importance for this reason. This demographic reality suggests that increased research attention to both steroid and non-steroidal anti-inflammatory medications is essential for the quality of life and general productivity of modern societies.

SEE ALSO *Biopharmaceuticals; Genetic Engineering; Patents and Other Intellectual Property Rights; Pharmacogenomics; Synthetic Biology; Translational Medicine*

BIBLIOGRAPHY

Books

Barnes, Linda P., ed. *New Research on Pharmacogenetics.* New York: Nova Biomedical Books, 2007.

Fischer, Jaános, and C. R. Ganellin, eds. *Analogue-Based Drug Discovery.* Weinheim: Wiley-VCH, 2006.

Garcia-Segura, Luis Miguel. *Hormones and Brain Plasticity.* New York: Oxford University Press, 2009.

Inoue, Yuuki, and Kouki Watanabe, eds. *Adverse Effects of Steroids.* New York: Nova Science Publishers, 2008.

Lednicer, Daniel. *Steroid Chemistry at a Glance.* Hoboken, NJ: John Wiley & Sons, 2010.

Vardanyan, R. S., and Victor J. Hruby. *Synthesis of Essential Drugs.* Boston: Elsevier, 2006.

AC Milan forward Marco Borriello, right, and Brescia defender Marius Stankevicius of Lithuania, jump for the ball during their Italian Cup eight-final first leg soccer match in Milan, Italy. Borriello was provisionally suspended after the Italian Olympic Committee found the banned substances prednisolone and prednisone, which are corticosteroids, in a urine sample in 2006. *© AP Images/Antonio Calanni.*

Web Sites

Mayo Clinic Staff. "Prednisone and Other Corticosteroids." *Mayo Clinic.* http://www.mayoclinic.com/health/steroids/HQ01431 (accessed November 6, 2011).

"Prednisone." *National Center for Biotechnology Information.* http://www.ncbi.nlm.nih.gov/pubmedhealth/PMH0000091/ (accessed November 6, 2011).

Bryan Thomas Davies

Cosmetics

■ Introduction

Cosmetics are products that are applied to the skin to enhance a person's appearance. Sometimes perfumes are also considered cosmetics. Cosmetics are closely related to toiletries, which are products such as shampoo and toothpaste used in personal grooming to keep the body clean.

The development and manufacture of cosmetics is a global business worth an estimated $200 billion per year. As the population of the United States ages, there is increasing interest in cosmetic products that may help delay skin aging, and the market in male cosmetics is also growing fast.

There are two broad categories of cosmetics. Makeup, which is applied to the skin, eyes, and lips, enhances facial features and disguises skin blemishes. Skin care products, such as moisturizers, enhance the appearance of skin. As cosmetic science has advanced, cosmetics have become more like pharmaceuticals, with companies making claims for the benefits of these cosmeceuticals that cannot always be backed up with scientific evidence. Another controversial area is the testing of cosmetics on animals, which has become banned in some countries: Cosmetic companies have to find other ways of ensuring the safety of product ingredients for the consumer. Labeling of cosmetics enables the consumer to avoid any ingredient that may cause harm such as an allergic reaction.

■ Historical Background and Scientific Foundations

People have long been interested in using cosmetics to beautify their faces and bodies. Women in ancient Egypt used kohl, a black powder containing lead sulfide, as eyeliner, and Greek women used lead carbonate, which is highly toxic, to make the skin look paler. In the early 2010s, cosmetics are discovered and developed in chemistry laboratories, and they are complex mixtures of well-defined ingredients of both synthetic and biologic origin.

The composition of cosmetic products varies with their intended function, but most contain a combination of water, emulsifier, preservative, thickener, color, and fragrance. Many cosmetics are based on emulsions. Without the presence of an emulsifier, the oil and water phases would separate, rendering the product unusable. Cosmetics are exposed to the air and the fingers of the user on application and could therefore become contaminated with microorganisms. Preservatives are added to prevent the growth of bacteria or fungi that might otherwise cause a skin or eye infection. Thickening agents are added to control cosmetics' consistency and texture,

A biologist checks dried seaweed (*Laminaria saccharina*) at Kiel, northern Germany. The biologist is the managing director of Germany's "o'well" company, which offers cosmetic products derived from algae.   © *AP Images/Heribert Proepper.*

making them easy to apply. Finally, color and fragrances are included to give the product a pleasant smell and appearance, which is crucial in its marketing.

Lipstick and moisturizer are among the most popular cosmetics. Lipstick is a combination of a water-insoluble dye, which gives the product its color, with wax and a non-volatile oil. Beeswax and castor oil is a common formulation. The resulting consistency is stiff, so it can be packaged in a stick, yet spreadable, so it can be applied to the lips. Moisturizer is intended to treat the dryness that results when water is lost from the epidermis by evaporation. First, the moisturizer prevents further moisture loss through the inclusion of oils such as isopropyl palmitate or stearyl alcohol, which form a waterproof layer on the epidermis. Second, inclusion of humectants such as glycerin in moisturizer attracts water to the skin.

■ Impacts and Issues

As with other consumer products, such as pharmaceuticals and foods, safety is an issue with cosmetics. Some people are allergic to cosmetic ingredients, particularly perfumes. Sodium lauryl sulfate, which is a common

emulsifier, can cause dermatitis. Even more concerning are reports that paraben preservatives found in many cosmetics may be carcinogenic. The parabens have estrogenic activity, which can stimulate the growth of tumors. They have been found in breast tumors, although there is no evidence that they actually cause cancer, and they are used at extremely low levels in cosmetics. Consumers may wish to avoid contact with parabens or allergens. In the United States and in many other countries, cosmetics are required to be labeled with a list of ingredients, so consumers can make an informed choice. There are also many fragrance-free and paraben-free ranges of cosmetics.

One way of testing the safety of cosmetics is by testing them on animals. This has always been a controversial practice. Pharmaceuticals are also tested on animals but here the suffering caused to the animal can perhaps be justified by balancing it against the potential benefit to human health from a new drug. With cosmetics the benefit is, at best, an increase in the consumer's self-esteem brought about by use of the product. Testing cosmetics on animals was banned in the European Union in 2009, but it continues to be allowed in the United States and in many other countries. However, the U.S. Food and Drug Administration (FDA) does strongly encourage companies to look at alternatives to animal testing wherever possible. Testing can be done on cell or tissue samples, and many such tests have been developed to assess the impact of cosmetics on human eyes or skin. Cosmetic testing can also be done on human volunteers. Some companies may outsource their animal testing to a country where regulations are weak or absent. It is also possible to use ingredients in cosmetics whose safety profile is well known from animal testing that has been done in the past.

Several cosmetics companies are using nanotechnology to improve the performance of their products, and government regulators are playing catch-up in order to insure their effectiveness and safety. Although initial regulations regarding nanotechnology have been enacted in Europe and the United States, few of the newer cosmetic products containing nanoparticles have received specific regulatory scrutiny. L'Oréal, La Prairie, Estée Lauder, Clinique, Lancôme, Avon, and Neutrogena all have at least one product manufactured using nanotechnology. Some of the benefits of nanotech-enhanced cosmetics include faster and deeper absorption into the skin, more vivid colors, and longer-lasting wear. Although a bill designed to examine the safety and regulate the use of nanotechnology in cosmetics, the Safe Cosmetics Act of 2010, stalled in the U.S. Congress, it was reintroduced in January 2011 and remained pending in November 2011. In the meantime, the market for nanotechnology-enhanced cosmetics is expected to top $150 million in 2012.

There is also concern over whether cosmetic companies make exaggerated claims about their products, particularly in the area of skin aging. Scientific knowledge of the processes responsible for the wrinkling and sagging of the skin with age is still incomplete, yet some products claim to slow or even reverse this process. Certain ingredients, such as peptides, are said to penetrate the dermis and carry out repair processes to restore a youthful appearance. However, carrying out proper clinical studies on cosmetics is costly and might require the product to be regulated as a pharmaceutical rather than a cosmetic. Therefore evidence on cosmetic claims is more likely to come from experiments on cells in test tubes that may not necessarily extrapolate to real human subjects, or from surveys of consumers, which are subjective. However, as dermatology research advances, it is likely that the number of cosmeceuticals on the market will grow, and it may be that in the future some of these products will be shown to do more for the skin than conventional cosmetics.

■ Primary Source Connection

Technology developed by the Hurel Corp. has the potential to offer cosmetic makers the ability to test for allergic reactions to their products without testing on

animals. The technology uses human skin cells grown in a laboratory rather than skin areas found on animals to stimulate an allergic response to foreign chemicals. Funded by L'Oréal, the technology may serve to replace tests currently performed on mice and guinea pigs, as a means to predict dermatological reactions from drugs and cosmetics in humans.

Technology Aims to Replace Animal Testing

WASHINGTON—Technology allowing cosmetic makers to test for allergic reactions to their products without controversial animal trials is in the works and could be in use by next year.

The technology developed by Hurel Corp., with funding from cosmetics maker L'Oreal, is designed to replace tests on mice and guinea pigs used to predict skin reactions from drugs and cosmetics. The device uses laboratory-grown human skin cells to simulate the body's allergic response to foreign chemicals. Preliminary experiments show promise, but rigorous tests are still needed to determine the technology's accuracy.

The standard method for testing allergic reactions involves applying chemicals to the ears of mice, which are later killed and dissected for study.

North Brunswick, N.J.-based Hurel said Thursday it hopes to eliminate the need for such tests, in an announcement with cosmetics giant L'Oreal, which provided funding for the test.

The product from Hurel consists of a glass chip with human skin cells and chemicals that simulate the body's immune system. When a foreign substance is dropped onto the chip, the cells and chemicals interact to mimic the human body's natural allergic response.

While the product is still in development, Hurel officials say a working prototype should be available by the second half of next year. In addition to cosmetics, the technology could be used to test household cleaners and pesticides.

Hurel Chief Executive Robert Freedman said it is too early to estimate the price or sales figures for the chip, but he pegs the market for a non-animal allergy test at $2 billion a year.

Like other companies in the cosmetics industry, L'Oreal is racing to develop alternatives for testing wrinkle creams and lipstick to comply with European Union laws. Regulators there have ordered companies to phase out animal skin testing by 2013.

L'Oreal has decreased its use of animal testing over the years, but still relies on the technique to test certain new chemicals.

"I give L'Oreal credit for being willing to explore these types of opportunities," said Dr. Charles Sandusky, of the Physicians Committee for Responsible Medicine. "This is the first thing I've ever seen where the immune system is being mimicked without using an animal component."

A spokeswoman for L'Oreal said the company has invested heavily in non-animal testing over 25 years, but declined to specify how much went into developing the Hurel chip.

Hurel will be free to license the technology to other companies once it has been proven effective, she said.

Sandusky, a former toxicologist at the Environmental Protection Agency, estimates Hurel's technology, if successfully applied, could eliminate the need for tens of thousands of test animals each year.

For that reason, the animal rights group People for the Ethical Treatment of Animals plans to grant the company an innovation award Thursday for "animal-friendly achievement in commerce." PETA's science policy adviser Dr. Kate Willett said the group has been following Hurel's research efforts.

Hurel was founded in 2005 and has one other product under development: a liver toxicity test. Given that regulators generally won't approve an experimental drug if there are signs it harms the liver, a liver toxicity test could be a boon to drugmakers who test their medicines in animals before submitting them to regulators.

Animal testing can be slow, and many researchers question how well an animal's response to a chemical predicts human reactions.

By eliminating the time, money and potential inaccuracies associated with animal testing, Chief Executive Robert Freedman estimates Hurel's test could shave $100 million off the roughly $1 billion cost of developing a new drug.

Matthew Perrone

PERRONE, MATTHEW. "TECHNOLOGY AIMS TO REPLACE ANIMAL TESTING." *U.S. NEWS & WORLD REPORT*, ASSOCIATED PRESS, JANUARY 14, 2010. AVAILABLE ONLINE AT HTTP:// WWW.USNEWS.COM/SCIENCE/ARTICLES/2010/01/14/ TECHNOLOGY-AIMS-TO-REPLACE-ANIMAL-TESTING

SEE ALSO *Biotechnology Products; Botox; Collagen Replacement; Nanotechnology, Molecular; Patents and Other Intellectual Property Rights*

BIBLIOGRAPHY

Books

Beck, Ruy, Silvia Guterres, and Adriana Pohlmann. *Nanocosmetics and Nanomedicines: New Approaches for Skin Care.* New York: Springer, 2011.

Betton, C. I., ed. *Global Regulatory Issues for the Cosmetics Industry.* Norwich, NY: W. Andrew, 2007.

Friends of the Earth. *Nanomaterials, Sunscreens and Cosmetics: Small Ingredients Big Risks.* Washington, DC: Friends of the Earth, 2006.

Garti, Nissim, and Idit Amar-Yuli, eds. *Nanotechnologies for Solubilization and Delivery in Foods, Cosmetics, and Pharmaceuticals.* Lancaster, PA: DEStech Publications, 2012.

Koslowski, Angela, ed. *Biotechnology in Cosmetics: Concepts, Tools, and Techniques.* Carol Stream, IL: Allured Pub, 2007.

Lad, Raj, ed. *Biotechnology in Personal Care.* New York: Taylor & Francis, 2006.

Morgan, Sarah E., Kathleen O. Havelka, and Robert Y. Lochhead, eds. *Cosmetic Nanotechnology: Polymers and Colloids in Cosmetics.* Washington, DC: American Chemical Society, 2007.

Naidu, B. David. *Biotechnology & Nanotechnology: Regulation under Environmental, Health, and Safety Laws.* New York: Oxford University Press, 2009.

Web Sites

Rinaldi, Andrea. "Healing Beauty? More Biotechnology Cosmetic Products That Claim Drug-Like Properties Reach the Market." *European Molecular Biology Organization (EMBO) Reports*, November 9, 2008. http://www.ncbi.nlm.nih.gov/pmc/ articles/PMC2581859/ (accessed November 8, 2011).

Schneider, Marc, et al. "Nanoparticles and Their Interactions with the Dermal Barrier." *Dermatoendocrinology*, July/August 2009. http://www.ncbi.nlm.nih. gov/pmc/articles/PMC2835875/?tool=pubmed (accessed November 8, 2011).

Susan Aldridge

Cotton, Genetically Engineered

■ Introduction

Cotton is among the world's most important fiber crops. Best known as a textile fiber, it is also a raw material for other products such as cottonseed oil, animal feed, and food additives. Many countries are now growing genetically engineered cotton carrying genes for pest resistance and herbicide tolerance, in order to boost yields and reduce farming costs.

Cotton fibers are derived from the seed hairs of the cotton plant *Gossypium hirsutum*, which develop in capsules called cotton bolls. When ripe, the bolls break open, revealing the fluffy white fibers. The plant grows in warm regions, including several of the southern states of the United States. Cotton is prone to attack by the bollworm, a type of moth larva that quickly can destroy an entire boll. However, genetic engineering of cotton can render it resistant to bollworm attack by transferring a gene for the Bt toxin (similar to the natural toxin produced by *Bacillus thuringiensis*). Genetic engineering can also make cotton tolerant of the herbicide glyphosphate, known as Roundup, so that weeds can be killed while leaving the crop unharmed. Both developments improve cotton production, which is why the amount of farmland planted with genetically engineered (GE) cotton has been increasing. However, there are concerns over whether GE cotton could encourage emergence of resistant pests and weeds.

■ Historical Background and Scientific Foundations

Cotton was being grown, spun, and woven into cloth in Pakistan and Egypt around 7,000 years ago. There is evidence of cotton planting in Mexico around the same time. Mechanization of cotton production in the nineteenth century ensured the place of cotton in the modern industrial world, and it remains an important commercial crop today. Genetically-modified (GM) cotton was first introduced into the United States in the mid-1990s and has been adopted around the world. The Philippines is the latest country to announce that it will commercialize GM cotton, with the planned introduction date being 2012.

The bollworm *Helicoverpa armigera*, which is sometimes known as the pink bollworm, is a major crop pest. It is the larval stage of a moth that damages the cotton boll, reducing fiber yield and also exposing the plant to bacterial and fungal attack. Bollworm also reduces yields in tomato, corn, soybean, tobacco, and alfalfa. The soil bacterium *Bacillus thuringiensis* produces a toxin called Cry1Ac, which kills bollworm larvae. This toxin is closely related to the Cry1Ab Bt toxin found in GM corn. All of the Cry toxins act by perforating the gut of the pest, which stops feeding and usually dies within a day or so. A spray containing *Bacillus thuringiensis* has been used for some years as a biological pesticide. In Bt cotton, the Cry1Ac gene has been transferred to the plant, conferring built-in resistance to the bollworm. Meanwhile, Roundup herbicide-tolerant cotton has been genetically modified to contain a different version of a plant enzyme that the herbicide would otherwise destroy. This means that Roundup can be sprayed on a crop, destroying weeds but leaving the cotton unscathed.

All genetic engineering experiments involve the transfer of a gene from one species to another in order to confer some beneficial trait, such as herbicide tolerance, on the recipient, thereby creating a transgenic organism. GM cotton can be produced in two ways. The first involves using the bacterium *Agrobacterium tumifaciens*, which naturally infects plants, as a vector to carry the gene into plant cells. In the second, a device called a gene gun is used to shoot particles coated with the DNA of the transgene into the plant cells. The transformed cells are then grown into mature plants under carefully controlled conditions. The resulting cotton plants are thoroughly tested for their characteristics and will be cultivated in small-scale outdoor field trials before moving to large scale cultivation.

■ Impacts and Issues

Grown in more than 50 countries around the world, cotton is a leading cash crop in the United States. Bollworm can destroy up 50 percent or more of a cotton crop, and pesticides to keep it under control account for a substantial part of cultivation costs. Heavy use of pesticides may also pollute water supplies and harm nonpest insects. Bt cotton promises to increase cotton yields and cut production costs for farmers. Field trials suggest these benefits can be realized. As of 2011, 73 percent of the cotton grown in the United States is herbicide tolerant, and 75 percent contains Bt toxin for pest resistance, while 95 percent of the cotton grown in China, the leading producer, is genetically engineered. Some plants have been engineered to have both traits.

Although genetically-modified cotton crops initially reduce pesticide use, there remain a number of environmental concerns associated with the use of GM crops, including cotton. Some scientists have long predicted that insects quickly would evolve resistance to Bt toxin, rendering the built-in pesticide useless. Researchers at the University of Arizona have, indeed, found resistant bollworm in some crop locations in Tennessee and Arkansas in experiments carried out between 2003 and 2006. However, thus far, given how many millions of hectares are planted with Bt cotton worldwide, bollworm resistance appears not to be a major issue. That could change in the long term. It looks as if it is important to plant refuges of non-Bt cotton around the genetically engineered plants to help keep resistance at bay.

A potentially more serious issue emerging with Bt cotton is the impact its presence is having on local insect ecosystems. GM cotton has been planted in China since 1997 after a period when bollworm pesticides were causing massive environmental damage. A team at the Chinese Academy of Agricultural Sciences has been monitoring pest populations around China over the last several years and has found a twelvefold increase in mirid bugs, previously known as only a minor pest. The surge has seen the bugs destroying cotton and other crops such as beans, cereals, and vegetables, and Chinese farmers are resorting to chemical pesticides, which is beginning to erode the economic benefits of Bt cotton.

Finally, if a GM plant transfers a herbicide resistance gene to a non-crop plant, there is the potential for creation of a so-called superweed. Such plants can be controlled only by the application of increasing amounts of chemical herbicides and may need a cocktail of different herbicides to kill them. Agricultural land might even be abandoned if superweeds crowd out crops and prove impossible to control. There is already evidence for the widespread existence of many species of herbicide-tolerant weeds. The long-term impact of their presence on agriculture remains to be seen.

■ Primary Source Connection

New types of insects are endangering the health of genetically-modified cotton currently grown by Chinese farmers. The genetically modified cotton, also known as Bt cotton, has thus far been resistant to pests such as bollworms. However, new pest populations, never before identified, are literally eating into profits of these farmers, who must use more equipment, labor, and additional sprays for the health of their crops. They are seeking assistance from scientists and governments to develop solutions, as secondary pest problems pose a potential threat in countries where the cotton has already been extensively planted across many regions.

Seven-year Glitch: Cornell Warns that Chinese GM Cotton Farmers are Losing Money Due to "Secondary" Pests

Although Chinese cotton growers were among the first farmers worldwide to plant genetically modified (GM) cotton to resist bollworms, the substantial profits they have reaped for several years by saving on pesticides have now been eroded.

The reason, as reported by Cornell University researchers at the American Agricultural Economics Association (AAEA) Annual Meeting in Long Beach, Calif., July 25, is that other pests are now attacking the GM cotton.

The GM crop is known as Bt cotton, shorthand for the *Bacillus thuringiensis* gene inserted into the seeds to produce toxins. But these toxins are lethal only to leafeating bollworms. After seven years, populations of other insects—such as mirids—have increased so much that farmers are now having to spray their crops up to

Former President of The Associated Chambers of Commerce and Industry of India Anil K Agarwal, left, with Director of Indicus Analytics Laveesh Bhandari, center, and Vice President and Executive Director IMRD international Nikhil Rawal, release a report on Bt Cotton Farming in India, in New Delhi. According to the report, the acreage of Bt cotton was expected to rise up to 3.5 million hectares by the end of 2008. Since the introduction of the Bt cotton in the country for the first time, there has been advent of a new varieties approved for the commercial cultivation.   © *AP Images/Manish Swarup.*

20 times a growing season to control them, according to the study of 481 Chinese farmers in five major cotton-producing provinces.

"These results should send a very strong signal to researchers and governments that they need to come up with remedial actions for the Bt-cotton farmers. Otherwise, these farmers will stop using Bt cotton, and that would be very unfortunate," said Per Pinstrup-Andersen, the H.E. Babcock Professor of Food, Nutrition and Public Policy at Cornell, and the 2001 Food Prize laureate. Bt cotton, he said, can help reduce poverty and undernourishment problems in developing countries if properly used.

The study—the first to look at the longer-term economic impact of Bt cotton—found that by year three, farmers in the survey who had planted Bt cotton cut pesticide use by more than 70 percent and had earnings 36 percent higher than farmers planting conventional cotton. By 2004, however, they had to spray just as much as conventional farmers, which resulted in a net average income of 8 percent less than conventional cotton farmers because Bt seed is triple the cost of conventional seed.

In addition to Pinstrup-Andersen, the study was conducted by Shenghui Wang, Cornell Ph.D. '06 and now an economist at the World Bank, and Cornell professor David R. Just. They stress that secondary pest problems could become a major threat in countries where Bt cotton has been widely planted.

"Because of its touted efficiency, four major cotton-growing countries were quick to adopt Bt cotton: the U.S., China, India and Argentina," said Wang. Bt cotton accounts for 35 percent of cotton production worldwide. In China, more than 5 million farmers have planted Bt cotton; it is also widely planted in Mexico and South Africa.

When U.S. farmers plant Bt crops, they, unlike farmers in China, are required by contracts with seed producers to plant a refuge, a field of non-Bt crops, to maintain a bollworm population nearby to help prevent the pest from developing resistance to the Bt cotton. The pesticides used in these refuge fields help control secondary pest populations on the nearby Bt cotton fields. Researchers do not yet know if a secondary pest problem will emerge in the United States and other countries, Pinstrup-Andersen said.

"The problem in China is not due to the bollworm developing resistance to Bt cotton—as some researchers have feared—but is due to secondary pests that are not targeted by the Bt cotton and which previously have been controlled by the broad-spectrum pesticides used to control bollworms," added Pinstrup-Andersen, who also is serving as president of AAEA for 2007.

Wang and her co-authors conclude, "Research is urgently needed to develop and test solutions."

Susan Lang

LANG, SUSAN. "SEVEN-YEAR GLITCH: CORNELL WARNS THAT CHINESE GM COTTON FARMERS ARE LOSING MONEY DUE TO 'SECONDARY' PESTS." *CORNELL UNIVERSITY CHRONICLE ONLINE* JULY 25, 2006. AVAILABLE ONLINE AT HTTP://WWW. NEWS.CORNELL.EDU/STORIES/JULY06/BT.COTTON.CHINA.SSL. HTML.

SEE ALSO *Agricultural Biotechnology; Agrobacterium; Bt Insect Resistant Crops; Corn, Genetically Engineered; Disease-Resistant Crops; Drought-Resistant Crops; Frost-Resistant Crops; Gene Banks; Plant Variety Protection Act of 1970; Salinity-Tolerant Plants; Transgenic Plants; Wheat, Genetically Engineered*

BIBLIOGRAPHY

Books

Kempken, Frank, and Christian Jung. *Genetic Modification of Plants: Agriculture, Horticulture and Forestry.* Berlin: Springer, 2010.

Moseley, William G., and Leslie Gray. *Hanging by a Thread: Cotton, Globalization, and Poverty in Africa.* Athens, Ohio: Ohio University Press, 2008.

U.S. National Research Council. *The Impact of Genetically Engineered Crops on Farm Sustainability in the United States.* Washington, DC: National Academies Press, 2010.

Web Sites

"Cotton." *GMO Compass.* http://www.gmo-compass. org/eng/grocery_shopping/crops/161.genetically_ modified_cotton.html (accessed October 4, 2011).

Susan Aldridge

Cryonics

■ Introduction

The premise of cryonics is that individuals (currently humans and some pets) can be cryopreserved shortly after clinical death, with the intention of restoration and reanimation at such time as future technology makes this a viable possibility. The goal of cryonics is the extension of life and restoration of health based on advanced medical, bio- and nanotechnologies. As of 2011, it is possible to induce a state of cryostasis but reanimation is not possible. Cryopreservation can be applied to the entire body, to the head only, or to sections of DNA. Tissue that has been cryostatically cooled and stored in liquid nitrogen is believed not to deteriorate or decompose.

The efficacy of cryonics cannot currently be evaluated, as the technology for reanimation and rejuvenation has not yet been developed and may never occur. In mainstream biotechnology, the efficacy of a protocol is measured through the clinical trial process; nothing similar exists for cryonics. Reanimation technology, if it occurs, may not advance to the point of efficacy for hundreds of years or more.

■ Historical Background and Scientific Foundations

Robert Ettinger (1918–2011), a college physics teacher, self-published *The Prospect of Immortality* in 1962; this was the first book describing the mechanics and logic of cryonics, although the term had not yet been created. Evan Cooper (1926–1982) independently published *Immortality: Physically, Scientifically, Now* (using the pseudonym Nathan Duhring), another tome describing the process and practice that would become cryonics, during the same year. In 1963 Cooper founded the Life Extension Society, in an effort to develop a network of like-minded individuals.

The term "cryonics" was developed by industrial designer Karl Werner in 1965, the year that he, Saul Kent, and Curtis Henderson founded the Cryonics Society of New York. In 1966 the Cryonics Society of California and the Cryonics Society of Michigan were established. The Cryonics Society of California cryopreserved its first patient, a 73-year-old psychology professor named James Bedford, in 1967. His cryogenic care was transferred to the Alcor facility in 1982, where he remains. In 1974 the New York State Department of Health closed down the Cryonics Society of New York's facility. The California facility was not properly maintained, and it too was closed in 1979.

The Bay Area Cryonics Society, later renamed the American Cryonics Society (1985), opened in 1969. It remains in operation in Cupertino, California, and is considered the longest continuously operating cryonics facility. Trans Time, founded in 1972 in Oakland, California, was originally intended as a cryonics provider for members of the Bay Area Cryonics Society; Trans Time created a cryostorage facility in 1974 and is a still operational as a for-profit cryonics suspension and maintenance provider. Currently there are just two patients stored at Trans Time.

In addition to Trans Time, there are three cryonics facilities currently using liquid nitrogen cryonic technology to store human and domestic animal patients: the Alcor Life Extension Foundation of Scottsdale, Arizona, created in 1974 by Linda and Fred Chamberlain; the Cryonics Institute in Clinton Township, Michigan, which Robert Ettinger opened in 1976; and KrioRus, located near Moscow, Russia, and created in 2006.

The Cryonics Institute developed a protocol for on-site fabrication of fiberglass cryostats, rather than using the larger dewars for provision of thermal insulation for cryonics patients. This allows storage of more patients in less space.

The goal of cryopreservation has been the preservation and protection of tissue at low temperatures while preventing or minimizing the formation of damage-causing ice crystals, particularly in patients' brains.

Glycerol was the most common cryoprotectant used until vitrification perfusion cryoprotection techniques were perfected in 2001 by Dr. Yuri Pichugin at the Cryonics Institute.

■ Impacts and Issues

Cryonicists assert that the current definition of clinical death as cessation of respiration/circulation and measurable brain activity do not define the end of life, they simply signify the state from which current medical practice fails to successfully revive patients. The preferred term in cryonics is "deanimation," with "reanimation" being the next desired state of being.

A central tenet of cryonics is that all material needed to reanimate an individual, including memories, knowledge, and personality traits/characteristics are enduringly encoded and remain relatively impervious to prolonged periods of cellular inactivity, such as what occurs with prolonged cryostats. Individuals who have experienced periods of hypothermia and been revived with intact long-term memory lend support to cryonic theory.

The burgeoning science of nanotechnology, in which molecules are intentionally created by manipulation of atoms and atomic structures, is expected to lead to rapid progress in reanimation techniques as it creates the blueprints for repair of damage occurring at the cellular level. In order for cryopreserved individuals to be reanimated, the initial cause of death must be reversed and the damage done as a result of cryopreservation repaired by nanotechnology.

The very last individuals placed into cryopreservation will be the first to be reanimated, as they will have experienced the most advanced technology. Because they will be the "newest" individuals, they will likely have living friends and relatives. Individuals involved in cryonics are often related to, or in relationship with, others committed to cryonics. As a result, they will probably know others who have been cryostored and will be highly motivated to facilitate their reanimation. A "LifePact" is when a group of individuals agree to assist one another to become reanimated. Cryonicists often encourage their close friends and family members to engage in cryonics so that they can all be together again in the future.

Individuals who engage in cryonics do so with the knowledge that the technology for reanimation might not be developed for hundreds of years, if ever. It is frequently stated that this is a gamble many consider worth taking, as the alternative entails no possibility of reanimation.

Currently, the cost of cryopreservation or cryostorage ranges from around $30,000 to ten times that. The cost is less for neuropreservation (brain only) than for full-body patients, and for those whose perfusion occurs through the use of a funeral director before the body is delivered to the cryostorage facility. The most expensive protocols are those in which a "standby team" of individuals waits at the bedside of the deanimating person and provides

cardiopulmonary support in order to minimize tissue damage and ischemia during transport, and the full body is cryostored. In all instances, full payment must occur before acceptance for cryostorage. Most facilities suggest use of life insurance monies as payment, so that the annual costs of maintenance are covered. Because relatively few individuals are cryostored, costs remain high. If a trend toward cryonics occurs, costs can be expected to decrease. In the United States and Russia, approximately 200 persons were in long-term cryostorage at the end of 2010.

Because cryostorage falls under the broad heading of experimental medical procedures and does not involve

WORDS TO KNOW

CRYONICIST: Cryonicists are those who believe cryonic preservation leading to reanimation and the restoration of health is a future certainty based on the current pace of bio- and nanotechnical advances. They actively support and may plan to participate in the cryonic preservation process. Per the January 1983 issue of *Cryonics*, cryonicists believe that individuals in cryostatic preservation are neither alive nor dead, but in a state of "uncertain waiting."

CRYOSTAT: Technically, this can be any device used to maintain cryostatic temperatures around −320°F (−196°C). In the context of cryonics, the term refers to a container for storing cryopreserved individuals, organs, or DNA in liquid nitrogen. It is also called a dewar. Individuals are stored head down in cryostats.

DNA/TISSUE CRYOPRESERVATION: The storage of DNA or tissue samples in special containers suspended in liquid nitrogen, with the intent of future reconstruction.

ISCHEMIA AND REPERFUSION DAMAGE: Refers to the potential for damage done to cells when deprived of oxygen and nutrients. Reperfusion damage occurs when blood is recirculated after a prolonged period of ischemia. The combination of ischemia and reperfusion can permanently, severely damage cells and organs.

SUSPENDED ANIMATION: An organism is said to be in suspended animation when all of its essential functions are temporarily interrupted.

VITRIFICATION: When the term is used in molecular biology, vitrification refers to the use of biotechnology to cool a cell to a glass-like state, removing water from the cell and preventing the formation of ice crystals that could cause cell damage. The process is reversible. As used in cryonics, vitrification refers to a process in which cooling and solidification occur with minimal tissue damage; this is typically done with a patient's brain. In cryonics, vitrification is not currently reversible through reanimation processes. In whole-body cryonics patients, the head/brain are often the only parts vitrified.

Head of Russian cryonics firm KrioRus Danila Medvedev and KrioRus customer look inside a low-temperature human brain storage unit just outside Moscow. Cryonics—or the freezing of humans in the hope of future resuscitation—is illegal in France and much of the world, but KrioRus has stored four full bodies and eight people's heads in liquid nitrogen-filled metal vats. © *Alexey Sazonov/AFP/Getty Images.*

death per se, it is generally not viewed by religious communities as controversial. Reanimation and rejuvenation are likened to reawakening after an extremely extended medical procedure involving general anesthesia and are not considered in any way related to the traditional religious concept of raising from the dead.

SEE ALSO *Bioethics; Biotechnology; Frozen Egg Technology; In Vitro Fertilization (IVF)*

BIBLIOGRAPHY

Books

Achrekar, Jayanto. *Concepts in Biotechnology.* New Delhi, India: Dominant, 2005.

Cavalier, Annie, Daniele Spehner, and Bruno M. Humbel. *Handbook of Cryo-Preparation Methods for Electron Microscopy.* Boca Raton, FL: CRC Press, 2009.

Web Sites

"Cryonics: the Chilling Facts." *The Independent,* July 27, 2011. http://www.independent.co.uk/ life-style/gadgets-and-tech/features/cryonics-the-chilling-facts-2326328.html (accessed September 16, 2011).

Glass, Ira. "This American Life: Mistakes Were Made: Act 1: You're as Cold as Ice." *National Public Radio,* April 18, 2008. http://www.thisamericanlife. org/radio-archives/episode/354/mistakes-were-made?act=1 (accessed September 16, 2011).

Howley, Kerry. "Until Cryonics Do Us Part." *The New York Times,* July 7, 2010. http://www.nytimes. com/2010/07/11/magazine/11cryonics-t. html?pagewanted=1 (accessed September 16, 2011).

Pamela V. Michaels

Dairy and Cheese Biotechnology

■ Introduction

Approximately 10,000 years ago, humans domesticated animals for milk production. The milk produced by goats, cattle, and other animals provides energy and nutrients for humans. The utilization of enzymes and acids to convert milk into yogurt and simple cheeses is one of the earliest applications of biotechnology in human history. Although archaeological evidence from around 2000 BC indicates that Egyptians made cheese, humans likely discovered these processes millennia earlier and used them to preserve dairy products.

Modern applications of biotechnology to dairy products continue the focus on food preservation, safety, and increasing yield. Refrigeration and pasteurization, the heating of milk to inhibit microbial growth, are common techniques to extend the shelf life of milk and other dairy products and reduce the transmission of diseases to humans. Over the last several decades, the use of genetically engineered hormones to increase milk production has created controversy over the safety of using such hormones.

■ Historical Background and Scientific Foundations

Neolithic farmers in modern day Iraq and Iran domesticated goats between 9000 and 8000 BC and began using the goats for milk production. Cattle were domesticated in Egypt and Sumer around 3000 BC, although some archaeological evidence indicates that humans may have domesticated cattle in Saharan grasslands as early as 4000 BC. Goats and cattle are ruminants, an order of mammals that chews cud regurgitated from the rumen, a compartment within the ruminant's four-part stomach. Sheep, yaks, water buffalo, camels, and the now-extinct auroch are ruminants that humans have used for milk production.

Whereas all mammals produce milk, ruminants have been favored for agricultural milk production for millennia because of their ability to convert cellulose, the structural component of plant cell walls, into food energy. Most mammals have only a limited ability to break down cellulose into energy. Anaerobic bacteria in the rumen produce enzymes that enable ruminants to break down cellulose and convert it to energy that may then be stored in the ruminant's muscle or milk. Ruminants, therefore, may graze on grasses and shrubs that humans cannot eat and convert the cellulose contained in these plants into a form of food energy, either meat or milk, that humans can use.

Humans domesticated ruminants for the production of raw milk and meat, and early farmers soon discovered, perhaps accidentally, that milk could be used for the production of cheese. Cheese is dairy product in which milk is separated in curds, a solid mass of milk proteins, and whey, the watery part of milk. The curd is separated from the whey through a process known as curdling. Curdling involves the addition of an acidifying agent, such as lemon juice, vinegar, rennet, or other enzymes. The use of only lemon juice, vinegar, or similar acidifying agents results in soft, fragile curds. Rennet is a set of enzymes, usually containing a combination of rennin, chymosin, protease, pepsin, and lipase, which is found in the stomachs of mammals. The use of rennet in the curdling process produces stronger curds and, therefore, firmer cheeses than the use of other acidifying agents.

The first cheese production, and, indeed, one of humankind's first forays into biotechnology, likely resulted accidentally from the storage of milk in the rennet-containing stomach of an animal. Earlier cheese makers realized that a simple cheese produced by making and salting curd would last longer than raw milk. Cheese making spread from the Middle East and North Africa to Europe, where the ancient Greeks and Romans became skilled cheese makers who employed aging to improve the flavor and other characteristics of cheese.

WORDS TO KNOW

PASTEURIZATION: The process of heating or irradiating milk, wine, or other products in order to partial sterilize them and, thereby, make them safer for human consumption.

PRECAUTIONARY PRINCIPLE: The principle that any product, action, or process that might pose a threat to public health or the environment should not be introduced in the absence of scientific consensus regarding its safety.

RECOMBINANT BOVINE GROWTH HORMONE (rBGH): A genetically engineered hormone given to cattle to increase milk production.

RUMINANT: A cloven-hoofed mammal that chews cud regurgitated from the rumen.

■ Impacts and Issues

Humans have long been concerned with the safe storage and transportation of dairy products. Nomads and traders in the Middle East made simple, acidified cheeses and yogurt for long journeys, because cheese and yogurt inhibit the growth of harmful bacteria. In the late-eighteenth and nineteenth centuries, economic activity in Europe and the United States became more specialized during the Industrial Revolution, and people became more concentrated in cities. More people came to rely on milk and other dairy products shipped into cities from remote farms, which increased the likelihood of spoilage.

In 1862 Louis Pasteur (1822–1895) and Claude Bernard (1813–1878) tested a process to preserve beer and wine by heating and rapidly cooling the liquid in order to inhibit microbial growth. This process, which became known as pasteurization, was adopted for the treatment of milk in the late-nineteenth century with the first commercial application in 1882. In 1908 the city of Chicago passed the first law requiring the pasteurization of milk prior to sale. Pasteurization became common for most commercially sold milk throughout the twentieth century.

Traditional batch pasteurization requires heating milk to 154.4°F (63°C) and holding the milk at that temperature for 30 minutes and then cooling quickly. In the 1950s Tropicana developed the High Temperature Short Time (HTST), or flash pasteurization, for treatment of orange juice, and the method quickly was adopted for milk processing. HTST involves heating milk to between 160 and 165°F (71.5 to 74°C) for between 15 and 30 seconds. Ultra-high temperature (UHT) pasteurization was developed in the 1960s and involves heating milk to 275°F (135°C) for one to two seconds.

Pasteurized milk typically is refrigerated during shipment and storage to further inhibit microbial growth, but UHT milk may be stored at room temperature for one year in hermetically sealed containers, such as Tetra Pak.

Natural animal rennet rarely is used in modern cheesemaking. Vegetable rennet produced from thistles, nettles, and other plants has been used for centuries, and some traditional cheeses have always been produced with vegetable rennet. The use of vegetable rennet produces cheeses that are suitable for vegetarians and traditionally was used in the production of kosher and halal cheeses. Microbial rennet produced from certain strains of mold also is used in cheesemaking. Microbial rennet, however, may produce bitter cheese, especially in cheese with long aging times. Most rennet, therefore, is vegetable rennet. In the early 2010s, genetically-engineered rennet often is used in industrial cheese production.

Much of the world's large-scale production cheese is made with genetically-engineered rennet, known as fermentation-produced chymosin rennet, which was introduced in the 1990s. Fermentation-produced chymosin rennet is made by inserting the rennin-producing gene from animal cells into bacteria, mold, or, most commonly, yeast. The rennin-producing genes cause the host organism to produce chymosin, one of the enzymes contained in natural animal rennet. Fermentation-produced chymosin rennet avoids the bitterness caused by rennet and is cost-effective. Fermentation-produced chymosin rennet is considered by some vegetarian, but it is animal-derived and is the result of genetic modification. Approximately 90 percent of the cheese produced in the United States is made using fermentation-produced chymosin rennet.

The United States also allows the use of recombinant bovine growth hormone (rBGH) in milk production. rBGH is a genetically-engineered version of bovine somatotropin, or bovine growth hormone (BGH), an amino acid complex that regulates metabolic processes in cattle, including increased milk production through the prevention of mammary cell death. The U.S. Food and Drug Administration (FDA) approved Posilac, the first rBGH, for use in dairy cattle in 1993. The FDA reviewed studies from Monsanto, the company that developed Posilac, and determined that milk from rBGH treated cattle is safe for human consumption. A later study by researchers associated with Health Canada, the Canadian government's public health department, raised concerns about the safety of rBGH.

The European Union, Canada, Japan, Australia, New Zealand, and other countries do not allow the use of rBGH on dairy cattle. These countries adhere to the precautionary principle with regards to food safety. The precautionary principle holds than any product, action, or process that might pose a threat to public or environmental health should not be introduced in

A sign greeting customers reading "rBST FREE FOREVER" at Perrydell Dairy Farms in York, Pennsylvania. The Pennsylvania Department of Agriculture approved rules in 2007 to prevent dairies from stamping milk containers with hormone-free labels in a precedent-setting decision watched by the industry. Following a public outcry, the state revised regulations to conform to USDA hormone labeling regulations. *© AP Images/Carolyn Caster.*

the absence of scientific consensus regarding its safety. Many of these countries also expressed concern about the concomitant increased use of antibiotics in rBGH-treated cattle due to increased incidence of mastitis, an inflammation of the mammary glands, and foot problems. Public concern over the safety of rBGH milk has decreased its use in the United States. According to the United States Department of Agriculture (USDA), only 17 percent of U.S. cattle were injected with rBGH in 2007.

SEE ALSO   *Agricultural Biotechnology; Bovine Growth Hormone; Cartegena Protocol on Biosafety; GE-Free Food Rights; Genetically Modified Food*

BIBLIOGRAPHY

Books

Ash, Rita. *Cheese Making.* New York: Skyhorse Pub, 2010.

Edgar, Gordon. *Cheesemonger: A Life on the Wedge.* White River Junction, VT: Chelsea Green, 2010.

Marchant, Gary Elvin, Guy A. Cardineau, and Thomas P. Redick. *Thwarting Consumer Choice: The Case against Mandatory Labeling for Genetically Modified Foods.* Washington, DC: AEI Press, 2010.

Weirich, Paul. *Labeling Genetically Modified Food: The Philosophical and Legal Debate.* New York: Oxford University Press, 2007.

Joseph P. Hyder

Designer Babies

■ Introduction

Designer babies refer to children whose genetics have been artificially selected or manipulated at the embryonic stage to exclude or produce certain traits. Designer baby technology or reproductive genetics combines genetic screening and engineering processes with in vitro fertilization (IVF).

The term designer babies was added to the Oxford English Dictionary in 2004. Despite its popular media use, some physicians and parents reject the term designer babies as offensive. They object to likening genetic screening or selection of certain traits—especially those relating to health—to commodities such as designer clothing. Critics feel the implied reference to fashion is fitting because genetic engineering could permit the selection of desired physical and aesthetic traits for non-medical reasons.

■ Historical Background and Scientific Foundations

English physician Walter Heape (1855–1929) established the scientific roots of IVF in the late-nineteenth century by transferring embryos from one rabbit to another. The first successful application of IVF in humans occurred almost a century later on July 25, 1978, when Louise Brown was born in Manchester, England. Brown was dubbed the world's first "test-tube baby," a reference to contemporary IVF's reliance on laboratory fertilization to create viable embryos for implantation. The British physiologist who pioneered IVF in humans, Robert G. Edwards (1925–), later won the 2010 Nobel Prize in Physiology or Medicine.

To create viable embryos, successful IVF requires healthy sperm to fertilize viable ova and a functional uterus into which embryos can be implanted for gestation. The ova and uterus used can belong to different women. At the time of implantation, embryos typically range in age from 2 to 5 days after fertilization though cryogenic freezing allows for long-term embryo storage.

The mapping of the human genome, completed in 2003, enabled researchers to study how each gene and gene sequence functioned and which genes were linked to certain traits or diseases. This opened the possibility for screening and diagnosis of certain conditions at the genetic level at early stages of cell development.

Genetic testing is a common component of many normal pregnancies. Amniocentesis is used to diagnose chromosomal abnormalities. These tests, however, occur during pregnancy. Pre-implantation screening processes are another component of the creation of designer babies. Pre-implantation genetic screening (PGS) identifies embryos at elevated risk of having genetic abnormalities. The process does not search for or identify specific diseases. Pre-implantation genetic diagnosis (PGD) can be coupled with IVF to screen oocytes and embryos for certain genetic markers prior to implantation. PGD can help identify the presence of genetic markers for certain hereditary diseases, allowing for the selection of embryos for implantation that are most likely to be free of such diseases. PGD is most often used to identify cystic fibrosis, Down syndrome, hemophilia A, and Tay-Sachs disease. PGD and PGS coupled with IVF offer parents with a high risk of passing on a fatal or painful genetic disorder to avoid waiting several months into a pregnancy for genetic testing and avoid more invasive medical procedures.

Germline therapy, the genetic manipulation of reproductive (germinal) cells, may allow future genetic scientists to insert almost any trait into sperm and ova used to create IVF embryos. The technology is used in animals but has not been applied to humans.

■ Impacts and Issues

Designer baby technology has been used to ensure that embryos are free from a hereditary genetic disease, to select the sex of a child, and to produce savior siblings

(children that are compatible stem cell, tissue, marrow, or organ donors for a critically ill sibling.)

Common public perception is that genetic screening and modification of human embryos is used to ensure offspring free of disease and composed of more desirable traits. However, there is considerable public debate over what qualifies as a desirable trait or what genetic selections and modifications should or should not be ethically permitted.

Most bioethicists assert that regulations on designer baby technology should not fetter its use to detect and prevent disease, but should prevent the technology from being used as a new means of promoting only a narrow set of socially admired physical or aesthetic traits. Some worry that genetic screening and modification could be used as the tools of new eugenics, the social theory popular during the early decades of the twentieth century that sought to cull from the population traits labeled undesirable. Traits considered undesirable by its largely educated, white, wealthy, and progressive proponents often coincided with negative stereotypes of cultural, economic, physical, and racial differences. Many of the traits that eugenics sought to eliminate were not—as proved by modern genetics—hereditary or even biologically determined. Those who support the technology dismiss concerns that a new form of eugenics will arise, asserting there is no automatic link between genetic manipulation and eugenics, and there is an ethical difference between targeting diseases or disorders and targeting traits such as height, intelligence, or eye color.

In general, supporters of designer baby technology support the widespread use of IVF technologies and approve of the use of genetic screening and selection for health and medical reasons. Some supporters also support selection for non-medical physical or aesthetic traits such as eye color; others see selection for such traits as unethical because they do not affect health.

Debate over designer baby technology in some ways mirrors the philosophical and moral debate over abortion or embryonic stem cell research. Some directly link the concept of a right to control one's own body to rights over their embryos and fetuses. Supporters of designer baby technology support the widespread use of IVF technology, whereas opponents may object to the destruction of unwanted or unused embryos. Such critics generally hold conception to be the determinative event that defines the beginning of human life. As such, some critics assert that genetic manipulation and screening are medical procedures to which embryos cannot give informed consent. Supporters of designer baby technology are more likely to have differing views on when human life begins or when human rights vest in a being, falling anywhere from conception to the moment of birth.

Access to designer baby technology is limited. IVF is expensive. In 2010, a cycle of IVF completed in the United States cost an average of $8,000 to $15,000 per

---

WORDS TO KNOW

EUGENICS: A social theory, popular in the early decades of the twentieth century, that sought to remove so-called socially undesirable hereditary traits from the population through institutionalization, segregation, and limits on childbearing, including sterilization, of people who were members of specific races or had certain medical conditions.

GENETIC ENGINEERING: The process by which the genes are manipulated or combined to produce desired traits that either may not or could not occur naturally.

IN VITRO FERTILIZATION (IVF): An assisted reproduction technology process involving the fertilization of ova by sperm outside of the body in a laboratory setting.

OOCYTE: The female gametocyte; the cell from which an ovum (egg) develops.

---

cycle. Multiple cycles of ova and sperm harvesting and implantation are often required to produce a successful pregnancy. Use of PGD further increases the U.S. average cost of IVF by $3,000 to $7,500.

Critics worry that the prohibitive costs of genetic screening and engineering and IVF make the possibility of designer babies available only to wealthy individuals. Poorer families may not have the financial resources to conceive savior siblings or children free of hereditary diseases; wealthier families may become more invested in creating designer children. Extreme criticisms of designer babies claim that this disparity could lead to genetically divided—not just economically divided—social classes. This dystopic vision of an emergent class of super-humans or genetic desirables is a common feature in science fiction that some fear could become reality.

Sex selection is another controversial aspect of designer baby technology. Whereas some parents may exercise sex selection of embryos because of cultural or personal preferences, others may employ the technique to reduce incidence of genetic diseases. For example, families with a history of hemophilia, a painful disease that primarily affects boys, may choose to have only female embryos implanted during IVF. However, when used to select sex merely for parental preference, potential parents overwhelmingly choose male embryos for implantation. Critics of unfettered sex selection of embryos note that it can promote gender imbalance in the population. As of 2009, 31 countries had banned sex selection—through termination of pregnancies based on sex of the fetus or through embryonic sex selection—for non-medical reasons. Five additional countries had banned sex selection outright, for any reason.

Dr. Jeffrey Steinberg, of the Fertility Institutes of Los Angeles and Las Vegas, is assisted by business manager Julia Vuille, right, who translates during a consultation with a couple from China about the gender selection process at his clinic in Los Angeles. Wealthy foreign couples are getting around laws in their home countries by traveling to the United States for medical procedures that can help them choose the sex of their next child.   © *AP Images/Phil McCarten.*

Disability advocates have spoken out against unregulated designer baby technology. Some fear that designer babies will help marginalize and further stigmatize those with disabilities by enforcing negative perceptions of disability. In 2007 deaf advocates in the United Kingdom opposed a bill restricting genetic selection of human embryos to selection against disease or disability; the advocates asserted that deaf or blind parents should have the right to select for deaf or blind children if hearing or sighted parents could select for hearing and sighted children.

■ Primary Source Connection

The question of whether it is possible versus whether or not it *should* be possible to choose ideal from non-ideal genes to make "the perfect baby" is addressed in this writing. Should embryos be determined and selected for use versus non-use? Scientists are making fast strides in the technology and development of genetic

manipulation; however, the ethics of what should and should not be done with this knowledge is still up for debate. Some proponents advocate for a future that may include the ability to virtually wipe out children being born with diseases and critical medical issues, whereas others counter with the critique stating that to do so is "playing God" and may lead to discriminatory practices based on questionable ethical foundations.

Building Baby From the Genes Up

The two British couples no doubt thought that their appeal for medical help in conceiving a child was entirely reasonable. Over several generations, many female members of their families had died of breast cancer. One or both spouses in each couple had probably inherited the genetic mutations for the disease, and they wanted to use in-vitro fertilization and preimplantation genetic diagnosis (PGD) to select only the healthy embryos for implantation. Their goal was to eradicate breast cancer from their family lines once and for all.

In the United States, this combination of reproductive and genetic medicine—what one scientist has dubbed "reprogenetics"—remains largely unregulated, but Britain has a formal agency, the Human Fertilization and Embryology Authority (HFEA), that must approve all requests for PGD. In July 2007, after considerable deliberation, the HFEA approved the procedure for both families. The concern was not about the use of PGD to avoid genetic disease, since embryo screening for serious disorders is commonplace now on both sides of the Atlantic. What troubled the HFEA was the fact that an embryo carrying the cancer mutation could go on to live for 40 or 50 years before ever developing cancer, and there was a chance it might never develop. Did this warrant selecting and discarding embryos? To its critics, the HFEA, in approving this request, crossed a bright line separating legitimate medical genetics from the quest for "the perfect baby."

Like it or not, that decision is a sign of things to come—and not necessarily a bad sign. Since the completion of the Human Genome Project in 2003, our understanding of the genetic bases of human disease and non-disease traits has been growing almost exponentially. The National Institutes of Health has initiated a quest for the "$1,000 genome," a 10-year program to develop machines that could identify all the genetic letters in anyone's genome at low cost (it took more than $3 billion to sequence the first human genome). With this technology, which some believe may be just four or five years away, we could not only scan an individual's—or embryo's—genome, we could also rapidly compare thousands of people and pinpoint those DNA sequences or combinations that underlie the variations that contribute to our biological differences.

With knowledge comes power. If we understand the genetic causes of obesity, for example, we can intervene by means of embryo selection to produce a child with a reduced genetic likelihood of getting fat. Eventually, without discarding embryos at all, we could use gene-targeting techniques to tweak fetal DNA sequences. No child would have to face a lifetime of dieting or experience the health and cosmetic problems associated with obesity. The same is true for cognitive problems such as dyslexia. Geneticists have already identified some of the mutations that contribute to this disorder. Why should a child struggle with reading difficulties when we could alter the genes responsible for the problem?

Many people are horrified at the thought of such uses of genetics, seeing echoes of the 1997 science-fiction film *Gattaca*, which depicted a world where parents choose their children's traits. Human weakness has been eliminated through genetic engineering, and the few parents who opt for a "natural" conception run the risk of

producing offspring—"invalids" or "degenerates"—who become members of a despised underclass. *Gattaca's* world is clean and efficient, but its eugenic obsessions have all but extinguished human love and compassion.

These fears aren't limited to fiction. Over the past few years, many bioethicists have spoken out against genetic manipulations. The critics tend to voice at least four major concerns. First, they worry about the effect of genetic selection on parenting. Will our ability to choose our children's biological inheritance lead parents to replace unconditional love with a consumerist mentality that seeks perfection?

Second, they ask whether gene manipulations will diminish our freedom by making us creatures of our genes or our parents' whims. In his book "Enough," the techno-critic Bill McKibben asks: If I am a world-class runner, but my parents inserted the "Sweatworks2010 GenePack" in my genome, can I really feel pride in my accomplishments? Worse, if I refuse to use my costly genetic endowments, will I face relentless pressure to live up to my parents' expectations?

Third, many critics fear that reproductive genetics will widen our social divisions as the affluent "buy" more competitive abilities for their offspring. Will we eventually see "speciation," the emergence of two or more human populations so different that they no longer even breed with one another? Will we re-create the horrors of eugenics that led, in Europe, Asia and the United States, to the sterilization of tens of thousands of people declared to be "unfit" and that in Nazi Germany paved the way for the Holocaust?

Finally, some worry about the religious implications of this technology. Does it amount to a forbidden and prideful "playing God"?

To many, the answers to these questions are clear. Not long ago, when I asked a large class at Dartmouth Medical School whether they thought that we should move in the direction of human genetic engineering, more than 80 percent said no. This squares with public opinion polls that show a similar degree of opposition. Nevertheless, "babies by design" are probably in our future—but I think that the critics' concerns may be less troublesome than they first appear.

Will critical scrutiny replace parental love? Not likely. Even today, parents who hope for a healthy child but have one born with disabilities tend to love that child ferociously. The very intensity of parental love is the best protection against its erosion by genetic technologies. Will a child somehow feel less free because parents have helped select his or her traits? The fact is that a child is already remarkably influenced by the genes she inherits. The difference is that we haven't taken control of the process. Yet.

Knowing more about our genes may actually increase our freedom by helping us understand the biological obstacles—and opportunities—we have to work with. Take the case of Tiger Woods. His father, Earl, is said to have handed him a golf club when he was still in the playpen. Earl probably also gave Tiger the genes for some of the traits that help make him a champion golfer. Genes and upbringing worked together to inspire excellence. Does Tiger feel less free because of his inherited abilities? Did he feel pressured by his parents? I doubt it. Of course, his story could have gone the other way, with overbearing parents forcing a child into their mold. But the problem in that case wouldn't be genetics, but bad parenting.

Granted, the social effects of reproductive genetics are worrisome. The risks of producing a "genobility," genetic overlords ruling a vast genetic underclass, are real. But genetics could also become a tool for reducing the class divide. Will we see the day when perhaps all youngsters are genetically vaccinated against dyslexia? And how might this contribute to everyone's social betterment?

As for the question of intruding on God's domain, the answer is less clear than the critics believe. The use of genetic medicine to cure or prevent disease is widely accepted by religious traditions, even those that oppose discarding embryos. Speaking in 1982 at the Pontifical Academy of Sciences, Pope John Paul II observed that modern biological research "can ameliorate the condition of those who are affected by chromosomic diseases," and he lauded this as helping to cure "the smallest and weakest of human beings . . . during their intrauterine life or in the period immediately after birth." For Catholicism and some other traditions, it is one thing to cure disease, but another to create children who are faster runners, longer-lived or smarter.

But why should we think that the human genome is a once-and-for-all-finished, untamperable product? All of the biblically derived faiths permit human beings to improve on nature using technology, from agriculture to aviation. Why not improve our genome? I have no doubt that most people considering these questions for the first time are certain that human genetic improvement is a bad idea, but I'd like to shake up that certainty.

Genomic science is racing toward a future in which foreseeable improvements include reduced susceptibility to a host of diseases, increased life span, better cognitive functioning and maybe even cosmetic enhancements such as whiter, straighter teeth. Yes, genetic orthodontics may be in our future. The challenge is to see that we don't also unleash the demons of discrimination and oppression. Although I acknowledge the risks, I believe

that we can and will incorporate gene technology into the ongoing human adventure.

Ronald M. Green is a professor of ethics at Dartmouth College. His most recent book is "Babies by Design: The Ethics of Genetic Choice."

Ronald M. Green

GREEN, RONALD M. "BUILDING BABY FROM THE GENES UP." *WASHINGTON POST* (APRIL 13, 2008).

SEE ALSO *Bioethics; Frozen Egg Technology; Genetic Discrimination; Genetic Engineering; Genetic Privacy; Genetic Technology; In Vitro Fertilization (IVF)*

BIBLIOGRAPHY

Books

Brinsden, Peter R. *A Textbook of In Vitro Fertilization and Assisted Reproduction*, 3rd ed. Boca Raton, FL: Taylor & Francis, 2005.

Gardner, David K., ed. *In Vitro Fertilization: A Practical Approach*. Boca Raton, FL: Informa Healthcare, 2007.

Gerris, Jan, F. Olivennes, and Petra De Sutter, eds. *Assisted Reproductive Technologies Quality and Safety*. Boca Raton, FL: Parthenon Publishing Group, 2004.

Jersild, Paul T. *The Nature of Our Humanity: Ethical Issues in Genetics and Biotechnology*. Minneapolis, MN: Fortress Press, 2009.

Kindregan, Charles P., and Maureen McBrien. *Assisted Reproductive Technology: A Lawyer's Guide to Emerging Law and Science*, 2nd ed. Chicago: American Bar Association Section of Family Law, 2011.

Schultz, Mark. *The Stuff of Life: A Graphic Guide to Genetics and DNA*. New York: Hill and Wang, 2009.

Web Sites

"Assisted Reproductive Technology (ART)." *Centers for Disease Control and Prevention (CDC)*. http://www.cdc.gov/art/ (accessed October 31, 2011).

"Family History and Genetics." *Centers for Disease Control and Prevention (CDC)*. http://www.cdc.gov/genomics/famhistory/famhist.htm (accessed October 31, 2011).

Adrienne Wilmoth Lerner

Designer Genes

■ Introduction

By definition, designer genes are those created or modified by genetic engineering, usually with a targeted or specific purpose. In humans, designer genes often are used for medical treatment or for specific research purposes, such as preventing or curing disease. Tacit gene design now occurs as a result of advances in human reproductive technology that make it possible to screen externally fertilized (in vitro) embryos for certain genetic diseases and to select for gender. Multiple eggs can harvested and fertilized; the embryos then are tested for the presence of specific genetic illnesses. Those that are free of the defective/undesirable genes and are otherwise healthy can be implanted in the female host. In the United Kingdom, a couple sought genetic counseling and fertility treatment due to high incidence of BRCA1 (highly heritable) breast cancer in the male partner's family. In vitro fertilization (IVF) was used, and two embryos free of the BRCA1 gene were implanted. A healthy female infant was born in late 2008; she will neither develop BRCA1 cancers nor transmit the gene to her progeny. Genetic engineering and IVF can be used to screen embryos for the presence of a number of potentially fatal genetic diseases, such as x-linked hydrocephalus, Fragile X syndrome, familial hypercholesterolemia, Diamond Blackfan Anemia, Down Syndrome, hemophilia, Fanconi's Anemia, and Duchenne's Muscular Dystrophy.

Moralists, ethicists, and religious groups express concern that advancing human reproductive technology, combined with ongoing specific gene mapping, will result in a future composed of humans whose genetic makeup has been carefully selected for specific, desired traits. The essential question is: Who decides which are desirable traits? Design likely will focus on eliminating disease, but could potentially also focus on what are commonly perceived in Western culture as less socially desired characteristics, such as imperfect vision or hearing, short stature, or limited cognitive efficiency.

■ Historical Background and Scientific Foundations

Records of agriculture in early civilizations provide evidence that farmers noted changes in plants as they naturally interbred with other species, producing new forms that came to be known as hybrids. Over time, crop manipulation became more deliberate, as farmers selectively saved and reused seeds from plants showing desired traits.

In the mid-1800s, Austrian monk Gregor Mendel (1822–1884) conducted experiments on pea plants, looking at the inheritance of specific traits in successive generations. Although his research was not given much scientific notice until the early 1900s, Mendel's research came to be considered foundational for the science of genetics. European scientists began to use Mendel's methods to alter, enhance, and create new plant species in the early decades of the twentieth century.

In 1944 it was determined that DNA is responsible for carrying genetic material from one generation to the next. In 1953, American James Watson (1928–) and Englishman Francis Crick (1916–2004), both molecular biologists, published their seminal paper on the double helix structure of DNA. This led to the development of recombinant DNA technology in the early 1970s. Using recombinant DNA technology, a section of DNA could be removed from one cell and transferred into the DNA of another. Recombinant DNA technology paved the way for modern genetic engineering techniques.

The 1980s heralded the era of cloning research. Techniques were developed by Scottish scientists Ian Wilmut (1944–) and his colleagues using adult animal cells to produce clones. In 1986 a sheep's embryo was cloned in the United Kingdom; that same year a cow's embryo was cloned in the United States. Mice were reproducibly cloned by Japanese and American researchers in 1998. Currently, cloning is still considered unreliable, because it typically takes dozens to hundreds of attempts in order to clone an organism successfully, and the cost

WORDS TO KNOW

DESIGNER BABY: A baby whose genetic material has been altered, either by adding or removing specific genes. The egg and sperm are combined through in vitro fertilization and the fertilized embryo undergoes germ line genetic manipulation. The genes are altered to make certain that unwanted characteristics are eliminated and desired ones are present. This primarily is done to select healthy embryos free of genetic diseases. Occasionally, this has been used to ensure that the resultant individual has healthy stem cells compatible with an older sibling with a life-threatening genetic condition that potentially could be cured via an infusion of cord blood/stem cells from a healthy sibling.

GENETIC ENGINEERING: Manipulating or changing DNA/genetic material in order to create a desired outcome, such as the elimination or removal of genes causing a specific syndrome or disease. The ability to manipulate specific genes and DNA has existed only since the 1970s, although selective breeding for desired traits and characteristics has existed since ancient times.

GERM LINE GENE THERAPY: Modification of genetic material at the pre-implantation level. This process involves making genetic changes in fertilized embryos by injecting new genes into an embryonic cellular nucleus. The embryo, with the new genetic material, is implanted into the female host with the hope that a healthy, full-term pregnancy will ensue and result in a live birth of a genetically enhanced/altered individual.

HUMAN GENOME PROJECT: An international research program designed to decode and base pair sequence all human DNA, to identify all genes in the human genome, and to create a relational database system for the storage of the information obtained. The research project began in 1990 in the United States of America and was completed in 2003.

SOMATIC GENE THERAPY: The process of embedding healthy or desired genes in an appropriate carrier, such as a virus, and injecting them into an existing individual with the hope of transmitting the new genetic material to the cells, which then will reproduce the "new" genes and incorporate them into the organism.

is extremely high. Because there have been relatively few successful clones reaching adulthood, there is insufficient scientific evidence regarding overall health and longevity of the individual organisms.

In 1980 the United States Supreme Court ruled that genetically modified organisms could be patented, leading to the patenting of an oil-consuming microorganism by Exxon (an organism intended for use cleaning oil spills). The U.S. Food and Drug Administration (FDA) approved the commercial use of Humulin, an insulin manufactured by Genentech, in 1982. This form of human insulin was the first produced via a bacteria using genetic engineering. In 1987 the United States Patent and Trademark office determined it appropriate to patent non-human, genetically-modified animals, and in 1988 the Harvard-DuPont Oncomouse, a genetically-engineered mouse that was highly predisposed to develop breast cancer and thus valuable to breast cancer research studies, was patented.

The FDA approved the first genetically-engineered vaccine for humans, used for the prevention of hepatitis B, in 1986. In 1996 the Genzyme Transgenics Corporation created a transgenic (genetically engineered) goat designed to carry a human, cancer-combating protein in her milk. Genzyme, in partnership with Tufts University, since has produced transgenic goats that carry substances in their milk shown to be effective in the treatment of some cancers, hemophilia, and kidney disease. By the year 2000, laboratories and research facilities were able to reliably produce other transgenic animals including chickens, pigs, cows, and sheep. Current research with these animals centers around creating new and more effective treatments for human diseases and in developing vaccines.

■ Impacts and Issues

In agriculture, designer genes are used in a variety of ways, from creating drought- and insect-resistant seed strains to the modification of traits and behaviors in animals that lead them to be more readily domesticated. In non-human genetic engineering (molecular genetics and agriculture, for example), the desired traits and characteristics of designer genes would not normally be found in the unaltered species. Genetic engineering in this realm involves a variety of plants such as canola, soybeans, corn/maize, wheat, tomatoes, grapes, and a variety of different types of trees. The animals normally targeted for agrarian genetic engineering are fish, shellfish, sheep, pigs, chickens, and cattle. Although much of the genetic engineering research occurs at universities and research facilities, large corporations also heavily invest in genetic engineering and the production of designer genes.

In humans, currently it is possible to produce an infant with selected genetic traits. The technology is commonly used to screen for the presence of a variety of potentially deadly genetically transmitted diseases through a process called pre-implantation genetic diagnosis (PGD). First, a woman is induced to produce multiple eggs in one ovarian cycle. The eggs (ova) can be harvested and fertilized, then those embryos found free of undesired genes can be implanted in the female host. In addition to producing an infant with selected genetic traits, a healthy infant's cord blood may be harvested at birth with the intent of treating an older sibling born with a deadly genetic disorder. Genetic screening also allows for sex selection of offspring, an important consideration in many cultures.

One form of genetic engineering, somatic gene therapy, impacts only the individual with a defective or undesirable gene. A corrected/healthy gene is introduced into the body of the host through a biological vector, usually in the form of a non-harmful virus. Ideally, the healthy gene becomes incorporated into the cells and then is replicated, creating a cure for the affected individual. One negative aspect of somatic therapy is that it is not always permanent: Most tissue cells eventually die and are replaced. It is possible for the unhealthy genes to be reproduced over time and the syndrome to require periodic treatment with an infusion of new genes. This has been shown to occur with somatic gene therapy of children with severe combined immunodeficiency syndrome. Somatic gene therapy results in change only in the affected individual; the genetic alterations are not heritable.

Germ line therapy, genetic engineering performed at the post-fertilization, pre-implantation stage, is still considered experimental, and as of 2011 is not legal for use on humans. Germ line gene therapy produces permanent changes in the DNA structure of the gene, and they may be continued through inheritance. The promise of this type of genetic engineering is that it can prevent and potentially eradicate fatal genetic disorders. The concern expressed among ethicists is that germ line therapy eventually could be broadened to include selection for specific, highly desirable traits and characteristics at the expense of human variation in the gene pool.

Proponents of genetic engineering and the creation of designer genes assert that this is the most effective and enduring means of eradicating deadly heritable diseases. Many diseases cannot be cured once expressed, but it is also argued that conditions such as some forms of muscular dystrophy likely could be eliminated by careful genetic screening before birth. The argument is that parents who know that they are potential carriers of serious or fatal genetically-transmitted diseases or syndromes should be allowed access to affordable genetic screening and the PGD process, eliminating current cost barriers. With access to genetic screening, over time, many of the most serious or fatal syndromes potentially could be eliminated from the world's population without bias toward socioeconomic status.

The first recorded birth of a baby born with "designer genes'" occurred in 2000, when Adam Nash was born both free of Fanconi's anemia and tissue-matched to supply stem cells for treatment of his older sister who has the disease. Jamie Whitaker was born in 2003 as another "savior sibling," for his older brother who was born with severe anemia. Jamie's genes were selected for freedom from the disease and tissue compatibility for stem cell treatment for his brother.

Critics of genetic engineering argue that humans will, if given the opportunity, genetically engineer so as to select for height, skin tone, IQ range, longevity, athletic prowess, personality type, body composition and myriad other traits and characteristics in addition to freedom from genetic diseases. Ethicists express concern that genetic engineering will create a homogenous species, selected for traits currently desirable, but also eliminating variation that evolution has shown essential to long-term survival.

■ Primary Source Connection

Scientists have been successful at creating and birthing the first genetically altered primate. ANDi, as he is called, is the product of a specific type of gene from jellyfish and viruses, and gene reproduction through rhesus monkey eggs. Scientists look at the future potential to model what they now know about rhesus monkey and human genes, to make advances in the alteration of genes that cause Alzheimer's disease, cancer, hereditary blindness, schizophrenia, and Parkinson's disease in humans. Although the creation of ANDi is considered a scientific breakthrough, questions and concerns arise surrounding bioethics and ethical responsibilities of creating genetically altered animals, and about eugenics with respect to the potential creation of designer babies.

Brave New Monkey

He's a Frisky little fellow, swinging from a ring in his doll-size white T-shirt with the black belt, clambering over and through an elaborate cat-scratching post, sucking his thumb and ducking for cover when playmates Sandy and Sammy ambush him. To all appearances, ANDi (we'll explain soon) is an ordinary rhesus monkey. But appearances deceive. Born by cesarean section last October, ANDi is the first genetically altered primate ever created. If he were human, he'd be called a designer baby. And that makes him the embodiment of the greatest hopes as well as the worst nightmares here at the dawn of the age of genetics: that desirable genes will be inserted into human eggs, producing "genetically enhanced" children. Although that was not the purpose of the research that produced ANDi, the little guy with the soulful eyes is a landmark proof of concept. "At some point in the future," admits Anthony Chan of the Oregon Regional Primate Research Center, who performed the manipulations that created ANDi, "it is conceivable that others may attempt this technique to enhance humans."

The researchers say they had no such goal in creating ANDi. Instead, they hope to create primate models of human diseases, and they had to start simply. They first retrieved a well-studied gene, called the green fluorescent protein gene, from jellyfish. True to its name, the gene makes a protein that, in blue light, glows green. They then put copies of the gene into viruses, since (as anyone with the flu knows) viruses are adept at penetrating cells. Each virus dutifully carried the green gene into 224 rhesus-monkey eggs, where it slipped into the monkey genes like a foreign spy hiding in a crowd. The eggs were then fertilized through microinjection of sperm. After 126 of the fertilized eggs grew and divided beyond the four-cell stage, Chan selected what looked like the 40 best embryos and transferred them into 20 surrogate mother monkeys. Five pregnancies resulted. One set of twins miscarried. One embryo failed to implant. Three monkeys were born. Sandy and Sammy show no sign of the green gene. But ANDi does. Hence his name: short for "inserted DNA, " backward.

The next step is to give rhesus monkeys human genes that play a role in Alzheimer's disease, cancer, hereditary blindness, schizophrenia, Parkinson's or other scourges. A primate version of such a disease, scientists believe, should lead more quickly to vaccines or treatments than the mouse models that currently exist. In addition, genetically altered monkeys carrying a gene "for" Alzheimer's or prostate cancer, say, might show whether the suspect gene always causes the disease, or whether environmental factors like diet or activity can cheat genetic destiny.

The researchers chose the green gene because its presence is so easy to detect, not because they wanted glow-in-the-dark monkeys. In fact, ANDi looks nothing like a hairy green Lava lamp; the gene, although present in every tissue the Oregon scientists tested, seems to be dormant. That's actually a red flag. Since ANDi shows no sign of the trait that was supposedly engineered into him, viruses—which insert the foreign gene randomly—may not be the way to produce genetically altered animals. For genetic enhancement, and perhaps even for making a monkey model of a human disease, you have to get the gene to the spot in the chromosomes where it will be properly controlled.

Until ANDi, genetic engineering had meant slipping a bit of healthy DNA into cells of a patient who is suffering from a genetic disease such as cystic fibrosis. ANDi breaks that mold, bringing us a step closer to tinkering with an individual's genetic endowment before birth, and with the genetic legacy of generations unborn. The Oregon scientists don't know whether ANDi's sperm contains the green gene, and they won't know for four years, when the little guy reaches sexual maturity. But if ANDi's sperm does carry the green gene, he will pass it on to all his offspring. And then the first genetically altered primate will claim another title: father to a genetically altered race.

Although Oregon's Gerald Schatten is emphatic that "we don't support the extrapolation of this work to people for genetic enhancement," some regard that step as an act of medical humanitarianism. Engineering eggs so that a healthy gene replaces a disease-causing one should help both the child-to-be and his or her descendants, lifting a family curse of cancer, atherosclerosis, schizophrenia or another disease with a strong genetic component. "If you could prevent future generations from having grave genetic disease, it would make the life of our species a little less terrible," says bioethicist Arthur Schafer of the University of Manitoba.

Even genetic enhancement for reasons that fall short of life and death has advocates. If parents want a tall, thin, hazel eyed, athletic, brainy kid, whose business is it of anyone else's? Today's well-off parents hire tutors, music teachers, private sports coaches and SAT advisers for their children, if they can afford to. No one calls for banning SAT tutors in the interest of egalitarianism. "But genes are different," argues bioethicist Margaret Somerville of McGill University in Canada. The human genome—the collection of some 80,000 genes carried by every human—is "the patrimony of the entire species, held in trust for us by our ancestors and in trust by us for our descendants. It has taken millions of years to evolve, should we really be changing it in a generation or two?" And Schafer warns of a new social division: in addition to the haves and have-nots, we will have the gene-rich and the gene-poor.

At infertility clinics, couples are not clamoring to choose their would-be child's traits the way they choose options on a car. At least not yet. "Some request a particular gender, but no one has asked for a specific gene," says Dr. Paul Gindoff of the George Washington University in vitro fertilization clinic in Washington, D.C. "The traits that couples might care about—eye color, hair color, height, intelligence and personality—don't come from a single gene anyway, but from a complex combination. We are orders of magnitude away from being able to offer traits to order." The closest couples come to their own private eugenics is embryo selection, says Dr. Martin Keltz of St. Luke's-Roosevelt Hospital Center in Manhattan. In this process, clinics screen the embryos they create to determine whether any carries a deleterious gene that runs in either partner's family. If it does, that embryo is discarded; only healthy-seeming ones go into the woman's uterus. But if today's patients aren't asking for custom-made babies, tomorrow's might. When a March of Dimes poll asked people if they would "improve" their child's appearance or intelligence through genetic tinkering, 42 percent answered yes.

Whether society allows that will likely depend on what might be called the "ick" factor—a sense of repugnance at treating children as a product, at robbing life of its sacredness. Society emitted a collective "ick" at the first test-tube baby, the first surrogate mother, the first baby born to its biological grandmother. But "ick" tends to dissipate; more than 300,000 test-tube babies have been born. Science now has the power not merely to create, but to manipulate creation as never before. In a 1944 essay, novelist C. S. Lewis warned that "the final stage is come when Man by eugenics, by prenatal conditioning . . . has obtained full control over himself." That day is now a little closer.

Sharon Begley

BEGLEY, SHARON. "BRAVE NEW MONKEY." *NEWSWEEK* (JANUARY 22, 2001).

SEE ALSO   *Bioethics; Biotechnology; Cloning; Designer Babies; DNA Sequencing; DNA Vaccines; Gene Banks; Gene Delivery; Gene Therapy; Genetic Bill of Rights; Genetic Discrimination; Genetic Engineering; Genetic Privacy; Genetic Technology; Genetic Use Restriction Technology; Genome; Insulin, Recombinant Human; In Vitro Fertilization (IVF); Molecular Biology; Oncogenomics; Pharmacogenomics; Recombinant DNA Technology*

BIBLIOGRAPHY

Books

Gessen, Masha. *Blood Matters: From Inherited Illness to Designer Babies, How the World and I Found Ourselves in the Future of the Gene.* Orlando, FL: Harcourt, 2008.

Harris, John. *Enhancing Evolution: The Ethical Case for Making Better People.* Princeton, NJ: Princeton University Press, 2007.

Henderson, Mark, Joanne Baker, and Anthony J. Crilly. *100 Most Important Science Ideas: Key Concepts in Genetics, Physics and Mathematics.* Buffalo, NY: Firefly Books, 2009.

McCabe, Linda L., and Edward R. B. McCabe. *DNA: Promise and Peril.* Berkeley: University of California Press, 2008.

Nuüsslein-Volhard, Christiane. *Coming to Life: How Genes Drive Development.* San Diego, CA: Kales Press, 2006.

Shanks, Pete. *Human Genetic Engineering: A Guide for Activists, Skeptics, and the Very Perplexed.* New York: Nation Books, 2005.

Wilmut, Ian, and Roger Highfield. *After Dolly: The Uses and Misuses of Human Cloning.* New York: W.W. Norton & Co, 2006.

Web Sites

Darnovsky, Marcy. "One Step Closer to Designer Babies." *Science Progress,* April 22, 2011. http://scienceprogress.org/2011/04/one-step-closer-to-designer-babies/ (accessed October 4, 2011).

Palca, Joe. "Screening Embryos for Disease." *National Public Radio (NPR),* December 20, 2006. http://www.npr.org/templates/story/story.php?storyId=6653837 (accessed October 4, 2011).

"Pre-implantation Genetic Testing." *Human Fertilisation & Embryology Authority.* http://www.hfea.gov.uk/PGD.html (accessed October 4, 2011).

Pamela V. Michaels

Diamond v. Chakrabarty

■ Introduction

Diamond v. Chakrabarty was a seminal biotechnology case decided by the Supreme Court of the United States in 1980. The Court examined the rejection of a patent claim on a genetically modified microorganism, a bacterium that could degrade crude oil. The Court addressed whether a living organism could be patented under U.S. patent laws. The Court ultimately ruled that the genetically modified microorganism could be patented, because it had been designed and manufactured for a specific use. The Court's decision gave rise to the rapid expansion of the biotechnology industry in the United States, which has resulted in the issuance of thousands of patents for genetically engineered organisms and related products.

■ Historical Background and Scientific Foundations

In 1972 Ananda Mohan Chakrabarty (1938–), a microbiologist at General Electric Company's Research and Development Center, genetically engineered a new species of *Pseudomonas* bacteria. The *Pseudomonas* bacteria created by Chakrabarty contained at least two stable, energy-generating plasmids, each of which was capable of degrading various components of crude oil. General Electric (GE) believed that Chakrabarty's genetically engineered bacteria held great promise for the treatment of oil spills. At the time, no naturally-occurring bacteria was known to have oil-degrading capabilities.

Chakrabarty applied for a patent on the genetically modified *Pseudomonas* with the U.S. Patent and Trademark Office (PTO). Chakrabarty's patent application made three categories of patent claims. First, Chakrabarty's application asserted process claims on the method for producing the genetically engineered, oil-degrading *Pseudomonas* bacteria. Second, the application made

claims for a material for culturing the bacteria that could float on water. Finally, Chakrabarty asserted claims to the modified bacteria themselves.

A U.S. PTO patent examiner granted a patent that included the claims related to the method for manufacturing the bacteria and the claims related to the culture, but the patent examiner rejected the claims related to patenting the bacteria themselves. The patent examiner stated that the bacteria were products of nature and as such were not patentable under U.S. patent law. The Patent Office Board of Appeals upheld the decision of the patent examiner in rejecting the claims on the bacteria themselves.

GE and Chakrabarty appealed this decision to the Court of Customs and Patent Appeals. After several complex legal maneuvers in which the Court of Customs and Patent Appeals issued, vacated, and then reissued a decision in the Chakrabarty case, the court denied the patent claims on the *Pseudomonas* bacteria themselves. Sidney Diamond (1915–1983), the U.S. Commissioner of Patents and Trademarks, then requested that the Supreme Court consider the Chakrabarty case and a companion case, *In re Bergy*, which was dismissed before the Supreme Court heard *Diamond v. Chakrabarty*.

■ Impacts and Issues

In *Diamond v. Chakrabarty*, the Supreme Court addressed whether a living organism could be patented under U.S. patent laws. Section 101 of the U.S. Patent Act (35 U.S.C. 101) states "Whoever invents or discovers any new and useful process, machine, manufacture, or composition of matter, or any new and useful improvement thereof, may obtain a patent therefor." The Court, therefore, had to consider whether Chakrabarty's genetically engineered bacteria constituted a "manufacture" or "composition of matter" for the purposes of U.S. patent law.

The Supreme Court has long held that natural phenomena, natural laws, and abstract ideas are not patentable under U.S. patent law. Likewise, a newly discovered, but naturally occurring, plant or animal is not patentable. The Court stated in *Funk Brothers Seed Co. v. Kalo Inoculant Co.* that naturally occurring substances, phenomena, and processes are "free to all men and reserved exclusively to none." The Court then examined Congressional records from 1952 when the U.S. Patent Act was recodified in an effort to understand the legislative intent behind the Patent Act. The Court noted that Congressional Committee Reports stated that the Patent Act was meant to "include anything under the sun that is made by man."

After analyzing Chakrabarty's genetically engineered bacteria, the Court, in a 5–4 decision, decided that the *Pseudomonas* bacteria were patentable. The Court stated that the oil-degrading *Pseudomonas* bacteria were not an undiscovered, naturally occurring bacteria for which Chakrabarty had merely found a use. Instead, Chakrabarty had invented a process for creating the human-made *Pseudomonas* bacteria

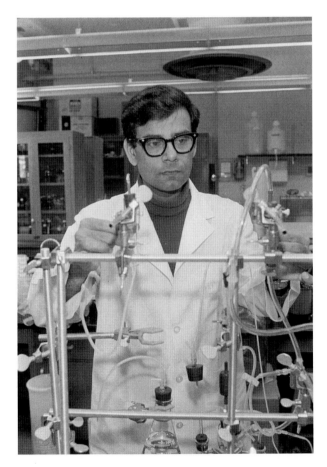

Ananda Chakrabarty (1938–) in his laboratory at the University of Illinois, Chicago, in 1980.   © *AP Photo/CEK*

and then manufactured the bacteria with a specific use in mind. The Court stated that Chakrabarty had manufactured "a new bacterium with markedly different characteristics from any found in nature and one having the potential for significant utility," and Chakrabarty, therefore, was entitled to a patent on his invention.

The dissenting justices asserted that Congress did not intend for living organisms to be patentable subject matter under Section 101 of the Patent Act. The dissent pointed to the fact that Congress had passed the 1930 Plant Protection Act and the 1970 Plant Variety Protection Act to allow for the patentability of certain plant varieties under those laws. The dissenting justices argued that had Congress contemplated patenting living organisms under Section 101 of the Patent Act, then Congress would not have needed to pass new laws to allow for the patentability of some living organisms under the plant patent laws. The dissent also noted that Congress had specifically excluded bacteria—the subject matter of Chakrabarty's patent claim—from patentability under the 1970 Plant Variety Protection Act.

The Supreme Court's decision in *Diamond v. Chakrabarty* settled the issue of patentability of modified single-celled organisms in the United States. The U.S. PTO had previously issued patents on modified organisms, including an 1873 patent issued to Louis Pasteur on a single-celled strain of yeast. The PTO's policy on the matter was unclear and unevenly applied before *Diamond v. Chakrabarty*.

The Supreme Court's decision in *Diamond v. Chakrabarty*, however, did not settle the issue of the patentability of multicellular organisms. In 1988 the PTO issued a statement claiming that the PTO considered all "non-naturally occurring non-human multicellular living organisms, including animals" to be patentable subject matter. Two weeks later the PTO issued its first patent on a multi-celled organism, known as "Harvard Mouse."

The Supreme Court's decision in *Diamond v. Chakrabarty* paved the way for the rapid expansion of the biotechnology sector in the United States after 1980. As of 2009, approximately 2,500 dedicated biotechnology companies were based in the United States and generated more than $60 billion in annual revenue, a number that is expected to continue to grow. Since the *Diamond v. Chakrabarty* decision and subsequent PTO rulings, the PTO has issued patents covering a wide variety of genetically-modified organisms, including herbicide- and drought-resistant plants, high-yield crops, and mice designed for use in specific medical tests.

SEE ALSO *Bioremediation; Bioremediation: Oil Spills; Cabilly Patents; Genetic Engineering; Patents and Other Intellectual Property Rights*

BIBLIOGRAPHY

Books

Bohrer, Robert A. *A Guide to Biotechnology Law and Business.* Durham, NC: Carolina Academic Press, 2007.

Chapman, Katherine L. *Understanding Biotechnology Law.* London: Taylor & Francis, 2007.

Ida Madieha bt Abdul Ghani Azmi. *Biotechnology Law and Policy: Searching for a Balance.* Gurgaon, India: Madhav Books, 2011.

Web Sites

"Genetics and Patenting." *U.S. Department of Energy Genome Programs.* http://www.ornl.gov/sci/techresources/Human_Genome/elsi/patents.shtml (accessed September 30, 2011).

Joseph P. Hyder

Digital-Biological Computer

Introduction

Digital-biological computing, also called DNA computing, evolved from research interfacing the disciplines of biotechnology, molecular biology, biochemistry, and DNA technology. Research scientists have been able to create DNA-based nano-sized processor systems capable of complex mathematical operations and storage of vast amounts of data. Digital-biological computers, able to detect and respond to cancerous changes on a molecular level, have been created in the laboratory. There are significant potential implications for future health monitoring and maintenance based on these emerging technologies.

The developing multidisciplinary field of synthetic biology combines the principles of engineering and biotechnology. Research scientists have successfully combined its principles with those of genetic engineering in order to modify biological organisms, creating electronic-like systems that could not occur in nature. Pioneering work has been done with yeast cells modified to communicate as if they are electronic circuits. There are great future implications for this technology in the fields of medicine and biodetection.

Historical Background and Scientific Foundations

The history of digital-biological computing using DNA began at the University of Southern California in 1994, when Leonard Adelman (1945–) created a nano-sized DNA-based computer capable of successfully solving a simplified version of a mathematical inquiry called "the traveling salesman problem." This calculation, rapidly accomplished by standard microprocessors, was solved by trial-and-error rather than the use of Boolean logic. The scientific significance of this experiment was the discovery that DNA could be used to perform calculations normally completed through the use of electronics.

In 1997 research scientists at the University of Rochester in New York constructed elementary Boolean logic gates made of DNA. This was scientifically important, because it paved the way for development of progressively more complex DNA-based processing systems.

Using the principles of nanotechnology, Ehud Shapiro (1955–) and his colleagues at the Weizmann Institute in Israel constructed their first protein and DNA-powered microprocessor in 2002. By 2003 they had produced a biological computer engineered from a strand of DNA, capable of processing speeds of 330 trillion operations per second. The next iteration of their DNA-powered biological computer in a test tube was created in 2004; it could identify molecular indicators of lung and prostate cancers and react to the presence of cancer cells by causing cell death. This biological computer was designed for medical applications and does not perform mathematical operations. The ultimate aim of this research program is advanced medical treatment. The ultimate hope is to develop technology to embed the biological computer systems in pharmaceuticals, which would then be metabolized, distributing the nano-sized biological computers via the circulatory system to the body's cells, where they would monitor cellular activity continuously for signs of disease, calculate the best means for eradicating it, and initiate targeted treatment.

In 2006 a group of research scientists at Jerusalem's Hebrew University constructed an enzyme-driven molecular computer capable of performing moderately complex calculations. Their goal, using nanotechnology, is to develop a minute computer capable of dwelling inside a human being. The function of the computer would be monitoring bodily systems for molecular indications of malfunction and reacting to eradicate disease by either causing cellular death or manufacturing and transmitting individual and cell-specific pharmacologic agents.

WORDS TO KNOW

BOOLEAN LOGIC: Also called Boolean algebra; a form of logical calculus derived by George Boole (1779–1848) in the mid-1800s. Boolean logic has two values, used as zero and one in the world of computers.

NANOTECHNOLOGY: Nanotechnology involves the engineering of molecules and atoms at sizes of less than 100 nanometers. One nanometer is equal to one billionth of a meter in length.

RIBONUCLEIC ACID: RNA is similar to DNA in that it is composed of long chains of nucleotides and encodes genetic information. In its messenger form (mRNA) it transmits genetic information utilized in protein synthesis.

SYNTHETIC BIOLOGY: An emerging science combining engineering and biology. The consortium of synthetic biologists at Harvard and Massachusetts Institute of Technology (MIT) defines the field as "the design and construction of new biological parts, devices, and systems and the re-design of existing, natural biological systems for useful purposes."

■ Impacts and Issues

Silicon-based microprocessor/computer chips have become steadily smaller during the past several decades; early computers were housed in large, sterile, cold rooms. Based on the tenets of Intel founder Gordon Moore's (1929–) mathematical derivation, called Moore's Law, microprocessors roughly double in capacity and processing speed and are halved in size about every 18 months; there is an absolute limit beyond which this cannot occur. DNA-based computer systems do not have similar constraints. DNA is organic and exists in all living organisms; it is both cost-effective and abundant. Traditional microprocessor-based computers require large facilities for manufacture. They engender toxic byproducts both during creation and after disposal, making them dangerous for the environment. This is not the case for DNA-based computers: they are generally far less expensive to create and produce no toxic waste. They are also extremely compact, created by nanotechnology. They can store vast amounts of data and can be made to solve highly complex mathematical equations by simultaneously considering a broad range of possible solutions, much like the parallel logic of traditional microprocessors.

Research scientists at the Lawrence Livermore National Laboratory in California combined biology with electronic technology in 2009 to create a device utilizing a lipid coating over nanowires to create an integrated bionanoelectronic device that could lead to the development of more precise biodiagnostic and biosensing devices. It is their expectation that this technology will lead to significant improvements in the effectiveness of neural prosthetics and biological computers.

In 2010, scientists at the University of Gothenburg in Sweden genetically modified yeast cells to function and to communicate with one another as synthetic electronic circuits that could be combined with genetically non-identical genetically modified yeast cells to form progressively more complex circuitry. They plan to advance this technology for the development of environmental biosensors able to detect toxins or pollutants. Anticipated medical uses would be on the level of intercellular monitoring and response to alterations in health status.

Researchers at the Imperial College in London created highly advanced modular biological logic gates composed of the bacteria *Escherichia coli* and DNA during 2011. Their eventual goal is to utilize the biological logic gates as the foundation for nanotechnology-based DNA computers designed to sense human bodily dysfunctions such as arterial plaque build-up or early cancer cell development. The computers would be designed not only to sense systemic disturbances with extreme precision but to evaluate rapidly all possible correction scenarios, select the best option, and initiate the most efficient and effective course of treatment.

SEE ALSO *Biocomputers; Biodetectors; Bioeconomy; Bioinformatics; Biomimetic Systems; Biorobotics; Nanotechnology, Molecular; Synthetic Biology*

BIBLIOGRAPHY

Books

Calude, Cristian, and Gheorghe Paun. *Computing with Cells and Atoms: An Introduction to Quantum, DNA, and Membrane Computing.* New York: Taylor & Francis, 2001.

Ignatova, Zoya, Israel Marck Martinez-Perez, and Karl-Heinz Zimmermann. *DNA Computing Models.* New York: Springer, 2008.

Paun, Gheorghe, Grzegorz Rozenberg, and Arto Salomaa. *DNA Computing: New Computing Paradigms.* New York: Springer, 2005.

Shasha, Dennis Elliott, and Cathy A. Lazere. *Natural Computing: DNA, Quantum Bits, and the Future of Smart Machines.* New York: W.W. Norton, 2010.

Web Sites

Bonsor, Kevin. "DNA Computers." *Engineering. com.* http://www.engineering.com/Library/ArticlesPage/tabid/85/articleType/ArticleView/articleId/86/DNA-Computers.aspx (accessed November 6, 2011).

Landweber, Laura. "DNA Computing: The Origin of Biological Information Processing." *Princeton University.* http://www.princeton.edu/~lfl/FRS.html (accessed November 6, 2011).

Pamela V. Michaels

Disease-Resistant Crops

■ Introduction

Farmers have long used a variety of breeding techniques to magnify desired traits, such as increased disease resistance, in crops. Representative members of pathogen-resistant strains have been bred selectively to increase their ability to withstand infection. Some plants are inherently immune, making them better able to resist infection; others acquire immunity by surviving specific pathogens. Some plants experience hypersensitivity reactions wherein they automatically shut down cells in regions attacked by pathogens, causing the infected areas to die off and preventing spread to the entire plant.

When there is a lack of disease resistance in crop plants, particularly to fungi, bacteria, and viruses, farmers frequently choose to employ chemical sprays to eradicate the pathogens. This raises public concerns of environmental impact as well as crop plant toxicity and is not considered a desirable way of protecting crop yields and marketability. Pathogens often mutate and transform, becoming immune to previously effective chemicals.

Marker-assisted selection is viewed as a means of combining modern biotechnology with traditional genetics, stopping short of actual genetic engineering, in order to strengthen desired gene pools. The drawback of this technique is its lack of specificity and absolute accuracy, because it involves selective breeding that combines the entire organism's DNA with that of another parent plant of the same species. It does not guarantee that the desired traits will be immediately evidenced.

Genetic engineering results in the creation of transgenic crops containing specific DNA sequences encoded for disease resistance. They may be drawn from the same plant variety or from another type of organism, possibly an animal-to-plant gene transfer. Public concerns have been expressed about environmental, economic, and human impact resulting from widespread use of this technology.

■ Historical Background and Scientific Foundations

Disease in crop plants accounts for 10 to 20 percent loss of yield per year in developed countries and 20 to 40 percent annual yield loss in developing countries. Biotic stress, caused by the presence of plant pathogens, weeds, and insect pests, accounts for roughly 40 percent of overall global crop losses. Many common pathogens have been identified, as have those strains or varieties of plant with either innate or secondary resistance to them. Innate resistance is already present in the plant; secondary resistance is acquired through infection with a particular disease and recovery from the infection.

Selective breeding has been used as a means of propagating disease-resistant crops for centuries. Generations of hardy, high-yield plants have been systematically cross-bred with equally disease-resistant members of the same species in order to strengthen breeding stock progressively. Breeding stocks have been fortified by using both cultivated and wild strains of the same plants. Through the use of biotechnology and gene mapping, it is possible to isolate genes associated with immune functions and reinforce them by selectively breeding through either traditional or genetically-engineered means.

In 2010 European Union (EU) researchers published a study in which they identified a plant-immune mediator called a pattern recognition receptor, or PRR. A PRR can recognize a pathogen and defend the plant against it. It is believed that reinforcing the activity of the PRR will markedly improve overall resistance to a variety of pathogens—a belief supported by study data. The EU researchers plan to use genetic engineering to transfer PRRs into a variety of important crop plants to determine whether they improve disease resistance across different types of crop plants.

Cisgenesis is another method used to increase disease resistance in crop plants. Genes for a specific, desired trait are isolated and then transferred to other members

WORDS TO KNOW

BIOTIC STRESS: When plants are negatively affected by the presence of pathogens such as fungi, bacteria, viruses, insects, or weeds, they are said to be impacted by biotic stress.

CISGENICS: A molecular biology technique in which genes associated with a particular trait, such as disease resistance, are identified in an organism. Those genes are reinforced and transferred into other members of the same species with the expectation that the trait will be evinced in the parent plants and magnified in successive generations.

GENE MAPPING: Creation of a genetic map by identifying and locating the genes in an organism's chromosomes or plasmid.

MARKER-ASSISTED SELECTION (MAS): A technique that combines molecular biology/biotechnology with classic genetics to select genes controlling specific desired traits and to systematically strengthen them across successive generations through various selective breeding processes.

PATHOGEN: A disease-causing organism.

TRANSGENIC: When genetic engineering, typically in the form of recombinant DNA techniques, is used to take a gene (or genes) from one type of organism and insert it into another, the resulting organism is said to be transgenic, containing the genes of more than one type of organism and possibly more than one species.

of the same species, magnifying the trait. This process has been implemented successfully with grape varieties in Florida. Dennis Gray, a researcher at the University of Florida, identified genes in Chardonnay grapes associated with resistance to fungal diseases; he then inserted them into Thompson seedless grapes and was able to demonstrate statistically significant increases in resistance to several different fungal and bacterial diseases, when compared to a control group. His goal is to improve disease resistance across many varieties of wine grapes and to impact the environment positively by significantly reducing need to spray fungicides on growing plants.

A multinational research group has been studying various mechanisms of disease resistance in crop plants since the first decade of the twenty-first century. One concern involves the ability of many pathogens to mutate or transform and overcome plant disease resistance or the chemical sprays designed to eradicate them (such as fungicides). Many plant pathogens act by injecting specific proteins designed to diminish the plant's ability to resist diseases. By studying those proteins, called effectors, researchers are able to determine which plant proteins they act on and then identify the genes associated with them. By identifying gene groups that impact

plants' immune systems, they can potentially engineer them to increase resistance.

Some plants exhibit a form of disease resistance called hypersensitivity reactions: They are genetically programmed to have cells in the area die off when attacked by specific pathogens, ending the disease process. Genetic engineering can transfer genes associated with hypersensitivity into non-resistant plants.

■ Impacts and Issues

Pathogens that attack and destroy crops, diminishing crop yields, also have an indirect impact on the environment. Farmers often use chemical sprays or heavy metal applications in order to eradicate crop diseases. Chemical spraying is most effective for fungi, less so for bacteria, and not at all effective for viruses. Many types of pathogens have evolved to develop resistance to agrochemicals, leading to use of progressively more toxic chemicals in an effort to maximize crop yields. In some cases, pathogen resistance has rendered spraying impossible due to the high levels of damage done to non-target areas by the toxic levels of agrochemicals necessary for effective eradication. By increasing disease resistance in crop plants, farmers will not need to spray or treat crops as much or as frequently. Crops affected by disease may have altered taste as well as changed appearance, making them unmarketable as food crops even if they are edible.

Marker-assisted selection (MAS) is a technique combining contemporary biotechnology with more traditional breeding techniques. Breeding plants that are used for study are selected based on desirable performance factors such as hardiness, yield, and disease-resistance. Plant DNA is mapped in order to identify major genes controlling traits of interest, such as resistance to specific pathogens. The gene is identified and associated with a molecular marker located near or attached to the gene of interest. The marker serves as a means of identifying the presence of the desired gene. When the plants are bred with others carrying marked genes for the same traits, it is possible to scan their offspring as seeds or seedlings and support the development of those carrying the desired genes long before the mature plant shows evidence of the trait. There has been significant research on the mapping of molecular markers as well as on uncovering the relationships between mapped markers and phenotypes. Phenotypes are the physical expression of genetic traits, such as flower color, leaf type, or resistance to disease. Molecular marker maps have been created for many different types of crops; the entire rice genome has been mapped. Because MAS is utilized during the course of selective breeding involving the entire organism, it remains approximate in its accuracy.

Genetic engineering involving recombinant DNA is often applied to tissue cultures to create transgenic plant

A wheat scientist examines a plant infected with Ug99 stem rust fungus at the Kenya Agricultural Research Institute in Njoro, Kenya. The disease has spread from Uganda and Kenya to the Mideast and endangers wheat crops worldwide. International researchers send wheat varieties to Njoro to test whether they resist the fungus. Along with climate change and stagnating yields, the rust is a threat to global wheat supplies. *© AP Images/Khalil Senosi.*

species with significantly improved resistance to numerous pathogens. An advantage of recombinant DNA technology, when compared to the use of marker-assisted selection, is the specificity of the gene changes. With genetic engineering, only the gene (or genes) specific to pathogen resistance is inserted into the target crop plant, leaving the remainder of the plant's original genome intact. In this way, a cultivated plant strain retains all of the desired genetics, without dilution caused by breeding with another gene pool in an effort to increase the likelihood of trait acquisition. Another advantage of genetic engineering for increased disease resistance in crop plants is that target genes for insertion need not come from the same variety or crop plant; they may be drawn from other species with desired resistance traits.

As of 2011, the most effective genetically engineered disease resistance in crop plants has involved transgenic virus resistance; less success has been obtained in increasing resistance to fungi and bacterial pathogens. Countries employing biotechnology for the production of transgenic disease-resistant crops are engaged in ongoing research on efficacy, cost-benefit relationships,

environmental impact, potential long-term human health effects, and public acceptance of genetically engineered crops.

■ Primary Source Connection

A research team of scientists in the United Kingdom are helping to minimize pesticide usage on crops by finding ways to help crops defend themselves against disease and invasive insects. In doing so, they are helping to minimize crop loss while at the same time working to increase and secure the production of food for a growing human population.

EU-Funded Study Yields Disease-Resistant Crops

Food security, the ability to produce the amount of food needed to sustain the ever-growing human population, is one of the priorities that drives the EU's research agenda. Through research, we are able to develop new ways to increase the amount of food we grow, while at the same time minimizing the impact food production can have on the environment. A breakthrough by a UK-led research team is a case in point: these scientists have found a way to help crops defend themselves against disease, ultimately reducing yield loss and minimizing pesticide usage. The research results are published in the journal *Nature Biotechnology.*

The study was supported by the ERA-NET ('European Research Area Network') on Plant Genomics (ERA-PG), funded under the EU's Sixth Framework Programme (FP6) with over EUR 2 million. ERA-PG's research programme consisted of a total of 41 transnational research projects.

In their paper, the authors of the current study highlight the immense crop losses incurred as a result of plant diseases. 'Microbial diseases and pests place major constraints on food production and agriculture,' they write. 'Agrochemical applications are the most common means of controlling these, but more sustainable methods are required. One way to improve plant disease resistance is to enhance the capability of the plants' own innate immune system.'

The majority of plants have in-built mechanisms to fight microbial pathogens, but the capacity to combat a particular pathogen varies from one species to the next.

Scientists from The Sainsbury Laboratory in the UK collaborated with an international team to focus on an immune receptor, known as the pattern recognition receptor (PRR), which is present in some plants (e.g. a wild species belonging to the mustard family).

PRRs can identify molecules that are key to keeping a pathogen alive. Since these essential molecules exist in many different microbes, if a plant is able to identify and defend itself against a given molecular pattern, it is likely to be able to fight off a host of other pathogens as well. The problem is that only a handful of PRRs have been identified in plants to date.

For their study, the team took a Brassica-specific PRR and transferred it into two plants, *Nicotiana benthamiana* and *Solanum lycopersicum* (tomato), to determine if adding new recognition receptors to the host arsenal would lead to better resistance.

The resistance of the transformed plants was then tested against many types of plant pathogens. Results showed a significantly enhanced resistance against many different bacteria, including some of the most hazardous. The team showed that PRRs can be successfully transferred

from one plant family to another, thus generating a new biotechnological solution to disease resistance.

'The strength of this resistance is because it has come from a different plant family, which the pathogen has not had any chance to adapt to,' explained Dr Cyril Zipfel of The Sainsbury Laboratory. 'We can now transfer this resistance across plant species boundaries in a way traditional breeding cannot.'

The team is now applying and testing their results to other crops that are highly susceptible to bacterial diseases such as potato, apple, cassava and banana.

"EU-FUNDED STUDY YIELDS DISEASE-RESISTANT CROPS."
CORDIS NEWS (MARCH 15, 2010): 1–2.

SEE ALSO *Agricultural Biotechnology; Agrobacterium; Bt Insect Resistant Crops; Corn, Genetically Engineered; Drought-Resistant Crops; Frost-Resistant Crops; Gene Banks; Plant Variety Protection Act of 1970; Salinity-Tolerant Plants; Transgenic Plants; Wheat, Genetically Engineered*

BIBLIOGRAPHY

Books

Bundgaard, Kristian, and Luke Isaksen, eds. *Agriculture Research and Technology.* New York: Nova Science Publishers, 2010.

Ciancio, A., and Krishna Gopal Mukerji, eds. *General Concepts in Integrated Pest and Disease Management.* Dordrecht, The Netherlands: Springer Verlag, 2007.

Copping, Leonard G., ed. *The GM Crop Manual: A World Compendium.* Alton, IL: BCPC, 2010.

Kidd, J. S., and Renee A. Kidd. *Agricultural versus Environmental Science: A Green Revolution.* New York: Chelsea House, 2006.

Lacey, Hugh. *Values and Objectivity in Science: The Current Controversy about Transgenic Crops.* Lanham, MD: Lexington Books, 2005.

Smith, Jeffrey M. *Genetic Roulette: The Documented Health Risks of Genetically Engineered Foods.* Fairfield, IA: Yes! Books, 2007.

Wesseler, Justus H. H., ed. *Environmental Costs and Benefits of Transgenic Crops.* Norwell, MA: Springer, 2005.

Pamela V. Michaels

DNA Databases

■ Introduction

DNA databases are government databases that contain genetic information on individuals that may be used for identification. Law enforcement personnel use DNA databases to identify criminal suspects by matching deoxyribonucleic acid (DNA) left at a crime scene with persons in a DNA database or to identify bodies or reunite family members during a natural disaster. Most national DNA databases originally contained only genetic information for certain classes of sex offenders, but most have expanded to include a wide variety of criminal offenses and other government purposes.

Typically, governments obtain DNA samples for inclusion in DNA databases from convicted criminals. Some jurisdictions, however, also retain DNA samples from suspects that are never convicted or, in some cases, not charged with a criminal offense. Civil libertarians have raised privacy right concerns about the use of DNA databases, particularly those that include DNA samples from innocent persons.

■ Historical Background and Scientific Foundations

The United Kingdom established the world's first national DNA database, the UK National Criminal Intelligence DNA Database, also known as the National DNA Database (NDNAD), in April 1995. New Zealand followed with its New Zealand National DNA Databank in 1996. In 1998 France unveiled its Fichier National Automatisé des Empreintes Génétiques (FNAEG), or Automated National File of Genetic Prints, and the United States launched the Combined DNA Index System (CODIS), a system of DNA databases. The national level of CODIS is known as the National DNA Index System (NDIS), which combines all of the CODIS databases from the national, state, and local levels. CODIS and NDIS are managed by the Federal Bureau of Investigation (FBI). Since 1998, most European nations, Australia, and Canada also have implemented DNA databases.

Most early DNA databases originated as databases of certain classes of sexual offenders. Since the year 2000, however, DNA databases have grown tremendously as additional crimes have been reclassified as DNA recordable offenses. As of September 2011, CODIS contained nearly 10.2 million offender profiles. As of April 2011, NDNAD contained DNA profiles of more than 5.7 million individuals, or approximately 9 percent of the UK population.

DNA databases record unique genetic markers, known as short tandem repeats, of individuals to create a genetic fingerprint for that individual. DNA databases do not store an individual's full genome. Short tandem repeats are short sequences of DNA, usually between two to five nucleobase pairs, that are repeated in sequence.

DNA databases may store information on different short tandem repeat regions, or loci. However, each database must utilize the same loci in each test in order to match a suspect's DNA with the sample on file in the database. CODIS, for example, uses 13 short tandem repeat loci, which differ from the loci used by the United Kingdom, European Union, and other countries. Most countries, however, include the seven loci set forth in the Interpol Standard Set of Loci, which allows countries to compare DNA samples taken in different countries.

■ Impacts and Issues

DNA databases have become important tools for law enforcement investigations. According to FBI statistics, as of September 2011, CODIS has produced more than 161,000 hits that have assisted in more than 155,000 law enforcement investigations. In the United Kingdom, more than 300,000 crimes have been detected using NDNAD. Despite the usefulness of DNA databases in assisting law enforcement investigations in the

WORDS TO KNOW

DEOXYRIBONUCLEIC ACID (DNA): A self-replicating, double-helix structure of nucleobases that is the main component of chromosomes and serves as the basis of heredity.

DNA DATABASE: A computerized database that stores DNA profiles of individuals, typically criminals.

NUCLEOBASE: A group of nitrogen-based molecules—adenine, cytosine, guanine, thymine, and uracil—that form nucleotides, the building blocks of DNA and RNA. The first four are found in DNA; in RNA, the nucleobase thymine is replaced by uracil.

SHORT TANDEM REPEATS: A repeated sequence of nucleobases. The use of multiple short tandem repeats contained at certain locations on chromosomes may be used to create an identifying genetic fingerprint for an individual.

United States and other countries, civil libertarians, policymakers, and others have raised numerous privacy and human rights issues related to the implementation and maintenance of DNA databases.

Civil libertarians consistently have criticized DNA databases as an invasion of privacy, especially in light of the continued expansion of groups covered by DNA databases. Originally, most DNA databases collected information only on certain subsets of the population but they have grown to include other groups over the years. When the FBI launched CODIS, the database was intended primarily to contain the DNA of convicted sex offenders. In the early 2010s, DNA samples submitted by U.S. states to CODIS include samples taken from all convicted felons, juveniles, and in some states, arrestees never convicted of a crime. FNAEG was also a database for the identification of child sex offenders but has expanded over the years to include numerous criminal offenses. The United Kingdom's NDNAD has also expanded to include juveniles and those arrested for petty crimes.

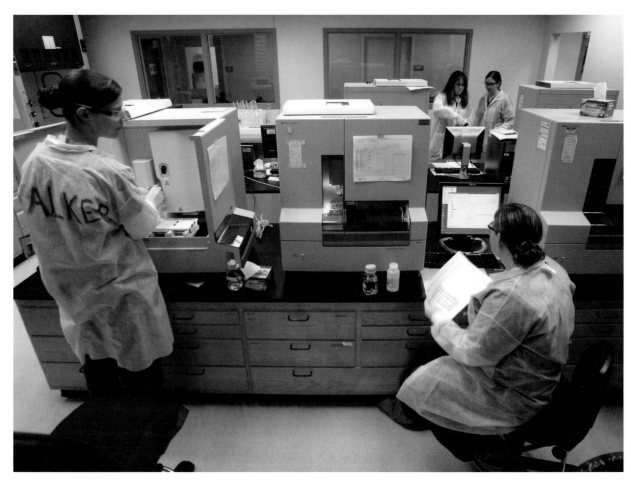

Forensic scientists process DNA samples at the New York State Police lab in Albany, New York. New York state law requires the collection of DNA samples from certain convicted criminals for inclusion in a DNA database. © *AP Images/Mike Groll.*

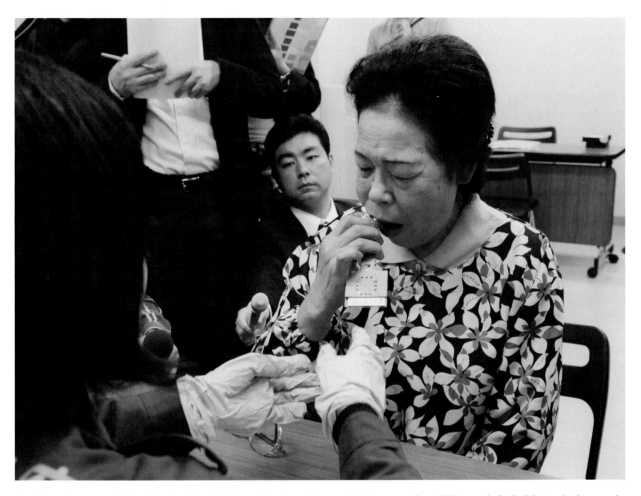

A relative of a person missing after the March 2011 earthquake and tsunami in Japan provides a DNA sample for building a database used to speed up the identification of human remains found after the disaster. Beyond their usefulness in crime-related situations, DNA databases can also serve additional purposes, such as missing person identification. © AP Image/Kyodo.

Civil libertarians assert that the permanent preservation of DNA samples raises significant privacy concerns, particularly when taken from arrestees, juveniles, or those convicted of minor offenses. In the United Kingdom, for example, begging and public intoxication are included in the offenses for which DNA samples may be taken and recorded. Furthermore, a BBC investigation in 2006 revealed that NDNAD contained more than 20,000 DNA samples taken from juveniles who were never convicted or even charged with a criminal offense. Many countries have far more stringent requirements for retaining DNA samples. In Germany, for example, a court order is required for the inclusion of a DNA sample in the national DNA database and will be granted only if a judge determines that the criminal is likely to reoffend.

The United Kingdom's retention of the DNA samples of an acquitted juvenile and an adult who was not prosecuted gave rise to the most prominent legal challenge to DNA databases. In 2008 a 17-judge panel of the European Court of Human Rights (ECHR) unanimously ruled in *S. and Marper v. United Kingdom* that the UK's DNA database policies violated Section 8 of the European Convention on Human Rights. Section 8 states, "Everyone has the right to respect for his private and family life, his home and his correspondence." The ECHR claimed that permanently retaining the DNA samples of persons not convicted of a crime "could not be regarded as necessary in a democratic society."

In 2011 the UK government announced revised policies for the retention of DNA samples in order to comply with the ECHR's ruling in *S. and Marper v. United Kingdom*. The UK government proposed destroying the DNA records of persons not convicted of a crime. The government later announced, however, that it would not be able to destroy the DNA records of innocent persons because those DNA samples were mixed with the DNA samples of convicted criminals.

Although criminal DNA databases are the most commonly employed type of DNA databases, a few more general DNA databases exist or have been proposed. Denmark maintains blood samples taken from every newborn born after 1981, known as the Danish Newborn Screening Biobank. Researchers at the biobank analyze the blood samples of Danish newborns to identify certain genetic disorders. Most developed countries analyze the blood of newborns for this purpose, but the samples are destroyed after testing in other countries. Denmark, in contrast, retains the samples and has used them for criminal DNA analysis. Parents, however, may request that the Danish Newborn Screening Biobank destroy their child's blood sample after initial screening.

SEE ALSO *Bioethics; Genetic Privacy; Genetic Technology; Individual Privacy Rights; Informed Consent*

BIBLIOGRAPHY

Books

Buckleton, John S., Christopher M. Triggs, and Simon J. Walsh, eds. *Forensic DNA Evidence Interpretation.* Boca Raton, FL: CRC Press, 2005.

Clarke, George. *Justice and Science: Trials and Triumphs of DNA Evidence.* New Brunswick, NJ: Rutgers University Press, 2007.

Hunter, William. *DNA Analysis.* Philadelphia: Mason Crest Publishers, 2006.

Krimsky, Sheldon, and Tania Simoncelli. *Genetic Justice: DNA Data Banks, Criminal Investigations, and Civil Liberties.* New York: Columbia University Press, 2011.

Web Sites

"Combined DNA Index System (CODIS)." *U.S. Federal Bureau of Investigation (FBI).* http://www.fbi.gov/about-us/lab/codis/codis (accessed November 4, 2011).

"Forensic DNA Databases: Linking Criminals to Crimes." *U.S. Department of Justice (DOJ), The DNA Initiative.* http://www.dna.gov/dna-databases/ (accessed November 4, 2011).

"Privacy Legislation and Regulations." *Centers for Disease Control and Prevention (CDC).* http://www.cdc.gov/od/science/integrity/privacy/ (accessed November 4, 2011).

Joseph P. Hyder

DNA Field-Effect Transistor

■ Introduction

A field-effect transistor (FET), or unipolar transistor, is one that uses an electrical field to affect the conductivity of one type of charge carrier in a semiconductor material. The most common semiconductor is currently silicon, although in 2008 IBM introduced the idea of a graphene transistor able to transmit at much higher frequencies.

A DNA field effect transistor, or DNAFET, is a type of field-effect transistor that uses a layer of single-stranded DNA (ssDNA) as the surface receptor of the gate, exploiting the finding that most biological polymers have an intrinsic charge. Complementary DNA strands hybridize with the DNA molecules on the gate, changing their charge and in turn affecting the current running back through the semiconductor transducer, enabling the DNAFET to act as a biosensor.

The use of DNAFETs as biosensors has applications wherever knowledge of the structure of a DNA strand is important, such as DNA sequencing, forensics, detection of hereditary diseases, and for monitoring in the food industry.

■ Historical Background and Scientific Foundations

In 1930 Julius Edgar Lilienfeld (1882–1963) was awarded a U.S. patent for the principle of the field-effect transistor, an idea also patented by Oskar Heil (1908–1994) in 1934 in the United Kingdom. In 1951 a research team at Bell Labs led by William Shockley (1910–1989) built the first JFET (junction gate field-effect transistor). In 1960, nearly a decade later, Dawon Kahng (1931–1992), also at Bell, created the metal-oxide-semiconductor transistor (MOSFET). The MOSFET has had an incalculable impact on the field of

electronics and modern communication systems, and the DNAFET is modeled on its structure, with the defining difference being the gate material.

The idea of combining these artificial transistors with biological elements really arose in 1962, when Professor Leland Clark (1918–2005) invented the enzyme electrode, the first biosensor. His idea evolved and eventually resulted in the glucose meter for blood testing in diabetics.

The Human Genome Project was an international research project, completed in 2003, the aim of which was to map the human genome and understand the functioning of the genes therein. Another element of the project involved looking at the tools and technologies related to gene research, improving them and licensing them to the private sector. The result was exponential growth of innovative research in both public and private industry, a major catalyst of the U.S. biotechnology industry. Significantly, the project also acted as a springboard for research into the use of DNA and RNA as biosensors.

■ Impacts and Issues

From the earliest experiments with transistors, and with the realization that the technology could be advanced to interact with biological targets, the concept of biosensors became seen as a way of creating hybrid systems from biological elements optimized through evolution in combination with human-crafted means of signal detection.

Optical detection and analysis of DNA structure via dye-based sequencing is the widespread method used currently. This requires labeling, for example with fluorescent dyes. The advantage of the DNAFET system is that it does not require labeling; the results are relayed via electrical current in near real time. This reduces

WORDS TO KNOW

BIOSENSOR: A sensor composed of two parts: a biological element and a human-made transducer.

DNA: Deoxyribonucleic acid; a self-replicating substance that carries genetic information. Its structure is that of two long chains in a double helix form.

POLYMORPHISM: A genetic variation with a frequency of at least 1 percent of the population, differentiating it from a mutation, and a causal factor in many hereditary diseases.

SEMICONDUCTOR: A solid material whose electrical conductivity varies between a conductor and insulator, depending on the addition of chemicals or differences in temperature.

TRANSDUCER: A device that converts variations in a physical quantity, such as pressure or brightness, into an electrical signal, or vice versa.

laboratory times and the possibility that the dyes will interact unevenly with the sample, making for an efficient, reliable, sensitive, and very specific biosensor.

The ability of DNAFETs to analyze DNA and also detect specific sequences has implications for many industries from medicine to the legal system. For example, DNA sequencing has a role to play in the investigation of genetic identity as well as forensics.

Deleterious mutations associated with single nucleotide polymorphisms such as sickle cell disease and osteoporosis can also be detected using DNAFETs. This capability has a far-reaching impact, not only on human health by aiding the design of new medications, but in a wider sense on the breeding of livestock and crops by enhancing the selection of favorable genetic traits.

Another interesting application, albeit a controversial one, is the use of DNA sequencing in stem cell research. Stem cell therapy relies on the theory that use of stem cells can regenerate damaged tissue or replace it by turning back its molecular clock to a more embryonic stage. Stem cell biologists have used DNA sequencing

A researcher for International Business Machines Corp. (IBM), works in a laboratory where development on a DNA transistor is ongoing at the company's Thomas J. Watson Research Center in Yorktown Heights, New York. *© Chris Ware/Bloomberg/Getty Images.*

to increase their understanding of how the DNA in embryonic stem cells replicates and organizes itself when those cells begin to differentiate and take on specialized functions.

In September 2011, stem cell research led to a groundbreaking experiment in which stem cell therapy was used in a trial on a group of subjects born blind in a bid to reverse their blindness. The potential success of such projects and ongoing research by other stem cell biologists offers hope for future cures and treatments for a wide range of diseases and disorders caused by the degeneration of genetic material, e.g., cancer and Parkinson's disease, or by external damage in the case of spinal cord injuries. In cancer diagnosis, DNAFETs and the identification of DNA sequences also allows for the detection of the gene CK20, a possible marker for many forms of the disease. Furthermore, the ability to detect variations in an individual's DNA may enable the design of patient-specific drugs for a wide variety of disorders.

DNAFETs represent an evolution in the analysis of DNA, and another example of technology and natural processes coming together to advance scientific progress.

SEE ALSO *Biochip; Biodetectors; DNA Sequencing; Forensic DNA Testing; Genome; Genomics; Microarrays; Nanotechnology, Molecular; Pharmacogenomics; Stem Cell Cloning; Synthetic Biology; Therapeutic Cloning*

BIBLIOGRAPHY

Books

Kumar, M. A. *Biosensors and Automation for Bioprocess Monitoring and Control.* Lund, Sweden: Department of Biotechnology, Lund University, 2011.

Ligler, Frances S., and Chris A. Rowe Taitt. *Optical Biosensors: Today and Tomorrow.* Boston: Elsevier, 2008.

Mutlu, Mehmet. *Biosensors in Food Processing, Safety, and Quality Control.* Boca Raton, FL: CRC Press, 2011.

Offenhausser, Andreas, and Ross Rinaldi. *Nanobioelectronics: For Electronics, Biology, and Medicine.* Dordrecht, The Netherlands: Springer, 2009.

Periodicals

Gu, Libo, JingHong Han, Hong Zhang, and Xiang Chen. "DNA Field-Effect Transistor." *Proceedings of the International Society for Optical Engineering (SPIE)* 4414 (2001): 47–49.

Nokhrin, Sergiy, Marcelo Baru, and Jeremy S. Lee. "A Field-Effect Transistor from M-DNA." *Nanotechnology* 18, no. 9 (2007): 95205–95500.

Kenneth Travis LaPensee

DNA Fingerprinting

Introduction

The genetic information carried by any two people is very similar. Even so, subtle genetic differences exist between each person, even between identical twins. The small differences in each individual's genetic makeup can be exploited in a technique called DNA fingerprinting. Just as people can be identified uniquely by their fingerprints, the use of DNA fingerprinting can be used to identify people to a high degree of confidence based upon patterns in their DNA that are unique to them. Other terms that sometimes are used interchangeably with DNA fingerprinting include DNA profiling, genetic fingerprinting, and DNA typing.

DNA fingerprinting has been used in many criminal cases to either eliminate individuals as suspects in a crime or implicate a person as being involved. The basic idea is that DNA from a crime scene often can be obtained, and from the collected evidence a DNA profile, or fingerprint, can be determined. That unique DNA fingerprint then can be compared to a suspect's DNA fingerprint to see if the two match. DNA fingerprinting has also proven helpful in many missing person cases.

The DNA fingerprinting technique can be applied to animals and plants as well. Materials from animals, such as fur, can be tested to see if they come from an endangered species. The same sort of technique can be applied to plants. Sports memorabilia, such as leather footballs, have had their DNA profiles catalogued as a unique method of identification in order to foil counterfeiters.

Historical Background and Scientific Foundations

The hereditary information of every known organism is carried in its DNA (and to a much lesser degree, in its RNA). DNA is known as a polymer because it consists of long strands of repeated molecular subunits called monomers. The DNA in cells is formed of two long strands that are wrapped around each other. The two DNA strands are linked together by short molecules called bases. There are four different types of bases, which join the two DNA strands together like the steps of a ladder. The bases linking the strands of DNA form pairs (called base pairs) all the way up and down the DNA ladder.

The long strands of DNA are coiled up into structures called chromosomes. Human cells possess 23 pairs of chromosomes that carry the DNA. Genes constitute a basic structure of heredity and are composed of stretches, or sequences, of DNA base pairs. The average gene length is 3,000 base pairs, but individual genes can be much smaller or larger. The main purpose of genes is to provide the information necessary for a cell to make proteins, which are essential for an organism to carry out its life functions.

In between the genes are long stretches of base pairs that are not used to code for proteins. Some of these base pair sequences on DNA strands can be used as a kind of marker that scientists can use to characterize an individual. These markers, known as short tandem repeats (STRs), are found in different locations on different chromosomes. The location of a particular marker on a chromosome is called its locus (loci refers to two or more such sites). By comparing markers from the same location (or locus) on the chromosomes of different people, scientists notice similarities and differences. It is not unusual that two individuals will both have an identical marker at a particular locus on a chromosome. However, the more markers that are identified between two individuals, the more likely it is that some markers will be different. For DNA fingerprinting, the U.S. Federal Bureau of Investigation (FBI) uses markers located at 13 distinct loci. The probability that two individuals will have identical markers at each of the 13 different loci (locations) within their DNA is one out of a billion. The exception is a case where the two individuals are monozygotic (identical) twins.

A set of loci, each corresponding to a different marker, is called a person's DNA fingerprint, or profile.

Different organizations and countries use different numbers of markers to form a DNA fingerprint. In Great Britain, 10 loci are required, whereas in the United States it is usually 13. Generally speaking, the greater the number of loci (individual markers), the higher the probability will be that no two people will share a given DNA profile. The concept of DNA profiling was developed by British scientist Alec Jeffreys (1950–) and first presented to the public by him in 1984.

■ Impacts and Issues

The availability of DNA fingerprinting techniques has had profound consequences in a number of areas, none more dramatic than in the criminal justice system. The ability to identify individuals uniquely in regards to various crimes has manifested itself in a number of ways, including connecting evidence at a crime scene to a particular individual's DNA fingerprint; identifying victims of foul play in which only limited remains of the victim are available; and exonerating people accused, or falsely convicted, of a crime.

The use of DNA fingerprinting has helped obtain the convictions of thousands of murder suspects. The very first use of DNA fingerprinting to solve a criminal case occurred in England in 1987. Two teenage girls had been raped and murdered on different occasions in nearby English villages, one in 1983, and the other in 1986. Semen was obtained from each of the two crime scenes. Ultimately, DNA fingerprinting led to the conviction of Colin Pitchfork (1960–) in 1988. Pitchfork confessed to committing the crimes after he was confronted with the evidence that his DNA profile matched DNA from the two crime scenes.

DNA fingerprinting has also helped exonerate several hundred people who had previously been convicted of murder. In many of those cases, the convictions had occurred prior to the advent of DNA profiling in criminal cases. The very first person in the world to be exonerated of a crime he did not commit was Englishman Richard Buckland (1969–), who was at one point the primary suspect in the aforementioned serial murder case involving the two slain English girls. Under police interrogation, Buckland confessed to one of the two murders, but not the other (he purportedly suffered from a learning disability, which may have had played a part in his confession). The use of DNA fingerprinting in the case showed that the same man had committed both crimes, and none of the DNA evidence collected at the two crime scenes matched Buckland. DNA fingerprinting did correctly identify the true murderer.

DNA databases are used in many countries to help the police solve crimes by matching the DNA collected at crime scenes to the genetic profiles contained in a DNA database. The information relating to each DNA profile is stored in such databases electronically. The largest DNA database in the world is the Combined

WORDS TO KNOW

CIVIL LIBERTIES: The fundamental rights of the individual, such as freedom of speech, guaranteed under the law against unwarranted infringement by the government or others.

DEOXYRIBONUCLEIC ACID (DNA): DNA appears in the cells of all known living organisms and embodies the information necessary for organisms to function. The information that DNA encodes is inherited. DNA is a polymer, meaning that it is composed of repeating subunits of smaller molecules. DNA forms a pair of long, intertwined strands linked together (like the steps of a ladder) by so-called base pairs.

FORENSICS: The application of science-based techniques and tests to the investigation of crime scenes.

GENE: The basic unit of inherited information in organisms necessary for them to develop and function. Genes are composed of sequences, or sections, of the base pairs found in DNA. They encode the information that a cell needs to produce amino acid sequences called proteins. It is estimated that humans possess from 20,000 to 25,000 genes.

DNA Index System (CODIS) in the United States, which is managed by the FBI. CODIS contains millions of DNA profiles that have been collected by law enforcement agencies at the national, state, and local levels. Many other countries have their own DNA databases. Such databases have helped to match perpetrators to the crimes they committed. They also are used extensively to match the DNA of missing persons, or the relatives of the missing, to unidentified human remains.

In addition to criminal and missing person cases, DNA fingerprinting has proved to be of great benefit in many other situations, such as the identification of human remains due to natural disasters or terrorism, including DNA fingerprinting that was used to identify remains of victims from the September 11, 2001, terrorist attacks. Some other ways DNA fingerprinting has been applied include matching donors with recipients for human organ transplants; determination of the lineage of plants or animals; and the identification of bacteria and other microorganisms associated with pollution.

Although DNA fingerprinting has provided benefits to society, the steady growth in the size of both federal and state DNA fingerprint databases has raised concerns among civil libertarians. DNA from convicted criminals is entered routinely into DNA databases. Even if such a procedure is deemed to be in the best interests of society, there are still opportunities for the abuse of an individual's rights. For instance, if a prisoner's conviction is overturned, that person may ask for his or her DNA profile to be removed from any databases containing it. But given the multiplicity of databases, how can one be sure that the

Professor Sir Alec Jeffreys, the man who discovered DNA fingerprinting, has called for a change in the law to remove the profiles of innocent people from the national database.   © *AP Images/Rui Vieira.*

profile was deleted properly? Another issue is whether or not a convicted criminal should continue to have a DNA profile stored in a database after the person has served out the prison term and returned to society.

Besides issues involving privacy and civil liberties, the application of DNA fingerprinting must be done carefully in order to prevent false leads in criminal investigations, and perhaps even wrongful arrests and convictions. In one particular case in Europe, evidence collected at numerous crime scenes was subjected to DNA fingerprinting, and the presence of the same individual—a female—was indicated at each crime. The case became known as the "phantom of Heilbronn." For many years police believed that they were dealing with a female serial killer, until it was discovered that the swabs used to gather evidence at each crime scene had all been contaminated at the point of production with human DNA by a female worker. So the DNA fingerprint of a supposed perpetrator was in reality the DNA profile of a factory worker handling the evidence swabs.

Probably a much more common and serious problem connected to DNA fingerprinting is the mismatching of DNA profiles with crime scene evidence. There are several scenarios in which erroneous or insufficient DNA fingerprints are more likely to be generated. The use of a degraded DNA sample might result in a

reduced DNA profile, meaning that there are less usable markers (loci) that can be compared with the DNA fingerprints in a database or the DNA fingerprint of a single suspect. An identification based on a reduced DNA profile is statistically less reliable. The same sort of problem can occur if the original DNA sample used to generate a DNA fingerprint is too small. In spite of degraded DNA profiles, there has been an overall trend of using smaller and smaller amounts of sample to derive such profiles. Another factor that can reduce the efficacy of DNA fingerprinting is the appearance of DNA from multiple people, which occurs quite often at crime scenes. The overlapping signals generated by different base pair sequences from different people can result in a hybrid DNA fingerprint that may not be usable.

The problems that can beset DNA fingerprinting can be curtailed to a large extent by proper procedures and training of the forensics personal conducting DNA fingerprint analysis. However, uniform standards, both in the United States and many other countries, have been lacking when it comes to DNA fingerprint analysis and to the training of forensic technicians. The widespread adoption of improved DNA fingerprint analysis techniques will strengthen the industry, while increasing the confidence of the public and criminal justice system in its reliability and accuracy.

■ Primary Source Connection

This article highlights the story of a woman who was found badly beaten in her Berlin apartment. Police officials ordered a number of laboratory tests for analysis of DNA found at the scene. A court order was obtained for Y-STR haplotyping in an attempt to gain information about the intruder's population origin. Discussion follows on the Y-STR Haplotype Reference Database and the usefulness of Y-STR haplotyping.

Male DNA Fingerprints Say More

In 2002, a woman was found with a smashed skull and covered in blood but still alive in her Berlin apartment. Her life was saved by intensive medical care. Later she told the police that she had let a man into her apartment, and he had immediately attacked her. The man was subletting the apartment next door. The evidence collected at the scene and in the neighboring apartment included a baseball cap, two towels and a glass. The evidence was sent to the state police laboratory in Berlin and was analyzed with conventional autosomal STR markers. Stains on the baseball cap and on one towel revealed a pattern consistent with that of the tenant, whereas two different male DNA profiles were found on a second bath towel and on the glass. The tenant was eliminated as a suspect because he was absent at the time of the offense, but two unknown men who shared the apartment were suspected. Unfortunately, the apartment had been used by many individuals of both European and African nationalities, so the initial search for the two men became very difficult.

The police obtained a court order for Y-chromosomal STR analysis (Y-STR haplotyping) to gain information about the unknown men's population affiliation. Y-STR haplotyping has been performed at the Institute of Legal Medicine, Humboldt University, Berlin, since 1991 to resolve male/female DNA admixtures in vaginal swabs and fingernails. In several high-profile cases, the Y-STR haplotyping method has been applied to gain a likelihood-based assessment of the population origin of a DNA sample. The specificity of Y-chromosome STR markers for specific populations is much higher than that of autosomal loci, and Y-STR haplotyping can often differentiate between closely related and admixed populations from the European continent. Prerequisites for such phylogeographic analyses are large reference databases containing Y-STR haplotypes from hundreds of different populations. The Y-STR Haplotype Reference Database (see www.yhrd.org), a collaborative project involving more than 100 forensic laboratories worldwide, proved useful to infer the population origin of the unknown male.

The search result indicated a patrilineage of European ancestry and African descent was unlikely. Also, none of the African, African-American or African-Caribbean haplotypes (n=1,587) matched the profile of the suspects; instead 7 matches were found in European populations.

The police were able to track down the tenant in Italy, and with his help, establish the identity of one of the unknown men, who was also Italian. When questioning this man, the police used the information that he had shared the apartment in Berlin with a paternal relative. This relative was identified as his nephew. Because of the close-knit relationship within the family, this information would probably not have been easily retrieved from the uncle without the prior knowledge. The nephew was suspected of the attempted murder in Berlin. He was later arrested in Italy, where he had committed another violent robbery. This actual case report illustrates the usefulness of the Y-STR haplotyping method, which was developed more than 10 years ago but rose to worldwide attention only recently with the development of highly sensitive multiplex PCR kits.

Lutz Roewer

ROEWER, LUTZ. "MALE DNA FINGERPRINTS SAY MORE." *PROFILES IN DNA* 7, NO. 2 (SEPTEMBER 2004): 14–15.

SEE ALSO *DNA Databases; DNA Testing: Post-Conviction; Electrophoresis; Forensic Biotechnology; Forensic DNA Testing; Individual Privacy Rights*

BIBLIOGRAPHY

Books

Butler, John M. *Fundamentals of Forensic DNA Typing.* Boston: Academic Press/Elsevier, 2010.

Echaore-McDavid, Susan, and Richard A. McDavid. *Career Opportunities in Forensic Science.* New York: Ferguson, 2008.

Embar-Seddon, Ayn, and Allan D. Pass, eds. *Forensic Science.* Pasadena, CA: Salem Press, 2009.

Kobilinsky, Lawrence F., Louis Levine, and Henrietta Margolis-Nunno. *Forensic DNA Analysis.* New York: Chelsea House, 2007.

Krimsky, Sheldon, and Tania Simoncelli. *Genetic Justice: DNA Data Banks, Criminal Investigations, and Civil Liberties.* New York: Columbia University Press, 2011.

Lynch, Michael. *Truth Machine: The Contentious History of DNA Fingerprinting.* Chicago: University of Chicago Press, 2008.

Read, M. M., ed. *Focus on DNA Fingerprinting Research.* New York: Nova Biomedical Books, 2006.

Web Sites

"About Forensic DNA." *U.S. Department of Justice (DOJ), The DNA Initiative.* http://www.dna.gov/basics/ (accessed November 7, 2011).

"DNA Forensics." *Human Genome Project Information.* http://www.ornl.gov/sci/techresources/Human_Genome/elsi/forensics.shtml (accessed November 7, 2011).

Philip E. Koth

DNA Sequencing

■ Introduction

DNA sequencing is a technology that uses chemical analysis to read the order, or sequence, of the four different bases in the nucleotide units that make up a DNA molecule. The sequence acts as a code in certain regions of the DNA molecule, the genes. The machinery of the cell uses this code to make protein molecules, with each gene specifying a different protein via its sequence.

DNA is a polymeric molecule (containing repetitions of simpler molecules), in which the bases are known as adenine, cytosine, thymine, and guanine (A, C, T, and G for short). Therefore, a small portion of a typical sequence might read ATTCGG. The sequence is obtained by first splitting the long DNA molecules up into smaller fragments, which are fed into an automated DNA sequencer that performs the necessary chemical reactions. Computer programs are then utilized to piece together the sequences of the fragments to obtain the sequence of the entire DNA molecule. DNA sequencing has many different applications. In medicine, it is used to identify mutations that cause diseases, which allows for more accurate diagnoses. Advances in DNA sequencing have also led to the sequencing of the human genome and the genomes of many other animal, plant, and microbial species, which allows a new understanding of biology on a genetic and molecular level.

■ Historical Background and Scientific Foundations

British biochemist Frederick Sanger (1918–), one of the pioneers of DNA sequencing, published the entire sequence of a virus, with around 5,400 bases, in 1977. He shared the Nobel Prize for Chemistry in 1980 with Walter Gilbert (1932–) of Harvard University. Both had developed different methods of sequencing DNA but Sanger's so-called dideoxy method eventually dominated the field. The sequence of the human genome

comprises around three billion bases, and the quest to decipher it, completed in 2000, drove many advances in DNA sequencing technology. Once a laborious manual procedure, DNA sequencing has become rapid, routine, and automated. In 2007 James Watson (1928–), who co-discovered the structure of DNA, and Craig Venter (1946–), a key player in the human genome project, both published the DNA sequences of their individual genomes, highlighting just how commonplace the technology has become.

In Sanger's dideoxy method, also known as the chain termination method, the DNA to be sequenced is used in its single stranded form, which acts as a template on which copies of the molecule can be made. Four separate copying reactions are set up. Each contains all the ingredients needed, which are the four nucleotides, the enzyme DNA polymerase, and a primer, which is a short stretch of DNA used to start the process off. Each of the four pots also contains a chain terminator molecule for A, C, G, or T, which is a dideoxy version of the nucleotide that cannot be incorporated into the growing sequence. When the new chain encounters the appropriate chain terminator, it ceases at that point. The result is a mixture of DNA fragments, each with a fluorescent tag marking a termination point for each base in the sequence. These can be separated and the position of each base read off. The dideoxy method readily lends itself to automation, which greatly improves the speed and accuracy of DNA sequencing. New methods known as next generation DNA sequencing carry out sequencing in parallel. This development enables hundreds of thousands, or even millions, of DNA fragments to be sequenced in one procedure.

■ Impacts and Issues

As DNA sequencing becomes easier and cheaper, it is beginning to play an important part in clinical diagnostics. The mutations underlying single-gene disorders, such as hemophilia and cystic fibrosis, are readily detectable within a patient sample, allowing for diagnosis at a molecular level. This has led to the development of pre-implantation diagnosis, in which DNA from embryos created for in vitro fertilization can be analyzed for specific mutations that create a risk of genetic disease. Only those embryos shown not to carry the mutations are implanted into the mother, and she can then continue the pregnancy, knowing the baby will be healthy. Sequencing of patient DNA is also becoming increasingly common in cancer, because certain mutations are known to influence response to specific cancer drugs and to prognosis, allowing a more individualized approach to diagnosis and treatment.

DNA sequencing technology has also led to the sequencing of more than 180 genomes from different organisms. Some of these organisms, such as rice, potato, and pig, are of great economic importance, whereas others, including HIV and various pathogenic bacteria, are of medical significance. Identifying genes within the DNA sequence of a genome can lead to new insights into the organism's functioning, which could have many applications, such as providing new targets for drugs. In the future, sequencing may become so cheap and accessible that each person could have his or her own genome sequenced as part of a routine medical screening. The target of the $1,000 genome, referring to the cost of sequencing, is one that is often mentioned.

Easy, low-cost DNA sequencing does, however, raise two challenges. The amount of sequence information has increased dramatically in recent years. The data needs to be processed and curated, and for this there is a need for bioinformatics tools and, more importantly, people who know how to use them. Much DNA sequence information is stored in public databases where it is available to scientists but they need to know how to access and make use of it. Much of a genome consists of non-coding DNA with genes making up only a small

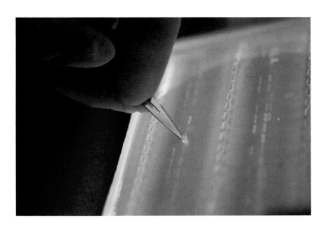

A lab officer cuts a DNA fragment under UV light from an agarose gel for DNA sequencing as part of research to determine genetic mutation in a blood cancer patient in Singapore. © AP Images/ Wong Maye-E.

> ## WORDS TO KNOW
>
> **BIOINFORMATICS:** The application of information technology to biological data, including DNA sequences.
>
> **DEOXYRIBONUCLEIC ACID (DNA):** DNA is found within every living cell and carries the organism's genetic information. DNA consists of two long chains of amino acid nucleotides; the chain is twisted into a double helix shape. The strands of the chain are joined by means of hydrogen bonds. The specific sequence of the DNA is what defines genetic characteristics.
>
> **GENOME:** The sum total of DNA in an organism, including its genes.
>
> **MUTATION:** A variation in the DNA sequence of a gene that may lead to a defect in or even absence of the protein the gene codes for. Mutations range from single base changes to more extensive changes in the DNA sequence. Mutation may or may not lead to disease.
>
> **NUCLEOTIDE:** The basic unit of the DNA polymer. A nucleotide is composed of three parts, which are a phosphate group, a sugar group, and a base that is either adenine (A), guanine (G), cytosine (C), or thymine (T).

proportion of the sequence. The function of non-coding DNA is not well understood, so most of the research focus is upon gene sequences.

It is possible to compare genes from different species to find out how similar organisms are and in what ways they differ from one another. For instance, there is about 95 percent similarity between human and chimpanzee DNA, according to a recent report, and it is possible that by concentrating on the differences that a better understanding of what it is that makes people uniquely human may be reached. However, there is still much to be discovered about what newly identified genes do and how they function in different environments. There is a danger of over-interpreting genetic information when understanding is still limited. It may be possible to sequence an individual's genome, but balancing the potential benefits and privacy issues generated from that information remains a challenge.

SEE ALSO *Biocomputers; Bioinformatics; Combinatorial Biology; DNA Databases; DNA Fingerprinting; DNA Testing: Post-Conviction; Forensic Biotechnology; Forensic DNA Testing; Gene Banks; Gene Therapy; Genetic Bill of Rights; Genetic Discrimination; Genetic Engineering; Genetic Privacy; Genetic Technology; Genetic Use Restriction Technology; Genome; Genomics; HapMap Project; Microarrays; Molecular Biology; Pharmacogenetics; Proteomics; Pyrosequencing; Recombinant*

DNA Technology; Selection and Amplification Binding Assay; SNP Genotyping; Yeast Artificial Chromosomes

BIBLIOGRAPHY

Books

Dale, Jeremy, and Malcolm von Schantz. *From Genes to Genomes: Concepts and Applications of DNA Technology*, 2nd ed. Chichester, UK: John Wiley & Sons, 2007.

Hunter, William. *DNA Analysis*. Philadelphia: Mason Crest Publishers, 2006.

Mitchelson, Keith R. *New High Throughput Technologies for DNA Sequencing and Genomics*, 2nd ed. Amsterdam and Boston: Elsevier, 2007.

Nunnally, Brian K. *Analytical Techniques in DNA Sequencing*. Boca Raton, FL: Taylor & Francis, 2005.

Periodicals

Hampshire, Andrew J., David A. Rusling, Victoria J. Broughton-Head, and Keith R. Fox. "Footprinting: A Method for Determining the Sequence Selectivity, Affinity and Kinetics of DNA-binding Ligands." *Methods* 42, no. 2 (June 2007): 128–140.

Web Sites

"Biology Animation Laboratory: Early DNA Sequencing." *Cold Spring Harbor Laboratory's DNA Learning Center*. http://www.dnalc.org/resources/animations/sangerseq.html (accessed September 27, 2011).

"DNA Sequencing." *National Human Genome Research Institute*. http://www.genome.gov/10001177 (accessed September 27, 2011).

Susan Aldridge

DNA Testing: Post-Conviction

■ Introduction

In use by authorities in the legal system since the mid–1980s, DNA testing of evidence collected at a crime scene to prove innocence or guilt has a short, but controversial, history. Crime scene evidence collection can yield organic human material that can be tested to produce the DNA sequence for the human to whom the tissue sample belongs. This forensic evidence can place a perpetrator at the scene of a crime, making DNA testing a gold standard for objective evidence in the legal system.

As forensic science evolves, and technology allows for more finely-honed testing, the question of post-conviction DNA testing raises questions about legal processes and prisoner rights. Technological advances drive legal change in this instance. DNA sample collection was not common before the 1980s. However, for older cases in which some sample—hair, semen, saliva, fingernail scrapings, skin tissue, or other organic matter—is available, the state's responsibility to provide access to this evidence, and access to DNA testing that could provide proof of guilt, or possibly innocence, comes into question.

■ Historical Background and Scientific Foundations

Using technology to analyze evidence that implicates a specific person stretches back to the late 1800s, when fingerprinting first came into vogue. The first use of fingerprint technology to implicate a person of a crime was in 1892, in Necochea, Argentina, where a matched fingerprint identified the murderer of two boys.

With fingerprint technology came fingerprint bureaus. In the late 1800s law enforcement officials in India, Argentina, and England created such bureaus, taking the fingerprints of known criminals and using newly-collected prints to compare.

DNA testing, or profiling, joined fingerprint technology to increase the accuracy of forensic science techniques for solving crimes. Developed in the early 1980s at the University of Leicester in England, reliable DNA testing became a viable option for law enforcement officials to pinpoint whether a human being was at a crime location based on testing of human tissue evidence. Gathering any evidence that may be DNA testable, from hair to semen to skin, and then testing for DNA sequence offers reliable evidence that is accurate 99.9 percent of the time.

The first criminal case in the United States in which DNA profiling was admitted as evidence was in the 1987 trial of Tommy Lee Andrews in the state of Florida. Andrews was convicted of rape after DNA testing showed that semen samples gathered from the rape victim matched Andrews's DNA. As with fingerprint records, creating records databases of known criminals' DNA became a goal for law enforcement. In 1989 the state of Virginia passed laws establishing a DNA laboratory for crime scene evidence, and as of 2011 all 50 states require convicted sex offenders to relinquish DNA samples for databases.

Technological advances allow for far more sophisticated data to be extracted from human tissue samples; DNA testing results from the mid–1980s, for instance, are less precise than those achievable in the early 2010s. During the decades that DNA testing has been available, prisoners have requested the right to post-conviction DNA testing to prove innocence. Throughout the 1980s, 1990s, and the first decade of the 2000s, lawsuits and legislation carved new territory as technology improved; by 2011, 48 states had passed laws requiring that convicted prisoners have access to evidence that could be tested for DNA sequencing, with Massachusetts and Oklahoma the only exceptions in the United States.

The Innocence Project, a nonprofit organization developed to address the use of technology in proving the innocence of wrongly-convicted prisoners, formed in 1992 at Cardozo School of Law. This project, a legal clinic that offers pro bono assistance to prisoners, works to provide prisoners with access to DNA-based evidence for post-conviction testing. As of October 2011, 273 prisoners had been exonerated via post-conviction DNA testing, 240 of them through the efforts of The Innocence Project.

■ Impacts and Issues

The 2009 United States Supreme Court case *District Attorney's Office v. Osborne* tackled the question of post-conviction DNA testing. In the case, convicted kidnapper and rapist William G. Osborne's DNA test, completed in 1994, showed that there was a one-in-six chance that he was the perpetrator in the beating, kidnapping, and rape of an Alaskan prostitute. Osborne's attorney chose to skip more detailed testing, fearing it would implicate him directly. Osborne requested the right to test the evidence after his conviction, and the case made its way to the U.S. Supreme Court. The court in a 5–4 decision, found that convicted prisoners do not have a constitutional right to access the evidence for post-conviction DNA testing, instead noting that the issue is one for individual states to decide.

In October 2011, Texas resident Michael Morton was exonerated in his murder conviction after serving nearly 25 years in prison for the 1987 murder of his wife, Christine Morton. Post-conviction DNA testing of a bandana found near the scene of the murder contained Ms. Morton's blood and blood from a man who whose DNA was listed in a national offender database. Michael Morton was freed on the basis of the new testing.

■ Primary Source Connection

This article gives an explanation and detailed account surrounding the conviction of a man for the murder of a 74-year-old woman on the night of June 3, 1997.

The woman, Alice Zolkowski, was savagely beaten and left to die at her home in Slavic Village. She eventually slipped into a coma and subsequently died two years later, never able to identify who was responsible for the brutal attack. Details follow of the physical evidence and witness accounts leading to the resulting conviction and the jail time that was served. Eventually, DNA testing performed post conviction led to a change in the original jail sentence previously put forth by the jury.

Call It Even

No one at the Barley House seems to notice the guy wearing jail-issued clothes on a Thursday night in late March. Not an orange jumpsuit—just denim blues, which from a distance look like something you might buy at the mall, and which easily blend in with the happy-hour crowd at the Warehouse District club. But there on the left pocket is his Trumbull County inmate number next to his name, T. Siller.

Just hours earlier, Thomas Siller was the property of Cuyahoga County, sitting in a cell, as he had for 13 years, for a murder he has maintained since day one that he had no part of, and for which new DNA evidence shows he is telling the truth.

Now he's sipping Guinness and Sam Adams, munching on a corned beef sandwich with a side of stuffed pretzel bites, and laughing. Asked how his first sandwich as a free man tastes, Siller responds, "Good, but the beer is better. How much is a Guinness these days, anyway?"

"Four or five dollars, depending on where you go," someone answers.

"Yours is still the most complicated case I've ever worked on," says Alba Morales, Siller's attorney from the Innocence Project in New York, and one among the gaggle of lawyers who not only helped secure his release, but also gathered to celebrate.

Faulty evidence made Siller an unlikely murderer all along. At the heart of Siller's case was the testimony of a single shady eyewitness and a forensics expert asleep on the job. But Cuyahoga County prosecutors haven't so much changed their minds as thrown up their hands in light of the overwhelming evidence that runs contrary to their 14-year-old stance.

Alice Zolkowski's nightgown had been ripped and used to tie her to a chair in her Slavic Village home on the night of June 3, 1997. The single 74-year-old had been gagged and savagely beaten, her blood spattered against the wall behind her and her home ransacked. When it was over, she was barely clinging to life.

Cops would find her in the early hours of June 4 after an anonymous 911 call placed at 3:49 a.m. That call was made by Thomas Siller.

318

Derrick Williams, at center, is released from Hardee Correctional Institution in Florida, on April 4, 2011, while accompanied by attorneys. Williams was released from prison after State Attorney Earl Moreland decided to end his fight to keep Williams in prison on a 1993 rape conviction. The 47-year-old man's conviction was overturned after tests showed his DNA was not a match to DNA found on clothing the rapist wore. At the time of his release, Williams was the 13th DNA exonerate in Florida and the 268th individual to be exonerated through the use of DNA testing nationwide. © *Paul Videla/Bradenton Herald/Getty Images.*

Fingerprints at the house included those of a then-42-year-old Siller, who had done work for Zolkowski, grown close to the woman, and borrowed money from her. Prints were also found matching 41-year-old Walter Zimmer, another worker at the house who had borrowed money, and 29-year-old Jason Smith, a drug dealer with a long rap sheet who never hung out with Siller or Zimmer or worked at Zolkowski's house. All three men had a thirst for crack cocaine.

Zolkowski slipped into a coma, then died two years later without having identified her attacker. But there are two accounts of what happened that night: one espoused by Siller and Zimmer, the other by Smith. The third man was offered a deal by prosecutors: just three years in prison in exchange for testimony against the other two. That testimony, it turns out, was a lie.

Life in Slavic Village led Siller to Wally Zimmer, who frequented the same bars and parties. Like Siller, Zimmer worked hard—driving trucks and doing other odd jobs—and numbed himself with crack. Also like Siller, he mostly steered clear of trouble. Despite their parallel paths, the two were never more than casual pals.

Siller knew of Jason Smith, but he didn't hang out with him. "I saw him a couple of times at a house or around, you know, but I would always get this bad feeling about him." He says he didn't buy from Smith and warned others to stay away. Zimmer, too, says he didn't buy from Smith.

Siller and Zimmer found work wherever they could get it. That's how they came to ask the old lady on Hosmer Avenue about her chimney.

"One day, me and Wally are walking down the street and I see this chimney. I can see daylight through it," says Siller. "So I knock on the door. This old lady very cautiously opens the door, and I say, 'I notice your chimney, it's not in good shape.'"

The woman, Alice Zolkowski, said she'd been looking for someone to repair it. Siller said he does chimneys.

"I looked at her, looked at the house, and say, 'Well, $175,' or something like that," he recalls. "That's like kibbles and bits compared to what I should have charged or would have charged for someone out in Beachwood. But I look at the house and I look at her, you know. So me and Wally tear it down and put up a new one."

Pleased with the work, Zolkowski began asking Siller about other projects around her house. *Could you fix the porch? Could you look at the thermostat?* Siller didn't know how to do all she needed, but he helped find others who did. He became a sort of general contractor, earning money to ensure that projects got done.

"One day, I see her giving one of the guys working on the porch some money," he says. "I ask her what's that about. She says, 'It's OK, it's OK,' and she says she knows the guy from before and she's just loaning him some money. She would then offer, *If you ever need anything. . . .*"

Siller eventually finished the porch job himself, and Zolkowski paid him extra. He says he told her not to. He says she insisted.

"I said, 'Why don't you put it down as a loan?' That's how the tally sheet got started."

The tally sheet eventually showed that Siller borrowed just over $12,000 from Zolkowski. Zimmer borrowed just over $7,000. It also showed money loaned to two other men. The ledger was dutifully kept by both Zolkowski and Siller, who would help her with the math.

Siller speaks of Zolkowski with great affection, often breaking down in stifled tears. Sometimes she paid him extra, and he told her to mark it as a loan. Some of it he asked for. Sometimes she just handed over checks.

"She started making it too easy," he says. "Worrying about where I was going to stay, how my girlfriend was. She worried about Wally's daughter. She was a lonely lady who wanted someone to be there." Siller tried to repay her various times, though she always refused.

"We did, I would say, take advantage of her," Wally Zimmer says today. "I'm just being honest. We borrowed money. She was nice—*real* nice. She'd ask me to fix a ceiling tile or something, and I would, and she'd say, 'Here, here, you did work,' and give me money. But I also had to feed my daughter and pay rent, and I'd ask sometimes."

The morning after Zolkowski was found beaten, neighbors told detectives that Siller and Zimmer were often seen around the house. So they went looking for the men on the tally sheet.

Siller and Zimmer have relayed essentially the same account of that night for 14 years: They weren't there. In voluntary statements given to detectives, the details remain the same save for some haziness on specific times. They were, after all, getting high that night.

Early in the evening on June 3, Zimmer and Siller say they went to a neighborhood bar called Chaulkie's. There, Siller cashed three checks from Zolkowski in the amounts of $240, $200, and $75. With more than $500 on him, he headed out for the night, driving to a friend's house in a car owned by Siller's sometime girlfriend, Rosie Crowder. The friend later told cops that Siller was there and that he arrived, nicely dressed, in Crowder's car.

Zimmer said he was with Crowder after leaving the bar; Crowder told cops the same; she would be charged with obstruction of justice, the prosecution claiming she lied. (Attempts to reach Crowder were unsuccessful.)

According to Zimmer's original statements to detectives, his testimony, and conversations after his release, he was walking by Zolkowski's house at around 2:30 a.m. and noticed the lights and TV on. Thinking that was strange, he went to knock on the door, but there was no answer. After peering in and seeing Zolkowski's feet sticking out from a chair, he tried the back door, which he found kicked in. He entered to find the house ransacked and Zolkowski tied to a chair, badly beaten.

Zimmer tried shaking her to elicit a response, then picked up the phone to call 911. At that moment, he considered an outstanding warrant he had for a traffic offense, as well as his state of impairment. So he returned to Crowder's house to tell her what had happened instead.

Crowder also didn't want to call 911 because she had drugs in the house. They began paging Siller, putting "911" in the message to convey urgency.

Siller figured Crowder was just bent out of shape and wanted her car back. He left the friend's house and headed down to Hosmer Avenue. When he arrived, Zimmer explained what had happened. Siller yelled at Zimmer for not calling for help, dropped him off at his home, then stopped at a pay phone on nearby Fleet Avenue to place an anonymous 911 call before heading back to Crowder's place.

All three told police what happened and paid it no mind. And for the better part of a year, all was quiet.

But in March 1998, Siller was working at a house in Shaker Heights when somebody from the neighborhood pointed at him and said, "There's the guy that's on TV!" and called the cops. Siller, it turned out, had been featured on an *America's Most Wanted* type show. He was deemed "armed and dangerous."

"I've shot a gun once in my life. I don't own a gun," he says. *"Armed and dangerous?"* Cops tracked him down, guns drawn. Zimmer had already been picked up. "I thought it was a case of mistaken identity. Or maybe we were being arrested for not calling 911. I didn't even think about going to jail. If you're innocent, how do you go to prison?"

Jason Smith's fingerprints were found on a dresser drawer in Zolkowski's home—a drawer where checks had been kept, but which had been grabbed and then dropped on her dining room table. Cops immediately zeroed in on Smith, who had never previously been to Zolkowski's home but had a long rap sheet.

And unlike the unchanging accounts of Siller and Zimmer, Smith's version of events proved to be much more malleable under scrutiny. This was the state's star witness.

Smith was arrested about a week after the death of Zolkowski. Officers had come looking for him at his girlfriend's home. They returned later, kicking down the door, to find Smith hiding in a closet. "I kind of had an idea of what they wanted," he would testify at trial.

Smith initially proclaimed his innocence, saying that he held down two jobs to stay out of trouble and that he was working that night at the Touch of Italy restaurant. He would later admit the claim was untrue.

When reminded at trial that he had already lied twice about the crime to detectives, Smith replied, "I lied, but I'm not a chronic liar."

While Smith was jailed, another man in the cell next to him was being held for a parole violation. He testified at trial that Smith told him that police had nothing on him, that he was the one who killed Zolkowski, and that he took off his gloves a couple of times while ransacking the house—and that's how a fingerprint ended up on the dresser. He went on to say that he had no partners; he just needed money to buy his girlfriend a house. Smith told the man he had an alibi ready—he was in Kentucky at the time—which is exactly what Smith would tell a detective the following day.

The jailhouse snitch wasn't the only one to say Smith had confessed. At the trials of Zimmer and Siller, another man said Smith called him at a motel in Independence where he was partying with two prostitutes on the night of the attack. Smith told him he "hit a lick" for $2,500, meaning he had stolen a substantial amount of money and was ready to party. Prosecutors questioned the man's word because he was a drug user.

Smith also admitted that he cut his leg on the night in question, and that it happened on the job at Touch of Italy. An odd detail to volunteer, but it came after Smith's girlfriend had warned him that cops confiscated a pair of his pants. Those pants, he knew, had blood on them.

Still, Smith maintained his innocence until March 1998. His girlfriend, who had been charged with obstruction of justice, agreed to testify against Smith, who faced burglary, kidnapping, and aggravated attempted murder charges. Now Smith was ready to talk to prosecutors. (Conveniently, Siller notes, it was clear by this time that Alice Zolkowski would not regain consciousness and thus would never identify her attacker.)

According to Smith, he stopped at Rosie Crowder's house around 9 on the evening of June 3, looking for Siller or Zimmer to help him cash a check to buy drugs. Smith, a black man in a predominantly white neighborhood, claimed he needed the help of Siller, who frequently cashed checks in the area. Instead, Smith claims, Zimmer suggested the trio just go to Zolkowski's house and borrow some money. Smith said he drove them there and waited while Siller and Zimmer went inside, then the three went to buy drugs and smoke them.

As the evening wore on, Smith said, the drugs and money ran out. He suggested they cash his check, then Zimmer suggested they return to Zolkowski's house. There, Zimmer and Siller went into the house, this time through the back door, while Smith waited in the car. Wondering what was taking so long, Smith went inside and found Siller rummaging through the house. He seized upon an opportunity to grab the checks, then came back to see Zimmer hovering over Zolkowski. He dropped the checks, left without saying a word, and went on in search of drugs.

The state could produce no witness who had ever seen Smith hanging out with Siller and Zimmer, let alone on the night in question.

In exchange for Smith's testimony, prosecutors dropped two pending drug charges, handed him just three years in a plea for a single burglary charge, and gave him immunity from further prosecution should Zolkowski die. Years later, Judge Christine McMonagle would call this a "breathtakingly favorable plea deal."

Smith had a history of those.

In March 1998, while setting up his deal to testify against Siller and Zimmer, Smith was double-dipping at the jailhouse: He wore a wire for the Sheriff's office and "entered into a sexual liaison" with a male nurse at the jail, according to court documents. He eventually testified at the nurse's trial.

One month after his release from prison on June 12, 2000, Smith was already back to using crack when he grabbed a purse from a lady at Kmart, racking up a fresh aggravated robbery charge. He then struck another deal: He told prosecutors another man had confessed to a murder in prison. Smith's testimony in that case whittled his sentence to one year.

In 2001, he was involved in an accident in the Warehouse District. After attempting to flee the scene, Smith struggled with an officer through the car window and was shot three times. He survived, but was rendered a quadriplegic.

Siller and Zimmer were tried together and convicted in 1998. Siller was given 20 years, Zimmer 40.

Alice Zolkowski died on April 26, 1999. Her death sent Siller and Zimmer back to trial, this time for murder.

Jason Smith explained the blood on his pants in three different ways.

At the time of his arrest in 1997, he volunteered that he had cut himself at work on the night Zolkowski was beaten. By the time of Siller and Zimmer's first trial, Smith claimed he had accidentally cut himself on broken glass after slamming a door at his girlfriend's house.

And when Siller and Zimmer faced murder charges a year later, Smith said he got hurt kicking in the door at his girlfriend's house because she had locked him out.

Besides the revolving testimony of Smith, the prosecution's case hinged mainly on the expertise of Joseph Serowik, a forensic specialist for Cleveland Police. In the first trial, Serowik testified that there was one spot of blood on Smith's pants, and that blood belonged to Smith.

Under cross examination in the second trial, however, Serowik buckled when asked about a discolored spot on the back of Smith's pants. Serowik could not say whether or not it had been tested. The judge ordered Serowik to retest the spot; Serowik returned six days later to testify that 1) the spot had been tested before the first trial and came up positive for blood, and 2) the new test showed that the blood belonged to the victim, Alice Zolkowski. Serowik also said every dark spot on the pants was retested, and that the circled spot was the only one that was blood.

This revelation sent the courtroom into a tizzy: The guy who said he was nowhere near the old lady suddenly had her blood on him. Prosecutors considered pulling Smith's deal on the spot, but instead went along with the case, giving Smith an opportunity to explain away the new evidence and for them to explain their case anew.

This time, Smith said he may have been close enough to Zolkowski that the blood flew and landed on him. Or maybe he brushed up against Zimmer.

Prosecutors, who had previously said that whoever beat Zolkowski would have her blood on his pants, now said that if Jason Smith was the one who beat her, "there would be *more* blood and it would be on the front of his pants." There would be a spatter pattern.

The jury bought Smith's revised story. Siller was convicted for murder and sentenced to 30 years without parole. Zimmer, whose trial was to begin after Siller's, faced the death penalty and had just watched a jury convict Siller. He took a plea bargain for involuntary manslaughter and was sentenced to ten years to run concurrently with his 40-year sentence from the first trial.

Siller filed for new DNA testing of Jason Smith's pants. The state opposed. After Siller acquired the counsel of the Innocence Project, new testing was approved, and 20 blood spatters were found on the front of Smith's pants. Nine were tested, seven were the blood of Zolkowski.

Siller filed a motion for a new trial, which was denied by Judge Steven Terry but successfully appealed. In June 2009, Siller was granted a new chance at justice: a retrial in September 2011.

But the Cuyahoga County Prosecutor's Office was not going to make it easy. Through the years, the office became known for unleashing mocking quips and puffy-chest declarations reaffirming its belief in Siller's guilt and support of Jason Smith.

Prosecutor Bill Mason opposed Siller's initial request for new DNA testing in 2004. In 2007, after the Innocence Project joined Siller's cause, Mason told *The Plain Dealer*, "I'll take a closer look at this case, but they lost their way on this one. They must be running out of innocent people to represent." (Mason's office declined via e-mail to comment for this story.)

When Mason's office decided to fight the appeal, Assistant County Prosecutor Matt Meyer told the paper, "This has been a battle from the get-go—and all it did was reconfirm what we already know. It is important to expose this fraud, because we've spent four and a half years of the public's money litigating a lie."

In late 2010, while preparing for Siller's retrial, the prosecutor's office shipped evidence—including Zolkowski's bindings and nightgown—for new DNA testing. The results: The only DNA on the bindings besides Zolkowski's belonged to the state's star witness, Jason Smith.

"He snookered us a little bit," Mason told *The Plain Dealer*. "We're working in the mud with defendants and witnesses all the time, and you make your best call on a person's credibility. But given all that has happened in this case, this was the right thing to do."

Mason also made a point to clarify that the DNA results did not rule out Siller or Zimmer as accomplices. In May, Smith was extradited from his new home in Atlanta and charged with two counts of perjury and three counts of obstruction of justice. A *Plain Dealer* editorial lauded Mason's "aggressive pursuit of justice."

"They [*The Plain Dealer*] went out of their way to point out that nothing happened until the prosecutors sent the bindings," says Alba Morales. "The bindings were initially sent off and with improvements we were able to find DNA this time. It was not the prosecution who initiated the testing of the bindings. When everything went out, the bindings went out."

Mason's office had no evidence against Siller and a mountain of new evidence against Smith. To Siller's camp, an aggressive pursuit of justice would have meant dropping the charges outright. Instead, prosecutors offered a deal: He could walk away with time served in exchange to pleading guilty to three felony counts of theft. Or they could take him to trial again.

"Everybody on our side held their nose," says Kevin Spellacy, an attorney who worked with the Innocence Project on Siller's case.

"Given the prosecutor's office we were dealing with, I knew better than to assume anything," says Alba Morales, another Siller attorney.

Siller took the deal, but only after dispatching his lawyers to ask his family whether they'd rather have him come home or continue the fight to clear his name. "Come home" was the resounding answer.

"Preferably, we'd like to have seen him vindicated," says Spellacy. "We like to say go to trial, make them prove it, and you're found not guilty. That's the ideal result. I told him point blank we'll try your case. Nothing would make me happier than to cross examine that idiot [Smith]. But when it comes with an enormous amount of risk . . . Tom had to decide the quickest way to go home."

"I had enough. I had enough of the madness," says Siller. "Let me throw myself under the bus, go home, and get out of the prosecutor's way so they can convict the right person in Alice's murder. That's what I was thinking. I always wanted to go back to trial. I always thought they'd eventually be stand-up enough to admit they're wrong."

Vince Grzegorek

GRZEGOREK, VINCE. "CALL IT EVEN." *CLEVELAND SCENE,* JULY 6, 2011.

SEE ALSO   *DNA Databases; DNA Fingerprinting; DNA Sequencing; Forensic Biotechnology; Forensic DNA Testing; Microarrays; Polymerase Chain Reaction (PCR)*

BIBLIOGRAPHY

Books

Acker, James R. and Allison D. Redlich. *Wrongful Conviction: Law, Science, and Policy.* Durham, NC: Carolina Academic Press, 2011.

Buckleton, John S., Christopher M. Triggs, and Simon J. Walsh, eds. *Forensic DNA Evidence Interpretation.* Boca Raton, FL: CRC Press, 2005.

Goodwin, William, Adrian Linacre, and Sibte Hadi. *An Introduction to Forensic Genetics,* 2nd ed. Chichester, UK: Wiley-Blackwell, 2011.

James, Stuart H., and Jon J. Nordby, eds. *Forensic Science: An Introduction to Scientific and Investigative Techniques,* 3rd ed. Boca Raton, FL: CRC Press, 2009.

Kobilinsk, Lawrence F., Thomas F. Liotti, and Jamel Oeser-Sweat. *DNA: Forensic and Legal Applications.* Hoboken, NJ: Wiley-Interscience, 2005.

Web Sites

"About Forensic DNA." *U.S. Department of Justice (DOJ), The DNA Initiative.* http://www.dna.gov/basics/ (accessed November 8, 2011).

"DNA Forensics." *Human Genome Project Information.* http://www.ornl.gov/sci/techresources/Human_Genome/elsi/forensics.shtml (accessed November 7, 2011).

"Post-conviction Exoneration of the Innocent." *U.S. Department of Justice (DOJ), The DNA Initiative.* http://www.dna.gov/postconviction/ (accessed November 8, 2011).

Melanie Barton Zoltan

DNA Vaccines

■ Introduction

DNA vaccines are based on genetic engineering technologies used to isolate, extract, and transfer a specific viral or bacterial gene encoding an antigen that can induce immune response or immunization in another organism. The viral or bacterial gene is transferred to an innocuous carrier such as a liposome capsule or is inserted into a plasmid before being inoculated into the host to induce immunization. DNA vaccines promise to be especially important in the prevention of those infectious diseases against which the conventional immunization techniques cannot provide long-term protection, such as tuberculosis, influenza, and malaria.

■ Historical Background and Scientific Foundations

The discovery of immunization against smallpox in the eighteenth century represented a giant leap in the history of medicine. As a consequence of that first success, several other new vaccines were developed in the following decades against a variety of infectious diseases such as whooping cough, rabies, yellow fever, tuberculosis, and poliomyelitis. Against certain diseases however, conventional vaccines are not uniformly effective, and show variations in degree of protection among individuals. Also, certain infections, including tuberculosis, malaria, and influenza, require repeated periodic vaccinations, because the period of immunity in available vaccines against these diseases varies. These diseases require highly specific protection, known as cell-mediated immunity. Diseases requiring cell-mediated immune response for development of immunity, such as the above mentioned, are those against which conventional vaccination is less effective.

Mammalian organisms have developed immune systems containing three major lines of defense against infectious microorganisms and toxins. The immune system is capable of two types of response: the innate immune response and the acquired immune response. The first line of defense is the innate immunity system that responds immediately after an invading microorganism enters the host. This line of defense is very effective against a wide variety of bacteria. Innate immune cells also interact with the second line of defense, the acquired immunity system, by triggering into action those cells involved in the formation of acquired immunity. Acquired immunity is not an immediate response; it takes several days to form and then mount a full attack against the new infectious or foreign agents.

Acquired immunity consists of two arms: cell-mediated immunity (T lymphocytes and T helpers cells) and humoral or antibody-mediated immunity (B lymphocytes and plasma cells). Antibodies are immunoglobulins, a class of proteins that react with specific antigens. Globulin proteins represent about 20 percent of all proteins present in the blood plasma. Antibody-mediated immunity is directed against toxins, certain viral infections, and bacteria such as pneumococci and meningococci, among others. Acquired immunity consists of the development of B or T cells' ability to recognize specific antigens and retain a long-term memory. This long-term memory enables these cells to recognize and respond against a given antigen many years after the first exposure. Another great advantage of cell-mediated and antibody-mediated immunity is that T and B cells are capable of forming a wide collection of specific memories against different infectious agents.

Cell-mediated immunity inhibits infections by parasites, intracellular bacteria, virus-infected cells, and fungi. Cell-mediated immunity can be effective against cancer cells as well. DNA vaccines, containing the genes for the antigens of specific microorganisms, have been under intensive investigation since the early 1990s because they can trigger both the humoral and cellular immune responses.

■ Impacts and Issues

Research in DNA vaccines is aimed at two different clinical situations: preventing infectious disease and treatment for existing conditions including autoimmune disease and cancer. DNA vaccines currently protect against, or are in various developmental and research stages, for influenza, tuberculosis, hepatitis type A and B, meningitis, and human immunodeficiency viruses (HIV).

After decades of being under control with the use of BCG vaccine and antibiotics, tuberculosis (TB) has become a major public health concern again. New and resistant strains of TB have appeared worldwide. BCG vaccine is made from *Mycobacterium bovis*, the bacteria that causes tuberculosis in cattle. However, BCG does not contain other gene sequences and antigens that are present in the genome of the tuberculosis bacteria that affects humans. New genetic studies aimed at developing a new and more effective vaccine against tuberculosis are under way.

AIDS (acquired immunodeficiency syndrome), caused by HIV, is another challenge that DNA-vaccine trials are hoping to overcome, with several different research groups experimenting with different approaches in animal models. As of late 2011, however, only two projects have completed clinical trials in humans to assess vaccine efficacy against HIV type 1 (HIV-1). One group used a protein isolated from the virus (monomeric HIV-1 glycoprotein 120), which initially triggered the production of specific antibodies, but could not elicit a wide reactive neutralizing antibody (Nabs) response. The second trial was designed to assess the efficacy of disabled adenoviruses as vectors. This trial was suspended when the first interim analysis showed that the vaccine failed to protect against infection or to reduce viral reproduction rates after infection.

The hurdles facing the success of anti-HIV vaccines are many, because HIV is a retrovirus and as such is highly prone to genetic mutation during replication of its genome, with each new population showing greater genetic diversity. This enables the virus to go unrecognized by some of the body's humoral and cellular immune responses. Furthermore, conventional immunization methods, such as the use of attenuated living pathogens, are often unsafe for humans, especially in the case of a nearly always fatal disease such as AIDS.

DNA vaccines for treating pre-existing diseases such as cancer have been under investigation since the early 1990s. Attenuated adenovirus vectors, plasmids, and liposome carriers were tried throughout the years as intracellular delivery agents for anticancer drugs, missing genes, and anti-proliferation factors, without much success. The rationale in utilizing vaccines against cancer is to teach the immune system to recognize mutated cancer cells as foreign bodies, eliciting an immune attack against the tumor cells. Although many tumor cell types are detected as non-self (e.g. alien),

WORDS TO KNOW

ANTIGEN: A protein fragment or the protein itself that is recognized by the infected host as an alien substance, thus triggering an immune response or attack against the antigen-producing source, the infectious microorganism.

DENTRITIC CELLS: Part of the immune defense with the function of capturing compounds, proteins, or any other foreign substance and presenting it to T cells or B cells for recognition. If the substance is alien, an immune response is prompted.

IMMUNITY: The ability by the host's immune system to recognize and kill invading microorganisms that cause infectious, contagious diseases.

IMMUNIZATION: Immunization or vaccination is the method of exposing the host's immune system to a dead or attenuated virus or bacterium or to an specific antigen in order to prompt recognition and attack by the host's immune cells against those infectious agents.

LIPOSOMES: Small lipids or fat particles able to cross cell membranes and deliver DNA fragments or a medication or another biologic compound into the targeted cells of a living organism.

PLASMID: A bacterial double-stranded circular DNA capable of replication independent of the bacterial chromosome. Plasmids also can be integrated into the chromosome or can be transferred or exchanged between bacteria.

VECTORS: Innocuous viruses or bacteria, able to cross cell membranes, used as carriers to transport a gene fragment or the antigen into the host's cells.

therefore eliciting an immune response, the tumor environment is a complex structure that shields most cancer cells from exposure to immune cells and the chemical armamentarium of the immune system. In spite of this and other huge challenges, some gradual success is being achieved with immunizations for cancer prevention and treatment.

The U.S. Food and Drug Administration (FDA) has approved two prophylactic cancer vaccines: HPV vaccines that prevent cervical and genital cancers caused by infection with Human Papilloma Virus (HPV type 16 and Type 18), and another vaccine to prevent liver cancer through immunization against hepatitis-B virus. DNA vaccines for treatment of existing cancer are approved for use only in treating metastatic prostate cancer in the United States.

In Brazil the first DNA vaccines against cancer were developed in a study of cell-fusion between dendritic cells and tumor cells extracted from patients with kidney cancer and melanoma. These therapeutic vaccines were

A research associate at the AIDS Vaccine Design and Development Laboratory collects bacteria transfected with DNA as part of research at the laboratory's campus in New York City. The laboratory, seeking a vaccine to prevent the spread of AIDS, is part of the International AIDS Vaccine Initiative (or IAVI), a global not-for-profit, public-private partnership working to accelerate the development of a vaccine to prevent HIV infection and AIDS. © *Chris Hondros/Getty Images.*

approved by the Brazilian regulatory agency, ANVISA, in 2004 as an adjuvant treatment for renal (kidney) cancer and melanoma.

SEE ALSO *Biotechnology Products; Genetic Engineering; Pharmacogenomics; Recombinant DNA Technology; Vaccination; Vaccines*

BIBLIOGRAPHY

Books

Bot, Adrian, and Mihail Obrocea, eds. *Cancer Vaccines: Challenges and Opportunities in Translation.* New York: Informa Healthcare, 2008.

Donnelly, Erin C., and Arthur M. Dixon, eds. *DNA Vaccines: Types, Advantages, and Limitations.* Hauppauge, NY: Nova Science, 2011.

Gabler, Maximilian. *DNA Replicons: Next Generation Vaccines against Allergy.* Saarbrucken, Germany: VDM Verlag, 2005.

Glick, Bernard R., Jack J. Pasternak, and Cheryl L. Patten. *Molecular Biotechnology: Principles and Applications of Recombinant DNA,* 4th ed. Washington, DC: ASM Press, 2010.

Saltzman, W. Mark, Hong Shen, and Janet L. Brandsma, eds. *DNA Vaccines Methods and Protocols,* 2nd ed. Totowa, NJ: Humana Press, 2006.

Schleef, Martin, ed. *DNA Pharmaceuticals: Formulation and Delivery in Gene Therapy, DNA Vaccination and Immunotherapy.* Weinheim, Germany: Wiley-VCH, 2005.

Stephenson, Frank Harold. *DNA: How the Biotech Revolution Is Changing the Way We Fight Disease.* Amherst, NY: Prometheus Books, 2007.

Thalhamer, Josef, Richard Weiss, and Sandra Scheiblhofer. *Gene Vaccines.* Vienna: Springer, 2011.

Periodicals

Munier, C. Mee Ling, Christopher R. Andersen, and Anthony D. Kelleher. "HIV Vaccines: Progress to Date." *Drugs* 71, no. 4 (March 5, 2011): 387–414.

Web Sites

"DNA Vaccines." *World Health Organization (WHO)* . http://www.who.int/biologicals/areas/vaccines/dna/en/index.html (accessed November 5, 2011).

"What are DNA Vaccines?." *Kenyon College.* http://biology.kenyon.edu/slonc/bio38/scuderi/partii.html (accessed November 5, 2011).

Sandra Galeotti

Drought-Resistant Crops

Introduction

Crop plants require a supply of water that enables them to survive and grow. The lack of rainfall that occurs in a drought can threaten the survival of crops, which can be a huge problem for the people and animals relying on the crops as a source of food. The problem of drought is becoming greater in some regions of the world, as the global climate changes with the warming of the planet's atmosphere.

The Intergovernmental Panel on Climate Change has estimated that the Earth's atmosphere warmed by 1.3°F (0.7°C) from 1906–2006. By 2106 the temperature of the atmosphere could increase by 2.8°F (1.5°C). The result could be more serious droughts in areas of the world already at risk, as well as the occurrence of drought in new regions. The World Bank has reported that the extensive cultivation of crops resistant to drought will be necessary if the ever-increasing global population is to be fed.

Adaptation to drought can occur naturally with some plants species. But this is not the case for all plants. Furthermore, natural adaptation can be a lengthy process, and it occurs at the local scale and not the global scale, which is needed to produce crop supply that is sufficient to feed the population of a region, country, continent, or even the planet.

Biotechnology and genetic engineering, in which genes that are crucial for survival in water-deficient conditions are added to the target crop, are being used to engineer plants that can continue to grow in drought conditions.

Historical Background and Scientific Foundations

The biotechnology technologies that can be used to make a drought-resistant plant are not new: They were developed in the 1970s and 1980s. More recently, the discipline of genomics has evolved. In genomics, the deoxyribonucleic acid (DNA) genetic material of organisms including plants can be compared to previously determined sequences to help pinpoint genes—regions that code for the production of proteins or other molecules.

Genes can be removed from the surrounding DNA using specific enzymes called restriction enzymes (as of 2011, hundreds of restriction enzymes are known) that recognize certain arrangements of the nucleotide building blocks of DNA and cut the DNA at those points.

Selection of other enzymes will enable the DNA of the target organism to be cut so that an introduced fragment of DNA can fit into and link up with the recipient DNA. When this occurs, the inserted gene is under the genetic control of the host, and whatever genes have been inserted will be expressed along with the normal host genes.

A different approach does not insert a new gene, but deliberately and specifically alters a gene that is already normally present. This can involve the replacement of one of the nucleotide bases that comprise the gene with a different nucleotide. The result will be a protein that has a different amino acid (amino acids are the building blocks of proteins) than the unaltered protein. A different amino acid will alter the three-dimensional structure of the protein, which can change the way the protein functions. The change can diminish or destroy the activity of the protein, but can also increase the protein. It is the latter that can be useful in designing plants resistant to drought.

Impacts and Issues

Research on drought resistance has identified several advantageous genes. If drought resistance genes are already in the plant, the plant might be able to be altered to increase their expression (thus producing more of the protein that is active in drought resistance). If not, the genes might transferred to the plant. A gene called

WORDS TO KNOW

DESICCATION: The removal of water from a living organism.

GLOBAL WARMING: The warming of Earth's atmosphere that is occurring due to the accumulation of compounds that increase the retention of heat.

SALINITY: The measure of the amount of salt dissolved in water.

erecta controls the formation of channels (pores) on the surface of some plants. The gene expression has been altered in *Arabidopsis* plants to reduce the amount of water vapor given off by the plants while maintaining the capability to take in the carbon dioxide that is essential for survival and growth. The resulting plants are hardier under drought conditions than their counterparts that lack erecta normally.

In another example, tomato plants have been engineered to increase their expression of a gene designated AVP1. The AVP1 protein functions in root development. The overproduction of AVP1 protein can lead to a more abundant root system, which increases the amount of moisture that a plant can obtain from the soil. Research has shown that the enhanced root development of engineered tomato plants greatly aids their growth under drought conditions, as compared to their counterparts that express AVP1 in normal amounts. Hopefully by identifying genes that perform a similar function, other plant species can be engineered for such drought resistance.

These examples are encouraging. As of 2011, drought-resistant plants remain in the developmental stage. Yet the prospects for the full-scale growth of such plants are good. For example, because rice and wheat plants contain gene sequences that are similar to the erecta gene in *Arabodopsis*, they may also be engineered for better growth in drought conditions.

Scientists are studying plants that grow in drought conditions to identify genes that are actively expressed

A hybrid tea crop is being grown at the Tea Research Institute to develop new types of drought-resistant tea seedlings in Kotmale, Sri Lanka. A severe drought in the first three months of 2009 sharply reduced the tea crop by nearly 50 percent while tea prices soared due to the shortfall in global production. © *AP Images/Lakruwan Wanniarachci*

IN CONTEXT: SEVEN BILLION PEOPLE ... AND COUNTING

Drought is the most common cause of all regional food shortages. Since 2001, recurring drought conditions in Ethiopia, Somalia, and Kenya have placed millions of people at risk from either death by starvation or poor health due to malnutrition.

Longer and more sustained droughts are an environmental phenomenon attributed by the vast majority of environmental experts to global warming, an aspect of climate change caused by increased levels of greenhouse gases such as carbon dioxide found in Earth's atmosphere. The warming planet has adverse consequences for humankind. The global population, which was estimated at around 7 billion in 2011, continues to grow at a rate that will add at least 1 billion more people by 2040. For this reason, the effects of drought on food production in affected regions are now greater than at any previous period in world history.

Many areas of the world are prone to low levels of precipitation; these regions are understood to be generally poor places to cultivate crops for food production. Increased soil salinity is an additional problem associated with soil exposed to drought conditions. Climate change concerns have stimulated the reassessment of traditional agricultural practices in many regions of the world.

The impact of drought on crop production is not a simple cause-and-effect relationship. Scientific studies confirm that where plants are subjected to heat stress induced by drought there are three likely consequences: The crop yield is reduced; the quality of the available crop for harvest is generally poor; and fewer seeds are produced by each affected plant, compromising the longer term sustainability of farming operations, especially those that depend on their own seed production year to year.

The study of global drought conditions engages four distinct agricultural sustainability issues. These are:

1. Food production. It is estimated that the global population will require twice as much food in 2050 as is produced today.
2. Environmental concern. Climate change has exerted its effects on food production, so it is imperative that the yield is increased from every unit of the finite resources of land, water, and energy that are directed to agricultural production.
3. Climate change. Crop yield stability must be increased in the face of the environmental stresses that result from climate change.
4. Economic growth. Improved economic circumstances of all persons involved in crop production will permit them to deal more effectively with the stresses imposed by drought.

This is the context in which drought-resistant crop development now occurs. Drought-resistant crops have been identified as a key component in the emerging strategy to counter the negative effects of climate change on agriculture. Given the acute drought-induced food shortages in Africa that affect upwards of 250 million people, significant research has been directed to improved drought resistance in indigenous crop species such as sorghum, pearl millet, chickpeas, and groundnuts. The ability of farmers to maximize crop yields may represent the difference between survival and death in developing countries where drought has been most prevalent.

---

during growth, and also those genes whose expression is diminished or turned off during growth. Some of these will prove to be crucial for drought survival. Already one such gene has been identified, which encodes for an enzyme called aldehyde dehydrogenase.

Drought-resistant plants are good for people in an increasingly dry planet. As well, the economic potential of such plants has been recognized by industry. Target crop plants such as corn, rice, wheat, and soybeans are being researched by companies including Bayer, Syngenta, Dow, Dupont, and especially Monsanto. Monsanto is on track to begin field testing of drought-resistant corn plants in the summer of 2012, with approval hoped for several years after that.

Despite the potential benefits of genetically engineered drought-resistant plants, concerns remain. A long-held concern with genetically modified plants involves the possibility that the engineered plants could spread their genetic traits to native plants living in the same environment. Although genetic transfer has occurred among some crops and their genetically modified varieties, transfer between GM crops and native non-crop plants is uncommon. Another worry is the design of plants that express certain genes only under certain conditions (such as when a particular nutrient is supplied). This approach, which is known as genetic use restriction technology (GURT) or "terminator technology," could place food in developing nations under the control of for-profit companies, critics argue, because the pollen or seeds of the crop plant would be sterile. Only by purchasing new seeds would drought-resistant plants be available in each growing season. This technology has been developed but is not currently in use in any crop plants.

SEE ALSO *Agricultural Biotechnology; Bt Insect Resistant Crops; Corn, Genetically Engineered; Cotton, Genetically Engineered; Disease-Resistant Crops; Frost-Resistant Crops; Genetically Modified Crops; Genetically Modified Food; Golden Rice; Herbicide-Tolerant Plants; Molecular Farming; Salinity-Tolerant Plants; Transgenic Plants*

BIBLIOGRAPHY

Books

Curtis, Ian S., ed. *Transgenic Crops of the World: Essential Protocols.* Boston: Kluwer Academic Publishers, 2004.

Lacey, Hugh. *Values and Objectivity in Science: The Current Controversy about Transgenic Crops.* Lanham, MD: Lexington Books, 2005.

Lichtfouse, Eric. *Alternative Farming Systems, Biotechnology, Drought Stress and Ecological Fertilisation.* Dordrecht, The Netherlands: Springer, 2011.

Wesseler, J. H. H., ed. *Environmental Costs and Benefits of Transgenic Crops.* Dordrecht, The Netherlands: Springer, 2005.

Web Sites

Bliss, Rosalie Marion. "Drought Hardy Soybean Lines Show Their Stamina." *Agricultural Research*, November/December 2008. http://ars.usda.gov/is/AR/archive/nov08/soybean1108.htm?pf=1 (accessed September 21, 2011).

"Coping with Water Scarcity in Developing Countries: What Role for Agricultural Biotechnologies? Summary Document to Conference 14 of the FAO Biotechnology Forum," March 5 and April 1, 2007. *Food and Agriculture Organization (FAO).* http://www.fao.org/biotech/logs/C14/summary.htm (accessed September 21, 2011).

The International Crops Research Institute for the Semi-Arid Tropics (ICRISAT). http://www.icrisat.org/ (accessed September 21, 2011).

Ledford, Heidi. "Researchers Engineer Drought-Resistant Plants." *Nature News*, November 26, 2007. http://www.nature.com/news/2007/071126/full/news.2007.289.html (accessed September 22, 2011).

Tran, Mark. "Small Farmers in Vanguard of Africa's Battle for Agricultural Development." *guardian.co.uk*, September 13, 2011. http://www.guardian.co.uk/global-development/2011/sep/13/africa-small-farmers-agricultural-development (accessed September 21, 2011).

"World Food Prize 2009: Ejeta." *The World Food Prize Foundation*, 2009. http://www.worldfoodprize.org/en/laureates/20002009_laureates/2009_ejeta/ (accessed September 21, 2011).

Brian Douglas Hoyle

Drug Price Competition and Patent Term Restoration Act

■ Introduction

The Drug Price Competition and Patent Term Restoration Act (DPC-PRA) is a U.S. law passed in 1984 to restructure the process by which generic drugs are approved and to modify the patent system for new, brand name drugs. The DPC-PRA, more commonly known as the Hatch-Waxman Act after its primary sponsors, U.S. Representative (D-CA) Henry Waxman (1939–) and U.S. Senator (R-UT) Orrin Hatch (1934–), sought to increase the availability of safe, low-cost generic drugs for American consumers by streamlining the U.S. Food and Drug Administration (FDA) approval process for generic drug. The Hatch-Waxman Act also exempted certain pre-sale activities by generic drug manufacturers from patent infringement claims. Additionally, the Hatch-Waxman Act modified the patent terms for new, brand name drugs in order to take into account the long clinical trial and FDA approval process for new drugs, which denies drug manufacturers the benefit of exclusive use under a patent.

■ Historical Background and Scientific Foundations

In 1984 the U.S. Court of Appeals for the Federal Circuit decided *Roche Products, Inc. v. Bolar Pharmaceutical Co.*, a patent infringement case between two pharmaceutical companies. In 1967 Roche obtained a patent on flurazepam hydrochloride, a medication for insomnia that Roche sold under the brand name Dalmane. Under existing patent laws, Roche's patent on flurazepam was set to expire in January 1984. In early 1983, Bolar Pharmaceutical began developing a generic version of Dalmane so they could rush the product to market shortly after the expiration of Roche's flurazepam patent.

In the summer of 1983, Bolar Pharmaceutical received a shipment of flurazepam from a foreign manufacturer and began conducting tests required for approval

of generic drugs by the FDA. Before the expiration of Roche's patent, Bolar Pharmaceutical did not produce a generic version of flurazepam for sale. In July 1983, Roche sued Bolar Pharmaceutical for patent infringement and sought to prevent Bolar Pharmaceutical from using flurazepam for any purpose, including conducting FDA required tests, before the expiration of the flurazepam patent.

Roche argued that the use of its patented drug by Bolar Pharmaceutical in order to gain a business advantage violated the exclusive rights granted to Roche under the flurazepam patent. A U.S. federal court sided with Bolar Pharmaceutical and decided that the use of flurazepam for the purpose of gaining FDA approval did not violate Roche's patent. On appeal, however, the United States Court of Appeals for the Federal Circuit decided that Bolar Pharmaceutical's actions infringed on Roche's patent, because the actions were undertaken purely for a business, and not a scientific, reason.

The Federal Circuit's decision meant that generic pharmaceutical manufacturers could not begin the process of seeking FDA approval for a generic drug until the patent on the brand name drug expired. At the time of the decision in *Roche Products, Inc. v. Bolar Pharmaceutical Co.*, FDA approval for a generic drug took more than two years. The court's decision, therefore, effectively extended the patent term on every patented drug.

■ Impacts and Issues

The U.S. Congress responded quickly to the Federal Circuit's decision in *Roche Products, Inc. v. Bolar Pharmaceutical Co.* and other issues that existed within the pharmaceutical patent framework and adopted the Drug Price Competition and Patent Restoration Act of 1984 (Hatch-Waxman Act). The Hatch-Waxman Act amended the Federal Food, Drug, and Cosmetics Act (FFDCA), which grants the FDA the power to regulate food, drugs, cosmetics, dietary supplements, medical

WORDS TO KNOW

FEDERAL FOOD, DRUG, AND COSMETIC ACT: The Food, Drug, and Cosmetic Act (FFDCA or FD&C Act) is a set of laws passed by Congress in 1938 and amended numerous times that grants the U.S. Food and Drug Administration the authority to regulate food, drugs, cosmetics, medical devices, food additives, dietary supplements, and related products.

GENERIC DRUG: A product that is pharmacologically similar to a brand name drug, including similar safety and efficacy, which is released following the expiration of the patent on the name brand drug.

PATENT: A government grant of the exclusive right to manufacture, sell, license, or use an invention or process for a limited period.

devices, and other products. The Hatch-Waxman Act modified the process by which generic drug manufacturers may seek FDA approval. The act also modifies the patent terms for companies that develop new drugs.

The Hatch-Waxman Act established a new, faster FDA approval process for generic drugs. The act allows the potential manufacturer of a generic version of a patented drug to file an Abbreviated New Drug Application (ANDA) if the generic version of the drug will be the bioequivalent of the original, patented drug. An ANDA allows the generic manufacturer to rely upon the safety and efficacy data submitted by the original drug manufacturer in conjunction with its initial application. The ANDA process saves generic drug manufacturers research costs and enables them to bring their products to the market shortly after the expiration of the original patent.

The Hatch-Waxman Act also included a provision that expressly overturned the Federal Circuit's ruling in *Roche Products, Inc. v. Bolar Pharmaceutical Co.* The act modified the 1952 Patent Act, exempting certain actions from patent infringement claims. The Hatch-Waxman Act states "It shall not be an infringement to make, use, offer to sell, or sell within the United States a patented invention…solely for uses reasonably related to the development and submission of information under a Federal Law which regulates the manufacture, use or sale of drugs or veterinary biological products." This provision allows a company or individual to use certain patented inventions in order to secure government approval for the new product.

The Hatch-Waxman Act also modified the patent term for pharmaceuticals to account for the reality that a portion of the patent term is lost to the FDA approval process. Typically, a patent term runs for 20 years from the date of application. This system allows an inventor a

period within which he or she has the exclusive right to manufacture, sell, or otherwise use the invention. Under the Hatch-Waxman Act, however, a pharmaceutical patent may be extended for a period equal to half the time required for clinical testing plus the entire time required for the FDA to approve the New Drug Application. This entire patent restoration period may not equal more than five years total. Furthermore, the remaining term of the restored patent may not exceed 14 years following FDA approval.

The Hatch-Waxman Act had the intended consequence of providing the American public with more low-cost generic drugs. Before the Hatch-Waxman Act, generic drugs accounted for less than 13 percent of the prescription drug market. Since 1984, however, more than 10,000 generic drugs have entered the U.S. marketplace, and generic drugs now account for more than 50 percent of the prescription drug market. Furthermore, the FDA estimates that the widespread availability of generic drugs will save Americans an estimated $35 billion between 2003 and 2013.

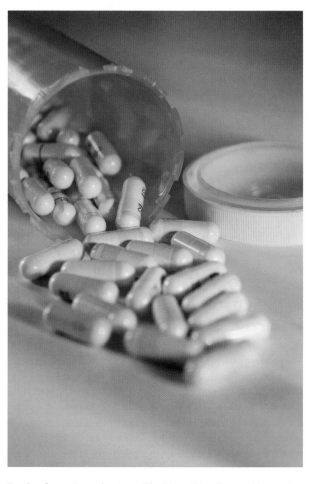

Bottle of generic medications. The Drug Price Competition and Patent Term Restoration Act, also known as the Hatch-Waxman Act, allowed for the current system of manufacturing and selling generic drugs. © Dgrilla/Shutterstock.com.

Protesters dressed as patients picket a trial court at the financial district of Makati city, east of Manila, Philippines, in November 2006 to coincide with the hearing on the lawsuit filed by drug company Pfizer against the Philippine government for patent infringement. Civic society groups, led by the international agency Oxfam, urged Pfizer to withdraw its lawsuit, which they claim will stifle the patient's option for cheaper but effective generic medicines.   © *AP Images/Bullit Marquez.*

SEE ALSO   *Antibiotics, Biosynthesized; Antirejection Drugs;* Ariad v. Lilly*; Bioethics; Biologics Control Act; Biopharmaceuticals; Cancer Drugs; Chemotherapy Drugs; Gene Therapy; Orphan Drugs; Penicillins; Pharmaceutical Advances in Quantitative Development; Pharmacogenomics;* Roche Products v. Bolar Pharmaceutical *(1984)*

BIBLIOGRAPHY

Books

Barr, David K., and Daniel L. Reisner, eds. *Pharmaceutical and Biotech Patent Law.* New York: Practising Law Institute, 2008.

Berry, Ira R., and Robert P. Martin. *The Pharmaceutical Regulatory Process,* 2nd ed. New York: Informa Healthcare, 2008.

Schacht, Wendy H., and John R. Thomas. *Follow-on Biologics Intellectual Property and Innovation Issues.* Washington, DC: Library of Congress, Congressional Research Service, 2007.

United States Federal Trade Commission. *Emerging Health Care Issues: Follow-on Biologic Drug Competition.* Washington, DC: U.S. Federal Trade Commission, 2009.

Web Sites

Crawford, Lester M. "Implementation of the Drug Price Competition and Patent Term Restoration Act of 1984." (Statement before the Subcommittee on Health House Committee on Energy and Commerce, October 9, 2002). *U.S. Food and Drug Administration (FDA).* http://www.fda.gov/ NewsEvents/Testimony/ucm115162.htm (accessed September 29, 2011).

Joseph P. Hyder

E. Coli: Protein Production Medium

■ Introduction

The development of recombinant proteins is one of the most important aspects of biotechnology research and related commercial applications. Whereas the further understanding of these proteins has been directed to numerous applications in biomedicine, recombinant protein science is also a vital component of modern agriculture. The ability to alter the genetic characteristics of crop plants to give them greater drought or pest resistance is an important contemporary research area given the demand for greater food production to support a growing global population.

Recombinant proteins are created through the application of recombinant DNA (rDNA) processes. Also known as molecular cloning, in its simplest terms rDNA is achieved when a strand taken from one DNA molecule is combined with the strand of a second sample. As with any technology that involves DNA modification, there are interesting ethical questions concerning the extent to which science ought to pursue cloning research that may lead to replicated human life forms or crops whose modified genetics might pose a threat to the health of the humans and animals that consume them.

Unlike the polymerase chain reaction (PCR) technologies widely employed by forensic scientists to replicate crime scene DNA sample analysis that may determine sample source identity, rDNA requires living cells as the medium in which DNA replication can occur. Bacterial *Escherichia coli* cells have been employed with considerable success in rDNA manufacture since the first rDNA was fashioned into protein successfully in 1973.

E. coli is renowned in the scientific research community as a model organism, as the features of *E. coli* organisms are similar to those observed in other bacteria that carry agricultural, medical, or environmental significance. *E. coli* are especially useful bacterial media for recombinant protein production because many strains of the bacteria have been studied extensively; it is a

relatively easy organism to grow from culture in the laboratory; and once rDNA has produced the desired proteins successfully, these are readily purified for research or biomedical applications, such as vaccines, pharmaceutical products, and blood clotting factors.

■ Historical Background and Scientific Foundations

The term recombinant DNA (rDNA) describes any product created in a process that joins two segments of DNA in a plasmid, the molecules usually present in bacteria. When rDNA is inserted into specific bacteria samples, a new protein composed of the rDNA will form in the bacteria. In biomedical processes, the proteins created are also known as therapeutic proteins. Examples include antibodies, antigens, hormones, and enzymes. Human insulin (developed in 1978) was the first drug created from recombinant proteins.

The production of recombinant protein (also known as protein expression) typically is achieved by either prokaryotic or eukaryotic systems. Prokaryotic cells such as *E. coli* are very small (diameter approximately 1–2 micrometers; length 10 micrometers), with simple physical structures that do not include the chromosomal DNA-carrying nuclei found in eukaryotic cells. Examples of eukaryotic protein expression systems are yeast and baculoviruses, the pathogens that attack various types of insects. *E. coli* as a recombinant protein medium is generally preferred to eukaryotic systems because its molecular properties are well understood, and it is a bacterium that can be cultured in the laboratory readily and safely.

Protein expression is achieved through a seven-stage process. The basic procedures do not differ between intended plant and human DNA applications. A DNA segment is selected for insertion into a cloning vector. The segment is cut from a DNA molecule using

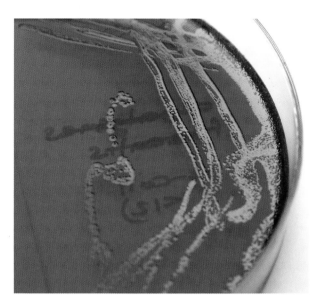

Escherichia coli on EMB agar.   © *Jarrod Erbe/Shutterstock*

bacteria-based restriction enzymes that recognize the specific places on the nucleotide where the DNA segments can be separated. The DNA inserted into the vector contains a marker that permits the recombinant molecules in the process to be identified. Where *E. coli* cells are used as the production medium, the vector prompts plasmid division within the cell. The plasmid divides and each part attaches itself to the cell wall. The cells divide quickly through binary fission to create two cells with identical genetic characteristics.

Following in the path of the numerous scientific advances associated with the discovery of the DNA double helix structure in the early 1950s, recombinant protein technology has emerged as an essential aspect of modern biotechnology research. The development of insulin through rDNA proteins carries potential benefits for millions of persons afflicted with diabetes. The ability to produce large protein quantities for experiments is important to the efficient function of a research laboratory. Protein therapy has emerged as an alternative to gene therapy in the treatment of mitochondrial disorders such as lactic acidosis, the potentially fatal condition in which the mitochondria, the "power house" of the human cell, malfunctions due to lactic acid accumulation. Protein therapeutics developed from rDNA technology have proved an effective counter to lactate accumulation.

■ Impacts and Issues

The advances in the understanding of how DNA can be manipulated into new protein forms have stimulated widespread debate. The issues generated by recombinant

WORDS TO KNOW

CLONING VECTOR: The selected DNA molecule that will replicate within a living cell during the recombinant process. Plasmids are a source used for these molecules.

ESCHERICHIA COLI: From the family Enterobacteriaceae, genus Escherichia, and species coli, *E. coli* is the widely recognized short form for the bacterium that is an abundant microorganism in the lower intestines of most mammals, including humans. First isolated in 1885, *E. coli* is the most commonly used bacterium as the medium for the production of recombinant proteins.

GENETICALLY-MODIFIED (GM) FOODS: GM foods include crop plants in which the original genetic traits are altered to improve a specific crop feature, such as drought resistance, herbicide tolerance, enhanced nutritional capacity, or protections against insects.

NUCLEOTIDES: The molecules that combine to form DNA. These are pentose (a sugar with 5 carbon atoms) and four nitrogenous bases (organic compounds essential to molecular structure): adenine (A), guanine (G), thymine (T) and cytosine (C).

PLASMIDS: DNA molecules that occur in bacteria. Plasmids can be replicated without the interaction of chromosomal DNA.

POLYMERASE CHAIN REACTION (PCR): PCR is one common method used to replicate DNA. The process is used in forensic biotechnology, where millions of copies of a DNA sample can be replicated from a single molecular strand. PCR is distinct from recombinant DNA processes in living cells, which are not required to replicate the desired sample.

PROTEIN EXPRESSION: The process to convert genetic information through synthesis into a protein.

protein technology have centered largely on the ethical issues associated with the apparent power to alter human, animal, or plant genetic structures fundamentally. In the early days of rDNA technology, the most prominent expressed concern was if the *E. coli* bacteria used in the rDNA procedures to create cancer-fighting drugs carried tumor-causing viruses; the failure to safeguard these materials properly by research facilities could precipitate a global cancer epidemic.

As recombinant protein science advanced into the twenty-first century, concern shifted from the nature of the *E. coli* media and associated rDNA processes to how the new protein forms will impact human societies and their environments. The development of genetically-modified (GM) foods has spurred intensive research as to their ultimate impact on human health, as well as the risk that GM crops will contaminate surrounding plants with herbicide-resistant genes and promote the growth of destructive invasive weed strains.

The overall impact of recombinant protein products in medicine remains an open question. The benefits associated with the development of potentially lifesaving rDNA products such as insulin must be weighed against the obvious potential that proteins could be designed to promote eugenics, the highly controversial concept of manipulating human genetic qualities to improve the human gene pool. Gene therapy, in which rDNA modified viruses are introduced into human cells to provide protection against specific diseases, is another medical advance of which the long term consequences are uncertain. It is equally clear that the prospect of gene therapy used to treat multiple sclerosis or other currently incurable diseases is a tantalizing one.

In an era in which international terrorism has dominated public discourse, recombinant protein technology could provide a means of creating active compounds from which different biological weapons could be developed. The prospect of rDNA being fashioned into virulent, treatment-resistant pathogens that terrorists could release into a target population remains an ongoing concern of military and emergency response leaders worldwide.

SEE ALSO *Genetic Engineering; Genetically Modified Crops; Industrial Genetics; Pharmacogenomics; Proteomics; Recombinant DNA Technology*

BIBLIOGRAPHY

Books

Lee, Sang Yup, ed. *Systems Biology and Biotechnology of Escherichia Coli.* Dordrecht, The Netherlands: Springer, 2009.

Maczulak, Anne E. *Allies and Enemies: How the World Depends on Bacteria.* Upper Saddle River, NJ: FT Press, 2011.

Williams, James A., ed. *Strain Engineering: Methods and Protocols.* New York: Humana, 2011.

Web Sites

"*E. coli* a Future Source of Energy?" *Biology News Net*, January 30, 2008. http://www.biologynews.net/archives/2008/01/30/ecoli_a_future_source_of_energy.html (accessed November 4, 2011).

"*E. Coli* to Store and Copy DNA." *National Human Genome Research Institute.* http://www.genome.gov/25019978 (accessed November 4, 2011).

Bryan Thomas Davies

Egg (Ovum) Donors

■ Introduction

Egg (ovum) donors donate either one or more eggs for the purpose of either assisted reproduction or medical research. The process allows an infertile female to have a baby. A donated egg is fertilized by sperm outside a woman's body, and then the resulting embryo is transferred to her womb where a pregnancy can be established.

Egg donation was shown to be possible in many other mammalian species before being tried in humans. It plays an important role in the cattle industry. The other application for egg donation is in cloning. When Dolly the sheep was cloned, an udder cell from one sheep was inserted into an egg from another sheep. The recipient egg was empty; its nucleus, with its genes, had been removed. Therefore, the resulting embryo had only the genes of the donor udder cell; Dolly was therefore a clone, or copy, of the cell donor. Cloning by nuclear transfer relies upon egg donation and could be used as a source of embryonic stem cells for repairing the body. The resulting embryo could also, like Dolly, be grown into a whole organism. Egg donation raises a number of safety and ethical issues, ranging from safety of hormonal treatments given to the donor to the status and fate of the resulting embryo.

■ Historical Background and Scientific Foundations

In the 1890s physiologist Walter Heape (1855–1929) of the University of Cambridge began to explore an idea called immunological privilege, which allows embryos to be transferred between individuals of the same species. By the early 1980s this had been done for several mammalian species including rodents, cattle, and sheep. Either the oocyte donor was fertilized in vivo and the embryo retrieved, or the egg was retrieved to be fertilized outside the body. In both cases, the resulting embryo was to be transferred to the womb of a second female. In 1983, teams in Los Angeles and in Melbourne, Australia, managed to achieve a human pregnancy using egg donation. The U.S. team used the in vivo technique and the Australian team the in vitro fertilization (IVF) approach. Healthy babies were born the following year. Egg donation has since become a mainstream process within assisted reproduction.

WORDS TO KNOW

ASSISTED REPRODUCTION: Assisted reproduction refers to a number of techniques, including in vitro fertilization, which are used to help achieve fertilization of an egg with a sperm in cases of infertility.

BLASTOCYST: A very early embryo, about five days old, containing hundreds of cells, organized into an inner and outer layer. A blastocyst is mature enough to be transferred to the uterus in IVF to establish a pregnancy.

CRYOPRESERVATION: A technique of preserving cells and tissues, by freezing them rapidly, usually in liquid nitrogen, and storing them at a temperature of −212°F (−100° C), or below.

IN VITRO FERTILIZATION (IVF): The word in vitro means "in glass" and IVF is a process in which a sperm fertilizes an ovum in a glass dish in a laboratory, instead of inside a woman's body. The term "test tube baby" is not accurate to describe a child born by IVF because the process does not occur in a test tube. The phrase "test tube" is probably being used as shorthand for "laboratory."

OOCYTE: The female mammalian germ cell, which matures into an ovum. The words oocyte and ovum are often used interchangeably, although this is not strictly accurate: An ovum can be fertilized, an oocyte cannot.

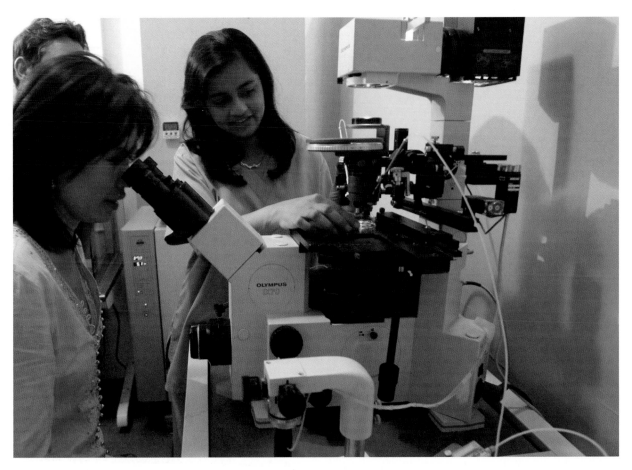

An Indian doctor shows a prospective mother slides of her two day old embryo before an implant procedure at her clinic in Mumbai, India. In about two weeks, the prospective mother will take a blood test in Miami that will tell her whether she and her husband are expecting their first child. They will not be the only ones waiting to hear the news. A 27-year-old woman in Vietnam—the egg donor—and the two doctors in this Indian city who implanted the embryo will be waiting eagerly for the results, too. © *Indranil Mukherjee/Getty Images.*

There are various types of egg donors. They may be designated by the recipient, who may feel more comfortable with the egg coming from someone she knows, like a sister. Alternatively, there are anonymous donors who may or may not be paid for their eggs. The procedure involves hormonal treatment to synchronize the menstrual cycles of the donor and recipient. This is to ensure that the recipient's womb is ready to receive the embryo shortly after the egg donation. The donor's ovaries are stimulated by hormones, to encourage more oocytes to mature, so there are more eggs to harvest. Finally, the oocytes and eggs must be retrieved from the follicles of the ovary, using a long needle that sucks the follicles out of the woman's body. Egg donation is becoming simpler and more widespread as cryopreservation techniques improve. It is no longer crucial to have live donation and embryo transfer occur at the same time. Women can donate eggs and have them safely stored for future use.

There are many reasons why a couple may choose to use egg donation. Some women do not produce their own oocytes because they were born without ovaries, or their ovaries do not function. The woman may have a severe genetic disorder and be afraid of passing it on if she uses her own eggs, or they may have been damaged by cancer treatment. Egg donation can also enable a woman who is past menopause to have a baby or for two gay men to have a child using the sperm of one man and a donor egg.

Nuclear transfer for cloning mammals, including humans, is the other main application of egg donation. A cell from any part of the body can be transferred to an enucleated, or empty, egg from a donor. The resulting embryo is a copy or a clone of the donor of the cell. It can be used in therapeutic cloning, as a source of embryonic stem cells, or the embryo could be grown into a whole organism, which is called reproductive cloning. There is much research going on in therapeutic cloning but reproductive human cloning is forbidden in countries that legislate cloning research. There have been no scientifically verifiable reports of human reproductive cloning to date.

■ Impacts and Issues

Egg donation has greatly expanded the scope of assisted reproduction. In-vitro fertilization was originally confined to the father and the biological mother. Egg donation creates an embryo with two mothers, one genetic, which is the egg donor, and one biological, the one who actually gives birth. This raises some interesting legal issues on which of the two women is the true mother; this needs to be clarified by a binding agreement between egg donor and recipient. There also need to be safeguards to ensure the donor is healthy and will not pass on any genetic or infectious disease, such as HIV, through the donation. Similarly, questions could be raised about the suitability of the recipient. Egg donation has enabled women in their 60s to give birth, which could not happen naturally. These children may well be deprived of a mother before reaching adult life and, indeed, the woman may be less physically able to meet the challenges of motherhood than a younger woman. There are also questions on the status of the embryos created by, but not used in, in vitro fertilization. These could be stored, destroyed, or used for medical research, but it is not clear who has the right to make the decision. Because each embryo is, potentially, a human being, their fate is naturally of concern for ethical reasons. In the United States, the American Society for Reproductive Medicine has issued guidelines for egg donation that are followed by reputable clinics.

In 2004 the South Korean scientist Hwang Woo Suk (1953–) claimed to have cloned human embryos by nuclear transfer, and in 2005 he published research on obtaining stem cells from these embryos for the treatment of disease. Although he claimed to have abided by high ethical principles, he later admitted to using eggs from his students and junior scientists, paying other women for oocytes, and having the women lie about the purpose of the donation. All of this is against the Declaration of Helsinki, which lays out ethical guidelines for medical research involving human subjects. It was also revealed that the Korean team had used hundreds of eggs in their experiments, far more than originally reported. This raises the question of how to obtain a supply of human eggs for therapeutic cloning in an ethical manner. Egg donation is not without risk to the donor, and if the end result is not a baby, the eggs may be used an experiment whose value is not known. One solution might be to use eggs from cows, instead of humans. A few laboratories do have a license to pursue this kind of work but many object to the fact that a cow-human hybrid embryo is thereby created.

SEE ALSO *Bioethics; Designer Babies; Designer Genes; Frozen Egg Technology; Helsinki Declaration (1964); Human Subject Protections; Human Subjects of Biotechnological Research; In Vitro Fertilization (IVF); Informed Consent; Twiblings*

BIBLIOGRAPHY

Books

Brinsden, Peter R. *A Textbook of In Vitro Fertilization and Assisted Reproduction.* Boca Raton, FL: Taylor & Francis, 2005.

Cassan, A. *Human Reproduction and Development.* Philadelphia: Chelsea House, 2006.

Gardner, David K. *In Vitro Fertilization.* Boca Raton, FL: Informa Healthcare, 2006.

Mundy, Liza. *Everything Conceivable: How Assisted Reproduction Is Changing Men, Women, and the World.* New York: Alfred A. Knopf, 2007.

Web Sites

"Assisted Reproductive Technology (ART)." *Centers for Disease Control and Prevention (CDC).* http://www.cdc.gov/art/ (accessed September 8, 2011).

Susan Aldridge

Electrophoresis

■ Introduction

Electrophoresis is a laboratory technique that allows molecules (including proteins or DNA fragments) to be sorted based on charge or size. A substance is placed in a liquid solution, a gel, or other media, to which an electric field is applied. Electrophoresis separates polar or ionic molecules based on their mobility through the media under a force supplied an electric field. For example, the charged particles move toward either an anode or cathode in what is known as moving boundary electrophoresis or, in zone or gel electrophoresis, larger molecules move more slowly than the smaller molecules through the media used and create distinct bands of particles.

Electrophoresis enables scientists to analyze numerous substances, including blood and DNA. The technology has many scientific uses, among which are DNA and protein fingerprinting, Southern and Western blot techniques, DNA sequencing, oncogene (genes associated with cancer) identification, and detecting the genes associated with several heritable diseases, such as Huntington's disease, sickle cell anemia, Duchenne's Muscular Dystrophy, and numerous viral infections. In addition to their popular use in laboratory and research settings, electrophoresis techniques are used frequently in hospitals, medical centers, and cancer treatment facilities.

Although there are mentions of foundational research beginning in the mid- to late-nineteenth century, the techniques used for the process of electrophoresis were developed primarily during the first half of the twentieth century. The research of Arne Tiselius (1902–1971), in which he separated proteins and performed chemical analyses, created the technology and methods for moving boundary electrophoresis, which used an electric charge in an aqueous solution to separate compounds based on the movement of charged particles. Rapid advances made in electrophoresis techniques, particularly in gel

methods, helped support the rise of molecular biology and biotechnology in the 1960s and 1970s.

■ Historical Background and Scientific Foundations

Descriptions of the concepts involved in the process of electrophoresis date back to research based on Faraday's Laws in the late eighteenth century. In the nineteenth century, Johann Hittorf (1824–1914), Friedrich Kohlrausch (1840–1910), and Walther Nernst (1964–1941) helped to develop the mathematical theories of electrochemistry by studying the movement of electrically charged particles through aqueous solutions when exposed to low level electrical voltage. Early in the twentieth century, Arne Tiselius pioneered sucrose-based gel electrophoresis methods. Tiselius won the Nobel Prize in Chemistry in 1948 for his research on electrophoresis and adsorption analysis of serum proteins. He developed the Tiselius apparatus used in moving boundary electrophoresis; it was considered the standard until zone electrophoresis methods were optimized in the 1940s and 1950s. Scientists working during the middle of the twentieth century realized that moving boundary electrophoresis was unable to separate compounds completely if they were very similar in terms of charge or molecular weight. With the advent of zone electrophoresis, first using starch-based gels and then moving to the use of polyacrylamide gels, it became possible to separate chemically complex macromolecules and protein compounds electrophoretically into stable zones for accurate analysis.

By 1955 starch gels has been introduced as media for electrophoresis; they were found to be less than optimally effective for clean separation. By the end of the 1950s, acrylamide gels were in use; Leonard Ornstein (1880–1941) and Baruch Joel (BJ) Davis were fine-tuning disk electrophoresis. Several aspects of the electrophoresis process became more standardized, including the ability to create consistent pore size and stability of gel media. Scientists discovered, in 1969, that the use of denaturing agents such as the detergent SDS (sodium dodecyl sulfate) greatly facilitated separation of protein subunits. In 1970 scientists pioneered the use of stacking gels in combination with SDS for separation of very small protein subunits. The use of sequencing gels was developed in 1977; agarose gel became routinely used in the late 1970s. The development of pulsed field gel electrophoresis in 1983 made it possible to separate large DNA molecules for the first time. Capillary electrophoresis, used extensively in hospitals and medical settings for the analysis of serum proteins and the detection of disease markers, was developed in the early 1980s.

■ Impacts and Issues

Electrophoresis is one of the most frequently used techniques in molecular biology. It is relatively simple, comparatively inexpensive, and quite accurate. For routine separation of nucleic acids and proteins, one-dimensional electrophoresis is used; the two-dimensional form is most commonly employed for DNA fingerprinting. Electrophoresis can be used to separate all of the proteins within highly complex molecules in order to facilitate accurate identification.

Electrophoresis is used for a number of purposes in biotechnology. It frequently is used to identify tissue and other biological materials (skin, semen, blood) found at a crime scene. In judicial and child welfare settings, electrophoresis often establishes (or excludes) paternity. It can be used to identify familial relationships within animal groups; this is especially helpful in selective breeding programs involving endangered or at-risk species. Electrophoresis can help scientists make the best decisions about selections of breeding pairs. It is useful in determining evolutionary relationships when comparing the genetics of different organisms or species. Electrophoresis is also employed to identify viral infections, diseased genes, and the presence of cancer-causing cells as well as to identify carriers of certain specific genetic diseases, such as Huntington's Disease, sickle cell anemia, and Duchenne's Muscular Dystrophy.

Macromolecular or DNA electrophoresis separates macromolecules or DNA segments according to size. DNA samples, purified and put into solution, are placed in the wells of an electrophoresis gel slab (or tube) and a low voltage current is passed across the medium for a specific period of time, generally several hours. The current is removed, and the pattern of movement across the surfaces of the gel (called the fragmentation gradient) undergoes analysis after application of DNA-specific fluorescent dye. The fragmentation gradient will show

WORDS TO KNOW

AGAROSE: A specialized chemical used in gel electrophoresis and made from agar, which is extracted from a type of seaweed called red marine algae. Although it is said to be non-toxic, it is best to wear gloves and use appropriate ventilation equipment when handling agarose. Agarose comes in powdered form for laboratory use; it is mixed with buffer solution, heated until it melts, then poured and allowed to set into a gel. The higher the concentration of agarose insulation, the firmer the gel will be.

DNA FINGERPRINTING: Often used in forensics or paternity testing, DNA fingerprinting involves sequencing an individual's DNA and comparing it to evidentiary DNA. It is also called DNA typing.

MOLECULAR WEIGHT: The weight of a molecule, which is the sum of the weights of all of its component atoms. The unit of weight is the dalton, which equals one-twelfth of the weight of a carbon 12 atom.

POLYACRYLAMIDE: A form of acrylamide that can be either a simple linear chain or a cross-linked structure. Polyacrylamide is very water absorbent and is used to form a soft gel in gel electrophoresis. It is a fairly strong neurotoxin, so it must be handled very carefully, using appropriate protective equipment.

POLYPEPTIDE: A chain composed of a long string of amino acids, joined by peptide bonds. Proteins are formed from polypeptides.

bands of DNA separated by molecular weight. They are compared to DNA markers of known lengths and molecular weights to facilitate identification.

Because DNA molecules have an overall negative charge resulting from their sugar-phosphate "backbones," they will be pulled across a gel field supplied with an electrical current toward the direction of the positive terminal (anode). Before undergoing electrophoresis, DNA often is treated with restriction enzymes in order to break it up into smaller segments. The larger the chain of DNA, the more slowly the segment will migrate.

Two types of gel are commonly used with DNA analysis: agarose and polyacrylamide. Agarose gel is most effective for moderate resolution of long-chain DNA segments, whereas polyacrylamide is best for very high resolution of small segments. Proteins are usually treated with heat and then with SDS in order to unfold them and make them linear before they are put through gel electrophoresis. When SDS-treated molecules are electrophoresed through polyacrylamide gel, it is called SDS-PAGE. When agarose gel or isoelectric focusing are used, proteins may be kept in their native (folded) state.

A University of Illinois student prepares soybean DNA samples for gel electrophoresis in the Agricultural Science laboratory at the University of Illinois. © *AP Images/Seth Perlman.*

■ Primary Source Connection

In the past, DNA fragments were separated by a high-speed centrifuge method known as velocity-sedimentation ultracentrifugation. Gel electrophoresis, electrophoresis using electricity (as opposed to gravitation), is a markedly more efficient method for analysis of DNA restriction patterns. Specifics on refinements and the benefits surrounding this technology are discussed in the excerpt below.

Gel Electrophoresis

DNA fragments of different sizes, such as those resulting from digestion with restriction endonucleases, were originally separated by a laborious method called velocity-sedimentation ultracentrifugation. DNA samples are spun in a high-speed centrifuge through a gradient of salt or sugar that would separate DNA fragments by size, largest to smallest, top to bottom. In 1970, Daniel Nathans used polyacrylamide gel

electrophoresis as a simple and rapid means to separate DNA fragments. Whereas centrifugation uses gravitational force to separate molecules, electrophoresis means literally to carry with electricity. Gel electrophoresis takes advantage of the fact that, as an organic acid, DNA is negatively charged. DNA owes its acidity to phosphate groups that alternate with deoxyribose to form the rails of the double helix ladder. In solution, at neutral pH, negatively charged oxygens radiate from phosphates on the outside of the DNA molecule. When placed in an electric field, DNA molecules are attracted toward the positive pole (anode) and repelled from the negative pole (cathode).

During electrophoresis, DNA fragments sort by size in the polyacrylamide gel. The porous gel matrix acts as a molecular sieve through which smaller molecules can move more easily than larger ones; thus, the distance moved by a DNA fragment is inversely proportional to its molecular weight. In a given period of time, smaller restriction fragments migrate relatively far from the origin compared to larger fragments. Because of its small pore size, polyacrylamide efficiently separates small DNA fragments of up to 1000 nucleotides. However, this level of resolution was inappropriate for isolating gene-sized fragments of several thousand nucleotides.

Although less time consuming than centrifugation, early polyacrylamide gel electrophoresis was still labor intensive and required the use of radioactively labeled DNA fragments. Following electrophoresis, the polyacrylamide gel was cut into many bands, and the amount of radioactivity in each slice was determined in a scintillation counter. The pattern of radioactivity was used to reconstruct the pattern of DNA bands in the gel.

A research team at Cold Spring Harbor Laboratory, led by Joseph Sambrook, introduced two important refinements to DNA electrophoresis that made possible rapid analysis of DNA restriction patterns. First, they replaced polyacrylamide with agarose, a highly purified form of agar. An agarose matrix can efficiently separate larger DNA fragments ranging in size from 100 nucleotides to more than 50,000 nucleotides. DNA fragments in different size ranges can be separated by adjusting the agarose concentration. A low concentration (down to 0.3%) produces a loose gel that separates larger fragments, whereas a high concentration (up to 2%) produces a stiff gel that resolves small fragments.

Second, they used a fluorescent dye, ethidium bromide, to stain DNA bands in agarose gels. Following a brief staining step, the fragment pattern is viewed directly under ultraviolet (UV) light. This technique is extremely sensitive; as little as 5ng (0.000000005g) of DNA can be detected. Thus, it is not difficult to understand why ethidium bromide staining quickly replaced radioactive

labeling for the routine analysis of DNA restriction patterns.

David A. Micklos
Greg A. Freyer

MICKLOS, DAVID A., AND GREG A. FREYER. "GEL ELECTROPHORESIS," EXCERPT FROM PAGES 113–114 IN *DNA SCIENCE: A FIRST COURSE.* COLD SPRING HARBOR, NY: COLD SPRING HARBOR LABORATORY PRESS, 2003.

SEE ALSO   *Biotechnologists; DNA Fingerprinting; DNA Sequencing; Nanotechnology, Molecular; Polymerase Chain Reaction (PCR); Recombinant DNA Technology; Selection and Amplification Binding Assay; Single Molecule Fluorescent Sequencing; SNP Genotyping*

BIBLIOGRAPHY

Books

Landers, James P., ed. *Handbook of Capillary and Microchip Electrophoresis and Associated Microtechniques.* Boca Raton, FL: CRC Press, 2008.

Sheehan, David, and Raymond Tyther, eds. *Two-Dimensional Electrophoresis Protocols.* New York: Humana, 2009.

Web Sites

"Electrophoresis." *National Human Genome Research Institute.* http://www.genome.gov/Glossary/index.cfm?id=56 (accessed November 2, 2011).

"Gel Electrophoresis Virtual Lab." *The University of Utah.* http://learn.genetics.utah.edu/content/labs/gel/ (accessed November 2, 2011).

Pamela V. Michaels

Eli Lilly & Co. v. Medtronic, Inc.

■ Introduction

Eli Lilly & Co. v. Medtronic, Inc. was a case decided by the Supreme Court of the United States in 1990 that addressed the acceptable use of patented technology in testing and marketing medical devices. The Supreme Court held (decided) that medical device manufacturers could use patented devices in testing and marketing new medical devices, even if such use constituted patent infringement. In essence, the Supreme Court ruled there was a special exception to patent infringement in similar cases.

The Supreme Court's decision in *Eli Lilly* turned on the application of the Drug Price Competition and Patent Term Restoration Act of 1984, also known as the Hatch-Waxman Act. The Hatch-Waxman Act provided a safe harbor provision (a form of legal exception), codified at 35 U.S.C. 271(e)(1), that exempted "uses reasonably related to the development and submission of information under a Federal law which regulates the manufacture, use, or sale of drugs" from patent infringement. Although the safe harbor provision of the Hatch-Waxman Act was designed to cover the testing and marketing of generic drugs, the Supreme Court held that the safe harbor also applies to the testing and marketing of medical devices.

■ Historical Background and Scientific Foundations

In 1983, a predecessor of Eli Lilly instituted proceedings to enjoin (prevent) Medtronic, Inc., a U.S. medical technology company, from testing and marketing an implantable cardiac defibrillator. Eli Lilly asserted that Medtronic's actions violated Eli Lilly's exclusive rights under two patents, which were utilized in Medtronic's testing and marketing. A jury in the federal district court that heard the initial suit found that Medtronic violated one of the patents held by Eli Lilly, and the judge ordered a finding of infringement on the second patent.

On appeal, the Court of Appeal for the Federal Circuit, which hears appeals involving patent issues, reversed the district court. The appellate court held that Medtronic's use of Eli Lilly's patented devices in testing and marketing its own product did not violate Eli Lilly's patents. The court stated that the safe harbor provision of the Hatch-Waxman Act protected the testing and marketing actions of Medtronic and other companies performing similar research, development, and marketing. This safe harbor provision allows for the use of otherwise infringing actions, so long as those actions are "reasonably related to the development and submission of information under a Federal law which regulates the manufacture, use, or sale of drugs."

■ Impacts and Issues

The Federal Food, Drug, and Cosmetics Act (FFDCA) grants the U.S. Food and Drug Administration the power to regulate the safety of food, drugs, cosmetics, medical devices, food additives, dietary supplements, and all related products. In 1984, Congress passed the Hatch-Waxman Act, which amended the FFDCA and modified existing drug regulations, particularly as they applied to the development and approval of generic drugs.

The safe harbor infringement exemption provision of the Hatch-Waxman Act was designed to reverse the decision of the Court of Appeals for the Federal Circuit's holding in *Roche Products, Inc. v. Bolar Pharmaceutical Co.* In *Roche*, the Federal Circuit held that a potential generic drug manufacturer could not even begin the testing required to seek U.S. Food and Drug Administration (FDA) approval based on the safety and efficacy of the generic drug without infringing on the patent. Such a system essentially granted the original patent holder additional years of exclusive rights due to the protracted FDA drug approval process, thereby affecting consumers who were faced with additional years of more expensive brand-name drugs. The safe harbor provision of the Hatch-Waxman Act, therefore, allowed potential generic drug manufacturers to begin testing generic drugs before the expiration of the patent.

The issue that faced the Supreme Court in *Eli Lilly v. Medtronic* centered on whether activities that would otherwise constitute patent infringement are permissible if they are undertaken for the purpose of developing and submitting information to the FDA in order to obtain marketing approval for a medical device under the FFDCA. Federal law normally "regulates the manufacture, use, or sale of drugs," which is covered under section 201(g) of the FFDCA, but the safe harbor provision clearly establishes noninfringing actions involving the development of products during which such drugs are used. The safe harbor provision, however, does not explicitly apply its requirements to the manufacture, use, or sale of medical devices, which are covered section 201(h) of the FFDCA. Eli Lilly argued the safe harbor provision's language, therefore, did not cover Medtronic's development of medical devices.

The Court stated that the Hatch-Waxman Act's use of the phrase "'a Federal law which regulates the manufacture, use, or sale of drugs' more naturally summons up the image of an entire statutory scheme of regulation," and, therefore, refers to the entire FFDCA. The FFDCA in its entirety authorizes the FDA to regulate

medical devices (along with many other items) in addition to drugs. Finding that the phrase "a Federal law which regulates the manufacture, use, or sale of drugs" referred to medical devices in addition to drugs, the Court held that Medtronic's actions fell under the safe harbor contemplated by Hatch-Waxman Act.

WORDS TO KNOW

FEDERAL FOOD, DRUG, AND COSMETIC ACT: The Food, Drug, and Cosmetic Act (FFDCA or FD&C Act) is a set of laws passed by Congress in 1938 and amended numerous times that grants the U.S Food and Drug Administration the authority to regulate food, drugs, cosmetics, medical devices, food additives, dietary supplements, and related products.

MEDICAL DEVICE: A medical device is a product used for the diagnosis or treatment of patients.

PATENT: A patent is a government grant of the exclusive right to manufacture, sell, license, or use an invention or process for a limited period.

The "Rising Man" symbol stands in front of the Fridley, Minnesota, based Medtronic. © *AP Images/Jim Mone.*

Eli Lilly & Co.'s headquarters in Indianapolis, Indiana. © *Tom Strickland/Bloomberg/Getty Images*

The Court's decision in Eli Lilly greatly changed the landscape of medical device development and manufacturing. Medical devices typically require years of research and development and often must rely on existing, and often patented, technology. If the Court had held that the use of existing, patented technology in the development of a medical device constituted patent infringement, then advances in medical devices would have been greatly curtailed or made more expensive to consumers.

■ Primary Source Connection

Eli Lilly & Co. v. Medtronic, Inc. is a case that was brought in front of the United States Supreme Court surrounding patent infringement in the medical device industry. Questions arose about patent laws relating to premarketing activity to gain approval of a device. The District Court initially entered judgment on verdicts for Eli Lilly, however the Court of Appeals reversed the judgment deciding that Medtronic's activities did not constitute infringement as related to the Federal Food, Drug, and Cosmetic Act.

Eli Lilly & Co. v. Medtronic, Inc.

Lilly & Co. v. Medtronic, Inc., 496 U.S. 661 (1990)

Eli Lilly and Company v. Medtronic, Inc.

No. 89–243

Argued Feb. 26, 1990

Decided June 18, 1990

496 U.S. 661

Justice SCALIA delivered the opinion of the Court.

This case presents the question whether 35 U.S.C. § 271(e)(1) renders activities that would otherwise constitute patent infringement noninfringing if they are undertaken for the purpose of developing and submitting to the Food and Drug Administration information necessary to obtain marketing approval for a medical device under § 515 of the Federal Food, Drug, and Cosmetic Act, 90 Stat. 552, 21 U.S.C. § 360e (FDCA)....

II

In 1984, Congress enacted the Drug Price Competition and Patent Term Restoration Act of 1984, 98 Stat. 1585 (1984 Act), which amended the FDCA and the patent laws in several important respects. The issue in this case concerns the proper interpretation of a portion of section 202 of the 1984 Act, codified at 35 U.S.C. § 271(e)(1). That paragraph, as originally enacted, provided:

"It shall not be an act of infringement to make, use, or sell a patented invention (other than a new animal drug or veterinary biological product (as those terms are used in the Federal Food, Drug, and Cosmetic Act and the Act of March 4, 1913)) solely for uses reasonably related to the development and submission of information under a Federal law which regulates the manufacture, use, or sale of drugs."

35 U.S.C. § 271(e)(1) (1982 ed., Supp. II). The parties dispute whether this provision exempts from infringement the use of patented inventions to develop and submit information for marketing approval of medical devices under the FDCA.

A

...The centrally important distinction in this legislation (from the standpoint of the commercial interests affected) is not between applications for drug approval and applications for device approval, but between patents relating to drugs and patents relating to devices. If only the former patents were meant to be included, there were available such infinitely more clear and simple ways of expressing that intent that it is hard to believe the convoluted manner petitioner suggests was employed would have been selected. The provision might have read, for example,

"It shall not be an act of infringement to make, use, or sell a patented drug invention . . . solely for uses reasonably related to the development and submission of information required, as a condition of manufacture, use, or sale, by Federal law...."

B

Under federal law, a patent "grant[s] to the patentee, his heirs or assigns, for the term of seventeen years, . . . the right to exclude others from making, using, or selling the invention throughout the United States."

35 U.S.C. § 154. Except as otherwise provided, "whoever without authority makes, uses or sells any patented invention, within the United States during the term of the patent therefor, infringes the patent."

35 U.S.C. § 271(a). The parties agree that the 1984 Act was designed to respond to two unintended distortions of the 17-year patent term produced by the requirement that certain products must receive premarket regulatory approval. First, the holder of a patent relating to such products would, as a practical matter, not be able to reap any

financial rewards during the early years of the term. When an inventor makes a potentially useful discovery, he ordinarily protects it by applying for a patent at once. Thus, if the discovery relates to a product that cannot be marketed without substantial testing and regulatory approval, the "clock" on his patent term will be running even though he is not yet able to derive any profit from the invention.

The second distortion occurred at the other end of the patent term. In 1984, the Court of Appeals for the Federal Circuit decided that the manufacture, use, or sale of a patented invention during the term of the patent constituted an act of infringement, see 35 U.S.C. § 271(a), even if it was for the sole purpose of conducting tests and developing information necessary to apply for regulatory approval. Since that activity could not be commenced by those who planned to compete with the patentee until expiration of the entire patent term, the patentee's de facto monopoly would continue for an often substantial period until regulatory approval was obtained. In other words, the combined effect of the patent law and the premarket regulatory approval requirement was to create an effective extension of the patent term.

The 1984 Act sought to eliminate this distortion from both ends of the patent period. Section 201 of the Act established a patent-term extension for patents relating to certain products that were subject to lengthy regulatory delays and could not be marketed prior to regulatory approval. The eligible products were described as follows:

"(1) The term 'product' means:"

"(A) A human drug product."

"(B) Any medical device, food additive, or color additive subject to regulation under the Federal Food, Drug, and Cosmetic Act."

"(2) The term 'human drug product' means the active ingredient of a new drug, antibiotic drug, or human biological product (as those terms are used in the Federal Food, Drug, and Cosmetic Act and the Public Health Services Act) including any salt or ester of the active ingredient, as a single entity or in combination with another active ingredient."

35 U.S.C. § 156(f). Section 201 provides that patents relating to these products can be extended up to five years if, inter alia, the product was "subject to a regulatory review period before its commercial marketing or use," and "the permission for the commercial marketing or use of the product after such regulatory review period [was] the first permitted commercial marketing or use of the product under the provision of law under which such regulatory review period occurred."

35 U.S.C. § 156(a).

The distortion at the other end of the patent period was addressed by § 202 of the Act. That added to the provision prohibiting patent infringement, 35 U.S.C. § 271, the paragraph at issue here, establishing that "[i]t shall not be an act of infringement to make, use, or sell a patented invention . . . solely for uses reasonably related to the development and submission of information under a Federal law which regulates the manufacture, use, or sale of drugs."

35 U.S.C. § 271(e)(1). This allows competitors, prior to the expiration of a patent, to engage in otherwise infringing activities necessary to obtain regulatory approval.

Under respondent's interpretation, there may be some relatively rare situations in which a patentee will obtain the advantage of the § 201 extension but not suffer the disadvantage of the § 202 noninfringement provision, and others in which he will suffer the disadvantage without the benefit. Under petitioner's interpretation, however, that sort of disequilibrium becomes the general rule for patents relating to all products (other than drugs) named in § 201 and subject to premarket approval under the FDCA. Not only medical devices, but also food additives and color additives, since they are specifically named in § 201, see 35 U.S.C. § 156(f), receive the patent-term extension; but since the specific provisions requiring regulatory approval for them, though included in the FDCA, are not provisions requiring regulatory approval for drugs, they are (on petitioner's view) not subject to the noninfringement provision of § 271(e)(1). It seems most implausible to us that Congress, being demonstrably aware of the dual distorting effects of regulatory approval requirements in this entire area—dual distorting effects that were roughly offsetting, the disadvantage at the beginning of the term producing a more or less corresponding advantage at the end of the term—should choose to address both those distortions only for drug products; and for other products named in § 201 should enact provisions which not only leave in place an anticompetitive restriction at the end of the monopoly term but simultaneously expand the monopoly term itself,

thereby not only failing to eliminate but positively aggravating distortion of the 17-year patent protection. It would take strong evidence to persuade us that this is what Congress wrought, and there is no such evidence here....

Antonin Scalia

SCALIA, ANTONIN. "ELI LILLY & CO. V. MEDTRONIC, INC." *UNITED STATES REPORTS (U.S.)* 496 (1990): 661–673.

SEE ALSO Ariad v. Lilly; Diamond v. Chakrabarty; Labcorp v. Metabolite, Inc.; Medimmune, Inc. v. Genentech, Inc.; Merck Kgaa v. Integra Lifesciences I, Ltd.; Monsanto Canada Inc. v. Schmeiser; *Patents and Other Intellectual Property Rights;* Roche Products v. Bolar Pharmaceutical

BIBLIOGRAPHY

Books

Amani, Bita. *State Agency and the Patenting of Life in International Law: Merchants and Missionaries in a Global Society.* Burlington, VT: Ashgate, 2009.

Bohrer, Robert A. *A Guide to Biotechnology Law and Business.* Durham, NC: Carolina Academic Press, 2007.

Ida Madieha bt. Abdul Ghani Azmi. *Biotechnology Law and Policy: Searching for a Balance.* Gurgaon, India: Madhav Books, 2011.

Mills, Oliver. *Biotechnological Inventions Moral Restraints and Patent Law,* Rev. ed. Farnham, VT: Ashgate Pub, 2010.

Naidu, B. David. *Biotechnology & Nanotechnology: Regulation under Environmental, Health, and Safety Laws.* Oxford: Oxford University Press, 2009.

Joseph P. Hyder

Environmental Biotechnology

Introduction

Environmental biotechnology refers to the use of biotechnology techniques to study or use natural environmental sources of material. The use of these materials can be therapeutic, such as the development of natural-source antibiotics. Environmental biotechnology is often a market-driven enterprise, in which the aim is to profit from the sale of the materials. As well, the approach can be used for social benefit; examples include the control of air and water pollution and the use of living organisms as energy sources (biofuels).

The benefits of environmental biotechnology can be enormous, for example in the cleanup of petroleum spills by bacteria that have been genetically engineered to prefer the petroleum compounds as a nutrient. However, this aspect of biotechnology has always created concerns over the release of the modified organisms to the natural environment. Even with stringent genetic restrictions on the environmental spread of the engineered organisms, the ongoing concern is that the genetically modified organisms will have a survival advantage and will outcompete the natural species.

Historical Background and Scientific Foundations

Sewage treatment is a type of environmental biotechnology that dates back more than 150 years. By the early twenty-first century, the technology included sophisticated treatment plants that harnessed oxygen-requiring (aerobic) bacteria and bacteria that grow in the absence of oxygen (anaerobes) to treatment sewage-laden water. The degradation of the waste compounds in the water by the bacteria is a crucial part of the process that renders contaminated water fit to drink.

Typically, the use of microorganisms in environmental biotechnology involves an extended process in which the particular species of microbe (bacteria are often used) is adapted to be able to use the target compound as a nutrient. The process involves growing the microbe in the presence of a low concentration of the target pollutant. Because the target compound is usually toxic, most of the microbial population will die. But some organisms that are capable of surviving in the presence of the noxious compound will survive. These survivors are then grown in the presence of a higher concentration of the pollutant. Again, most of the microbial population of this first wave of survivors will die but a few will survive. Repeated cycles with increasing concentrations of the pollutant will yield a microbial population that is capable of growing in the presence of a high concentration of the pollutant. Most often, growth occurs because the organisms are capable of using the pollutant as a nutrient source (typically as source of carbon and/or nitrogen).

This process can be taken one step further; the microbe can be designed to be absolutely dependent on the presence of the pollutant for survival. If other sources of carbon and/or nitrogen are present, for example, the microbe will preferentially use the pollutant.

The engineered microbe can be used to degrade spills involving the target pollutant and, when the spill has been degraded and the pollutant is no longer present as a nutrient source, the organism will die off. This die off is ensured by additional genetic engineering to create the need for one or more compounds as nutrients that are not found naturally, but which are supplied during the degradation process.

Plants can also be used in environmental cleanup. Similar to the above process, plants can be bred to be very resistant to high concentrations of compounds that would otherwise be toxic, such as arsenic. As the plants grow, they will accumulate the noxious compound in their cells. The plants can then be harvested and safety disposed of, which removes much of the pollutant from the site of contamination. As well, the gas that is produced during the bacterial degradation (biogas, typically

WORDS TO KNOW

BIOREMEDIATION: The use of living organisms to degrade noxious compounds that have been accidently released into water or soil.

EXXON VALDEZ: The name of an oil tanker that ruptured in Prince William Sound, Alaska, in March 1989, spilling upwards of 750,000 barrels of oil. Part of the cleanup involved the use of a genetically-modified bacterium that was the first patented living organism.

GENE: A sequence of deoxyribonucleic acid that codes for a molecule that has a function.

methane) can be recovered and ignited, and the released energy harnessed to perform work.

Species of fungi are also used as "scrubbers" to remove pollutants from the water used in the pulp and paper industry. The various types of degradation rely on enzymes—proteins that can accelerate the speed of chemical reactions but which are not altered by the reaction. In technical language, the microbes function as catalysts. Plants can also be biofuels—alternate sources of energy to fossil fuels.

■ Impacts and Issues

Environmental biotechnology is extremely valuable in a number of ways. The design of organisms capable of remediating spills of compounds such as oil, gas, airline fuel, and even radioactive waste has enabled the cleanup of contaminated soil and water that would otherwise be impossible. As well, the use of plants as a fuel source could be prudent in the face of dwindling supplies of oil that some models have predicted could begin within the next century. In addition, biofuels could help curb emission of atmosphere-warming gases.

Although the use of plants such as corn and sugarcane as biofuels has been demonstrated and is practiced in countries including the United States, the approach has its critics. The critics contend that the energy required to grow and harvest the crops, the use of environmentally noxious fertilizers and pesticides during growth of the crops, and the exorbitant use of water during irrigation and processing is too great to justify the widespread use of biofuels.

Another concern with environmental biotechnology, especially the use of deliberately genetically-engineered organisms, is the potential that the natural balance of the various types of organisms will be upset by the overgrowth of the organism being used in the remediation. Genetic safeguards make it theoretically very unlikely that engineered organisms will persist in

Cliff Leeper, overseeing the cleanup of the Wyckoff/Eagle Harbor Superfund site, a former wood-treating plant on Bainbridge Island near Seattle, walks across tanks filled with water contaminated with creosote and other chemicals that has been mixed with a naturally occurring oil-eating microbe to help break down the pollution. Contaminated water passes through the microbe treatment tanks and two other cleaning stages for a total of approximately 36 hours of cleaning before being released back into the harbor. © *AP Images/Ted S. Warren.*

the environment very long after the target compound has been degraded. Yet, studies have demonstrated persistence of months to years. Furthermore, adaptation of the released organisms to the natural environment can never be absolutely ruled out.

■ Primary Source Connection

Researchers in Ireland are studying a link between methane production and ties with chemicals found in the stools of livestock such as cows and sheep. One may question the importance of this research and if so, one will find that the research findings offer insight into the environment and how the estimation of methane emissions by animals may contribute to global warming.

Measuring Ruminant Emissions Through Biomarkers Found in Stool

Livestock is a significant contributor to greenhouse gases. The ruminant digestive system creates ample amounts of methane which is released into the atmosphere. It is difficult to measure the amount of methane produced by cows because unlike emission stacks, ruminant exhaust cannot be controlled or monitored. However, researchers from the University of Bristol and the Teagasc Animal and Grassland Research Centre in Ireland have made the connection between methane production and a certain chemical found in the stool of cows, sheep, and other animals. This link may be used to more accurately estimate methane emissions by animals and assess their contribution to global warming.

Through various methods, people are able to calculate the emissions they produce from generators, turbines, boilers, and furnaces. Many governments demand an annual statement of emissions by every company with a regulated system. Some companies have to install continuous emissions monitors which constantly measure what comes out of the stack. But livestock operators have been exempt from such rules even though their animals produce methane, a potent greenhouse gas.

"We're quite good at measuring man-made CO2 emissions, but techniques to measure the animal production of methane—a much more potent greenhouse gas—have serious limitations," stated co-author, Dr. Fiona Gill, postdoctoral researcher at Bristol who is now at the University of Leeds. "If we can identify a simple biomarker for methane production in animal stools, then we can use this along with information on diet and animal population numbers to estimate their total contribution to global methane levels."

The chemical found in stool is called archaea, which are symbiotic microbes living in the foregut of ruminant animals. The product of their metabolism is methane which is expelled from the body through burping or flatulence. The microbe was initially found inside the animals guts, and the research expanded to locate it in the feces. The scientists then compared the microbe concentrations in feces of animals which were on different diets.

Principal investigator, Dr. Ian Bull said, "Two groups of cows were fed on different diets and then their methane production and faecal archaeol concentration were measured. The animals that were allowed to graze on as much silage as they wanted emitted significantly more methane and produced faeces with higher concentrations of archaeol than those given a fixed amount of silage, supplemented by concentrate feed. This confirms that manipulating the diet of domestic livestock could also be an important way of controlling methane gas emissions."

Ruminants are thought to produce about one fifth of all methane gas worldwide. In the past, researchers had tried to measure methane production directly from the animals using respiration chambers, but this proved to be too cumbersome for the researchers and the animals. This research is significant for more accurately measuring greenhouse gas production.

David A. Gabel

GABEL, DAVID A. "MEASURING RUMINANT EMISSIONS THROUGH BIOMARKERS FOUND IN STOOL." *ENVIRONMENTAL NEWS NETWORK*, JUNE 12, 2011. AVAILABLE ONLINE AT HTTP://WWW.ENN.COM/WILDLIFE/ARTICLE/42802.

SEE ALSO *Agricultural Biotechnology; Biodegradable Products and Packaging; Biodiversity; Biodiversity Agreements; Bioremediation; Bioremediation: Oil Spills; Biotechnology; Genetically Modified Crops; Solid Waste Treatment; Wastewater Treatment*

BIBLIOGRAPHY

Books

Agarwal, S. K. *Advanced Environmental Biotechnology.* New Delhi, India: A.P.H. Publishing Corp, 2005.

Ahmad, Wan Azlina, Zainoha Zakaria, and Zainul Akmar Zakariab. *Bacteria in Environmental Biotechnology: The Malaysian Case Study-Analysis, Waste Utilization and Wastewater Remediation.* New York: Nova Science Publishers, 2011.

Evans, Gareth, and Judith C. Furlong. *Environmental Biotechnology: Theory and Application.* Chichester, West Sussex, UK: Wiley-Blackwell, 2011.

Saran, V. K. *Environmental Biotechnology: Commercial Applications.* Delhi, India: Medhashri Publications, 2011.

Scragg, Alan H. *Environmental Biotechnology.* Oxford, UK: Oxford University Press, 2005.

Web Sites

"Environmental Remediation & Water Resources Program." *Lawrence Berkley National Laboratory, Earth Sciences Division.* http://esd.lbl.gov/research/programs/erwr/ (accessed September 23, 2011).

"Environmental Technologies Action Plan: Eco-Innovation for a Sustainable Future." *European Commission: Environment.* http://ec.europa.eu/environment/etap/index_en.htm (accessed September 23, 2011).

"Environmental Technology Verification Program." *U.S. Environmental Protection Agency.* http://www.epa.gov/etv/ (accessed September 23, 2011).

Brian Douglas Hoyle

Enzyme Replacement Therapy

■ Introduction

In many persons with single-gene disorders, an enzyme needed for correct biochemical functioning is lacking because of the existence of a mutation in the gene that codes for the enzyme. Enzyme replacement therapy involves administering the missing enzyme in its genetically engineered form. Thus far, enzyme replacement therapy has been successfully applied to a group of rare genetic disorders called the lysosomal storage diseases. The lysosome is an organelle whose role is to break down lipids, glycoproteins, and mucopolysaccharides. In these diseases, enzymes involved in these processes are lacking, and the molecules build up in the cell, leading to potentially fatal cellular damage. With the supply of the correct enzyme, the lysosome can function normally.

Genetic engineering technology has made it possible to make synthetic versions of many human proteins from the corresponding genes. The drugs are administered by injection and, as of 2011, four lysosomal storage diseases are treatable by enzyme therapy. The first enzyme replacement therapy made available was for Type I Gaucher's disease, which is caused by a lack of the enzyme glucocerebrosidase. The disorder causes a buildup of lipids in various organs, which can lead to anemia, bruising, and bone disease. It affects around one person in every 50,000 to 100,000 of the population. The treatment is expensive, and the diseases are rare, so patients, government, and biotechnology companies have had to work together to ensure the treatments are developed and made available.

■ Historical Background and Scientific Foundations

In 1991 researchers at the National Institute of Neurological Diseases and Stroke, part of the National Institutes of Health (NIH), announced that enzyme replacement therapy could reverse symptoms in 12 patients with Type 1 Gaucher's disease. In this disorder, the missing enzyme is glucocerebrosidase, which is needed for breakdown of lipid molecules. In its absence, lipid accumulates, causing a range of serious symptoms such as liver and spleen swelling, bone fractures, and increased bruising and bleeding. The 12 patients were treated with infusions of a modified version of glucocerebrosidase called alglucerase (Ceredase), and the drug was approved in the same year.

The biotechnology company Genzyme, which is based in Cambridge, Massachusetts, has pioneered enzyme replacement therapy. Following the launch of alglucerase, the company has developed a second-generation therapy for Gaucher's disease called imiglucerase. The other Genzyme drugs for lysosomal storage disease are laronidase for mucopolysaccharidosis, agalsidase-beta for Fabry disease, and alglucosidase-alfa for Pompe disease. All of these enzyme therapies are manufactured by transferring the appropriate gene to mammalian cells. These are then grown in a bioreactor, where they produce enzyme molecules. After harvesting the protein from the host cells, it is purified using filtration and chromatography. Because the enzyme is to be infused into human subjects, quality control demands from the regulatory authorities must be carefully followed.

■ Impacts and Issues

There are more than 40 different lysosomal storage diseases, and each is classified as a rare disease. Taken together the diseases do, however, affect around one in 5,000 people, and they are potentially life threatening, with no treatment available prior to the advent of enzyme replacement therapy. However, developing a new drug can cost at least a billion dollars, and pharmaceutical companies can afford this investment only

WORDS TO KNOW

FABRY DISEASE: A lysosomal storage disease caused by deficiency of the enzyme alpha-galactosidase A, which is essential to the breakdown of molecules known as glycosphingolipids. Without the enzyme, these molecules accumulate in the nerves, kidneys, and heart.

LYSOSOME: Organelles that are involved in the degradation of proteins and other substances in the cell. Lysosomal storage disorders lead to accumulation of lysosomes because there is a lack of an enzyme involved in normal lysosomal functioning.

POMPE DISEASE: A lysosomal storage disease caused by deficiency of the enzyme alpha-glucosidase. The deficiency leads to buildup of the sugar glycogen in the cells, which impairs functioning of muscle and nerve cells throughout the body.

RARE DISEASE: A disease affecting fewer than 200,000 people in the U.S. population.

if they know there is a large enough market to recoup the money. With a rare disease, the market is limited. The development of genetic engineering technologies in the 1970s opened up the possibility of treating lysosomal storage diseases by replacement of the missing enzyme. However, the orphan drug regulations introduced into the United States and Europe were a necessary step in persuading biotechnology companies such as Genzyme that it was worthwhile to develop treatments for rare diseases.

The United States Orphan Drug act was introduced in 1983. There are similar regulations in Europe, Singapore, Australia, and Japan. Drug companies developing drugs for rare diseases benefit from tax incentives, an accelerated pathway to approval, and extended patent life. The legislation has enabled the development of around 350 drugs, including the enzyme replacement therapies. It arose from intense lobbying from patient groups keen to promote the interests of those suffering from rare diseases and ensuring they have the same rights as those with more common disorders.

A patient receives his weekly four-hour infusion, called enzyme replacement therapy, in Chicago, Illinois. The patient has a rare genetic disorder called Maroteaux-Lamy Syndrome, also known as MPS, type 6. *© AP Images/M. Spencer Green.*

Enzyme replacement therapies are classed as therapeutic proteins, which also include drugs such as human insulin for treatment of diabetes and erythropoietin for anemia. Protein drugs are broken down by stomach acid, so they usually have to be given by injection or infusion. There is much interest in trying to deliver these drugs orally by developing tablet versions of the enzymes or alternative drugs. An oral drug may be more acceptable to patients because it is so much easier to take. Genzyme has an oral drug for Gaucher's disease in clinical trials. It is not an enzyme, but knowledge about the disease derived from developing alglucerase has helped in its discovery and development.

Another option for patients with rare genetic disorders is gene therapy. Instead of administering the enzyme, which has to be taken regularly to maintain therapeutic effect, the gene for the missing enzyme could be delivered into the patients' cells. This may provide a longer-lasting therapy because the patient has a built-in source of the missing enzyme. However, various technical challenges remain before gene therapies are proven safe and effective for patients, such as how to best deliver them to cells and how to make the gene produce enough of the desired enzyme.

SEE ALSO *Bioreactor; Gene Therapy; Pharmacogenomics*

BIBLIOGRAPHY

Books

Colavito, Mary C. *Gene Therapy.*, San Francisco: Pearson, 2007.

Connock, M. *The Clinical Effectiveness and Cost-Effectiveness of Enzyme Replacement Therapy for Gaucher's Disease: A Systematic Review.* Tunbridge Wells, Kent, UK: Gray Publishing, 2006.

El-Metwally, Tarek H. *Medical Enzymology: A Simplified Approach.* New York: Nova Science, 2011.

Elstein, Deborah, Gheona Altarescu, and Michael Beck. *Fabry Disease.* New York: Springer, 2010.

Web Sites

"Genes and Gene Therapy." *National Institutes of Health (NIH).* http://www.nlm.nih.gov/medlineplus/genesandgenetherapy.html (accessed September 29, 2011).

"Lipid Storage Diseases Fact Sheet." *National Institute of Neurological Disorders and Stroke, National Institutes of Health (NIH).* http://www.ninds.nih.gov/disorders/lipid_storage_diseases/detail_lipid_storage_diseases.htm (accessed September 29, 2011).

Susan Aldridge

Enzymes, Industrial

■ Introduction

An enzyme is a biological catalyst that speeds up essential biochemical reactions, such as the breakdown of glucose, in living organisms. Without enzymes, these reactions would occur at only a very slow rate. Enzymes are very specific in the reactions they catalyze, and generally work best at low temperatures and in aqueous solutions. Enzymes are being used increasingly in the detergent, food, beverage, and pharmaceutical industries. Such industrial enzymes open up the possibility of achieving reactions and processes under milder conditions than are possible with conventional chemical catalysts.

Industrial enzymes enable manufacturers to reduce waste and energy, and to cut down on their use of toxic solvents. For drug manufacturers, the application of a biocatalyst in manufacturing has the added advantage of producing a chirally pure drug, which may be safer for the patient and is preferred by regulatory authorities. Industrial enzymes are produced by genetic engineering and cell culture, with a host cell carrying the gene that codes for the enzyme. Discovering and developing new enzymes for industrial applications is a major focus in so-called white biotechnology. There is great interest in discovering new sources of industrial enzymes with microbes, particularly those found in extreme conditions, which are being explored as sources of novel industrial enzymes.

■ Historical Background and Scientific Foundations

Enzymes in wild yeasts have been used for thousands of years in brewing and wine making. But modern enzyme technology began with Danish chemist Christian Hansen's (1843–1916) isolation of rennet from dried calves' stomachs in 1874. Rennet is actually a mixture of enzymes that is applied in the coagulation of milk proteins to make cheese. The twentieth century saw increasing application of enzymes in the food industry, particularly in Japan, where mold fungi are used to make a range of products based on soy protein, such as miso and tempeh. Enzymes also proved useful for desizing textiles. Bacterial amylase, derived from *Bacillus subtilis*, was derived for this purpose in 1917. German industrialist Otto Röhm (1876–1939) studied the role of enzymes in the bating of leather, which traditionally was carried out with dog and pigeon excrement. He argued that it was the digestive enzymes in the excrement that helped prepare the leather and began to use pancreatic extracts instead, with successful results. He began marketing it as Oropon in the early 1900s. The first purified industrial enzyme was a bacterial protease, first marketed in 1959 and employed by the major detergent manufacturers beginning in 1965.

There are around 50 commercially available industrial enzymes. They are mainly applied in the detergent, textile, starch, baking, and animal feed industries, but are used increasingly in drug manufacture. Enzymes are classified according to the reaction that they catalyze. A protease breaks down protein into amino acids, whereas a hydrolase catalyzes the breaking of a chemical bond with the addition of water. The majority of industrial enzymes are hydrolases. Most are protein molecules, and they can be used either in isolated form or within microbial cells. Often they will be used in immobilized form, attached to polymer beads, because they are more stable in this format and can be recycled for use in the next manufacturing batch.

The availability of industrial enzymes in amounts needed to make commercial batches has been driven by the success of genetic engineering. Previously, enzymes could be extracted only in relatively small amounts from animal and plant tissue. Currently most industrial enzymes are produced by cultured microorganisms in bioreactors. The gene for the desired enzyme is inserted into the host microorganism, using a vector, to produce a high-producing strain. The culture medium and production conditions are optimized by small-scale experiments. Commercial quantities are obtained subsequently by harvesting the cell culture and isolating and purifying the enzyme.

■ Impacts and Issues

Because the natural environment of enzymes is the living cell, they catalyze reactions in physiological conditions, which usually means near neutral pH, aqueous solution, and at temperatures in a narrow range around 98.6°F (37°C). Replacing more conventional processes with one based on a biocatalyst often makes it more environmentally friendly by reducing energy inputs and waste. For instance, use of biological detergent enables laundry to be washed at a much lower temperature than with a soap-based detergent. However, it is in the chemical and pharmaceutical industries that biocatalysts have most to offer over traditional catalysts. Production processes requiring several steps often can be streamlined, which cuts down on raw materials and waste. This may go hand in hand with the elimination of one or more reactions that require a toxic solvent, replacing it with one that can be utilized in aqueous solution.

In drug manufacturing, the use of biocatalysts has the advantage that it can produce a chiral drug molecule, in which all the carbon atoms are in the correct orientation in space. The three-dimensional structure of an enzyme drives its specificity in a way that does not apply to simple metal catalysts. The enzyme will produce the correct chiral form of the drug from intermediate molecules in a very selective way. The importance of having the correct chiral form of a drug was revealed by the thalidomide tragedy of the 1960s. Babies born to mothers taking this drug suffered from congenital birth defects, and the drug had to be withdrawn. It was later discovered that one of the chiral forms of thalidomide caused the damage, while the other form was the one that had the therapeutic impact on the patient. Increasingly, regulators demand that manufacturers minimize harm to the patient by making the correct chiral form of the drug only.

The introduction of enzymes into industrial processes is limited by their availability. Around 3,000 enzymes are known, but only a tiny proportion is available in amounts large enough and cheap enough to allow them to be used on a large scale. Genetic engineering processes are complex and costly to develop, so most enzymes have not yet been exploited in this way. Even when an enzyme is available, it may not be appropriate for an industrial process. Enzymes are proteins and depend upon their three-dimensional structure for correct function. They rapidly unfold if subject to harsh conditions such as high temperatures or high salt concentration, both of which are prevalent in industrial processes. It is not always easy to replace a conventional process with one that can be carried out in mild conditions suitable for enzymes.

WORDS TO KNOW

BIOCATALYST: A substance that initiates or accelerates the rate of a biochemical reaction. Most biocatalysts are enzymes, and the two terms are sometimes used interchangeably, but hormones may also be considered as biocatalysts.

BIOLOGICAL DETERGENT: A detergent containing enzymes including proteases, amylases, and lipases, which break down protein, carbohydrate, and fat stains respectively.

CHIRAL: A carbon atom in an organic molecule that is attached to four different atoms or groups of atoms is said to be chiral. A molecule with a chiral carbon atom exists in space as two non-superimposable mirror images that may have different biological properties. Enzymes, amino acids, and other biological molecules are chiral.

DESIZING: A sizing material is added to a textile to protect its fibers during weaving. Desizing is the removal of this material from the textile after the weaving process.

WHITE BIOTECHNOLOGY: Also known as industrial biotechnology, white biotechnology is the use of biological products in industrial products that are normally made from petroleum, such as fuel, polymers, bulk chemicals, and specialty chemicals.

The biodiversity of nature, however, may provide many more industrial enzymes in the future. There are thousands of enzyme genes in genome databases just waiting to be exploited commercially. Some are more able to tolerate harsh conditions than enzymes currently available. A particularly rich source of enzymes is microbial life existing in extreme environments. For example, researchers at the Idaho National Laboratory have isolated an enzyme from a microbe that exists in the hot springs of Yellowstone National Park. The temperature range of this catalyst is 86°F to more than 212°F (30°C to more than 94°C), and its pH range is 6–10. Experiments show that this catalyst is very stable compared to other catalysts at higher temperatures and pH. Its half life was 15 days at 176° (80°C) and pH 10, whereas a conventional enzyme from *Aspergillus niger* had a half life of just 15 seconds under the same conditions. This novel enzyme has potential application for the removal of hydrogen peroxide in processes involving bleach, such as those used in the paper and pulp, textiles, and food pasteurization industries. The advantage would be cheaper treatment of wastewater and lower energy costs.

SEE ALSO *Algae Bioreactor; Antimicrobial Soaps; Biopharmaceuticals; Bioplastics (Bacterial Polymers); Biopreservation; Bioreactor; Bioremediation; Biosurfactants; Food Preservation; Forensic Biotechnology; Laundry Detergents; Solid Waste Treatment; Wastewater Treatment*

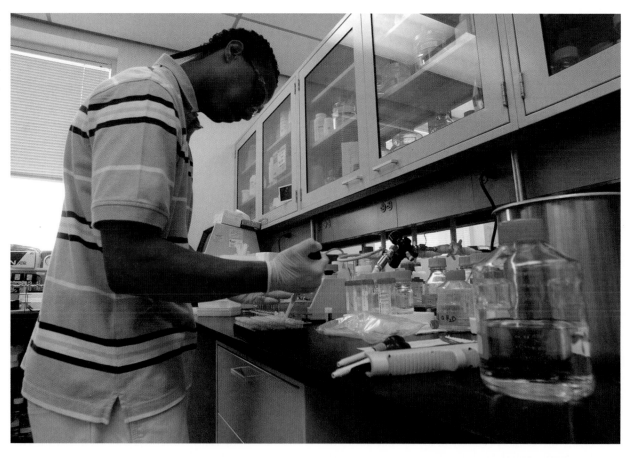

A researcher works in the fermentation and recovery lab inside Genencor International, Inc. in Palo Alto, California. Genencor has the genetic material of tens of thousands of strains of microbes stored in deep-freeze refrigerators in Palo Alto and the Netherlands. It has numerous industrial products on the market and is using living material—enzymes and proteins, rather than fossil fuels—to develop cleaner and cheaper ways of making industrial chemicals. *© AP Images/Marcio Jose Sanchez.*

BIBLIOGRAPHY

Books

Polaina, Julio, and Andrew P. MacCabe. *Industrial Enzymes.* Dordrecht, The Netherlands: Springer, 2007.

Sijbesma, Feike, and Hugo Schepens. *White Biotechnology: Gateway to a More Sustainable Future.* Brussels: Europabio, 2003.

Ulber, Roland, and Dieter Sell. *White Biotechnology.* London: Springer, 2007.

Varfolomeev, Sergei D., G. E. Zaikov, and Larisa P. Krylova. *Biotechnology and the Ecology of Big Cities.* New York: Nova Science, 2011.

Periodicals

Frazzetto, Giovanni. "White Biotechnology." *European Molecular Biology Organization (EMBO) Reports* 4, no. 9 (2003): 835–837. Available online at http://www.nature.com/embor/journal/v4/n9/full/embor928.html (accessed October 1, 2011).

Web Sites

"Third Time Lucky." *The Economist*, June 4, 2009. http://www.economist.com/node/13725783 (accessed October 1, 2011).

Susan Aldridge

Ethanol

■ Introduction

Ethanol refers to ethyl alcohol. The chemical formula is C_2H_5OH. Ethanol is important in biotechnology in two main aspects: It is the active component of alcoholic beverages, and it is a biologically-derived energy source (biofuel) for vehicles equipped to combust liquid ethanol. The global use of ethanol as a biofuel has increased markedly beginning in the early 1990s. However, this use is not without controversy, because it requires the growth and pollution-producing and water-consuming processing of crops such as corn that could otherwise be a food source. Nonetheless, the use of ethanol biofuel contributes an estimated 85 percent less greenhouse gas emissions into the atmosphere than the burning of gasoline as vehicle fuel.

■ Historical Background and Scientific Foundations

Ethanol is produced by the microbial degradation of lignocellulose containing plants such as corn and sugarcane, as well as wood chips. The lignocellulose makes up a large proportion of the plant support structure. It is inedible to humans, who lack the necessary enzyme to degrade the compound. This form of ethanol is known as cellulosic ethanol.

The history of ethanol production from wood dates back over 150 years to Germany. In the United States, Henry Ford (1863–1947) designed a car in 1896 that operated on ethanol, and the Model T was capable of running purely on ethanol or a gasoline-ethanol mixture. By the time of World War I (1914–1918), several ethanol production facilities were operating in the United States. Thereafter, the abundant availability and inexpensive (at the time) purchase price of gasoline reduced interest in ethanol fuel. Another reason for the decline in interest was due to Prohibition (1920–1933), which discouraged the manufacture of ethanol for drinking purposes.

With the increasing realization beginning in the 1970s of the influence of gasoline-related emissions on the warming of Earth's atmosphere, ethanol was again considered as a fuel source. Embracing the use of ethanol as a biofuel took longer in some countries, with the United States lagging behind other nations. But in a State of the Union address on January 31, 2006, U.S. President George W. Bush (1946–) proposed to expand use of cellulosic ethanol to 35 billion gallons annually by 2017. Because it has been estimated that the maximum U.S. yield of cellulosic ethanol from corn starch is only 15 billion gallons per year, the presidential mandate necessitated the increasing use of corn that would otherwise be used for food, as well as new sources of ethanol.

Ethanol can be produced from cellulose in two ways. The first involves the use of specific enzymes that degrade the cellulose into simple sugar molecules, which can then be used as nutrient by microorganisms. The microbial breakdown of the sugars is done in the near-absence or complete absence of oxygen. This process, which is called fermentation, yields ethanol. The second process is called gasification. This involves the change of lignocellulose to carbon monoxide and hydrogen, with the subsequent production of ethanol by microbial fermentation or chemical reactions.

The production of ethanol by yeasts such as *Saccharomyces cerevisiae* has been used in the production of alcoholic beverages including beer for centuries.

■ Impacts and Issues

The use of ethanol as a fuel source is rapidly increasing globally, from 17 billion liters in 2000 to about 52 billion liters in 2007 to 87 billion liters in 2010. The United States has become the world's greatest producer of ethanol fuel, with almost 58 percent of global production in 2010.

Until recently, the fermentative production of ethanol was done by bacteria, fungi, algae, and yeasts that were naturally available, albeit selected based on their

WORDS TO KNOW

COMBUSTION: The chemical destruction of a compound in the presence of oxygen, which produces energy.

LIGNIN: A rigid polymer found in the cell walls of many types of plants.

PHOTOSYNTHESIS: The use of sunlight as an energy source for the manufacture of nutrients from carbon dioxide and water.

effectiveness over time. Since 2006, reports have described the use of a genetically-altered strain of yeast and of the bacterium *Escherichia coli* that are able to directly degrade cellulose, bypassing the step in which cellulose is converted to simple sugars. The use of yeast is especially attractive, because it is active at a low pH that tends to prevent the growth of bacteria, reducing the potential of bacterial contamination that could detract from the efficiency of ethanol production. While the use of genetically-modified species would save time and money if successfully scaled-up to the volumes required for industrial purposes,

their use is controversial and is opposed by those who are concerned over the consequences of the release of the organisms to natural settings and by the potential for industrial monopolization of a food crop such as corn. Also, in the United States, ethanol production is highly subsidized by the government. If the cash subsidy was not in place, the price of ethanol fuel would be very high.

The use of algae to produce ethanol is being refined and may represent a potentially important source in the future. Algae that can use the energy of sunlight can be grown in sun-exposed ponds; ethanol is produced directly and can be recovered without harming the algae. The production of algae-derived ethanol has been estimated as 6,000 gallons (22,712 liters) per acre per year, as compared with 400 gallons (1,514 liters) per acre per year for corn.

SEE ALSO *Agricultural Biotechnology; Biofuels, Gas; Biofuels, Liquid; Biofuels, Solid; Bioreactor; Biotechnology; Corn, Genetically Engineered*

BIBLIOGRAPHY

Books

Biofuels, Biorefinery & Renewable Energy: Issues & Developments. Hauppauge, NY: Nova Science, 2011.

People march to protest recent price increases in tortillas in Mexico City, on January 31, 2007. Tens of thousands of trade unionists, farmers, and leftist groups marched through downtown Mexico City on Wednesday to protests price increases for basic food items like the corn in tortillas. Corn has risen in price due in part to its increased use in biofuels like ethanol. © *AP Images/Gregory Bull.*

An E85 Ethanol gas fuel pump in Washington state.   © Carolina K. Smith, M.D./Shutterstock.com.

Buckeridge, Marcos Silveira, and Gustavo Henrique Goldman. *Routes to Cellulosic Ethanol*. New York, NY: Springer, 2011.

Goettemoeller, Jeffrey, and Adrian Goettemoeller. *Sustainable Ethanol: Biofuels, Biorefineries, Cellulosic Biomass, Flex-Fuel Vehicles, and Sustainable Farming for Energy Independence*. Maryville, MO: Prairie Oak Publishing, 2007.

Kukathas, Uma. *The Global Food Crisis*. Detroit, MI: Greenhaven Press, 2009.

Mousdale, David M. *Introduction to Biofuels*. Boca Raton, FL: CRC Press, 2010.

Zuurbier, P. J. P., and J. van de Vooren. *Sugarcane Ethanol: Contributions to Climate Change Mitigation and the Environment*. Wageningen, The Netherlands: Wageningen Academic Publishers, 2008.

Periodicals

Cunningham, A. "Going Native: Diverse Grassland Plants Edge Out Crops as Biofuel." *Science News* 170 (December 9, 2006): 372.

Farrell, Alexander E. "Ethanol Can Contribute to Energy and Environmental Goals." *Science* 311 (2006): 506–508.

Scharlemann, Jörn P. W., and William F. Laurance. "How Green Are Biofuels?" *Science* 319 (2008): 43–44.

Searchinger, Timothy. "Use of U.S. Croplands for Biofuels Increases Greenhouse Gases through Emissions from Land-Use Change." *Science* 319 (2008): 1598–1600.

Wald, Matthew L. "Is Ethanol for the Long Haul?" *Scientific American* 296 (January 2007): 42–49.

Web Sites

"Biofuels." *National Geographic Society*. http://environment.nationalgeographic.com/environment/global-warming/biofuel-profile.html (accessed September 8, 2011).

Clayton, Mark. "The Politics of Ethanol Outshine Its Costs." *Christian Science Monitor*, November 15, 2007. http://www.csmonitor.com/2007/1115/p02s02-uspo.html (accessed September 12, 2011).

"Renewable Energy: Solar, Wind, Biomass Energy, Heat Pumps, and Biofuels." *The Energy Savings Trust*. http://www.energysavingtrust.org.uk/generate_your_own_energy/how_renewable_energy_works (accessed September 8, 2011).

"Survey of Energy Resources 2007: BioEnergy: Biofuels." *World Energy Council*. http://www.worldenergy.org/publications/survey_of_energy_resources_2007/bioenergy/714.asp (accessed September 8, 2011).

Brian Douglas Hoyle

Fat Substitutes

■ Introduction

A fat substitute is a substance that can replace some or all of the fat that naturally occurs in a food. Fat plays an important role in adding a creamy texture to foods and makes fried foods crispy. It also helps to dissolve flavor components so they can be experienced by the consumer.

However, fat contains 9 calories per gram, compared to 4 calories per gram for carbohydrate and protein. A high-fat diet can easily lead to weight gain whereas saturated and trans fats are associated with a risk of heart disease. With two-thirds of American adults, and one-third of children, now being overweight or obese, and heart disease being the leading cause of death, cutting back on fat in the diet may play an important part in improving the health of the nation. Developing a fat substitute that mimics the texture and flavor properties of real fat is challenging. There are carbohydrate, protein, and fat-derived fat substitutes, of which the best known is Proctor and Gamble's Olestra (Olean®). It is not yet known what the long-term risks and benefits of fat substitutes are, and most health and nutrition experts still recommend opting for a low-fat diet, rather than substituting fat in a high-fat diet, as a better way of losing weight.

■ Historical Background and Scientific Foundations

The search for fat substitutes goes back more than a century. In 1911 Crisco Shortening introduced an all-vegetable shortening that contained only 27 percent saturated fat, compared to the 51 percent of ordinary lard. In the 1960s a number of carbohydrate-based fat substitutes were introduced, such as FMC Corp's Avicel, which is a cellulose gel, and the National Starch and Chemical Company's dextrin-based N-Oil.

Carbohydrate-based fat substitutes are distributed widely in foods and can be seen listed on food labels as dextrins, modified starch, guar gum, xanthan gum, and polydextrose. They can simulate the creamy texture of fat. However, they often contain a lot of sugar, and therefore calories. Carbohydrate-based fat substitutes are heat-stable for baking purposes but they do not melt and cannot be used for frying.

Protein-based fat substitutes are made of egg white, dairy or whey protein, or soy in the form of microscopic particles. Their texture and appearance make them well suited for use in dairy products. They can be baked into cheesecake and heated in pizza but cannot be used in frying. Protein-based substitutes entered the fat substitute market in the early 1990s.

Different varieties of chips containing the fat substitute Olestra (also known by the brand name Olean) are displayed for a promo in 1998. Side effects included gastrointestinal problems in some people, which made several food manufacturers discontinue its use. However, it is still used in several "light" versions of potato chips. © *John Barr / Getty Images.*

A real breakthrough came with the discovery of Olestra by scientists at Proctor and Gamble. Olestra, trade name Olean, is a synthetic fat that is not broken down by digestive enzymes and was approved by the U.S. Food and Drug Administration (FDA) for use in the manufacture of potato chips and similar snack foods in 1996. Olestra is a molecule designed to mimic the structure of triglycerides, with a sucrose backbone instead of a glycerol backbone. This means that it can bind 6–8 fatty acids instead of the three found in triglycerides. Digestive enzymes are unable to break down Olestra, so it passes unchanged through the digestive tract and is therefore calorie-free. An ounce of potato chips made with Olestra contains 70 calories, which come from the potatoes, compared to an ounce of conventional potato chips that contain 160 calories. Olestra has the same taste and texture as fat and retains this during deep frying. The FDA looked at 150 studies on Olestra covering 20,000 people before approving its use in food.

■ Impacts and Issues

The FDA received more than 20,000 consumer complaints about Olestra, mostly within the first five years of its launch, making it the most controversial food

additive the agency has ever approved. Olestra can cause diarrhea, abdominal cramping, incontinence, and a distressing oily leakage into toilet bowls and underwear. But not everyone experiences these side effects, and the product is still in use. There are also many more fat substitutes to choose from for those who want to reduce the fat content of their diet, including fiber-based Z-Trim and Oatrim. No doubt, as food science progresses, there will also be new and improved fat substitutes.

The body needs fat to help with the absorption of essential nutrients, to build cell membranes, and to protect immunity. However, because fat is calorie-dense and saturated fats and trans fats are bad for the heart, the government offers advice on limiting the fat content of the diet. According to the 2010 Dietary Guidelines for Americans, total fat intake should be limited to between 20 and 35 percent of total calorie intake. There is an emphasis on the type of fat that should be included in the diet, with a recommendation to keep trans fat intake to a minimum, and saturated fat to less than 10 percent of total daily calories.

Thus far, fat substitutes seem not to have made any significant contribution to reducing levels of overweight and obesity among the general population. Indeed, a study from the American Psychological Association suggests they may even be making the problem worse. Laboratory rats were fed either a high-fat or a low-fat diet. Half of the rats in each group also received high-fat potato chips and the other were given high-fat chips on some days and low-fat chips, made with Olestra, on other days. In the high-fat group, those eating both types of chips ate more food, gained more weight, and developed more fatty tissue than those eating only high-fat chips. Moreover, they did not lose this extra weight when the low-fat chips were withdrawn. This did not happen to the rats on a low-fat diet. The researchers suggest that consumption of fat substitutes confuses the brain because fatty tastes are usually accompanied by a large number of calories. If these are absent, food-seeking behavior may be triggered, leading to overeating.

There is still much to be learned about eating behavior, obesity, and losing weight. It may be that fat substitutes have a role to play in helping people reach and maintain a healthy weight. However, it is likely wise not to rely on low-fat or reduced-fat products for weight loss. A balanced diet combined with physical activity and a psychological motivation to reach a healthy weight remains the most commonly prescribed weight-loss approach.

SEE ALSO *Agricultural Biotechnology; Lipidomics; Patents and Other Intellectual Property Rights; Synthetic Biology*

BIBLIOGRAPHY

Books

Chen, Nancy N. *Food, Medicine, and the Quest for Good Health: Nutrition, Medicine, and Culture.* New York: Columbia University Press, 2009.

Eskin, N. A. Michael, and Snait Tamir. *Dictionary of Nutraceuticals and Functional Foods.* Boca Raton, FL: Taylor & Francis Group/CRC Press, 2006.

Hughes, Holly, Helen Arrowsmith, and Liz Mulvey. *Reducing the Fat Content of Meat Products: A Review of Fat Replacement Ingredients.* Chipping Campden, UK: Campden BRI, 2010.

Shetty, Kalidas, et al., eds. *Functional Foods and Biotechnology.* Boca Raton, FL: CRC/Taylor & Francis, 2007.

Wilson, Rachel. *Fat Substitutes*, 2nd ed. Leatherhead, UK: Leatherhead Publishing, 2008.

Periodicals

"The Fad for Functional Foods: Artificial Success." *Economist* 392, no. 8650 (September 24, 2009): 84.

Web Site

"Fat Replacers." *Calorie Control Council.* http://www.caloriecontrol.org/sweeteners-and-lite/fat-replacers (accessed November 7, 2010).

"Position of the American Dietary Association: Fat Replacers." *Journal of the American Dietary Association,* February 2005. http://www.adajournal.org/article/S0002-8223(04)01853-X/fulltext (accessed November 7, 2010).

Susan Aldridge

Fermentation

Introduction

The exploitation of microorganisms and their biochemical products has led to the development of the fermentation industry. Humankind has used yeasts such as *Saccharomyces cerevisiae* and microorganisms such as lactobacilli for millennia to produce dairy products, wine, and beer, as well as other fermented beverages, such as the Japanese sake. Fermentation also is essential for cheese and yogurt, and for producing food flavor ingredients, such as soy sauce and vinegar. The complete DNA sequencing of microorganisms involved in fermentation and the development of more accurate genetic engineering techniques have led to the commercial-scale production of more stable and efficient yeast microorganisms for wine, beer, and other fermented food items. Fermentation plays a significant part in other industrial processes such as biofuel production, life sciences basic research, drug and vaccine development, diagnostic assays, production of vitamins and essential amino acids, and promotion of biodegradation of garbage and waste during recycling.

Historical Background and Scientific Foundations

The term fermentation originates from the Latin verb *fervere* meaning to boil, because ancient wine and beer-makers observed that substances undergoing fermentation resembled a boiling process. Fermentation consists of the breaking down of sugars (glucose, maltose, and lactose) to lactic acid, in the absence of oxygen (an anaerobic process).

Essentially, fermentation is an energy-generating natural process mediated by enzymes produced by anaerobic microorganisms. However, the term fermentation is also used in a broader connotation by the food industry and food engineers to refer to the products obtained by mass culturing of microorganisms—chiefly, several types of enzymes. Enzymes derived from genetically-modified (GM) yeasts first were introduced for making antibiotics and vaccines. Subsequently, GM yeasts were approved for use in manufacturing fermented alcoholic beverages and fruit juices, flavor-enhancing food additives, and amino acids and vitamins used to enrich animal feed. They also are used in some foods as the flavor-enhancer glutamate, the sweetener aspartame, or the flour-treating agent cysteine. Numerous GM-yeast derived enzymes are used to make cheeses, bread, and baked foods, as well as to produce corn syrup and other starch products.

Almost all pioneering investigations in microbiology and biochemistry were conducted using yeast as a biological model for fermentation, especially *S. cerevisiae*. The same rationale behind the use of biotechnology

WORDS TO KNOW

AMINO ACIDS: Simple small peptides (protein units) that serve as the building blocks of more complex proteins. Amino acids also play essential regulatory roles in physiological systems such as basal metabolism, cell functioning, brain chemistry, and functional regulation.

ANAEROBIC MICROORGANISMS: Bacteria that cannot grow in the presence of oxygen (obligatory anaerobes) or those that do not require oxygen to produce energy (facultative anaerobes).

BIODEGRADATION: Microbial process of breaking complex molecules and digesting both organic and inorganic matter through fermentation.

ENZYMES: Small proteins capable of activating or enhancing chemical reactions among other biomolecules or inorganic chemical compounds.

ENZYMOLOGY: The study of enzymes, their different types and functions in biological systems, and applications in therapeutics or in industrial chemical reactions.

Kimchi pots sit in the sun in South Korea. The heavy pots contain kimchi, a delicacy from Korea. The kimchi ferments in the pots while the lid keeps air from inside.   © *Keith Brooks/Shutterstock.com.*

for mass cultivation of microorganisms was at the root of the choice of *S. cerevisiae* and other yeasts for initial research: yeasts grow rapidly and are easy to cultivate. Yeasts were especially important for the studies of animal and human metabolism and of enzymology.

The first genetic engineering experiment was performed by Jean Beggs (1950–) in 1978 using *Escherichia coli* and *S. cerevisiae*, which enabled cloning in yeast. *S. cerevisiae* was also the first organism to have its genome completely sequenced in 1996, thanks to an international public consortium that involved 600 scientists worldwide. This achievement not only facilitated the comparative study of DNA similarities between species, but also paved the way for new molecular technologies that were used for genomic studies and for sequencing multicellular organisms, including humans.

■ Impacts and Issues

Genetic engineering has evolved significantly since the early 1980s, making it possible to modify bacteria and fungi to cause them to produce enzymes and other substances used by the pharmaceutical and food industry or other industries, such as biofuels, paints, and biodegradable plastics. Fermentation processes are conducted inside airproof stainless steel tanks used to reproduce the controlled optimum environment where yeast microorganisms can thrive and produce the desired product in large quantities. When the growth and production phase

is completed, the product (whether a vitamin or amino acid or other specific enzyme) is isolated and purified, to eliminate completely the microorganisms that produced them from the final product.

The U.S. Food and Drug Administration's (FDA) position on the safety of *S. cerevisiae* and other yeasts found in bread, dairy products, wine, and beer fermentation, is that they generally are recognized as safe, because humans and animals have been exposed to them in food products and in the environment for at least 7,000 years.

Genetically-modified yeasts have part of their genome either removed or enhanced in order to eliminate undesired effects during fermentation, or to increase flavors and aromas that improve the taste of wine or food, or to diminish or augment alcohol content in the final product. Safety standards and purity requirements for vitamins and food additives are set by regulatory agencies such as the FDA in the United States, the EFSA in the European Union, and their counterparts elsewhere. Products containing genetically-modified microorganisms must satisfy additional requirements: They must be thoroughly tested in compliance with laws governing the facilities, and so must the microorganisms used for genetic engineering.

GM yeasts also are designed to increase the speed at which sugarcane or other plants used in biofuel production ferment. By reducing the fermentation time required for ethanol production, energy costs to the final consumer are reduced as well. The genetically-modified yeasts used to increase ethanol production are

Grains ferment in a fermenting cellar in a brewery. Genetic modification can be used to create yeasts that can give beers and other food products enhanced tastes.   © Martin D. Vonka/Shutterstock.com

recombinant *Saccharomyces cerevisiae* containing genes from another yeast, *Pichia stipitis*.

Ethanol (alcohol used as automotive fuel) can be produced naturally through fermentation from renewable resources, and it is responsible for more than 40 percent of all automotive fuels consumed in countries such as in Brazil. Ethanol is obtained easily through the fermentation of sugarcane juice and has lower levels of toxicity and a higher degree of biodegradability than methanol.

SEE ALSO *Agrobacterium; Biofuels, Liquid; Dairy and Cheese Biotechnology; Enzymes, Industrial; Ethanol; Genetically Modified Food; Metabolic Engineering; Recombinant DNA Technology*

BIBLIOGRAPHY

Books

Glick, Bernard R., Jack J. Pasternak, and Cheryl L. Patten. *Molecular Biotechnology: Principles and Applications of Recombinant DNA*, 4th ed. Washington, DC: ASM Press, 2010.

Hutkins, Robert W. *Microbiology and Technology of Fermented Foods*. Chicago: IFT Press, 2006.

Lens, Piet Nicholas Luc, ed. *Biofuels for Fuel Cells: Renewable Energy from Biomass Fermentation*. Seattle: IWA Publishing, 2005.

McNeil, Brian, and Linda M. Harvey. *Practical Fermentation Technology*. Chichester, UK, and Hoboken, NJ: John Wiley & Sons, 2008.

Miesfeld, Roger L. *Molecular Biotechnology*, 2nd ed. New York: Wiley, 2006.

Srivastava, Manish. *Fermentation Technology*. Oxford: Alpha Science International, 2008.

Periodicals

Barnett, James A. "A History of Research on Yeasts: Foundations of Yeast Genetics." *Yeast* 24, no. 10 (October 2007): 799–845.

Web Sites

Colin, Veronica L., Analia Rodriguez, and Hector A. Cristobal. "The Role of Synthetic Biology in the Design of Microbial Cell Factories for Biofuel Production." *Journal of Biomedicine and Biotechnology*, August 2, 2011. http://www.ncbi.nlm.nih.gov/pmc/articles/PMC3197265/ (accessed November 13, 2011).

Saccharomyces Genome Database. http://www.yeastgenome.org/ (accessed November 13, 2011).

Sandra Galeotti

Fertilizers

■ Introduction

The fundamental relationship between improved soil quality and increased crop yields has been understood since the practice of agriculture became an essential part of virtually all human societies. The earliest fertilizers were readily available animal and human wastes that were spread over the soil. Through natural decomposition, these organic materials released their available macronutrients and micronutrients directly into the soil. These substances also contributed to soil density essential to moisture retention and the reduced risk of runoff that could damage farm fields. In addition, the biodegradation of these organic waste products introduces microorganisms into the soil that contribute to plant growth.

In the twentieth century, fertilizer science evolved from the localized use of organic waste methods to the sophisticated manufacture and processing of inorganic fertilizers capable of delivering larger amounts of nutrients to the soil in a shorter period than achieved through organic methods. Inorganic fertilizers usually are manufactured through ammonia synthesis processes that alter the chemical characteristics of nitrogen to render it reactive with its immediate environment. Synthetic controlled-release nitrogen fertilizers are used throughout the world to increase crop yields; slower nitrogen release rates also reduce the contamination of rivers, lakes, and subsurface water tables often caused by inorganic fertilizer use.

The resolution of the serious environmental concerns associated with inorganic fertilizer use has emerged as a key twenty-first century agricultural science research and development objective. As the world population continues to grow at a rate of approximately 1.5 percent per year, the need to increase crop yields to sustain human populations is a global imperative. The production of inorganic fertilizers involves significant amounts of non-renewal resources, including natural gas required for ammonia production, and mined micronutrient additives such as zinc.

■ Historical Background and Scientific Foundations

The objective of all fertilizer applications is increased crop yield and harvest quality. The positive effect on plant growth in soils that have been mixed with organic

---

WORDS TO KNOW

AMMONIUM NITRATE: A fertilizer with high nitrogen content produced through the combination of ammonia and nitric acid. Ammonia nitrate is also a powerful explosive; its sale and distribution has been restricted in many countries due to its potential use by terrorists.

EUTROPHICATION: The process that leads to the depletion of dissolved oxygen in bodies of water (hypoxia). Excessive nutrient levels in water promote the growth of algae, reduce water quality, and eliminate habitats for aquatic species that require colder, oxygen-rich waters. Fertilizer runoff is a primary contributor to eutrophication.

MACRONUTRIENTS: The essential elements required for healthy crop plant development. Nitrogen, sulfur, and phosphorous are macronutrients typically found in most fertilizer formulations.

MENHADEN: A herring fish species found in North America Atlantic coastal regions. Too oily for regular human consumption, menhaden are used to produce fish oils, animal feed, and organic fertilizer.

MICRONUTRIENT: The trace elements required for healthy crop plant development. Boron, chlorine, and zinc are important fertilizer micronutrients.

UREA: A nitrogen-based compound used in solid fertilizers. The first organic compound to be synthesized (1828); human and animal urine are the most common sources of organic urea.

A man prepares to spread fertilizer in a tea field.   © *Radu Razvan/Shutterstock.com.*

wastes has been observed since ancient times. Emigrants who settled near the Chesapeake Bay region of America in the early eighteenth century adopted the indigenous practice of adding menhaden fish to their fields to improve soil quality.

Terra preta, the dark soil found in the Amazon basin where charcoal from cooking fires significantly increased its fertility, is an example of a soil improved from an anthropogenic (human-made) source.

Crop rotation is a related agricultural practice that improves soil quality. Farmers grew different crops on the same fields because they appreciated that the pests and fungal diseases associated with a specific crop were less likely to reoccur if a different crop followed. In the mid-eighteenth century, the four crop rotation methods promoted by British scientist Charles Townshend (1674–1738) became the accepted agricultural standard in Northern Europe. The rotation of the field crops clover, barley, wheat, and turnips achieved healthier, more nitrogen-rich and productive soils. Clover later became known as green manure, because when it is plowed into

the soil, clover will convert atmospheric nitrogen into ammonia that is subsequently broken down into soil nutrients.

The increased demand for cereal crops worldwide prompted the development of inorganic fertilizers. Ammonia is the essential component of the Haber-Bosch process developed in the early twentieth century. Methane extracted from natural gas is the crucial raw material; repeated exposure of the gas to an iron catalyst permits its component nitrogen and hydrogen molecules to form ammonia, because the process reduces the usual intermolecular strength of nitrogen atoms that renders the element inert in the atmosphere.

The first controlled-release, nitrogen-based fertilizers were sold in 1955. These were designed to prolong the nutritional benefits achieved through the exposure of nitrogen to depleted soils. Sulfur was the typical ingredient used to coat the fertilizer for slow nitrogen release; sulfur provides an additional macronutrient. Ammonia-based fertilizers have become the most widely used variety in the world.

■ Impacts and Issues

The development of synthetic fertilizers demonstrates how two important food sustainability objectives can come into direct conflict. The ability of agricultural science to develop technologies that promote increased crop yields is of paramount importance to the survival of many global populations. The accumulated pressure imposed on food supplies is generated by the combined effects of these factors:

- Steady increases in the global population from 2 billion persons in 1930 to 7 billion in 2011.

- Lands removed from agricultural production and converted to housing or industrial uses.

- Impact of droughts, frosts, and other extreme weather attributed to climate change; all impact the length and quality of crop growing seasons.

- Traditional agricultural regions where the soils have been nutrient-depleted. Overworked soils suffer from macronutrient depletion due to poor agricultural practices such as the failure to rotate crops.

- Insect infestation and fungal diseases that impair or destroy crop harvest quality.

The positive impacts of synthetic fertilizers on world agriculture are considerable. It is doubtful that there are sufficient sources of organic fertilizers available to sustain current and projected global agricultural needs. Modern slow-release nitrogen products, especially those that are further enhanced with additional micronutrients, contribute to increased yields for corn, wheat, and sorghum, three key crops in the food supply for the developing world.

The ease of handling and the powerful impact on crop yields realized through the use of synthetic fertilizers also carries a range of significant environmental risks. When soils are overexposed to nitrogen-based fertilizers, the soils will tend to become acidic and less capable of supporting crop growth. The runoff experienced in the ecosystems adjacent to fields where fertilizers are used extensively often leads to eutrophication of adjacent watercourses, where water quality and the diversity of aquatic life are diminished. Further, the manufacture of synthetic fertilizers requires the consumption of non-renewable fossil fuels, particularly natural gas, and minerals such as zinc and calcium that are mined from open pits using significant energy resources.

Organic fertilizers are recognized as having far fewer negative environmental impacts than the inorganic synthetic products. The principal drawbacks associated

Anhydrous ammonia transport tanks outside Basin Chemical and Fertilizer company in Klamath Falls, Oregon. Fertilizer prices have risen as the result of several factors, including the booming demand for fertilizer to produce animal feed for rapidly developing nations like India and China, where people are adopting diets richer in meat. This has an increasing effect on the environment. © TFoxFoto/Shutterstock.com.

with these natural sources are their greater bulk than the synthetics, which poses additional handling and transportation costs. Further, organic wastes that are used as fertilizers may contain pathogens than can pose threats to human and animal health.

Biotechnology's role in modern fertilizers is a limited one in the early 2010s, but biotechnology likely will change the way fertilizers are used in agriculture in the future. Scientists are working to enable cereal plants to fix their own nitrogen levels, lessening the need for fertilizers, although this development is many years away, because nitrogen fixation requires the input of more than a dozen genes in most organisms. Genetically-engineered microorganisms within fertilizers that efficiently deliver nitrogen to crops and help them retain it are likely to happen sooner, which would also reduce the need for fertilizers. For the near future, biotechnology is having a quick impact in heartier species of plants that require less water or fewer insecticides, or have other qualities that make them desirable crops. Farmers are likely to justify increased use of fertilizers on these premium crops.

■ Primary Source Connection

Crops need certain chemical forms of nitrogen (called reactive nitrogen) found in Earth's atmosphere to thrive and grow. Reactive nitrogen is not always found at optimum levels in soil for productivity; therefore, it is often necessary to add it to soil. The added nitrogen is well known as a major source of pollution to the environment, and as such, the practice raises concerns about resulting water and air pollution secondary to crop fertilization. Farmers seek to grow abundant, healthy crops while at the same time reduce the impact of nitrogen pollution. Options to achieve this goal are discussed.

No Sure Fix: Prospects for Reducing Nitrogen Fertilizer Pollution through Genetic Engineering

Executive Summary

Nitrogen is essential for life. It is the most common element in Earth's atmosphere and a primary component of crucial biological molecules, including proteins and nucleic acids such as DNA and RNA—the building blocks of life.

Crops need large amounts of nitrogen in order to thrive and grow, but only certain chemical forms collectively referred to as reactive nitrogen can be readily used by most organisms, including crops. And because soils frequently do not contain enough reactive nitrogen (especially ammonia and nitrate) to attain maximum productivity, many farmers add substantial quantities to their soils, often in the form of chemical fertilizer.

Unfortunately, this added nitrogen is a major source of global pollution. Current agricultural practices aimed at producing high crop yields often result in excess reactive nitrogen because of the difficulty in matching fertilizer application rates and timing to the needs of a given crop. The excess reactive nitrogen, which is mobile in air and water, can escape from the farm and enter the global nitrogen cycle—a complex web in which nitrogen is exchanged between organisms and the physical environment—becoming one of the world's major sources of water and air pollution.

The challenge facing farmers and farm policy makers is therefore to attain a level of crop productivity high enough to feed a growing world population while reducing the enormous impact of nitrogen pollution. Crop genetic engineering has been proposed as a means of reducing the loss of reactive nitrogen from agriculture. This report represents a first step in evaluating the prospects of genetic engineering to achieve this goal while increasing crop productivity, in comparison with other methods such as traditional crop breeding, precision farming, and the use of cover crops that supply reactive nitrogen to the soil naturally.

The Importance of Nitrogen Use Efficiency (NUE)

Crops vary in their ability to absorb nitrogen, but none absorb all of the nitrogen supplied to them. The degree to which crops utilize nitrogen is called nitrogen use efficiency (NUE), which can be measured in the form of crop yield per unit of added nitrogen. NUE is affected by how much nitrogen is added as fertilizer, since excess added nitrogen results in lower NUE. Some agricultural practices are aimed at optimizing the nitrogen applied to match the needs of the crop; other practices, such as planting cover crops, can actually remove excess reactive nitrogen from the soil.

In the United States, where large volumes of chemical fertilizers are used, NUE is typically below 50 percent for corn and other major crops— in other words, more than half of all added reactive nitrogen is lost from farms. This lost nitrogen is the largest contributor to the "dead zone" in the Gulf of Mexico—an area the size of Connecticut and Delaware combined, in which excess nutrients have caused microbial populations to boom, robbing the water of oxygen needed by fish and shellfish. Furthermore, nitrogen in the form of nitrate seeps into drinking water, where it can become a health risk (especially to pregnant women and children), and nitrogen entering the air as ammonia contributes to smog and respiratory disease as well as to acid rain that damages forests and other habitats. Agriculture is also the largest human-caused domestic source of nitrous oxide, another reactive form of nitrogen that contributes to global warming and reduces the stratospheric ozone that protects us from ultraviolet radiation.

Nitrogen is therefore a key threat to our global environment. A recent scientific assessment of nine global environmental challenges that may make the earth unfavorable for continued human development identified nitrogen pollution as one of only three—along with climate change and loss of bio-diversity—that have already crossed a boundary that could result in disastrous consequences if not corrected. One important strategy for avoiding this outcome is to improve crop NUE, thereby reducing pollution from reactive nitrogen.

Can Genetic Engineering Increase NUE?

Genetic engineering (GE) is the laboratory-based insertion of genes into the genetic material of organisms that may be unrelated to the source of the genes. Several genes involved in nitrogen metabolism in plants are currently being used in GE crops in an attempt to improve NUE. Our study of these efforts found that:

- Approval has been given for approximately 125 field trials of NUE GE crops in the United States (primarily corn, soybeans, and canola), mostly in the last 10 years. This compares with several thousand field trials each for insect resistance and herbicide tolerance.

- About half a dozen genes (or variants of these genes) appear to be of primary interest. The exact number of NUE genes is impossible to determine because the genes under consideration by companies are often not revealed to the public.

- No GE NUE crop has been approved by regulatory agencies in any country or commercialized, although at least one gene (and probably more) has been in field trials for about eight years.

- Improvements in NUE for experimental GE crops, mostly in controlled environments, have typically ranged from about 10 to 50 percent for grain crops, with some higher values. There have been few reports of values from the field, which may differ considerably from lab-based performance.

- By comparison, improvement of corn NUE through currently available methods has been estimated at roughly 36 percent over the past few decades in the United States. Japan has improved rice NUE by an estimated 32 percent and the United Kingdom has improved cereal grain NUE by 23 percent.

- Similarly, estimates for wheat from France show an NUE increase from traditional breeding of about 29 percent over 35 years, and Mexico has improved wheat NUE by about 42 percent over 35 years.

Available information about the crops and genes in development for improved NUE suggests that these genes interact with plant genes in complex ways, such that a single engineered NUE gene may affect the function of many other genes. For example:

- In one of the most advanced GE NUE crops, the function of several unrelated genes that help protect the plant against disease has been reduced.

- Another NUE gene unexpectedly altered the output of tobacco genes that could change the plant's toxicological properties.

Many unexpected changes in the function of plant genes will not prove harmful, but some may make it difficult for the crops to gain regulatory approval due to potential harm to the environment or human health, or may present agricultural drawbacks even if they improve NUE. For the most advanced of the genes in the research pipeline, commercialization will probably not occur until at least 2012, and it will likely take longer for most of these genes to achieve commercialization—if they prove effective at improving NUE. At this point, the prospects for GE contributing substantially to improved NUE are uncertain.

Other Methods for Reducing Nitrogen Pollution

Traditional or enhanced breeding techniques can use many of the same or similar genes that are being used in GE, and these methods are likely to be as quick, or quicker, than GE in many cases. Traditional breeding may have advantages in combining several NUE genes at once.

Precision farming—the careful matching of nitrogen supply to crop needs over the course of the growing season—has shown the ability to increase NUE in experimental trials. Some of these practices are already improving NUE, but adoption of some of the more technologically sophisticated and precise methods has been slow.

Cover crops are planted to cover and protect the soil during those months when a cash crop such as corn is not growing, often as a component of an organic or similar farming system. Some can supply nitrogen to crops in lieu of synthetic fertilizers, and can remove excess nitrogen from the soil; in several studies, cover crops reduced nitrogen losses into groundwater by about 40 to 70 percent.

Cover crops and other "low-external-input" methods (i.e., those that limit use of synthetic fertilizers and pesticides) may also offer other benefits such as improving soil water retention (and drought tolerance) and increasing soil organic matter. An increase in organic matter that contains nitrogen can reduce the need for externally supplied nitrogen over time.

With the help of increased public investment, these methods should be developed and evaluated fully, using an ecosystem approach that is best suited to determine how reactive nitrogen is lost from the farm and how NUE can be improved in a comprehensive way. Crop breeding or GE alone is not sufficient because they do not fully address the nitrogen cycle on real farms, where nitrogen loss varies over time and space, such as those times when crops—conventional or GE—are not growing.

Conclusions

GE crops now being developed for NUE may eventually enter the marketplace, but such crops are not uniquely beneficial or easy to produce. There is already sufficient genetic variety for NUE traits in crops, and probably in close relatives of important crops, for traditional breeding to build on its successful track record and develop more efficient varieties.

Other methods such as the use of cover crops and precision farming can also improve NUE and reduce nitrogen pollution substantially.

Recommendations

The challenge of optimizing nitrogen use in a hungry world is far too important to rely on any one approach or technology as a solution. We therefore recommend that research on improving crop NUE continue. For traditional breeding to succeed, public research support is essential and should be increased in proportion to this method's substantial potential.

We also recommend that system-based approaches to increasing NUE—cover crops, precision application of fertilizer, and organic or similar farming methods—should be vigorously pursued and supported. These approaches are complementary to crop improvement because each addresses a different aspect of nitrogen use. For example, while breeding for NUE reduces the amount of nitrogen needed by crops, precision farming reduces the amount of nitrogen applied. Cover crops remove excess nitrogen and may supply nitrogen to cash crops in a more manageable form.

Along with adequate public funding, incentives that lead farmers to adopt these practices are also needed. Although the private sector does explore traditional breeding along with its heavy investment in the development of GE crops, it is not likely to provide adequate support for the development of non-GE varieties, crops that can better use nitrogen from organic sources, or improved cover crops that remove excess nitrogen from soil. We must ensure that broad societal goals are addressed and important options are pursued nevertheless.

In short, there are considerable opportunities to address the problems caused by our current overuse of synthetic nitrogen in agriculture if we are willing to make the necessary investments. The global impact of excess reactive nitrogen will worsen as our need to produce more food increases, so strong actions—including significant investments in

technologies and methods now largely ignored by industrial agriculture—will be required to lessen the impact.

Doug Gurian-Sherman
Noel Gurwick

GURIAN-SHERMAN, DOUG, AND NOEL GURWICK. "NO SURE FIX: PROSPECTS FOR REDUCING NITROGEN FERTILIZER POLLUTION THROUGH GENETIC ENGINEERING." *UNION OF CONCERNED SCIENTISTS* (DECEMBER 2009): 1–4.

SEE ALSO *Agricultural Biotechnology; Agrobacterium; Biofuels, Liquid; Bt Insect Resistant Crops; Corn, Genetically Engineered; Cornstarch Packing Materials; Cotton, Genetically Engineered; Disease-Resistant Crops; Drought-Resistant Crops; Environmental Biotechnology; Frost-Resistant Crops; Genetically Modified Crops; Golden Rice; Phytoremediation; Salinity-Tolerant Plants; Soybeans; Transgenic Plants; Wheat, Genetically Engineered*

BIBLIOGRAPHY

Books

Chrispeels, Maarten J., and David E. Sadava. *Plants, Genes, and Crop Biotechnology*, 2nd ed. Boston: Jones and Bartlett, 2003.

Gurian-Sherman, Doug, and Noel Gurwick. *No Sure Fix: Prospects for Reducing Nitrogen Fertilizer Pollution through Genetic Engineering*. Cambridge, MA: UCS Publications, 2009.

Heldman, Dennis R., Matthew B. Wheeler, and Dallas G. Hoover, ed. *Encyclopedia of Biotechnology in Agriculture and Food*. Boca Raton, FL: CRC Press, 2011.

Lei, Xingen. *Biotechnological Approaches to Manure Nutrient Management*. Ames, IA: Council for Agricultural Science and Technology, 2006.

Lichtfouse, Eric. *Alternative Farming Systems, Biotechnology, Drought Stress and Ecological Fertilisation*. Dordrecht, The Netherlands: Springer, 2011.

Murphy, Denis J. *Plant Breeding and Biotechnology: Societal Context and the Future of Agriculture*. New York: Cambridge University Press, 2007.

Web Sites

"Biotechnology No Sure Fix for World's Nitrogen Fertilizer Pollution Problem, New Report Finds." *Union of Concerned Scientists*, December 9, 2009. http://www.ucsusa.org/news/press_release/biotechnology-no-sure-fix-0321.html (accessed November 7, 2011).

"Nutrient Management and Fertilizer." *U.S. Environmental Protection Agency (EPA)*. http://www.epa.gov/agriculture/tfer.html (accessed November 7, 2011).

Bryan Thomas Davies

Food Preservation

Introduction

Food preservation has been a preoccupation of the human species throughout its entire existence. Increased abilities to maintain food quality for longer periods enabled various peoples to range further from their original settlements. Success with basic preservative methods such as dried fish and cured meats inspired the pursuit of better systems, especially in climates where fruits and vegetables quickly spoiled and their consumption posed a threat to human health. For more than 5,000 years, raw milk has been fashioned into more durable cheeses and yogurts through the cumulative effect of preservative bacterial action.

Biopreservation is the modern scientific term that also encompasses many traditional food preservation methods. Biopreservation is defined as any process in which naturally occurring microbiota are utilized to extend the useful life of a food. Fermentation is a common form of biopreservation used throughout the ancient world that remains a prominent modern commercial food and beverage process. Microorganisms (primarily lactic acid bacteria such as yeast) act on available carbohydrates to produce alcohols that are toxic to other microorganic pathogens but preserve food stability and quality.

Advances in biotechnology have led to the widespread use of more sophisticated food preservation methods that prolong shelf life. These include pasteurization; food irradiation with gamma rays; vacuum packaging to create an oxygen and moisture-free environment less likely to sustain harmful microorganism growth; and recombinant proteins that preserve frozen dairy products.

Historical Background and Scientific Foundations

Food preservation techniques have been a feature of every human culture. The earliest methods were directed at keeping the food away from air and water, the two natural agents that contribute most to natural food spoilage. Sealed clay pots were used for this purpose in Egyptian and Roman societies.

Drying foods was another popular preservation method in ancient times, as most bacteria that promote food deterioration thrive in moist environments. North American indigenous peoples dried venison, fish, and fruits such as cranberries to provide sustenance during the winter seasons. Salt (sodium chloride) was the first chemical used to preserve foods. Plasmolysis occurs on any surface where salt is applied; the health risks associated with excessive sodium consumption were not fully understood until the later twentieth century. In potentially dangerous strains of bacteria such as *Salmonella*, growth is inhibited by as little as 3 percent salt in the curing solution; *Listeria* will survive in concentrations of up to 12 percent. Brine and other forms of pickling solutions have high salt concentrations that tend to preserve foods such as pickles and cabbage for periods as long as two years without risk of deterioration.

If the water in a specific food is bound to other compounds, preservation characteristics are enhanced. Water activity (aw) is determined on a scale that ranges from 0 to 1, with 0 indicating a total water absence and 1 as 100 percent water composition. The degree to which a food preservation process inhibits water activity is also influenced by the pH level of the compound—the greater the pH value (more acidic), the greater the likelihood of water activity and speedier food deterioration.

Baking is a simple example of a process in which the combination of water and the desired ingredients at high temperatures reduces the amount of water available to promote other bacterial growth and food deterioration. The addition of sugar and salt to a mixture also immobilizes water and leaves the food stable at room temperature. Freeze-dried foods are the modern extension of the knowledge associated with the ability of dried foods to ward off deterioration. Instant coffee is an example.

In addition to moisture, the commonly understood threats to food quality are: contamination by

Octopus dry on a line in Greece. Drying is one of the oldest methods of food preservation. It removes the moisture from foods, discouraging bacterial growth. © *Graham Prentice/Shutterstock.com.*

microorganisms such as fungus, mold, or bacteria; enzyme activity, in which protein-based chemicals speed decomposition caused by oxidation; and insects and vermin. Pests consume available unprotected food, any they also heighten the risk of foodborne disease through the droppings left on the food surface.

Meat and most vegetables have a pH value greater than 4.6. Heat greater than 50°C (121°F) in combination with pressure is necessary to sterilize these foods from the effects of pathogens and enzyme activity. Canned foods are produced using these methods.

Flash-frozen foods are a product of the technological advances made in refrigeration. Powerful commercial freezers provide a storage environment of approximately –4°F (–20°C), temperatures that significantly inhibit deteriorative microbial activity. Blanching, the preliminary treatment of preserved foods by brief immersion in boiling water followed by freezing, deactivates the enzymes that stimulate initial food decomposition at room temperatures.

Another modern preservation technique is gamma ray irradiation, high frequency electromagnetic radiation that is used to limit decomposition in certain foods, such as the sprouts that grow in stored potatoes and onions; insect infestations in powdered grains; and microbes that occur in ground spices and seasonings. When arranged in sequences, a hurdle effect is created, and the likelihood of effective preservation is very high.

Biopreservation has a counterintuitive quality given the often negative impact of microorganisms on food quality and the heightened risk posed to human health. Current uses of lactic acid bacteria (LAB) are an extension of ancient food preservation practices in which friendly microbiota introduced to the food in a

Pickling and canning vegetables allows them to be preserved for extended periods of time. The salt and vinegar used in pickling kills bacteria in that method of preservation, but for canning, the food must be cooked to kill bacteria before sealing the lids.
© Alaettin Yildirim/Shutterstock.com.

controlled process neutralize potential pathogens. Probiotics, chemicals similar to the microorganisms found in the human intestinal tract, are added to milk products such as yogurt to promote better health.

■ Impacts and Issues

The primary concern associated with all food preservation processes is that the further the process takes the food from its natural state, the greater the health risks to a consumer. Although increasingly rare in the highly-regulated commercial food sectors of developed nations, food poisoning caused by microorganisms introduced to the food through the improper handling, storage, or packaging of foods can lead to widespread illness or death. The U.S.-based Center for Disease Control (CDC) estimates that foodborne illness affects 48 million Americans each year, leading to 128,000 hospitalizations and 3,000 deaths. Peanut butter processed by Peanut Corporation of America was linked to 714 illnesses and 9 deaths between late 2008 and early 2009. It triggered the largest food recall up to that date, affecting products including peanut butter crackers, cookies, granola bars, and dog biscuits. The CDC confirmed in November 2011 that, in the deadliest foodborne disease outbreak in a decade, 139 people had become sickened and 29 people had died from listeriosis traced back to cantaloupe from a farm in Colorado. The U.S. Food and

Drug Administration reported that *Listeria monocytogenes* was found in the farm's cold storage, processing, and packing areas.

The biotechnological applications used in food preservation reveal a common theme—no single process is likely to guarantee food quality, stability, and shelf life. It is also recognized that the cost to achieve perfect food preservation in modern societies where food distribution networks extend across continents is perceived as prohibitive. The object of all commercial food preservation is not to satisfy a present need, such as the challenge faced by earlier human societies to ensure their food supply lasted long enough to survive the challenges of the winter months. Modern preservation systems are premised on the principle that food quality should be achieved in a cost-effective fashion.

The future of food preservation will be driven by the greater use of biotechnology and nano-biotechnological applications. For consumers who express concern that foods are overly processed through chemical preservatives, methods including the use of recombinant proteins that act as antifreeze to extend frozen dairy products and fruit shelf life will become more common. Fermented whey with higher acidity will preserve cheese more effectively.

■ Primary Source Connection

A mixture of six viruses, added to foods, aims to kill deadly bacteria that are often present in uncooked meats and poultry. The virus mixture, developed by Intralytix Inc., is designed to target bacteria that cause food-borne illnesses such as listeriosis. Federal food regulators granted permission for use of the bacteria-killing virus mixture, which food manufactures spray on meats and poultry before final processing. Food related sickness due to strains of *Listeria* bacteria, causing flu-like symptoms, can be reduced with the use of the virus mixture.

Baltimore Firm Wins FDA Nod for Mix of Beneficial Viruses

WASHINGTON—A Baltimore company received yesterday the first permission that federal food regulators have ever granted for killing a common but sometimes deadly bacteria with a mixture of viruses added to foods.

The mixture of six viruses, developed by Intralytix Inc., aims to sharply reduce the 500 deaths and 2,500 illnesses caused in Americans each year by exposure to the bacteria often present in some uncooked meats and poultry. After four years of review, the Food and Drug Administration said the antimicrobial combination was safe and works in deli meats and other ready-to-eat foods.

John Vazzana, chief executive officer of Intralytix, described the approval of the mixture as a "huge milestone" in the fight against bacteria and antibiotic-resistant bacteria that cause food-borne illnesses. The viruses, he said, "are very specific, and they won't kill or destroy any other organism that is there. The only thing they will do is kill their target bacteria."

The combination of viruses that Intralytix developed kills various strains of the Listeria monocytogenes bacteria, a widely occurring microbe that especially sickens pregnant women, their fetuses and adults with weakened immune systems.

Before final processing, food manufacturers would spray the mixture on sliced ham, turkey and other foods that usually aren't cooked or reheated before eaten. Cooking and reheating, as well as processing, kills the Listeria bacteria, but foods can become contaminated after processing or even while sitting in a refrigerator.

Judged safe, effective

Consumers shouldn't notice any difference in the taste or color of foods sprayed with the mixture of bacteriophages, as the bacteria-killing viruses are called, the FDA said. Also, the agency found Intralytix's recipe safe and effective even among men in their 20s, who eat the largest quantities of ready-to-eat foods and consequently would ingest the largest amounts of the viruses.

Doug Gurian-Sherman, a senior scientist at the Center for Food Safety, a nonprofit public advocacy group based in Washington, said bacteriophages were safely used in the Soviet Union to kill bacteria during surgeries and other medical treatments. He said the only possible harm he could envision from the viruses' use as a food additive was allergic reactions in some people.

"But that's always an issue, and we are exposed to these things all of the time," he said. "I generally wouldn't be concerned about it."

In its application for FDA approval, Intralytix said it would purify the viruses during manufacture to reduce any potential for allergic reactions. An FDA review of studies on the company's combination of viruses, completed earlier this year, found that they were safe and effective, including for children. The U.S. Department of Agriculture will provide additional regulation, monitoring its actual use in foods.

"As long as it is used in accordance with the regulations, we have concluded it's safe," said Andrew J. Zajac, of the FDA's office of food additive safety. The illness caused by the Listeria bacteria carries flu-like symptoms, such as fever, muscle aches and sometimes, stomach pains. It can lead to severe headaches, a stiff neck, loss of balance and convulsions.

Food manufacturers have been searching for additives that would target Listeria, Salmonella and other bacteria that sicken consumers. They have relied on antibiotics to kill bacteria, but the microbes have developed resistance to some of those drugs.

Although the incidence of listeriosis is rare among the 76 million food-borne illnesses contracted each year, it's responsible for a disproportionately large percentage of hospitalizations and for many deaths.

Since 1987, regulators have been sampling ready-to-eat foods for the bacteria, but the sampling process destroys the product and thus can't be widely applied, according to the American Meat Institute, an industry association.

Perdue's part

Julie DeYoung, a spokeswoman for Perdue Farms in Salisbury, said the chicken processor would consider using Intralytix's mixture. "The industry is always looking for more effective ways to control pathogens in the processing environment," she said....

Jonathan D. Rockoff
Hanah Cho

ROCKOFF, JONATHAN D., AND HANAH CHO. "BALTIMORE FIRM WINS FDA NOD FOR MIX OF BENEFICIAL VIRUSES." *BALTIMORE SUN*, AUGUST 19, 2006. HTTP://ARTICLES.BALTIMORESUN. COM/2006-08-19/NEWS/0608190338_1_BACTERIA-CENTER-FOR-FOOD-VIRUSES (ACCESSED DECEMBER 30, 2011.)

SEE ALSO *Agricultural Biotechnology; Biopreservation; Dairy and Cheese Biotechnology; Fermentation; Genetically Modified Food; Nanotechnology, Molecular; Security-Related Biotechnology*

BIBLIOGRAPHY

Books

Gutierrez-Lopez, Gustavo F., and Gustavo V. Barbosa-Canovas. *Food Science and Food Biotechnology.* Boca Raton, FL: CRC Press, 2003.

McElhatton, Anna, and Richard J. Marshall, eds. *Food Safety: A Practical and Case Study Approach.* New York: Springer, 2007..

Rahman, Shafiur, ed. *Handbook of Food Preservation*, 2nd ed. Boca Raton, FL: CRC Press, 2007.

Shepard, Sue. *Pickled, Potted, and Canned: How the Art and Science of Food Preserving Changed the World.* New York: Simon & Schuster, 2006.

Varfolomeev, Sergei Dmitrievich, Gennady E. Zaikov, and Larisa P. Krylova, eds. *Biotechnology in Medicine, Foodstuffs, Biocatalysis, Environment, and Biogeotechnology.* New York: Nova Science Publishers, 2010.

Zaikov, Gennadii Efremovich, ed. *Biotechnology in Agriculture and the Food Industry.* New York: Nova Science Publishers, 2004.

Periodicals

Lee, Sun-Young. "Microbial Safety of Pickled Fruits and Vegetables and Hurdle Technology." *Internet*

Journal of Food Safety 4 (2004): 21–32. Available online at http://www.internetjfs.org/articles/ijfsv4-3.pdf (accessed November 3, 2011).

Web Sites

"Estimates of Foodborne Illness in the United States." *U.S. Centers for Disease Control and Prevention (CDC).* http://www.cdc.gov/foodborneburden/ (accessed November 3, 2011).

"Extended Shelf Life Refrigerated Foods: Microbiological Quality and Safety." *Food Technology*, February 1998. http://www.ift.org/Knowledge-Center/ Read-IFT-Publications/Science-Reports/Scientific-Status-Summaries/Extended-Shelf-Life-Refrigerated-Foods.aspx (accessed November 3, 2011).

"Functional Materials in Food Nanotechnology." *Journal of Food Science*, November/December 2006. http://www.ift.org/Knowledge-Center/Read-IFT-Publications/Science-Reports/Scientific-Status-Summaries/Functional-Materials-in-Food-Nanotechnology.aspx (accessed November 3, 2011).

Bryan Thomas Davies

Forensic Biotechnology

■ Introduction

Since the development of the first conclusive DNA matching techniques in the early 1980s, the public perception of forensic science has undergone a remarkable transformation. The role of trained specialists whose laboratories are as important to solving crime as the police interview room has been popularized with relative accuracy in the media. When DNA was sequenced successfully to permit ever smaller biological evidentiary fragments to be analyzed to determine the identity of the sample source, the technological developments that took DNA science from nuclear analysis to new mitochondrial bases moved at a pace that in scientific terms has been breathtakingly swift.

Crime scene procedures, particularly those that ensure the continuity of the biological sample used for forensic analysis, provide the most common avenue of attack taken by defense attorneys in DNA cases. It is an indication of the power of the DNA-based identifications that defense arguments to rebut an apparently compelling conclusion about the defendant's involvement in a crime have come to be largely directed at the investigative procedures used to gather and store the sample, as opposed to the DNA result itself. The suggestion that the sample may have been contaminated through improper handing is one made in many contemporary prosecutions.

The power of DNA to exonerate the wrongly charged or convicted is a feature of modern forensic biotechnology that reveals its importance to the pursuit of criminal justice, particularly in cases where otherwise compelling independent eyewitness evidence or circumstantial evidence appears to confirm the guilt of a particular defendant. DNA evidence is also an essential component of paternity testing and other legal procedures such as the tracing of legal heirs to an estate, where proof of identity is at the heart of the case to be determined, as well as identification of victims in the aftermath of large natural disasters like tsunamis and earthquakes.

■ Historical Background and Scientific Foundations

Forensic science is defined broadly as the specific scientific disciplines in which the evidence produced or revealed through scientific inquiry is admissible to determine questions raised in civil or criminal law proceedings. The modern associations made in the media between crime scene investigations and forensic biotechnological applications, especially DNA technology, might suggest that forensics is a modern scientific advance. Modern forensic applications have roots that extend to the ancient law courts of the Roman Empire, where many of the modern rules pertaining to the admissibility of criminal evidence originate.

Prior to the development of DNA technologies to determine the identity of a person through the analysis of a specific biological sample, the forensic sciences were generally understood to encompass the physical or chemical analyses of evidence seized at crime scenes or as revealed in a subsequent investigation, such as:

- Ballistics, especially the trajectory or physical characteristics of firearms and ammunitions
- Blood or other biological samples
- Blood spatter (or "splatter") patterns
- Chemical analyses conducted to determine the properties of a seized item, or to determine the presence of alcohol, drugs, poisons, or other chemicals in the subject's bloodstream
- Fingerprints
- Gunshot residue
- Wounds, especially the apparent angle of entry, depth, or other features that might be associated with specific weapon use.

The scientific understanding that the molecules present in every human carry unique genetic identifiers was first

WORDS TO KNOW

ALLELOMORPH (ALLELE): The generic term that describes the type of genes found in a biological sample; a common example is the dominant and recessive alleles that contribute to variable human characteristics such as hair color.

BLOOD PATTERN ANALYSIS: Also known as "blood spatter" analysis, the branch of forensic science devoted to the relationship between the patterns left by blood at a crime scene and type of injury sustained or the manner in which a weapon might have been used to produce the particular pattern.

DNA PROFILING: The omnibus term used to describe the forensic processes that determine identity based on the match between a biological sample and a known DNA source.

INNOCENCE PROJECTS: A term that began with the Innocence Project started at the Benjamin N. Cardoza School of Law in 1992, it is now a generic term used to describe legal organizations that pursue justice for persons who assert that they were wrongly convicted. The use of DNA evidence discovered after conviction has been an important feature of all innocence projects.

MITOCHONDRIAL DNA: The DNA located within the mitochondria, the center for energy production located within all human cells. Mitochondrial DNA is the smallest unit of the genome, the hereditary information stored throughout the chromosomes found in every human cell.

NUCLEOTIDES: The three-part molecules that combine to form DNA. The parts are pentose (a sugar with five carbon atoms) and four nitrogenous bases (organic compounds essential to molecular structure): adenine (A), guanine (G), thymine (T), and cytosine (C).

POLYMERASE CHAIN REACTION (PCR): A biological process essential to modern forensic DNA applications that permits millions of copies of a DNA sample to be replicated from a single molecular strand. PCR was developed by a team led by American molecular biologist Kary Mullins (1944–) in 1983, an achievement that resulted in Mullins receiving a share of the Nobel Prize for Biology.

RESTRICTION FRAGMENT LENGTH POLYMORPHISM (RFLP): An early and labor-intensive method of DNA analysis. A restriction enzyme is the chemical used to "cut" the two strands of the DNA double helix structure at a predetermined point. The length of the resulting fragments is used to determine whether a potential match exists between the biological sample and a known source.

SHORT TANDEM REPEATS (STR): The process that employs polymerase chain reaction (PCR) principles in combination with known population variability to determine the probability that there is a match between two compared DNA sources. STR analysis is directed to specific known regions of the nuclear DNA. In this way, DNA samples can be distinguished, enabling scientists to differentiate one DNA sample from another. For example, the likelihood that any two individuals (except identical twins) will have the same 13-loci DNA profile can be as high as 1 in 1 billion or greater.

crystallized through the work of Russian biologist Nikolai Koltsov (1872–1940) in 1927. Identical twins are the only people on Earth who share the same DNA. The platform on which all future DNA technologies were built was the 1953 identification of the DNA double helix structure confirmed by molecular biologists James Watson (1928–) and Francis Crick (1916–2004). Once subsequent research revealed the physical structure and properties of the DNA molecule, the restriction fragment length polymorphism (RFLP) techniques that permitted DNA matches to be made with greater precision and confidence were the next stage in the evolution of forensic biotechnology.

The utilization of DNA technology in criminal investigations during the late 1980s and early 1990s established the ability of biotechnology to provide definitive answers to legal questions that had previously depended on human recollection or other variables. Whereas an accused person could previously challenge visual identification evidence as untrustworthy due to the inherent weaknesses in human recollection, a DNA match was established as a virtually certain identifier. Matches were made on the basis of ratios of 1 to several hundred thousand to reveal that the alleged perpetrator was, or was not, the identified subject. STR technology represented another scientific breakthrough, because any available DNA could be copied for the most thorough matching analysis possible, no matter how minute the biological samples obtained in the crime scene investigations. Short tandem repeat (STR) analysis ensured that all available DNA could be extracted efficiently from a given sample, with each allele identified in the process.

STR analysis has been accepted in most Western jurisdictions as a means by which to make a positive identification. There are 13 DNA locations on a human cell used in STR analysis that are universally accepted as the sites at which the analysis will yield accurate results. The laws of probability determine that where a match between two samples is made at three of the locations on the examined samples, the probability of a match can be calculated against the entire human population. It is for this reason that the match probabilities given as evidence in trial proceedings are often stated in ratios of 1 to several billion or more.

Mitochondrial DNA analysis represents the latest scientific frontier in forensic biotechnology applications to have been successfully conquered through concerted research efforts. Unlike nuclear DNA, which is extracted and analyzed using STR technology, mitochondrial DNA is found in the eukaryotic cells within the body that convert stored glucose into adenosine triphosphate (ATP), the energy that powers all human physical activity. The smallest chromosome yet identified within the body, mitochondrial DNA is known to be inherited solely from the mother, making it a very valuable investigative tool. Mitochondrial DNA also can be replicated in the same fashion as in polymerase chain reaction (PCR) process, which has been used since the later 1980s to reproduce sufficient nuclear DNA samples necessary to perform a range of comparative forensic investigations.

A famous application of DNA technology outside of its common use in criminal and paternity investigations is the case of Anna Anderson (1896–1984), who had claimed that she was the sole surviving member of the Romanov family, the rulers of Imperial Russia prior to the 1917 Revolution. Anderson's claim that she was the Grand Duchess Anastasia was disproved without doubt by a DNA analysis performed after her death that compared her stored tissue samples with known biological samples traced to Anastasia.

The rapid rise of DNA testing as a criminal investigative tool has prompted the development of accreditation standards for laboratories that perform these analyses. In North America, the American Society of Crime Laboratory Directors/Laboratory Accreditation Board (ASCLD/LAB) generally is regarded as the most authoritative forensic DNA laboratory accreditation program.

■ Impacts and Issues

The combined effect of nuclear and mitochondrial DNA research has advanced forensic science beyond its traditional physical/chemical science boundaries. Concepts that were only research theories until the early 1980s have come to occupy the mainstream of the forensic sciences. The power of properly conducted DNA testing to identify a biological sample as the product of a specific individual human being has become so certain as to be undoubted. The importance of DNA analysis has placed significant pressure on law enforcement agencies to ensure that their crime scene investigators have the

A poster encourages relatives of Srebrenica massacre victims to give a blood sample at the International Commission on Missing Persons' Podrinje Identification Project Center in Tuzla, Bosnia-Herzegovina. © *Marco Di Lauro/Getty Images.*

IN CONTEXT: IDENTIFICATION OF VICTIMS

The fate of a missing person strikes a powerful chord in all human societies. Throughout history, genocide and similar atrocities have proved a tragic feature of armed conflict. The Rwandan civil war was a prominent example of how the mass murder of a specific population caused immense pain for the survivors, who were forced to rebuild their society with the knowledge that many thousands of people were not unaccounted for when the conflict ended. The ability to provide closure to the families of genocide victims is argued to be a humane and necessary step in national reconstruction. DNA technology provides the means to identify the bodies of persons indiscriminately buried in the mass graves that are often a common feature of genocide.

The International Commission on Missing Persons (ICMP) was established in 1996 to locate missing persons from the conflicts in the former Yugoslavia. The ICMP was one of the first international agencies to use DNA technology to resolve the fate of those found in various mass graves located in Croatia, Bosnia and Herzegovina.

Natural disasters are also instances in which DNA technology can be used to identify victims. In 2006 the ICMP assisted American authorities in the identification of victims of Hurricane Katrina, the massive storm that struck the U.S. Gulf Coast in August 2005. DNA tests were performed on 300 bone samples in an effort to match known biological samples to remains of approximately 170 Katrina victims that could not otherwise be identified.

The March 2011 tsunami that struck northeastern Japan also illustrates how DNA identification procedures promote orderly recovery and assist in achieving closure for the survivors. Japanese police established DNA databases using biological samples taken from known sources such as the clothing of missing persons. The databases were used by investigators to identify some of the thousands of bodies found in the tsunami's aftermath.

---

requisite training to handle biological samples located at a scene.

DNA science has advanced from its earlier reliance on the more expensive, time-consuming RLFP analysis to the modern PCR methods to make conclusive identifications from nuclear and mitochondrial DNA. Rather than the relatively large DNA samples required by RLFP, even minute amounts of DNA can be amplified and analyzed to achieve a definitive conclusion concerning identification of the sample source.

PCR requires the following materials: a minimum single DNA strand taken from the biological sample found at the investigation scene; primers that prepare the samples for analysis, polymerase, and the DNA-specific enzyme; and the DNA nucleotides adenine, guanine, cytosine, and thymine. PCR is completed through a three-stage process. In stage one, the target DNA, primers, polymerase, and nucleotides are combined at a temperature of approximately 195°F (90°C) for a 30-second period. The DNA double helix structure separates at this temperature. In the second PCR stage, the mixture is cooled to approximately 130°F (55°C), a process that lasts 20 seconds, sufficient to encourage the DNA molecule to reform in its double strand configuration. The primers are used to isolate the precise section of the DNA molecule that will be copied for forensic analysis. In the final stage, the temperature of the mixture is increased to 166°F (75°C) to optimize the effectiveness of the polymerase as it generates an exact copy of the target DNA desired for analysis.

Through the repetition of the three stage PCR cycle, the second cycle produces four copies, the fourth cycle generates eight copies, and through simple mathematical progressions of the cycle millions of copies are produced after 20 repetitions. PCR technology combines scientific precision in the copy quality and massive savings of time for analysts who require large sample numbers for comprehensive DNA matching analysis.

Cases involving the analysis of biological materials that contain evidence of more than one DNA source pose special challenges for analysts. This issue most often arises in cases of alleged sexual assault when the forensic examination of the victim's clothing yields more than one DNA type, confirming the possibility of more than one possible perpetrator. PCR techniques have been used successfully in numerous cases to distinguish the DNA found in mixed biological samples.

There are four DNA handling errors that are most commonly encountered in criminal prosecutions that carry the potential for a faulty analysis that will compromise the legal proceeding. In order of their frequency of occurrence, the errors are:

- Sample confusion. Samples that have been erroneously labeled or otherwise mixed up in the lab due to human error.

- Sample contamination. The most common form is the accidental touching of the sample by a crime scene investigator or laboratory analyst that introduces their DNA into the sample. Sample contamination also occurs when bacteria or other materials come into contact with the sample.

- DNA degradation. DNA may degrade due to the failure to preserve or store the crime scene sample in a cool or dry receptacle. If the sample is not properly stored, the possibility of an inaccurate DNA analysis is increased (mitochondrial analysis is less susceptible to errors of this kind).

- Faulty statistical analysis. The accuracy of DNA analysis may depend in some cases on the application of probabilities and statistics principles that go beyond the standard software packages now used to provide the results used in court.

In criminal prosecutions, paternity cases, genealogy, disaster victim identification, and other instances in which the ability to provide a definitive connection between the known and sample is essential, DNA technology provides the evidentiary backbone to the proceeding. It is obvious from the hundreds of cases in which DNA has provided pivotal evidence to establish the identity of perpetrators of crimes that DNA technologies are now essential to the proper administration of justice.

DNA plays an equally important role in the determination of factual innocence in criminal cases. In hundreds of documented cases in the United States, Canada, Australia, the United Kingdom, and other Western jurisdictions where there is ready access to DNA technology, DNA technology has provided the basis for the subsequent exoneration of persons who were convicted of crimes, particularly when the evidence was solely or largely based on that of eyewitnesses. These exonerations have included many instances in which the defendant faced the death penalty upon conviction.

Eyewitness evidence is in many respects the antithesis of the precision associated with DNA results. The courtroom witness is providing a subjective account of what they saw that led them to conclude that the particular accused person committed or was otherwise involved in the crime in question. Lighting, personal bias, the influence of fatigue, alcohol, drugs, apprehension about involvement as a witness, or various perceptual limitations all may influence the quality of an eyewitness account. Numerous studies conducted by various international innocence projects have confirmed that in some cases, third party independent witnesses to an alleged crime will add details subconsciously to their recollection that fill in gaps in what they remember, or are otherwise based on erroneous assumptions of what the witness believes must have occurred.

DNA has the decided advantage of providing an objective, rational determination of identification issues, assuming that the analysis of the relevant biological sample was conducted in accordance with accepted scientific standards. Hundreds of convicted persons have been exonerated subsequently through forensic biotechnological methods that did not exist at the time of their conviction.

Falsified DNA is an emerging issue in crime scene investigations that involve the collection of biological samples for subsequent DNA analysis. The chemical characteristics of DNA, especially its ability to provide a unique personal identifier, mean that a crime scene can be contaminated deliberately by a manufactured DNA sample that disguises the identity of the true perpetrator of a crime.

■ Primary Source Connection

A victim found shot on May 10, 2002, subsequently died at the hospital. A cigarette butt found at the scene was examined as evidence involved in the crime. Questions arose around findings of the DNA on the cigarette butt. The prosecutor and lead detective sought an explanation as to how DNA findings could show both the suspect's as well as the victim's DNA on this piece of evidence. A study ensued to find out the answer to this question.

When is a Kiss Just a Kiss?

In the early morning hours of May 10, 2002, a male was found shot, bleeding and in need of medical attention on a Richland County roadway. The victim expired while at the hospital, and although he could speak, he did not say who shot him. The autopsy revealed the victim suffered a single gunshot wound to the chest. The evidence collected at the scene included two live rounds, one spent casing, a cell phone and a cigarette butt. The crime scene investigator noted the clean condition of the cigarette butt and collected it. The casing and live rounds were .380 caliber; no weapon was recovered.

In January of 2003, the cigarette butt was submitted to the Mansfield Police Laboratory for analysis for an upcoming court date. In addition to the cigarette butt, a buccal swab sample from the female suspect, an acquaintance of the victim's girlfriend, was submitted. Analysis of the cigarette butt revealed a readily distinguishable major/minor mixture consistent with an unknown male and the suspect. The victim's blood was submitted in February for analysis. The victim could not be eliminated as the major donor of the male portion of DNA recovered from the evidence. The Prosecutor and lead Detective could not believe that DNA from both the suspect and victim could be on the cigarette butt from the crime scene. They felt they had the only evidence they needed to convict their suspect. We considered the possibility that the suspect lit a cigarette for the victim before she killed him. Perhaps they shared a cigarette? My favorite, "the made for TV" version, is "the perpetrator took the last puff of the cigarette that she removed from his dying hand and then simply flicked it away," not knowing she just placed herself at the death scene. To address whether there was a reasonable explanation of how this evidence wound up at the scene, we decided to use the PowerPlex 16 System to explore an alternate explanation of how both profiles had been deposited on the cigarette butt. Was it possible that the donor of the female DNA was never at the scene? This was a thought the Prosecutor and lead Detective did not want to hear.

I was unable to locate any kissing studies, and my neighbors have pretty much had it with my intimate requests, so I needed help from my colleagues. The study involved minimal supplies, an appropriate location and an occasion that would make the collection feasible. The participants were required to kiss for a period of 5 seconds and then each smoke a cigarette. The process was repeated after 1 hour, but the kiss was extended to 10 seconds. The cigarettes were air-dried and sealed in a paper envelope for transportation to the laboratory. Analysis of the cigarette butts revealed readily distinguishable major/minor profiles in 67% of the samples using the PowerPlex 16 System. The profiles were extremely similar to the profile obtained in the case sample.

At the end of March, the DNA evidence was introduced during the trial. Cross-examination included discussions of primary and secondary transfer theories associated with biological fluids and specifically cigarette butts. The research results were discussed during this period.

The jury found the defendant guilty of murder. In addition to the DNA evidence, statements of the defendant's associates and her cell phone call, which relayed off the closest cell phone tower, ruined her alibi.

Anthony J. Tambasco

TAMBASCO, ANTHONY J. "WHEN IS A KISS JUST A KISS?" *PROFILES IN DNA* 7, NO. 1 (MARCH 2004): 14.

SEE ALSO *DNA Databases; DNA Fingerprinting; DNA Sequencing; DNA Testing: Post-Conviction; Forensic DNA Testing; Microarrays; Polymerase Chain Reaction (PCR)*

BIBLIOGRAPHY

Books

Buckleton, John S., Christopher M. Triggs, and Simon J. Walsh, eds. *Forensic DNA Evidence Interpretation.* Boca Raton, FL: CRC Press, 2005.

Goodwin, William, Adrian Linacre, and Sibte Hadi. *An Introduction to Forensic Genetics*, 2nd ed. Chichester, UK: Wiley-Blackwell, 2011.

James, Stuart H., and Jon J. Nordby, eds. *Forensic Science: An Introduction to Scientific and Investigative Techniques*, 3rd ed. Boca Raton, FL: CRC Press, 2009.

Kobilinsk, Lawrence F., Thomas F Liotti, and Jamel Oeser-Sweat. *DNA: Forensic and Legal Applications.* Hoboken, NJ: Wiley-Interscience, 2005.

Web Sites

"About Forensic DNA." *DNA Initiative.* http://www.dna.gov/basics/ (accessed October 27, 2011).

"DNA Forensics." *National Human Genome Research Institute.* http://www.ornl.gov/sci/techresources/ Human_Genome/elsi/forensics.shtml (accessed October 27, 2011).

Innocence Project. http://www.innocenceproject.org/ (accessed October 27, 2011).

van Oorschot, Roland A. H., Kaye N. Ballantyne, and R. John Mitchell. "Forensic Trace DNA: A Review." *Investigative Genetics*, December 2010. http://www.ncbi.nlm.nih.gov/pmc/articles/PMC3012025/ (accessed October 27, 2011).

Bryan Thomas Davies

Forensic DNA Testing

■ Introduction

A person's identity is written into his or her genome in the form of its DNA sequence. Advances in DNA analysis have allowed the link between DNA and identity to be applied in forensic testing. Samples of tissue, such as skin flakes, blood, or hair, left behind at the scene of a crime, can be compared to the DNA of a suspect, helping to establish innocence or guilt. In addition, forensic DNA testing may be used to determine the identities of individuals who are victims of large natural disasters or buried in mass graves.

Forensic DNA testing does not involve sequencing the DNA of a suspect. This would be prohibitively expensive and is unnecessary. The distribution of certain features within DNA differs sufficiently between individuals to establish identity. Although DNA analysis is a powerful forensic tool that has made a huge contribution to the justice process, it does raise a number of technical, statistical, and ethical issues. DNA analysis is highly dependent on strictly controlled laboratory conditions; where these are not met, data may be compromised. There is also some debate on whether a person's DNA profile, as currently measured, is really as unique to that individual as is generally believed. Finally, the storage of DNA samples in databases is felt by some to raise serious civil liberty issues.

■ Historical Background and Scientific Foundations

Professor Alec Jeffreys (1950–) at Leicester University in the United Kingdom developed DNA fingerprinting, or DNA profiling as it is now more commonly known. In 1985, he became the first to use the method to establish identity on the basis of an individual's DNA sample. The world's first DNA profiling success in the forensic area was the solving of a double rape and murder in Leicestershire, United Kingdom, in 1986. In this case, a suspect had confessed to the second murder, but was cleared because his DNA did not match that found at the scene of the crime. The guilty man, Colin Pitchfork (1960–), persuaded another man to stand in for him when DNA testing of all men in the area was organized. The police were tipped off about the deception, and Pitchfork was arrested, tried, and given a life sentence for his crimes.

WORDS TO KNOW

ADMISSIBILITY: Refers to evidence, such as forensic evidence, that is accepted by a court of law.

DNA FINGERPRINTING: Also known as DNA profiling, DNA fingerprinting is a technology that generates a pattern of specific repeated DNA sequences in a DNA sample and is used to establish a person's identity.

ELECTROPHORESIS: A technology for separating a mixture of DNA, or other biological molecules, on a gel by subjecting it to an electric field. Fragments move at different speeds under the influence of the field, and this achieves a separation effect. The fragments can be visualized by fluorescent or radioactive tags.

LOCUS: The location of a gene, or other DNA sequence, on a specific chromosome within the genome (plural, loci).

MINISATELLITE DNA: Short repetitive stretches of DNA within the genome that are used as a basis for DNA fingerprinting.

POLYMORPHISM: Variations in DNA sequence at a specific locus in the genome.

RESTRICTION ENZYME: An enzyme that cuts a DNA molecule at a known base pair sequence.

SHORT TANDEM REPEAT: A short repeating sequence of DNA that is found at a specific locus on a genome.

VARIABLE NUMBER TANDEM REPEAT: A base pair sequence that repeats throughout a specific locus on a genome.

IN CONTEXT: POST-CONVICTION TESTING

The 2009 U.S. Supreme Court case *District Attorney's Office v. Osborne* highlighted the state-by-state variability regarding post-conviction DNA testing rights for convicted prisoners. As of October 2011, Massachusetts and Oklahoma are the only states that do not have post-conviction DNA testing access laws, and yet Massachusetts does require fair compensation for wrongly-convicted individuals. A patchwork of state laws creates variability regarding access to evidence, preservation of evidence, compensation for wrongful conviction, and eyewitness interrogation processes. This variability leads organizations such as The Innocence Project to claim that the system is inherently unfair from a legal standpoint. A convicted felon in California, for instance, has a set of evidence access rights that differs for the same criminal conviction in Nevada, giving unequal protection. According to the

Innocence Project, as of November 2011, 275 people had been exonerated post-conviction based on DNA evidence, including some who had been on death row.

In March 2011, the U.S. Supreme Court, in *Skinner v. Switzer*, determined that a convicted prisoner could file a claim that federal civil rights had been violated in cases where state law prevented prisoners from access to evidence that could be tested for DNA typing. The case concerns Texas law; Hank Skinner was convicted of triple homicide in 1993 and filed the suit from death row. In the suit, he claimed that the Texas law barring his access to evidence that had never been tested before his trial constituted a violation of federal civil rights. The court concurred in a 6-3 decision, paving a narrow channel for other convicted prisoners to argue against state laws that might violate their federal rights.

In the United States, DNA evidence was first introduced into the trial of a sexual assault case in Florida in 1987.

The idea of detecting the identity of a perpetrator through some kind of biological signature left at the scene of a crime goes back to fingerprint analysis, first used in forensics in the late nineteenth century. No two people have the same fingerprints, so fingerprinting analysis remains a powerful forensic technology. However, criminals have long since learned to use gloves or to wipe away fingerprint evidence, so fingerprinting analysis has its limitations. It is far more difficult for a criminal to remove all evidence of his or her DNA from the scene, particularly since the advent of the polymerase chain reaction (PCR), which allows DNA analysis of the most minute amounts of tissue. Traces of blood, saliva, hair follicles, skin, and semen are commonly left at the scene. All can provide DNA for analysis to determine if there is a match to DNA from a blood sample or swab taken from a suspect. Even a stamp, toothbrush, or a bite wound can yield enough perpetrator DNA for analysis, as can a single hair follicle.

DNA profiling depends on looking at various loci within the genome that show polymorphism and therefore differ between individuals. These regions are located in so-called junk DNA, which does not contain genes. Certain bits of sequence in these regions, known as minisatellites and microsatellite DNA, are repeated many times, and it is the analysis of these features that forms the scientific basis of DNA profiling. There are two main approaches to DNA profiling. The first is based on analyzing variable number tandem repeats (VNTRs), which are minisatellites that are hundreds of base pairs long, repeating a variable number of times along the region. The second involves analysis of short tandem repeats (STRs), which are microsatellites, consisting of only three to seven base pairs, repeating over segments of DNA up to 400 base pairs long. Forensic scientists

use fluorescent or radioactive probes based upon sets of VNTRs and STRs mixed with the DNA sample to create a barcode-like pattern reflecting the variability of the repeats in an individual.

Prior to analysis, the DNA is chopped up into smaller fragments using restriction enzymes, and these are separated by electrophoresis. It is the analysis of these fragments that provides the DNA profile. The forensic scientist compares the profiles of the DNA from the crime scene side by side with that from the suspects to see if there is a match. The prosecutor will generally have forensic DNA testing carried out in a state laboratory, whereas the defense will use a commercial laboratory if they wish to challenge the DNA evidence.

■ Impacts and Issues

The introduction of forensic DNA testing into the courts led to several years of legal arguments over its admissibility as evidence. Many courtroom challenges were based upon lack of standards and guidelines in DNA testing. Therefore, the U.S. Federal Bureau of Investigation (FBI) helped set up a technical working group that produced a set of guidelines in 1993. These have been regularly updated to keep pace with developments in DNA testing technology.

There are many ways in which DNA evidence may be compromised, and it can be challenged in the court on any of these points. DNA from a crime scene must be either dried or frozen. It is then stored in a protective container before analysis to prevent its degradation. Degradation breaks the DNA into fragments that are too small to work with, essentially destroying the evidence. Great care must be taken to secure the chain of custody at every stage, from the crime scene to the laboratory. As well as being prone to degradation, DNA is also easily contaminated with extraneous DNA, either from the

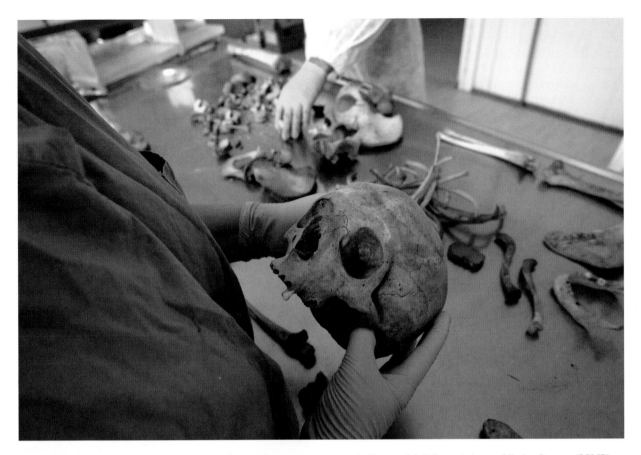

A forensic anthropologist examines a cranium exhumed from a mass grave at the International Commission on Missing Persons (ICMP) Lukavac re-association center in Bosnia-Herzegovina. ICMP is using DNA samples to help identify remains from mass graves in Srebrenca in the former Yugoslavia. Forensic pathologists make the final determination on the identification of remains, comparing DNA samples with samples taken from family members of the missing. ICMP pioneered the use of DNA for identifying missing persons on a mass scale. © *Marco Di Lauro/Getty Images.*

crime scene or from the skin of anyone who handles the DNA sample along the chain of custody without using appropriate precautions. The risk of contamination is particularly great when PCR is involved, because the procedure will amplify the contaminating DNA alongside that of the evidence sample.

The defense can challenge forensic DNA evidence on a number of other technical issues. Mishandling, sample mix-ups, laboratory equipment failure, human error, and departure from standard testing protocols can all form the basis of a successful challenge. The defense can also question the qualifications of the forensic scientist, for DNA analysis and interpretation demand particular skills and training. The laboratory itself must also be accredited for this kind of forensic work. Independent testing, or second opinions, can raise other issues, because different laboratories may obtain different results when testing the same DNA sample. Indeed, different results may occur in the same laboratory with the same sample tested on different days. These difficulties underline the importance of adhering very strictly to guidelines and protocols when it comes to DNA work,

because the testing is expensive and the outcome of a case may depend upon the quality of the DNA evidence.

There is also some controversy over just how good an indicator a DNA profile is of an individual's unique identity. A person's genome is unique, unless that person happens to be an identical twin. However, a DNA profile is not a genome. It samples only about 1,000 of the estimated 3 billion nucleotides in the genome. A profile is a pattern of DNA taken from a number of different loci on a genome. Although many individuals may share a pattern at one or two loci, the chances of them sharing the same pattern at 13 loci, the number generally examined in forensic DNA testing, are vanishingly small because the number of possible profiles vastly exceeds the world's population. It is upon this kind of statistical analysis that the reliability of DNA evidence rests. On this basis, it would be safe to assume that matches between individuals whose profiles are stored in DNA databases must be impossible.

However, in 2001, forensic scientist Kathyrn Troyer found two individuals in the Arizona criminal DNA database who had remarkably similar profiles. They matched at nine of the 13 loci, but they were unrelated, with one being black and the other white. According to the FBI, the chances of unrelated people sharing a profile to this extent would be one in 113 billion. Critics of Troyer's findings point out that she was searching for any matches among thousands of profiles, whereas in forensic investigations a match to a specific profile is sought. Those who believe her work raises some questions about the statistics put out by the FBI would like further searches carried out on criminal databases. The FBI argues that this would violate the privacy of offenders whose profiles are stored there. However, such searches have been carried out in at least two other states, and nearly 1,000 matches were found, according to a report in the *Los Angeles Times*. The implication of such searches remains unclear but does cast some doubt over whether a person's DNA profile is really unique. As the sizes of DNA databases grow all over the world, there will be opportunities to explore this issue further. DNA profiling is indeed a powerful forensic tool, but it needs to be used, and set in context with, other forensic evidence.

Another important issue is the storage of DNA samples in huge databases. The FBI runs a program of support for criminal justice known as the Combined DNA Index System (CODIS), which includes the National DNA Index System, a database containing DNA profiles contributed from federal, state, and local forensic laboratories. The DNA Identification Act of 1994 gave the FBI the authority to set up the DNA database, and it became fully operational in 1998. At present, the database contains more than 10 million offender profiles. State laws vary as to which crimes qualify for needing the deposition of a DNA profile in the database, but generally this is required at least for homicide and sexual offenses. Sometimes people have their DNA collected and deposited on arrest, before conviction of a crime, to aid a forensic investigation. The database also contains DNA of missing persons and of unidentified human remains.

The CODIS software is used by more than 40 law enforcement agencies in more than 25 countries. The development of DNA databases for forensic purposes is growing. The United Kingdom Police National DNA database, established in 1995, is the world's largest in terms of the percentage of the population whose samples are held on it. Around 10 percent of the U.K. population has samples in the database, compared to just 1 percent in Austria, whose database is the second largest in this respect. There is much controversy over whose DNA should be entered into such databases, especially whether DNA samples should remain stored if a person is found to be innocent or has been convicted of a less serious crime. In the United Kingdom, samples from those who are acquitted are now to be destroyed.

It is still unclear whether expanding forensic DNA databases will help solve more crimes. As the collection and storage of DNA has become easier, some have argued that a DNA database may as well be expanded to

cover the whole of the population. This would be a valuable resource for research applications other than forensics. There are many such databases around the world being used for medical purposes, for instance. Others say that a forensic DNA database should be as small as possible compatible with the purpose of fighting crime and seeing justice done, with only the only DNA from those convicted for very serious crimes being there.

There are civil liberties arguments against DNA databases, mainly because of fears that the information could be accessed by those outside the criminal justice system, although police systems do have security measures built in. Those who argue for such databases state that forensic DNA databases are not to be used for purposes other than forensic ones. Personal information is generally not stored in the databases and the samples are to come from accredited laboratories only. Ideally, security is tight,

A U.S. Army soldier walks his guard post before ceremonies May 14, 1998, at the Tomb of the Unknowns at Arlington National Cemetery, Virginia, in which the remains of the Vietnam War unknown were exhumed. The remains were taken to Walter Reed Army Medical Center for DNA testing in an attempt to make a positive identification. They were confirmed by mitochondrial DNA to be those of Air Force pilot Michael Blassie. © *Tim Sloan/ AFP/Getty Images.*

with only authorized personnel having access. However, as with all computer systems, a forensic DNA database is open to malfunction and abuse. Ongoing vigilance is needed, therefore, to ensure that forensic DNA helps justice be sought without harming civil liberties.

■ Primary Source Connection

Questions have arisen regarding the accuracy of some DNA database results in identifying a trustworthy match to one person, and one person only. Similar genetic profiles between more than one person, even non-related individuals, have scientists and legal experts questioning the accuracy of some FBI statistics, as they look at the validity and accuracy of profiles in the Combined DNA Index System (CODIS), the national database system. New findings of matching locations on chromosomes and genetic markers used to distinguish individuals have raised questions about DNA findings, leading to legal fights between authorities over the possibility of broadening access to the national database for scrutiny.

FBI resists scrutiny of "matches"

A crime lab's findings raise doubts about the reliability of genetic profiles. The bureau pushes back.

State crime lab analyst Kathryn Troyer was running tests on Arizona's DNA database when she stumbled across two felons with remarkably similar genetic profiles.

The men matched at nine of the 13 locations on chromosomes, or loci, commonly used to distinguish people.

The FBI estimated the odds of unrelated people sharing those genetic markers to be as remote as 1 in 113 billion. But the mug shots of the two felons suggested that they were not related: One was black, the other white.

In the years after her 2001 discovery, Troyer found dozens of similar matches—each seeming to defy impossible odds.

As word spread, these findings by a little-known lab worker raised questions about the accuracy of the FBI's DNA statistics and ignited a legal fight over whether the nation's genetic databases ought to be opened to wider scrutiny.

The FBI laboratory, which administers the national DNA database system, tried to stop distribution of Troyer's results and began an aggressive behind-the-scenes campaign to block similar searches elsewhere, even those ordered by courts, a Times investigation found.

At stake is the credibility of the compelling odds often cited in DNA cases, which can suggest an all but certain link between a suspect and a crime scene.

When DNA from such clues as blood or skin cells matches a suspect's genetic profile, it can seal his fate with a jury, even in the absence of other evidence. As questions arise about the reliability of ballistic, bitemark and even fingerprint analysis, genetic evidence has emerged as the forensic gold standard, often portrayed in courtrooms as unassailable.

But DNA "matches" are not always what they appear to be. Although a person's genetic makeup is unique, his genetic profile—just a tiny sliver of the full genome—may not be. Siblings often share genetic markers at several locations, and even unrelated people can share some by coincidence.

No one knows precisely how rare DNA profiles are. The odds presented in court are the FBI's best estimates.

The Arizona search was, in effect, the first test of those estimates in a large state database, and the results were surprising, even to some experts.

Lawyers seek searches.

Defense attorneys seized on the Arizona discoveries as evidence that genetic profiles match more often than the official statistics imply—and are far from unique, as the FBI has sometimes suggested.

Now, lawyers around the country are asking for searches of their own state databases.

Several scientists and legal experts want to test the accuracy of official statistics using the 6 million profiles in CODIS, the national system that includes most state and local databases.

"DNA is terrific and nobody doubts it, but because it is so powerful, any chinks in its armor ought to be made as salient and clear as possible so jurors will not be overwhelmed by the seeming certainty of it," said David Faigman, a professor at UC Hastings College of the Law, who specializes in scientific evidence.

FBI officials argue that, under their interpretation of federal law, use of CODIS is limited to criminal justice agencies. In their view, defense attorneys are allowed access to information about their specific cases, not the databases in general.

Bureau officials say critics have exaggerated or misunderstood the implications of Troyer's discoveries.

Indeed, experts generally agree that most—but not all—of the Arizona matches were to be expected statistically because of the unusual way Troyer searched for them.

In a typical criminal case, investigators look for matches to a specific profile. But the Arizona search looked for any matches among all the thousands of profiles in the database, greatly increasing the odds of finding them.

As a result, Thomas Callaghan, head of the FBI's CODIS unit, has dismissed Troyer's findings as "misleading" and "meaningless."

He urged authorities in several states to object to Arizona-style searches, advising them to tell courts that the probes could violate the privacy of convicted offenders, tie up crucial databases and even lead the FBI to expel offending states from CODIS—a penalty that could cripple states' ability to solve crimes.

In one case, Callaghan advised state officials to raise the risk of expulsion with a judge but told the officials that expulsion was unlikely to actually happen, according to a record of the conversation filed in court.

In an interview with The Times, Callaghan denied any effort to mislead the court.

The FBI's arguments have persuaded courts in California and other states to block the searches. But in at least two states, judges overruled the objections.

The resulting searches found nearly 1,000 more pairs that matched at nine or more loci.

Jason Felch
Maura Dolan

FELCH, JASON, AND MAURA DOLAN. "FBI RESISTS SCRUTINY OF 'MATCHES'" *LOS ANGELES TIMES* (JULY 20, 2008): 1–3.

SEE ALSO   *DNA Databases; DNA Fingerprinting; DNA Testing: Post-Conviction; Electrophoresis; Forensic Biotechnology; Individual Privacy Rights*

BIBLIOGRAPHY

Books

Butler, John M. *Fundamentals of Forensic DNA Typing.* Boston: Academic Press/Elsevier, 2010.

Echaore-McDavid, Susan, and Richard A. McDavid. *Career Opportunities in Forensic Science.* New York: Ferguson, 2008.

Embar-Seddon, Ayn, and Allan D. Pass, eds. *Forensic Science.* Pasadena, CA: Salem Press, 2009.

Kobilinsky, Lawrence F., Louis Levine, and Henrietta Margolis-Nunno. *Forensic DNA Analysis.* New York: Chelsea House, 2007.

Krimsky, Sheldon, and Tania Simoncelli. *Genetic Justice: DNA Data Banks, Criminal Investigations, and Civil Liberties.* New York: Columbia University Press, 2011.

Lynch, Michael. *Truth Machine: The Contentious History of DNA Fingerprinting.* Chicago: University of Chicago Press, 2008.

Read, M. M., ed. *Focus on DNA Fingerprinting Research.* Hauppage, NY: Nova Biomedical Books, 2006.

Web Sites

"About Forensic DNA." *DNA.gov: DNA Initiative.* http://www.dna.gov/basics/ (accessed October 4, 2011).

"DNA Forensics." *U.S. Department of Energy Human Genome Program.* http://www.ornl.gov/sci/techresources/Human_Genome/elsi/forensics.shtml (accessed October 4, 2011).

Susan Aldridge

Frost-Resistant Crops

■ Introduction

The steady increase observed in global human population growth rates has stimulated intense scientific research into ways to improve food production. An important research area is the development of frost-resistant crops. The utilization of crop species that can withstand colder temperatures represents an important breakthrough, as the result is improved crop yields realized from the same available arable land. Where crop yields are increased, the need to clear land of trees and natural vegetation to create more farmland is reduced; deforestation contributes to the increased levels of greenhouse gases associated with global warming and climate change. Frost-resistant crops will also reduce the need for farmers to resort to expensive and time-intensive protective measures to save their crops, such as heaters and sprinkler systems, in colder weather or when sudden weather changes produce frost conditions.

Frost damage to crops, including fruits and vegetables, has a significant limiting effect on food production capacity in regions that have shorter growing seasons or those with higher altitudes. Frost-resistant crops are being developed for commercial application through the study of the genetic composition of plants with cellular structures that are able to withstand the extreme temperatures in global polar regions.

■ Historical Background and Scientific Foundations

Agriculture is a prominent example of the elemental battle that has been waged by humankind against Earth's environmental forces throughout the entire history of the human species. The cultivation of plant crops has been an essential component of human food production activities. The first human societies that practiced agriculture were generally established in temperate climate regions that are located between 20° and 50° latitude in each hemisphere. As the global population increased over time, people migrated into regions that experienced colder weather and correspondingly shorter growing seasons. The ability to identify hardier crop species that could better resist frost has been an element of all successful agricultural practices in colder climates throughout recorded history.

Frost is a term that has scientific and colloquial meanings. For scientists, frost is the phenomenon observed when ice crystals form on surfaces. The crystals form either when dew freezes or when water in the air changes from vapor to ice. Frost is also widely used to describe any metrological event that causes damage to crops or plants due to freezing. Most plants tend to die when frost forms on their surfaces. Plant cells are destroyed when the water they contain becomes ice due to frost contact, and the ice expands inside the plant. Frost is most commonly formed when cold air radiates outwards to reduce the dew point to freezing (32°F / 0°C). Advection is the less common but equally destructive process in which frost forms when cold winds instantly freeze water vapor directly to plant surfaces.

Plants such as tomatoes, cherries, peaches, and other tender fruits are especially vulnerable to frost damage. These crop species tend to have a higher water content and a cellular structure that promotes freezing and irreversible cell death when the plant is exposed to frost. Root vegetables such as radishes and turnips have a denser cell structure that better withstands the destructive effects of frost.

Otherwise productive agricultural regions are often most vulnerable to the economic consequences of frost damage. Farmers will seek to extend the growing season by planting crops at the earliest possible opportunity or by delaying the harvest to the last possible moment to maximize crop yields. The rich prairie lands of central North America, Australia, and Eastern Europe regularly experience sudden spring or fall frosts that can destroy an entire crop. Sudden frosts are a feature of the citrus growing seasons in Florida and California.

WORDS TO KNOW

ARABIDOPSIS: *Arabidopsis thaliana* is a plant that is widely used by research biologists as the model organism in many frost-resistant crop experiments. *Arabidopsis* is a member of the Brassicaceae plant family, and it is genetically similar to radish and cabbage plants. The genetic structure of *Arabidopsis* is well understood. This plant provides researchers with numerous investigative avenues to advance the understanding of plant genetics.

CRYOBIOLOGY: The study of how different components in living organisms react when exposed to low temperatures. The contemporary study of frost tolerance is primarily directed to the cell and protein structures of the crop plant species that make the greatest contribution to global food supplies. These species include wheat and other grain crops.

DEW POINT: The temperature at which water vapor will condense into liquid form. The condensation is dew; the dew point varies in accordance with the prevailing barometric pressure.

FROST PROTECTION: Many crops are protected from frost by artificial means; the objective is to maintain the crop ambient air temperature above 32°F (0°C). In some instances, wind turbines are used to redirect warm air from the atmospheric inversion layer to the crop. Water sprinkler systems are also employed to moisten the soil surface. This procedure tends to prevent the surface temperature of the crop and soil from falling below freezing.

■ Impacts and Issues

The impact of frost damage on global agricultural production is profound. Approximately 15% of all projected annual crop harvests throughout the world are lost to frost. In the United States, there are greater economic losses that result from frost damage than those caused by any other environmental phenomenon, including droughts, violent storms, or insect attack. Crops such as the spring wheat varieties grown in North America and Australia are especially vulnerable to frosts that can strike in the early part of growing season. In a world where the continual need to increase food production is driven by the pressures imposed by population growth, the development of frost-resistant crops that permit farmers to maximize their yields is a research imperative.

The research directed to the development of frost-resistant crops has largely centered on furthering the understanding of how certain plants are able to thrive in otherwise inhospitable climatic regions. Since 2006, different varieties of Antarctic haregrass that will grow at temperatures as low as −8°F (−22°C) have been the research focus in several studies. The grass contains a protective protein that is activated when the surrounding air temperature falls below 41°F (5°C). As the water content in the grass reaches its freezing point, the protein creates a barrier around the minute particles of ice formed inside the plant structure to prevent further expansion that would damage the surrounding cells. Experiments using the *Arabidopsis* plant as the model confirm that when these haregrass proteins are introduced into its structure, *Arabidopsis* acquires significantly greater frost resistance.

There are significant long term benefits associated with the development of frost-resistant crops. The ability to plant crops that will better withstand frost makes all arable land in colder agricultural regions more productive. Greater numbers of crop species can be planted with the confidence that the effort to cultivate the crop is less likely to be compromised by frost damage. The expenses associated with the equipment currently used to prevent frost damage would be largely eliminated.

The research of genetic modifications to improve frost resistance is far more benign than other modern plant biology research that has sought to make crops more drought or insect resistant. These latter research initiatives have generated considerable academic controversy. The benefits associated with these enhanced resistance qualities may be offset by the risk that the genetically modified plants will ultimately mix with native plants to produce highly invasive "super weeds" that will destroy croplands.

■ Primary Source Connection

Frost resistant crops are the focus of this article. A plant found in Antarctica holds important information at the cellular level that may help farmers throughout the world save billions of dollars while at the same time increase their grain harvests. The discovery of an anti-freeze gene found in Antarctic hairgrass may be a key component in the scientific evolution of preventing cereals and grasses from being destroyed by frost. The genes are being studied with the hope that application of their frost-resistant characteristics may be better understood and someday translate to other crops as a solution to minimize frost-related losses that farmers withstand every year.

Frost Tolerant Crops In Reach

VICTORIAN scientists have made a major breakthrough that promises to save farmers worldwide billions of dollars a year and dramatically boost grain harvests.

The discovery of an anti-freeze gene in a plant in Antarctica by State Government scientists is expected to substantially improve farm incomes and strengthen rural economies by preventing cereals and grasses being destroyed by frost.

The Australian team isolated the gene sequence responsible for the Antarctic plant's frost resistance and inserted it into a test plant in Victoria.

The Department of Primary Industries team headed by Professor German Spangenberg and Dr Ulrik John, of the Victorian AgriBiosciences Centre at La Trobe University, then found that under certain conditions the test plant had the same frost-resistant characteristics.

The discovery has huge implications for agriculture in view of crop losses to frost each year. Globally, 15 percent of agricultural production is lost to frost and in the US there are more economic losses due to frost than from any other weather-related phenomenon.

Farmers in Victoria and South Australia have barley crops that are damaged by frost, while wheat belts in NSW and Western Australia also suffer significant damage.

The findings will be announced at the Biotechnology 2006 conference in Chicago this week. A State Government team headed by Premier Steve Bracks and Innovation Minister John Brumby is in Chicago for the conference, which is attended by 20,000 delegates from 60 countries.

"Over the next few years, we should see the development and application of technologies for frost tolerance in crops based on the knowledge gained from the functional analysis of these antifreeze genes," Mr. Brumby said.

The search for a solution to the devastating frost-related losses suffered by Victorian farmers started with a Department of Primary Industries program to study the genes of plants and animals that grow in harsh environments, including freezing conditions.

The department's manager of biotechnology policy, Robert Sward, said: "We are interested because we know that often there are particular genes associated with the survival of plants and organisms in these incredibly harsh environments.

"The question was: how can they withstand that kind of freezing when plants and grasses at those low temperatures go to mush and die?"

"Often it will kill the whole plant, or the growing centre of the plants where the cereal will come from. It is one of the major destructive forces for Victorian cereal growers."

"Farmers have desperately hoped over the decades for a solution to these great losses and have tried where possible to breed varieties less susceptible to frosts. But no major genes had ever been identified in terms of breeding."

Dr Sward is presenting the findings in Chicago on behalf of the scientists.

The search for a solution led the scientists to Antarctic hairgrass, a plant that thrives "on the edge of survival" on the Antarctic Peninsula, according to Dr Sward. It has the ability to inhibit the growth of ice crystals.

Samples of hairgrass were brought to Melbourne, and the scientists discovered a protective protein in the plant that is activated at a temperature of about five degrees.

The hairgrass has been growing for possibly millions of years but scientists did not know why it survived until the Victorian team extracted the genes responsible for preventing freezing.

Dr Sward believes it will be seven to 10 years before the breakthrough has practical application on farms. He says there's excitement about the discovery but that the real

excitement comes "when you hand a farmer this new seed and say, 'Here, plant these and you're not going to have any frost damage this year.' The real benefits are yet to come. It will happen. But these steps we have made are some of the very crucial ones."

Martin Daly

DALY, MARTIN. "FROST TOLERANT CROPS IN REACH." *THE AGE* (APRIL 10, 2006): 1–2.

SEE ALSO *Agricultural Biotechnology; Corn, Genetically Engineered;* Diamond v. Chakrabarty; *Genetically Modified Crops;* Monsanto Canada Inc. v. Schmeiser; *Patents and Other Intellectual Property Rights; Plant Patent Act of 1930; Plant Variety Protection Act of 1970*

BIBLIOGRAPHY

Books

Benkeblia, Noureddine. *Sustainable Agriculture and New Biotechnologies.* Boca Raton, FL: CRC Press, 2011.

Ferry, Natalie, and A. M. R. Gatehouse. *Environmental Impact of Genetically Modified Crops.* Cambridge, MA: CABI, 2009.

Gordon, Thomas T., and Arthur S. Cookfair. *Patent Fundamentals for Scientists and Engineers*, 2nd ed. Boca Raton, FL: Lewis Publishers, 2000.

Heldman, Dennis R., Matthew B. Wheeler, and Dallas G. Hoover. *Encyclopedia of Biotechnology in Agriculture and Food.* Boca Raton, FL: CRC Press, 2011.

Herren, Ray V. *Biotechnology: An Agricultural Revolution.* Clifton Park, NY: Delmar Learning, 2005.

Kimbrell, Andrew, and Joseph Mendelson. *Monsanto vs. U.S. Farmers.* Washington, DC: Center for Food Safety, 2005.

Snyder, Richard L., J. Paulo de Melo-Abreu, and Scott Matulich. *Frost Protection: Fundamentals, Practice and Economics.* Rome: Food and Agriculture Organization (FAO), 2005. Available online at http://www.fao.org/docrep/008/y7223e/y7223e07.htm#TopOfPage.

Thomson, Jennifer A. *Seeds for the Future: The Impact of Genetically Modified Crops on the Environment.* Ithaca, NY: Comstock Publishing Associates, 2007.

Web Sites

Kish, Stacy. "Researchers Identify Gene to Improve Wheat Frost Tolerance." *U.S. Department of Agriculture*, February 9, 2009. http://www.csrees.usda.gov/newsroom/impact/2009/nri/02091_wheat_frost.html (accessed September 23, 2011).

The Research Council of Norway. "Research for Frost-resistant Strawberry Plants." *Science Daily*, September 2, 2011. http://www.sciencedaily.com/releases/2011/09/110902104743.htm (accessed September 23, 2011).

Bryan Thomas Davies

Frozen Egg Technology

■ Introduction

Reliable oocyte cryopreservation, or egg freezing, is a recently successful biotechnological advance. Although sperm has been frozen and thawed since the mid-1950s, and fertilized embryos have been retrieved and viably implanted since the late 1970s, egg-freezing techniques were not considered effective until the dawn of the twenty-first century.

Although women preparing to undergo chemotherapy, radiation, or other medical procedures with high likelihood of impacting future reproductive potential could be offered egg harvesting and freezing in order to preserve future reproductive options, those eggs were often unusable after thawing. The egg is the largest cell in the human body, and it contains a high percentage of water that must be removed to ensure freezing without ice crystal formation. Lab methods used to dehydrate the eggs and replace water with cryoprotectants could result in irreparable damage to the egg; incomplete dehydration led to formation of ice crystals, making the egg unviable. Biotechnological advances have made it possible to efficiently dehydrate and cryoprotect eggs. Vitrification (the process by which all water is removed from the egg and replaced with a cryoprotectant substance) has been reported effective as well.

Women seek to freeze their eggs for three primary reasons: impending medical procedures with significant likelihood of eliminating reproductive potential; desire to postpone reproduction until later in life while minimizing age-related infertility issues; and religious or cultural objections to traditional IVF's production of excess embryos per single implantation cycle.

■ Historical Background and Scientific Foundations

At birth, most human females have about 2 million eggs contained in follicles within their ovaries. There is no further egg production after birth. By the time a girl reaches adolescence, half to three-fourths of those eggs have degraded and become unusable. From puberty onward, varying numbers of follicles mature, and eggs are shed every month until menopause. For human females, peak fertility occurs between the onset of puberty and the late 20s, with a sharp decline occurring after the mid-30s. During peak fertility, the odds of achieving a viable

WORDS TO KNOW

AGE-RELATED INFERTILITY: Female mammals are born with their lifetime supply of eggs, no more are manufactured after birth. Over time, the viability of healthy eggs decreases, with a dramatic decline in human fertility after the age of 35.

CRYOGENIC: Relating to freezing and storage of an entity at temperatures far below zero. Cryogenics are typically used for longer-term storage of substances that are fragile or unstable at traditional freezing temperatures.

IN VITRO FERTILIZATION (IVF): The process in which eggs are removed from the female and placed in a culture medium containing a large quantity (typically) of sperm. Fertilization occurs outside the body, and fertilized embryos are implanted in the female host several days after successful fertilization occurs. This is typically used when there are fertility problems or there are concerns regarding genetically-related diseases.

INTRA-CYTOPLASMIC SPERM INJECTION (ICSI): An assisted reproductive process in which sperm are injected directly into an egg in order to facilitate fertilization as part of an IVF process.

VITRIFICATION: In oocyte cryogenics, vitrification refers to the process by which all water is removed from the egg and replaced with a cryoprotectant substance, after which the egg is flash-frozen to a temperature of around −392°F (−200°C). Vitrified eggs have a glassy appearance.

pregnancy without biotechnological assistance maximize at about 1 in 5. Women in their late 30s and 40s are least likely to achieve successful pregnancies without external reproductive assistance.

Although human sperm have been reliably cryo-frozen since 1955, and frozen human embryos have been thawed to result in successful pregnancies since the late 1970s (the first recorded live birth from in vitro fertilization occurred in 1978 in the United Kingdom), there was only marginal success in freezing and thawing viable human eggs (oocytes) to create full-term pregnancy until recently.

Some cultures and religions decry freezing or disposing of the unused fertilized embryos resulting from IVF. Freezing or discarding excess sperm or eggs is considered non-controversial as they are not potentially viable. For persons who choose to postpone conception and have religious convictions about freezing or disposing of embryos, egg cryopreservation may be the most practical option.

For the past few decades, women diagnosed with cancer or life-threatening conditions for which treatment could cause infertility have been offered the option of harvesting and freezing eggs for possible later use. The previously frozen eggs resulted in few reported pregnancies.

Since the advent of effective birth control, some women have waited until their late 30s or 40s before attempting pregnancy. Many were unsuccessful as a result of age-related infertility. The use of in vitro fertilization and assisted reproductive technologies has steadily increased as a result. Simultaneously, egg-freezing processes have become progressively more successful, making oocyte cryopreservation an increasingly reasonable option.

Sperm banks have been used for several decades by those wishing to achieve pregnancy but who lack access to appropriate sperm. More recently, egg banks have been established, and female donors have provided oocytes for use by others.

Fertility researchers have noted a statistically significant rise among women in their early- to mid-30s

Rachel Lehmann-Haupt, 40, decided to have her eggs frozen at the age of 37. As more women delay childbearing into their 30s and 40s, they find they may need to utilize assisted reproductive technologies to bear children.   © *Cynthia E. Wood/Chicago Tribune/MCT/Getty Images.*

seeking voluntary egg-harvesting and freezing in order to increase the likelihood of achieving successful pregnancies later in life while decreasing the impact of age-related infertility.

■ Impacts and Issues

During in vitro fertilization, a woman's ovaries are hormonally stimulated to produce multiple mature follicles and eggs, which are then harvested and inseminated with either fresh or thawed sperm. The fertilized embryos are allowed to undergo several days of cell division before assessing for health and viability, and a specific number are selectively implanted in the female's uterus. The additional embryos are often frozen for possible later use. IVF is utilized when pregnancy is the immediate goal. This is not an immediate consideration for those needing or desiring to postpone reproduction. For those with religious or cultural prohibitions concerning freezing or disposing of unused embryos, IVF is not deemed appropriate. Egg cryopreservation involving a single cell that is not viable before fertilization can provide an alternative for such individuals. A significant amount of

research and development of egg freezing techniques has occurred in Italy and Spain, countries that discourage the freezing or disposal of embryos.

Human sperm and embryo cryopreservation for later use have been successful for several decades. This has not been the case for egg freezing until recent biotechnological advances improved outcomes of egg dehydration and cryopreservation. Many physicians and scientists still consider egg freezing an experimental process, citing a lack of sufficient research outcome data to consider it a mainstream practice. Critics of the experimental label posit that the designation is self-perpetuating. They assert that few reproductive specialists will recommend a procedure considered relatively untested, and fewer still will be inclined to learn and employ the technology, resulting in decreased numbers of individuals engaging in a potentially useful alternative to more mainstream options. Thus, the end result is that the experimental label is retained due to lack of use of the oocyte cryopreservation techniques.

The human egg is the largest cell in the female body and contains a large portion of water. Using traditional cryogenic methods, ice crystals often formed, rendering the eggs unusable. With many "slow-freezing" methods, the egg was moved through a succession of cryopreservative

French doctor Rene Frydman poses in his office at Antoine Beclere hospital in Clamart, a Paris suburb, after announcing the birth of children resulting from a donor of frozen eggs for the first time in France. © *Fred Dufour/AFP/Getty Images.*

baths, gradually removing the water in the egg and replacing it with sugars and chemical cryoprotectants. The egg would then be deep frozen for later use. Because the process involved allowing water and the chemical solutions to perfuse the cell wall at a comparatively slow rate, damage to the egg could occur due to prolonged exposure. Some water might also remain in the cell, causing ice crystals to form and destroy the egg. Advances in slow-freezing technology have significantly increased the efficiency and effectiveness of the process. With the newer vitrification method, eggs can be flash-frozen, retaining stability while preventing development of ice crystals. Vitrified eggs have a rigid, glass-like consistently. After thawing, vitrified eggs are fertilized using ICSI technology in order for the sperm to pass through rigid cell walls.

In addition to human use of oocyte cryopreservation, the technology has potential implication for the preservation of endangered animal species. Several laboratories, such as the Frozen Zoo affiliated with the San Diego Zoo, collect and cryopreserve eggs from endangered species in anticipation of preventing their disappearance by using assisted reproductive technology.

SEE ALSO *Designer Babies; Designer Genes; Egg (Ovum) Donors; Human Embryonic Stem Cell Debate; In Vitro Fertilization (IVF); Stem Cells; Stem Cells, Embryonic; Twiblings*

BIBLIOGRAPHY

Books

Brinsden, Peter R. *A Textbook of In Vitro Fertilization and Assisted Reproduction*, 3rd ed. Boca Raton, FL: Taylor & Francis, 2005.

Carlson, Bruce M. *Human Embryology & Developmental Biology*, 3rd ed. Philadelphia: Mosby/Elsevier, 2004.

Dudek, Ronald W. *Embryology*, 5th ed. Philadelphia: Wolters Kluwer Health/Lippincott Williams & Wilkins, 2011.

Masters, J. R. W., Bernhard Palsson, and James A. Thomson. *Embryonic Stem Cells*. Dordrecht: Springer, 2007.

Mundy, Liza. *Everything Conceivable: How Assisted Reproduction Is Changing Men, Women, and the World*. New York: Alfred A. Knopf, 2007.

Web Sites

"Assisted Reproductive Technology (ART)." *Centers for Disease Control and Prevention (CDC).* http://www.cdc.gov/art/ (accessed September 10, 2011).

Pamela V. Michaels

Fuel Cell Technology

■ Introduction

The quest for safe and sustainable energy sources is epitomized by the research that has been directed to fuel cell technology since the mid-nineteenth century. "Fuel cell" is a generic term that describes any device that is capable of producing electricity through a self-contained chemical reaction that involves oxygen. No matter how a fuel cell is devised, the basic operating principles are constant—a chemical reaction between the fuel (most often hydrogen gas) and an oxidizing agent produces electricity at the anode and cathode electrodes. The chemical reaction is facilitated by a catalyst, and an electrolyte transports the charged particles created through the reaction between the electrodes. The thermal energy generated by the fuel cell is available for other energy applications.

The relatively simple chemistry at the heart of fuel cell operation has made this technology an attractive commercial energy application. Fuel cells produce limited environmental impacts, because water is the only significant emission generated in their operation. When compared with conventional internal combustion engines, fuel cells are more efficient and have a longer life span; the cell mechanisms have fewer parts and thus are less likely to require extensive repair. Stationary fuel cells are used in numerous commercial applications as either primary or backup power systems. The large scale use of fuel cells in motor vehicles is problematic when compared with conventional vehicles due to the relatively high cost of the cell components and the lack of a fuel station infrastructure.

■ Historical Background and Scientific Foundations

The underlying chemical principles that govern the generation of fuel cell electricity were understood in the European scientific community after 1800. These early theorists confirmed that when hydrogen and oxygen where combined, the resulting reaction should produce electricity and water.

In 1838, British physicist William Grove (1811–1896) was the first scientist to build a working fuel cell that proved the theory. The Grove "wet cell battery" generated a constant electrical current of 12 amps at a rate of 1.8 volts. Its two platinum electrodes were each aligned with one end immersed in a container of sulfuric acid. One electrode had its opposite end placed in a sealed tube of oxygen gas; the second electrode had its opposite end sealed in an identical tube that contained hydrogen gas. The Grove battery operated as predicted in the earlier theory: As current was generated, the water level rose in both tubes.

The Grove cell sparked an intense scientific debate concerning the precise nature of the current generated across the electrodes. Was the current the product of the physical contact between the materials in the cell? Alternatively, did a chemical reaction produce the desired current? By 1900 scientific knowledge had coalesced around a conclusion that combined contact and chemical concepts, as science accepted that the chemical reaction produced in a fuel cell will only occur in a precise contact zone formed by the fuel, electrolyte, and catalyst.

The primitive Grove fuel cell also identified the problem that continues to bedevil modern researchers. The expensive metal platinum was the catalyst used by Grove; the quest to find an inexpensive catalyst that would be suitable for commercial energy production, especially the mass produced fuel cell technology essential for motor vehicle power, has continued into the twenty-first century.

The five most prominent modern fuel cell designs are similar in their fundamental construction to the simple Grove cell, in that each uses a hydrogen fuel source. The name of each type is taken from the electrolyte used in its operation:

- Alkali
- Molten carbonate
- Polymer electrolyte membrane
- Phosphoric acid
- Solid oxide

■ Impacts and Issues

A global society in which all energy needs are satisfied by devices that are efficient and reliable, fuel sources are virtually limitless, and environmental impacts approach zero would be the foundation for a modern-day utopia. The relative lack of progress in the development of fuel cell technology from the Grove model between the mid-1850s and the 1970s can be attributed to several causes. Petroleum, coal, natural gas, and hydroelectricity were plentiful and very cheap. Nuclear power technology had evolved from the immense resources expended in the World War II (1939–1945) race to unleash the perceived unlimited power of the atom. Finally, the incentives to innovate in the energy field were centered on greater ability to extract or harness readily accessible fuel sources as opposed to the fuel cell research necessary to take the technology into the commercial mainstream.

The U.S. Gemini space program in the 1960s provided an important research stimulus for commercial fuel cell development. The alkali fuel cells developed by NASA supplied energy and necessary drinking water for the use of astronauts during flight. The energy crisis that struck the Western world in the early 1970s was the second spur to consider how fuel cells could be utilized in a broader range of commercial applications. A rapid rise in global petroleum prices was precipitated by the co-ordinated action of the Organization of Petroleum Exporting Countries (OPEC). Crude oil prices trebled and placed pressure on all industrialized nations that relied on imported oil to reevaluate their domestic energy policies in terms of supply security, available reserves, and sustainability.

Twenty-first century fuel cell advocates point to the proven energy efficiency, reliability, and environmentally friendly aspects of the technology. Their arguments are cited in favor of the greater infusion of public and private sector resources into further improvements to fuel cells function. The detractors do not question the soundness of the scientific concepts first proven persuasively by Grove. The opposing position is rooted in commercial pragmatism: If fuel cell technology is not cost-effective for business and consumers, it must be relegated to the historical slag heap of great ideas that were ultimately impossible to achieve in practice, such as perpetual motion machines and internal combustion engine carburetors, which in theory could increase automotive fuel efficiency to 150 miles per gallon (34 km/L).

Molten carbonate and solid oxide fuel cells require the electrolyte to be heated to temperatures in excess of 1250°F (600°C), a factor that makes these systems best suited to use in stationary power plants. The otherwise wasted thermal energy provides opportunities to implement cogeneration systems in which steam generated from the excess heat is used for space heating, or it is directed through turbines to generate additional electricity.

WORDS TO KNOW

ANODE: The fuel cell electrode where oxidation occurs, resulting in a loss of electrons in the particles involved in the chemical reaction. All fuel cells have an anode as the negative terminal.

CATHODE: The fuel cell electrode where reduction occurs and the particles gain electrons. In a fuel cell, the cathode is the positive terminal.

COGENERATION: Fuel cell systems generally convert to electricity between 40 to 50 percent of the chemical energy in the hydrogen fuel source. Gasoline powered engines perform at approximately 20 percent efficiency, and conventional electrical battery powered systems achieve 65 percent efficiency. Cogeneration is the process used to harness system thermal output to improve overall efficiency of the fuel cell system; energy efficiency rates approach 80 percent when the fuel cell systems incorporates thermal cogeneration.

FUEL CELL STACK: The stack is composed of individual fuel cells that are arranged in circuits to increase voltage.

PEMFC: Polymer electrolyte membrane fuel cells (PEMFC) are a common type of fuel cell employed in many commercial applications. A solid polymer membrane is the electrolyte used to transport the protons generated by the chemical reaction within the cell. The PEMFC will typically have an operating temperature of 140°F (60°C) to 212°F (100°C).

Microbial fuel cells (MFCs) represent a separate but equally exciting advance in fuel cell technology. Also called bio-electrochemical systems (BESs), they produce electrical energy directly from the conversion of organic matter, such as the bacteria found in all wastewater. In an oxygen-free fuel cell chamber, any available bacteria attach to the electrode where the electrons flow to the anode. The cathode is exposed to oxygen gas, where it combines with the available electrons and protons to form water. The constant addition of fresh bacteria to the system enables the cell to generate a sustained flow of electricity. In this way MFCs represent a dual scientific breakthrough—readily available electrical energy, and improved sanitation because pure water is the only by-product.

It has been suggested that if the energy infrastructures of the United States and other developed nations were reconstructed to make hydrogen easily available for motor vehicle use, the natural supply and demand features of consumer markets would drive greater technological innovation and consumer acceptance to the point at which fuel cells would achieve economic critical mass. Battery powered vehicles, and their variants that rely upon a combined battery or conventional gasoline

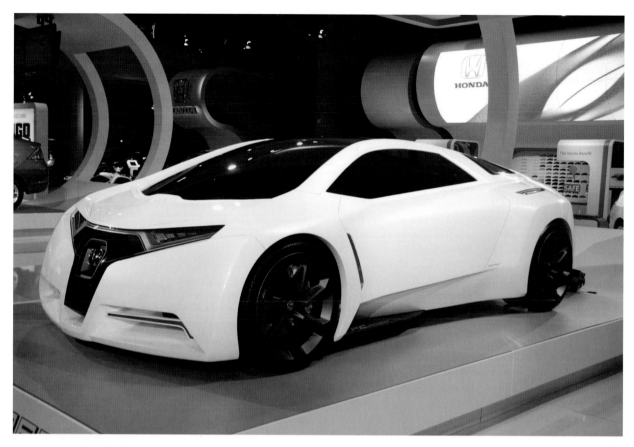

The Honda FC Sport concept car using fuel cell technology is presented at the Canadian International AutoShow 2009. It was one of six concept cars unveiled at CIAS2009. © *KSPhotography/Shutterstock.com.*

engine have gained greater traction in the public consciousness than fuel cell alternatives. This development is largely attributed to the ability of consumers to recharge the vehicle battery from conventional power supplies. As was first observed by Groves, the components necessary to make the most efficient fuel cell catalysts remain an attractive yet more expensive electrical energy option.

SEE ALSO *Algae Bioreactor; Biochemical Engineering; Environmental Biotechnology; Green Chemistry*

BIBLIOGRAPHY

Books

Li, Xianguo. *Principles of Fuel Cells.* New York: Taylor & Francis, 2006.

Logan, Bruce E. *Microbial Fuel Cells.* Hoboken, NJ: Wiley-Interscience, 2008.

Web Sites

Logan, Bruce E. "Research-Bioenergy: Check Out the MFC-Cam." *Pennsylvania State University, Department of Civil and Environmental Engineering.* http://www.engr.psu.edu/ce/enve/logan/bioenergy/research_mfc.htm (accessed September 27, 2011).

"Fuel Cells." *National Geographic Society.* http://environment.nationalgeographic.com/environment/global-warming/fuel-cell-profile.html (accessed September 26, 2011).

Bryan Thomas Davies

GE-Free Food Rights

■ Introduction

The principle underlying food or pharmaceutical consumer rights is that people have the right to know what they are buying and the exact compositional content of foods and medications that are for sale on the market. In other words, people have the right to make informed decisions when choosing what they are buying or eating.

Uncertainties about the long-term effects of human and animal consumption of genetically-engineered (GE) grains or processed foods containing GE products have been an object of controversy since the early 1990s. Previously existing consumer-rights legislation in some countries was used as reference by regulatory agencies to require mandatory labeling of both *in natura* GE foods and processed ones, whereas in other nations, the struggle for the consumers' right to know and to choose continues.

■ Historical Background and Scientific Foundations

Consumer rights activists, concerned citizens, and scientists have formed an organized political front to face the intense lobby of biotechnology companies upon local and federal lawmakers and regulatory agencies in the United States, Brazil, the United Kingdom, Australia, and many other countries. These critics of GE foods believe that the "precautionary principle" should be used in the absence of scientific consensus about the safety of GE foods. Biotechnology companies often see mandatory labeling of products as "GE" or "GE-Free" as unnecessary, stating that GE foods have "substantial equivalence" with non-GE foods and pose no risks.

Brazil is the second-largest world producer of GE grains (soy, corn, beans, rice, etc.) and some GE plants have been developed by Brazilian scientists at the state-owned biotechnology company, EMBRAPA. In turn, ANVISA is the Brazilian regulatory governmental agency for drugs and food. ANVISA requires that GE foods in Brazil be labeled as such, per a presidential act issued in 2003. The law ensures that Brazilian consumers will have the right to information about GE ingredients in food products and GE foods (fruits, vegetables, grains) intended for human or animal consumption.

Mandatory labeling of GE products was first introduced in the European Union in the 1990s, followed by other countries around the world including New Zealand, Norway, Japan, Russia, Saudi Arabia, South Korea, Switzerland, Taiwan, Brazil, Chile, Croatia, Ecuador, El Salvador, Indonesia, Mauritius, Serbia, Sri Lanka, Ukraine, and Vietnam. Enforcement of mandatory labeling, however, has yet to be completely implemented in some countries of Eastern Europe, Asia, and Latin America.

In Brazil, for instance, mandatory labeling reinforcement suffered some setbacks between 2007 and 2010, due to disagreements between the regulatory agency ANVISA and CTNBio, the Council of the Brazilian National Commission for Biosafety (which is responsible for assessment and approval of commercial GE crops in the country). As a result, only partial enforcement of mandatory labeling was implemented during that period.

The European Food Safety Agency (EFSA) was established to provide scientific advice on matters directly or indirectly impacting food and animal feed safety in the European Union, including animal health, plant and environment protection, and consumer information. EFSA assesses and establishes regulations for products containing genetically-modified organisms (GMOs) as ingredients and for GMOs themselves, based on two sources of information: the worldwide scientific published literature and the opinions of its internal scientific panel. EFSA's GMO Panel adopted its guidance document for the risk assessment of genetically-modified (GM) plants and derived food and feed in 2004. In 2006 the Panel enhanced the document by adding a chapter on surveillance of unanticipated effects of GM foods.

EFSA also has adopted regulations designed to prevent non-compliance, such as occurred in the Starlink maize case, when a type of GM maize that was authorized in the United States only for animal feed was found in human food products. According to EFSA regulations, food and feed in the European Union must carry a label that refers to the presence of GMOs.

■ Impacts and Issues

Labeling food provides information for consumers in the ingredients lists of products and allows them to make an informed choice. In the European Union and other countries that adopted mandatory labeling policies, pre-packaged products consisting of or containing GMOs must indicate in the list of ingredients the phrases "genetically modified" or "produced from genetically modified [name of the organism]." However, this information in many labels is barely discernible, due to the small size of the printed letters. As a consequence, some countries adopted regulations that call for more visible mandatory signs on labels such as a black "T" (for transgenic) inside a yellow triangle. In the case of products without packaging, those words or sign still must be clearly displayed in close proximity to the product, such as a sign on the supermarket shelf.

In the United States, the Food and Drug Administration (FDA) decided in 1992 that there is no basis for concluding that bioengineered foods differ from

An anti-biotech demonstrator marches through downtown San Diego, California. About 1,000 demonstrators, some dressed as genetically engineered tomatoes, staged a colorful but peaceful protest on the opening day of a biotechnology trade show. Many of the demonstrators said they are concerned that biotech businesses are introducing genetically modified crops and seeds into the food supply without knowing the long-term consequences. *© AP Images/Joe Cavaretta.*

other foods in any meaningful or uniform way, or that as a class, foods developed by the new techniques present any different or greater safety concern than foods developed by traditional plant breeding. The FDA also provided guidance for voluntary GE-product labeling by the food industry.

Arguments against labeling GE foods in the United States are as follows:

- Labels on GE foods could imply to the consumer that they pose health risks or adverse effects;

- According to the FDA, there is no significant difference between GE and conventional foods;

- Mandatory labeling in Europe, Japan, the European Union, and New Zealand fomented fear and prejudice against GE foods, leading many retailers to stop selling them;

- Mandatory labeling of GE foods could increase product prices;

- The FDA already requires a mandatory label or warning stating the type of allergen present in products, whether conventional or GE;

- Consumers desiring to avoid GE foods already have as an alternative in the certified organic foods, which by definition cannot be produced with GE ingredients.

Arguments in favor of mandatory labeling for GM foods include the following:

- Consumers have the right to know what is in their food, especially in those products for which health and environmental issues have been raised;

- Mandatory labeling would allow consumers to avoid food products that they do not trust or that they believe may cause them health problems;

- Surveys indicate that a majority of Americans support mandatory labeling for GE foods;

- For ethical reasons, many Americans want to avoid eating animal products, including animal DNA transferred to plants;

- Many countries have already established mandatory labeling without experiencing any logistical problems for storage, transport ,and distribution, or significant increase in food prices;

- There has been too little time to know for certain that new allergies or immunogenic disorders will not originate from GE foods and feed after years of human and animal repeated exposure to modified transgenic proteins.

Among the many organizations fighting for consumers rights and mandatory labeling of GE foods, Greenpeace and Friends of the Earth (FoE) are perhaps the most pervasive movements, with successful chapters in more than 50 countries around the world. In 2010, FoE issued surveillance reports produced as part of its "Feeding and Fuelling Europe" project, that criticized current EU legislation, which does not require labeling meat from animals that were fed GE/GM feed.

In the FoE report, EFSA is criticized because Current EU legislation requires only the labeling of GM animal feed: animals reared on GM animal feed do not have to be identified. This has led to consumers unknowingly consuming animals fed on a GM diet. The report also relates that in Germany, animal products derived from animals not fed with GE feed can be labeled "without biotechnology" and that supermarket chains are also adopting this approach. Similar "without GMO" or "GE-free" labeling legislation is being proposed in France and Ireland, according to the same report. Although not mandatory, "Free of Transgenic" labeling was adopted in Brazil by many food industries. "Transgenic Free" labeling requires certification (through genetic tests performed by companies certified by ANVISA) of the grains, animal products, and plants used by the food industry.

Many food companies in the United States carry labels on their products that state they do not contain GMOs. The FDA has become a battlefield between those demanding that the FDA take actions against companies using these labels and others who support companies for informing their customers. For example, the phrases "GMO-free" and "not genetically modified" are not recommended by the FDA because the first would be almost impossible to verify, and almost all plants are genetically modified through selective breeding. Some special interest groups and politicians regard these comments as examples of undue influence that the biotechnology lobby could have on the FDA decision-making process.

Most major cities around the world have seen organized protests against GE/GM foods since the beginning of the twenty-first century. Some industry analysts blame the biotechnology companies for the resultant prejudice and suspicion raised in the public eye concerning genetically-modified products, stating such fears are caused by a lack of proactive information and forthright communication. Doubters suggest that the biotechnology industry has relied on the concept of substantial equivalence, embedded in regulatory requirements, to justify ignoring the concerns of consumers for most of the GM foods currently on the market. Therefore, GM foods have entered some markets without offering choice to consumers. Current scientific evidence holds that GM foods developed so far are safe, but many sectors of the public remain unconvinced. Until public confidence in GM foods is built, largely by providing the public with information and choice, the market share of GM foods, as well as their potential role in alleviating world hunger, will likely be diminished.

SEE ALSO *Agricultural Biotechnology; Bioethics; Corn, Genetically Engineered; Genetic Engineering; Genetically Modified Crops; Genetically Modified Food; Government Regulations; Informed Consent; Transgenic Animals; Transgenic Plants; Wheat, Genetically Engineered*

BIBLIOGRAPHY

Books

Bansal, Sangeeta, and Bharat Ramaswami. *The Economics of GM Food Labels: An Evaluation of Mandatory Labeling Proposals in India.* Washington, DC: International Food Policy Research Institute, 2007.

Weirich, Paul, ed. *Labeling Genetically Modified Food: The Philosophical and Legal Debate.* Oxford: Oxford University Press, 2007.

Periodicals

Khoury, Lara, and Stuart Smyth. "Reasonable Foreseeability and Liability in Relation to Genetically Modified Organisms." *Bulletin of Science, Technology & Society* 27, no. 3 (June 2007): 215–232.

Web Sites

Byrne, P. "Labeling of Genetically Engineered Foods." *Colorado State University,* September 2010. http://www.ext.colostate.edu/pubs/foodnut/09371.html (accessed November 8, 2011).

"Genetically Modified Foods and Organisms." *U.S. Department of Energy Genome Program.* http://www.ornl.gov/sci/techresources/Human_Genome/elsi/gmfood.shtml (accessed October 20, 2011).

"GM Food and Feed - Labeling." *European Commission, EUROPA.* http://ec.europa.eu/food/food/biotechnology/gmfood/labelling_en.htm (accessed November 8, 2011).

Sandra Galeotti

GenBank

Introduction

GenBank is a repository of all publicly available nucleotide sequences and their protein translations, along with associated bibliographic data. The National Center for Biotechnology Information (NCBI) at the National Library of Medicine in the National Institutes of Health maintains the database. GenBank is operated in partnership with the European Molecular Biology Laboratory (EMBL) and the DNA Databank of Japan (DDBJ), with which it shares information daily. The three systems make up the International Nucleotide Sequence Database Collaboration (INSDC). Data submitted to GenBank by an individual laboratory or research group is subjected to a series of quality assurance checks before being assigned an accession number and incorporated into the public database.

In addition to the international collaboration with EMBL and DDBJ as means of continuously updating the population of the database system, staff members from all three groups meet at least annually in order to keep the systems running smoothly. There is an international advisory board available for additional technical or administrative assistance.

Most scientific journals require that nucleotide sequences be entered into GenBank, EMBL, or DDBJ prior to publication so that accession numbers can also be published, making the data retrievable. Authors can request that the sequences not be made public until the research is published. Data from the Human Genome Project is also contained in GenBank.

Historical Background and Scientific Foundations

In 1965 Margaret Dayhoff (1925–1983) published the *Atlas of Protein Sequence and Structure*, representing the world's largest collection of nucleic acid and protein sequences. Her expertise included combining computer science, chemistry, mathematics, and biology; she is considered the founder of bioinformatics. Sequencing proteins and DNA was initially an arduous and expensive process, and there was little standardization across research facilities regarding data presentation, making information sharing almost impossible.

As the disciplines of molecular biology and computer science burgeoned, scientists recognized a pressing need to create a database system for collecting and systematizing the growing body of DNA sequence data while making it available for research and analysis. Prior to the availability of EMDL, DDBJ, and GenBank, research groups were collecting and analyzing genetic sequences primarily for organisms of particular interest. It was cost-prohibitive to do large-scale sequencing research. There was little standardization across facilities, rendering data-sharing inefficient.

The global scientific community recognized the need for an international database to contain base pair and nucleic acid sequence information. The U.S. National Institutes for Health (NIH) convened several workshops and study groups from 1979 to 1981, the focus of which was to refine and specify the parameters for such a system. A Request for Proposals (RFP) was generated, and a five-year contract for the creation and development of a nucleic acid sequence database was awarded to the private company of Bolt, Beranek and Newman in conjunction with the Los Alamos National Laboratory (LANL). The data was originally housed at Los Alamos. GenBank was designed to become a repository for all known, publicly-available nucleotide sequences, along with their associated bibliographic and supporting scientific information. After the collaboration with Bolt, Beranek and Newman ended in 1987, LANL was issued another five-year contract for GenBank's management in partnership with Intelligenetics at Stanford University. In 1989 Congress mandated that the GenBank database be moved to the National Center for Biotechnology Information (NCBI) on the Bethesda, Maryland, campus of NLM/NIH, although

WORDS TO KNOW

BASE PAIR: Two chemical bases bound together by chemical bonds to form a rung of the DNA double helix. The amino acid pairs in DNA are adenine, cytosine, guanine, or thymine. The ways in which the bases are paired identifies and defines the organism.

DNA: The scientific abbreviation for deoxyribonucleic acid. Found within every living cell and carrying the organism's genetic information, DNA consists of two long chains of amino acid nucleotides; the chain is twisted into a double helix shape. The strands of the chain are joined by means of hydrogen bonds. The specific sequence of the DNA is what defines genetic characteristics.

HUMAN GENOME PROJECT: A multinational bioinformation project, the goal of which was to identify all of the genes existing in human DNA by determining all of the base pair relationships. All of this information has been gathered and stored in computer databases. The project commenced in the United States in 1990 and was completed in 2003.

SEQUENCE DATABASE: In this context, a sequence database refers to a relational computerized database containing an ever-increasing body of information about DNA, RNA, protein, and nucleic acid sequences.

the actual transition did not occur until the end of the LANL/Stanford contract in 1992.

In 1986 GenBank had begun sharing data with the EMBL and extended the collaboration to the DDBJ in 1987. In the mid-1990s GenBank became part of the International Nucleotide Sequence Database Collaboration (INSDC), along with EMBL and DDBJ. The three exchange information on a daily basis as part of the INSDC in order to ascertain that all sequence data is comprehensive and up-to-date. Although each database has unique ways of managing the data, each record must contain the same parameters in order to be easily searchable. GenBank is publicly available through the Internet, and there is no cost to search or data-mine. GenBank is considered the first stop for biomedical and biological researchers studying potential new organisms. The collaboration accepts thousands of submissions daily; the database doubles in size roughly every 18 months. In order to assure data consistency, data entries can be updated only by the database that was the original recipient.

■ Impacts and Issues

Although each database has a unique formatting style, the Collaboration created a comprehensive sequence-based taxonomy and annotation system through the development of a Feature Table Definition affording a standard outline for the legal features and syntax required in DDBJ's, EMBL's, and GenBank's feature tables.

Most GenBank submissions are composed of a single DNA or RNA sequence segment, accompanied by annotation representative of the biological information for the sample. Each completed database entry contains the organism's scientific name and taxonomy, the collaboration-specified feature table, a brief description of the sequence, and all requisite bibliographic information. The files are categorized into divisions corresponding to taxonomy.

Each GenBank record is assigned a permanent accession number that serves to identify the sequence and its annotations across the collaboration systems. If the sequence or annotation is changed or updated, the new information is added to the original record. Each version of the sequence within the record is also assigned an NCBI identifier called a gi. The numbering system and nomenclature for the GenBank entries utilize standard procedures for incorporating updates into existing records, which assures that those needing the information have continuous access to the entire history of the sequences being researched.

Per NCBI statistics, as of April 2011 GenBank contained 126,551,501,141 bases in 135,440,924 sequence records in the standard GenBank divisions and 191,401,393,188 bases and 62,715,288 sequences in the Whole Genome Shotgun (WGS) Division. Updated release notes are available every two months on the NCBI file transfer protocol (ftp) site.

SEE ALSO *Bioinformatics; DNA Databases; DNA Sequencing; Genomics; HapMap Project*

BIBLIOGRAPHY

Books

Batiza, Ann Finney. *Bioinformatics, Genomics, and Proteomics: Getting the Big Picture.* Philadelphia: Chelsea House, 2006.

Savolainen, Vincent. *DNA and Tissue Banking for Biodiversity and Conservation: Theory, Practice and Uses.* Richmond, Surrey, UK: Kew Publishing, 2006.

Periodicals

Benson, Dennis A., Ilene Karsch-Mizrachi, David J. Lipman, James Ostell, and Eric W. Sayers. "GenBank." *Nucleic Acids Research* 37, Supplement 1 (January 2009): D26–D31.

Web Sites

"GenBank Overview: What Is GenBank?" *National Center for Biotechnology Information.* http://www.ncbi.nlm.nih.gov/genbank/ (accessed September 21, 2011).

Pamela V. Michaels

Gene Banks

■ Introduction

A gene bank, or DNA bank, is a repository of stored DNA or genetic information that is normally used for species preservation or research purposes.

GenBank, located at the U.S. National Institutes of Health (NIH) in Bethesda, Maryland, is the cumulative American data storage facility for all publicly available nucleotide sequences, their protein translations and bibliographic information, as well as the data from the Human Genome Project. GenBank collaborates with the European Molecular Biology Laboratory (EMBL) and the DNA Databank of Japan (DDBJ), in order to continuously update all three systems. GenBank, EMBL, and DDBJ, comprise the International Nucleotide Sequence Database Collaboration (INSDC).

The first seed gene bank was established by the Russian scientist Nikolai Vavilov (1887–1943) in Leningrad in 1920, out of a strong belief in the importance of conserving biodiversity. Currently, there are approximately 1,500 seed gene banks operating around the globe. These banks are primarily of two varieties: frozen and ambient temperature banks. Frozen seed gene banks dehydrate and store seeds for several decades at conventional freezer temperatures, or indefinitely at cryonic temperatures (below −238°F / −150°C). Frozen banked seeds are meant to be stored until needed as a result of disaster or climate change; they are generally housed in large facilities operated as laboratories and research facilities. Ambient temperature banks are typically locally owned and operated by agrarian communities. They are used to increase biodiversity or to bring back strains that are in short supply (or experiencing increased demand due to environmental changes). Because they are readily accessible, ambient banks are susceptible to natural disasters or human interference.

Large international gene banks often focus on crops with food or medicinal importance. The Global Crop Diversity Trust seeks to maintain biodiversity necessary for nutritiously feeding the world's burgeoning population, in part by helping to fund many gene banks around the world. The trust prioritizes seed genes for food crops considered to be the most critically important to the healthy survival of the world's population.

The Svalbard Global Seed Vault is supported by the Government of Norway as well as the Global Crop Diversity Trust. It was designed to serve as a back-up for every seed gene bank on the planet.

■ Historical Background and Scientific Foundations

Early in the 1970s, the global scientific communities determined that it was necessary to create an international database containing base pair and nucleic acid sequence information. In 1982, the NIH entered into a 5-year contract for the creation and development of a nucleic acid sequence database (GenBank) with the private company of Bolt, Beranek and Newman in conjunction with the Los Alamos National Laboratory (LANL). The data was originally housed at Los Alamos. In 1987, a new contract for GenBank's continued development was issued to LANL in partnership with Intelligenetics at Stanford University. In 1989, Congress mandated that the GenBank database be moved to the National Center for Biotechnology Information (NCBI) on the campus of NIH, and the relocation was completed in 1992.

INSDC accepts thousands of submissions daily; the database doubles in size roughly every 18 months. According to NCBI statistics published in April 2011, GenBank contained 126,551,501,141 bases in 135,440,924 sequence records in the standard GenBank divisions and 191,401,393,188 bases and 62,715,288 sequences in the Whole Genome Shotgun (WGS) Division. Updated release notes are available every two months on the NCBI file transfer protocol (ftp) site.

The Food and Agriculture Organization of the United Nations (FAO) has utilized programs for the preservation of animal genetic resources for more than

WORDS TO KNOW

CLONING: Growing an individual organism from a single cell and producing a genetically identical copy of the individual from whom the cell was removed. This has successfully been done with animals. Human cloning is currently illegal.

DEOXYRIBONUCLEIC ACID (DNA): DNA is found within every living cell and carries the organism's genetic information. DNA consists of two long chains of amino acid nucleotides; the chain is twisted into a double helix shape. The strands of the chain are joined by means of hydrogen bonds. The specific sequence of the DNA is what defines genetic characteristics.

ENDANGERED SPECIES: A species at risk of extinction throughout all or most of its habitat. There are two risks with endangered species: first, that the species will disappear completely from the planet; second, that the surviving gene pool will be weakened or threatened by the low numbers of organisms available for breeding.

GENETICALLY MODIFIED ORGANISM (GMO): Also called genetically engineered organism (GEO). An entity in which the genetic make-up has been permanently altered by the insertion of a gene or genes from a different variety of the same species or a completely different species through the use of genetic engineering. Such modifications are likely to be heritable, i.e., they are passed on to offspring.

GLOBAL CROP DIVERSITY TRUST: A partnership between the private and public sectors, dedicated to the assurance of funding for important food crop genetic preservation in perpetuity. They seek to create and guide efficient and effective sustainable conservation methods in order to provide a means for preserving and protecting diverse food crop genetic material.

SVALBARD GLOBAL SEED VAULT: A global seed storage vault located below the permafrost level in a remote region (Svalbard) of Norway. The vault stores duplicate copies of nearly every important food crop seed from around the world. It has been built to withstand natural and human-created disasters and is designed as a resource for repopulating the world's food supply, if necessary. Because the seeds are stored at cryonic temperatures, they are able to survive indefinitely.

to prevent significant loss of biodiversity, the FAO established a program for the preservation of animal genetic resources through the establishment of Regional Animal Gene Banks as well as a Global Animal Genetic Data Bank. The Gene Banks utilize cryogenic storage methods for the preservation of germplasm from endangered breeds and species of animals. Semen, fertilized embryos, oocytes, tissue samples, chromosomes, and DNA segments/sequences are all being preserved in the gene banks. Scientists have also developed a "World Watch List" for endangered species and those at risk of endangerment so that their germplasm can be prioritized for collection.

The San Diego Zoo's Institute for Conservation Research has been cryopreserving skin and tissue cells from rare and endangered animal species since 1972, with the hope of extinction prevention. Their gene bank is called The Frozen Zoo.

Plant seed saving and storage have been practiced since the earliest agrarians noticed that some plants and food crops possessed more desirable traits, superior taste, or hardiness than others. It has been common farming practice to take the seeds from the best of the harvest to clean, dry, and store for use in the following year's planting.

Seed gene banks have become an essential part of most food-growing countries. By the end of the first decade of the twenty-first century, nearly 1,500 seed gene banks were in operation around the globe. Such banks catalogue, store, and preserve as many varieties of crop seeds as possible, in part to assure that there are enough varieties to assure ongoing food supplies and in part to preserve history by saving seeds from plants that are no longer being cultivated but might have future use due to progressive climate change or as a result of human- or nature-caused disasters. Some plants may have traits that have not yet been discerned or cultivated and should be saved for future research or for possible medicinal use. Many plant varieties disappeared (became extinct) before their seeds could be saved. It has been estimated that nearly 75 percent of all crop seed gene varieties have disappeared since the year 1900.

The Consultative Group on International Agricultural Research (CGIAR) is tasked with global oversight and maintenance of seed gene banks. By the end of the first decade of the twenty-first century, there were nearly three-quarters of a million different varieties of plants that had seed genes stored in banks. CGIAR is attempting to assure that back-up copies of every cultivated plant seed gene are stored in the Svalbard Global Seed Vault. The Svalbard Global Seed Vault, located under permafrost on the Norwegian island of Spitsbergen on an archipelago called Svalbard, is sometimes referred to as the "Doomsday Vault." It was designed to withstand global climate change and natural disasters and is home to duplicate copies of seeds from gene banks worldwide. The

a half century. Demands for specific strains of livestock based on changing market conditions have led to the use of selective breeding techniques that have caused a significant narrowing of breeds, as well as significant culling and cross-breeding within those strains. As a result, many local breeds (indigenous and adapted to the specific environment) are in danger of extinction. In order expectation is that seeds stored at Svalbard will remain viable for hundreds, possibly thousands, of years because they are stored at extremely low, cryonic temperatures. Currently, there are more than 20 million seeds stored in the vault, representing more than 500,000 plant varieties. The cost of Svalbard's construction was borne by the Norwegian Government; operating costs are shared by Norway and the Global Crop Diversity Trust.

There are two main types of seed gene banks located around the world: long-term cold and shorter-term zero energy gene seed banks. Banks that utilize cold storage methods describe, catalogue, dry and preserve the seed and store it in air- and water-resistant containers at temperatures at or below zero degrees Fahrenheit. The genes can be successfully stored for several decades before they must be planted and re-cultivated in order to maintain genetic viability. Zero-energy seed gene banks use desiccation methods to dry and preserve seeds, which are then prepared for short- or medium-term at ambient temperatures. These banks are located in agrarian communities, and local farmers set up, run and use the banks. The seeds stored represent locally cultivated crops and are available to all who need them.

■ Impacts and Issues

With the emergence of modern biotechnology and the increasing sophistication of computerized relational database systems, American scientists expressed the need for a comprehensive, publicly available database system to be used as a repository for the rapidly expanding body of DNA sequence data. Prior to the creation of GenBank, it was cost-prohibitive for laboratories to collect and analyze genetic sequences other than for organisms they were studying. Data-sharing was hampered by the lack of standards for classification and storage.

Most GenBank submissions consist of a single DNA or RNA sequence segment with annotation representative of the sample's biological information. Each finalized database entry contains scientific name and taxonomy of the organism, a brief description of the sequence, and all necessary bibliographic information. The files are categorized into divisions by taxonomy. Each GenBank record is assigned a permanent accession number that serves to identify the sequence and its annotations across the collaboration's systems. If the sequence or annotation is changed or updated, the new information is added to the original record. The standardization of GenBank records assures that those needing the information have continuous access to the entire history of the sequences being researched.

As of 2006, the FAO's Global Data Bank for Animal Genetic Resources contained 14,017 samples representing 10,512 national breed populations of mammals and 3,505 national breed populations of avian species, spanning 182 countries across the Southwest Pacific, North America, the Near and Middle East, Latin America and the Caribbean, Europe and the Caucasus, Asia, and Africa. More than 90 percent of the breed samples come from domesticated livestock. The mammalian genes were drawn from buffalo, cattle yak, goat, sheep, pig, ass, horse, camel, dromedary, alpaca, llama, guanaco, vicuna, rabbit, guinea pig, and dog. The avian species were comprised of ostrich, nandu, emu, cassowary, swallow, pigeon, peacock, quail, pheasant, partridge, guinea fowl, duck, domestic goose, turkey, domestic duck, and chicken. Roughly 20 percent of the breeds with preserved samples are considered to be at risk for depopulation.

A challenge for the Global Data Bank involves its reporting status; much population data is missing, limiting the knowledge of the risk status of large numbers of reported breeds. Difficulties also exist in measuring breed erosion due to cross breeding between species. The FAO seeks to broaden the usefulness of the data collected by fine-tuning information about local breeds' geographic distribution, as well as collecting information about importation and exportation of live animals and genetic materials for each country sampled.

Geneticists at San Diego's Scripps Research Institute have successfully transformed animal skin cells into induced pluripotent stem cells, or IPS, using techniques based on the genetic engineering research of Shinya Yamanaka (1962–) at Kyoto University in Japan. Stem cells are not specialized and can therefore be transformed into virtually any cell type through the use of biotechnology. As a result, it may be possible some day to use them to clone new members of endangered or rare species.

In addition to seed-saving, researchers around the world seek to preserve the genes of rare, wild, or exotic plants that may prove beneficial in the future. Many locally owned and operated seed gene banks use principles of selective breeding to cultivate the most useful seed genes, or to attempt to recreate crops prevalent in ancient times for food or possible medicinal uses. Scientists also seek to preserve seed genes from threatened or nearly extinct plants because they may have future usefulness.

Seeds conventionally frozen can only be stored for several decades before they need to be thawed and cultivated in order to remain viable. Seeds stored at cryonic temperatures can remain in dormant storage indefinitely.

Banked seeds have been used to recover food sources after natural disasters: after the tsunami in Malaysia and Sri Lanka destroyed rice paddies in 2004, farmers were able to obtain salt-water tolerant rice seeds from banks in other areas.

Local seed banks can be impacted by natural disasters or unusual weather occurrences related to climate change; they are also susceptible to human intervention in times of war or civil unrest. Local seed banks in both

The Svalbard Global Seed Vault in Longyearbyen, Norway, built to protect millions of food crops from climate change, wars and natural disasters and to preserve the genetic diversity of plants. © *Arcticphoto/Alamy.*

Iraq and Afghanistan were destroyed during conflicts in each country. This underscores the critical nature of back-up seed gene banks such as the Svalbard vault, which opened in 2008.

The Global Crop Diversity Trust, a primary support for the Svalbard vault, prioritizes food crop seed genes for banking based on the Articles of the International Treaty on Plant Genetic Resources for Food and Agriculture. The overarching goal is to conserve and protect seed genes for numerous varieties of those diverse food crops considered critical for the feeding and survival of Earth's growing population, now and for the future.

A primary seed gene bank goal is to protect biodiversity, both currently and for the future. As the world grows warmer and drier, some crops may not continue to be viable in their traditional areas. Seed gene banks help assure that it will be possible to substitute one crop variety for another when changing conditions dictate a need to shift.

Gene banks are also used to store human eggs and sperm for use with assisted reproductive technology. Additional human genetic material is stored in laboratories for current and future medical research. Non-human animal gene banks serve several purposes: to provide material for genetic research, to save genes representative of endangered species, and to protect biodiversity in livestock.

■ Primary Source Connection

According to the program officials, "The Global Crop Diversity Trust is an independent international organization with a mission to ensure the conservation, availability and use of crop diversity collections, forever. Since its founding in 2004, the Trust has carried out many projects to restore, maintain and access collections of crop diversity in gene banks, most of which are located in developing countries. Most famously, the Trust has been involved with the Svalbard Global Seed Vault, which was officially opened in 2008. Nicknamed the "Doomsday Vault" by popular media, the Svalbard Global Seed Vault now holds over 650,000 crop samples—making it the most diverse collection of crop diversity anywhere in the world." Additional information about the current status of the project may be found at www.croptrust.org.

At the time of publication, the author, Cary Fowler, was the Executive Director of the Global Crop Diversity Trust.

The Svalbard Global Seed Vault: Securing the Future of Agriculture

Executive Summary

This report combines the historical view and a unique moment in the story of agriculture. The formal opening of the Svalbard Global Seed Vault deep inside an Arctic mountain on February 26, 2008 marks a turning point toward ensuring the crops that sustain us will not be lost. It follows millennia of haphazard forms of protecting crop diversity, and decades of catch-up preservation efforts to save more than a million different varieties of crops. With growing evidence that unchecked climate change could seriously threaten agricultural production and the diversity of crops around the world, the opening of the Seed Vault also represents a major step toward finishing the job of protecting the varieties now held in seed banks. A quiet rescue mission is underway. It will intensify in the coming years, as thousands of scientists, plant breeders, farmers, and those working in the Global Crop Diversity Trust identify and save as many distinct crop varieties as possible.

The story of agriculture dates to some 13,000 years ago, when human societies began the transformation from hunting and gathering to forms of growing food. But the story of systematically saving varieties of crops didn't begin until less than 100 years ago. In the 1920s, plant breeders assembled collections of seeds to breed new varieties. Gradually, scientists began to sample and collect more generally in an attempt to assemble the complete diversity of each crop before distinct varieties were lost.

These scientists delved into the makeup of these varieties. Plant breeders created variety upon variety. Today, the documented pedigrees of modern crop varieties are longer than those of any monarchy. One type of wheat, for instance, has a pedigree that runs six meters long in small type on paper, recording hundreds of crosses, using many different types of wheat from many countries. A number of crops could not be produced on a commercial scale if not for genes obtained from their botanical wild relatives and used in breeding programs.

Around the world, countries and institutions created seed banks, also called genebanks. Today, there are some 1400 collections of crop diversity, ranging in size from one sample to more than half a million. These seed banks now house about 6.5 million samples. About 1.5 million of these are thought to be distinct samples. And within each crop, the diversity of varieties is stunning. Experts, for instance, estimate 200,000 types of wheat, 30,000 types of corn, 47,000 types of sorghum, and even 15,000 types of groundnut.

Some of the more popular varieties are widely distributed in seed banks, occurring in literally hundreds of collections, while others are in just a single facility. Information systems will eventually aid in identifying unintended duplication. About half of the stored samples are in developing countries, and about half of all samples are of cereals.

The Global Crop Diversity Trust is working with the Consultative Group on International Agricultural Research (CGIAR) and seed banks from around the world to assist in preparing and shipping seeds to the Seed Vault in Svalbard. The Trust has assembled leading experts in all of the major crops to identify priority collections. Some 500 scientists from around the world have been involved. The rescue and regeneration effort is under way, and will result in a steady flow of samples being sent to Svalbard in coming years as the genebanks produce fresh new seed. For the February 26 opening of the Seed Vault, workers will load shipments from 21 seed banks, which have sent 268,000 samples that contain about 100 million seeds.

When fully stocked, the Seed Vault will contain samples deposited by large and small genebanks, by those in developed and developing countries as well as international institutions, by those that have state-of-the-art facilities, and by those whose facilities fall far short of international standards. They will share a common desire to use the Seed Vault to insure against losses in their own facility.

Why do they want a backup? Put simply, without the diversity represented in these collections, agriculture will fail. This diversity is vital in guaranteeing a successful harvest and in satisfying our needs for variety. On one level, consumers want diversity within crops because they need wheat for pasta and wheat for bread (for which they need two types of wheat), or they want tomatoes for eating fresh and for making sauce (again, two types of tomato.) On another, farmers want diversity not just to supply consumer demands, but because different farming and environmental conditions require crop varieties with different characteristics.

Plant breeders help consumers and farmers. They have to produce varieties that are productive and popular. This is a moving target. Pest and diseases evolve, the climate changes and so do consumer preferences, and the plant breeder has to incorporate the appropriate characteristics into the variety he or she breeds. And so a farmer's field, over time, is a study of change. One has to run fast just to stay in the same place, just to beat back the pests and diseases and other constantly evolving challenges.

Three partners are overseeing the Seed Vault: the Nordic Gene Bank, the Norwegian Ministry of Agriculture and Food, and the Global Crop Diversity Trust.

They have a simple purpose: provide insurance against both incremental and catastrophic loss of crop diversity held in traditional seed banks around the world. The Seed Vault offers "fail-safe" protection. It serves as an essential element in a global network of facilities that conserve crop diversity and make it available for use in plant breeding and research. Its genesis lies primarily in the desire of scientists to protect against the all-too-common small-scale loss of diversity in individual seed collections. With a duplicate sample of each distinct variety safeguarded in the Seed Vault, seed banks can be assured that the loss of a variety in their institution, or even the loss of the entire collection, will not mean the extinction of the variety or varieties and the diversity they embody.

Svalbard, in the northern reaches of Norway, was chosen for a variety of reasons: The permafrost in the ground offers natural freezing for the seeds; the vault's remote location enhances the security of the facility; the local infrastructure is excellent; Norway, a global player in many multinational efforts, is a willing host; and the area is geologically stable.

In the case of a large-scale regional or even global catastrophe, it is quite likely that the Seed Vault would prove indispensable to humanity. Still, we need not experience apocalypse in order for the Seed Vault to be useful and to repay its costs thousands of times over. If the Seed Vault simply re-supplies genebanks with samples that those genebanks lose accidentally, it will be a grand bargain.

Cary Fowler

FOWLER, CARY. "THE SVALBARD GLOBAL SEED VAULT: SECURING THE FUTURE OF AGRICULTURE." THE GLOBAL CROP DIVERSITY TRUST, FEBRUARY 26, 2008.

SEE ALSO   *Bioinformatics; Cloning; Combinatorial Biology; DNA Sequencing; Genbank; Gene Delivery; Gene Therapy; Genetic Engineering; Genetic Technology; Genome; Genomics; Industrial Genetics; Molecular Biology; Recombinant DNA Technology; Stem Cells*

BIBLIOGRAPHY

Books

Dworkin, Susan. *The Viking in the Wheat Field: A Scientist's Struggle to Preserve the World's Harvest.* New York: Walker, 2009.

Fry, Carolyn, Sue Seddon, and Gail Vines. *The Last Great Plant Hunt: The Story of Kew's Millennium Seed Bank.* Richmond, Surrey: Kew, 2011.

Glasner, Peter E., Paul Atkinson, and Helen Greenslade. *New Genetics, New Social Formations.* London: Routledge, 2007.

Savolainen, V. *DNA and Tissue Banking for Biodiversity and Conservation: Theory, Practice, and Uses.* Richmond, Surrey: Kew, 2006.

Watson, P. F., and W. V. Holt. *Cryobanking the Genetic Resource: Wildlife Conservation for the Future.* London: Taylor & Francis, 2001.

Web Sites

"Gene Banks." *Biodiversity International.* http://www .bioversityinternational.org/research/conservation/ genebanks.html (accessed September 26, 2011).

Seabrook, John. "Sowing For Apocalypse: The Quest for a Global Seed Bank." *The New Yorker*, August 27, 2007. http://www.newyorker.com/ reporting/2007/08/27/070827fa_fact_seabrook (accessed September 26, 2011).

"Svalbard Global Seed Vault." *The Global Crop Diversity Trust.* http://www.croptrust.org/main/ arcticseedvault.php?itemid=842 (accessed September 26, 2011).

"Svalbard Global Seed Vault." *Ministry of Agriculture and Food, Norway.* http://www.regjeringen.no/ en/dep/lmd/campain/svalbard-global-seed-vault .html?id=462220 (accessed September 26, 2011).

"UK Biobank — What Is It?" *UK Biobank.* http:// www.ukbiobank.ac.uk/about/what.php (accessed September 26, 2011).

Pamela V. Michaels

Gene Delivery

■ Introduction

Gene delivery refers to the introduction of deoxyribonucleic acid (DNA) containing one or more genes—sequences that code for functional products—of one organism into a different organism (termed the host). If done correctly, the introduced DNA will be transcribed and the messenger ribonucleic acid (mRNA) translated along with the host genes to produce the encoded functional product(s). The routing of genes into the host cell can be done in a number of different ways, which, in general, either involve the use of virus as the transmission vehicle or the use of techniques that do not involve viruses.

Without the ability to add genes to cells of a target organism, it would be impossible to modify cells to produce compounds of medical and industrial importance.

■ Historical Background and Scientific Foundations

In general, gene delivery relies on the transport of the target gene(s) into a cell as part of the genetic material of a virus, or by other means that do not use viruses.

The use of viruses exploits the natural course of events that occur when viruses infect host cells. This process involves the movement of genetic material from the virus into the host cell and usually the integration of the viral genetic material into the host genetic material. After this integration, the viral material is transcribed and translated along with the host genetic material. The viral gene products that are produced are then used to assemble new virus particles, beginning another cycle of infection, or are used to maintain the presence of the viral genetic material in the host's genome (latent infection).

Retroviruses (including the human immunodeficiency virus, HIV) are especially adept at the introduction and integration of the viral genetic material. Such viruses can be modified genetically to render them incapable of progressing further in the infection cycle. The gene(s) of interest can be incorporated into the viral genome; the crippled virus then becomes useful as a vehicle to move the added genes into host cells.

There are a number of non-viral gene delivery methods. One method is known as microinjection. In this method, a very thin glass tube termed a micropipette is carefully advanced towards a cell. This is not done by hand: The micropipette is attached to an apparatus called a micromanipulator that can be advanced in very small increments (i.e., one or several micrometers at a time). With care and operator skill, the micropipette can pierce the cell wall, allowing the DNA contents to be released into the cell. The micropipette is then withdrawn without lasting damage to the cell wall. The introduced DNA can then integrate into the host genome or can be independently used for transcription and translation. Microinjection is commonly used in the genetic modification of plant cells and eggs of transgenic animals.

Another technique utilizes an apparatus called a gene gun. In this approach, DNA that has been linked to small beads of a material such as gold is propelled into cells growing on a nutrient gel. Although many of the cells will be destroyed by the impact of the beads, some beads will pass into the cells without destroying them. The DNA can then be available for transcription and translation.

A popular method of gene delivery in plant cells (particularly tobacco) is the use of genes that have been inserted into a bacterium called *Agrobacterium tumefaciens*. The inserted gene(s) incorporates into genetic material that is separate from the main genome, known as the Ti plasmid. The plasmid is very mobile (can move from the bacterium to the host cells infected by the bacterium), which provides a ready means of transporting the genes of interest into the host cells.

Gene delivery can also be performed by impaling the host cells on a nanoarray that contains needle-like projections. The genes of interest are attached to the projections and so are thrust into the cells as they are impaled.

WORDS TO KNOW

HOST: The cell that receives the added genetic material.

PLASMID: A circular section of DNA located independent of the main genome, which is able to move more easily from cell to cell than the genomic DNA.

TRANSFORMATION: The alteration of the genetic material of a cell by the introduction of DNA from another organism.

Another technique is called electroporation. In this approach, an electrical field is created, and the DNA of interest is electrically driven across the host cell wall into the cell interior. Similarly, this can be done using chemicals.

■ Impacts and Issues

The ability to deliver target genes into host cells is the basis of genetic modification of cells, and so it is one of the fundamentally important approaches in biotechnology. Without gene delivery, cell modification would be virtually impossible.

Even though gene delivery methods were developed decades ago, the approach remains challenging. Targeting the correct host cells is important, especially if the process is to be done in an intact multi-organ system such as a human when the intent is to treat a disease. Successful delivery of genes to the brain will be useless at best and could have dire consequences at worst—if the therapy is directed at correcting a problem with liver cells, for example. Ensuring that the introduced gene becomes active is another issue; if an introduced gene does not produce a functional product, the process is futile. Finally, the introduced gene or its manufactured product should not stimulate an immune response by the host or harm the host.

SEE ALSO *DNA Sequencing; Gene Therapy; Genetic Technology; Genome; RNA*

BIBLIOGRAPHY

Books

Atkinson, Paul, Peter E. Glasner, and Margaret M. Lock. *Handbook of Genetics and Society: Mapping the New Genomic Era.* London: Routledge, 2009.

Colavito, Mary, and Michael Palladino. *Gene Therapy,* Old Tappan, NJ: Benjamin Cummings, 2006.

Dale, Jeremy, and Malcolm von Schantz. *From Genes to Genomes: Concepts and Applications of DNA Technology,* 2nd ed. Chichester, UK: John Wiley & Sons, 2007.

Rose, Nikolas S. *Politics of Life Itself: Biomedicine, Power, and Subjectivity in the Twenty-First Century.* Princeton: Princeton University Press, 2007.

Web Sites

"Gene Delivery: The Key to Gene Therapy." *Genetic Science Learning Center at the University of Utah.* http://learn.genetics.utah.edu/content/tech/genetherapy/gtdelivery/ (accessed September 23, 2011).

"Gene Therapy." *U.S. Department of Energy Genome Programs.* http://www.ornl.gov/sci/techresources/Human_Genome/medicine/genetherapy.shtml (accessed September 23, 2011).

"Gene Therapy for Cancer: Questions and Answers." *National Cancer Institute at the National Institutes of Health (NIH).* http://www.cancer.gov/cancertopics/factsheet/Therapy/gene (accessed September 23, 2011).

"Genetics Home Reference: Gene Therapy." *National Institutes of Health (NIH).* http://ghr.nlm.nih.gov/handbook/therapy (accessed September 23, 2011).

Brian Douglas Hoyle

Gene Police

■ Introduction

The earliest farmers noted that some plants provided greater yields than others; some resulted in better-tasting, higher-quality crops, and some were unsuccessful. As a result, ancient agrarians began the process of selecting seeds from the best, most useful, and tastiest producers to use for the next planting season. In more modern times, the process was deliberate: Differing strains were crossbred, or hybrids were created in order to reinforce the most desirable plant and animal traits. The rubric became: Save the seeds from the best crops to use next growing season, feed the poorest in appearance to the livestock, and sell or eat the rest.

With the advent of genetic engineering, the worldwide practice of agriculture changed. Large corporations such as Monsanto, Syngenta, AgriPro, and Pioneer have developed genetically modified seed types that are specific for certain traits, including resistance to particular herbicides and pesticides, imperviousness to crop-specific insects and pests, drought hardiness, and ability to withstand very low temperatures. Each of those trait-bearing seeds has been patented by the developers; farmers or agribusinesses purchasing those seeds are required to enter into contracts with the companies, signing stewardship agreements requiring that they not sell or reuse the seed after a single growing season. They also sign agreements allowing representatives from the seed-production companies (colloquially referred to as gene police or seed police) onto their land for a period of several years after the end of the contract agreement in order to verify that seeds are not being saved or pirated. Another form of seed piracy occurs when farmers sell their seeds to other farmers or agribusinesses without labeling the seed as genetically modified or trait bearing. Because all of the genetically modified seed is patented, farmers who are believed to violate their contracts with seed developers or who purchase/use pirated seed are actively investigated, penalized, sued, and/or prosecuted by the patent-holders.

■ Historical Background and Scientific Foundations

The precursors to genetic engineering in agriculture were hybridization and selective breeding. Plants were progressively cultivated on the basis of desirable traits such as high crop yield, rapid growth, resistance to various pests, and general hardiness. Farmers have traditionally saved the seeds from the best and most desirable performers for use with successive planting seasons. Such breeding practices have been utilized for as long as farming has occurred.

WORDS TO KNOW

AGRIBUSINESS: A large commercial farm operation, often encompassing development, production, marketing, and branding of food or other farmed products.

SEED PIRACY: The process of gathering, reusing, or selling genetically modified, patented seeds from one season to the next.

SEED SAVING: Cleaning and reconditioning seeds gathered from one year's harvest for planting during the following growing season.

SEED TRAITS: Desired characteristics such as insect and pest resistance, imperviousness to pesticides and herbicides, and ability to withstand drought or extreme cold weather conditions. The traits are not naturally occurring; they are intentionally created through the use of biotechnology and molecular biology techniques involving isolating the desired traits in other plants or animals and inserting specific segments of their DNA into the seed to be genetically modified. Seed traits are patented by the companies that create or develop them.

With the advent of the science of genetics and the work of Gregor Mendel (1822–1884) during the 1800s, plant breeding became more fine-tuned as the mechanisms for identifying and reinforcing specific traits and characteristics were progressively defined. In 1944 Oswald Avery (1877–1955), Colin McLeod (1909–1972), and Maclyn McCarty (1922–2005) discovered that DNA was responsible for the transmission of genetic information. James Watson (1928–) and Francis Crick (1916–2004) uncovered the helical structure of DNA in 1953. These discoveries paved the way for recombinant DNA technology, in which a section of DNA can be removed from one cell and inserted into another. Recombinant DNA technology was foundational for the inception of genetic engineering.

In 1982 scientists working for the Monsanto corporation genetically modified plant cells. The first genetically engineered crops developed by Monsanto and grown in the United States were tomato and tobacco; they were tested in 1987. At that time, it was reported by the National Academy of Sciences that genetic engineering did not substantially alter the nature of the plants being modified, nor did it pose any significant health or environmental risks.

In 1992, the commercial production and distribution of seeds used to produce a tomato engineered for longer shelf-life, Calgene's Flavr Savr, was approved by the U.S. Department of Agriculture. That same year, the U.S. Food and Drug Administration reported that genetically engineered foods did not require regulation because they did not pose specific risks for production and consumption.

Companies producing genetically modified seed routinely patent their products and the proprietary methods used to create the specific genetic modifications, called traits or trait stacks (when more than one trait is engineered into a type of seed, such as resistance to a particular pesticide and imperviousness to specific pests and insects). When the seeds are made available for purchase, the purchaser signs agreements regulating the ownership and use of the seed.

This aerial photograph by Greenpeace shows a huge circle made by local farmers and Greenpeace volunteers on a corn farm planted with a genetically-modified Bt corn in Isabela province, 186 miles (300 kilometers) north of Manila, Philippines. The crop circle, with a slash over the letter "M" symbolizes farmer rejection of genetically-modified Bt corn crops from Monsanto corporation, which many feel destroys small farming communities with its business practices of not allowing seed saving. The protest coincides with Global Day of Action to protect corn, one of the world's most important staple foods, against contamination from genetically-engineered (GE) varieties. *© Melvyn Calderon/AFP/Getty Images.*

■ Impacts and Issues

When farmers contract to use genetically modified seed, they are required to sign a stewardship agreement and agree that they are leasing a technology rather than purchasing a product. They make commitments not to save and reuse the seed, not to sell the seed to others, and to abide by a variety of conditions relating to the protection of the proprietary, patented seed technologies. One of the areas of agreement concerns conferring permission for agents of the seed manufacturer to enter the farmer's land in order to ascertain that the seed is properly used, to provide technical assistance as necessary, and to examine crops and fields for proper use of seed technology for a period of several years after the expiration of the contract. This allows the seed manufacturer to ascertain that seed piracy is not occurring either during or after the stewardship period.

There is a significant difference between selective breeding and genetic engineering. With the former, all changes in DNA are natural and occur across generations of the same species. In genetic engineering, DNA representative of a desired trait is removed from another type of plant or animal and inserted into the cells of the organism to be modified. The new trait is immediately expressed. Engineered genes do not occur in nature; they are created by humans through scientific processes. The theory as well as the laboratory methods used to create those genes are patented and deemed the intellectual property of the company employing the seed developers.

Companies who develop and patent genetic engineering methods in agriculture state that the technology is essential in order to feed the planet's growing population under often inhospitable growing conditions. They assert that the cost of the seeds and the need for annual purchase are necessary in order to pay for the developing technology as well as funding ongoing research and development.

Opponents of genetic engineering in agriculture assert that engineered seeds are "infecting" other farms when transported by wind, insects, and airborne pollen; this impinges on some farmers' ability to meet stringent organic farming requirements. Some non-organic growers report that the unwanted genetically engineered seeds are getting mixed in with their saved, cultivated, non-engineered seeds. There are reports in the popular and agricultural press concerning the appearance of "seed police" or "gene police" at establishments owned by individuals not contracted to use the patented seeds. Those individuals report being unfairly targeted, investigated, and sued by the patent holders.

The large genetically-engineered seed-producing companies utilize several methods to assure fair business practices and "a level playing field" for all users of their patented seeds and technology. Monsanto reports that from 1995 to 1997, it engaged in intensive farmer education on patent law for engineered seeds. In 1998 the company began enforcing patent infringement. One aspect of enforcement involves staffing a toll-free line used by concerned individuals to anonymously report suspicious seed- or technology-related activity by neighboring farmers. When they have credible information, Monsanto researches the individual's file. If warranted, it engages the services of contracted field investigators. Those personnel employ a variety of off-site surveillance techniques to make preliminary assessments of possible patent infringement. If a preponderance of evidence does not support infringement, the investigation ends. If there is a likelihood of patent infringement, investigators meet directly with the identified farmer.

Monsanto's investigators state that the majority of farmers engaging in patent infringement admit it during interview and will request an out-of-court settlement. When that does not occur, and depending on the

---

IN CONTEXT: SEEDS OF DISPUTE

Large agricultural corporations sell genetically modified seeds to farmers under license. A common provision in seed license agreements requires farmers to destroy or return all of their seed after harvest. Traditionally, farmers would retain seed for planting during the next season. Under the seed license agreement, farmers must purchase new seed each year from the genetically modified seed manufacturer.

In 2004, Monsanto Canada, Inc., sued Canadian farmer Percy Schmeiser (1931–) for patent infringement after Roundup Ready Canola, a genetically modified (GM) crop produced by Monsanto, was discovered in Schmeiser's field. Schmeiser discovered the GM plants on his property and saved some of the seeds for the next planting season. The Supreme Court of Canada ruled in favor of Monsanto and determined that Schmeiser had violated Monsanto's right to exclusive use under Canadian patent law.

Since the mid-1990s, Monsanto, one of the largest producers of genetically modified seed, has sued over 150 farmers for violating the company's patents on genetically modified seeds.

However, such practices have been controversial. It is not uncommon for genetically modified plants from one farmer's field to cross-pollinate or drift onto an adjacent farmer's field of conventional crops without the conventional farmer's knowledge or consent. Studies indicate that this form of cross-pollination, known as outcrossing, can occur over great distances.

In 2007 the U.S. Department of Agriculture (USDA) fined Scotts Miracle-Gro Company when modified genetic material from a new variety of genetically modified grass was discovered in the genetic material of conventional grasses located miles from the company's testing area.

outcome of the investigation and interviews, investigators request access to the farmer's records and permission to take crop samples. Large agribusinesses such as Monsanto, Pioneer, AgriPro, and Syngenta, all of whom hold genetic engineering patents, conduct business with hundreds of thousands of farming operations around the world. Of the hundreds of farmers investigated since the patent enforcement began, less than 150 lawsuits have been filed, and fewer than a dozen cases have proceeded to trial. The large agribusinesses rarely lose in court.

■ Primary Source Connection

The Monsanto Company of St. Louis, Missouri, touts its challenge as "Meeting the Needs of Today While Preserving the Planet for Tomorrow." Products from this company include agricultural and vegetable seeds, plant biotechnology traits, and crop protection chemicals. In 1999 the company accused a farmer of using saved seed from one harvest and replanting it the following season, a practice that the company considers unlawful. Monsanto cited accusations of technology piracy, stating that its seeds are genetically engineered for use in farming and that the practice of saving seeds for reuse is a violation of the company's contract with farmers. To protect its company biotech-related patents, Monsanto filed several similar suits to follow, with ensuing fines and criminal prosecution as a consequence to contract violations.

Seed Company Cracks down on Recycling Farmers

SAN FRANCISCO—Monsanto Co.'s "Seed Police" snared soybean farmer Roman McFarling in 1999, and the company is demanding he pay it hundreds of thousands of dollars for alleged technology piracy.

McFarling's sin? Resaved seed from one harvest and replanted it the following season, a revered and ancient agricultural practice.

"My daddy saved seed. I saved seed," said McFarling, 62, who still grows soybeans on the 5,000-acre family farm in Shannon, Miss., and is fighting the agribusiness giant in court.

Saving Monsanto's seeds, genetically engineered to kill bugs and resist weed sprays, violates provisions of the company's contracts with farmers.

Since 1997, Monsanto has filed similar lawsuits 90 times in 25 states against 147 farmers and 39 agriculture companies, according to a report issued Thursday by The Center for Food Safety, a biotechnology foe.

In a similar case a year ago, Tennessee farmer Kem Ralph was sued by Monsanto and sentenced to eight months in prison after he was caught lying about a truckload of cotton seed he hid for a friend.

Ralph's prison term is believed to be the first criminal prosecution linked to Monsanto's crackdown. Ralph also has been ordered to pay Monsanto more than $1.7 million.

The company itself says it annually investigates about 500 "tips" that farmers are illegally using its seeds and settles many of those cases before a lawsuit is filed.

In this way; Monsanto is attempting to protect its business from pirates in much the same way the entertainment industry does when it sues underground digital distributors exploiting music, movies and video games.

In the process, it has turned farmer on farmer and sent private investigators into small towns to ask prying questions of friends and business acquaintances.

Monsanto's licensing contracts and litigation tactics are coming under increased scrutiny as more of the planet's farmland comes under genetically engineered cultivation.

Some 200 million acres of the world's farms grew biotech crops last year, an increase of 20 percent from 2003, according to a separate report released Wednesday.

Many of the farmers Monsanto has sued say; as McFarling claims, that they didn't read the company's technology agreement close enough. Others say they never received an agreement in the first place.

The company counters that it sues only the most egregious violations and is protecting the 300,000 law-abiding U.S. farmers who annually pay a premium for its technology. Soybean farmers, for instance, pay a "technology fee" of about $6.50 an acre each year.

Some 85 percent of the nation's soybean crop is genetically engineered to resist Monsanto's herbicide Roundup, a trait many farmers say makes it easier to weed their fields and ultimately cheaper to grow their crops.

"It's a very efficient and cost effective way to raise soybeans and that's why the market has embraced it," said Ron Heck, who grows 900 acres of genetically engineered soybeans in Perry, Iowa.

Heck, who also is chairman of the American Soybean Association, said he doesn't mind buying new seed each year and appreciates Monsanto's crackdown on competitors who don't pay for their seed.

"You can save seed if you want to use the old technology;" Heck said.

The company said the licensing agreement protects its more than 600 biotech-related patents and ensures a return on its research and development expenses, which amount to more than $400 million annually.

"We have to balance out obligations and our responsibilities to our customers, to our employees and to our share holders," said Scott Baucum, Monsanto's chief intellectual property protector.

Still, Monsanto's investigative tactics are sowing seeds of fear and mistrust in some farming communities, company critics say.

Monsanto encourages farmers to call a company hot line with piracy tips, and private investigators in its employ act on leads with visits to the associates of suspect farmers.

Baucum acknowledged that the company walks a fine line when it sues farmers. "It is very uncomfortable for us," Baucum said. "They are our customers and they are important to us."

The Center for Food Safety established its own hot line Thursday where farmers getting sued can receive aid. It also said it hopes to convene a meeting among defense lawyers to develop legal strategies to fight Monsanto.

The company said it has gone to trial five times and has never lost a legal fight against an accused pirate. The U.S. Supreme Court in 1980 allowed for the patenting of genetically engineered life forms and extended the same protections to altered plants in 2001. Earlier this year, a Washington D.C. federal appeals court specifically upheld Monsanto's license.

"It's sad. It's sickening. I'm disillusioned," said Rodney Nelson, a North Dakota farmer who settled a Monsanto suit in 2001 that he said was unfairly filed. "We have a heck of an uphill battle that I don't think can be won."

Paul Elias

ELIAS, PAUL. "SEED COMPANY CRACKS DOWN ON RECYCLING FARMERS." *THE ASSOCIATED PRESS* (JANUARY 15, 2005).

SEE ALSO *Agricultural Biotechnology; Corn, Genetically Engineered;* Diamond v. Chakrabarty; Monsanto Canada Inc. v. Schmeiser*; Patents and Other Intellectual Property Rights; Plant Patent Act of 1930; Plant Variety Protection Act of 1970*

BIBLIOGRAPHY

Books

Agriculture and Environment Biotechnology Commission. *What Shapes the Research Agenda?* Great Britain: Department of Trade and Industry, 2005.

Aoki, Keith. *Seed Wars: Controversies and Cases on Plant Genetic Resources and Intellectual Property.* Durham, NC: Carolina Academic Press, 2008.

Australian Farm Institute. *Can Agriculture Manage a Genetically Modified Future?* Surrey Hills, New South Wales: Australian Farm Institute, 2011.

Becker, Geoffrey S. *Agricultural Biotechnology Background and Recent Issues.* Washington, DC: Congressional Information Service, Library of Congress, 2005.

Gordon, Thomas T., and Arthur S. Cookfair. *Patent Fundamentals for Scientists and Engineers*, 2nd ed. Boca Raton, FL: Lewis Publishers, 2000.

Heldman, Dennis R., Matthew B. Wheeler, and Dallas G. Hoover. *Encyclopedia of Biotechnology in Agriculture and Food.* Boca Raton, FL: CRC Press, 2011.

Herren, Ray V. *Biotechnology: An Agricultural Revolution.* Clifton Park, NY: Delmar Learning, 2005.

Kessler, Rob. *Seeds: Time Capsules of Life*, 2nd ed. Richmond Hill, Ontario: Firefly Books, 2009.

Kimbrell, Andrew, and Joseph Mendelson. *Monsanto vs. U.S. Farmers.* Washington, DC: Center for Food Safety, 2005.

Web Sites

"Agricultural Giant Battles Small Farmers." *CBS News,* January 4, 2011. http://www.cbsnews.com/stories/2008/04/26/eveningnews/main4048288.shtml (accessed September 19, 2011).

Barlett, Donald L., and James B. Steele. "Monsanto's Harvest of Fear." *Vanity Fair,* May 2008. http://www.vanityfair.com/politics/features/2008/05/monsanto200805 (accessed September 19, 2011).

Freeman, E. "Seed Police? Part 4." *Monsanto.com,* 11/10/2008. http://www.monsanto.com/newsviews/Pages/Seed-Police-Part-4.aspx (accessed September 19, 2011).

"Percy Schmeiser's Battle." *CBC News Online,* May 21, 2004. http://www.cbc.ca/news/background/genetics_modification/percyschmeiser.html (accessed September 19, 2011).

Pamela V. Michaels

Gene Therapy

■ Introduction

Gene therapy refers to the replacement of a defective gene by a non-defective, identical gene. The replacement, which is accomplished using an approach called gene delivery, enables the normally functioning gene product to be produced. If a body malfunction or disease is caused exclusively by the abnormal gene (sometimes other conditions contribute to the problem), then gene therapy potentially can correct the problem.

Gene therapy is far from a tried and true approach. The procedure has been responsible for the deaths of several people, usually due to immune reactions to the virus used to transport the therapeutic genes into the cells of the host. More recent experiments using animal models have produced more promising results, but routine human use of gene therapy still faces many hurdles.

WORDS TO KNOW

DNA: The scientific abbreviation for deoxyribonucleic acid. Found within every living cell and carrying the organism's genetic information, DNA consists of two long chains of amino acid nucleotides; the chain is twisted into a double helix shape. The strands of the chain are joined by means of hydrogen bonds. The specific sequence of the DNA is what defines genetic characteristics.

HOST: The cell that receives the added genetic material.

PLASMID: A circular section of DNA located independent of the main genome, which is able to move more easily from cell to cell than the genomic DNA.

TRANSFORMATION: The alteration of a cell's genetic material by the introduction of DNA from another organism.

■ Historical Background and Scientific Foundations

The ability to alter selectively the arrangement of deoxyribonucleic acid (DNA) by introducing genes was first developed in the 1970s. Even then, the knowledge that some diseases were primarily caused by a defect in one or several genes prompted the suggestion that replacing the defective gene with its normal counterpart could alleviate the disease. In the 1980s researchers began to explore the use of viruses as a vehicle to transport foreign DNA into host cells.

Since then, several methods of gene therapy have been developed. One involves the introduction of a normal gene at any point in the host genome. The idea is that the production of the normal product will override the defect associated with the abnormal gene located elsewhere in the genome. In another, more specific approach, the defective gene is removed from the chromosome using enzymes that recognize DNA patterns at either end of the target gene and cut the DNA. Then the normal gene can be introduced in place of the excised defective gene, and the normal gene product produced. In a third approach, a gene that is different from the defective gene is introduced; the product of the introduced gene reverses the defect in the host gene. Finally, if the gene defect is related to the regulation of a gene (i.e., how the gene is activated or activity is stopped), then manipulations that correct the regulation, rather than the actual structure of the gene, can be effective.

Most often, gene therapy relies on the introduction of a normal form of the target gene into host cells using a virus as the carrier system. This approach exploits the normal mechanism used by viruses in infection (i.e., introduction of viral DNA or RNA into the host and subsequent use of the host's transcription and translation machinery to produce the viral products). Essentially, the virus unloads its payload of genetic material into the target host cell.

of the particular target host cell. The idea is that injection of the liposomes will lead to their specific binding to the target host cells, blending of the liposome membrane with the host membrane, and releasing the genetic material into the host cells.

■ Impacts and Issues

The theoretical potential of gene therapy has so far not translated into clinical reality. In many cases, the normal gene product is produced only for a relatively short time before gene activity ceases or the gene is lost. Maintaining stability of introduced genes as cells undergo many cycles of division is a challenging issue. The use of viral vectors is a risk. Indeed, several human deaths such as 18-year-old Jesse Gelsinger in 1999 and Jolee Mohr in 2007 were attributed to a massive immune reaction to the viruses. Gelsinger was the first person known to have died in a gene therapy clinical trial. Preventing an immune reaction from occurring is a very daunting challenge.

Optimistically, more recent studies using animal models have achieved success in treating ailments including blindness due to the progressive death of photoreceptors. As well, research with human volunteers with Parkinson's disease reported short-term (months) relief from some symptoms and in a single case of a form of anemia. Other successes have been reported, but all are far from routine use as of 2011.

Even if gene therapy could be advanced to the point of clinical use, the high cost of the approach is problematic. Without some financial support, potentially life-altering or even life-saving therapy could be restricted to those with the economic means to pay.

SEE ALSO *Bioethics; Biotechnology; Gene Delivery; Genetic Privacy; Genetic Technology; Genome; Genomics*

BIBLIOGRAPHY

Books

Atkinson, Paul, Peter E. Glasner, and Margaret M. Lock. *Handbook of Genetics and Society: Mapping the New Genomic Era.* London: Routledge, 2009.

Colavito, Mary, and Michael Palladino. *Gene Therapy,* Old Tappan, NJ: Benjamin Cummings, 2006.

Dale, Jeremy, and Malcolm von Schantz. *From Genes to Genomes: Concepts and Applications of DNA Technology,* 2nd ed. Chichester, UK: John Wiley & Sons, 2007.

Merkeley, Ward. *A Model for Gene Therapy.* Philadelphia: Xlibris Corporation, 2005.

Panno, Joseph. *Gene Therapy: Treating Disease by Repairing Genes.* New York: Facts on File, 2004.

In this photo taken May 16, 2011, Timothy Brown, the only man ever known to have been cured of HIV, poses with his dog, Jack, on Treasure Island in San Francisco. Brown has been called "The Berlin patient" because that's where he was treated. It seemed like a fluke for several years, but doctors continued to have early signs of success using gene therapy to do the same thing. *© AP Images/ Eric Risberg.*

Several types of virus have been explored and tried in gene therapy, including retroviruses such as the human immunodeficiency virus (HIV), adenoviruses, adeno-associated viruses, and herpes simplex viruses. Normally these viruses cause disease; two examples are the several known types of HIV, which cause acquired immunodeficiency syndrome (AIDS), and herpes simplex virus, which causes cold sores. Therefore it is absolutely essential that the viruses be genetically crippled to ensure that while they maintain their ability to transfer DNA or RNA to host cells, the infectious process goes no further.

Non-viral approaches include the use of spheres of lipid (liposomes) that encapsulate DNA or RNA. The liposomes can even be equipped with proteins on the surface that recognize and bind to proteins on the surface

Rose, Nikolas S. *Politics of Life Itself: Biomedicine, Power, and Subjectivity in the Twenty-First Century.* Princeton: Princeton University Press, 2007.

Ryan, Allen F. *Gene Therapy of Cochlear Deafness: Present Concepts and Future Aspects.* Basel: Karger, 2009.

Web Sites

"Gene Therapy." *U.S. Department of Energy Genome Programs.* http://www.ornl.gov/sci/techresources/Human_Genome/medicine/genetherapy.shtml (accessed September 23, 2011).

"Gene Therapy for Cancer: Questions and Answers." *National Cancer Institute at the National Institutes of Health (NIH).* http://www.cancer.gov/cancertopics/factsheet/Therapy/gene (accessed September 23, 2011).

"Genetics Home Reference: Gene Therapy." *National Institutes of Health (NIH).* http://ghr.nlm.nih.gov/handbook/therapy (accessed September 23, 2011).

Brian Douglas Hoyle

Genetic Bill of Rights

■ Introduction

The Genetic Bill of Rights is a statement of principles related to the ethical, social, and legal implications of biotechnology. Promulgated in 2000 by the Council for Responsible Genetics (a nonprofit organization), the Genetic Bill of Rights employs the precautionary principle to address the unforeseen effects that genetic modification could have on humans and the environment. The bill also contains principles designed to protect human rights and autonomy. To date, no government has adopted the Genetic Bill of Rights into law.

■ Historical Background and Scientific Foundations

In the 1970s scientists devised recombinant DNA (rDNA) technologies, including molecular cloning and polymerase chain reactions, which allow genetic material from multiple organisms to be combined to create a genetic sequence that does not exist in nature. Recombinant DNA technology enables scientists to create organisms that express desired traits that are not naturally found in that organism's genetic code. Scientists have used rDNA technologies to create a variety of useful products, including recombinant human insulin and vaccines, herbicide- and pest-resistant crops, and human growth hormones.

Despite the advantages of rDNA technologies, many scientists, bioethicists, and policymakers have expressed concern about potential harm to humans or other organisms posed by rDNA technologies. In 1974 the scientific community even imposed a voluntary moratorium on rDNA research pending the adoption of a set of ethical and safety guidelines for rDNA research. In 1975 scientists at the Asilomar Conference on Recombinant DNA Molecules adopted a set of guidelines for the use of appropriate biological and physical controls to minimize any potential threats posed by rDNA research.

In 1983 concerned scientists, attorneys, and others formed the Council for Responsible Genetics, a nonprofit public interest group focused on biotechnology issues. The Council for Responsible Genetics advocates the responsible use of biotechnology. The organization also seeks to engage the public on biotechnology issues and encourages public participation in policymaking on biotechnology issues. Following the 1997 announcement of the cloning of Dolly the sheep, the first mammal cloned from an adult somatic cell, the Council for Responsible Genetics began devising a set of principles to protect humans and the environment from possible risks posed by biotechnology. In 2000 the Council for Responsible Genetics proposed the Genetic Bill of Rights, a document that embodies the organization's ideals that all people have the right to participate in evaluating the social, legal, and environmental impacts of biotechnology.

■ Impacts and Issues

The Genetic Bill of Rights adopts a precautionary approach to addressing issues raised by genetic engineering. The precautionary principle states that any product, action, or process that might pose a threat to public or environmental health should not be introduced in the absence of scientific consensus regarding its safety. Under the precautionary principle, the entity that desires to introduce a new technology bears the burden of proving that the technology does not pose a threat to the public or environment. The Genetic Bill of Rights states that commercial, scientific, medical and government entities have a "profound ignorance" of how genetic engineering affects humans and other living organisms. Based on this premise, the Genetic Bill of Rights endorses 10 principles for protecting the human rights, privacy, health, and dignity of humans and the health of the environment from any unforeseen consequences of genetic engineering.

WORDS TO KNOW

MOLECULAR CLONING: A technique that assembles and replicates recombinant DNA (rDNA) molecules with a host.

POLYMERASE CHAIN REACTION: A laboratory technique used to amplify (copy) small segments of DNA.

PRECAUTIONARY PRINCIPLE: A principle stating that any product, action, or process that might pose a threat to public or environmental health should not be introduced in the absence of scientific consensus regarding its safety.

RECOMBINANT DNA (RDNA): A form of DNA containing two or more genetic sequences normally not occurring together, which are spliced together.

Several provisions of the Genetic Bill of Rights address potential human health and reproductive issues posed by genetic engineering. The Genetic Bill of Rights states that all humans have the right to protection from toxins or other substances that could harm their genetic makeup or that of their children. Other provisions claim that humans have the right to freedom from forced sterilization or other eugenics measures and the freedom from mandatory screening of fetuses. The Genetic Bill of Rights also asserts that each human has the right "to have been conceived, gestated, and born without genetic manipulation." Finally, the Genetic Bill of Rights states that everyone has the right to genetic privacy. This provision has been codified in the laws of the United States and other countries.

The Genetic Bill of Rights also promotes social justice and environmental principles. The first right listed in the Genetic Bill of Rights states "All people have the right to preservation of the earth's biological and genetic diversity." Other provisions state that living organisms cannot be patented and that indigenous peoples have the right to manage their biological resources. In addition, people have the right to a food supply that is free of genetically engineered food. Finally, the Genetic Bill of Rights asserts that people have the right to DNA evidence to defend themselves in criminal proceedings.

Despite being introduced in 2000, no nation has passed the Genetic Bill of Rights into law. In January 2011, legislators in the General Court of Massachusetts, the state legislature, introduced the Genetic Bill of Rights. As of October 2011, the General Court has not passed the Genetic Bill of Rights. Massachusetts has been at the forefront of public protection on biotechnology issues in the past. A Massachusetts law, advocated by the Council for Genetic Rights, served as the basis for the Genetic Information and Nondiscrimination Act of 2008 (GINA), a U.S. law that prohibits the use of genetic information in health insurance and employment.

■ Primary Source Connection

The Genetic Bill of Rights was introduced in response to a belief that ethical, legal, social, and environmental implications of biotechnology need to be determined. It was written as a foundational document highlighting fundamental values deemed to be put at risk by new applications of genetics. The tenets relate to human integrity, individual liberty, and the health of Earth's environment as it pertains to genetic technologies. It has been made available as a pamphlet with the hope that by making it available to the public, it will become a vehicle for public education and open dialogue.

Genetic Bill of Rights

PREAMBLE

Our life and health depend on an intricate web of relationships within the biological and social worlds. Protection of these relationships must inform all public policy.

Commercial, governmental, scientific and medical institutions promote manipulation of genes despite profound ignorance of how such changes may affect the web of life. Once they enter the environment, organisms with modified genes cannot be recalled and pose novel risks to humanity and the entire biosphere.

Manipulation of human genes creates new threats to the health of individuals and their offspring, and endangers human rights, privacy and dignity.

Genes, other constituents of life, and genetically modified organisms themselves are rapidly being patented and turned into objects of commerce. This commercialization of life is veiled behind promises to cure disease and feed the hungry.

People everywhere have the right to participate in evaluating the social and biological implications of the genetic revolution and in democratically guiding its applications.

To protect our human rights and integrity and the biological integrity of the earth, we, therefore, propose this Genetic Bill of Rights.

1. All people have the right to preservation of the earth's biological and genetic diversity.

2. All people have the right to a world in which living organisms cannot be patented, including human beings, animals, plants, microorganisms and all their parts.

3. All people have the right to a food supply that has not been genetically engineered.

A member of Greenpeace displays sacks of U.S. long grain rice that made it to the shelves of a high-end supermarket chain in the Philippines. The rice allegedly was grown using GMOs (genetically-modified organisms) in Manila, Philippines. Part of the Genetic Bill of Rights calls for all people to have access to food that has not been genetically engineered. © *AP Images/Bullit Marquez.*

4. All indigenous peoples have the right to manage their own biological resources, to preserve their traditional knowledge, and to protect these from expropriation and biopiracy by scientific, corporate or government interests.

5. All people have the right to protection from toxins, other contaminants, or actions that can harm their genetic makeup and that of their offspring.

6. All people have the right to protection against eugenic measures such as forced sterilization or mandatory screening aimed at aborting or manipulating selected embryos or fetuses.

7. All people have the right to genetic privacy including the right to prevent the taking or storing of bodily samples for genetic information without their voluntary informed consent.

8. All people have the right to be free from genetic discrimination.

9. All people have the right to DNA tests to defend themselves in criminal proceedings.

10. All people have the right to have been conceived, gestated, and born without genetic manipulation.

BOARD OF THE COUNCIL FOR RESPONSIBLE GENETICS. "GENETIC BILL OF RIGHTS," SPRING 2000.

SEE ALSO *Bioethics; Genetic Discrimination; Genetic Engineering; Genetic Privacy; Genetic Technology; Government Regulations; Individual Privacy Rights*

BIBLIOGRAPHY

Books

Albertson, Leana J. *Genetic Discrimination.* Hauppage, NY: Nova Science, 2008.

Berry, Roberta M. *The Ethics of Genetic Engineering.* London: Routledge, 2007.

Jersild, Paul T. *The Nature of Our Humanity: Ethical Issues in Genetics and Biotechnology.* Minneapolis, MN: Fortress Press, 2009.

Willett, Edward. *Genetics Demystified.* New York: McGraw-Hill, 2006.

Web Sites

Centers for Disease Control and Prevention (CDC). "Privacy Legislation and Regulations." http:// www.cdc.gov/od/science/regs/privacy/index.htm (accessed October 21, 2011).

"Genetic Testing." *Centers for Disease Control and Prevention (CDC).* http://www.cdc.gov/ genomics/gtesting/ (accessed October 21, 2011).

World Health Organization (WHO)). "Genetics." http://www.who.int/topics/genetics/en (accessed October 21, 2011).

Joseph P. Hyder

Genetic Discrimination

■ Introduction

Genetic discrimination refers to a form of discrimination based on the genetic information of an individual, typically related to that person's health. Genetic information may be determined through genetic testing or through analyzing family health information. Genetic discrimination occurs when a third party, typically an employer or health insurance company, learns of an individual's genetic predisposition towards certain diseases or disorders and consequently treats that person differently. Employers may use genetic information to deny or terminate employment. Health insurers may use genetic information to deny or revoke an individual's or family's health insurance. Many governments have adopted legislation that prohibits genetic discrimination.

■ Historical Background and Scientific Foundations

Genes are hereditary units that control and define all human characteristics. Genes are distinct sequences of nucleotides, molecules that comprise deoxyribonucleic acid (DNA) and ribonucleic acid (RNA), which form part of a chromosome. Each sequence of nucleotides codes for a specific trait, which may be inherited by offspring. The entire complement of genes within an organism is known as the genome.

Mutations, variations within the DNA of a gene, may lead to disease or indicate an increased risk of disease. Genetic testing involves detecting these mutations, typically through a blood, skin, or hair test that measures the presence and amount of key proteins or metabolites or observed mutations in chromosomes or DNA. More than 1,000 genetic tests are currently available and check for the presence or an increased risk of numerous diseases or genetic disorders, including breast cancer, colon cancer, cystic fibrosis, and sickle cell anemia.

A genetic test may test for a single gene disorder, in which the presence of a mutation is highly predictive of developing that particular disease or disorder. A person carrying a Huntington's chorea mutation is certain to develop the disease. A genetic test may also test for susceptibility genes, in which the presence of a mutated gene only plays a small role in whether that person develops the disease or disorder. Additional factors, such as lifestyle and environment, also will influence whether a person with a susceptible gene ultimately develops that particular disease or disorder. The presence of susceptible genes may increase a person's risk of developing a disease or disorder, but other genes and factors usually play a more important role.

In addition to undergoing genetic testing to reveal a potential disease or disorder, people may take a genetic test to determine whether they carry genes for particular hereditary diseases or disorders that may be passed on to children. Although a carrier may not have a particular disease, and may never develop that disease, the carrier may pass on the genes for that disease to a child. If the child inherits those particular genes, then the child will have an increased risk of disease. The child, however, may also be merely a carrier that never develops that disease. Parents may also choose to have genetic testing performed on a

---

WORDS TO KNOW

DEOXYRIBONUCLEIC ACID (DNA): A self-replicating, double-helix structure of nucleobases that is the main component of chromosomes and serves as the basis of heredity.

GENE: The functional unit of genetic inheritance that is composed of specific sequence of nucleotides in DNA or RNA.

RIBONUCLEIC ACID (RNA): Nucleic acid that carries protein synthesis instructions from DNA.

fetus or fertilized embryos prior to in-vitro fertilization to assess the risk of genetic diseases or disorders.

The first genetic test, conducted in the early 1960s, tested newborns for phenylketonuria (PKU), a metabolic genetic disorder that may lead to mental retardation without treatment. The PKU test was quick, painless, and could lead to the successful treatment of PKU. Over the following decades, researchers developed hundreds of genetic tests, many of which detected diseases that are not as easily treatable as PKU.

■ Impacts and Issues

As genetic testing entered mainstream medicine, public interest groups and concerned citizens raised concerns that the dissemination of genetic test results could lead to discrimination by employers and insurers against people carrying mutations. Public interest groups documented numerous cases of genetic discrimination throughout the 1990s and first decade of the twenty-first century in the Australia, the United Kingdom, the United States, and other countries.

Nations responded to genetic discrimination concerns in a variety of ways. The European Union (EU) considers genetic discrimination to fall under the EU's broad personal data protections laws. Genetic discrimination is prohibited under the EU Charter of Fundamental Rights and the Council of Europe Convention on Human Rights and Biomedicine. The EU also considers the Universal Declaration on the Human Genome and Human Rights to protect individuals from genetic discrimination under human rights law.

After more than a decade of failed attempts to pass a genetic antidiscrimination law, the U.S. Congress passed the Genetic Information Nondiscrimination Act of 2008 (GINA) with only one dissenting vote. GINA, the most comprehensive U.S. antidiscrimination law since the 1980s, went into effect on November 21, 2009. GINA prohibits employers from requesting genetic testing or genetic information, which may be derived from family history, in hiring, promoting, or firing employees. GINA also prohibits insurance companies from requesting genetic testing or family history to make decisions to deny or terminate coverage or to increase insurance premiums.

Pamela Fink, shown at her home in 2010, filed complaints against her employer that alleged she was dismissed after making it known that genetic testing showed she carried the BRCA2 breast cancer gene and then underwent a double mastectomy to prevent cancer. © AP Images/Douglas Healey.

Prior to the passage of GINA, the U.S. National Institutes of Health's National Human Genome Research Institute (NHGRI) noted that many patients avoided genetic testing out of fear that employers or insurers could learn the results and use them against individuals. The Council for Responsible Genetics (CRG), a U.S. non-profit organization that promotes the public interest on biotechnology issues, documented numerous cases of employers firing employees and insurers denying or revoking insurance coverage, including coverage for children, after learning the results of genetic tests. Both the NHGRI and CRG called for laws to protect employees and insured people. The NHGRI stated that the absence of genetic nondiscrimination laws could inhibit biomedical research.

GINA, however, does not protect all employees or insured persons from genetic discrimination. Whereas GINA applies to all individual and group insurance plans, the legislation applies only to employers with 15 or more employees. GINA also has several additional employment exceptions. Although GINA prohibits employers from asking employees about family medical history, the employer may collect information about employees' family medical history if such information is overheard by managers or other employees or if the employee volunteers the information. Employers may also collect family medical history about employees from public records, such as obituaries of family members. In addition, GINA permits group health insurance plans to request family medical information if the information is collected in association with a health or wellness plan as long as no financial penalty or incentive is attached to that plan.

■ Primary Source Connection

The Genetic Information Nondiscrimination Act of 2007 details efforts to establish protections for the public against genetic discrimination. The Introductory Remarks of Senator Edward Kennedy (D-MA) are presented, in which he urges the Senate to approve the bill.

Introductory Remarks of Senator Edward Kennedy (D-MA) on the Genetic Information Nondiscrimination Act

Mr. President, it is a privilege to introduce the Genetic Information Nondiscrimination Act of 2007. It is an honor to join Senator Snowe, Senator Enzi, Senator Dodd, Senator Harkin, Senator Gregg, and other members of our committee in support of this needed legislation.

I especially commend Senator Snowe for her leadership in this effort to establish protections for the public against genetic discrimination. It is now over a decade since Senator Snowe first introduced legislation on the issue. It passed the Senate 98–0 in the last Congress, and I am very hopeful we can work with our colleagues in the House and enact it into law, so that our people will finally have the protections they need against the misuse of genetic information.

In this century of the life sciences, much of what we learn through biomedical research is being translated into new treatments and cures, and nowhere is the explosion of scientific progress more apparent than in the field of genetics. Four years after the remarkable achievement of discovering the sequence of the human genome, clinical testing is now possible for over a thousand genetic diseases. It has led to rapid growth in the field of personalized medicine, in which patients' treatment and care is individualized according to their genetic makeup.

In the absence of federal protections, however, patients fear that undergoing genetic tests may lead to disqualification from future insurance coverage, or that an employer will fire them or deny a promotion based on the results of a genetic test. The consequence is that many Americans are choosing not to be tested, and are declining to participate in clinical trials so important for the development of new treatments.

Discrimination based on genetics is just as wrong as discrimination based on race or gender. Our bill provides specific protections for citizens against genetic discrimination. It prohibits health insurers from picking and choosing their customers based on genetics. Employers cannot fire or refuse to hire persons because of their genetic characteristics. It enables Americans to benefit from better health care through the use of genetic information, without the fear that it will be misused against them.

It is difficult to imagine information more personal or more private than a person's genetic makeup. It should not be shared by insurers or employers, or be used in making decisions about health coverage or a job. It should only be used by patients and their doctors to make the best diagnostic and treatment decisions they can.

In the near future, genetic tests will become even cheaper and more widely available. If we don't ban discrimination now, it may soon be routine for employers to use genetic tests to deny jobs to employees, based on their risk for disease.

If Congress enacts clear protections against genetic discrimination in employment and health insurance, all Americans will be able to enjoy the benefits of genetic research, free from the fear that their personal genetic information will be misused. If Congress fails to make sure that genetic information is used only for legitimate purposes, we may well squander the vast potential of genetic research to improve the nation's health.

The bill that we are considering today has been unanimously approved by the full Senate in the past two Congresses. We passed it 95–0 in the 108th Congress, and 98–0 in the 109th Congress. It had over 240 cosponsors in the House in both Congresses, but the leadership refused to bring it to a vote.

As President Bush himself has said, "Genetic information should be an opportunity to prevent and treat disease, not an excuse for discrimination. Just as our nation addressed discrimination based on race, we must now prevent discrimination based on genetic information."

We are closer than ever to enactment. I urge the Senate to approve the bill, and this time, I think we will finally see it become law.

Edward Kennedy

KENNEDY, EDWARD. "INTRODUCTORY REMARKS OF
SENATOR EDWARD KENNEDY (D-MA) ON THE GENETIC
INFORMATION NONDISCRIMINATION ACT,"
JANUARY 22, 2007.

SEE ALSO   *Bioethics; Genetic Bill of Rights; Genetic Privacy; Genetic Technology; Individual Privacy Rights; Informed Consent*

BIBLIOGRAPHY

Books

Atkinson, Paul, Peter E. Glasner, and Margaret M. Lock, eds. *Handbook of Genetics and Society: Mapping the New Genomic Era.* New York: Routledge, 2009.

Avise, John C. *Inside the Human Genome: A Case for Non-Intelligent Design.* Oxford: Oxford University Press, 2010.

Web Sites

"Privacy Legislation and Regulations." *Centers for Disease Control and Prevention (CDC).* http://www.cdc.gov/od/science/integrity/privacy/ (accessed November 1, 2011).

"Public Health Genomics." *Centers for Disease Control and Prevention (CDC).* http://www.cdc.gov/genomics/ (accessed November 1, 2011).

Joseph P. Hyder

Genetic Engineering

■ Introduction

Genetic engineering involves the transfer of genes from one species to another. It is used mainly to manufacture proteins that have medical and other applications. The gene coding for the protein of interest is transferred to a host cell, which could be a bacterium, insect, plant, or mammalian cell. This host cell is cultured in an industrial-scale fermenter, where it turns out copies of the desired protein, treating the foreign gene as one of its own. With widespread impact in medicine and agriculture, genetic engineering is one of the most significant technologies to emerge from the 1953 discovery of the structure of DNA by Francis Crick (1916–2004) and James Watson (1928–).

When an organism, be it an animal, plant, or bacterium, receives a foreign gene, it is said to be genetically-modified (GM). Sometimes the GM organism itself, rather than the product it makes, is of interest. For instance, genes for pest resistance can be transferred to important crop plants such as soy and cotton to make agriculture more productive. Because GM organisms do not occur in nature, their existence raises concern over whether they pose a health hazard if consumed or whether they are a threat to the environment.

■ Historical Background and Scientific Foundations

Genetic engineering was developed in the early 1970s by Paul Berg (1926–) and Herbert Boyer (1936–) of Stanford University and by Stanley N. Cohen (1935–) of the University of California at Berkeley. Patents on genetic engineering technology have earned the two universities millions of dollars since they were issued in 1981, coming from worldwide sales of products such as human insulin and hepatitis B vaccine. Another milestone was the discovery of restriction enzymes, which earned Werner Arber (1929–), Hamilton Smith (1931–), and Daniel

Nathans (1928–) of Johns Hopkins University the Nobel Prize for Physiology or Medicine in 1978. Recombinant insulin, manufactured by genetic engineering, was first approved by the United States Food and Drug Administration in 1976 for the treatment of diabetes.

Genetic engineering is a cut-paste-copy process. Restriction enzymes are used to cut the gene of interest out of the donor organism. Alternatively, the relevant stretch of DNA can be synthesized. It is then pasted, or inserted, into a vector, which carries the gene of interest into the host cell. This stage creates a GM version of the host cell containing recombinant DNA. Growing the host cell in a fermenter then allows it to manufacture

---

WORDS TO KNOW

DNA: The scientific abbreviation for deoxyribonucleic acid. Found within every living cell and carrying the organism's genetic information, DNA consists of two long chains of amino acid nucleotides; the chain is twisted into a double helix shape. The strands of the chain are joined by means of hydrogen bonds. The specific sequence of the DNA is what defines genetic characteristics.

RECOMBINANT PROTEIN: A protein produced by recombinant deoxyribonucleic acid (rDNA), which is itself DNA produced by transferring a gene from one species to another.

RESTRICTION ENZYME: Bacterial enzymes that can recognize specific parts of a DNA sequence and snip the sequence at this point, thereby cutting out the sequence of a particular gene, in effect acting as a pair of very precise biochemical scissors.

TRANSGENIC: An organism containing DNA from another species.

VECTOR: In genetic engineering, an intermediary DNA molecule, typically within a plasmid, virus, or bacterium, which is used to carry a foreign gene into a host.

many copies of the protein for which the foreign gene codes. Alternatively, the foreign gene may be inserted into a multicellular organism, creating a transgenic plant or animal. The bacterium *Agrobacterium tumiefaciens* is often used as a vector in plant genetic engineering, while viruses can be used to carry foreign genes into a range of animals including mice and sheep.

Although genetic engineering has become an almost routine operation, there are a number of technical hurdles involved. Host cells or organisms that take up a foreign gene and express it as one of their own genes are said to be transformed. The aim is to find a vector that can transform the host with high efficiency, which will maximize the yield of the desired product. To distinguish which host cells, or organisms, have been transformed and which have not, a marker gene is usually transferred along with the gene of interest. This is often a gene for antibiotic resistance. If a host cell has been transformed, then it contains the foreign gene and a new antibiotic resistance gene. These cells will grow into colonies on medium containing the antibiotic. The non-transformed cells do not have an antibiotic resistance gene, so they will not grow into colonies on the same medium.

■ Impacts and Issues

Genetic engineering has significant applications in medicine, food, and agriculture. It has led to the manufacture and clinical use of many new recombinant proteins, vaccines, and monoclonal antibodies. Proteins include enzymes and hormones, and they are used to treat patients with deficiency diseases. For instance, Type 1 diabetes is characterized by the lack of the hormone insulin, and hemophilia is defined by lack of Factor VIII, the blood clotting protein. There are also a number of genetic disorders involving deficiency of a vital enzyme. One example is Gaucher's disease, in which a lack of acid-beta-glucosidase leads to an enlarged liver and spleen, as well as neurological problems. Previously, these therapeutic proteins were either unavailable or had to be extracted from animal or human tissue, which carries a risk of contamination. Thus many patients with hemophilia became infected with HIV when their Factor VIII was extracted from infected blood. Similarly, others have contracted variant Creutzfeldt-Jakob disease, a fatal brain disorder, from treatment with growth hormone extracted from infected pituitary glands. Recombinant proteins eliminate this risk of contamination. Furthermore,

Leaves from genetically engineered poplar trees. The leaf on the left is an average leaf from an unmodified control tree and the one on the right was genetically engineered to be larger and will be used to identify genes that affect gene growth and productivity.
© *AP Images/John Gress.*

genetic engineering offers completely new treatments for some diseases. Beta-interferon is used in multiple sclerosis, whereas several monoclonal antibodies are being used in the treatment of rheumatoid arthritis and cancer. The main drawback of recombinant drugs is that they are very expensive, compared to their small molecule counterparts such as aspirin or penicillin. This has led to some health systems restricting their use on cost ground, meaning that these drugs do not reach all of those who could benefit from them.

Plant genetic engineering generally aims to produce plants with improved characteristics, such as pest resistance or resistance to harsh conditions such as drought. For example, the bacterium *Bacillus thuringiensis* (Bt) produces a toxin that kills certain species of insect, including mosquitoes and blackfly. The gene for this toxin has been transferred to many crop plants, including cotton, corn, and potato, and these have been widely planted around the world. Plants can also be genetically engineered for resistance to herbicides. Thus crops can be safely sprayed, killing weeds but leaving the crop plant unharmed. GM plants, with fewer losses to pests and disease, could increase the world's food supply. Meanwhile, genetic engineering is also having an impact in the food industry. One product is GM vegetarian cheese. Chymosin, an enzyme used in cheese making, is traditionally produced from the stomach of a calf and is therefore unacceptable to vegetarians. It is possible, however, to use a synthetic version of the chymosin gene to create a recombinant version of the enzyme for use in the manufacture of a cheese with no animal-derived ingredients.

Genetic engineering has always been a controversial technology. GM foods are particularly unpopular and many stores and restaurants declare that they will not sell or use them. There is concern that the antibiotic resistance genes used in selecting transformed cells could enter the human gut when GM food is consumed. This might encourage the growth of antibiotic resistance, which is already a major public health problem. Plant genetic engineering has led to increasing dominance of big companies in agriculture, possibly at the expense of small farmers and organic cultivation methods. There are also environmental implications in introducing GM plants. If they are made fitter by introducing disease resistance genes, and if these genes spread to weeds, it could make these unwanted species increasingly hard to kill. It could take many years before it is known what impact GM plants have on an ecosystem, by which time it would be hard to remove them.

SEE ALSO *Agricultural Biotechnology; Bioethics; Biotechnology; Drought-Resistant Crops; Genetic Technology; Genetically Modified Crops; Genome*

BIBLIOGRAPHY

Books

Atkinson, Paul, Peter E. Glasner, and Margaret M. Lock. *Handbook of Genetics and Society: Mapping the New Genomic Era*. London: Routledge, 2009.

Australian Farm Institute. *Can Agriculture Manage a Genetically Modified Future?* Surrey Hills, New South Wales, Australia: Australian Farm Institute, 2011.

Berry, Roberta M. *The Ethics of Genetic Engineering*. London: Routledge, 2007.

Dale, Jeremy, and Malcolm von Schantz. *From Genes to Genomes: Concepts and Applications of DNA Technology*, 2nd ed. Chichester, UK: John Wiley & Sons, 2007.

Harding, Stephen E. *Biotechnology & Genetic Engineering Reviews*. Chicago: Nottingham University Press, 2011.

Hayes, Lynn A., Karen R. Krub, and Jill E. Krueger. *Farmers' Guide to GMOs*. Saint Paul, MN: Farmers' Legal Action Group, 2008.

Hodge, Russ. *Genetic Engineering: Manipulating the Mechanisms of Life*. New York: Facts On File, 2009.

Peacock, Kathy Wilson. *Biotechnology and Genetic Engineering*. New York: Facts on File, 2010.

Shanks, Pete. *Human Genetic Engineering: A Guide for Activists, Skeptics, and the Very Perplexed*. New York: Nation Books, 2005.

Web Sites

United States Department of Energy (DOE) Human Genome Program. "PRIMER: Genomics and Its Impact on Science and Society: The Human Genome Project and Beyond (2008)." *Biological and Environmental Research Information System BERIS)*. http://www.ornl.gov/sci/techresources/Human_Genome/publicat/primer/index.shtml (accessed September 7, 2011).

Susan Aldridge

Genetic Privacy

Introduction

Genetic privacy involves maintaining the confidentiality of a person's genetic information. Genetic testing or an evaluation of a family's medical history may reveal genetic information about existing or potential diseases or disorders. People may desire to keep genetic information confidential for a variety of reasons, including avoiding discrimination by employers, insurers, or others; avoiding the stigma associated with certain diseases or disorders; or general privacy concerns. The collection of genetic information in government-controlled databases raises additional privacy concerns among civil libertarians.

Historical Background and Scientific Foundations

Genes, distinct sequences of nucleotides, form part of a chromosome and define the characteristics of living organisms. Each gene codes for a specific trait and may be passed on to offspring. Mutations, variations within the deoxyribonucleic acid (DNA) of a gene, may result in a disease or disorder or indicate an increased risk of disease.

Genetic privacy concerns center on maintaining confidential information about a person's genetic makeup in order to avoid discrimination, stigma, or other unwanted attention based on genetic information. Information about a person's genetic makeup may be determined from genetic testing or through an evaluation of family health history.

Genetic testing involves detecting genetic mutations and usually involves a simple blood, skin, or hair test. Most genetic tests measure the presence and amount of key proteins or metabolites or determine mutations in chromosomes or DNA. As of 2011, researchers have developed more than 1,000 genetic tests that may check for numerous diseases or genetic disorders, including breast cancer, colon cancer, cystic fibrosis, and sickle cell anemia. A genetic test may test for a single gene disorder—the presence of which is highly predictive of developing that particular disease or disorder. A genetic test may also test for susceptibility genes, which only indicate an increased risk of disease.

Impacts and Issues

Concerns over genetic privacy may have an adverse effect on an individual's health. In 2007 Francis Collins (1950–), then serving as director of the National Institutes of Health (NIH) National Human Genome Research Institute (NHGRI), stated that genetic privacy concerns led many patients to seek genetic testing under fictitious names or ask physicians to lie about test results so insurers would not discover the results. Collins stated that many other individuals did not seek genetic testing at all out of fear that the information may be disclosed to insurers or employers.

Many of the people who choose not to submit to genetic testing may miss discovering an increased susceptibility to a disease for which medical treatment may reduce their risk of developing the disease. Early genetic testing may enable physicians to prevent or treat a disease better. People with a genetic test that shows an increased likelihood of developing colon cancer, for example, may greatly reduce their risk of developing colon cancer by undergoing annual colon exams and having any polyps surgically removed.

Maintaining genetic privacy, however, often requires doing more than keeping an individual's genetic test results private. Because a person's genetic predisposition to certain diseases or disorders may be deduced from that person's family history, maintaining genetic privacy often requires maintaining the genetic privacy of family members, too. If an insurer obtains genetic information from a parent that identifies the parent as a carrier of a genetically inheritable disease, then the insurer could refuse to insure that person's children because they are at an increased risk of developing the disease.

European Union (EU) laws and regulations treat genetic privacy as a human rights issue. The EU Charter of Fundamental Rights and the Council of Europe Convention on Human Rights and Biomedicine guarantee genetic privacy and prohibit genetic discrimination. The EU Universal Declaration on the Human Genome and Human Rights also asserts that genetic privacy is a human right. Additionally, EU courts have interpreted the EU's broad personal data privacy laws as guaranteeing genetic privacy.

In the United States, the Genetic Information Nondiscrimination Act of 2008 (GINA) prohibits genetic discrimination in some circumstances. GINA does not guarantee genetic privacy per se: The legislation prohibits employers with 15 or more employees from requiring genetic testing or requesting genetic information, such as family medical history, for use in hiring, promoting, or firing employees. GINA also prohibits insurance companies from requiring genetic testing or requesting family medical history for use in making decisions regarding coverage or premium amounts. GINA does not prohibit collecting genetic information on employees or insured

Horacio Pietragalla shows a photo of himself with his mother Liliana Corti in Buenos Aires, Argentina. Pietragalla was kidnapped from his mother when he was a few months old and given to another family who discovered his true identity in May 2003 at age 27. Valuing the right to truth over the right to privacy, Argentina's Congress has authorized the forced extraction of DNA from suspected orphans of Argentina's "Dirty War" who refuse to help identify their birth parents. © *AP Images/Natacha Pisarenko.*

WORDS TO KNOW

DEOXYRIBONUCLEIC ACID (DNA): A self-replicating, double-helix structure of nucleobases that is the main component of chromosomes and serves as the basis of heredity.

DNA DATABASE: A computerized database that stores DNA profiles of individuals, typically criminals.

GENE: The functional unit of genetic inheritance that is composed of a specific sequence of nucleotides in DNA or RNA.

---

persons when the information is volunteered by an individual or gleaned from public sources, such as obituaries.

In addition to privacy issues related to genetic testing, the use of government-controlled DNA databases also raises privacy concerns. The United Kingdom established the first national DNA criminal database in 1995. New Zealand, France, and the United States followed shortly thereafter. In the early 2010s most developed countries maintain DNA databases, and their use is a cause of a special privacy concern with regards to taking and maintaining the DNA of persons arrested but never convicted of crimes as well as juveniles. The use of DNA criminal databases to conduct familial searches to link the family members of suspects to a crime is also controversial. To date, most countries maintain databases with DNA samples of criminals and suspects only. In 2005 the incoming government of Portugal proposed a DNA database containing samples from every Portuguese citizen. Following public objections, the government limited the database to criminals.

SEE ALSO *Bioethics; DNA Databases; Genetic Bill of Rights; Genetic Discrimination; Genetic Technology; Individual Privacy Rights; Informed Consent*

BIBLIOGRAPHY

Books

Atkinson, Paul, Peter E. Glasner, and Margaret M. Lock, eds. *Handbook of Genetics and Society: Mapping the New Genomic Era.* London: Routledge, 2009.

Avise, John C. *Inside the Human Genome: A Case for Non-Intelligent Design.* Oxford: Oxford University Press, 2010.

Web Sites

"Privacy Legislation and Regulations." *Centers for Disease Control and Prevention (CDC).* http://www.cdc.gov/od/science/integrity/privacy/ (accessed November 1, 2011).

"Public Health Genomics." *Centers for Disease Control and Prevention (CDC).* http://www.cdc.gov/genomics/ (accessed November 1, 2011).

Joseph P. Hyder

Genetic Technology

■ Introduction

Genetic technology represents the synthesis of genetics, molecular biology, information technology, biotechnology, and environmental sciences, with the combined goal of improving human life. At the most fundamental level, genetic technology seeks to discern the expression of each gene in the genome; to understand and expand upon naturally occurring genetic variations; to identify genetic malfunctions and prevent or correct their expression; to identify, isolate, expand, and modify or engineer desired genes and genetic sequences; and to insert those novel genes into new hosts for the purpose of creating targeted improvements.

Humans have been creating and exploiting genetic technology since the first farmers selectively bred crop plants to increase yields. Animal breeding pairs have been chosen based on size, strength, or work capacity. With rapid advances in genetic technologies, scientists are able to identify genes expressing desired traits, amplify and recombine them, and insert them into new hosts to serve a multitude of purposes ranging from the detection and prevention of heritable genetic disorders to diagnosis and genetic treatments for cancer and other debilitating diseases. An ever-expanding set of instruments and techniques are available to research scientists working in the field of genetics, enabling them to isolate and express individual genes and facilitate their combination with the DNA of unrelated organisms, often from differing species, to create beneficial outcomes and life-enhancing technologies.

■ Historical Background and Scientific Foundations

Austrian monk Gregor Mendel (1822–1864) published founding genetic research on trait inheritance in pea plants in 1866. In 1869 Johann Friedrich Miescher (1844–1895) discovered a weak acidic substance in cell nuclei he termed nuclein. In 1902 Walter Sutton (1877–1916) described the chromosome theory of inheritance and Theodor Boveri (1862–1915) noted differences between chromosomes. Boveri and Sutton share credit for the chromosome theory of inheritance.

In 1944 Oswald Avery (1877–1955), Maclyn McCarty (1911–2005), and Colin MacLeod (1909–1972) discovered that DNA transmitted genetic inheritance. In 1953 James Watson (1928–) and Francis Crick (1916–2004) reported that DNA has a double helix shape containing base pairs encoding genetic information. In 1966 the genetic code was discovered. Hamilton Smith (1931–) purified the first restriction enzyme in 1970.

Paul Berg (1926–) created the first recombinant DNA molecule in 1970. In 1973 Stanley N. Cohen (1935–), Annie Chang, and Herbert Boyer (1936–) used recombinant DNA technology to create the first transgenic organisms. Fred Sanger (1918–) developed DNA sequencing technology in 1977; in 1980 the United States Supreme Court ruled that genetically engineered organisms could be patented.

The first commercially available genetically engineered pharmaceutical was a form of human insulin produced by bacteria, publicly marketed in 1982, after approval by the U.S. Food and Drug Administration (FDA). The first genetically engineered vaccine for humans, used for the prevention of hepatitis B, was approved by the FDA in 1986. In 1987 the United States Patent and Trademark office determined that non-human genetically modified animals could be patented.

In 1985 polymerase chain reaction (PCR) technology was discovered by Kary Mullis (1944–). It created the ability for researchers to amplify millions of copies of a minute sample of DNA. PCR technology is used widely for the diagnosis of human genetic disorders and diseases such as HIV/AIDS. PCR has significant forensic utility for connection of suspected perpetrators to victims or crime scenes through DNA comparisons of hair, skin, blood, semen, or other tissue samples.

The Human Genome Project (HGP) began in 1990; its goals were to identify all genes in human DNA; to determine the sequence of each of the three billion human DNA base pairs and to establish a relational database for the storage of this information; to create tools for advanced analysis of data generated; to make the newly developed technologies available to the private sector; and to manage ethical, legal, and social issues (ELSI) stemming from the HGP. A working draft of the human genome was announced in 2000. The human genome was fully sequenced in 2003. The HGP was an international effort funded by the U.S. National Institutes of Health and the U.S. Department of Energy. In addition to sequencing the human genome, those of the bacterium *Escherichia coli*, the fruit fly, the rat, the mouse, and the nematode *Caenerhabditis elegans* were sequenced.

The first somatic gene transfer experiments were initiated in 1991; W. French Anderson (1936–) was awarded a patent for human *ex vivo* gene manipulations in 1995. In 1998 Anderson's group submitted a proposal to conduct experiments on human germline manipulation to the National Institutes of Health's Recombinant DNA Advisory Committee. By 1999 development of the first artificially engineered human chromosomes was underway. Researchers successfully genetically engineered human stem cells in 2003.

In 2004 the births of children designed through the use of pre-implantation genetic diagnosis (PGD) to donate stem cells for the treatment of older siblings with inherited genetic diseases was announced. PGD involves screening several-day-old in vitro fertilized embryos for the presence of a variety of potentially deadly genetically transmitted diseases. Healthy embryos are implanted in the female parent. The healthy newborn's cord blood is harvested at birth in order to treat an older sibling born with the deadly genetic disorder. PGD also allows for sex selection of offspring.

■ Impacts and Issues

Electrophoresis is a frequently-used genetic technology. It is relatively simple, comparatively inexpensive, and quite accurate. For routine separation of nucleic acids and proteins, one-dimensional electrophoresis is used; the two-dimensional form is used most commonly for DNA fingerprinting. Electrophoresis can be utilized to separate all of the proteins within highly complex molecules in order to facilitate accurate identification. It is used to identify tissue and other biological materials found at crime scenes. In addition, electrophoresis often is used to establish (or exclude) paternity as well as to identify viral infections, diseased genes, and the presence of cancer-causing cells, and to identify carriers of genetic diseases such as Huntington's Disease, sickle cell anemia, and Duchenne's Muscular Dystrophy.

WORDS TO KNOW

BIOINFORMATICS: From the National Center for Biotechnology Information (NCBI), "bioinformatics is the field of science in which biology, computer science, and information technology merge to form a single discipline. The ultimate goal of the field is to enable the discovery of new biological insights as well as to create a global perspective from which unifying principles in biology can be discerned."

ELECTROPHORESIS: A technique used to separate various molecules, such as proteins (RNA or DNA segments), according to size and molecular weight by moving them across a field, usually a gel, carrying an electrical charge.

GENE: Genes are made up of specific sequences of bases that contain the code determining how each protein must be made. The bases determine the function performed by the gene.

GENOME: This constitutes the entirety of an organism's DNA. All human cells, with the exception of red blood cells, contain the complete human genome. The simplest bacterial organisms have genomes containing about 600,000 base pairs; the human and mouse genomes each contain more than three billion base pairs.

RECOMBINANT DNA (rDNA) TECHNOLOGY: This constitutes the entire set of processes utilized in order to combine a specific DNA sequence from one organism with the DNA of a specific vector, usually a bacteria or virus.

Bioinformatics, which bridges biological sciences, information technology, and computer science, is a critical adjunct to the growth and development of genetic technology. Rapid developments in genetics and molecular biology have magnified the need for comprehensive, real-time data storage and retrieval systems capable of processing and organizing vast quantities of information. Researchers must be able to mine discrete pieces of information readily and accurately and compare large arrays and sequences. The National Center for Biotechnology Information (NCBI) was created in 1988 as a comprehensive computerized literature resource and data warehouse for the disciplines of bioinformatics and biomedicine. Research scientists require access to genetic, biological, and information technologies to better understand the roles of proteins and DNA in health and disease processes. Bioinformatics facilitates the analysis and interpretation of ever-increasing biological information, giving researchers critical tools for the understanding of the healthy functioning of organisms, which elucidates the processes leading to biological system malfunctions. Scientists then can create disease models to aid in the development of pharmaceutical and therapeutic protocols.

Recombinant DNA technology (rDNA) is critical to nearly every aspect of genetic technology. The process begins with the selection and localization of a desired gene, which is placed into a specific vector, typically a bacteria plasmid or a viral phage, and then cloned in order to make multiple copies of the desired sequence. When the desired DNA sequence is integrated into the DNA of the vector cell, it is said to have recombined, or joined together, with the DNA of the vector. The synthesized DNA then is inserted along with the vector into the cells of a new host organism. The host organism synthesizes new proteins, incorporating the recombined DNA.

Although rDNA techniques are an essential genetic technology research tool, they are also very important in biomedical diagnosis and treatment. Recombinant DNA techniques are used to create biologically based pharmaceuticals such as human insulin, human growth hormone, and follicle-stimulating hormone; they aid in detection of the presence of genetic mutations associated with specific diseases such as neurofibromatosis, retinoblastoma, and several types of cancer. rDNA techniques are used for identification of hereditary genetic disease carriers, such as Huntington's Disease, early onset Alzheimer's Disease, Duchenne's Muscular Dystrophy, sickle cell disease, and Tay-Sachs Disease, as well as a variety of fatal hereditary anemias. rDNA techniques are central to the evolving field of gene therapy for the treatment of vascular diseases, various forms of cancer, the arthritis disease spectrum, and cystic fibrosis.

SEE ALSO *Agricultural Biotechnology; Asilomar Conference on Recombinant DNA; Bayh-Dole Act of 1980; Biodiversity; Bioethics; Bioinformatics; Biotechnology; Biotechnology Organizations; Biotechnology Products; Biotechnology Valuation; Cell Therapy (Somatic Cell Therapy); Cloning; Corn, Genetically Engineered; Cotton, Genetically Engineered; Diamond v. Chakrabarty; DNA Sequencing; Gene Delivery; Gene Therapy; Genetic Bill of Rights; Genetic Engineering; Genetic Use Restriction Technology; Genetically Modified Crops; Genetically Modified Food; Genewatch; Genome; Genomics; Germ Line Gene Therapy; GloFish; Government Regulations; Human Embryonic Stem Cell Debate; Industrial Genetics; Microarrays; Molecular Biology; Molecular Farming; Oncogenomics; Patents and Other Intellectual Property Rights; Pharmacogenomics; Proteomics; Replication; Stem Cells; Terminator Technology; Therapeutic Cloning; Tissue Engineering; Transgenic Animals; Transgenic Plants; Translational Medicine; Vaccines; Viral Transfer; Yeast Artificial Chromosomes*

BIBLIOGRAPHY
Books

Atkinson, Paul, Peter E. Glasner, and Margaret M. Lock, eds. *Handbook of Genetics and Society: Mapping the New Genomic Era.* New York: Routledge, 2009.

Australian Farm Institute. *Can Agriculture Manage a Genetically Modified Future?* Surrey Hills, New South Wales: Australian Farm Institute, 2011.

Dale, Jeremy, Malcolm von Schantz, and Nick Plant. *From Genes to Genomes: Concepts and Applications of DNA Technology,* 3rd ed. Chichester, UK: John Wiley & Sons, 2011.

DeSalle, Rob, and Michael Yudell. *Welcome to the Genome: A User's Guide to the Genetic Past, Present, and Future.* Hoboken, NJ: Wiley-Liss, 2005.

Glick, Bernard R., Jack J. Pasternak, and Cheryl L. Patten. *Molecular Biotechnology: Principles and Applications of Recombinant DNA,* 4th ed. Washington, DC: ASM Press, 2010.

Hayes, Lynn A., Karen R. Krub, and Jill E. Krueger. *Farmers' Guide to GMOs,* 2nd ed. Saint Paul, MN: Farmers' Legal Action Group, 2008.

Hodge, Russ. *Genetic Engineering: Manipulating the Mechanisms of Life.* New York: Facts on File, 2009.

Peacock, Kathy Wilson. *Biotechnology and Genetic Engineering.* New York: Facts on File, 2010.

Shanks, Pete. *Human Genetic Engineering: A Guide for Activists, Skeptics, and the Very Perplexed.* New York: Nation Books, 2005.

Solway, Andrew. *Using Genetic Technology.* Chicago: Heinemann Library, 2009.

Watson, James D. *Recombinant DNA: Genes and Genomes: A Short Course,* 3rd ed. New York: W. H. Freeman, 2007.

Web Sites

"Celebrating a Milestone: FDA's Approval of First Genetically Engineered Product." *U.S. Food and Drug Administration (FDA).* http://www.fda.gov/AboutFDA/WhatWeDo/History/ProductRegulation/SelectionsFromFDLIUpdateSeriesonFDAHistory/ucm081964.htm (accessed November 13, 2011).

"DNA Manipulation." *Cold Spring Harbor Laboratory.* http://www.dnai.org/b/ (accessed November 13, 2011).

Pamela V. Michaels

Genetic Use Restriction Technology

■ Introduction

Genetic Use Restriction Technology (GURT), also called terminator gene technology, involves the insertion of a particular genetic sequence into the DNA of a plant seed. There are two types of GURT: v-GURT and t-GURT. For v-GURT to function, the inserted gene must be activated by a chemical catalyst. The mature plant produces non-viable seeds and cannot reproduce. In t-GURT, the inserted gene will not express the desired trait unless externally activated, typically by exposure to a particular chemical substance.

Proponents of GURT report that it could prevent leakage of engineered plant genes into nature or into non-genetically-modified (non-GM) crops and neighboring farms. GURT is also proposed as a means of preventing seed piracy or contract violation through the process of saving and replanting seeds from one year to the next. Seed trait patent holders also describe the technology as a means of creating business relationships with developing countries. Critics of terminator technology, also called the Technology Protection System, assert that GURT research has just one purpose: to increase profit margins for seed trait developers and patent holders by forcing farmers to buy new seed each year rather than saving seeds.

The United Nations Food and Agriculture Organization (FAO) has recommended an ongoing ban on deployment of terminator technology, particularly with food crops. The largest global seed trait developer and marketer, Monsanto, has agreed not to pursue GURT in developing food crops at present.

■ Historical Background and Scientific Foundations

In 1873 French bacteriologist Louis Pasteur (1822–1895) was awarded a patent for yeast. This was the first time that a living organism used for food production

was patented. The Plant Patent Act was passed in 1930 in the United States, making it possible to patent novel varieties of asexually reproduced plants. The patent was modified in 1954 to include hybrids, newly discovered wild seedlings, mutants, and cultivated plants, so long as they reproduced asexually. In 1975 Earl Patterson (1923–1999) of the University of Illinois filed the first seed utility patent on maize. In 1980 the United States Supreme Court ruled that genetically-altered life forms could be patented, and in 1982 a genetically altered bacterium was patented. In 1985 the Patent and Trademark Office ruled that plants could be patented.

In 1982 scientists working for the Monsanto Company successfully genetically-modified plant cells; in 1996 Monsanto began marketing seeds created through the use of biotechnology. Monsanto and most other genetically engineered seed developing companies, particularly Bayer, Syngenta, Pioneer, and AstraZeneca, have all patented the technology that produces their genetically modified seeds. In 1986 the first genetically modified crop, tobacco, was field tested in Belgium, and in 1987 genetically-engineered tomatoes and tobacco were field tested in the United States.

As of 2011 there no legally-sanctioned GURTs in commercial use. GURTs are experimental technologies that either prevent reproduction of a plant variety or cancel/allow the expression of a specific engineered trait in a plant variety by deactivating or activating a genetic switch. The former is called v-GURT and involves restriction or elimination of plant reproductive capability. The latter is t-GURT, and it allows for the expression of the engineered trait solely by applying a specific process (such as spraying the plant with a designated chemical). If the trait is not catalyzed, the plant still grows to maturity but does not express the engineered trait.

Because there is not yet sufficient data on the safety and efficacy of GURTs, the FAO has recommended against the use of this technology. In 2000 the United Nations' Convention on Biological Diversity recommended that GURT should not be used "until

WORDS TO KNOW

AGRIBUSINESS: When farming is run as a large-scale industry; that is, crops or livestock are produced, processed, stored, and marketed, and equipment and supplies are manufactured and distributed.

PATENT: When intellectual property is created, such as the technology involved in the creation of the process for alteration of a plant by genetic engineering to make it resistant to a particular pathogen, the inventor protects the technology by patenting it. A patent guarantees exclusive rights for the manufacture and public distribution of the specific technology.

SEED PIRACY: Farmers engage in seed piracy if they save and replant patented seeds in violation of their lease or stewardship agreements with agribusiness companies.

SEED POLICE: The euphemistic term applied to employees of the Monsanto Corporation whose task it is to investigate, at the local level, possible allegations of seed piracy or seed saving.

TERMINATOR SEEDS: Refers to seeds that will grow into genetically-modified plants that have been made to be sterile and unable to reproduce.

appropriate, authorized and strictly controlled scientific assessments with regard to, inter alia, their ecological and socioeconomic impacts and any adverse effects for biological diversity, food security and human health have been carried out in a transparent manner and the conditions for their safe and beneficial use validated." This stance was reaffirmed in 2006, with another five-year ban recommended for GURT.

■ Impacts and Issues

Between 1994 and 2000, applications were made for more than 30 patents, spanning more than 80 countries, for gene use restriction technologies, with the greatest number submitted by Syngenta and Delta & Pine Land/Monsanto. The most commonly listed reasons for the development of GURT are protection of gene technology, prevention of genetically modified seeds from entering the general crop population, and creation of new markets for seed development companies, especially in developing countries.

In 1999 Monsanto published an open letter from its Chairman and CEO, Robert Shapiro (1938–), to the Rockefeller Foundation President, Gordon Conway, making a commitment not to develop and market gene protection systems involving either sterile seed technologies or trait specific deactivation systems.

Critics of GURT technologies have termed them a potential economic threat to small farmers engaged solely in production of food crops, particularly those in developing countries where seed-saving and traditional selective breeding are considered important business practices.

For many decades, a substantial percentage of farmers have utilized hybridized, non-genetically engineered seeds to produce higher yielding crops; such seeds are sterile and cannot be used for replanting. Hybridized seeds need to be repurchased annually.

A variety of lobbying, popular press, and public-interest groups have expressed opposition to GURT and have suggested that frequent mergers of large biotechnology companies could create monopolies exerting power and control over farmers. The implication is that genetically-modified seeds and those with so-called terminator genes could be produced and marketed in ways that might become unaffordable for small farmers, thereby supporting only agribusiness and corporate farms.

Opponents of sterile seed (v-GURT) technology express concern that traits from those genetically-engineered plants could be transmitted to wild plant varieties or onto neighboring farms, inadvertently rendering affected plants unable to reproduce and possibly resulting in significant crop yield losses.

Biotechnology companies including Monsanto, Syngenta, Pioneer, Bayer and Dow AgroSciences engaged in the development and marketing of genetically-engineered seeds, obtain intellectual property patents for the seeds' technology. When they market the seeds to growers, they engage in stewardship agreements in which the seeds and the patented biotechnology that they represent are leased as part of a contract. The purchaser agrees not to reuse the seeds past one growing season and commits to allowing representatives of the producing corporation, sometimes referred to as seed police, onto the farmed land in order to ascertain that the seeds are not saved and replanted. Monsanto reports that more than one-quarter of a million farm businesses contract with them yearly.

One way of assuring that seeds cannot be reused and intellectual patents cannot be violated is through the use of GURT. The major biotechnology seed development companies continue to state that they will not develop and commercialize sterile seed technologies for use in food crops, but they do not rule out using t-GURT to limit the propagation of specific engineered traits in otherwise fertile plant materials. The technology for activating the traits could be restricted to those growers specifically contracting and paying for their use. T-GURT technology may be helpful in preventing the movement of genetically engineered traits into other crops because it requires specific procedures for activation.

SEE ALSO *Agricultural Biotechnology; Biodiversity Agreements; Bioethics; Biological Confinement; Corn, Genetically Engineered; Environmental Biotechnology; Gene Banks; Genetic Engineering; Genetically Modified Crops; Monsanto Canada Inc. v. Schmeiser; Patents and Other Intellectual Property Rights; Plant Patent Act of 1930; Plant Variety Protection Act of 1970; Terminator Technology*

BIBLIOGRAPHY

Books

Danforth, Arn T, ed. *Corn Crop Production: Growth, Fertilization and Yield.* New York: Nova Science Publishers, 2009.

Kesan, Jay P. *Agricultural Biotechnology and Intellectual Property Protection Seeds of Change.* Wallingford, UK: CABI, 2007.

Somsen, Han. *The Regulatory Challenge of Biotechnology Human Genetics, Food and Patents.* Northampton, MA: Edward Elgar, 2007.

Swanson, Timothy M. *Biotechnology, Agriculture, and the Developing World: The Distributional Implications of Technological Change.* Northampton, MA: Edward Elgar, 2002.

Periodicals

Bustos, Keith. "Sowing the Seeds of Reason in the Field of the Terminator Debate." *Journal of Business Ethics* 77, no. 1 (January 2008): 65–72.

Web Sites

"Gene Use Restriction Technologies." *Foundation for Biotechnology Awareness and Education.* http://fbae.org/2009/FBAE/website/our-position-gene-use-restriction-technologies.html (accessed November 4, 2011).

"Is Monsanto Going to Develop or Sell 'Terminator' Seeds?" *Monsanto Company.* http://www.monsanto.com/newsviews/Pages/terminator-seeds.aspx (accessed November 4, 2011).

Pamela V. Michaels

Genetically Modified Crops

■ Introduction

Genetically modified (often known as GM) crops are plants grown for food or other useful purposes that contain a genome altered by technology, usually including beneficial genes inserted from another species. The transgene is transferred into the plant by genetic engineering. The transgenes confer some desirable trait upon the crop plant, such as insect resistance or herbicide resistance, which helps to maximize yields and so increase the world's food supply. GM plants may also be created for increased nutritional value, which can improve the quality of food. Another application of genetic engineering is to use plants as factories for drugs or vaccines, by transferring the appropriate genes to them.

The main GM crops in the early 2010s are soy, maize, cotton, and rapeseed (also known as canola). In 2010 GM crops were planted on 148 million hectares of agricultural land, up from 1.7 million hectares in 1996. GM crops have led to significant benefits such as reduced farm costs, increased profits, and a decrease in crop losses to pests and weeds, whereas newer crops, such as golden rice, offer more direct benefit to the consumer. However, GM crops are unpopular with the public, who fear potential health risks associated with transgenes. There are also concerns that GM crops may have adverse effects on the environment.

■ Historical Background and Scientific Foundations

Farmers have always tried to exchange genes between two plants to create plants that have desirable traits such as high yield or disease resistance. But this conventional plant breeding is limited to exchange of genes between the same, or very closely related, species. Genetic engineering techniques, invented in the early 1970s at Stanford University, opened up the possibility of transferring genes from very different species, such as bacteria,

into a plant. The resulting GM plants are truly novel, in that they could not exist in nature. The first commercially grown example of a GM plant was the Flavr Savr tomato, developed by the Californian company Calgene in the early 1990s. The tomato had been genetically modified to take longer to soften and decompose once picked, so it could safely be left to ripen on the vine. A similar GM tomato was used to make a tomato paste that was marketed in Europe shortly afterwards.

One of the most significant of the early developments in GM crops was the introduction of the toxin gene from the soil bacterium *Bacillus thuringiensis* (Bt) into plants. This toxin destroys the gut of many species of insect pest, and has been used on its own as a spray for biological pest control. Transfer of the Bt toxin gene led to the creation of GM tomato, potato, cotton, and other crops. Another landmark was the creation of GM plants resistant to the herbicide glyphosphate or Roundup, by Monsanto, which is the main company in plant genetic engineering. Herbicides have a tendency to kill crop plants as well as weeds. A field of herbicide-resistant crop plants, however, can be sprayed and the weeds will die, while the crop remains unscathed. The first generation of GM crops was developed for these improved agronomic characteristics. The focus in research has shifted to creating crops that have clear consumer benefit. Examples include golden rice (with its higher vitamin A content), allergen-free nuts, maize with enhanced amino acid content, and edible vaccines in bananas.

There are two main methods for introducing a transgene into a crop plant. The first involves using the bacterium *Agrobacterium tumifaciens*, which naturally infects plants, as a vector to carry the gene into plant cells. In the second, a gene gun is used to shoot particles coated with the DNA of the transgene into the plant cells. The cells that have taken up the transgene generally are identified through the presence of an antibiotic resistance marker gene, transferred alongside the transgene. The transformed cells then are grown into GM plants under very carefully controlled conditions. The resulting plants are tested thoroughly for their characteristics and cultivated in small-scale outdoor field trials before moving to large-scale cultivation.

■ Impacts and Issues

GM crops have had a huge impact on world agriculture. They are grown by 14 million farmers in 25 countries around the world, both developed and developing. In 17 of these countries, more than 50,000 hectares of GM crops are being grown, with the United States, Brazil, Argentina, and India leading the way. Less developed countries can benefit from any new technology that can increase food production and quality and lower food prices. GM crops promise these benefits and therefore could play a leading role in alleviating malnutrition in these countries. However, it could take some time for these benefits to be realized. In the early 2010s, much of the GM planting in less developed countries is used for animal feed. Furthermore, substantial amounts of GM crops are exported, depriving the population of land for local food production. Many less-developed countries lack the scientific expertise to assess the safety of GM crops and the legal and regulatory framework to

control them. There is therefore concern that a technology gap may develop, with those not able to plant GM crops being left behind. GM technology is currently in the hands of a few powerful multinational companies, such as Monsanto, Syngenta, and Bayer CropScience, who have control of the supply of GM seeds and methods of cultivation. It is not clear what impact this dominance may have on small farmers' livelihoods, and more traditional farming methods.

WORDS TO KNOW

GENE GUN: A device that shoots tungsten or gold particles coated with DNA into plant cell tissue in genetic engineering experiments.

GOLDEN RICE: A GM rice that contains genes that make vitamin A, thereby increasing its nutritional value.

TRANSFORMED: A cell that has taken up a transgene in a genetic engineering experiment.

TRANSGENE: The gene that is transferred into a host crop plant in genetic engineering.

A Burkinabe farming technician of the National Environment and Research Institute (INERA) inspects transgenic cotton in a single field in Fada Ngourma, eastern Burkina Faso. The state has started a test program of transgenic cotton in Fada Ngourma.   *© Issouf Sanogo/AFP/ Getty Images.*

In 1998 Arpad Pusztai (1930–) of the Rowett Research Institute, Aberdeen, Scotland, published research suggesting that GM potatoes were toxic to rats. The report was apparently a factor in destroying public confidence in GM foods and many safety concerns and questions still surround their consumption. First, food allergies seem to be on the increase; it is possible that genetic modification could compound the problem by introducing new allergens, or other harmful factors, into food. Second, if the GM food still contains the antibiotic resistance genes used in the selection of the transformed cells, then these might enter the human gut. Antibiotic resistance is already a growing public health issue, and it is possible that exposure through GM food could make the problem worse. There are many other questions to resolve, such as whether and how GM food should be labeled, whether imports should be restricted or banned, and how to stop GM foods from being mixed with non-GM, especially organic, food. Of course the public may, in time, grow used to the idea of GM food, and start to accept it. This is more likely if GM foods that have some direct benefit to them are introduced.

There are also some potential environmental risks associated with GM crops. When pests are exposed to a toxin, some inevitably evolve a resistance to it. Already, a species of Diamondback moth resistant to Bt cotton has been identified. There is also the possibility that transgenes could escape from cultivated crops to wild plants. Suppose, for instance, that the Roundup resistance gene was to transfer to a weed. This would create a superweed that could not be killed by Roundup and would then compete successfully with crop plants for light and nutrients. The magnitude of these risks is still not completely understood. It is important to have strict monitoring of GM crops in place so that any problems, such as transgene spread, can be avoided or dealt with promptly. There is strong anti-GM campaigning in many countries. Often supermarkets and restaurants ban GM foods and ingredients. In some instances, campaigners have sabotaged GM field trials, pulling up and destroying plants. This sometimes makes companies involved in GM reluctant to be wholly transparent about their work, particularly where the location of trials is concerned.

■ Primary Source Connection

Biotechnology offers powerful tools and techniques that can be useful in agriculture and food production. The Food and Agriculture Organization published this statement relating to the use, methods, and application of biotechnology during the "Codex Alimentarius Ad Hoc Intergovernmental Task Force on Foods Derived from Biotechnology" meeting that took place in Chiba, Japan in 2000. Potential health and environmental risks associated with biotechnology are discussed as well.

FAO Statement on Biotechnology

Biotechnology provides powerful tools for the sustainable development of agriculture, fisheries and forestry, as well as the food industry. When appropriately integrated with other technologies for the production of food, agricultural products and services, biotechnology can be of significant assistance in meeting the needs of an expanding and increasingly urbanized population in the next millennium.

There is a wide array of "biotechnologies" with different techniques and applications. The Convention on Biological Diversity (CBD) defines biotechnology as: *"any technological application that uses biological systems, living organisms, or derivatives thereof, to make or modify products or processes for specific use"*.

Interpreted in this broad sense, the definition of biotechnology covers many of the tools and techniques that are commonplace in agriculture and food production. Interpreted in a narrow sense, which considers only the new DNA techniques, molecular biology and reproductive technological applications, the definition covers a range of different technologies such as gene manipulation and gene transfer, DNA typing and cloning of plants and animals.

While there is little controversy about many aspects of biotechnology and its application, genetically modified organisms (GMOs) have become the target of a very intensive and, at times, emotionally charged debate. FAO recognizes that genetic engineering has the potential to help increase production and productivity in agriculture, forestry and fisheries. It could lead to higher yields on marginal lands in countries that today cannot grow enough food to feed their people. There are already examples where genetic engineering is helping to reduce the transmission of human and animal diseases through new vaccines. Rice has been genetically engineered to contain pro-vitamin A (beta carotene) and iron, which could improve the health of many low-income communities.

Other biotechnological methods have led to organisms that improve food quality and consistency, or that clean up oil spills and heavy metals in fragile ecosystems. Tissue culture has produced plants that are increasing crop yields by providing farmers with healthier planting material. Marker-assisted selection and DNA fingerprinting allow a faster and much more targeted development of improved genotypes for all living species. They also provide new research methods which can assist in the conservation and characterization of biodiversity. The new techniques will enable scientists to recognize and target quantitative trait loci and thus increase the efficiency of breeding for some traditionally intractable agronomic problems such as drought resistance and improved root systems.

However, FAO is also aware of the concern about the potential risks posed by certain aspects of biotechnology. These risks fall into two basic categories: the effects on human and animal health and the environmental consequences. Caution must be exercised in order to reduce the risks of transferring toxins from one life form to another, of creating new toxins or of transferring allergenic compounds from one species to another, which could result in unexpected allergic reactions. Risks to the environment include the possibility of outcrossing, which could lead, for example, to the development of more aggressive weeds or wild relatives with increased resistance to diseases or environmental stresses, upsetting the ecosystem balance. Biodiversity may also be lost, as a result of the displacement of traditional cultivars by a small number of genetically modified cultivars, for example.

FAO supports a science-based evaluation system that would objectively determine the benefits and risks of each individual GMO. This calls for a cautious case-by-case approach to address legitimate concerns for the biosafety of each product or process prior to its release. The possible effects on biodiversity, the environment and food safety need to be evaluated, and the extent to which the benefits of the product or process outweigh its risks assessed. The evaluation process should also take into consideration experience gained by national regulatory authorities in clearing such products.

Careful monitoring of the post-release effects of these products and processes is also essential to ensure their continued safety to human beings, animals and the environment.

Current investment in biotechnological research tends to be concentrated in the private sector and oriented towards agriculture in higher-income countries where there is purchasing power for its products. In view of the potential contribution of biotechnologies for increasing food supply and overcoming food insecurity and vulnerability, FAO considers that efforts should be made to ensure that developing countries, in general, and resource-poor farmers, in particular, benefit more from biotechnological research, while continuing to have access to a diversity of sources of genetic material. FAO proposes that this need be addressed through increased public funding and dialogue between the public and private sectors.

FAO continues to assist its member countries, particularly developing countries, to reap the benefits derived from the application of biotechnologies in agriculture, forestry and fisheries—through, for example, the network on plant biotechnology for Latin America and the Caribbean (REDBIO), which involves 33 countries. The Organization also assists developing countries to participate more effectively and equitably in international commodities and food trade. FAO provides technical information and assistance, as well as socio-economic and environmental analyses, on major global issues related to new technological developments. Whenever the need arises, FAO acts as an "honest broker" by providing a forum for discussion.

For instance, together with the World Health Organization, FAO provides the secretariat to the Codex Alimentarius Commission which has just established an ad hoc Intergovernmental Task Force on Foods Derived from Biotechnologies, in which government-designated experts will develop standards, guidelines or recommendations, as appropriate, for foods derived from biotechnologies or traits introduced into foods by biotechnological methods. The Codex Alimentarius Commission is also considering the labeling of foods derived from biotechnologies to allow the consumer to make an informed choice.

Another example is the FAO Commission on Genetic Resources for Food and Agriculture, a permanent intergovernmental forum where countries are developing a Code of Conduct on Biotechnology aimed at maximizing the benefits of modern biotechnologies and minimizing the risks.

The Code will be based on scientific considerations and will take into account the environmental, socio-economic and ethical implications of biotechnology. As in applications in medicine, these ethical aspects warrant responsible consideration. Therefore the Organization is working towards the establishment of an international expert committee on ethics in food and agriculture.

FAO is constantly striving to determine the potential benefits and possible risks associated with the application of modern technologies to increase plant and animal productivity and production. However, the responsibility for formulating policies towards these technologies rests with the Member Governments themselves.

FOOD AND AGRICULTURE ORGANIZATION. "FAO STATEMENT ON BIOTECHNOLOGY." *BIOTECHNOLOGY IN FOOD AND AGRICULTURE* (MARCH 2000).

SEE ALSO *Agricultural Biotechnology; Biodiversity; Bt Insect Resistant Crops; Corn, Genetically Engineered; Cotton, Genetically Engineered; Drought-Resistant Crops; Environmental Biotechnology; Fertilizers; Frost-Resistant Crops; GE-Free Food Rights; Gene Banks; Genetic Engineering; Genetic Technology; Genetic Use Restriction Technology; Genetically Modified Food; Genewatch; Golden Rice; Herbicide-Tolerant Plants;* Monsanto Canada Inc. v. Schmeiser; *Patents and Other Intellectual Property Rights; Salinity-Tolerant Plants; Terminator Technology; Transgenic Plants; Wheat, Genetically Engineered*

BIBLIOGRAPHY

Books

Ferry, Natalie, and Angharad M. R. Gatehouse, eds. *Environmental Impact of Genetically Modified Crops.* Cambridge, MA: CABI, 2009.

Fiechter, Armin, Christof Sautter, et al, eds. *Green Gene Technology: Research in an Area of Social Conflict.* New York: Springer, 2007.

Gordon, Susan, ed. *Critical Perspectives on Genetically Modified Crops and Food.* New York: Rosen Publishing Group, 2006.

Halford, Nigel, ed. *Plant Biotechnology: Current and Future Applications of Genetically Modified Crops.* Hoboken, NJ: John Wiley, 2006.

Liang, George H., and Daniel Z. Skinner, eds. *Genetically Modified Crops: Their Development, Uses, and Risks.* New York: Food Products Press, 2004.

Lurquin, Paul F. *High Tech Harvest: Understanding Genetically Modified Food Plants.* Boulder, CO: Westview Press, 2002.

Thomson, Jennifer A. *Seeds for the Future: The Impact of Genetically Modified Crops on the Environment.* Ithaca, NY: Comstock Publishing Associates, 2007.

Weirich, Paul, ed. *Labeling Genetically Modified Food: The Philosophical and Legal Debate.* Oxford: Oxford University Press, 2007.

Wesseler, Justus H. H., ed. *Environmental Costs and Benefits of Transgenic Crops.* Norwell, MA: Springer, 2005.

Young, Tomme. *Genetically Modified Organisms and Biosafety: A Background Paper for Decision-Makers and Others to Assist in Consideration of GMO Issues.* Gland, Switzerland: World Conservation Union, 2004.

Periodicals

Abbott, Alison. "European Disarray on Transgenic Crops." *Nature* 457, no. 7232 (February 19, 2009): 946–947.

Charles, Dan. "U.S. Courts Say Transgenic Crops Need Tighter Scrutiny." *Science* 315, no. 5815 (February 26, 2007): 1069.

Moeller, Lorena, and Kan Wang. "Engineering with Precision: Tools for the New Generation of Transgenic Crops." *Bioscience* 58, no. 5 (May 2008): 391–401.

Web Sites

Food and Agricultural Organization of the United Nations (FAO). "The Gene Revolution: Great Potential for the Poor, but No Panacea." *FAO Newsroom,* May 17, 2004. http://www.fao.org/newsroom/en/news/2004/41714/index.html (accessed November 5, 2011).

"Food, Genetically Modified." *World Health Organization (WHO).* http://www.who.int/topics/food_genetically_modified/en (accessed November 5, 2011).

"Oxfam International's Position on Transgenic Crops." *Oxfam International.* http://www.oxfam.org/en/campaigns/agriculture/oxfam-position-transgenic-crops (accessed November 5, 2011).

Susan Aldridge

Genetically Modified Food

■ Introduction

Foods are called genetically modified if biotechnology is used to alter the genetic makeup of an animal or plant that is used for food. A genetically-modified (GM) food has had its genetic makeup changed by any available method, including selective breeding, hybridization, or through recombinant DNA technology. A genetically-engineered (GE) food has genes that were directly transferred or removed through recombinant DNA methods. Transgenic foods contain genes added from other organisms.

The population of the world continues to increase and is expected to reach 9 billion by the year 2050. Climate change and growing population place increasing demands on the world's ability to produce enough food. Genetic engineering provides a potential solution to increasing food demands based on global climate change and increasing world population. Critics of genetic engineering assert that there has not been a sufficient accumulation of long-term data concerning the impact of GM foods on human health and on the environment.

■ Historical Background and Scientific Foundations

Genetic engineering is based on the science of genetics pioneered by the Austrian monk Gregor Mendel (1822–1884) in the mid-1800s. He selectively crossbred pea plants to determine heritability of a variety of traits. By studying plant offspring across successive generations, he was able to generate detailed hypotheses about the nature and process of genetic inheritance. His work did not achieve scientific notice for more than three decades.

In 1944, Colin McLeod (1909–1972) and Maclyn McCarty (1911–2005) determined that DNA is responsible for transmission of genetic data. James Watson (1926–) and Francis Crick (1916–2004) discovered the double helix structure of DNA in 1953; they determined that DNA consists of paired amino acids (called base pairs) arranged in specific sequences.

Recombinant DNA technology was pioneered in 1973 by Stanley N. Cohen (1935–) and Herbert Boyer (1936–); they removed a DNA segment from one *E. coli* bacterium and implanted it into the DNA of another. In 1980 the U.S. Supreme Court handed down a decision permitting patenting of GE organisms, including GM crops.

In 1986 GE tobacco was field tested in Belgium, and in 1987 the first GE crops in the United States, tobacco and tomatoes, were field tested. The U.S. Food and Drug Administration (FDA) reported in 1992 that genetically modified foods did not require special regulation because they were "not inherently dangerous." In 1994 the first GE crop in the European Union, tobacco, was approved for commercial production in France.

In 1990 the FDA approved the commercial production and distribution of chymosin, the first GE food substance made by altering bacteria. Chymosin mimics the activity of rennet, an enzyme used in the making of cheese. In nature, it is produced in the stomachs of cows. In the United States, between 80 and 90 percent of all hard cheeses are currently made with chymosin, as the commercial demand for cheese greatly exceeds the availability of rennet.

In 1993 the FDA approved the use of Bovine Growth Hormone (BGH) to stimulate milk production in dairy cows. BGH is a GE drug. The FDA does not require specific labeling on milk produced by cows injected with BGH. The U.S. Department of Agriculture (USDA) approved the commercial production of Calgene's Flavr Savr tomato, engineered to ripen and decompose more slowly than traditional tomatoes, in 1994.

Two years later, Genzyme Transgenics Corporation reported the birth of a transgenic goat created to transmit a human cancer-fighting protein through her milk. Genzyme partnered with Tufts University to produce goats carrying heart-disease, hemophilia- and

WORDS TO KNOW

CODEX ALIMENTARIUS: According to the World Health Organization, the Codex Alimentarius was created in 1963 for the purpose of developing food standards, guidelines, and food-related codes of practice. The primary purpose of the Codex is to protect the health and safety of consumers as well as to ensure that global fair trade practices are maintained throughout the food industry.

CROSS POLLINATION: Cross pollination occurs when plants of two different varieties are bred together. The offspring will have a gene pool that combines characteristics of both varieties, creating a new strain. Cross pollination can occur intentionally or through pollination by wind or insect.

GENE TRANSFER: Gene transfer occurs when a DNA sequence is transferred from one organism to another, unrelated organism. This may involve two members of the same family or organism, or transfer between different types of organisms.

GENETIC ENGINEERING: When recombinant DNA or other biotechnology techniques are used to intentionally alter the genetic material of an organism in order to introduce a new trait or characteristic that can then be transmitted to future generations.

LIVING MODIFIED ORGANISMS: Living organisms whose genetic make-up has been intentionally altered through the use of biotechnology/genetic engineering and that are capable of reproduction and passing the altered genes to successive generations.

OUTCROSSING: Outcrossing occurs when two organisms from different varieties of the same species are bred. Outcrossing serves as a means of introducing new genetic material into the breeding pool.

PLASMID: A segment of DNA, either linear or double-stranded, that can replicate itself independently of the DNA found in chromosomes. Some plasmids can be inserted into chromosomes, making them useful in recombinant DNA technology. Plasmids may be used to transfer genes in genetic engineering techniques.

RECOMBINANT DNA MOLECULES: Recombinant DNA sequences result when genes from two different organisms, whether the same or different species, are combined through the use of biotechnology to create new genetic sequences that could not otherwise occur in nature.

TRANSGENIC: An organism in which DNA from an unrelated organism has been added through a genetic engineering process.

Traditional selective breeding as a means of genetically modifying food crops is a lengthy and inexact method for creating desired changes. Seeds are sown, and the resulting plants are observed for desired traits. Those that show strongest evidence of the trait are bred progressively and selectively until a strain of food crop that strongly represents the characteristic is developed. This process can take many years, because the entire gene pool of the crop plant is involved, adding a significant degree of chance to the process. The trait can be magnified only by repetition across successive generations. Contemporary biotechnology enables the specific genes carrying the desired trait, whether from the crop type being improved or within another type of organism (or species), to be mapped, targeted, and inserted directly into the food to be changed. The effect is both immediate and heritable; that is, the trait is expressed in the current crop and transmitted to successive offspring generations.

■ Impacts and Issues

The world's climate is changing and becoming more unpredictable, significantly impacting traditional farming practices. When there is drought, flooding, increased incidence of catastrophic events such as earthquakes and tsunamis, or prolonged freezing temperatures, crops can be destroyed, creating food insecurity for entire populations. Pathogens such as bacteria, nematodes, fungi, and viruses can infect and markedly diminish crop yields. With advances in biotechnology, it is possible to engineer crops so as to increase resistance to pathogens, magnify hardiness under changing weather conditions, and cultivate higher yielding crops. It is also possible to use genetic engineering to increase nutrient value, inject vaccines, and add essential vitamins to food crops.

The world's population is expected to reach 9 billion by 2050. It is imperative that farms be able to produce enough food crops to meet the demands of the growing population in order to assure food security. The changing climate is creating prolonged high temperatures and drought conditions in some areas, extended freezes and longer rainy periods or flooding in others. Increasing incidences of catastrophic events such as earthquakes, cyclones, typhoons, tsunamis, hurricanes, floods, and wildfires have also been reported. These conditions can change growing seasons or alter soil conditions, rendering traditional crop seeds and cultivation methods inadequate.

The two most widely produced transgenic crop types are herbicide- and insect-resistant varieties, most frequently in the form of soybeans, cotton, corn, and canola. In addition to creating herbicide and insect-resistant plants, food crops are being engineered to improve taste, magnify nutrient content, increase hardiness in harsh weather conditions, and improve disease

cancer-fighting drugs in their milk. PPL therapeutics created a genetically-engineered hen designed to transmit an anti-neoplastic protein in her eggs in 2000. As of late 2010, no GM animals were commercially or legally available for human consumption.

A tomato plant that has been genetically modified to improve the quality of the fruit by slowing the ripening process. Between 1996 and 1999 puree made from genetically modified (GM) tomatoes such as these were sold in British supermarkets. The strain had been made by Zeneca seeds (later part of Syngenta). It was popular at first but was a victim of more generalised public concerns over GM foods. © *SSPL/Getty Images.*

resistance, thereby improving yield and viability. Research is underway to produce GM foods containing essential vaccines for use in developing countries. Rice strains containing added iron and vitamins are being cultivated in regions of Asia.

Proponents of GM foods assert that the methods used can produce increased crop yields, create foods that mature and produce more quickly than traditional crop plants, and result in foods that are tastier, more nutritionally dense, and may contain added vitamins or vaccines. Engineered foods have greater resistance to pathogens, insect pests, and herbicides. The biotechnology used to create engineered foods results in the advancement of agricultural techniques and technology. Overall, cultivation of genetically-modified food, whether plant or animal, is designed to decrease hunger and food insecurity and increase nutrition and health on a global scale.

Genetically-modified livestock typically have increased hardiness and produce larger amounts of meat, milk, or eggs. They also mature and reproduce more rapidly than traditionally bred animals, and have increased disease resistance.

Those supporting GM food crops hold that their cultivation is beneficial for the environment as they may decrease the need to use chemical fertilizers, herbicides and insecticides. Foods that require less energy, water, or fertilizer in order to grow successfully result in greater conservation of resources.

The Monsanto Corporation is one of the largest producers of transgenic seeds in the world. Monsanto reports that it takes at least 10 years and costs approximately $300 million dollars to create a single strain of successful GE seeds. Roughly one third of all crops grown by farmers in the United States utilize GM seeds. The majority of those seeds are purchased through lease or stewardship agreements with the companies that modify and develop the seed technologies. The lease agreements allow the farmers to use each batch of seeds for one growing season; they may not save, sell, trade, or reuse them. Part of the agreement gives the manufacturer's representatives the right to enter the farmland for a period of up to three years after the end of the agreement in order to ascertain that no part of the contract has been violated.

Opponents of genetic engineering express concern that the world's food supply could be controlled by a few large corporations, rendering developing countries financially and agriculturally dependent on industries located in wealthier and more developed nations. Those who oppose GM foods also assert that cross-pollination of GM seeds with traditionally created seed strains may occur without the knowledge or consent of farmers. When a farmer is cultivating both GM and non-GM crops, or is traditionally farming land near a GM crop, it is possible for cross pollination to take place via wind or insects. For those who choose not to raise GM crops, this can create problems, particularly if their crops are genetically tested and found to contain patented seeds. Because GM seed creators hold patents on their creations, farmers who use them, whether intentionally or not, are subject to potential legal action for patent or contract violation. In addition, organic farmers can lose their designation if crops become "contaminated" by inadvertent cross breeding with engineered crop seeds.

Some animal and other organism rights activists express opposition to the concept of genetic engineering. They assert that selective breeding, particularly when augmented by induced genetic changes, violates the rights of the involved animals or crop plants because it produces physiological alterations designed to benefit humans, not the hosts. It is asserted that genetic engineering of animals causes them to be treated as property rather than as living beings. Activists consider the patenting of laboratory and other animals as further evidence of ignoring the rights of animals and treating them as human property. Similarly, mixing of genes between species and the creation of transgenic organisms is considered "tampering with the natural order" by activists. Others assert that combining plant and animal genes, particularly when there is no labeling of this occurrence, violates the rights of vegans and vegetarians to choose not to ingest animal products.

GM foods have been significantly incorporated into the American food supply. There is no legal requirement to label GM foods, so many Americans might be unaware of the presence of GM foods in their diet. Many organic and natural food producers have chosen to state specifically on their labels that their food products are not genetically engineered. In the

Workers sort through apples at Crunch Pak, an apple slicing company in Cashmere, Washington. A Canadian biotechnology company has asked the United States to approve a genetically modified apple that doesn't turn brown when sliced. © *AP Images/ Shannon Dininny.*

European Union, there has been some opposition to the cultivation and marketing of GM foods. European Union critics of GM foods have asserted that there is no long-term data on the impact of cultivating GM food on the environment or on the physical health of humans.

Concerns have been expressed that genetic engineering of foods may increase the expression of allergies in humans when genes from potential allergens are incorporated into typically non-allergic foods. In addition, it is possible that antibiotic resistance could increase due to transfer of marker genes for antibiotic resistance.

The World Health Organization (WHO) and the Food and Agriculture Organization (FAO) of the United Nations have convened Expert Consultation Groups to define and oversee world food and food ingredient safety. They provide scientific and technical advice to Member States of FAO and WHO. The Codex Alimentarius Commission advises and assesses the safety of foods created or modified through the use of biotechnology. WHO/FAO reports that GM foods currently produced and distributed have been successfully assessed for risk to human health and to the environment.

■ Primary Source Connection

Simon Barber, Director of the Plant Biotechnology Unit at EuropaBio, discusses the framework surrounding the assessment process in place for the development and approval of making genetically modified plants, and for the cultivation of genetically modified plants grown in Europe. He offers insight as to why European consumers are so wary about genetically modified foods and what steps are being taken to change their perceptions about the safety of genetically modified foods.

Modifying GM Food Perception

Issue

Approval of genetically modified foods in Europe. Europeans have been far more nervous about the safety of GM foods than North Americans, essentially halting the approval of new varieties since 1998. Are skittish regulators and consumers finally warming up to the technology?

TECHNOLOGY REVIEW (TR): After almost six years in which no genetically modified food or crop had been approved for sale in Europe, a few varieties of corn finally made it through the regulatory process this year. Where do things stand now?

SIMON BARBER: *(Director of the Plant Biotechnology Unit at EuropaBio, the European biotech industry association.)* There is a complete regulatory framework in place for assessment and approval of genetically modified plants that are going to be grown or imported for food or food ingredients or animal feed. We have seen two approvals through that process for imports for food and animal-feed use of maize, for instance. So that system seems to be beginning to work. Getting approvals to grow new GM crops here, that's a different matter. That doesn't seem to be moving yet.

TR: Why not? Isn't there a process to approve new GM crops for cultivation?

BARBER: The framework is there. What is under discussion, though, is the concept of coexistence: once a crop has approval, how can I, as a farmer, choose to grow one of these varieties while minimizing any pollen and gene movement into my neighbor's crops? At the moment, I don't think we anticipate having new EU legislation on coexistence; we see the European Commission having its guidelines and then the member states making their own legislation around those guidelines. Some member states are developing their rules in a way that might well prohibit their farmers from ever choosing genetically modified seed, but others are being more pragmatic. The fact that the coexistence rules are in development doesn't give a very strong incentive for people to go for authorization to cultivate just at the moment.

TR: Why have European consumers been so wary about GM foods?

BARBER: There are groups that have made a huge amount of noise about it. They raise the question of the precautionary principle and say that we're not absolutely certain of safety—which actually we can say about everything. If we're honest, no science will say that anything is 100 percent safe. But there have been food scares here, such as mad cow, which means that our citizens are concerned about the safety of their food supply. There isn't an awful lot of what I would call very balanced debate; the debate tends to be very antagonistic, so you would have people very much "for" talking to people very much "against." If people don't have things explained to them well, there's room there for them to have concerns, and they're legitimate concerns.

TR: Has the European biotech industry done its share to explain the technology?

BARBER: They have recently made more efforts in that direction, but at the outset perhaps not as much as they ought. But it's not just the job of the industry. If you look at the industry, it's very small compared to the others that it supports. Plant variety developers and people who produce seed—that's our industry—support the farmers, which is a larger industry; the farmers then support food processors, and the value gets bigger and bigger. At the top, one U.K. supermarket chain probably has the same annual turnover as the whole international seed trade. So in some ways, we are a limited resource to be able to teach everybody in the world about modern biology and its uses. It's something that I think everybody has to be involved in. It's easy for people, once this had become an issue, to say, well, industry didn't do a good job, but before anything can be imported into Europe and used as animal feed or as an ingredient as food for us humans, it had to go through a safety approval process. The governments of the EU and the EU itself have institutions that did all this. Well, how were they explaining to their citizens what was going on? It's something that has to be shared across the board.

TR: But biotech companies would seem to have the most to gain from consumer acceptance of GM foods, so shouldn't they bear most of the educational responsibility?

BARBER: They should bear some responsibility, and in more recent years, they have put effort into this. There is something called Agricultural Biotechnology in Europe, which is a program that some of the companies have put money into to try to provide materials for outreach into the food chain and to citizens, and in some schools. It's not all-encompassing, but we are making real efforts to do that.

TR: Some agricultural biotech companies, such as Syngenta, have reduced or halted research in Europe due to consumer resistance and regulatory inactivity. How does the industry perceive current regulations?

BARBER: I don't think that the regulatory machine in the EU is running consistently yet. If you're using this technology to develop a product and you want to have it on the market in 10 years, if the machine isn't running consistently, you never know whether you'll get your product to market. So perhaps one would move one's research somewhere else, where there's a history of consistent application of the regulation.

TR: Will some of these companies eventually return?

BARBER: Until the question of how the regulatory system is going to run—is it going to run at an even speed, or is it going to be run in a discriminatory way in some countries?—is sorted out, people will probably think very carefully about that. For instance, it's very difficult to do field trials here now. In the interest of transparency, researchers make the locations of field trials known, and many of them are destroyed every year by people with a conviction against anybody using the technology. So that is part of the judgment a company that is interested in using these technologies has to make about where it does its basic research. Europe is one of the centers of origin of this technology, in Belgium at Ghent University, 25 years ago. And I think it's a sad thing that not just industry but also the public institutions have been much reduced in their plant science activity because of the way things have gone here. Universities are finding people don't want to get involved in plant science because they don't see a future for it. But we still have a commitment from companies here to continue. Bayer CropScience, which is a German company, recently opened a new facility for plant science research in Belgium, for instance. The industry does want to see this move forward, and they really do think that plant science in Europe is important.

TR: Will GM foods and crops ever enjoy the acceptance level in Europe that they have in the United States?

BARBER: I would like to think so. It may be a good many years away. But if you look to see how the technology is being used to date, it's provided benefits to farmers. I think it's a very, very sad thing that a lot of people in the West living in urban areas don't perceive a benefit to a farmer as a benefit to themselves, because they are benefits to us. But this is also an opportunity to diversify the way we use plants to meet some other needs in a more environmentally sustainable way. I hope eventually we'll see Europe embrace the technology and move down these roads. It's a tool. We say "GM," and we think of one or two crops. But it's a tool that we can use to do a multitude of useful things.

Erika Jonietz

JONIETZ, ERIKA. "MODIFYING GM FOOD PERCEPTION." *MIT'S TECHNOLOGY REVIEW* (DECEMBER 1, 2004).

SEE ALSO *Agricultural Biotechnology; Agrobacterium; Asilomar Conference on Recombinant DNA; Biodiversity Agreements; Bioethics, Biological Confinement; Breeding, Selective; Bt Insect Resistant Crops; Corn, Genetically Engineered; Cotton, Genetically Engineered; Disease-Resistant Crops; Environmental Biotechnology; Fertilizers; Food Preservation; Frost-Resistant Crops; Gene Banks; Genetic Engineering; Genetically Modified Crops; Golden Rice; Herbicide-Tolerant Plants; Patents and Other Intellectual Property Rights; Phytoremediation; Recombinant DNA Technology; Salinity-Tolerant Plants; Terminator Technology; Transgenic Animals; Transgenic Plants; Wheat, Genetically Engineered*

BIBLIOGRAPHY

Books

Baram, Michael S., and Mathilde Bourrier. *Governing Risk in GM Agriculture.* New York: Cambridge University Press, 2011.

Brunk, Conrad G., and Harold G. Coward, eds. *Acceptable Genes?: Religious Traditions and Genetically Modified Foods.* Albany, NY: SUNY Press, 2009.

Levidow, Les, and Susan Carr. *GM Food on Trial: Testing European Democracy.* New York: Routledge, 2010.

Lurquin, Paul. *High Tech Harvest: Understanding Genetically Modified Food Plants.* Boulder, CO: Westview Press, 2002.

Marchant, Gary Elvin, Guy A. Cardineau, and Thomas P. Redick. *Thwarting Consumer Choice: The Case against Mandatory Labeling for Genetically Modified Foods.* Washington, DC: AEI Press, 2010.

Pollack, Mark A., and Gregory C. Shaffer. *When Cooperation Fails: The International Law and Politics of Genetically Modified Foods.* New York: Oxford University Press, 2009.

Walters, Reece. *Eco Crime and Genetically Modified Food.* New York: Routledge, 2011.

Weasel, Lisa H. *Food Fray: Inside the Controversy over Genetically Modified Food.* New York: Amacom-American Management Association, 2009.

Weirich, Paul. *Labeling Genetically Modified Food: The Philosophical and Legal Debate.* New York: Oxford University Press, 2007.

Web Sites

"Food Safety: 20 Questions on Genetically Modified Foods." *World Health Organization.* http://www.who.int/foodsafety/publications/biotech/20questions/en/ (accessed October 20, 2011).

"Genetically Modified Foods and Organisms." *U.S. Department of Energy Genome Program.* http://www.ornl.gov/sci/techresources/Human_Genome/elsi/gmfood.shtml (accessed October 20, 2011).

Pamela V. Michaels

Genewatch UK

Introduction

GeneWatch UK is a not-for-profit non-governmental organization (NGO) that monitors developments in genetic technologies from a public interest, human rights, and environmental perspective. GeneWatch UK should not be confused with *GeneWatch*, the magazine of the Council for Responsible Genetics. GeneWatch UK has mounted a number of campaigns across the board in a wide range of different genetics technologies. For instance, the organization has focused on differing opinions on the usefulness of current genetic testing methods in identifying the risk of some diseases. Another campaign is focused upon public DNA databases.

GeneWatch UK joined with its American counterpart, the Council for Responsible Genetics, and the campaigning group Privacy International to call on the UK Government to scrap contracts it had set up under which the Forensic Science Service would build a database of the entire population of the United Arab Emirates. Other campaigning programs deal with genetically modified (GM) insects, GM fish, and GM contamination, as well as whether biomarkers are good indicators of disease risk.

Historical Background and Scientific Foundations

Environmental scientist Dr. Helen Wallace, who has worked in academia, in industry, and for Greenpeace, joined GeneWatch UK in 2001 and became the organization's Executive Director in 2007. Most of GeneWatch UK's funding comes from charitable trusts and individual donations. The European Commission also gives some money to GeneWatch UK. A smaller amount of funding comes from commissioned work.

GeneWatch UK operates according to three principles. First is the belief that genetic technologies are too often promoted as the solution to society's problems, such as cancer and climate change, which masks the underlying social, economic, and environmental issues. Second, there is often a lack of transparency on decisions on funding genetic technology and related issues. Finally, consideration of the impact of genetic technologies on human rights, animal welfare, and the environment ought to be more at the center of making big decisions. GeneWatch UK aims to ensure that genetic technologies are developed with human interest, animal welfare, and the environment in mind. The organization also supports public involvement in decision making on genetic technologies, and works to increase public understanding of genetics.

GeneWatch UK studies and analyzes the latest developments in genetic technology and clarifies new findings and makes them accessible to the public. It also communicates key issues to decision makers, the public, and other stakeholders, such as farmers and doctors. Furthermore, GeneWatch UK looks for practical methods to minimize the impact of genetic technologies on human interest, animal welfare, and the environment. It has a strong focus upon networking and alliance building with other organizations. Finally, it has a watching brief (taking on the responsibility to observe and monitor developments) in order to challenge the biotechnology industry if it is thought they are producing misleading information.

Impacts and Issues

Biotechnology, including genetics, has a very wide scope of applications, ranging from human health to agriculture and transgenic technology. The range of issues that GeneWatch UK has addressed over recent years is impressive. For instance, early in 2011 DuPont and Syngenta each announced the development of drought-resistant corn varieties created with conventional breeding techniques, not GM technology. While welcoming this advance, Dr. Wallace criticized Syngenta for saying

WORDS TO KNOW

23ANDME: A company that provides genetic testing for over 100 traits and diseases as well as DNA testing for ancestry studies.

GREENPEACE: A leading organization campaigning on environmental issues.

PRIVACY INTERNATIONAL: An organization that campaigns against privacy intrusions on individuals by government and business.

their variety of the conventionally bred corn would also be genetically modified for pesticide and herbicide resistance. GeneWatch UK has also campaigned against DNA databases and their applications. The police national DNA database in the United Kingdom is the second biggest in the world. While it plays an important role in fighting crime, GeneWatch UK believes the database infringes upon people's human rights and privacy because many people who are on the database have no criminal record. The concern is that this genetic information could find its way into the hands of potential employers and insurance companies, and it could used to discriminate against individuals because of their genetic background.

GeneWatch UK has also tackled the complex subject of patenting genes. The organization points out that allowing patents on genes restricts access to basic research, and that a gene is a discovery and not an invention that a biotechnology company can own. Finally, GeneWatch UK has also campaigned against biological weapons, arguing for general and institutional awareness of possible abuses of biotechnology that could lead to development of a weapon system. GeneWatch UK provides a source of basic information about genetic technology through its extensive series of briefing sheets.

SEE ALSO *Agricultural Biotechnology; Bioethics; Biotechnology; Biotechnology Organizations; DNA Databases; Genetic Technology; Genetically Modified Crops; Genome; Patents and Other Intellectual Property Rights*

BIBLIOGRAPHY

Books

Atkinson, Paul, Peter E. Glasner, and Margaret M. Lock. *Handbook of Genetics and Society: Mapping the New Genomic Era.* London: Routledge, 2009.

Everson, Ted. *The Gene: A Historical Perspective.* Westport, CT: Greenwood Press, 2007.

Hartwell, Leland. *Genetics: From Genes to Genomes.* Boston: McGraw-Hill Higher Education, 2008.

Hughes, Sally Smith. *Genentech: The Beginnings of Biotech.* Chicago: University of Chicago Press, 2011.

Peacock, Kathy Wilson. *Biotechnology and Genetic Engineering.* New York: Facts on File, 2010.

Web Sites

"Genetics." *World Health Organization (WHO).* http://www.who.int/topics/genetics/en (accessed September 8, 2011).

"Genomic Resources." *Centers for Disease Control and Prevention (CDC).* http://www.cdc.gov/genomics/resources/index.htm (accessed September 13, 2011).

"Welcome to GeneWatch UK." *GeneWatch UK.* http://www.genewatch.org/ (accessed September 28, 2011).

Susan Aldridge

Genome

■ Introduction

The genome is the sum total of the DNA in an organism. It consists of genes, which code for proteins. There is also a large amount of non-coding DNA in a genome. Often this DNA is called junk DNA, but it may have some as yet undiscovered function. All organisms on Earth exist as either eukaryotes or prokaryotes. A eukaryotic genome contains much more junk DNA than a more compact prokaryotic genome.

From the 1980s, a number of mapping tools have been developed so that the features of an organism's genome can be investigated. Chief among this is DNA sequencing, which enables the DNA code to be deciphered, letter-by-letter. Genes have a special coding signal in their sequences that allows them to be distinguished from non-coding DNA. Because of these new technologies, the genomes of many organisms, including humans, have been sequenced in the last several years. Such studies show how many genes an organism is likely to have. Sequencing has become almost routine, and there has been a shift in emphasis towards functional genomic studies, so that applications can be developed in medicine, agriculture, and the environment.

■ Historical Background and Scientific Foundations

The first genome to be sequenced was that of a bacteriophage, a virus that infects bacteria, in 1977. Since then, efforts to map and sequence other genomes were carried out in parallel. The first bacterium to be sequenced was *Haemophilus influenza* in 1995, and the first eukarote to have its genome deciphered was a type of yeast, in 1996. More complex genomes followed with the first multicellular organism, the worm *Caenorhabditis elegans* being sequenced in 1998, and the first plant, *Arabidopsis thaliana*, in 2000. These particular organisms were chosen because they are often seen as models for genetic

studies, and the genome adds to the large amount of information already accumulated on them. The complete sequencing of the human genome, announced in 2003, was the culmination of more than 10 years of international scientific effort. The project was led by the genetics company Celera and the U.S. National Institutes of Health, with significant contributions from the United Kingdom, Japan, Pakistan, Germany, France, and China.

Genome mapping and sequencing involves first chopping up the DNA, using enzymes, into fragments of a manageable size. (It would not be practical to start sequencing, base-by-base, from one end of the whole genome and finishing at the other.) Each fragment is then scanned with a number of DNA probes, which determines some sections of the sequence. Then these scanned sections are put back in order by looking for overlapping regions. The human genome was found to have 3.2 gigabytes of DNA. At first, it was expected that this would correspond to the presence of about 100,000 genes. However, analysis revealed that the number is only 25,000 to 30,000. This is probably sufficient to specify the complexity of the human body, because genes can mix and match, each one producing several different versions with different functions. Nearly 200 genomes have been sequenced, enabling scientists to begin looking at the similarities and differences between organisms at the DNA level. This might shed some light on the intriguing question of the C-value paradox, in which the amount of DNA an organism contains bears no relation to its complexity. For instance, the common house plant *Tradescantia*, the water lily, the salamander, and the South American lungfish all contain 20 to 70 times as much DNA as a human.

■ Impacts and Issues

The discovery of so many new genes through genome sequencing projects opens up many applications. Each person, except identical twins, has a unique genome, and

WORDS TO KNOW

C-VALUE: The total amount of DNA in an organism, measured in pictograms.

DNA PROBE: A short length of DNA tagged with a fluorescent or radioactive atom that binds to a complementary sequence in the genome.

EUKARYOTE: An organism whose cells have a nucleus, such as yeast, plants, and animals.

FUNCTIONAL GENOMICS: The branch of genomics concerned with the function of genes.

PROKARYOTE: An organism whose cells lack a nucleus. Bacteria are prokaryotes.

mutations in genes. Whereas the identities of genes that cause rare genetic disorders such as hemophilia have been known for some time, the search is now on for genes that influence more common diseases such as heart disease and cancer. These genes are known as susceptibility genes, and it is believed that there are several, maybe even hundreds, of mutations of these genes. This could lead to whole genome scans being carried out to assess someone's risk of getting such diseases. DNA testing was formerly very expensive, but the cost has come down dramatically, opening up the possibility of the whole genome scan being a part of routine health care. Nobel Prize winner and co-discoverer of the structure of DNA, James Watson (1928–), had his own genome sequenced in 2007.

The discovery of several thousand new genes offers drug companies many new targets to study. If they could synthesize a drug molecule that binds to a target, then it may be possible to block or interrupt a disease pathway. However, because of the time taken to develop a new drug, there have not yet been any new medicines that have arisen directly from the Genome Project.

Meanwhile, genome sequencing offers benefits in agriculture. Understanding which genes important crops such as rice (which was sequenced in 2005) use to resist

this can be used for identification. For example, the FBI uses probes based on a set of repeated DNA sequences to create a pattern that is indicative of a person's identity. People convicted of a crime have their DNA placed in a huge database to which traces of DNA in forensic samples can be matched. The sequencing of the human genome has also accelerated the search for disease-causing

Celera Genomics President J. Craig Venter's shadow is cast against a map of human genes during a Washington news conference on February 12, 2001, to discuss genome research. *© AP Images/Ron Edmonds.*

disease, insects, and drought could help to produce plants in which the effect of these genes is enhanced. The same is true of farm animals. The genomes of the cow and pig have been sequenced, and identifying the relevant genes could help to produce healthier, higher yielding, and more disease-resistant animals. Microbial genomes are easier to sequence than those of animals and plants. Biofuels use bacteria to generate energy from biomass, and the U.S. Department of Energy has therefore set up a project to gather together microbial genome information for application in energy production, environmental remediation, toxic waste production, and industrial processing.

However, before any of these applications of genomics can become a reality, there needs to be much more research into how genes function, rather than just their identification. Obtaining an individual's genome could pose real dangers if it is not clear what should be done with the information. It could be used to discriminate against someone in the workplace, or for insurance purposes, if an individual is found to be carrying a disease susceptibility gene. If the gene is not well understood, then the person is exposed to all the disadvantages of this knowledge, without any benefit in terms knowing whether the disease will occur or if a new treatment will be offered for it. Although many genomes have been sequenced, understanding lags far behind, which is why there needs to be careful discussion of how this information should be used.

SEE ALSO *Bioethics; Biotechnology; Gene Delivery; Gene Therapy; Genetic Privacy; Genetic Technology; Genome; Genomics; Individual Privacy Rights*

BIBLIOGRAPHY

Books

Acharya, Tara, and Neeraja Sankaran. *The Human Genome Sourcebook*. Westport, CT: Greenwood Press, 2005.

Atkinson, Paul, Peter E. Glasner, and Margaret M. Lock. *Handbook of Genetics and Society: Mapping the New Genomic Era*. London: Routledge, 2009.

Dale, Jeremy, and Malcolm von Schantz. *From Genes to Genomes: Concepts and Applications of DNA Technology*, 2nd ed. Chichester, UK: John Wiley & Sons, 2007.

Rose, Nikolas S. *Politics of Life Itself: Biomedicine, Power, and Subjectivity in the Twenty-First Century*. Princeton: Princeton University Press, 2007.

Periodicals

Esposito, Joseph J., et al. "Genome Sequence Diversity and Clues to the Evolution of Variola (Smallpox) Virus." *Science* 313 (2006): 807–812.

McElheny, Victor K., and Francis S. Collins. "The Human Genome Project +5." *The Scientist* 20 (2006): 42–48.

Web Sites

"Genes and Gene Therapy." *National Institutes of Health (NIH)*. http://health.nih.gov/topic/GenesandGeneTherapy (accessed September 7, 2011).

"Human Genome Epidemiology Network (HuGENet)." *Centers for Disease Control and Prevention (CDC)*. http://www.cdc.gov/genomics/hugenet/whatsnew/current.htm (accessed September 7, 2011).

U.S. Department of Energy (DOE) Genomes Programs. "Primer: Genomics and Its Impact on Science and Society: The Human Genome Project and Beyond (2008)." *genomics.energy.gov*. http://www.ornl.gov/sci/techresources/Human_Genome/publicat/primer/index.shtml (accessed September 7, 2011).

Susan Aldridge

Genomics

Introduction

The genome is the entire genetic content of an organism. In the early twentieth century, biologists began to investigate the relationship between chromosomes and the molecular structure of human cells. Genetics assumed greater importance as a distinct research avenue as the relationship between genetic structure and the incidence of various hereditary diseases became clearer. These findings ultimately led to the discovery of the double helix deoxyribonucleic acid (DNA) molecule that has provided the foundation for modern genetics research.

An international scientific movement emerged in the 1980s to pursue the development of a catalog that would contain the data necessary to understand the entire human genetic structure. The Human Genome Project (HGP) was initiated in 1990. When completed in 2003, the HGP provided a comprehensive mapping of the human genome. As a consequence, genomics completed its advance from a theoretical pursuit to a foundation for many practical scientific applications.

The HGP established the baseline data that has been the impetus for a number of research initiatives that span molecular biology, genetics, and information technologies. Future approaches to questions as diverse as disease prevention and treatment, pharmaceutical development, the understanding of human evolution, and crop plant cultivation are each significantly influenced by current genome research. Genomics also engages a number of important ethical, moral, and social issues; an example is the use of genetics to determine whether the conception of a child is likely to lead to one born with a congenital disease.

Historical Background and Scientific Foundations

The history of genomics confirms the importance of this discipline to the study of human health. The rise of genomics to prominence has been driven by parallel breakthroughs in various aspects of computational and computer science.

The term genome was introduced into the scientific lexicon in 1920. It combines the words chromosome and gene. Genomics had its point of commencement in the pioneering work undertaken at the Carnegie Institute in Washington, DC, after World War II (1939–1945). Radioisotopes were successfully employed to further the understanding of how human metabolic pathways function. During this period, the *Escherichia coli* bacterium was established as the model organism that would provide the foundation for the research later undertaken to understand the enzymes transported along each pathway.

These developments led to the discovery of how macromolecules such as proteins and nucleic acids were formed within the body. In 1953, the discovery by James Watson (1928–) and Francis Crick (1916–2004) of the DNA double helix would prove to be the inspiration for several generations of geneticists whose work collectively revolutionized the fundamental understanding of human biology. DNA research, and in particular the ability to replicate specific segments of DNA molecules, enabled scientists to develop important hypotheses concerning protein synthesis, genetic structure, and how genetic material tends to undergo different exchanges and recombinations that influence all aspects of human evolution. The ability to clone genes in both humans and animals is an outcome of this early research.

Technological advances have been instrumental to the growth and prominence of genomics. There are two methodologies that were especially important. The polymerase chain reaction (PCR) is the automated process that enables researchers to quickly reproduce millions of exact copies of the desired DNA sequence for study. PCR eliminated the need to use bacteria as a host where DNA could be replicated. Further, minute amounts of available DNA can be amplified with PCR methodology. For this reason, PCR became a prominent feature of criminal investigations. The available DNA extracted from a small sample of bodily fluids or human tissue discovered at a crime scene could be copied efficiently to permit a virtually unlimited number of forensic tests to be performed to match the crime scene evidence against known samples. Subsequent advances in the identification of mitochondrial DNA allowed even more minute samples to be successfully matched to their donor.

Automated sequencing technologies now permit scientists to access millions of DNA base pairs in a period of days, as opposed to the weeks or months that were previously needed to assemble DNA for analysis using physical means. The HGP was an important outcome of these advances driven by the combined forces of molecular biology and computational science.

The HGP began in 1990 as an expression of the collective resolve of the international scientific community to better understand human genetic structure. The HGP sought to sequence the entire genome. Research identified the importance of the relationship between the enzyme RNA and the genome, as it was determined that only 3 percent of the genome structure contained ribonucleic acid (RNA). This finding was a significant impetus to genome research, as molecular biologists and geneticists accelerated their efforts find genes using RNA as the mechanism to better understand the processing machinery contained in all human cells.

The speed with which the HGP reached its genome sequencing objective of is one of the remarkable feats of modern science. In 2001 the HGP working draft was published; it confirmed that approximately 90 percent of the genome had been successfully mapped. The final 2003 project report provided conclusions that are the foundation for all future genomics research:

- Humans have approximately 30,000 genes, a number that is roughly similar to those determined in other mammals.

- The key distinction noted between the human genome and comparable animal structures is the greater number of almost identical DNA segments found in the human genome.

- Proteins comprise between 1.1 percent and 1.4 percent of the sequence codes found in the genome; only 7 percent of the proteins identified in the genome are ones specific to vertebrates.

■ Impacts and Issues

The present and anticipated future benefits associated with genomics research are profound. The fields of medicine and biotechnology are the most obvious beneficiaries of the scientific knowledge gained from the HGP. The ongoing analysis of the HGP data contributes to an ever-increasing body of scholarship and practical applications. The relationship between specific genes and the onset of particular diseases is better understood as a

WORDS TO KNOW

DNA: An abbreviation for deoxyribonucleic acid, the genetic material that is the basis of life for most organisms. Found within every living cell and carrying the organism's genetic information, DNA consists of two long chains of amino acid nucleotides; the chain is twisted into a double helix shape. The strands of the chain are joined by means of hydrogen bonds. The specific sequence of the DNA is what defines genetic characteristics.

FUNCTIONAL FOODS: Foods that are enhanced by an additive or in which the quantity of a present bioactive compound is increased to enhance specific physiological benefits.

HUMAN GENOME PROJECT (HGP): The international genomics initiative was established in 1990 with two primary objectives. The project sought to identify all of the genes present in human DNA. Once identified, the HGP assembled a map of the entire genome and the sequences of the estimated three billion DNA base pairs. The HGP was completed in 2003 with these objectives achieved.

MICROBIAL GENOMICS: The study of the genetic structure of microorganisms. This field includes important research into the genetic structure of the pathogens such as *Salmonella* that cause food poisoning.

MITOCHONDRIAL DNA: Mitochondrial DNA (mtDNA) is the smallest human chromosome. These molecules were the first important element of human genome sequencing that led to the success of the Human Genome Project. In humans, the mother is the source of all mtDNA, a finding that influenced the course of the studies of inherited traits. The ability to identify definitively the smaller sources of DNA is an important aspect of modern forensic science.

SEQUENCING: The generic term for the processes used to determine the order in which the nucleotide bases are arranged in a DNA molecule. The bases are adenine (A), guanine (G), cytosine (C), and thymine (T). The bases form pairs (e.g. AT, GC) that comprise the individual components of the strands of the double helix DNA molecule. Each base pair carries a specific genetic coding.

result of genomics research. Breast cancer diagnosis and prevention techniques are examples of how persons predisposed to cancers can be the subjects of earlier identification through genetic screening processes.

Enhanced human genome knowledge, in combination with the study of other animal genomes, will enable plant crops and animal food sources to be specifically engineered with the inclusion of nutritional components that benefit human health. The combination of genomics and the expanding information technologies now employed by agriculture will promote more specialized crop production. Microbial genomics contributes to the reduction

President Bill Clinton and Dr. Craig Venter, left, President of Celera Genomics Corporation, and Dr. Francis Collins, right, Director of the National Institutes of Health, talk Monday, June 26, 2000, prior to a teleconference of a joint announcement that the International Human Genome Project and Celera Genomics Corporation have both completed an initial sequencing of the human genome, the working blueprint for human beings. The discovery is seen as one of history's great scientific milestones, the biological equivalent of the moon landing. © AP Images/Ron Edmonds.

of food pathogens and the impact of foodborne diseases. Consumers may be equipped with a genetic passport generated by the analysis of their own genetic structure. With such information, consumers could access how the cultivation and processing of specific functional foods are best suited to their personal genetic needs.

Genome research generally and the HGP specifically have engaged a variety of complex ethical and social issues. The readily appreciated benefits derived from deeper knowledge of all biological process are offset in some respects by concerns that science may become an instrument that promotes equally profound ethical errors. A prominent example is the use of genetic information to predict the likelihood of conditions such as Down's syndrome or spina bifida in a fetus. The use of genomics in prenatal screening is closely linked to concerns that genomics will ultimately provide a basis for genetic engineering and eugenics, the science associated with improving the genetic composition of a population through selective breeding.

In the United States, the National Human Genome Research Institute's (NHGRI) Ethical, Legal and Social Implications (ELSI) Research Program was established in 1990. The planners of the HGP determined that the research undertaken within its domain had potentially far-reaching effects on all of human society. The NHGRI and other ethics organizations strive to ensure that genome research is not undertaken in a vacuum in which ethical questions or social impacts are detached from scientific applications. The NHGRI regularly publishes articles and position papers that consider these complex issues.

A related objective of the HGP was the desire to ensure that the genome map and all DNA sequence research is freely accessible to anyone. As bio-patents and the commercialization of scientific research have gained prominence since 1990, challenges have arisen for those

A poster of the Archon X PRIZE for Genomics is seen during a news conference in Washington, DC, in 2006, that announced the global competition to develop technology that could dramatically reduce the time and cost of sequencing human genomes. The first team that builds a device that can successfully sequence 100 human genomes in 10 days will be awarded with $10 million. The competition was on-going as of June 2011. *© Alex Wong/Getty Images.*

who advocate open access. One of their biggest concerns is that genomic research does not become the exclusive preserve of a single institution, commercial enterprise, or nation.

SEE ALSO *Bioinformatics; Biotechnology; Designer Genes; DNA Databases; DNA Sequencing; DNA Vaccines; Gene Banks; Gene Delivery; Gene Therapy; Genetic Bill of Rights; Genetic Discrimination; Genetic Engineering; Genetic Privacy; Genetic Technology; Genetic Use Restriction Technology; Genetically Modified Crops; Genetically Modified Food; Genome; Molecular Biology; Oncogenomics; Pharmacogenomics; Recombinant DNA Technology*

BIBLIOGRAPHY

Books

Campbell, A. Malcolm, and Laurie J. Heyer. *Discovering Genomics, Proteomics and Bioinformatics*, 2nd ed. San Francisco: CSHL Press, 2007.

Choudhuri, Supratim, and David B. Carlson. *Genomics: Fundamentals and Applications*. New York: Informa Healthcare, 2009.

Lesk, Arthur M. *Introduction to Genomics*. New York: Oxford University Press, 2007.

Web Sites

"Genomics." *World Health Organization (WHO).* http://www.who.int/topics/genomics/en (accessed September 20, 2011).

"Genomics and Health." *Centers for Disease Control and Prevention (CDC).* http://www.cdc.gov/genomics/default.htm (accessed September 29, 2011).

"PRIMER: Genomics and Its Impact on Science and Society: The Human Genome Project and Beyond (2008)." *U.S. Department of Energy (DOE) Genome Program.* http://www.ornl.gov/sci/techresources/Human_Genome/publicat/primer/index.shtml (accessed September 29, 2011).

"Public Health Genomics." *Centers for Disease Control and Prevention (CDC).* http://www.cdc.gov/genomics/ (accessed September 20, 2011).

Bryan Thomas Davies

Germ Line Gene Therapy

■ Introduction

The Human Genome Project, completed in 2003, successfully mapped the 25,000 genes in human DNA and gave scientists and physicians the tools to determine each patient's genetic strengths and vulnerabilities to potential illnesses. Scientists and engineers predict that the detailed laying out of all of the elements that go into humanity's genetic constitution will transform medicine from a reactive mode in which people get sick and then are treated into an activity that is predictive, personalized, preventive, and participatory (termed the "four Ps").

Genetic engineering can take many forms. Gene therapy, in broad terms, involves the introduction of a therapeutic gene to replace a defective one. This can take place in two ways, first with somatic therapy, in which the result of the modification is restricted to that organism, and secondly with germ line gene therapy, in which the sperm or eggs (germ cells) are genetically altered. In this case, the genetic modification would be passed on to future generations and would be a huge boon in the treatment of genetic disorders. The ethical considerations of such therapy are so vast that it is a controversial topic, and any potential benefits of genetic engineering for humanity must be weighed against potential criticisms.

■ Historical Background and Scientific Foundations

In 1971 Donald Munro and others undertook an experiment to introduce a viral vector (in this case ultraviolet-irradiated herpes simplex virus or HSV) into thymidine kinase-deficient mouse cells. The HSV virus contains coding for the deficient enzyme and restored the mice's ability to synthesize it. Munro's work was cited in a 1972 paper in the journal *Science* titled "Gene Therapy for Human Genetic Disease?" In it, the authors expressed their view that gene therapy could eventually conquer some human genetic diseases.

A fertility treatment that treats the egg of an infertile woman with healthy ooplasm from a donor egg resulted in what the researchers claimed was the first case of human germ line genetic modification resulting in normal healthy children in 2001. The process resulted in 30 births, and two of those babies were found to have DNA from three parents. Mitochondrial DNA, which was carried in the donor ooplasm, had been incorporated into the genetic makeup of the children. Some debate still occurs as to whether the incorporation of ooplasmic DNA as opposed to nuclear DNA can be considered germ line gene therapy, as well as if this type of modification should actually be seen as simply an inadvertent side effect of the fertility treatment.

One major issue of gene therapy is the risk that the implanted gene will be rejected by the immune system as "foreign," similar to the rejection that can occur after organ transplants. In 2006 a team from the San Raffaele Telethon Institute for Gene Therapy (HSR-TIGET) discovered a way of using microRNA molecules to "switch off" the identity of the "normal" gene being inserted. The process will be of great significance in the continuing work to treat disease and disorders with both somatic and germ line gene therapy.

■ Impacts and Issues

Germ line gene therapy is the power of creation through science. In most cases this creation involves the formation of new normal cells to replace defective ones that can cause disease, but it can also be seen as the creation of new phenotypes within a species. Despite the case of the children born with three types of DNA mentioned above, germ line gene therapy is currently focused on animal and plant subjects as opposed to humans. This focus is largely due to the decreased ethical concerns related to gene therapy in other species, and also the positive impact that gene therapy on those species can result in for humans.

Malaria is caused by the parasitic protozoa *Plasmodium*, which requires both mosquito and vertebrate hosts to complete its life cycle. It kills around 1 million people every year, and germ line therapy has provided one method of fighting back. By injecting foreign DNA into the mosquito egg, scientists aim to disrupt the life cycle of the *Plasmodium* in various ways with the intention of controlling the spread of the disease. One way would involve a gene to increase the resistance of the mosquitoes to the parasite, another to produce sterile males, and another would modify the mosquito's sense of smell, preventing it from identifying human hosts. In experiments, 50 percent of those that hatched showed strong expression for the therapeutic gene.

A lot of current research involves the idea of transgenic animals as bioreactors, producing desired proteins in their milk (in the case of sheep and rabbits, producing an alpha-1 antitrypsin to treat emphysema or human interleukin to fight cancer, respectively) or in the albumin of their eggs (in the case of chickens producing the drug interferon). Advanced Cell Technology, a Massachusetts-based company, has taken the process further by aiming to produce not only proteins in milk, but also specific organs and cells within the bodies of cows. Modified DNA is injected into cow egg cells from which the original nucleus has been removed. The resulting calves would not have genes from their mother but those chosen for them in the lab.

Germ line gene therapy can be used to change the DNA of eggs or sperm, and also the blastomere (the very earliest grouping of cells after fertilization has taken place.) There are various gene delivery tools (vectors) used, each with its own advantages and disadvantages. The most common is the use of a virus as vector. A virus is in some ways the perfect vector, given that it is naturally designed to invade cells. However the fact of introducing a foreign body raises the problem of an immune response, toxicity, and also the risk that the virus might somehow become reactivated at some point and cause disease. Another real concern is the risk of inadvertent germ line transfer even with somatic gene therapy. The issue that arises with somatic gene transfer that is being used as a therapy with some diseases is that it needs to take place early in development to produce a higher rate of success. At this early stage, it is more likely, some would say inevitable, that germ line transfer would take place, changing the genetic makeup of the infant, and possibly being handed down to future generations. This would contradict the Weismann barrier principle that states that genetic information passes in one direction from germ to somatic, but soma-germline transfer has already been shown to occur with some retroviruses.

Other vectors include DNA in a lipid and protein envelope and the idea of a 47th (artificial human)

WORDS TO KNOW

GERM LINE: The lineage of cells resulting in eggs and sperm.

OOPLASM: Cytoplasm of an ovum.

SOMATIC: Pertaining to the corporeal; the cells of the body that make up tissues and organs, distinct from those involved in reproduction.

TRANSGENIC: An organism or cell into which the genes of another species have been combined.

chromosome to carry new genetic material. In this case, the big advantage would be that the new genetic material could exist alongside current genes, would have the ability to carry a large amount of genetic code, and would also be less likely to invoke an immune response from the body.

In the United States, the processes of germ line gene therapy are overseen by two agencies, the Food and Drug Administration (FDA) and the National Institutes of Health (NIH), both of which have strict protocols in place. Currently the use of human eggs or sperm for experimentation is deemed impermissible. In a 2002 report, "Guidelines for Research Involving Recombinant DNA Molecules," the NIH stated:

> Significant additional preclinical and clinical studies addressing vector transduction efficacy, biodistribution, and toxicity are required before a human in utero gene transfer protocol can proceed. In addition, a more thorough understanding of the development of human organ systems, such as the immune and nervous systems, is needed to better define the potential efficacy and risks of human in utero gene transfer.

For religious groups, politicians and bioethical think tanks, the moral and ethical issues pertaining to germ line gene therapy are myriad. Apart from the medical problems, many have also expressed concerns about a slippery slope of genetic tinkering. If livestock can be manipulated for physical characteristics, will mankind one day end up with "designer babies," not only selecting the sex but also physical appearance and abilities? Changing the genetics to wipe out genetic mutation might seem clear cut to some, but those who are benefiting are often not the current patients but their unborn children and future generations, none of whom can give informed consent.

The potential benefits of this technology are tempting, a way to control living organisms to human advantage, to overcome disease and hunger. With this vision, germ line therapy could transform the world, but the pitfalls of manipulating the genetic code are also to be weighed.

SEE ALSO *Agricultural Biotechnology; Artificial Organs; Bioethics; Cell Therapy (Somatic Cell Therapy); Cloning; Designer Babies; Designer Genes; Egg (Ovum) Donors; Genbank; Gene Delivery; Gene Therapy; Genetic Bill of Rights; Genetic Discrimination; Genetic Engineering; Genetic Technology; GeneWatch; Genome; Genomics; HapMap Project; Human Cloning; Human Embryonic Stem Cell Debate; Human Subject Protections; In Vitro Fertilization (IVF); Nuremberg Code (1948); Pharmacogenomics; Recombinant DNA Technology; Stem Cell Cloning; Stem Cell Laws; Stem Cell Lines; Stem Cells; Stem Cells, Adult; Stem Cells, Embryonic; Therapeutic Cloning*

BIBLIOGRAPHY

Books

Battler, Alexander, and Jonathan Leor, eds. *Stem Cell and Gene-Based Therapy Frontiers in Regenerative Medicine.* London: Springer, 2006.

Greenwell, Pamela, and Michelle McCulley. *Molecular Therapeutics: 21st-Century Medicine.* Chichester, UK, and Hoboken, NJ: J. Wiley, 2007.

Herzog, Roland, and Sergei Zolotukhin. *A Guide to Human Gene Therapy.* Hackensack, NJ: World Scientific, 2010.

Niewöhner, Jörg, and Christof Tannert, eds. *Gene Therapy: Prospective Technology Assessment in Its Societal Context.* Amsterdam and Boston: Elsevier, 2006.

Nusslein-Volhard, C. *Coming to Life: How Genes Drive Development.* San Diego, CA: Kales Press, 2006.

Walther, Wolfgang, and Ulrike Stein, eds. *Gene Therapy of Cancer: Methods and Protocols*, 2nd ed. New York: Humana Press, 2009.

Web Sites

"Genetics Home Reference: Gene Therapy." *Lister Hill National Center for Biomedical Communications.* http://ghr.nlm.nih.gov/handbook/therapy?show=all (accessed October 19, 2011).

"Types of Gene Therapy." *Gene Therapy Net.* http://www.genetherapynet.com/types-of-gene-therapy.html (accessed October 19, 2011).

Kenneth Travis LaPensee

GloFish®

■ Introduction

GloFish® are a species of zebrafish, and are the first transgenic pets sold in the United States. The distinguishing feature of GloFish is their ability to emit light, or fluoresce. There are several distinct types of GloFish, each fluorescing in a different color. The available colors of GloFish include green, red, blue, and purple. The use of particular types of aquarium lighting, especially blue or black lights, will highlight and enhance the fish's fluorescence. Black lights emit mostly ultraviolet light that is invisible to humans (hence the name), but which can cause certain substances to glow or fluoresce.

GloFish are a type of zebrafish genetically modified to fluoresce. Like other zebrafish, GloFish are relatively small—typically a maximum length in captivity of a little less than two inches—that are well-suited for home aquariums. Zebrafish, including GloFish, have long been a popular pet fish because of their hardiness, appearance, and easy maintenance.

■ Historical Background and Scientific Foundations

The first fluorescing ancestors of GloFish were successfully produced by scientists at the National University of Singapore in 1999. The appropriate gene from a fluorescing type of jellyfish was implanted into a fertilized egg of a tropical fish commonly known as the zebrafish (so-called because of its long stripes). The resulting mutated (genetically-modified) zebrafish was able to fluoresce, and it passed the new genetic trait to its offspring. The original motivation of the scientists involved was to produce a fish that would fluoresce selectively in the presence of particular waterborne pollutants.

In the early twenty-first century, a company in the United States called Yorktown Technologies entered into an agreement to commercialize specific types of fluorescing zebrafish developed by the University of Singapore

scientists. As zebrafish have long been a favorite of aquarium owners throughout much of the world, the American investors saw the advantage of having an established type of pet fish that could also fluoresce. The financial agreement obtained by Yorktown Technologies included the worldwide rights to market and sell GloFish.

Before being allowed to sell GloFish in the United States and other countries, permission would have to be obtained from the appropriate regulatory bodies within each country. In the United States the Food and Drug Administration (FDA) regulates the commercial uses of genetically engineered (GE) animals. The FDA's purpose in regulating GE organisms is to ensure the safety of both people and the environment. For example, if a GE crop is intended for human consumption, the FDA must ensure that the introduction of such plants into the public food supply is safe.

In submitting their GloFish for approval from the FDA, the petitioners could point out that GloFish are simply a special type of zebrafish, intended solely for use as pets. Because zebrafish are a freshwater tropical fish, and are not adapted to either seawater or the cooler climate found throughout most of the continental United States, it is difficult for them to survive if released into the wild. In addition, studies funded by Yorktown Technologies showed that, except for their fluorescence, GloFish are identical to other zebrafish in terms of their lifespan, fertility, pet-care, and behavior. Taking these factors into consideration, the FDA released an announcement in late 2003 approving the sale of GloFish in the United States.

■ Impacts and Issues

The fact that the fluorescing trait of GloFish is due to genetic modification has meant that its market is more limited than would otherwise be the case. GE organisms have met with resistance in many countries, such as the nations of the European Union (EU), as well as Canada and Australia. In all those countries the sale of GloFish is banned. In the United States, only the State

WORDS TO KNOW

FLUORESCENCE: The property of certain substances to emit electromagnetic radiation, most often visible light, in reaction to the absorption of radiation (such as visible or ultraviolet light) from another source.

GENETIC MODIFICATION: Also called genetic engineering, genetic modification is the deliberate modification of one or more parts of an organism's genome, often to produce some new and desirable trait. An organism's genome is the entirety of its hereditary information, embedded within its DNA.

PATENT: A form of intellectual property related to a specific invention. A patent is bestowed by a government upon an invention's creator, conferring the exclusive right to produce, use, and sell said invention.

TRANSGENIC ORGANISM: An organism that has been genetically modified by the insertion of genes from another species.

ZEBRAFISH: A type of freshwater minnow indigenous to certain parts of Asia.

of California prevents their sale. Regulations in California require that the environmental safety of GloFish be demonstrated by performing certain environmental impact studies; however, the company that owns the rights to GloFish has declined to pursue the process required to lift the California ban. They cite cost concerns, as well as the length of time involved in the approval process.

The name GloFish is trademarked, and the fish itself is considered a form of intellectual property, which has led to it being patented. In the United States, trademarks and patents are issued by the U.S. Patent and Trademark Office (PTO). Having a U.S. patent or trademark imparts certain rights to the owner(s), especially the right to bring a lawsuit against those who violate the patent or trademark. The trademark on the GloFish name basically means that the owner retains the rights to the use of the name, especially for commercial purposes. The GloFish patent prevents others from legally breeding the fish for the purpose of selling them. This means that if a pet owner has legally purchased male and female GloFish, and the fish reproduce, the owner cannot legally sell, trade, or barter the offspring because of the GloFish patent. Some people have raised objections to such practices, both to the idea that an organism can be patented (even if some of the organism's traits are the result of genetic modification),

GloFish are seen swimming around a fish tank. The fish, which were genetically engineered to glow, were originally intended to help scientists study pollution but are now being marketed as the first genetically altered house pet. © *GloFish/Getty Images.*

as well as to the prohibition on selling the offspring of animals that have been legally purchased.

It is interesting to note that even though GloFish are limited to the pet market, the technology used to give them their fluorescence is being used in a variety of scientific research. For instance, efforts continue to try to produce fish that can fluoresce in the presence of specific pollutants, while refraining from fluorescing when such pollutants are absent. The fluorescent trait is also being used in medical research.

Despite the controversy surrounding the breeding and selling of genetically modified creatures for purely aesthetic reasons, GloFish have found a niche in the retail pet market. The future of GloFish, as well as other genetically modified organisms that have been created by biotechnologists for people's enjoyment, seems to be inextricably bound to the broader question of how much genetic engineering of living things is acceptable to society as a whole.

SEE ALSO *Genetic Engineering; Genetically Modified Crops; Genetically Modified Food; Patents and Other Intellectual Property Rights; Recombinant DNA Technology; Transgenic Animals; Transgenic Plants*

BIBLIOGRAPHY

Books

Bonnicksen, Andrea L. *Chimeras, Hybrids, and Interspecies Research: Politics and Policymaking.* Washington, DC: Georgetown University Press, 2009.

Wall, Robert. *Animal Productivity and Genetic Diversity: Cloned and Transgenic Animals.* Ames, IA: Council for Agricultural Science and Technology, 2009.

Web Sites

"GFP: Green Fluorescent Protein." *Connecticut College.* http://www.conncoll.edu/ccacad/zimmer/ GFP-ww/GFP-1.htm (accessed October 10, 2011).

Pray, Leslie. "GloFish Are the First Transgenic Animals Available to the American Public. But What's the Biotechnology behind Them?" *Nature Education,* 2008. http://www.nature.com/ scitable/topicpage/recombinant-dna-technology-and-transgenic-animals-34513 (accessed October 10, 2011).

Philip Edward Koth

Golden Rice

■ Introduction

The term Golden Rice refers to rice (*Oryza sativa*) that has been genetically altered to contain carotenoids, mainly beta carotene, in the portion of the rice that is eaten. The presence of the carotenoids imparts a golden color to the edible part of the rice grains (the endosperm), in contrast to the usual white or brown color, hence the name of the rice.

In the body, carotenoids such as beta carotene can be converted to a precursor of vitamin A (pro-vitamin A), which is further converted to the vitamin. This supplementation of the vitamin can be beneficial to health and in the prevention of some diseases related to vitamin A deficiency including blindness in children and increased risk of death from measles and infections that produce diarrhea.

■ Historical Background and Scientific Foundations

Golden Rice was developed by the research team of Ingo Potrykus (1933–) at the Swiss Federal Institute of Technology, Zurich, Switzerland and Peter Beyer (1952–) at the University of Freiburg, Germany. Their research began in the early 1980s and culminated almost two decades later with a publication in a 1999 issue of the journal *Science*. Their work identified a chemical pathway in rice that, when supplemented with the addition of two genes, could be converted into a pathway for vitamin A synthesis.

The identification of the key genes in the biosynthetic pathway, the genes needed to drive the synthesis of beta carotene, the successful transfer and continued maintenance of the two genes to rice seeds, and the successful operation of the genetically modified biosynthetic pathway took more than 15 years of work to achieve.

A superior version of the rice, dubbed Golden Rice 2, was introduced by a biotechnology company called Syngenta in 2005. This variety, which was further genetically modified to increase the biosynthetic efficiency of beta carotene, produces about 23 times more beta carotene per rice grain than the original variety.

Further research funded by the Bill and Melinda Gates Foundation has explored the additional supplementation of the rice varieties with other important nutrients including vitamin E, zinc, and iron.

During the first decade of the twenty-first century, both Golden Rice varieties have been grown, first in the greenhouse and then in real field conditions. In contrast to other plants, where the potential shown in the greenhouse does not translate to the real world, beta carotene production by Golden Rice cultivated in the field is four to five times higher than that produced by rice cultivated in the greenhouse.

■ Impacts and Issues

Vitamin A deficiency is especially severe in underdeveloped countries, where the United Nations Children's Emergency Fund (UNICEF) has estimated that nearly 125 million children suffer from vitamin A deficiency. This deficiency accounts for almost 25 percent of the global burden from malnutrition. The use of rice as the vehicle for vitamin A supplementation reflects that fact that rice is the most important source of food providing the bulk of daily caloric intake for hundreds of millions of people in underdeveloped and developing countries, and for almost half of the world's population.

Although Golden Rice may certainly help alleviate vitamin A deficiency in those most affected, critics argue that the amount of rice necessary to achieve the desired effect would be so large as to be unattainable. Time has proven this criticism wrong, with great benefits resulting from consumption of about 75 grams of Golden Rice each day.

As with all genetically-modified foods, there is a concern by some over the possible, and as yet unknown, adverse health effects that might not become apparent until the product is in wide circulation and is being consumed by many people. Also, there is concern over the for-profit control of the access to a food supply, especially a food with health benefits. In addition, there is concern about the potential for the loss of crop diversity as one species becomes dominant.

As of 2011, Golden Rice has not been released for distribution. It may be approved for distribution as early as 2012. The development of Golden Rice by a network of researchers and regulatory authorities is being directed by the International Rice Research Institute. Under the guidance of Syngenta and the Golden Rice Humanitarian Board, Golden Rice will be made available free of charge to developing countries.

SEE ALSO *Agricultural Biotechnology; Biotechnology; Breeding; Selective; Corn, Genetically Engineered; Cotton, Genetically Engineered; GE-Free Food Rights; Genetic Engineering; Genetically Modified Crops; Genetically Modified Food*

BIBLIOGRAPHY

Books

Australian Farm Institute. *Can Agriculture Manage a Genetically Modified Future?* Surrey Hills, New South Wales, Australia: Australian Farm Institute, 2011.

Bodiguel, Luc, and Michael Cardwell. *The Regulation of Genetically Modified Organisms: Comparative Approaches.* Oxford: Oxford University Press, 2009.

Ferry, Natalie, and A. M. R. Gatehouse. *Environmental Impact of Genetically Modified Crops.* Wallingford, UK: CABI, 2009.

> # WORDS TO KNOW
>
> **BIOFORTIFICATION:** The supplementation of a biological material with a compound or compounds that enhance health.
>
> **GENE:** A sequence of DNA that codes for a functional product.
>
> **GENETIC MODIFICATION:** Alteration of the genetic content and/or activity of an organism to achieve a desired outcome.

Golden Rice has been bioengineered to contain increased amounts of beta-carotene, which the body uses as vitamin A.
© *zhuda/Shutterstock.com.*

Gordon, Susan, ed. *Critical Perspectives on Genetically Modified Crops and Food.* New York: Rosen Pub. Group, 2006.

Liang, G. H., and Daniel Z. Skinner, eds. *Genetically Modified Crops: Their Development, Uses, and Risks.* Binghamton, NY: Food Products, 2004.

Wolf, Timm, and Jonas Koch. *Genetically Modified Plants.* New York: Nova Science Publishers, 2008.

Web Sites

"Food, Genetically Modified." *World Health Organization (WHO).* http://www.who.int/topics/food_genetically_modified/en (accessed September 9, 2011).

"Vitamin and Mineral Nutrition Information System (VMNIS)." *World Health Organization (WHO).* http://www.who.int/entity/vmnis/en (accessed September 9, 2011).

Brian Douglas Hoyle

Government Regulations

■ Introduction

Government regulation of biotechnology generally seeks to ensure that modern biotechnology products are safe for humans and will not harm the environment. Some biotechnology regulations, such as prohibitions on human cloning and embryonic stem cell research, are grounded in ethical or moral considerations. Governments may choose to regulate modern biotechnology through laws that govern which biotechnology procedures may be used or by regulating only the final product without regard to the production processes involved. In addition to government regulations, biotechnology products and processes may be regulated by international treaties or trade agreements.

■ Historical Background and Scientific Foundations

Humans have used biotechnology processes to manipulate agriculture, medicine, and other endeavors for millennia through fermentation, hybridization and selective breeding of plants and animals, the development of vaccines and antibiotics, and numerous other natural processes. In the second half of the twentieth century, however, scientists developed the field of modern biotechnology, which relies on genetic engineering techniques to manipulate deoxyribonucleic acid (DNA) or other genetic material. The development of modern biotechnology raised ethical concerns among scientists, policymakers, and the public about the limits and applications of modern biotechnology. Policymakers responded to these concerns about modern biotechnology and began to implement laws and procedures to regulate modern biotechnology.

The U.S. National Institutes of Health (NIH) was the first U.S. agency to regulate modern biotechnology in the United States. The NIH established the Recombinant DNA Advisory Committee in 1974 to address ethical, legal, and scientific concerns about the safety of recombinant DNA (rDNA) technology. rDNA technology involves the use of one or more laboratory methods, such as molecular cloning or polymerase chain reactions, to manipulate genetic material from one or more sources to create a new genetic sequence. The new genetic sequence then is inserted into the genetic material of another organism in order to force the modified organism to express desired genetic traits.

The NIH devised the Guidelines for Research Involving Recombinant DNA Molecules based upon voluntary guidelines for researchers adopted at the Asilomar Conference on Recombinant DNA Molecules. The NIH Guidelines outline safety procedures for rDNA research, including biosafety and containment procedures. The NIH Guidelines have been revised numerous times in order to keep pace with technological developments. Although the NIH Guidelines are mandatory only for projects receiving NIH funding, the guidelines have become an important safety standard followed by most companies and institutions involved in rDNA research.

By the mid-1980s advancements in rDNA technology expanded the field beyond the biomedical purview of the NIH. The application of modern biotechnology techniques in agriculture and other areas prompted the U.S. government to reevaluate its approach to biotechnology regulation. In 1984 the White House Office of Science and Technology Policy (OSTP) proposed a new policy for regulating the development of modern biotechnology products. The government finalized the policy in 1986 as the Coordinated Framework for Regulation of Biotechnology. The Coordinated Framework for Regulation of Biotechnology relied on existing health and safety laws to divide responsibility for regulating modern biotechnology products among three government agencies: the U.S. Department of Agriculture (USDA), the U.S. Environmental Protection Agency (EPA), and the U.S. Food and Drug Administration (FDA).

WORDS TO KNOW

LIVING MODIFIED ORGANISM (LMO): Any living organism that has been genetically modified through the use of biotechnology.

PRECAUTIONARY PRINCIPLE: A principle that any product, action, or process that might pose a threat to public or environmental health should not be introduced in the absence of scientific consensus regarding its safety.

RECOMBINANT DNA (RDNA): A form of artificial DNA that contains two or more genetic sequences that normally do not occur together, which are spliced together.

SOMATIC CELLS: All human cells except egg or sperm cells.

THERAPEUTIC CLONING: The process in which cells are created through somatic cell nuclear transfer (SCNT) from embryonic stem cells. Implanted stem cells in the DNA donor used for SCNT ensure that the embryonic stem cells created are identical to those present in the surrounding tissues to provide immunocompatiblity between new and existing cells.

■ Impacts and Issues

Biotechnology laws may regulate either modern biotechnology processes or final products. The NIH Guidelines for Research Involving Recombinant DNA Molecules are an example of government guidelines that regulate modern biotechnology processes. All rDNA research receiving NIH funding must follow the biosafety and containment requirements of the NIH Guidelines regardless of the proposed final product.

Most government biotechnology regulations focus on ensuring the safety of biotechnology products for humans and animals. In the United States the USDA, EPA, and FDA regulate agricultural and other biotechnology products under the Coordinated Framework for Regulation of Biotechnology and all related laws and regulations. The regulatory regime in the United States focuses on the characteristics and safety of the end product and not the biotechnology processes involved in its production. Biotechnology regulation in the United States adheres to the principle of substantial equivalence, which holds that a novel food that has the same characteristics and composition as a conventional food is considered as safe as the conventional variety. The safety of totally new foods that do not have a conventional counterpart, however, must be evaluated on their own composition and properties.

The European Union (EU) eschews the principle of substantial equivalence in favor of a precautionary principle approach to genetically modified foods and livestock feed. The precautionary principle states that any product, action, or process that may pose a threat to public or environmental health should not be introduced in the absence of scientific consensus regarding its safety. Novel foods and other biotechnology products, therefore, are not treated as the substantial equivalent of conventional products, even if the novel product shares a similar composition and characteristics. The novel product must be proven safe on its own merits.

The EU also allows member states to prohibit the use of sale of genetically-modified (GM) foods within that nation under the EU's GM foods safeguard clause. Each nation seeking to restrict the use or sale of GM foods must state justifiable reasons for believing that a GM food poses a threat to human health or the environment. The EU and many other countries also have labeling requirements for food containing GM products. In the EU all food and feed containing 0.9 percent or higher of GM products must be labeled as a GM food. U.S. biotechnology regulations do not require labeling of GM foods.

International agreements, such as Agenda 21, the Convention on Biological Diversity, and the Cartagena Protocol on Biosafety, also influence biotechnology regulation. Agenda 21, a United Nations (UN) action plan on sustainable development adopted in 1992, called for the international community to agree upon principles to be applied to modern biotechnology regulation and development. The Convention on Biological Diversity, a legally binding international treaty on preserving biological diversity, requires parties to establish national biotechnology regulatory frameworks to ensure the safe use of biotechnology. The Cartagena Protocol on Biosafety, a supplementary agreement to the Convention on Biological Diversity, regulates the trade of living modified organisms (LMOs) among countries. LMOs are genetically-modified living organisms. The Cartagena Protocol authorizes countries to prohibit the import of LMOs if, based on scientific information, that nation believes that imported LMOs could threaten that country's biodiversity.

Biotechnology products are also regulated under the terms of international trade agreements and organizations, including the World Trade Organization (WTO). Some countries use the issue of biotechnology safety as a reason to restrict the import of GM goods. In 2006 the WTO issued a seminal ruling, "Dispute Settlement 291" (DS291), in a trade dispute in which the United States, Canada, and Argentina accused the European Community (EC) of blocking GM agricultural products in violation of WTO rules. The EC imposed a de facto moratorium on GM food and feed in 1998 and restricted the import of GM products from the United States, Canada, Argentina, and other countries. In DS291, a WTO panel decided that the EC blocked the import of GM products for political reasons, rather than scientific reasons. Whereas the EU

President Barack Obama, left, speaks during an event on the South Lawn of the White House in Washington, D.C. With Obama is John P. Holdren, Director of the White House Office of Science and Technology Policy (OSTP). The OSTP formulates policy regarding biotechnology in the United States. *© AP Images/Pablo Martinez Monsivais.*

could block the import and sale of GM products for scientific reasons under WTO rules and the Cartagena Protocol, the WTO stated that the EC could not do so on political grounds.

In addition to human health and safety, governments have legislated certain areas of modern biotechnology research considered unethical or immoral by some individuals. Reproductive human cloning, which involves creating a genetic copy of a human being, has been banned almost universally. Governments are divided on the use of therapeutic cloning, which involves cloning adult cells through somatic-cell nuclear transfer. Most nations, however, allow therapeutic cloning. Government laws and regulations on embryonic stem cell research are more evenly split. Since 2009 the United States has permitted the use of federal funds for some forms of embryonic stem cell research. The EU does not fund research that results in the destruction of embryos, and some European nations have complete bans on embryonic stem cell research. Brazil, China, India, Japan,

and South Korea have few restrictions on embryonic stem cell research. Most African and Latin American countries, however, have severe restrictions or complete bans on embryonic stem cell research.

SEE ALSO *Bayh-Dole Act of 1980;* Diamond v. Chakrabarty; *Genetic Discrimination; Individual Privacy Rights; Plant Patent Act of 1930; Plant Variety Protection Act of 1970; Stem Cell Laws*

BIBLIOGRAPHY

Books

Bodiguel, Luc, and Michael Cardwell, eds. *The Regulation of Genetically Modified Organisms: Comparative Approaches.* Oxford: Oxford University Press, 2009.

Coleman, Carl H. *The Ethics and Regulation of Research with Human Subjects.* Newark, NJ: LexisNexis, 2005.

Naidu, B. David. *Biotechnology & Nanotechnology: Regulation under Environmental, Health, and Safety Laws.* Oxford: Oxford University Press, 2009.

Pak, Un-jong. *Bioethics, Research Ethics and Regulation.* Seoul, South Korea: Seoul National University Press, 2005.

Termini, Roseann B. *Life Sciences Law: Federal Regulation of Drugs, Biologics, Medical Devices, Foods, and Dietary Supplements.* Wynnewood, PA: FORTI Publications, 2010.

Web Sites

"Legislation, Regulations, and Policies." *Centers for Disease Control and Prevention (CDC).* http://www.cdc.gov/stltpublichealth/Policy/legislation-regulation-policies.html (accessed November 12, 2011).

"United States Regulatory Agencies Unified Biotechnology Website." *Center for Biological Informatics of the U.S. Geological Survey.* http://usbiotechreg.nbii.gov/ (accessed November 12, 2011).

Joseph P. Hyder

Green Chemistry

Introduction

Green chemistry is the term given to a sustainable approach to chemical synthesis and related processes. While not a formal branch of chemistry such as organic chemistry, green chemistry is being used increasingly in sectors of industry that rely on chemistry, such as pharmaceuticals.

Many conventional chemical reactions create by-products. Green chemistry aims to design syntheses so that by-products are reduced or eliminated. Such reactions often will join two reagents to form a single product. Another important focus is the nature of the solvent used. For example, chlorinated solvents such as dichloromethane traditionally play an important role in industrial chemistry. However, in recent years, concern has been growing over the potential health impact of pollution of the water supply by such solvents. Green chemistry tries to carry out reactions and processes in aqueous solvent wherever possible, or to use alternative safer solvents such as ionic liquids. Enzymes play a central role in green chemistry because they can carry out chemical reactions with high efficiency and under mild conditions. The main challenges in the more widespread adoption of the green chemistry approach lie in lack of education and awareness. These barriers will be overcome once the benefits of green chemistry in terms of waste and cost reduction are more widely recognized.

Historical Background and Scientific Foundations

From its early days in the nineteenth century and throughout most of the twentieth century, the chemical industry paid little attention to the impact its activities had on the environment. This aggressive approach may have been necessary to build the industry, which has provided many of the products that are seen as essential in modern life. However, the growth of the green movement in the 1960s turned the spotlight on chemicals and their polluting effects on the water supply and other aspects of the environment.

Paul Anastas (1962–), a Yale University professor working with the United States Environmental Protection Agency (EPA), pioneered green chemistry in the early 1990s. Anastas described its goal as promotion of innovative chemical technologies that reduce or eliminate the use or generation of hazardous substances in the design, manufacture, and use of chemical products. In 1995 President Bill Clinton (1946–) established the Presidential Green Chemistry Challenge in recognition of key developments in green chemistry. Two years later Anastas went on to found the Green Chemistry Institute at the American Chemical Society and now leads Yale's Center for Green Chemistry and Green Engineering. In 2011 he was serving as the Assistant Administrator for EPA's Office of Research and Development and the Science Advisor to the EPA.

Anastas and John C. Warner, also considered a pioneer of green chemistry, laid down the 12 principles of green chemistry in their classic text *Green Chemistry: Theory and Practice*. They urge chemists to aim for high atom economy in a reaction to reduce waste and increase efficiency. Processes should be designed with energy efficiency in mind and carried out at the lowest temperature and pressure. Renewable feedstocks are preferred as raw materials. Enzymes are important in this context. Industrial enzymes are made in cell culture and can be used in tiny quantities. A major principle of green chemistry is to avoid toxic raw materials, solvents, and products. This last has been strengthened in practice by the development of legislation such as the Control of Substances Hazardous to Health in the United Kingdom, which requires risk assessment of chemicals used in both laboratory work and manufacturing.

Impacts and Issues

The green chemistry approach has had a significant impact upon the way the chemical industry functions. For example, the Green Chemistry Institute set up a

WORDS TO KNOW

ATOM ECONOMY: A measure of how much starting material in a chemical reaction is turned into the desired product. Wasteful processes, with by-products, have low atom economy, whereas efficient processes, with high product yield, have high atom economy.

IONIC LIQUID: An ionic compound that is liquid below 212°F (100°C), the boiling point of water. Conventional ionic compounds have much higher melting points. Many ionic liquids are powerful solvents, without the toxic properties of organic solvents, and so play a role in green chemistry.

SUSTAINABLE: Definitions vary, but broadly speaking sustainable refers to a process, such as drug manufacture or energy production, that can be continued in the long-term with only minimal impact on the environment and minimal depletion of natural resources.

Pharmaceutical Roundtable in 2005 to discuss how best to integrate green chemistry into drug discovery and development. The major pharmaceutical companies now participate in this forum. The Presidential Green Challenge provides many examples of how the pharmaceutical industry has begun to change its ways. For instance, Merck and Codexis are now using enzymes in the manufacture of Januvia®, a diabetes treatment. Codexis also uses a designer enzyme in the manufacture of the cholesterol-lowering drug Lipitor®, which is one of the world's best selling medications.

A growing awareness of the impact of global warming and higher energy prices is one of the major factors driving green chemistry, with its reduced energy consumption and costs. The green approach also promises to improve public perceptions of the chemical industry, which is often seen as a major polluter of the environment. Major chemical companies have implemented programs committed to social responsibility, and green chemistry fits well with this agenda.

However, there are a number of challenges remaining to the more widespread adoption of green chemistry. Companies may be reluctant to change processes, such as those for purifying a drug molecule, that have already been approved by regulatory authorities. There will be short-term costs involved in process changes, and there are no guarantees that long-term savings will be achieved or that new processes will work as well. Strengthening of environmental legislation may help here, with more of

Cambridge College professor James Stephen Lee uses specimens, including a brood of fish affected by growth hormones, during a class in green chemistry at the school in Cambridge, Massachusetts. The American Chemical Society, which certifies more than 600 college chemistry programs, lists only about a dozen that teach green chemistry, though the number is growing. © *AP Images/Josh Reynolds.*

an emphasis on pollution prevention rather than requiring the polluter to clean up. At present, manufacturers reluctant to change can outsource processes to countries where environmental legislation is weaker.

Green chemistry also needs to become more central to chemical education. Many universities have already made an excellent start in this area. However, more should be done to convince chemists to look for simple, rather than complex, chemicals and processes. In conventional chemistry, water is not seen as an effective solvent compared to tried and tested organic solvents, and enzymes are perceived as being unstable and difficult to work with compared to conventional heavy metal catalysts. Eventually, however, green chemistry will create a new body of knowledge that will help the chemical industry to a more economic and sustainable future.

SEE ALSO *Algae Bioreactor; Biofuels, Gas; Biofuels, Liquid; Biofuels, Solid; Bioplastics (Bacterial Polymers); Biopreservation; Environmental Biotechnology*

BIBLIOGRAPHY

Books

Ahluwalia, V. K. *Green Chemistry: Environmentally Benign Reaction.* Boca Raton, FL: CRC, Taylor & Francis, 2009.

Anastas, Paul, and Warner, John C. *Green Chemistry: Theory and Practice.* New York: Oxford University Press, 1998.

Anastas, Paul T., Irvin J. Levy, and Kathryn E. Parent, eds. *Green Chemistry Education: Changing the Course of Chemistry.* Washington, DC: American Chemical Society, 2009.

Lacey, Hugh. *Values and Objectivity in Science: The Current Controversy about Transgenic Crops.* Lanham, MD: Lexington Books, 2005.

Nelson, William M., ed. *Oxford Handbook of Green Chemistry: From Philosophy to Industrial Applications.* New York: Oxford University Press, 2010.

Pearlman, Jeffrey T., ed. *Green Chemistry Research Trends.* New York: Nova Science Publishers, 2009.

Web Sites

"Green Chemistry." *United States Environmental Protection Agency (EPA).* http://www.epa.gov/greenchemistry/ (accessed October 8, 2011).

Goodman, Sarah. "'Green Chemistry' Movement Sprouts in Colleges, Companies." *New York Times,* March 25, 2009. http://www.nytimes.com/gwire/2009/03/25/25greenwire-green-chemistry-movement-sprouts-in-colleges-c-10287.html (accessed October 8, 2011).

"What Is Green Chemistry?" *California Department of Toxic Substances Control.* http://www.dtsc.ca.gov/PollutionPrevention/GreenChemistryInitiative/index.cfm (accessed October 8, 2011).

Susan Aldridge